PLANTS AND SOCIETY

PLANTS AND SOCIETY

Estelle Levetin
University of Tulsa

Karen McMahon
University of Tulsa

WCB Wm. C. Brown Publishers

Dubuque, IA Bogota Boston Buenos Aires Caracas Chicago
Guilford, CT London Madrid Mexico City Sydney Toronto

Book Team

Editor *Margaret J. Kemp*
Developmental Editor *Kathleen R. Loewenberg*
Production Editor *Kay J. Brimeyer*
Designer *K. Wayne Harms*
Art Editor *Miriam J. Hoffman*
Photo Editor *Lori Hancock*
Permissions Coordinator *Gail I. Wheatley*

President and Chief Executive Officer *Beverly Kolz*
Vice President, Publisher *Kevin Kane*
Vice President, Director of Sales and Marketing *Virginia S. Moffat*
Vice President, Director of Production *Colleen A. Yonda*
National Sales Manager *Douglas J. DiNardo*
Marketing Manager *Thomas C. Lyon*
Advertising Manager *Janelle Keeffer*
Production Editorial Manager *Renée Menne*
Publishing Services Manager *Karen J. Slaght*
Royalty/Permissions Manager *Connie Allendorf*

Photo research by *Kathy Husemann*

Cover photo: © Bryan Haynes/The Image Bank

The credits section for this book begins on page 429 and
is considered an extension of the copyright page.

Library of Congress Catalog Card Number: 94–73161

ISBN 0–697–14064–4

Printed in the United States of America

10 9 8 7 6 5 4 3 2

In loving memory of our mothers,
Pauline Levetin and Dorothy McMahon

A personal library is a lifelong source of enrichment and distinction. Consider this book an investment in your future and add it to your personal library.

Brief Contents

Contents

Unit I

Plants and Society: The Botanical Connections to Our Lives

Chapter 1

Unit II

Introduction to Plant Life: Botanical Principles

Chapter 2

Chapter 3

The Plant Body 32

Chapter 4

Plant Physiology 49

Chapter 5

Plant Life Cycle: Flowers 70

Chapter 6

Plant Life Cycle: Fruits and Seeds 84

Chapter 7

Genetics 96

Chapter 12

The Grasses 175

Chapter 13

Legumes 194

Chapter 14

Starchy Staples 204

Chapter 15

Feeding A Hungry World 217

Unit IV

Commercial Products Derived From Plants

Chapter 16

Stimulating Beverages 236

Chapter 17

Herbs and Spices 249

Chapter 18

Materials: Cloth, Paper, and Wood 265

Unit V

Plants and Human Health

Chapter 19

Medicinal Plants 286

Chapter 20

Psychoactive Plants 302

Chapter 21

Poisonous and Allergy Plants 319

UNIT VI

FUNGI: THE IMPACT OF FUNGI ON HUMAN AFFAIRS

CHAPTER 22

CHAPTER 23

CHAPTER 24

UNIT VII

PLANTS AND THE ENVIRONMENT

CHAPTER 25

ECOLOGY 396

Preface

As we approach the 21st century, plant science is once again assuming a prominent role in research. Renewed emphasis on developing medicinal products from native plants has encouraged ethnobotanical endeavors. The destruction of the rain forests has made the timing for this imperative, and also spurred on efforts to catalog the plant biodiversity in these environments. Efforts to feed the growing populations in developing nations have also positioned plant scientists at the cutting edge of genetic engineering with the creation of transgenic crops. However, in recent decades botany courses have seen a decline in enrollment, and some courses have even disappeared from the curriculum in many universities. We have written *Plants and Society* in an effort to offset this trend. By taking a multidisciplinary approach to studying the relationship between plants and people, we hope to stimulate interest in plant science and encourage students to further study. Also, by exposing students to society's historical connection to plants, we hope to instill a greater appreciation for the botanical world.

Audience

Recently, general botany courses have placed greater emphasis on the impact of plants on society. In addition many institutions have developed Plants and Civilization courses devoted exclusively to this topic. This has invigorated the traditional Economic Botany from a dry statistical treatment of "bushels/acre" to an exciting discussion of "botanical marvels" that have influenced our past and will change our future. *Plants and Society* is intended for use in this type of course which is usually one semester or one quarter in length. There are no prerequisites since it is an introductory course. The course covers basic principles of botany with strong emphasis on the economic aspects and social implications of plants and fungi.

Students usually take a course of this nature in their freshman or sophomore year to satisfy a science requirement in the general education curriculum. Typically they are not biology majors. Although most students enroll to satisfy the science requirement, many become enthusiastic about the subject matter. Students, even with a limited science background, should not encounter any problems with the level of scientific detail in this text.

As indicated, the primary market for this text would be a Plants and Civilization course; however it would certainly be suitable for an introductory general botany course as well.

Design

We feel that *Plants and Society* is a textbook with a great deal of flexibility for course design. It offers a unique balanced approach between basic botany and the applied or economic aspects of plant science. Similar texts emphasize either the basic or applied material, making it difficult for instructors who wish to provide better balance in an introductory course. Another distinctive feature is the unit on fungi. While other texts cover certain aspects of this topic, we have an expanded coverage of fungi and their impact on society.

Plants and Society is organized into 25 chapters that are grouped into seven units. The first nine chapters cover the basic botany found in an introductory course. However, even in these chapters we have included many applied topics; some in the boxed essays but others directly in the chapter text.

Unit I—Plants and Society: The Botanical Connections to Our Lives. Chapter one stresses the overall importance of plants in everyday life. The properties of life and an introduction to chemistry are also included.

Unit II—Introduction to Plant Life: Botanical Principles. This unit addresses basic botany. Chapters cover plant structure, from the cellular level through the mature plant. Reproduction, including mitosis and meiosis and the life cycle of flowering plants, is discussed in a separate chapter. Other chapters cover genetics, plant physiology, plant taxonomy, and plant diversity. Among the economic topics in this unit are vegetables, fruits, sugar, and perfumes.

Unit III—Plants as a Source of Food. This unit describes the major food crops. It begins with a chapter on the requirements for human nutrition and continues with a chapter on the origin of agriculture. Other chapters cover the grasses, the legumes, and starchy staples. The unit ends with a chapter on the Green Revolution, the loss of genetic diversity, the search for alternative crops, and development of new crops.

Unit IV—Commercial Products Derived from Plants. This unit covers other crops that provide us with consumable products such as beverage plants, herbs and spices, and materials such as cloth, paper, and wood. The origin and historical impact of these crops is explored.

Unit V—Plants and Human Health. Unit five introduces students to the historical foundations of western medicine, the practice of herbal medicine, and the chemistry of secondary plant products. Descriptions of those plants that provide us with medicinal products as well as psychoactive drugs are discussed. The unit also covers the ubiquitous poisonous and allergy plants in our environment.

Unit VI—Fungi. This unit describes the impact of fungi including their biology, their role in the environment, and the plant diseases they cause. Fermented beverages and foods from fungi are discussed, as is the medical importance of fungi as sources of antibiotics, toxins, poisons, and human disease.

Unit VII—Plants and the Environment. Chapter 25 is an introduction to the principles of ecology: the ecosystem, niche, food chains, biogeochemical cycles, and ecological succession. Th e major biomes of the world are discussed, with the economic value of certain desert plants and the strategy of extractive reserves emphasized.

Approach

This textbook is written at the introductory level suitable for students with little or no background in biology. Like any introductory book, this is a broad-brush treatment. The nature of the course dictates an applied approach, with the impact of plants on society as the integrating theme, but the theoretical aspects of basic botany are thoroughly covered.

Learning Aids

In addition to the textual material, each chapter begins with a chapter outline and chapter objectives. Key terms are in boldface throughout the text, and each chapter ends with a summary, review questions, and suggested readings. Topics of special interest are included as boxed essays. The text also has appendices on the metric system and plant taxonomy as well as a glossary.

Supplements

An instructor's manual and test bank accompany the text. Also available is a slide set containing 75 slides, many of which are images not found in the book.

Acknowledgments

We want to acknowledge the continued help and encouragement we received from Kathy Loewenberg, Kay Brimeyer, Kathy Husemann, Lori Hancock, Miriam Hoffman, Gail Wheatley and others at Wm. C. Brown Publishers. We especially want to thank Meg Johnson who believed in our concept from the very beginning.

We want to recognize the Harriet G. Barclay Slide Collection at the University of Tulsa and Dr. Paul Buck who curates the collection. The slides have been an exceptional resources for both this book and for our Plants and Civilization classes.

Finally, we wish to extend our sincere appreciation to our families for their patience and encouragement throughout the development and writing of this book.

We also want to thank the reviewers whose constructive criticism and comments helped in the development of the manuscript.

List of Boxed Readings

List of Reviewers

C. Gerald Van Dyke, *North Carolina State University*

Beth A. Gaydos, *San Jose City College*

Marshall D. Sundberg, *Louisiana State University*

Robert D. Bergad, *Lakewood Community College*

R. Gordon Perry, *Fairleigh Dickinson University*

David B. Czarnecki, *Loras College*

Richard E. Hall, *Cypress College*

Grant L. Pyrah, *Southwest Missouri State University*

Michael F. Gross, *Georgian Court College*

Dr. L. Elliot Shubert, *University of North Dakota* and *The Natural History Museum, London, U.K.*

B. Dwain Vance, *University of North Texas*

Ronald C. Doney, *SUNY—Cortland*

Lawrence C. Matten, *Southern Illinois University at Carbondale*

UNIT

I

PLANTS AND SOCIETY: *THE BOTANICAL CONNECTIONS TO OUR LIVES*

The botanical connections to our lives are many: food, medicines, materials, and beverages are just a few of the ways plants serve humanity.

1

Plants in Our Lives

Chapter Outline

Chapter Concepts

1. Green plants, especially angiosperms, are more than just landscaping for the planet, since they supply humanity with all the essentials of life: food, and oxygen, as well as other products that have shaped modern society.
2. Fungi are also an economically important group of organisms impacting society in numerous ways, from fermentation in the brewing process to the use of antibiotics in medicine, to their role as decomposers in the environment, and as the cause of plant and animal diseases.
3. All living organisms share certain characteristics: growth, ability to respond, reproduction, metabolism, organized structure, and organic composition.
4. The processes of life are based on the chemical nature and interactions of carbohydrates, lipids, proteins, and nucleic acids.

Much of modern society is estranged from the natural world; people living in large cities often spend over 90% of their time indoors and have little contact with nature. Urbanized society is far removed from the source of many of the products that make civilization possible: most food is purchased in large supermarkets, most medicines are purchased at pharmacies, and most building supplies are purchased at lumber yards. Society's dependence on nature, especially plants, is forgotten (table 1.1).

In less urbanized environments, life-styles are more attuned to nature. The farmer's existence is dependent on crop survival, and the farmer's work cycle is timed to the growing season of the crops. The few hunter-gatherer cultures that remain in isolated areas of the world are even more dependent on nature as they forage for wild plants and hunt wild animals. These foragers know that without grains there would be no flour or bread; without plant fibers there would be no cloth, baskets, or rope; without medicinal herbs there would be no relief from pain; without wood there would be no shelter; without firewood there would be no fuel for cooking or heat; and without vegetation there would be no wild game.

Plants and Human Society

Whether forager, farmer, or city dweller, humans have four great necessities in life: food, clothing, shelter, and fuel. Of the four, an adequate food supply is the most pressing need, and, directly or indirectly, plants are the source of virtually all food through the process of **photosynthesis.** Through photosynthesis, plants use solar energy to convert carbon dioxide and water into sugars and, as such, are the **producers** in the **food chain.** They are the base of all food chains, whether eaten directly by humans as **primary consumers,** or indirectly as **secondary consumers** when eating beef (that comes from grain-fed or pasture-fed cattle). In addition to the food produced by photosynthesis, oxygen is given off as a by-product and is the Earth's only continuous supply of oxygen. As sources of food, oxygen, lumber, fuel, paper, rope, fabrics, beverages, medicines, and cosmetics, plants support and enhance life on the planet.

The Flowering Plants

The word *plant* means different things to different people: to an ecologist a plant is a producer, to a forester it is a tree, to a home gardener a vegetable, and to an apartment dweller a houseplant. Although there are many different types of plants, the most abundant and diverse plants in the environment are the flowering plants or **angiosperms.** These are also the most economically important plants and are the primary focus of this book. From the more than 250,000 known **species*** of angiosperms, an overwhelming diversity of products has been obtained and utilized by society. The food staples of civilization, wheat, rice, and corn, are all angiosperms; in fact, with minor exceptions, all food crops are angiosperms. The list of other products from angiosperms is considerable and includes cloth, hardwood, herbs and spices, beverages, many drugs, perfumes, vegetable oils, gums, and rubber.

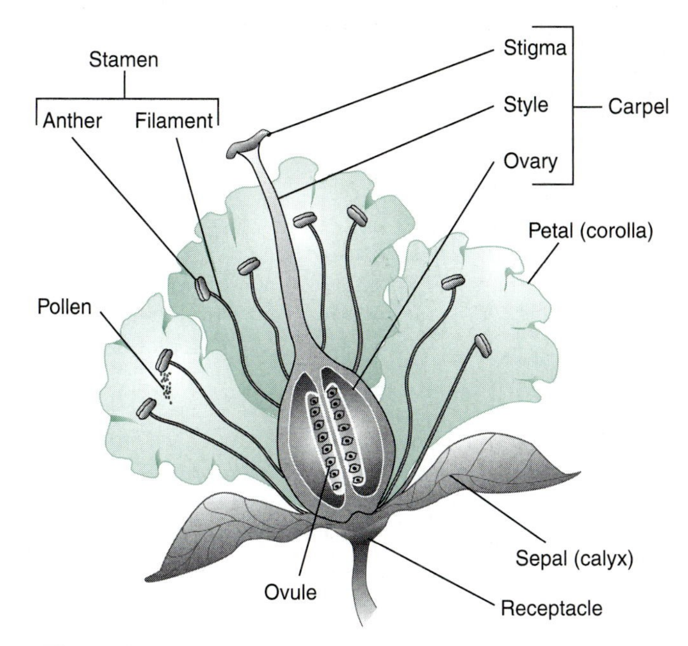

Figure 1.1 *Angiosperm flower. All the stamens comprise the androecium, and the carpels comprise the pistil, or gynoecium.*

All angiosperms are characterized by flowers that contain the sexual reproductive structures and fruits that contain seeds. A typical angiosperm flower consists of four whorls of parts: **sepals, petals, stamens,** and **carpels** (fig. 1.1). It is from the carpel that the fruit and its seeds will develop. The angiosperms are divided into two groups, the **monocots** and the **dicots,** on the basis of structural and anatomical differences. Among the most familiar monocots are lilies, grasses, palms, and

*Each kind of organism, or species, has a two-part scientific name consisting of a genus name and a specific epithet; for example, white oak is known scientifically as *Quercus alba.* After the first mention of a scientific name, the genus name can be abbreviated, *Q. alba.* When referring to oaks in general it is acceptable to use the genus name, *Quercus,* alone. Sometimes an abbreviation for species, "sp." or plural "spp.", stands in for the specific epithet; i.e., *Quercus* sp. or *Quercus* spp. Both common and scientific names are used throughout this book; details on this topic are found in Chapter 8.

TABLE 1.1
How Much Do Plants Affect Society?

_____ 1.	*True or False* - Most of the world's population depends mainly on plants for calories and protein.
_____ 2.	*True or False* - Although plants were important as medicinals in the past, they are seldom used today.
_____ 3.	*True or False* - The search for cinnamon led to the discovery of North America.
_____ 4.	*True or False* - Coffee is native to the mountains of Colombia and Brazil.
_____ 5.	*True or False* - The introduction of the potato to Europe in the sixteenth century initiated events that led to a devastating famine in Ireland.
_____ 6.	*True or False* - One hundred fifty acres of forest are cut down for each *New York Times* Sunday edition.
_____ 7.	*True or False* - Although a good source of protein for livestock, soybeans are inadequate for the human diet.
_____ 8.	*True or False* - The Salem Witchcraft Trials in the 1690s might have resulted from a case of fungal poisoning.
_____ 9.	*True or False* - Tomatoes were once considered to be an aphrodisiac.
_____10.	*True or False* - A well-balanced meal must consist of a meat entree with vegetable side dishes.

Answers

1. True	In nations like the United States or those in Western Europe, approximately 65% of the total caloric intake and 35% of the protein is obtained directly from plants, while in developing nations close to 90% of the calories and over 80% of the protein is from plants (Chapters 10, 15).
2. False	Approximately 25% of all prescription drugs in Western society contain ingredients derived from plants; however, 80% of the world's population does not use prescription drugs but relies exclusively on herbal medicine (Chapter 19).
3. True	Columbus was one of many explorers trying to find a sea route to the rich spicelands of the Orient. Cinnamon and other spices were so valued in the fifteenth century that a new, faster route to the East would bring untold wealth to the explorer and his country (Chapter 17).
4. False	Although premium coffee is grown today in the mountains of South America, the coffee tree is actually native to Ethiopia and had a long history of use in the Arab world before its introduction into European society in the seventeenth century. Plantations in the New World were started in the eighteenth century (Chapter 16).
5. True	The potato, native to South America, became a staple food for the poor in many European countries, especially Ireland. The widespread dependence on a single crop led to massive starvation when a fungal disease, late blight of potato, destroyed potato fields in the 1840s. Over one million Irish died from starvation or subsequent diseases; another 1.5 million emigrated (Chapter 14).
6. True	Most of the world's paper comes from wood pulp. In the United States, each person uses an average of 275 kilograms (600 pounds) of paper per year, with less than 40% being recycled. Recycling a 1.2 meter (4 foot) stack of newspapers would save a 12 meter (40 ft) tree (Chapter 18).
7. False	Like most legumes, soybeans are high in protein; in fact, soybeans have a higher protein content than lean beef. Although soybeans have a long history of food use in Asia, in the United States they have largely been used for livestock feed until recently (Chapter 13).
8. True	Searching for the cause of the hysteria that led to the accusations of witchcraft in Salem, Massachusetts, some historians have suggested ergot poisoning. Caused by a fungal disease of rye plants, an ergot forms in place of a normal grain and produces hallucinogenic toxins. Consumption of contaminated rye flour can lead to hallucinations, neurological symptoms, or even death (Chapter 24).
9. True	When tomato plants were first introduced to Europe, they were viewed with suspicion by many people, since poisonous relatives of the tomato were known. Neither poisonous nor an aphrodisiac, it took centuries for the tomato to fully overcome its undeserved reputation (Chapter 6).
10. False	It is possible to have a well-balanced meal from vegetarian dishes only. In fact, nutritionists have come to realize that a diet rich in animal fat is connected to increased risk of heart disease and certain forms of cancer. Increasing plant fiber and reducing animal products in the diet should be the dietary goals of every person (Chapter 10).

orchids. A few common dicots are geraniums, roses, tomatoes, dandelions, and most broad-leaved trees. The structure and reproduction of the angiosperms will be described in detail in later chapters.

The Fungi

One other group of organisms that has had a significant impact on society is the fungi, including the molds, mildews, yeast, and mushrooms (fig. 1.2). Although once considered a type of simple plant, biologists today classify the fungi as neither plants nor animals but in their own kingdom. Second only to the angiosperms in economic importance, the fungi provide many beneficial items such as penicillin, beer, wine, cheese, edible mushrooms, and leavened bread. A negative aspect of their economic importance is the impact of fungal disease and spoilage. The most serious diseases of our crop plants are caused by fungi, resulting in billions of dollars in crop losses each year.

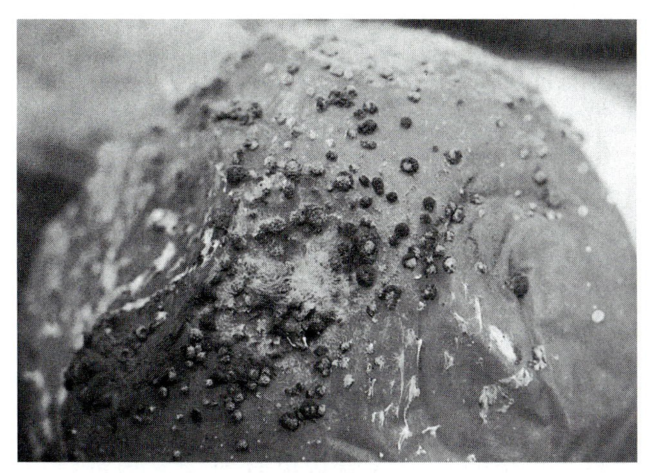

(a)

(b)

Figure 1.2 *Diversity of fungi. (a) Rotting apple with the common mold, Penicillium; (b) A mushroom in the genus* Boletus *found in a woodland area.*

Fungi generally have a thread-like body, the **mycelium,** and propagate by reproductive structures called **spores.** Fungi are nonphotosynthetic organisms, obtaining their nourishment from decaying organic matter as **saprobes** or as **parasites** of living hosts. Ecologically the fungi play an essential role as **decomposers,** recycling nutrients in the environment. Because of their traditional ties to botany (the study of plants), the fungi and their impact on humanity will be considered in this book and are presented in Chapters 22–24.

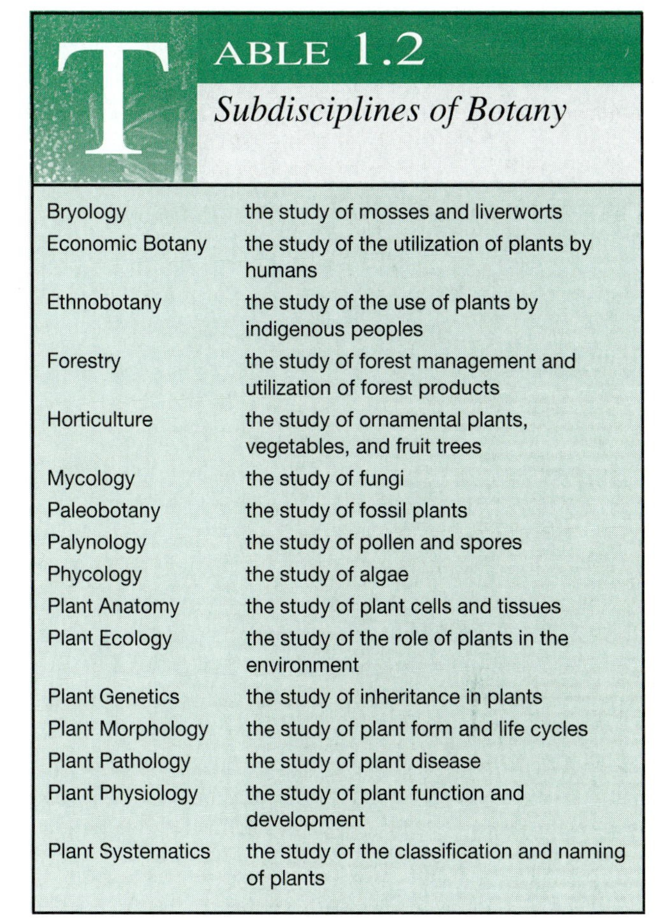

TABLE 1.2

Subdisciplines of Botany

Bryology	the study of mosses and liverworts
Economic Botany	the study of the utilization of plants by humans
Ethnobotany	the study of the use of plants by indigenous peoples
Forestry	the study of forest management and utilization of forest products
Horticulture	the study of ornamental plants, vegetables, and fruit trees
Mycology	the study of fungi
Paleobotany	the study of fossil plants
Palynology	the study of pollen and spores
Phycology	the study of algae
Plant Anatomy	the study of plant cells and tissues
Plant Ecology	the study of the role of plants in the environment
Plant Genetics	the study of inheritance in plants
Plant Morphology	the study of plant form and life cycles
Plant Pathology	the study of plant disease
Plant Physiology	the study of plant function and development
Plant Systematics	the study of the classification and naming of plants

Plant Sciences

When humans began investigating the uses of plants for food, bedding, medicines, and fuel, the beginnings of plant science were evident. Early peoples were skilled regional botanists and passed on their knowledge to succeeding generations. This folk botany gradually amassed a great body of knowledge, laying the foundation for scientific botany that began in ancient Greece. As the body of knowledge expanded over the centuries, areas of specialization developed within botany, and today many of them are recognized as disciplines in their own right (table 1.2).

Fundamental Properties of Life

Although living organisms can be as different as oak trees, elephants, and bacteria, they share certain fundamental properties common to all life. These properties include the following:

1. **Growth and Reproduction** Living organisms have the capacity to grow and reproduce. Growth is defined as an irreversible increase in size and should not be confused with simple expansion. Although balloons or

A Closer Look 1.1

Biological Mimic: Velcro™

The architecture of nature far surpasses any design developed by modern technology. In fact, engineers and inventors often appropriate their best ideas directly from the natural world. Perhaps the best example of this is the development of Velcro™ fasteners. Today, Velcro™ has hundreds of uses in diapers, running shoes, space suits, even in sealing the chambers of artificial hearts. But it started from observations during the tedious task of removing cockleburs from clothing. In 1948, a Swiss hiker, George de Mestral, observed the manner in which cockleburs clung to clothing and thought that a fastener could be designed using the pattern. Cockleburs (box fig. 1.1a) have up to several hundred curved prickles that function in seed dispersal. These tiny prickles tenaciously hook onto clothing or the fur of animals and are thus transported to new areas. De Mestral envisioned a fastener with thousands of tiny hooks, mimicking the cocklebur prickles on one side, and on the other side, thousands of tiny eyes for the hooks to lock onto (box fig. 1.1b). It took ten years to perfect the original concept of the "locking tape" that has become Velcro™, so common in modern life.

(a)

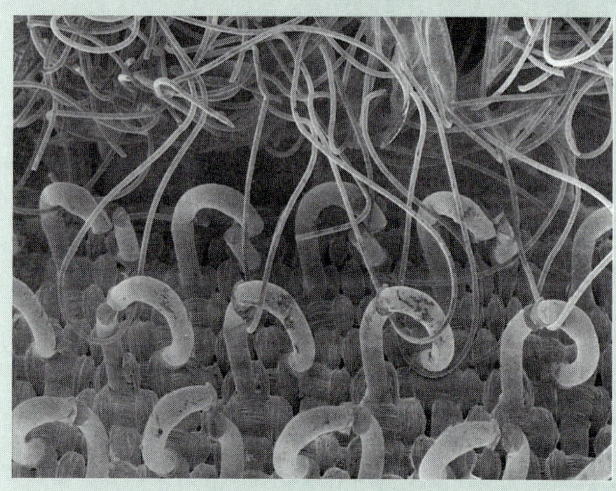

(b)

Box Figure 1.1 *Biological mimic is seen in (a) the prickles on cocklebur and (b) hooks on Velcro.™*

crystals can enlarge, this is not true growth. The ability to reproduce, produce new individuals, is common to all life. Reproduction can be **sexual,** involving the fusion of **sperm** and **egg** to form a **zygote,** or **asexual,** in which the offspring are genetic clones of a single parent.

2. **Ability to Respond** The environment is never static; it is always changing, and living organisms have the capabilities to respond to these changes. These responses can be obvious, such as a stem growing toward the light (fig. 1.3) or an animal hibernating for the winter. Sometimes, however, the responses are subtle, such as changes in the chemical composition of leaves in trees under attack by insects. The chemical composition of intact leaves is altered, making the leaves unpalatable to the insects.
3. **Ability to Evolve and Adapt** All life constantly changes or evolves. This process has been going on for billions of years, as evidenced by the fossil record. Sometimes changes promote survival because the altered species is better adapted to its environment. Many desert plants have evolved water-storing tissue, an adaptation that helps them survive their arid environment.
4. **Metabolism** Metabolism is the sum total of all chemical reactions occurring in living organisms. Two of the most important metabolic reactions are **cellular**

Figure 1.3 *A field of sunflowers with their heads all pointed in the same direction as the flowers track the sun from dawn to dusk.*

respiration and **photosynthesis.** Respiration is a metabolic process where food is chemically broken down to release energy; this occurs in all living organisms. Photosynthesis occurs in green plants, algae, and some bacteria. It is the process that links the energy of the sun with life on Earth. In this process, photosynthetic organisms utilize solar energy to manufacture sugars.

5. **Organized Structure** All living organisms are composed of one or more cells, the basic structure of life. From the smallest **unicellular organism** to the largest **multicellular organism,** all show a high degree of organization and coordination. The simplest level of organization is seen in bacteria that are **prokaryotic cells** (fig. 1.4a). These cells are the most primitive type of cells known. All other organisms are composed of **eukaryotic cells**. In a eukaryotic cell, the **nucleus,** containing the hereditary material, is clearly visible (fig. 1.4b), and different metabolic activities are compartmentalized into specialized membrane-bound structures called **organelles.** Prokaryotic cells lack a discernible internal organization. Prokaryotes have no organized nucleus or other obvious membrane-bound structures, but they have hereditary material and carry out all the activities of life.
6. **Organic Composition** All living organisms are composed mainly of four types of compounds: **carbohydrates, proteins, lipids,** and **nucleic acids.** These are the chemicals of life.

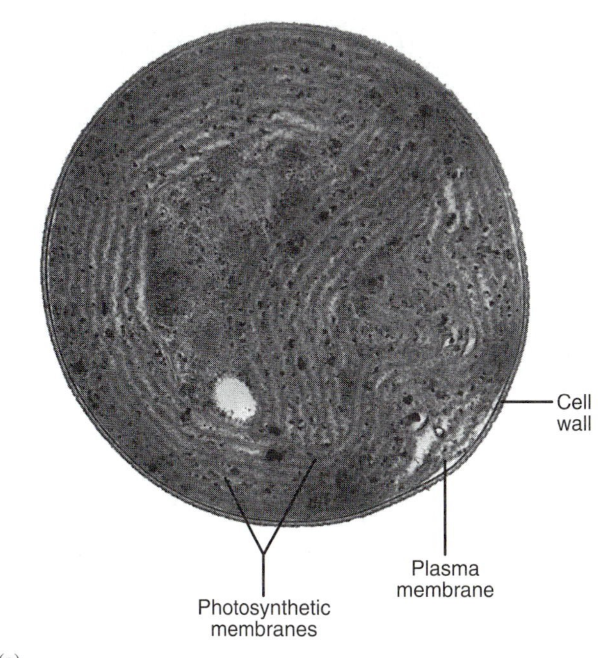

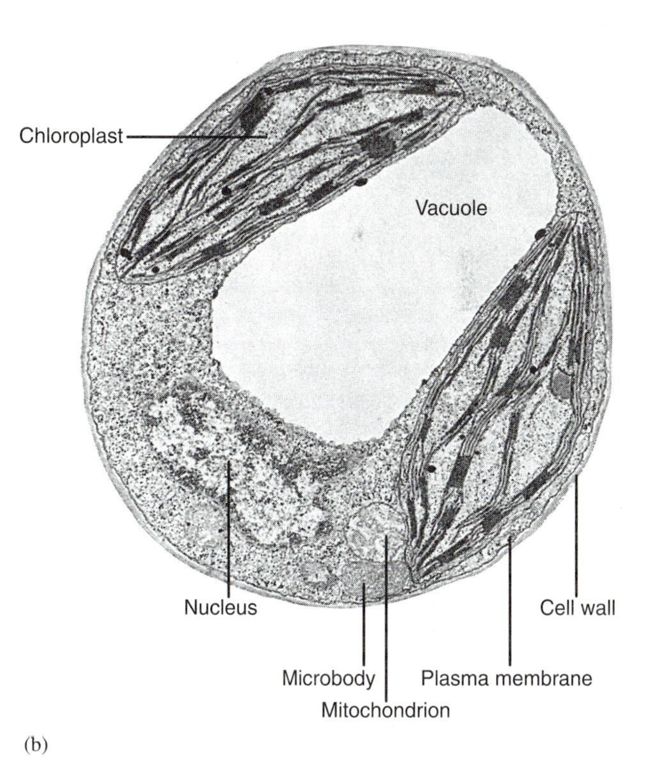

Figure 1.4 *Cellular organization. (a) Prokaryotic cell. Although this cyanobacterial cell does contain internal photosynthetic membranes, there are no membrane-bound organelles or nucleus. (b) Eukaryotic cell. This* Coleus *leaf cell shows a prominent nucleus and such membrane-bound organelles as chloroplasts, mitochondria, vacuole, and microbodies.*

Table 1.3
Essential Elements in Plants

Element	Symbol	Approximate Percent of Dry Weight
Macronutrients		
Carbon	C	45
Oxygen	O	45
Hydrogen	H	6
Nitrogen	N	1.5
Potassium	K	1.0
Calcium	Ca	0.5
Magnesium	Mg	0.2
Phosphorus	P	0.2
Sulfur	S	0.1
Micronutrients		
Chlorine	Cl	0.01
Iron	Fe	0.01
Manganese	Mn	0.005
Zinc	Zn	0.002
Boron	B	0.002
Copper	Cu	0.0006
Molybdenum	Mo	0.00001

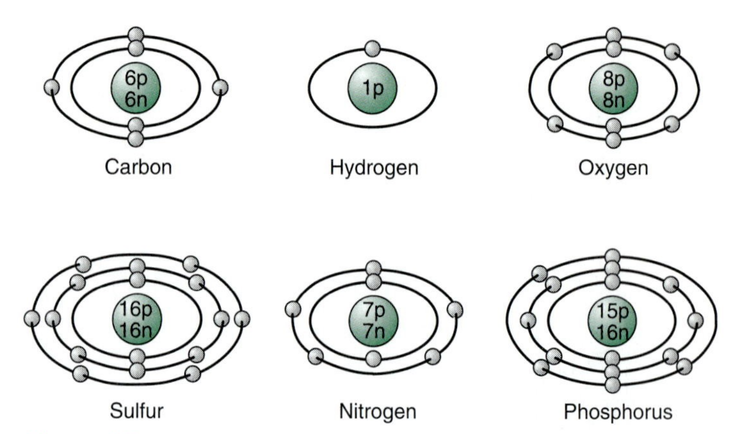

Figure 1.5 *Atomic structures of six atoms commonly found in organic molecules.*

Chemistry of Life

Atoms

All matter in the universe is composed of a limited number of substances called **elements;** there are 92 naturally occurring elements and about a dozen more that have been synthesized in the laboratory. Each element is assigned a symbol consisting of one or two letters. Of all the elements, nine make up approximately 99.5% of living matter: **carbon, oxygen, hydrogen, nitrogen, potassium, magnesium, calcium, phosphorus,** and **sulfur.** These are not the only elements required by life, but the others, known as **trace elements,** or **micronutrients** are required in extremely small amounts (table 1.3).

An **atom** is the smallest part of an element that still retains the chemical and physical characteristics of that element. Each atom is made up of subatomic particles; the three most important of these are **protons, neutrons,** and **electrons.** Protons and neutrons occupy the central position in atoms, a region referred to as the **atomic nucleus,** while electrons are in orbit around the nucleus. Protons and electrons are both electrically charged particles, with protons carrying a positive (+) charge and electrons a negative (–) one. Neutrons, however, have no electrical charge. Because of the presence of protons, the atomic nucleus has an overall positive charge, and it is the attraction of unlike charges that helps keep the negatively charged electrons in orbit.

Each element is assigned an **atomic number,** which refers to the number of protons in the atomic nucleus, ranging from the smallest atom hydrogen with an atomic number of 1 to the largest naturally occurring atom, uranium, with an atomic number of 92. All atoms of an element have the same atomic number. Atoms are also characterized by their **atomic mass** or **mass number,** which is determined by the number of protons and neutrons in the atomic nucleus. Protons and neutrons have approximately the same mass, each equal to about 1 **dalton (atomic weight unit),** while the mass of electrons is negligible and not usually considered. Thus, the mass of an atom is concentrated in the nucleus (fig. 1.5).

While the atomic number is constant for an element, the mass number may vary, since the number of neutrons may differ among atoms of the same element. **Isotopes** are atoms of an element with different mass numbers. For example, the most common isotope of carbon is ^{12}C with 6 protons and 6 neutrons; however, another isotope of carbon is ^{14}C with 6 protons and 8 neutrons.

Some isotopes, such as ^{14}C, are **radioactive;** their atomic nuclei are unstable and decay, emitting radiation. Radioactive isotopes are useful in research and have been used in dating fossils and artifacts. The time period of some early agricultural societies has been determined by this method. Radioactive isotopes have other applications in biology as tracers in metabolic pathways. For example, ^{14}C has been used to trace the path of carbon dioxide in photosynthesis.

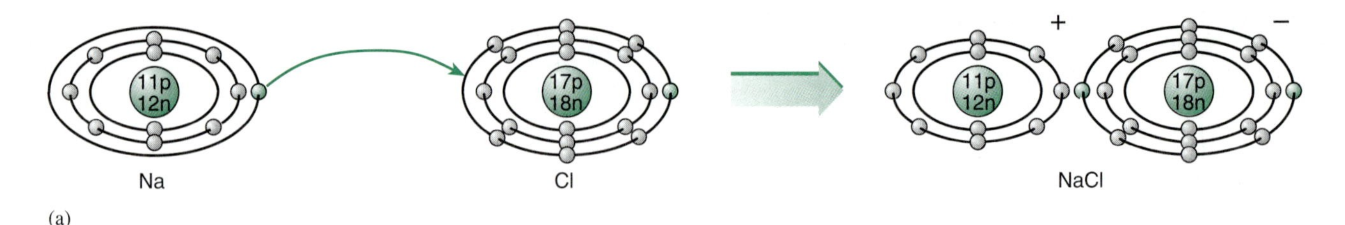

(a)

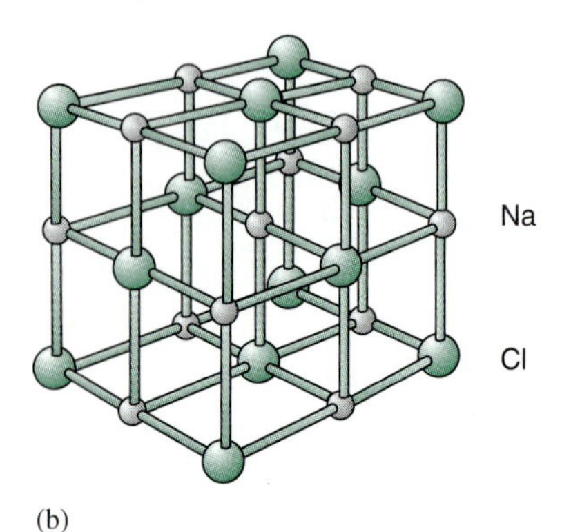

(b)

Figure 1.6 *Ionic bonding in table salt, NaCl. (a) The sodium ion donates one electron to the chlorine atoms. The two oppositely charged atoms bond to form the compound NaCl. (b) The ions are arranged in a repeating pattern.*

Molecules and Bonding

Most atoms do not occur singly in nature but exist combined with like or different atoms as **molecules.** A molecule of oxygen contains two atoms of oxygen chemically bonded together and represented by the symbol O_2. A molecule of water, H_2O, contains two atoms of hydrogen and one of oxygen. A **compound** like water is any substance formed by two or more elements in a definite proportion.

Chemical bonds are forces that hold atoms together; three common types of bonds are **ionic, covalent,** and **hydrogen.** Atoms are naturally electrically neutral, having an equal number of positively charged protons and negatively charged electrons. When an atom gains or loses one or more electrons, it becomes a charged particle known as an **ion.** The number of electrons lost or gained is specific for each element and is known as **valence.** If an atom loses electrons, it becomes positively charged because the number of protons exceeds the number of electrons. Conversely, if an atom gains one or more electrons it develops a negative charge because the number of electrons is greater. Ions are shown with the charge indicated as a superscript next to the chemical symbol; for example, sodium atoms can lose an electron, leaving 11 protons and 10 electrons, resulting in a sodium ion with a single positive charge (symbolized by Na^+). Chlorine atoms can gain an electron and thereby have an overall negative charge with 17 protons and 18 electrons (Cl^-). Calcium ions (Ca^{++}) have an overall positive charge of 2 since 2 electrons are lost, leaving 20 protons and 18 electrons.

When two oppositely charged ions are in contact, they are strongly attracted to each other. This attraction is referred to as an ionic bond. Common table salt, sodium chloride (NaCl), forms as a result of an ionic bond between oppositely charged Na^+ and Cl^- (fig. 1.6). In water, ionic compounds tend to dissociate, separating into the component negative and positive ions. The minerals that plants require for growth and development are absorbed as ions from soil water. In addition, many biologically important molecules occur as ions in living tissues.

Covalent bonds form when two atoms share electron pairs. Covalent bonds are strong bonds that are not easily broken. In the oxygen molecule, two electron pairs are shared by each atom making a double covalent bond; the covalent bonds between the atoms are represented by O=O (fig. 1.7a). Most biologically important molecules are formed from various combinations of the elements carbon, hydrogen, and oxygen joined by covalent bonds. The number of covalent bonds formed by an atom is usually specific for an element; for example, each carbon atom must form four covalent bonds (fig. 1.7c).

When electron pairs are shared between atoms of the same element such as O_2, H_2, and N_2, they are shared equally. These covalent bonds are referred to as **nonpolar covalent bonds.** When electrons are shared between two atoms that differ in their ability to attract electrons, the electrons are shared unequally. This creates a **polar covalent bond** and the atom with the greater attraction for the electrons develops a slight negative charge. Polar covalent bonds are found in water molecules (fig. 1.8a). The electrons are pulled more closely to the oxygen atom, giving the oxygen a slight negative charge and each hydrogen atom a slight positive charge.

Attractions form between each weakly (–) charged oxygen of a water molecule and weakly (+) charged hydrogens from adjacent molecules (fig. 1.8b). These attractions are known as **hydrogen bonds.** Hydrogen bonds

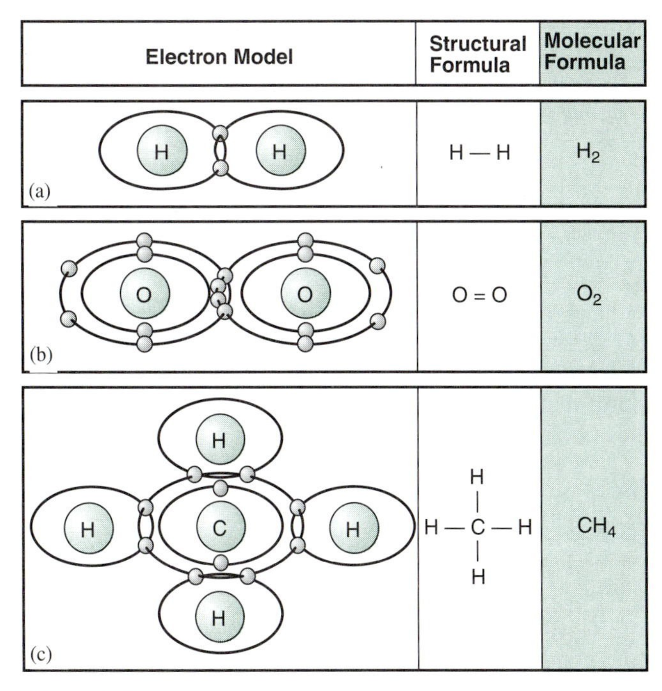

Figure 1.7 *Covalently bonded molecules share electrons. (a) A molecule of hydrogen is formed when two atoms share one pair of electrons. (b) A molecule of oxygen is formed when two atoms share two electron pairs. (c) A molecule of methane consists of one carbon atom covalently bonded to four hydrogen atoms.*

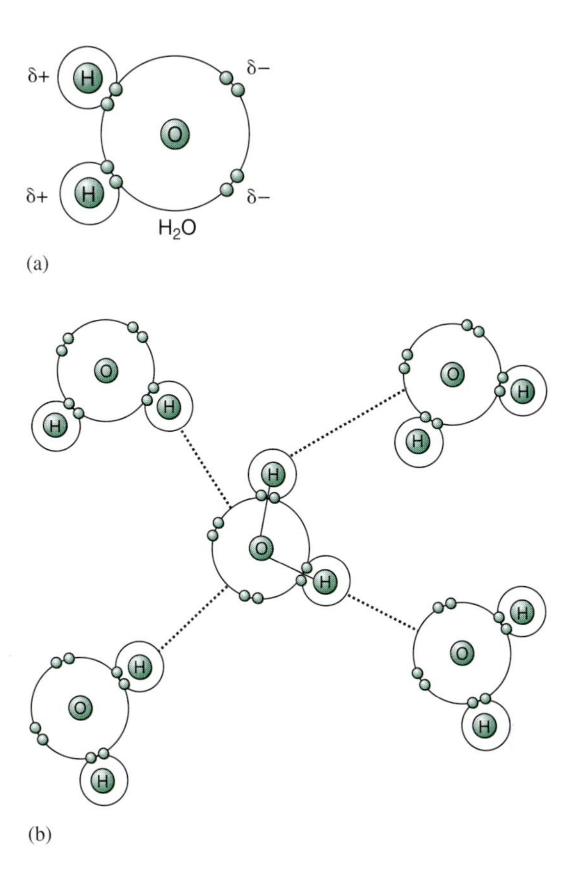

Figure 1.8 *Water molecules. (a) The shared electrons are pulled toward the oxygen atom creating a polar covalent bond, giving oxygen a slight negative charge and each hydrogen a slight positive charge. (b) The weakly charged regions of the water molecule are attracted to the opposite charges on adjacent water molecules. These attractions are known as hydrogen bonds.*

occur in other molecules also, forming between the weakly charged oxygen or nitrogen of one molecule and the weakly charged hydrogen of another. Hydrogen bonds are weak bonds. They are easily made and easily broken; but when they occur in number, they add stability to a molecule. Hydrogen bonds are important for the structure, and therefore the function, of both proteins and nucleic acids.

Molecules of Life

The chemical composition of life is based on the element carbon, and the classes of carbon compounds known as carbohydrates, lipids, proteins, and nucleic acids. Carbon is covalently bonded to other carbon atoms to create carbon chains that form the skeletons of these molecules. These four classes of compounds are the most important molecules in living organisms and often exist as large complex **macromolecules.** Carbohydrates, lipids, and proteins also constitute the major nutrients in the human diet and are discussed in detail in Chapter 10.

Carbohydrates

Carbohydrates, which include **sugars** and **starches** as well as **cellulose,** are composed of carbon, hydrogen, and oxygen (fig. 1.9). Many carbohydrates, especially **glucose,** are sources of energy for cells, while other carbohydrates, such as cellulose, are structural materials. The smallest carbohydrates are the **monosaccharides,** or the simple sugars. These contain only one sugar molecule; the most familiar examples of monosaccharides are glucose and **fructose.** The general formula for monosaccharides is $C_nH_{2n}O_n$, with n equal to 3, 4, 5, 6, or 7. Both glucose and fructose have the same general formula $C_6H_{12}O_6$ but have different arrangements of the atoms and react differently.

Two sugar molecules chemically bonded together are known as **disaccharides.** Common table sugar, **sucrose,** is a disaccharide composed of one glucose molecule and one fructose molecule. Although most plants transport carbohydrates from one part of the plant to another in the form of sucrose, only a few plants actually store this molecule. Most sucrose for table use comes from either sugar cane or

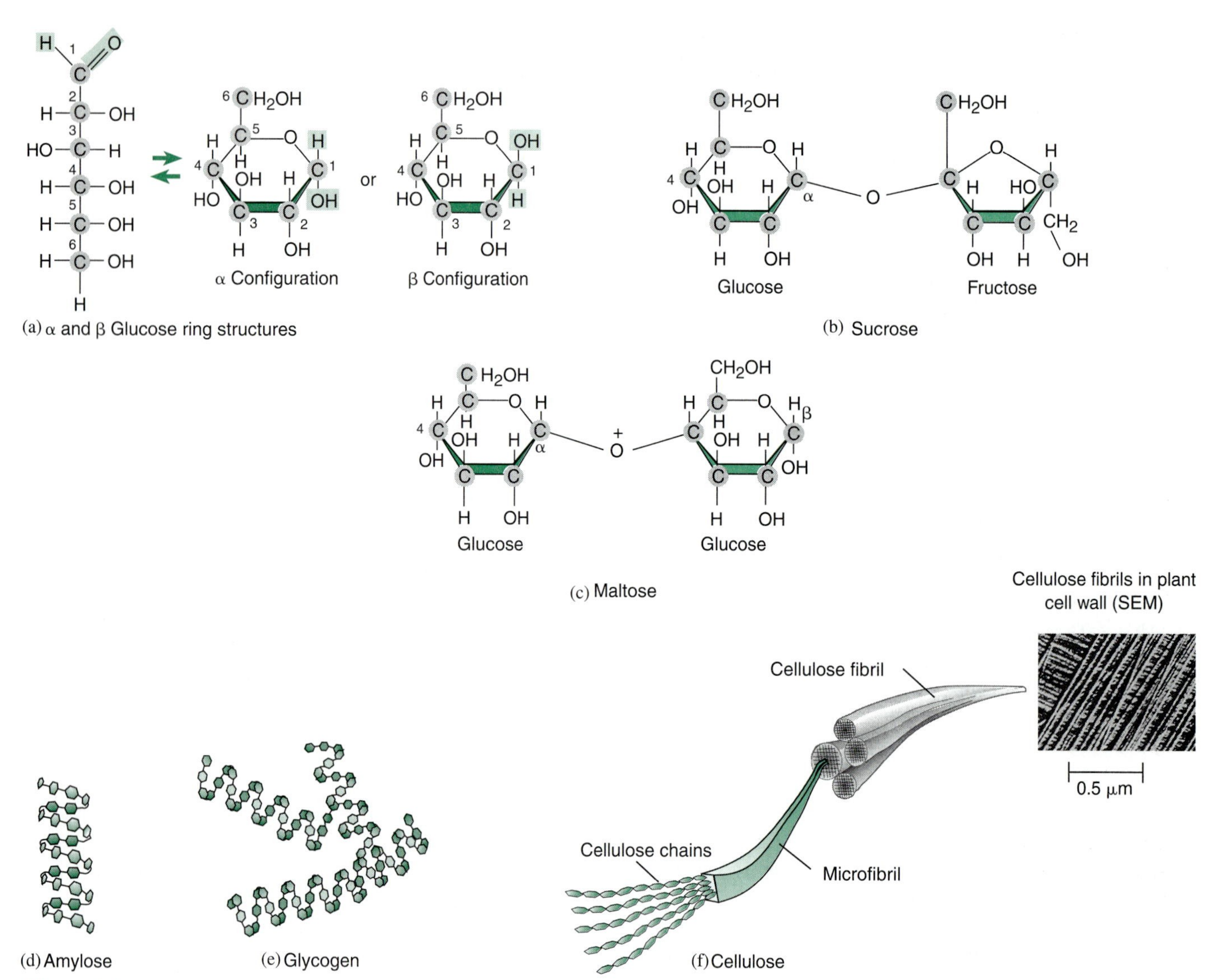

Figure 1.9 *Carbohydrates. (a) Monosaccharides are known as simple sugars. Glucose, the most abundant monosaccharide, can exist in a straight chain or ring configuration. (b) Sucrose, a disaccharide, composed of a molecule of glucose and a molecule of fructose bonded together. (c) Maltose, another disaccharide, forms from two glucose molecules. (d)–(f) Polysaccharides. All three molecules are made from thousands of glucose molecules but have different bonding arrangements. (d) Starch or amylose found as a storage molecule in green plants; (e) Glycogen found as a storage molecule in animals, bacteria, and fungi; (f) Cellulose, a structural component of plant cell walls, scanning electron micrograph.*

sugar beet (fig. 1.9). **Maltose,** another disaccharide, contains two glucose molecules. This sugar is seldom found free in plants, but is a breakdown product of starch and an important ingredient in the brewing of beer.

Polysaccharides consist of many thousands of sugar molecules bonded together. The three most common polysaccharides are starch, glycogen, and cellulose. These three are all composed of repeating glucose molecules but have different chemical bonding and arrangements (fig. 1.9). Both glycogen and starch are storage molecules; starch occurs in green plants, while glycogen is found in fungi, bacteria, and animals. Cellulose is a structural component of plant cell walls while chitin, a more complex molecule, is the major structural component in fungal cell walls.

Proteins

Carbon, hydrogen, oxygen, nitrogen, and sulfur are the elements found in proteins. Proteins are large complex molecules composed of long chains of **amino acids.**

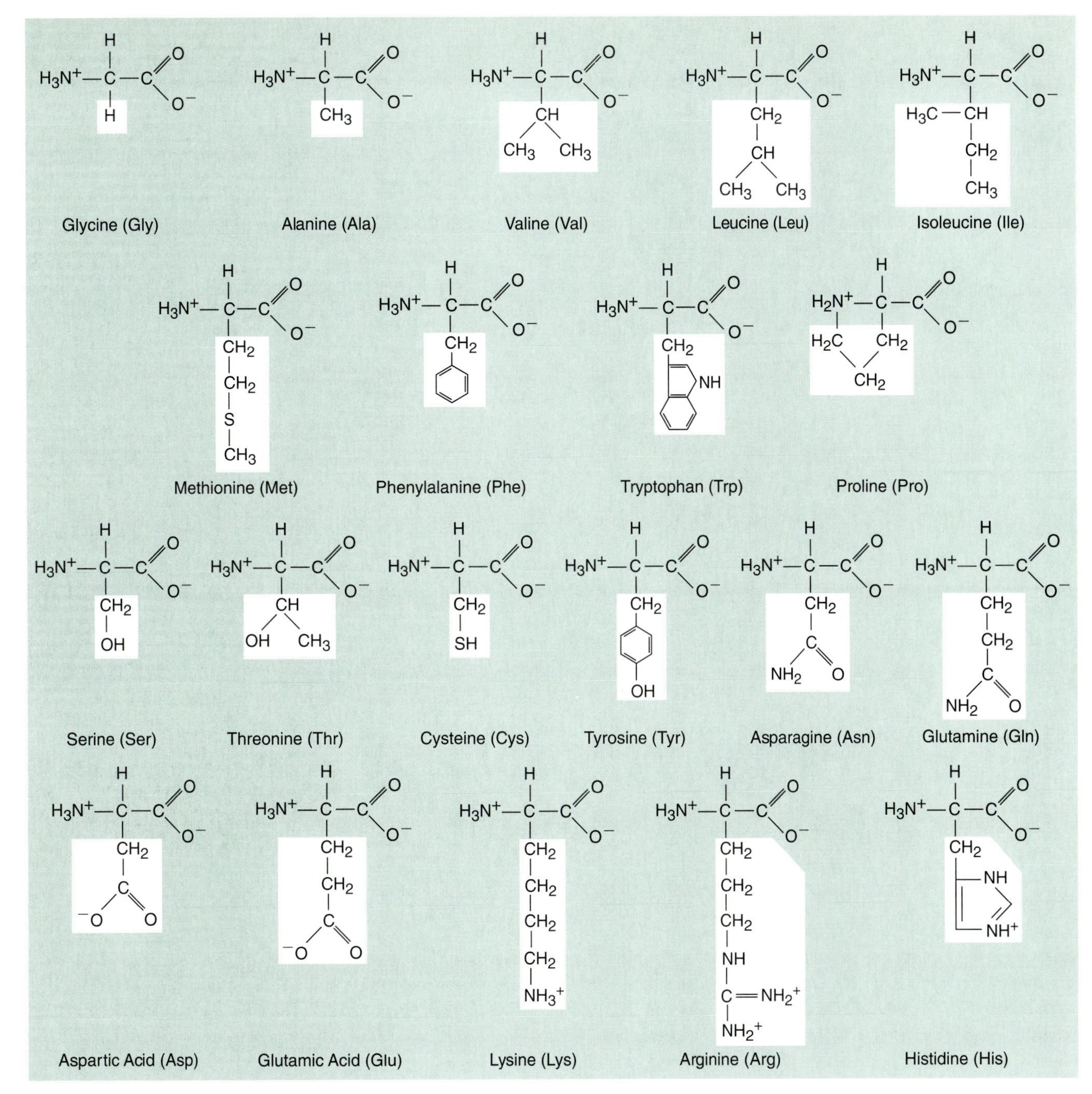

Figure 1.10 *Amino acids. There are 20 naturally occurring amino acids. All have the same backbone (N-C-C) and differ only in the side group attached to the center carbon.*

There are 20 different amino acids that are common to all life forms (fig. 1.10). The number and arrangement of these 20 amino acids result in an infinite variety of proteins. In the complete protein structure, the amino acid chain is twisted and folded into a specific three-dimensional shape (fig. 1.11). Proteins have many functions; they can serve as enzymes (biological catalysts), structural materials, regulatory molecules, or transport molecules, to name a few of their many roles.

Lipids

Lipids are a diverse group of substances largely composed of only carbon and hydrogen. Small amounts of oxygen may occur in some lipids. There are many

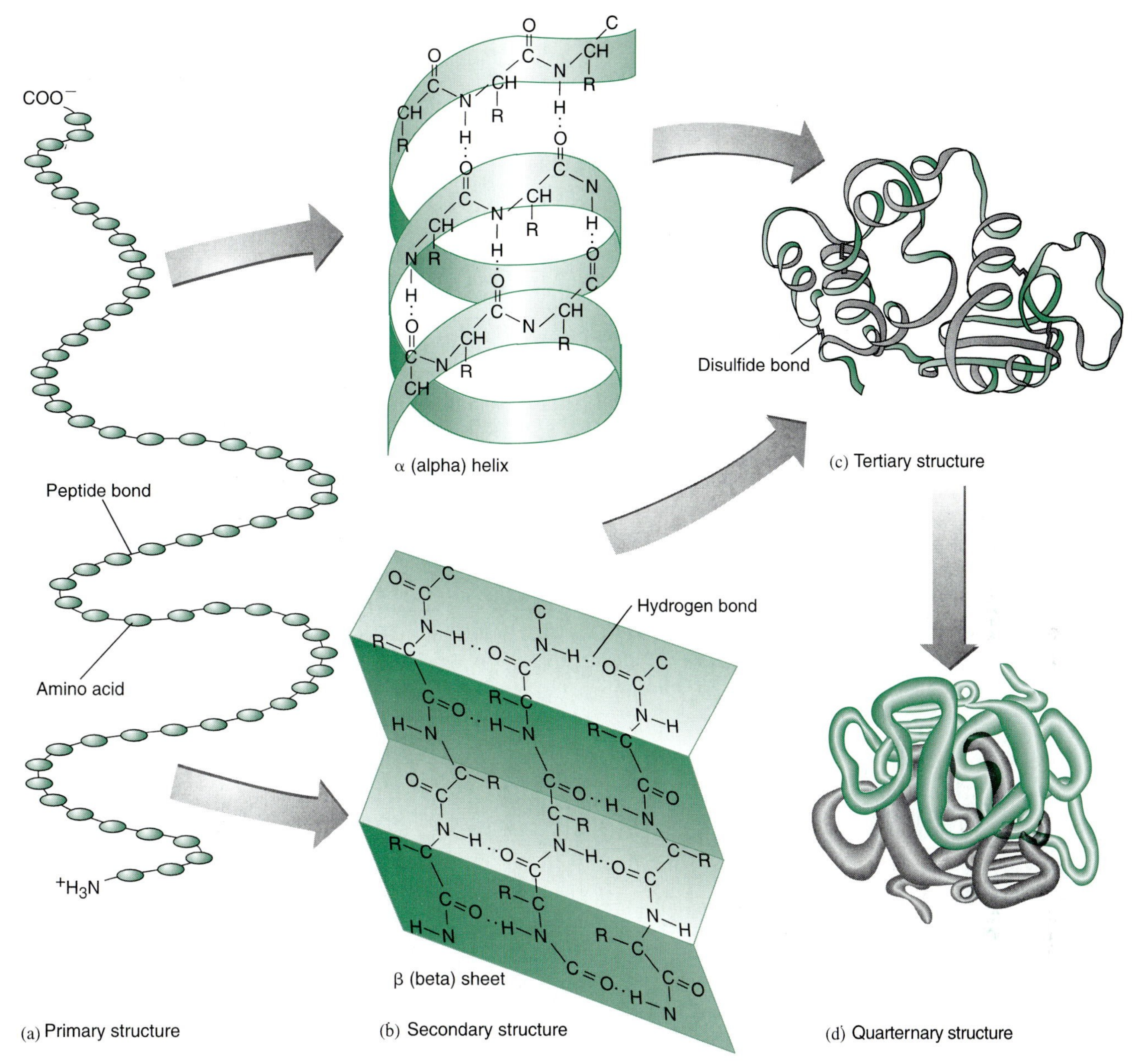

Figure 1.11 *Protein structure. (a) Primary protein structure consists of the sequence of amino acids bonded together by peptide bonds to make a polypeptide chain. (b) Secondary protein structure consists of a helix or pleated sheet that spirals or folds the polypeptide chain. This is stabilized by hydrogen bonds. (c) Tertiary structure is a twisting and folding of the molecule. (d) Quaternary structure contains more than one polypeptide chain, each with its own tertiary structure.*

different types of lipids; what they have in common is that they are insoluble in water. Lipids include such compounds as **triglycerides, phospholipids, waxes,** and **steroids** (fig. 1.12). Different types of lipids have different functions. They can be important as sources of energy (triglycerides), as structural components of cell membranes (phospholipids and cholesterol), or as hormones (steroids). Triglycerides, better known as fats and oils, function as food reserves in many organisms. Fats are the usual energy reserves in animals, while seeds of certain plants store appreciable amounts of oil.

Nucleic Acids

Nucleic acids contain carbon, hydrogen, oxygen, nitrogen, and phosphorus. Nucleic acids are composed of repeating units called **nucleotides,** which consist of a sugar (either **ribose** or **deoxyribose**), a **phosphate** group (PO_4), and a **nitrogenous base** (either a **purine** or a

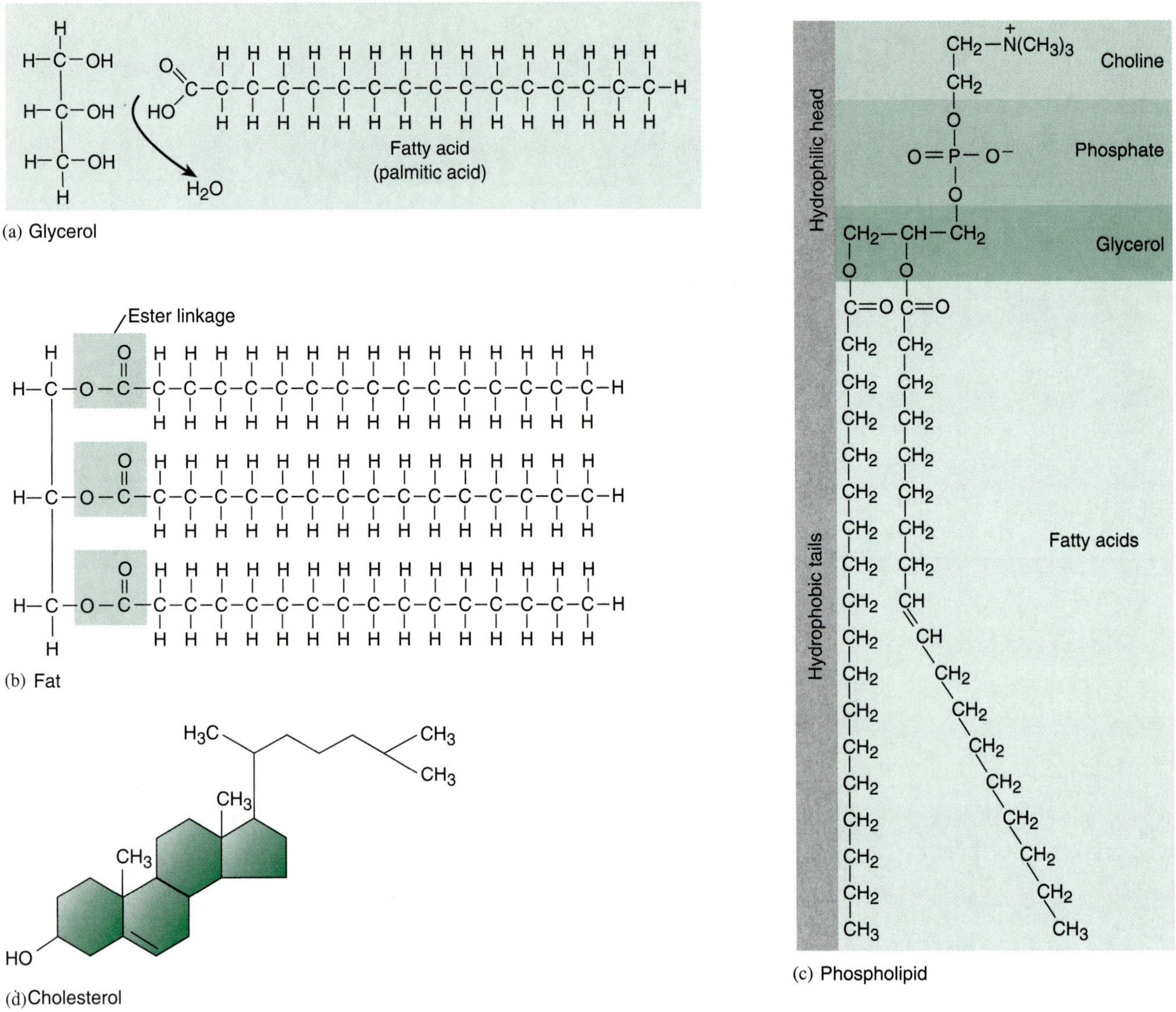

Fatty acid (palmitic acid)

(a) Glycerol

Ester linkage

(b) Fat

(d) Cholesterol

Choline

Phosphate

Glycerol

Fatty acids

Hydrophilic head

Hydrophobic tails

(c) Phospholipid

Figure 1.12 *Lipids. (a) The building blocks of fats or oils consist of glycerol and fatty acids. (b) A fat or triglyceride formed from glycerol and three fatty acids. (c) A phospholipid is formed from glycerol, two fatty acids, a phosphate group, and choline. (d) Cholesterol, one of many steroids, is a more complex lipid. The four-ring steroid backbone is shaded.*

pyrimidine base) (fig. 1.13). Five different types of nucleotides occur, depending on the type of base. There are two purine bases, **adenine** and **guanine,** and three pyrimidine bases, **thymine, cytosine,** and **uracil. Deoxyribonucleic acid (DNA)** and **ribonucleic acid (RNA)** are the two types of nucleic acids. Nucleotides containing adenine, guanine, and cytosine occur in both DNA and RNA. Thymine nucleotides only occur in DNA, while uracil replaces thymine in RNA. Thus, both DNA and RNA contain four types of nucleotides, two purines and two pyrimidines.

It is the sequence of these nucleotide bases in the DNA molecule that is the essence of the genetic code. DNA is the hereditary material of life, unique in its ability to replicate itself and, thus, pass on the genetic code from one generation to the next. DNA, often called the **double helix** (fig. 1.13), exists as a double-stranded molecule that is twisted into a helix. The sides of the helix are made of alternating sugar (deoxyribose) and phosphate groups, and the nitrogenous bases are found as purine-pyrimidine pairs (adenine always pairs with thymine and guanine with cytosine) in the interior of the helix.

Unlike DNA, RNA consists of a single strand, with ribose as part of the sugar-phosphate backbone. RNA is involved in the manufacture of proteins by transcribing and then translating the instructions coded on the DNA molecule. The sequence of bases in DNA is ultimately translated into the sequence of amino acids in proteins.

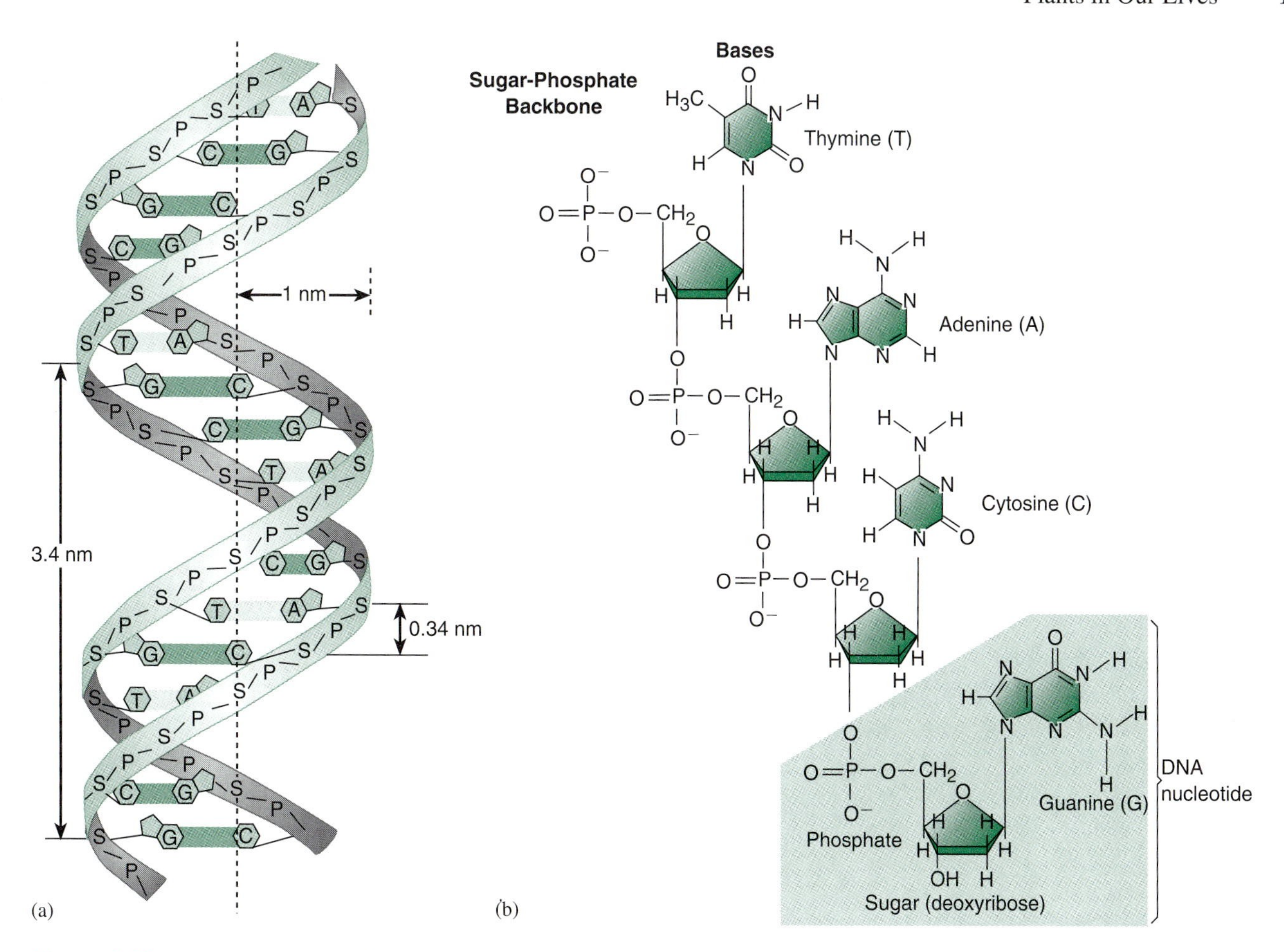

Figure 1.13 *DNA molecule. (a) The double helix with the sugar-phosphate backbone making up the sides and the paired nitrogenous bases the interior. (b) Structures of the nucleotides (sugar, phosphate, and base) that make up the DNA molecule.*

Chapter Summary

1. Angiosperms, also called flowering plants, supply humanity with the essentials of life. The food staples of civilization, wheat, rice, and corn, are all angiosperms, as are almost all other food crops. Other angiosperm products that have shaped modern society include cloth, hardwood, herbs and spices, beverages, drugs, perfumes, vegetable oils, gums, and rubber.
2. Fungi are also an economically important group of organisms that include molds, mildews, yeast, and mushrooms. These organisms are second only to the angiosperms in their significance to society. Fungi provide many beneficial items such as penicillin, beer, wine, cheese, edible mushrooms, and leavened bread. A negative aspect of their economic importance is the impact of fungal disease and spoilage. The most serious diseases of our crop plants are caused by fungi and result in billions of dollars in crop losses each year.
3. All living organisms have the capacity to grow and reproduce, the ability to respond, the ability to evolve and adapt, a metabolism based on ATP, an organized structure, and organic composition.
4. Nine elements (carbon, oxygen, hydrogen, nitrogen, potassium, magnesium, calcium, phosphorus, and sulfur) make up 99.5% of living matter. The chemical nature of living matter is based on the element carbon and its ability to covalently bond to other carbon atoms to form the skeletons of carbohydrates, lipids, proteins, and

nucleic acids, the molecules of life. Monosaccharides, especially glucose molecules, serve as sources of energy for cells, while polysaccharides have storage and structural functions. Proteins, composed of long chains of amino acids, have many functions as enzymes, structural molecules, regulatory molecules, and transport molecules. Lipids are a diverse group of compounds that are insoluble in water. Some serve as energy reserves, others as structural materials or hormones. DNA serves as the hereditary material of life by encoding information in the sequences of bases. RNA functions in the manufacture of proteins by transcribing and translating information encoded in the DNA molecule.

Review Questions

1. What are the characteristics of the angiosperms? the fungi?
2. What characteristics are common to all forms of life?
3. In what ways are plants essential to life on Earth?
4. Distinguish the three types of chemical bonds. Give examples.
5. Describe the levels of protein structure.
6. What are the differences between monosaccharides, disaccharides, and polysaccharides?
7. How do triglycerides and phospholipids differ in structure and function?
8. Define the following terms: atom, proton, electron, neutron, molecule, and isotope.
9. Can you find other examples of materials that mimic biological designs?

Further Reading

Campbell, Neil A. 1993. *Biology,* 3rd Edition, Benjamin Cummings Publishing, Redwood City, CA.

Mader, Sylvia. 1993. *Biology,* 4th Edition. Wm. C. Brown Publishers, Dubuque, IA.

Mauseth, James D. 1991. *Botany: An Introduction to Plant Biology,* Saunders College Publishing, Philadelphia, PA.

Moore, Randy and W. Dennis Clark. 1995. *Botany: Plant Form and Function.* Wm. C. Brown Publishers, Dubuque, IA.

Raven, Peter H. and George B. Johnson. 1992. *Biology,* 3rd Edition, Mosby, St. Louis, MO.

Raven, Peter, Ray F. Evert, and Susan E. Eichhorn. 1986. *Biology of Plants,* 4th Edition. Worth Publishers, Inc., New York, NY.

Stern, Kingsley R. 1994. *Introductory Plant Biology,* 6th Edition. Wm. C. Brown Publishers, Dubuque, IA.

UNIT

II

Introduction to Plant Life: *Botanical Principles*

All plants, including this magnificent Victoria water lily, display a highly organized structure and complex physiology.

CHAPTER

2

THE PLANT CELL

Chapter Outline

Chapter Concepts

1. The Cell Theory establishes that the cell is the basic unit of life, that all living organisms are composed of cells, and that cells arise from preexisting cells.
2. Plant cells are eukaryotic, having an organized nucleus and membrane-bound organelles.
3. Substances can move in and out of cells by diffusion and osmosis.
4. Mitosis, followed by cytokinesis, results in two genetically identical daughter cells. Growth, replacement of cells, and asexual reproduction all depend on the process of cell division.

All plants (and every other living organism) are composed of cells. In some plants, the whole organism consists of a single cell, but angiosperms are complex multicellular organisms composed of many different types of cells. Plant cells are microscopic in size and typically range from 10 to 100 µm in length. This means that there would be between 254 to 2,540 of these cells to an inch (fig. 2.1). In Chapter 3 we will be looking at the variety of cells, but in this chapter we will focus on a composite angiosperm plant cell.

Early Studies of Cells

The first person to describe cells was the Englishman Robert Hooke in 1665. Hooke was examining the structure of cork with a primitive microscope and noticed that it was organized into small units that resembled the cubicles in monasteries where monks slept (fig. 2.2). These rooms were called "cells." He gave that name to the little compartments in cork and the term was eventually applied to mean the basic unit of life. Although the cork was not living, Hooke later looked at living plants and identified cells there also.

Other scientists in the late seventeenth and eighteenth centuries continued the microscopic examination and study of a variety of organisms. It was not until the midnineteenth century, however, that Matthias Schleiden and Theodor Schwann, and later Rudolf Virchow, firmly established the Cell Theory that recognizes the cell as the basic unit of life. The Cell Theory further states that all organisms are composed of cells and all cells arise from preexisting cells. This theory is one of the major principles in biology.

Although these early scientists were unable to identify many structures within a cell, today it is possible to magnify extremely small details of the cell using an electron microscope. This has greatly expanded our knowledge of cellular structure and function. The structures in a eukaryotic plant cell that are visible with an electron microscope are illustrated in Figure 2.3.

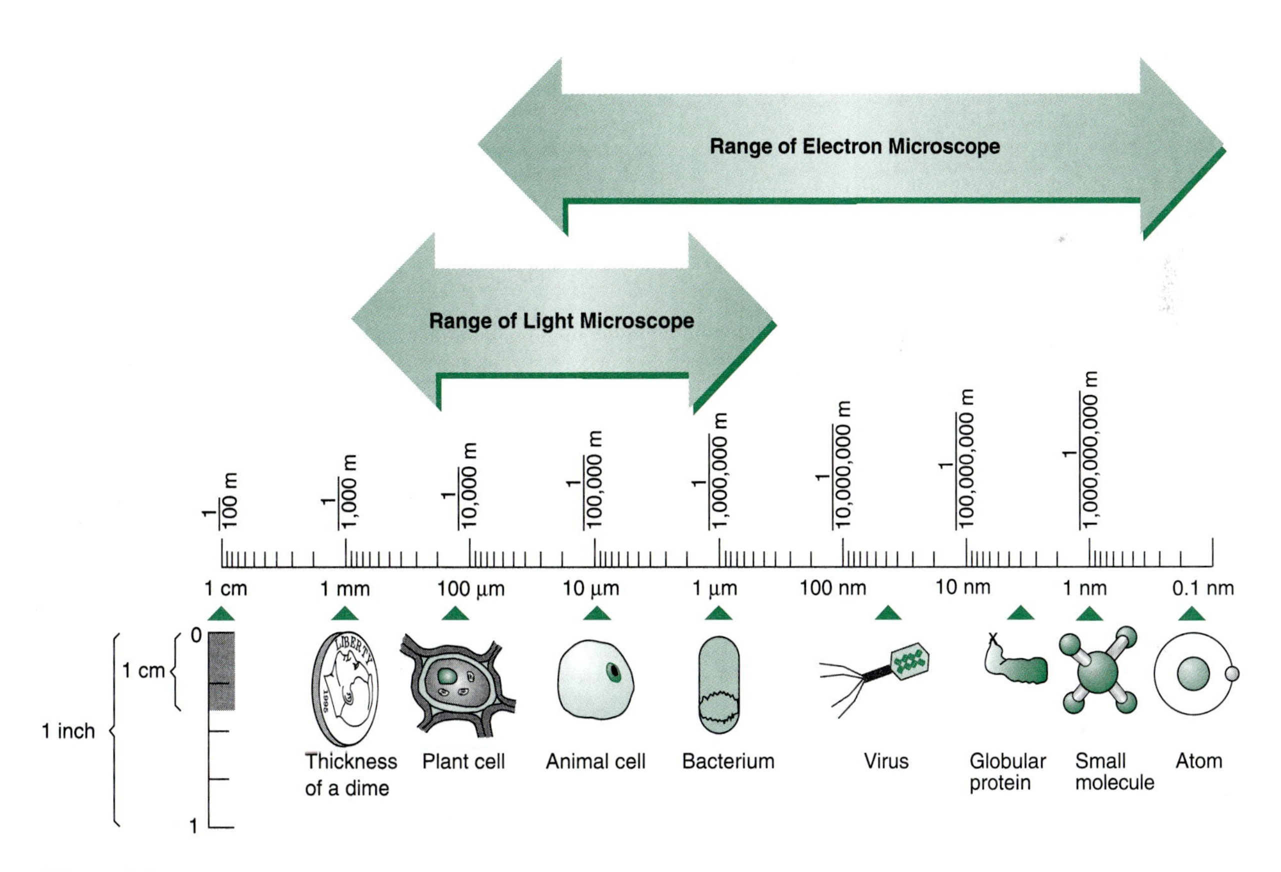

Figure 2.1 *Biological measurements. The scale ranges from 1 centimeter (0.01 meter) down to 0.1 nanometer (0.0000000001 meter).*

(a)

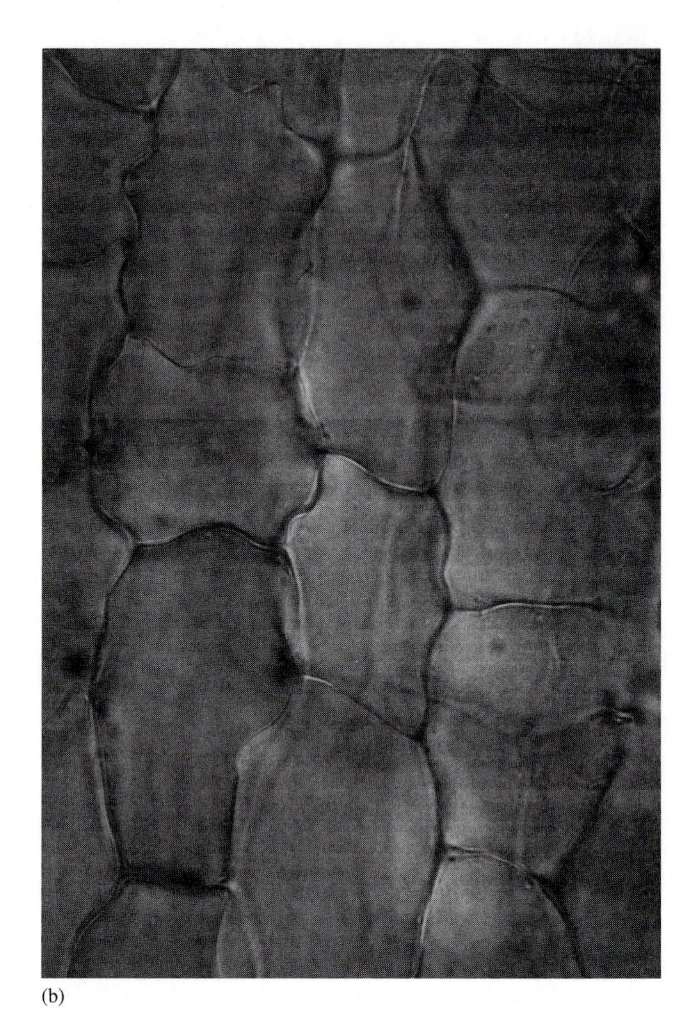

(b)

Figure 2.2 *(a) Robert Hooke's microscope; (b) Cork cells, × 1500.*

CELL WALL

The plant cell is composed of a **cell wall** that surrounds all the other parts of the cell, collectively called the **protoplast.** The cell wall material is formed by the protoplast. Plant cell walls may consist of one or two layers. The first layer, or the **primary wall,** is formed early in the life of a plant cell. It is composed of a number of polysaccharides, principally cellulose. The cellulose is in the form of fibrils, extremely fine fibers, which can be seen with an electron microscope (fig. 2.4). These fibrils are embedded in a matrix of other polysaccharides.

The **secondary wall** is laid down internal to the primary wall. In cells with secondary walls, lignin, a very complex organic molecule, is a major component of the walls, in addition to the cellulose and other polysaccharides. Considering all the plant material on Earth, it is not surprising that cellulose is the most abundant organic compound, with lignin a close second.

Only certain types of plant cells have secondary walls, usually just those specialized for support, protection, or water conduction. Lignin is known for its toughness; it gives wood its characteristic strength and also provides protection against attack by pathogens (disease-causing agents) and consumption by herbivores. To compare the characteristics of primary and secondary walls, imagine a chair made of lettuce leaves instead of wood!

Although the cell wall is one or two layers thick, it is not a solid structure. Minute pores, or **pits,** exist; most of these are large enough to be seen with the light microscope. Pits allow for the transfer of materials through cell walls. Cytoplasmic connections between adjacent plant cells often occur. These are called **plasmodesmata** and pass through the pits in the cell wall. These allow for the movement of materials from cell to cell (fig. 2.5).

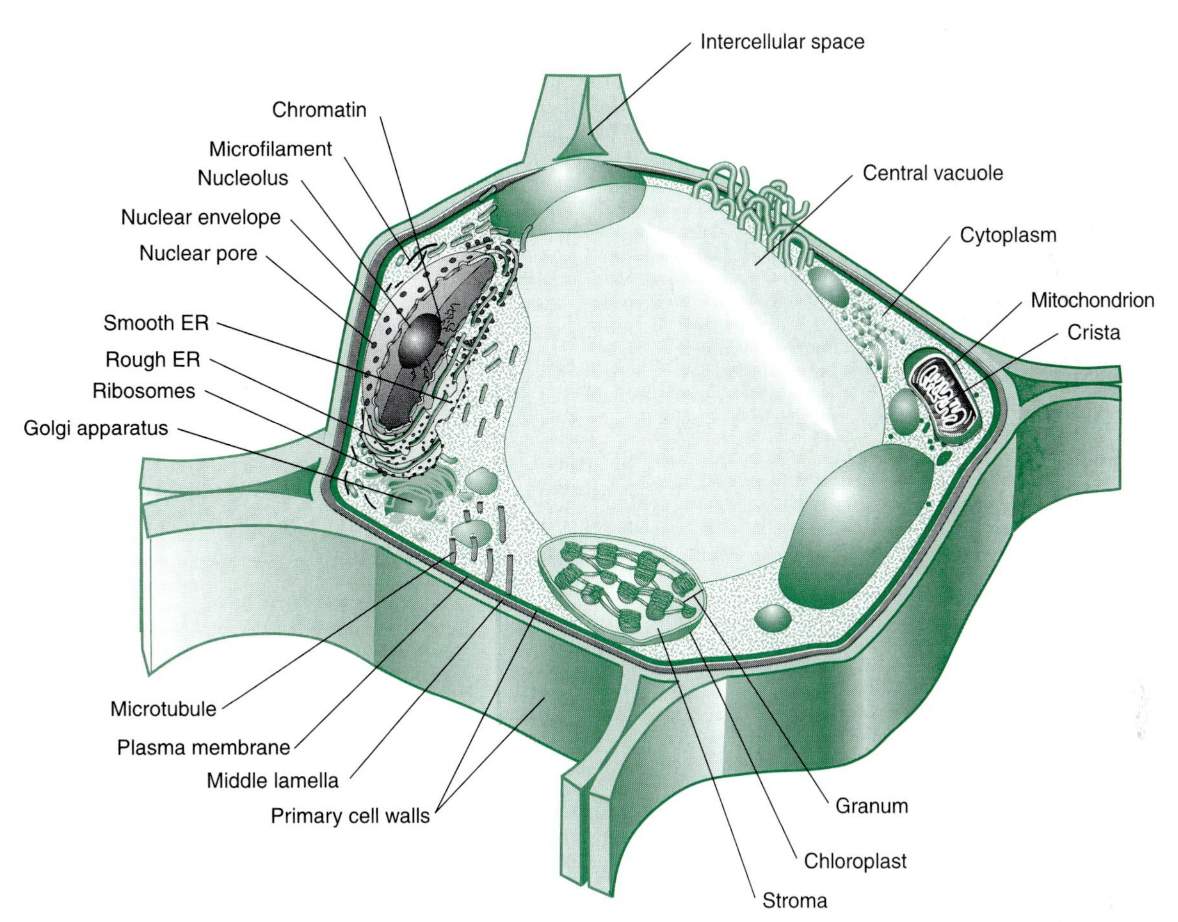

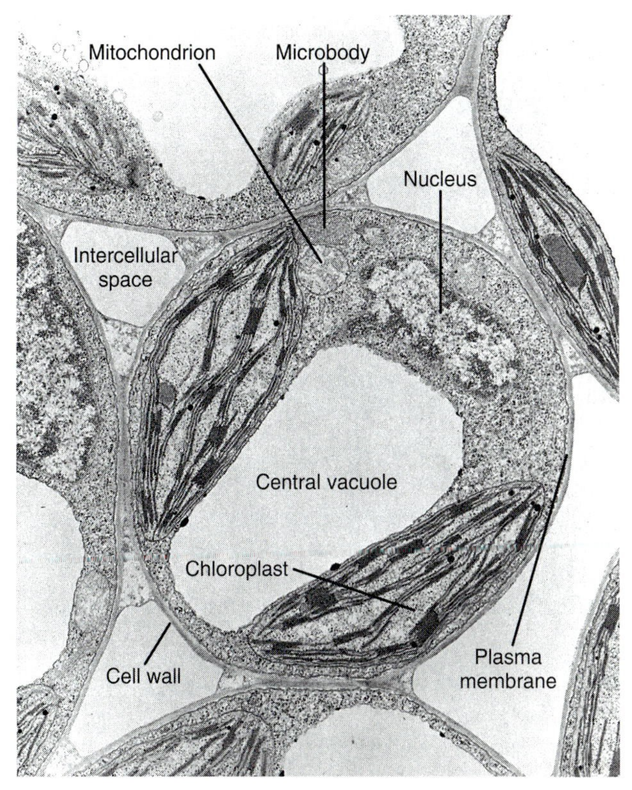

Figure 2.3 *Plant cell structure. (a) Diagram of a generalized plant cell as seen under an electron microscope; (b) Electron micrograph of a plant cell, ×20,000.*

A sticky substance called the **middle lamella** (*see* fig. 2.3) can be found between the walls of adjacent plant cells. This acts similar to a cellular cement, glueing cells together. It is composed of **pectins,** the additive often used in making fruit jellies.

The Protoplast

The **protoplast** is defined as all of the plant cell enclosed by the cell wall. It is composed of the nucleus plus the **cytoplasm.** The cytoplasm consists of various **organelles** (cellular structures) distributed in the **cytosol,** a matrix consisting of large amounts of water (in some cells up to 90%), proteins, other organic molecules, and ions. Also found in the cytoplasm is a network of proteinaceous **microtubules** and **microfilaments** that make up the **cytoskeleton,** a cellular scaffolding that helps support and shape the cell (*see* fig. 2.3).

Figure 2.4 *Electron micrograph of a cell wall. Note the cellulose fibrils.*

Membranes

The outermost layer of the protoplast is the plasma membrane, which is composed of phospholipids and proteins. The **fluid mosaic model,** the currently accepted idea of membrane structure, is shown in Figure 2.6. This model consists of a double layer of phospholipids with scattered proteins. Some of the proteins go through the lipid bilayer **(integral proteins),** while others are either on the inner or outer surface **(peripheral proteins).** Some of the membrane proteins and lipids have carbohydrates attached; they are called glycoproteins and glycolipids, respectively. The carbohydrates are usually short chains of about five to seven monosaccharides in size. Some have described this membrane model as "protein icebergs in a sea of lipids." The plasma membrane serves as a permeability barrier allowing some molecules (like water) to pass through but not others.

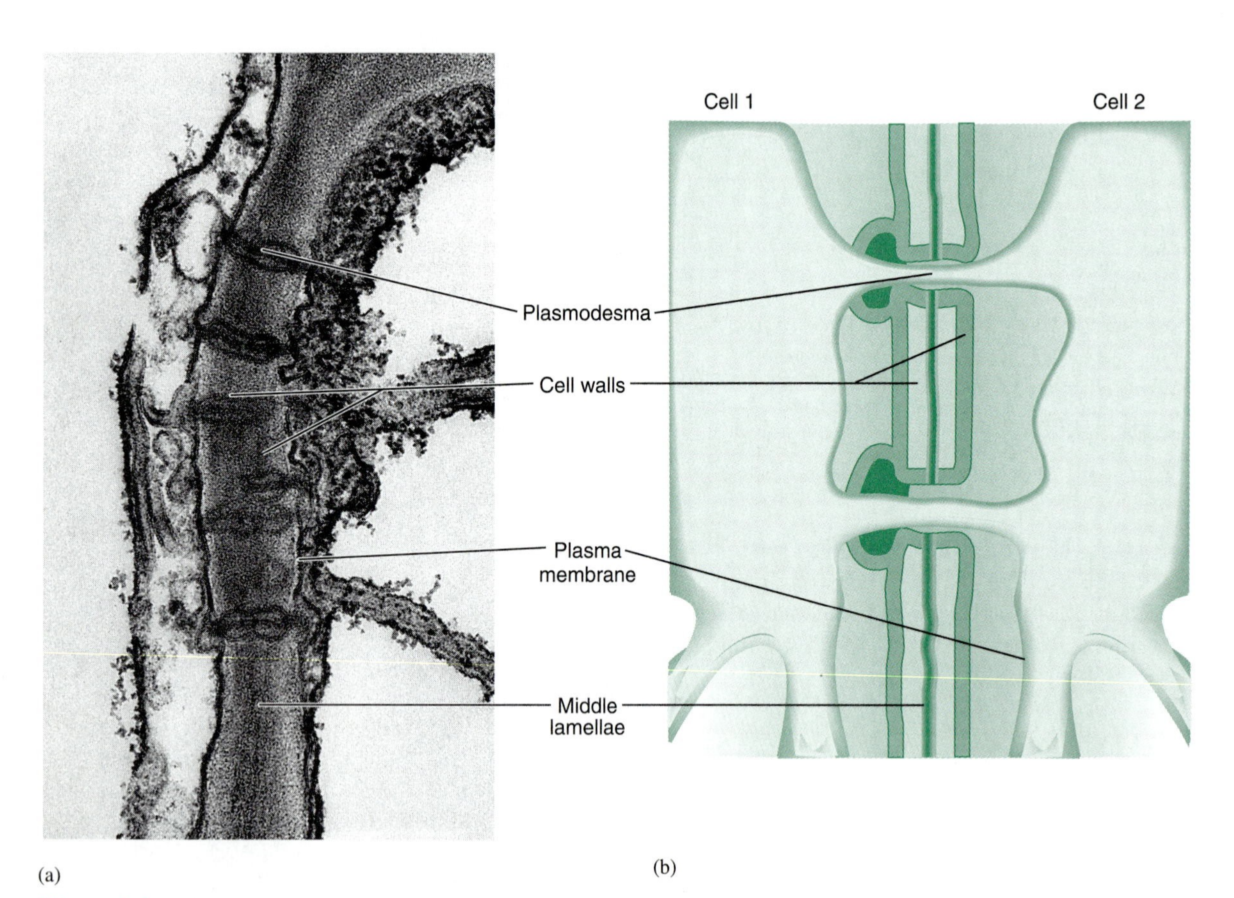

Figure 2.5 *Plasmodesmata permit the passage of materials from cell to cell. (a) Electron micrograph; (b) drawing.*

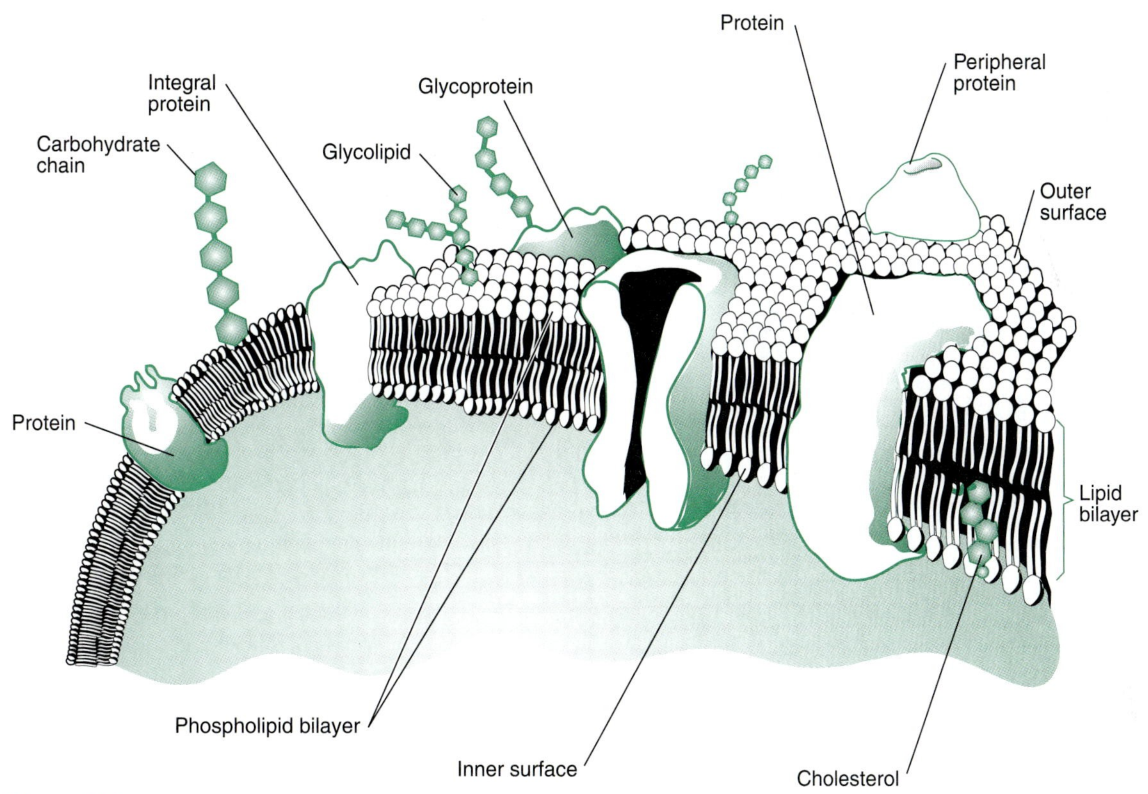

Figure 2.6 *Fluid mosaic model of the plasma membrane. The plasma membrane is composed of a phospholipid bilayer with embedded proteins.*

Organelles

A variety of organelles can be found in the plant cell (*see* fig. 2.3). Most of these are membrane-bound, with the membrane similar in structure and function to the plasma membrane. In leaf cells, the most distinctive organelles are the disk-shaped **chloroplasts,** which, in fact, are double membrane-bound. These organelles contain several pigments; the most abundant pigments are the chlorophylls, making leaves green. Carotenes and xanthophylls are other pigments present; these orange and yellow pigments are normally masked by the more abundant chlorophylls but become visible in the autumn when the chlorophylls break down before the leaves are shed. The pigments are located within the internal membranes of the chloroplasts and are most concentrated in membranous stacks called **grana** (sing., **granum**). The individual grana are interconnected and embedded in the **stroma,** a protein-rich environment. Although chloroplasts are easily seen with a light microscope, the internal organization, or ultrastructure, is only visible with an electron microscope. Photosynthesis occurs in the chloroplasts; this process allows plants to manufacture food from carbon dioxide and water using the energy of sunlight. More details on the ultrastructure of chloroplasts and the photosynthetic process are covered in Chapter 4.

Two other organelles that may be found in plant cells are **leucoplasts** and **chromoplasts** (fig. 2.7). Leucoplasts are colorless organelles that can store various materials, especially starch. The starch grains filling the cells of a potato are found in a type of leucoplast called an **amyloplast.** Chromoplasts contain orange, red, or yellow pigments and are abundant in colored plant parts like petals and fruits. The orange of carrots, the red of tomatoes, and the yellow of marigolds result from pigments stored in chromoplasts. Chloroplasts, leucoplasts, and chromoplasts are collectively called **plastids.**

A Closer Look 2.1

DMSO and the Lignin Connection

An interesting use of lignin is in the manufacturing of DMSO (dimethyl sulfoxide), a compound that has been promoted as a potent pain-reliever and treatment for a number of ailments such as arthritis, burns, toothaches, headaches, and muscle strains. DMSO was originally derived as a by-product of the wood pulping industry, but attracted little attention until 1950 when it became widely used as an industrial solvent. In the early 1960s its pharmaceutical applications were investigated when it was noticed that it rapidly penetrated the skin and cell membranes. The scientific community was excited by the possibilities of DMSO as a carrier for transporting drugs through the skin. Soon the pain-killing applications were discovered; applied to the site of pain or inflammation, DMSO could alleviate discomfort.

One drawback to DMSO usage is that patients rapidly develop garliclike breath that may last for several hours. Despite its potential, DMSO was banned for human use in the United States by the Food and Drug Administration (FDA) in 1965 when a study reported that high dosage resulted in eye damage and the death of several experimental animals. Disregarding this ban, people suffering from arthritis and skeletal-muscular injuries continue their unsanctioned use of DMSO while claiming relief from pain.

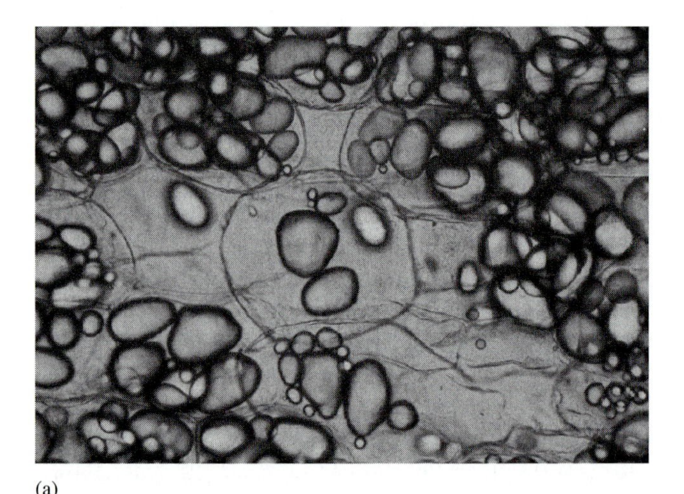

(a)

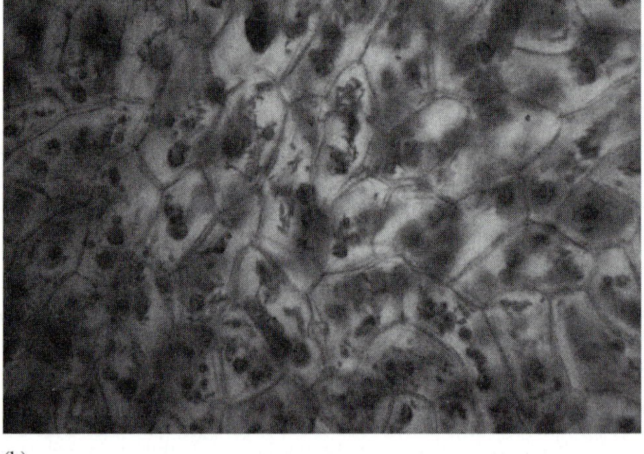

(b)

Figure 2.7 *Various leucoplasts and chromoplasts; (a) amyloplasts in white potato and (b) chromoplasts in bittersweet fruit.*

Another organelle bound by a double membrane is the **mitochondrion** (pl., **mitochondria**), which is the site of many of the reactions of cellular respiration in all eukaryotic cells (*see* fig. 2.3). Recall that cellular respiration is the metabolic process where glucose is chemically broken down to release energy in a useable form, ATP. Mitochondria are not easily studied with a light microscope; the electron microscope has made the study of their ultrastructure possible. The size, shape, and numbers of mitochondria vary between different types of cells, but all have a smooth outer membrane and an inner membrane with numerous infoldings called **cristae.** The compartment enclosed by the inner membrane is called the **matrix;** the matrix contains enzymes that are used in cellular respiration, while the cristae are the sites of ATP formation. Chapter 4 contains additional information on the role of mitochondria in cellular respiration.

Most mature plant cells (*see* fig. 2.3) are characterized by a large **central vacuole** that is separated from the rest of the cytoplasm by its own membrane. In some cells, the vacuole takes up 90% of the cell volume, pushing the cytoplasm into a thin layer against the plasma membrane. The vacuole contains the cell sap, a watery solution of sugars, salts, amino acids, proteins, and crystals, all separated from the cytoplasm by the vacuolar membrane. The cell sap is often acidic; the tartness of lemons and limes is due to their very acidic cell sap. Some of the substances in the vacuole are waste products, while others can be drawn upon when needed by the cell. The concentrations of these materials in the vacuole may become so great that they precipitate out as crystals. The leaves of the common house plant dumb cane (*Dieffenbachia* sp.) are poisonous due to the presence of large amounts of calcium oxalate crystals (*see* Chapter 21). If consumed, the crystals can injure the tissues of the mouth and throat, causing a temporary inability to speak—hence the common name dumb cane. Pigments can also be found in the vacuole; these are called anthocyanins and are responsible for deep red, blue, and purple colors of many plant organs, such as red onions and red cabbage. Unlike the pigments of the chloroplasts and chromoplasts, the anthocyanins are water soluble and are distributed uniformly in the cell sap.

An internal membrane system also occurs in plant cells (*see* fig. 2.3). This consists of the **endoplasmic reticulum (ER),** the **Golgi apparatus,** and **microbodies.** These structures are all involved in the synthesizing, packaging, and transporting of materials within the cell. The ER is a network of membranous channels throughout the cytoplasm. In some places the cytoplasmic side of the ER is studded with minute bodies called **ribosomes.** Ribosomes, composed of RNA and protein, are not membrane-bound and are the sites of protein synthesis. Portions of the ER, with ribosomes attached, are referred to as **rough ER,** while portions without ribosomes are called **smooth ER.** Due to the presence of ribosomes, rough ER is active in protein synthesis, while smooth ER functions in the transport and packaging of proteins. Ribosomes are also found free in the cytoplasm.

The **Golgi apparatus** is a stack of flattened hollow sacs with distended edges; small vesicles are pinched off the edges of these sacs (*see* fig. 2.3). The Golgi apparatus functions in the storage, modification, and packaging of proteins that are produced by the ER. Once the proteins are transported to the Golgi sacs, they are modified in various ways to form complex biological molecules. Often carbohydrates are added to proteins to form glycoproteins. The vesicles that are pinched off contain products that will be secreted from the cell. Some of the polysaccharides (not cellulose) found in the cell wall are also secreted by these Golgi vesicles.

Microbodies are small, spherical organelles in which various enzymatic reactions occur. Plant cells can contain two types of microbodies: **peroxisomes,** which are found in leaves and play a limited role in photosynthesis under certain conditions, and **glyoxysomes,** which are involved in the conversion of stored fats to sugars in some seeds.

Nucleus

One of the most important and conspicuous structures in the cell is the **nucleus,** the center of control and hereditary information (*see* fig. 2.3). The nucleus is surrounded by a double membrane with small openings called **nuclear pores,** which lead to the cytoplasm. In places, the nuclear membrane is connected to the ER. Contained within the nucleus is granular-appearing **chromatin,** which consists of DNA, the hereditary material, and proteins. Another structure within the nucleus is the **nucleolus;** one or more dark-staining nucleoli are always present. The nucleolus is not membrane-bound and is roughly spherical in shape; it is involved in the formation of ribosomes. Table 2.1 is a summary of the functions of the cellular components.

Cell Division

The cell, with its organelles just described, is not a static structure but dynamic, continually growing, metabolizing, and reproducing. Inherent in all cells are the instructions for cell reproduction or **cell division,** the process by which one cell divides into two.

The Cell Cycle

The life of an actively dividing cell can be described in terms of a cycle, which is the time from the beginning of one division to the beginning of the next (fig. 2.8). Most of the time is spent in the nondividing or **interphase** stage. This is a metabolically active stage and consists of three phases: G_1, S, and G_2. The G_1, or the first gap phase, is a period of intense biochemical activity; the cell is actively growing, enzymes and other proteins are rapidly synthesized, and organelles are increasing in size and number. The S, or synthesis phase, is crucial to cell division, for this is the time when DNA is duplicated; other chromosomal components such as proteins are also synthesized in this phase. Following the S phase, the cell enters the G_2, or second gap, phase where there is an increase in protein synthesis and the final preparations for cell division take place. The G_2 phase ends as the cell begins division.

ABLE 2.1

Plant Cell Structures and Their Functions

Structure	Description	Function
Cell Wall	Cellulose fibrils	Support and protection
Plasma Membrane	Lipid bilayer with embedded proteins	Regulates passage of materials into and out of cell
Central Vacuole	Fluid-filled sac	Storage of various substances
Nucleus	Bounded by nuclear envelope, contains chromatin	Control center of cell, directs protein synthesis and cell reproduction
Nucleolus	Concentrated area of RNA and protein within the nucleus	Ribosome formation
Ribosomes	Assembly of protein and RNA	Protein synthesis
Endoplasmic Reticulum	Membranous channels	Transport and protein synthesis (rough ER)
Golgi Apparatus	Stack of flattened membranous sacs	Processing and packaging of proteins; secretion
Chloroplast	Double membrane-bound; contains chlorophyll	Photosynthesis
Leucoplast	Colorless plastid	Storage of various materials, especially starch
Chromoplast	Pigmented plastid	Imparts color
Mitochondrion	Double membrane-bound	Cellular respiration
Microbodies	Vesicles	Various metabolic reactions
Cytoskeleton	Microtubules and microfilaments	Cell support and shape
Plasmodesmata	Cytoplasmic bridges	Movement of materials between cells

During cell division, two exact copies of the nucleus result from a process known as **mitosis. Cytokinesis,** the division of the cytoplasm, usually occurs during the later stages of mitosis.

Chromatin, consisting of DNA and protein, is prominent in the nucleus of a nondividing or interphase cell. Although in photographs the chromatin appears granular, it is actually somewhat threadlike (fig. 2.9). The chromatin has already been duplicated during the S phase prior to mitosis. The events of mitosis are described as four intergrading stages: **prophase, metaphase, anaphase,** and **telophase** (fig. 2.9).

Prophase

During prophase, the appearance of the nucleus changes dramatically. The chromatin begins to condense and thicken, coiling up into bodies referred to as chromosomes. Each **chromosome** is double, composed of two identical **chromatids,** which represent the condensed duplicated strands of chromatin. The chromatids are joined at a constriction known as the **centromere** (*see* fig. 2.8). By the end of prophase, the chromosomes are fully formed. Also during prophase, the nuclear membrane and the nucleoli disperse into the cytoplasm and are no longer visible. This leaves the chromosomes free in the cytoplasm.

Metaphase

The chromosomes arrange themselves across the center of the cell during metaphase, the second stage of mitosis. The **spindle,** which begins forming in prophase, is evident during this stage. Spindle fibers, composed of

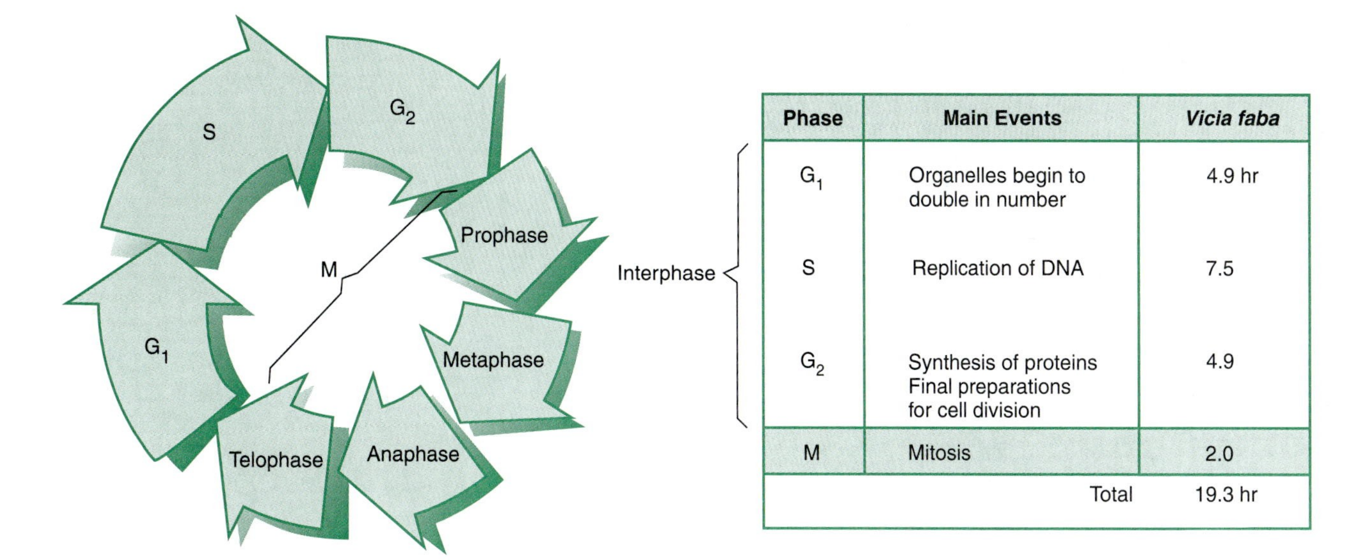

Phase	Main Events	*Vicia faba*
G_1	Organelles begin to double in number	4.9 hr
S	Replication of DNA	7.5
G_2	Synthesis of proteins Final preparations for cell division	4.9
M	Mitosis	2.0
	Total	19.3 hr

(a)

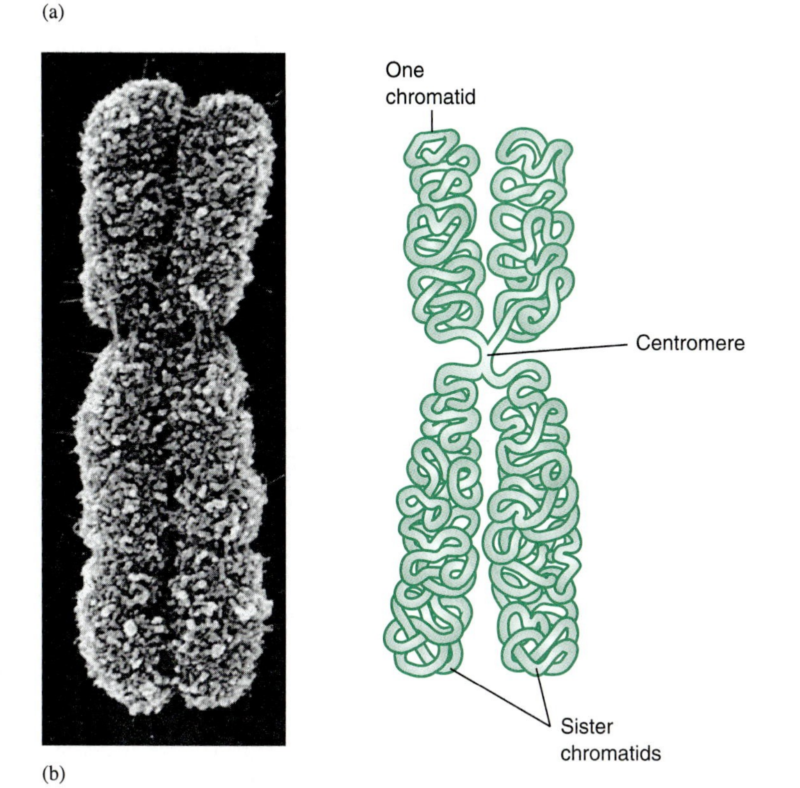

(b)

Figure 2.8 *(a) The cell cycle consists of four stages (G_1, S, G_2, and M for mitosis). The events that occur in each stage, and the length of each stage using the broad bean* (Vicia faba) *as an example, are depicted in the accompanying table. (b) A duplicated chromosome consists of two sister chromatids held together at the centromere, ×9,600.*

microtubules, stretch from each end or pole of the cell to the centromeres of the chromosomes. Other spindle fibers stretch from pole to pole (fig. 2.9).

Anaphase

In anaphase, chromatids of each chromosome separate, pulled by the spindle fibers to opposite ends of the cell. This step effectively divides the genetic material into two identical sets, each with the same number of single chromosomes. At the end of anaphase the spindle is less apparent (fig. 2.9).

Telophase

During telophase the chromatin appears again as the chromatids, at each end of the cell, begin to unwind and lengthen. At each pole a nuclear membrane reappears around the chromatin. Now two distinct nuclei are evident. Within each nucleus, nucleoli become visible (fig. 2.9).

Cytokinesis

Cytokinesis, the division of the cytoplasm, separates the two identical daughter nuclei into two cells. Cytokinesis begins during the latter part of anaphase and is completed

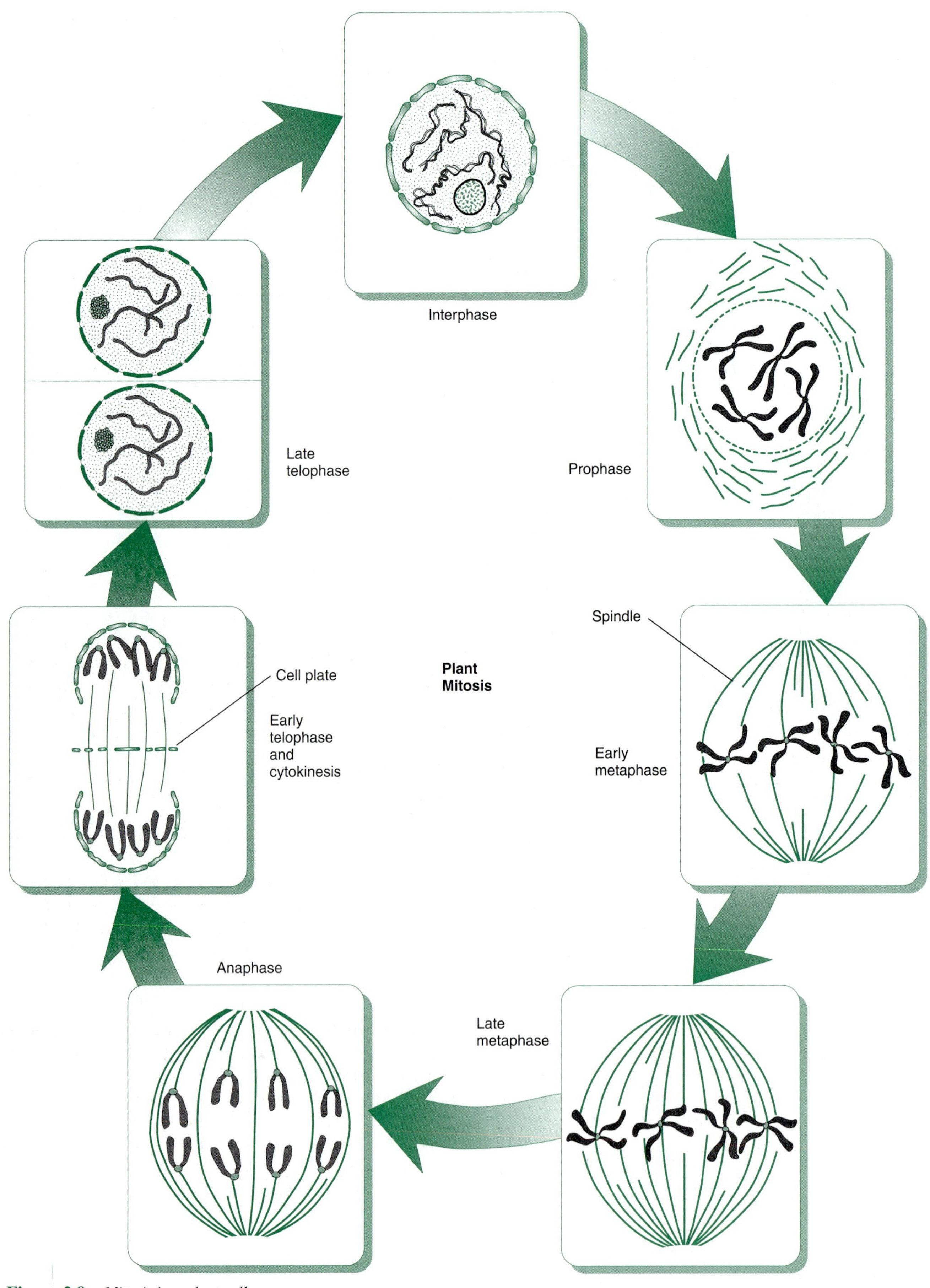

Figure 2.9 *Mitosis in a plant cell.*

A Closer Look 2.2

Osmosis and Diffusion: How Things Move in and out of Cells

Cells constantly exchange materials with their environment. One way this occurs is by **diffusion.** Diffusion is the spontaneous movement of particles or molecules from areas of higher concentration to lower concentration. Examples of diffusion occur everywhere. Open a bottle of perfume; soon the scent spreads throughout the room. Try the same thing with a bottle of ammonia. In both cases, the molecules have diffused from where they were most concentrated. Diffusion can also be easily demonstrated in liquids. Place a sugar cube in a cup of hot tea; eventually the sugar will diffuse and be distributed, even without stirring.

Diffusion also occurs within living organisms, but the membranes present barriers to this movement of molecules. Membranes like the plasma membrane can be described as **differentially permeable.** They permit the diffusion of some molecules, but present a barrier to the passage of other molecules. Many molecules are simply too large to diffuse through membranes.

The diffusion of water across cell membranes is called **osmosis.** Water can move freely through membranes. The direction the water molecules move is dependent on the relative concentrations of substances on either side of the membrane. If you place a cell in a highly concentrated solution of salt or sugar, water will leave the cell. The water is actually diffusing from an area of higher concentration in the cell to an area of lower concentration. On the other hand, if you place a cell in distilled (pure) water, water will enter the cell. Again the water is moving from higher (outside the cell) to lower concentration (box fig. 2.2).

If a plant cell is left in a highly concentrated, or **hypertonic,** solution for any length of time, so much water will leave that the protoplast actually shrinks away from the cell wall. When this happens the cell is said to be **plasmolyzed.** In a wilted leaf, many of the cells would be plasmolyzed.

When a plant cell is in pure water or a very weak, **hypotonic,** solution, water will enter until the vacuole is fully extended, pushing the cytoplasm up against the cell wall. Such cells look plump, or **turgid.** This is the normal appearance of cells in a well-watered plant. The crispness and crunch of fresh celery is due to its turgid cells. When the cell is placed in a solution of the same concentration, **isotonic,** there is no net movement of water and the cell is not turgid.

Diffusion and osmosis take place when molecules move along a **concentration gradient,** from higher to lower concentrations. However, cells can also move substances against a concentration gradient; sometimes sugars are accumulated this way. This type of movement is called **active transport** and requires the expenditure of energy by the cell. Membrane proteins are involved in transporting these substances across the membrane.

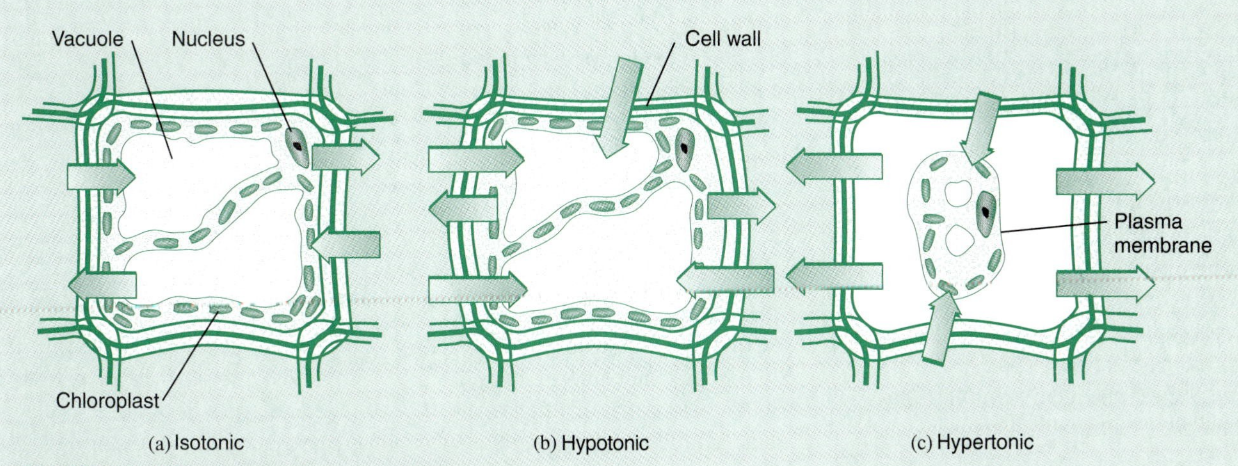

Box Figure 2.2 *Osmosis in plant cells. The arrows indicate the net movement of water. (a) In an isotonic solution, the cell neither gains nor loses water; (b) in a hypotonic solution, the cell gains water; and (c) in a hypertonic solution, the cell loses water.*

A Closer Look 2.3

Origin of Chloroplasts and Mitochondria

As stated in Chapter 1, prokaryotes were the first organisms on Earth. Evidence indicates that prokaryotes first appeared approximately 3.5 billion years ago, while eukaryotes only appeared around 1.5 billion years ago. One question that has intrigued biologists for many years is how did the eukaryotic cell evolve? Dr. Lynn Margulis of the University of Massachusetts is one of the main proponents of a possible answer to this question, the Endosymbiont Theory. This theory states that the organelles of eukaryotic cells are the descendants of once free-living prokaryotes that took up residence in a larger cell, establishing a symbiotic relationship (**symbiosis:** two or more organisms living together). This association evolved into the well-studied eukaryotic cell.

Chloroplasts and mitochondria provide the best examples of this theory. Both organelles resemble free-living prokaryotes. In fact, as long ago as the 1880s some biologists observed that chloroplasts of eukaryotic cells resembled cyanobacteria (then called blue-green algae). Both chloroplasts and mitochondria have structures that are associated with free-living cells; for example, they contain DNA as well as ribosomes, which allows them to synthesize some of their own proteins. Both chloroplasts and mitochondria can divide to produce new chloroplasts and mitochondria in a manner very similar to prokaryotic cell division. The inner membranes of both organelles closely resemble the plasma membrane of prokaryotes. These features, as well as additional biochemical similarities, provide support for the validity of the Endosymbiont Theory.

by the end of telophase. The **phragmoplast,** which consist, of vesicles, microtubules, and portions of ER, accumulates across the center of the dividing cell. These coalesce to form the **cell plate,** which becomes the cell wall separating the newly formed daughter cells (fig. 2.9).

The production of new cells through cell division enables plants to grow, repair wounds, and regenerate lost cells. Cell division can even lead to the production of new, genetically identical individuals or **clones.** This type of reproduction is known as asexual. When you make a leaf cutting of an African violet and a whole new plant develops from the cutting, you are facilitating asexual reproduction and seeing the results of cell division on a large scale. Many of our crops are actually propagated through asexual methods that will be discussed in detail in future chapters. Also, in a later chapter, you will learn of another type of cell division called meiosis, which is involved in sexual reproduction.

Chapter Summary

1. All life on Earth, including plant life, has a cellular organization. The plant cell shares many characteristics in common with other eukaryotic cells. The plant protoplast includes the cytoplasm with the embedded organelles and nucleus. Within the nucleus is DNA, the genetic blueprint of all cells.
2. The plasma membrane, composed of phospholipids and proteins according to the fluid mosaic model, regulates the passage of materials into and out of the cell. Numerous mitochondria can be found within the cytoplasm; they are the sites of cellular respiration. The endoplasmic reticulum, Golgi apparatus, and microbodies make up an internal membrane system that functions in the synthesizing, packaging, and transporting of materials.
3. Some features of a plant cell are unique. The primary cell wall containing cellulose surrounds a plant protoplast providing protection and support. In certain specialized plant cells, a secondary cell wall, impregnated with the toughening agent lignin, imparts extra strength. Chloroplasts are the site for photosynthesis; they are one of several types of plastids. Other plastids are the food storing leucoplasts and the pigment-containing chromoplasts. A large

central vacuole may take up approximately 90% of the mature plant cell and act as a storage site for many substances.

4. The life of a cell can be described in terms of a cycle. Most cells spend the majority of the time in interphase, a nondividing stage. But at certain times in the life of a cell, it may undergo division whereby one cell divides into two. Mitosis is the duplication of the nucleus into two exact copies. There are four intergrading stages in mitosis: prophase, metaphase, anaphase, and telophase. The division process is complete when, in the process of cytokinesis, the cytoplasm is split, and two identical daughter cells are formed.

Review Questions

1. What is the significance of the Cell Theory to biology?
2. List the parts of a plant cell, and for each part describe its structure and function.
3. Describe the events occurring during G_1, S, and G_2 stages of interphase.
4. Describe the stages of mitosis.
5. Describe the similarities and differences between chloroplasts and mitochondria.
6. Differentiate between osmosis and diffusion.

Further Reading

Mauseth, James D. 1991. *Botany: An Introduction to Plant Biology.* Saunders College Publishing, Philadelphia, PA.

Moore, Randy and W. Dennis Clark. 1995. *Botany: Plant Form and Function.* Wm. C. Brown Publishers, Dubuque, IA.

Raven, Peter, Ray F. Evert, and Susan E. Eichhorn. 1986. *Biology of Plants,* 4th Edition. Worth Publishers, Inc., New York, NY.

Stern, Kingsley R. 1994. *Introductory Plant Biology,* 6th Edition. Wm. C. Brown Publishers, Dubuque, IA.

CHAPTER

3

THE PLANT BODY

Chapter Outline

Chapter Concepts

1. Tissues are groups of cells that perform a common function and have a common origin and structure.
2. Flowering plants are made up of three basic tissue types: dermal, ground, and vascular.
3. The vegetative organs of higher plants are roots, stems, and leaves.

The earliest life forms were unicellular, with that single cell capable of carrying out all the necessary functions of life. When multicellular organisms evolved, certain cells became specialized in structure and function, leading to a division of labor. These specialized cells performing specific functions are usually referred to as **tissues.** In flowering plants, various tissues compose the familiar organs: roots, stems, and leaves.

Plant Tissues

Meristems

All flowering plants are multicellular, with the cells all originating from regions of active cell division. These regions are known as **meristems.** The cells arising from meristems give rise to the various tissue types that make up a plant, such as the cells of the epidermis that form the protective layer in a plant.

Apical meristems are located at the tips of all roots and stems and contribute to the increase in length of the plant. Tissues that develop from these apical meristems are part of the **primary growth** of the plant and give rise to the leaves and nonwoody stems and roots. Some plants have additional meristematic tissues that contribute to increases in diameter. These are the **vascular cambium** and **cork cambium.** Tissues developing from them are considered part of the plant's **secondary growth** (fig. 3.1).

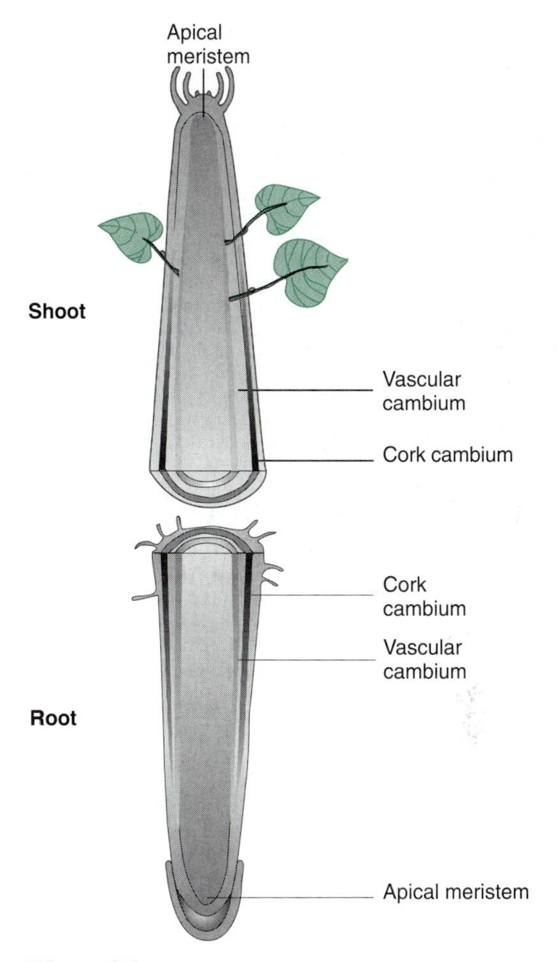

Figure 3.1 *Plant meristematic tissues in a diagram of a shoot tip and root tip. Apical meristems contribute to increases in the length of the plant or primary growth. Vascular cambium and cork cambium are present in those plants that have secondary growth, an increase in the girth of the plant body.*

Dermal Tissue

The three basic tissue types in higher plants are **dermal, ground,** and **vascular.** Dermal tissues are the outermost layers in a plant. In young plants and nonwoody plant parts, the outermost surface is the **epidermis.** It is usually a single layer of flattened cells. Epidermal cells in leaves and stems secrete **cutin,** a waxlike substance that makes up the **cuticle** on the external surface. The cuticle prevents evaporative water loss from the plant by acting as a waterproof barrier. In many leaves the cuticle is so thick that the leaf has a shiny surface; this is especially true in succulents such as the jade plant and tropical plants such as philodendron.

In some plants, **hairs (trichomes)** may be present on the epidermis (fig. 3.2). Although usually microscopic, they may be abundant enough to give a fuzzy appearance and texture to leaves or stems. Trichomes may also be glandular, often imparting an aroma when they are brushed, as you can experience by rubbing a geranium or tomato leaf.

Scattered through the leaf epidermis are pores known as **stomata** (sing., **stoma**). Gases such as carbon dioxide, oxygen, and water vapor are exchanged through these stomata. A pair of sausage-shaped cells, **guard cells,** occur on either side of the pore and regulate the opening and closing of the stomata (fig. 3.2). The guard cells are the only epidermal cells with chloroplasts. Stomata and guard cells can also be found in the epidermis of some stems.

In those plant parts that become woody, the epidermis cracks and is replaced by a new surface layer, the **periderm,** which is continuously produced by the cork cambium as the tree increases in girth. The periderm, which consists of cork cells, the cork cambium, and sometimes other cells, makes

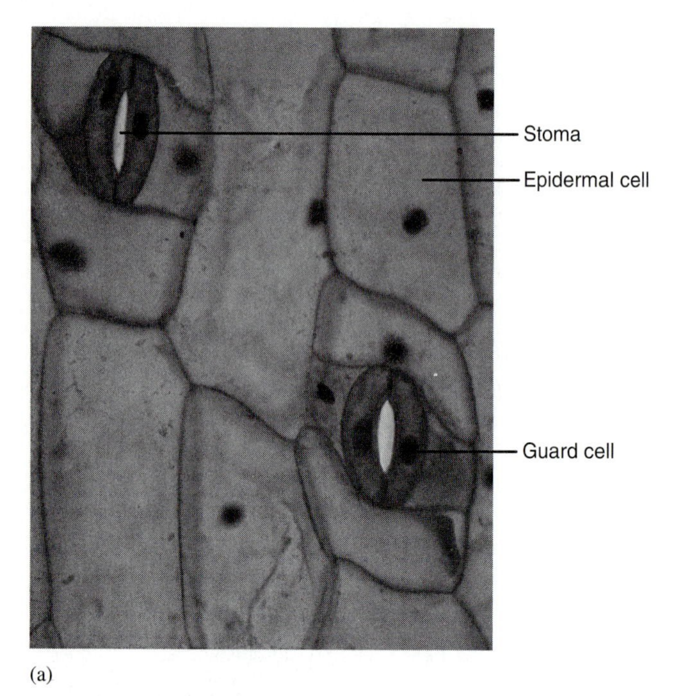

(a)

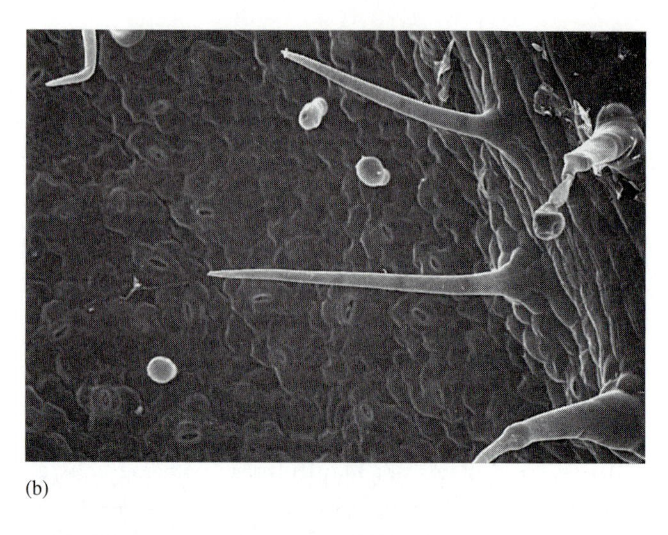

(b)

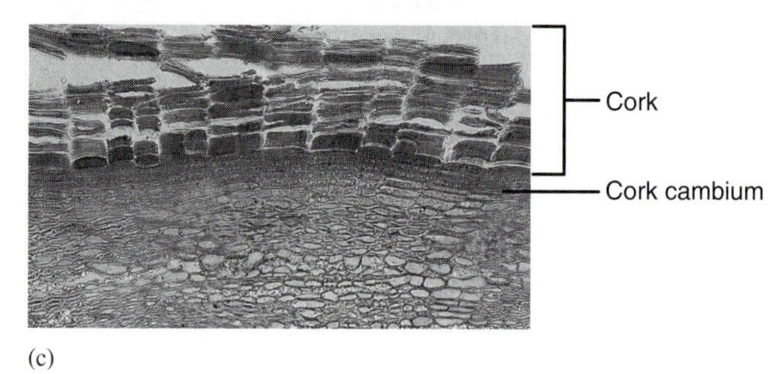

(c)

Figure 3.2 *Dermal tissues. (a) Leaf epidermis contains stomata for gas exchange, ×100; (b) Trichomes (×150); (c) Periderm is a complex tissue consisting of a thick outer layer of cork cells that arise from the cork cambium, ×300.*

up the outer bark seen on mature trees (fig. 3.2). In fact, the cork in wine bottles is the periderm from *Quercus suber,* the cork oak tree native to the western Mediterranean. Cork is principally made up of dead cells whose walls contain **suberin,** another waterproofing fatty substance. It prevents water loss and protects underlying tissues (see chapter 18).

Ground Tissue

Ground tissues make up the bulk of nonwoody plant organs and perform a variety of functions. The three categories of ground tissue are **parenchyma, collenchyma,** and **sclerenchyma.** The most versatile of these is parenchyma. Although often described as a thin-walled 14-sided polygon, parenchyma cells can be almost any shape or size. Usually parenchyma tissue is loosely arranged, with many intercellular spaces. Parenchyma cells are capable of performing many different functions (fig. 3.3a). They are the photosynthetic cells in leaves and green stems and the storage cells in all plant organs. The starch in potato tubers, the water in cactus stems, and the sugar in sugar beet roots are all stored in parenchyma cells.

Collenchyma cells are the primary support tissue in young plant organs. They can be found in stems, leaves, and petals. Collenchyma cells are elongated cells with unevenly thickened primary cell walls, with the walls thickest at the corners (fig. 3.3b). They are found tightly packed together just below the epidermis. The tough strings in celery are actually strands of collenchyma cells.

Sclerenchyma tissue has two cell types: fibers and sclereids. Like collenchyma cells, the fibers are elongated cells that function in support. Unlike collenchyma, they are nonliving at maturity and have thickened secondary walls (fig. 3.3c). For centuries people have used leaf and stem fibers from many plants in the making of cloth and rope. Sclereids have many shapes but are seldom elongate like fibers. The major function of these cells is to provide mechanical support and protection. The extremely thick secondary walls of sclereids account for the hardness in nut shells and the grit of pear fruit .

Vascular Tissue

Vascular tissues are the conducting tissues in plants. You can readily see the vascular tissues in a leaf; they are the **veins.** The vascular tissues form a continuum throughout the plant, allowing the unrestricted movement of materials. There are two types: **xylem,** which conducts water and minerals from the roots upward, and **phloem,** which transports organic materials synthesized by the plant. Both xylem and phloem are complex tissues composed of several cell types.

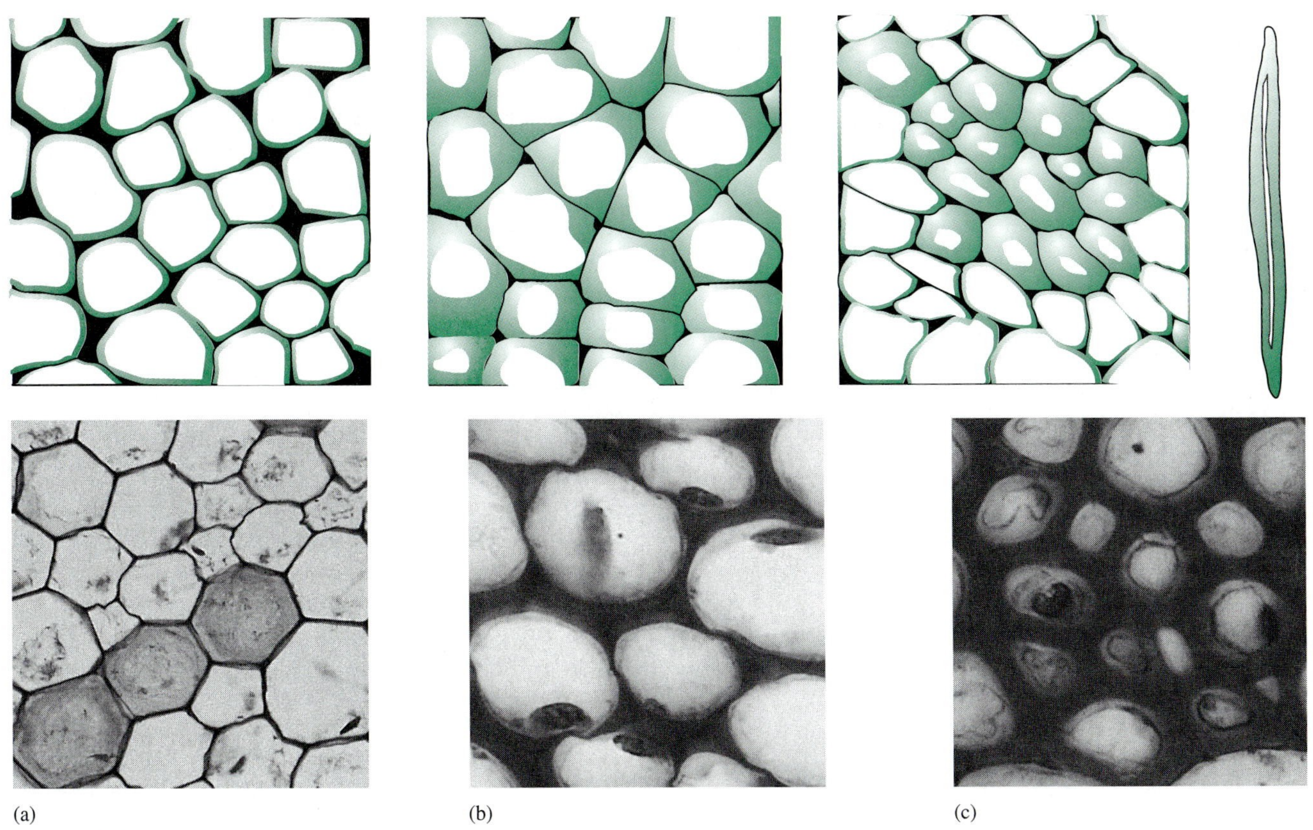

Figure 3.3 *Ground tissues. (a) Parenchyma cells are the most abundant plant tissue type and have characteristically thin cell walls, ×250. (b) Collenchyma cells have primary cell walls that are thickest at the corners, ×450. (c) Sclerenchyma cells have very thick secondary cell walls and are nonliving, ×560.*

Tracheids and **vessel elements** are the water-conducting cells in the xylem. Both cell types have secondary walls and, at maturity, these cells are dead and consist only of cell walls. Tracheids are long thin cells with tapering ends and numerous pits in the walls; these cells also function in support. Vessel elements are usually shorter and wider, and often have horizontal end walls with large openings. Like the tracheids, the side walls have numerous pits. Vessel elements are attached end to end to form a long, pipelike **vessel** (fig. 3.4a).

Tracheids and vessel elements are found in angiosperms, while only tracheids occur in other vascular plants. Fibers are present in the xylem, where they provide additional support. Parenchyma cells, which also occur in the xylem, are the only living and metabolically active cells in this tissue.

Xylem can be either primary or secondary; primary xylem originates from the apical meristem, while secondary xylem comes from the vascular cambium. In trees, secondary xylem is very extensive; it is what we call wood.

The cells involved in the transport of organic materials in the phloem are the **sieve tube members.** Unlike the conducting cells in the xylem, the sieve tube members are living cells with only primary walls. But they are unusual living cells, since the nucleus and some organelles degenerate as the sieve tube member matures. The end walls of these cells have several-to-many large pores and are called **sieve plates.** They allow plasmodesmata, cytoplasmic connections, to occur between adjacent sieve tube members and provide channels for conduction. The column of connected sieve tube members is referred to as a **sieve tube** (fig. 3.4b).

Adjoining each sieve tube member is a **companion cell,** which is physiologically and developmentally related to its sieve tube member. Numerous plasmodesmata connect the two cells. The companion cells are involved in the loading and unloading of organic materials for transport. As in the xylem, both fibers and parenchyma cells are found in the phloem. Both primary

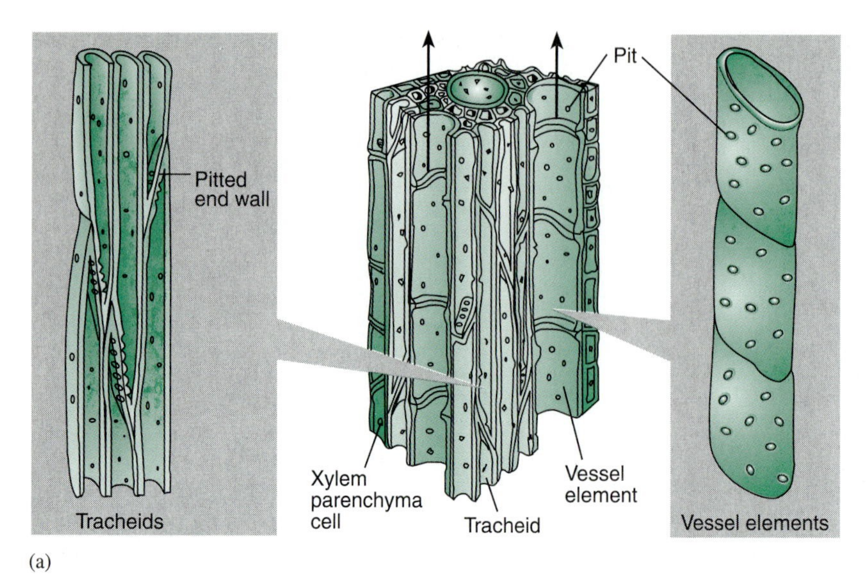

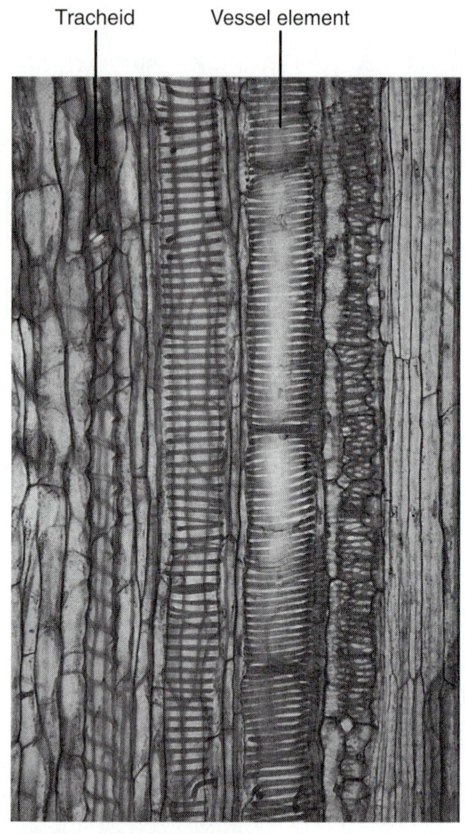

(a)

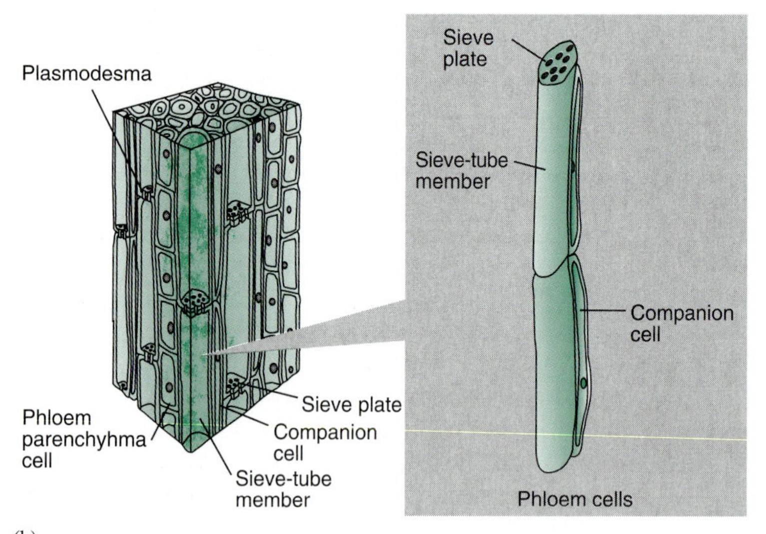

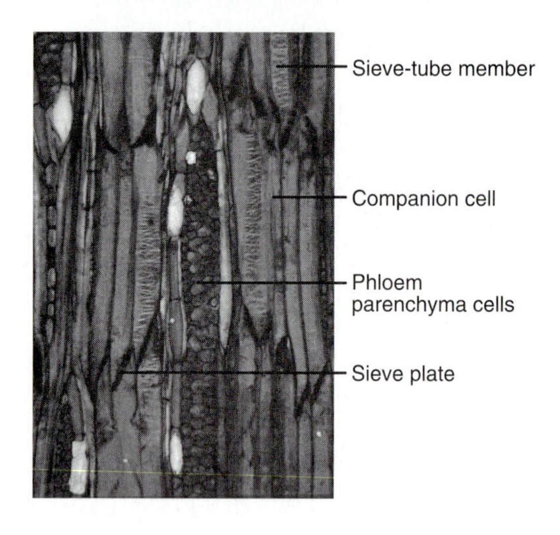

(b)

Figure 3.4 *(a) Xylem. The conducting cells of xylem are tracheids and vessel elements. (b) Phloem. Sugars are loaded by companion cells into sieve tube members for transport.*

TABLE 3.1
Plant Tissues

Tissue Type	Cell Types	Function
Dermal		
Epidermis	Epidermal cells	Protection
Periderm	Cork cells	Protection
Ground		
Parenchyma	Parenchyma cells	Storage, photosynthesis
Collenchyma	Collenchyma cells	Support
Sclerenchyma	Sclereids, fibers	Support, protection
Vascular		
Xylem	Tracheids, vessel elements, fibers, parenchyma	Water conduction, support
Phloem	Sieve tube members, companion cells, fibers, parenchyma	Food transport

and secondary phloem occur; again the primary phloem is produced by the apical meristem and the secondary by the vascular cambium. Table 3.1 is a summary of these plant tissues.

Plant Organs

The principal vegetative organs of flowering plants are stems, roots, and leaves. Roots anchor the plant and absorb water and nutrients from the soil; stems support the plant and transport both water and organic materials; and leaves are the main photosynthetic structures.

Stems

Recall that angiosperms are divided into two classes of plants, the dicots and the monocots. Although the major differences between these classes are in the flower and seed, anatomical differences can also be seen in stems, roots, and leaves.

A monocot stem is best exemplified by examining a cross section of a corn stem. The outermost tissue is a single layer of epidermis. Beneath the epidermis are two to three layers of sclerenchyma for support. **Vascular bundles** are scattered throughout the stem. These vascular bundles are composed of both xylem and phloem and are usually surrounded by a **bundle sheath** of fibers. Parenchyma cells or tissues fill in the rest of the stem (fig. 3.5).

Dicot stems can be either **herbaceous** (nonwoody) or **woody.** In herbaceous dicots the vascular tissue occurs as a ring of separate vascular bundles. Again each vascular bundle contains both xylem and phloem, with the xylem toward the center of the stem and the phloem toward the outside. This ring of vascular bundles surrounds the **pith,** a central area of ground tissue composed of parenchyma cells. On the other side of the ring of vascular bundles, toward the outside of the stem, is the **cortex,** another region of ground tissue. Although the cortex consists mainly of parenchyma cells, fibers often occur in this region. Between the vascular bundles, the ground tissue of the pith and cortex is continuous. The outermost layer of the stem is the epidermis. In some plants, support tissue, either sclerenchyma or collenchyma, can be found beneath the epidermis (fig. 3.5).

In woody dicots the vascular tissue, especially the xylem, is much more extensive and makes up the bulk of the stem. Like the herbaceous dicots, the pith occupies the center of the stem. Surrounding the pith are rings of secondary xylem. Each ring represents the xylem formed by the vascular cambium during one growing season and is called an **annual ring.** The rings, which are easily visible to the naked eye, are due to the different sizes of cells formed through the growing season. In the spring, when water is more abundant, the cells, **springwood,** are larger than those produced during the late summer, **summerwood.** The portion of each ring with springwood appears lighter than the area with the smaller, densely packed cells of summerwood. Since each ring typically represents one growing season, in temperate regions the age of the tree can be determined by counting the annual rings (fig. 3.6).

Surrounding the outermost ring of xylem is the vascular cambium, the meristematic tissue that produces both secondary xylem toward the inside and secondary phloem toward the outside. The amount of secondary phloem produced each year is very small when compared to the xylem. No annual rings are evident in the phloem, although bands of fibers occur in some plants (fig. 3.6 and *see* A Closer Look: Studying Ancient Tree Rings) on page 42.

Vascular rays, resembling spokes of a wheel, are seen crossing both xylem and phloem. Composed of parenchyma cells, these rays are involved in radial transport of materials.

A small band of cortex can be found outside the phloem. In older trees, however, the cortex is completely replaced by the periderm or cork (fig. 3.6). In fact, even the older, outermost layers of phloem are replaced by the periderm. The thickness and texture of the periderm depends on the type of tree and varies from thin and papery in cherry or paper birch to extremely thick in cork oak.

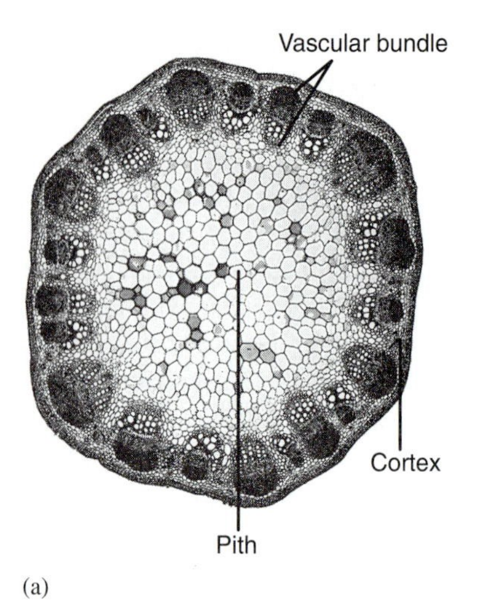

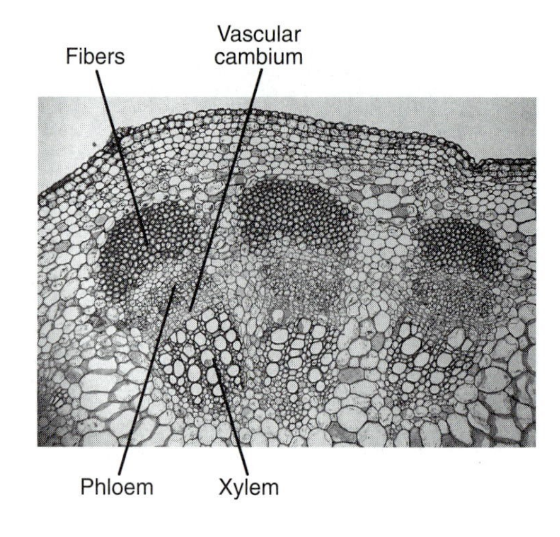

(a)

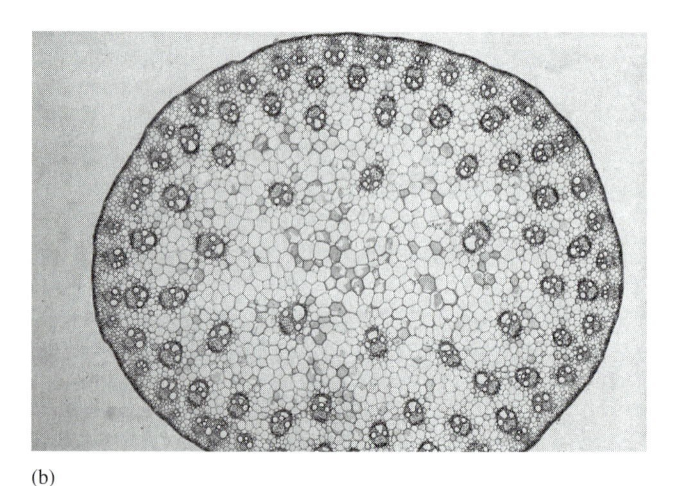

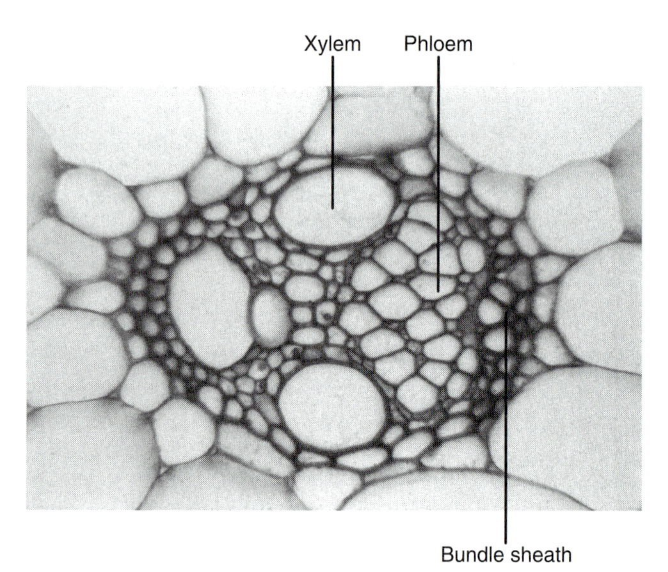

(b)

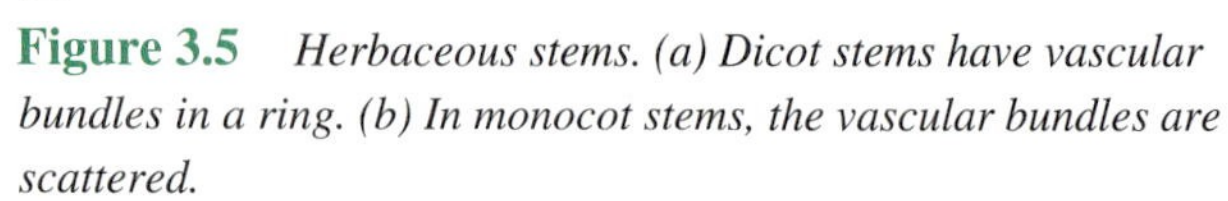
Figure 3.5 *Herbaceous stems. (a) Dicot stems have vascular bundles in a ring. (b) In monocot stems, the vascular bundles are scattered.*

Roots

Two major types of root systems can be found in flowering plants: **taproots** and **fibrous roots.** Taproots have one large main root with small lateral or branch roots. Taproots can be enlarged for storage as evident in carrots, turnips, and beets. Fibrous roots are highly branched, and lack a central main root, as in many grasses (fig. 3.7).

At the tips of all main roots or branch roots are thimble-shaped **root caps,** which protect the root meristems as the roots grow through the soil. The meristem **(zone of cell division)** accounts for primary growth in roots. Just behind the meristem, the newly formed cells elongate considerably **(zone of elongation)** before they begin to differentiate into the various tissues that compose the root (fig. 3.8a).

A cross section of a dicot root in the region where cells have differentiated **(zone of maturation)** is seen in Figure 3.8. The vascular tissue is found in the center of the root making up the **stele** or **vascular cylinder.** In the very center of the stele is the xylem, usually in a star-shaped configuration. The number of arms of this star is variable with bundles of phloem found between the arms of xylem. In monocot roots, a pith is present that is encircled by alternating bundles of xylem and phloem

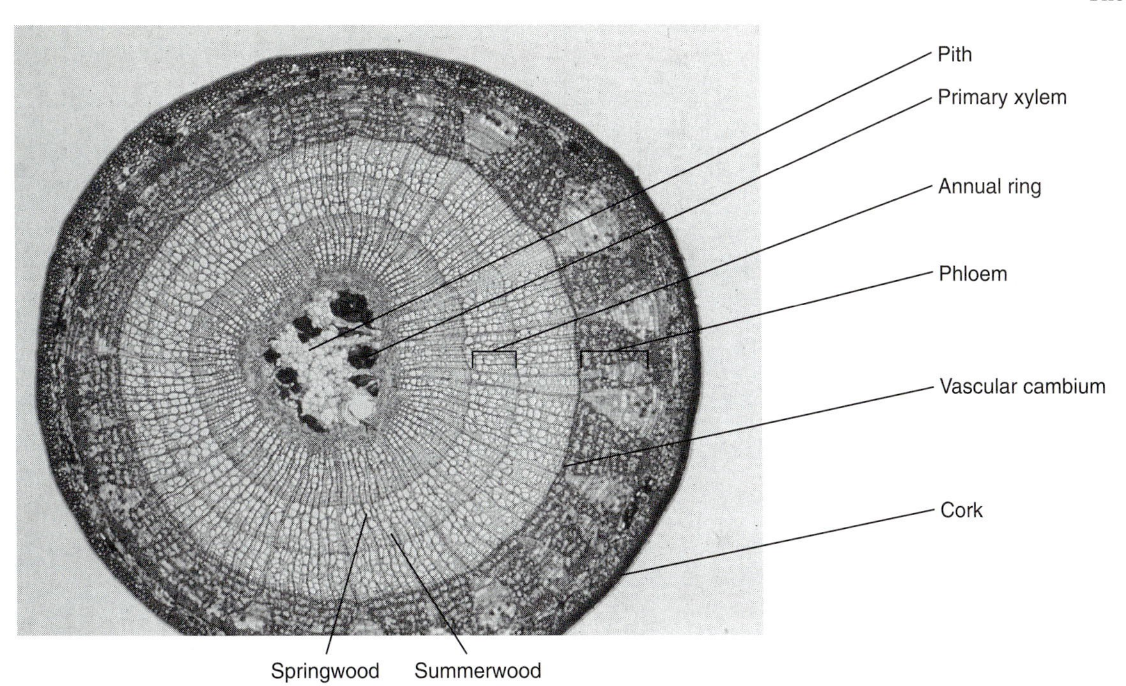

Figure 3.6 *Anatomy of a woody stem, ×35.*

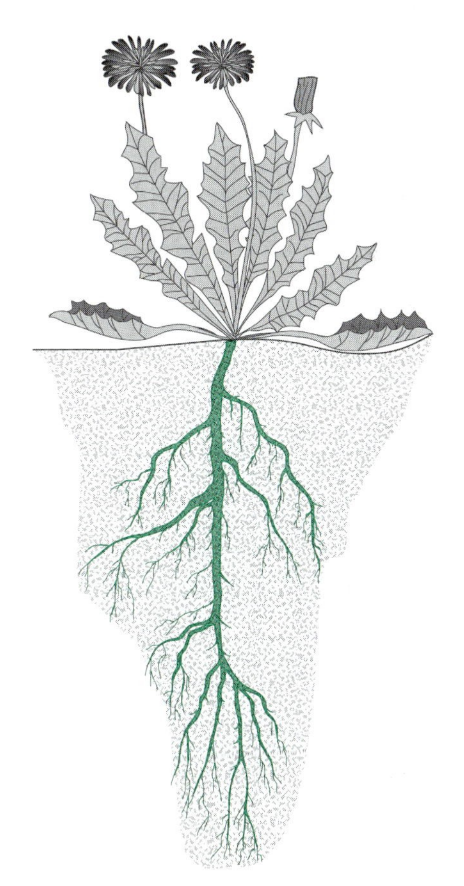

Figure 3.7 *Root systems. (a) The fibrous root system of barley; (b) The taproot of a dandelion.*

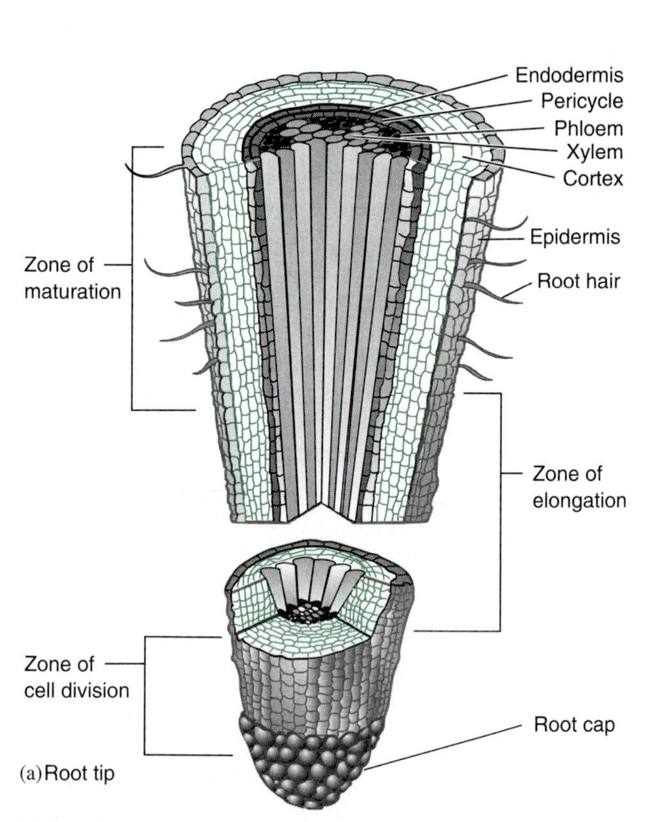

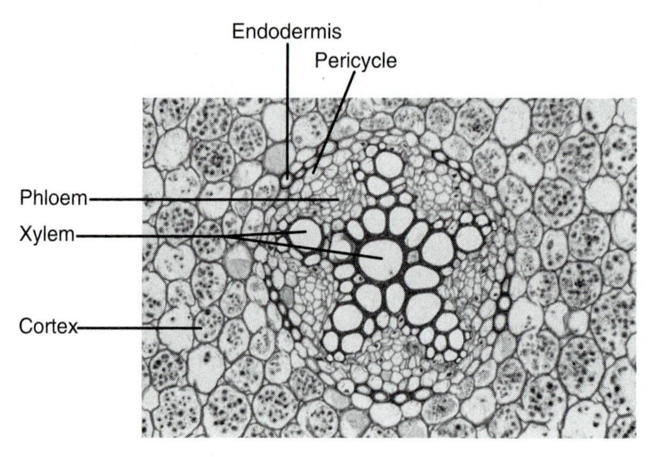

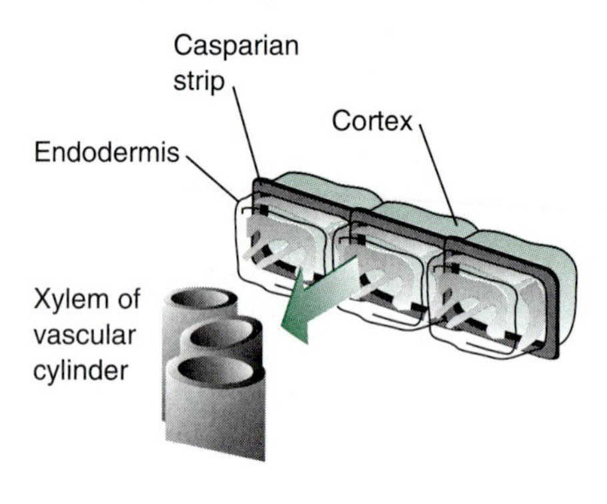

Figure 3.8 *Dicot root tip. (a) A root tip is divided into four zones as seen in this longitudinal section. (b) The vascular cylinder or stele seen in cross section typically shows the xylem in a star-shaped pattern, ×200. (c) The Casparian strip in the wall of the endodermis directs the passage of water and minerals into the xylem.*

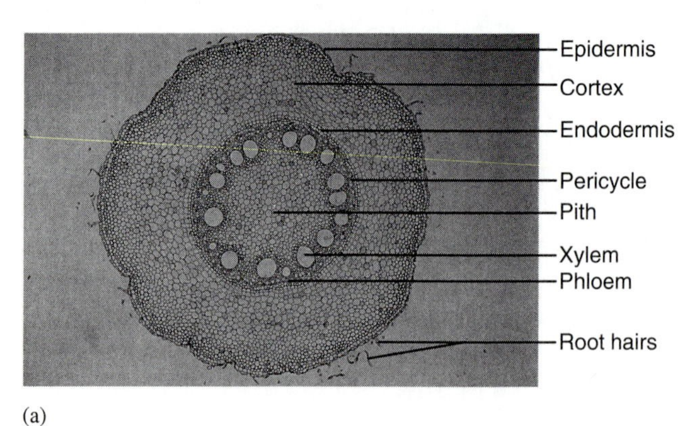

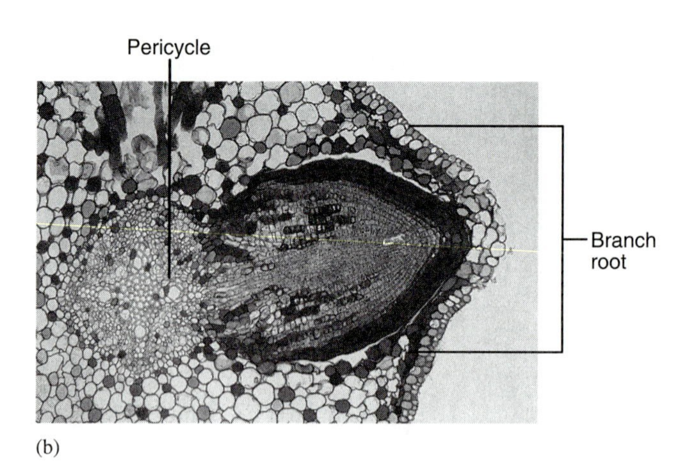

Figure 3.9 *(a) The vascular cylinder of monocot roots typically surround a pith, ×20. (b) Branch roots originate from the pericycle, ×40.*

(fig. 3.9). The outermost layer of cells in the stele is known as the **pericycle,** which is a meristematic layer that can give rise to branch roots (figs. 3.8 and 3.9).

Surrounding the stele is the cortex composed of parenchyma cells, which are sites of storage. The innermost layer of the cortex (just outside the pericycle) is known as the **endodermis.** Endodermal cells are characterized by the presence of a **Casparian strip,** a waxy material ringing each endodermal cell. The faces of the cell wall next to the cortex and stele do not have a Casparian strip. Because of this strip, water and minerals must pass through the endodermal cells, not between them (*see* Chapter 4).

The cortex is usually quite large, making up the bulk of the root. The outermost layer of cells is the epidermis. Extensions of the epidermal cell are called **root hairs;** these greatly increase the surface area and are the sites of maximum water and mineral absorption.

In those plants that have woody stems, extensive secondary xylem and annual rings occur in roots as well as stems. One major difference between a woody root and woody stem is that no pith occurs in the woody root.

Leaves

Leaves have often been called the photosynthetic factories of the plant since this is their major function. The flat, expanded **blade** of the leaf is ideally suited for this photosynthetic process. The **petiole** or leaf stalk connects the leaf blade to the stem and transports materials to and from the blade. In some leaves the petiole may be absent; in those cases the blade is attached directly to the stem. Small paired appendages called **stipules** may be present at the base of the leaf (fig. 3.10a). Stipules are varied in form; in some plants they are leaflike, while in others they are thornlike.

The place where the petiole is attached to the stem is called the **node.** The areas of the stem between adjacent nodes are **internodes.** There are three patterns of leaf arrangement on stems. If only one leaf is present at a node the arrangement is known as **alternate.** If two leaves occur at a node, the arrangement is **opposite;** with three or more, the arrangement is **whorled** (fig. 3.10b).

There is a great variety of leaf forms and shapes, ranging from small, **simple leaves** as in elm where the blade is undivided, to large, **compound leaves** as in pecan and buckeye where the blade is divided into **leaflets.** When the leaflets occur in a featherlike pattern, it is called **pinnately compound,** and **palmately compound** when the leaflets have a common attachment. Figure 3.10 illustrates the varieties of simple and compound leaves.

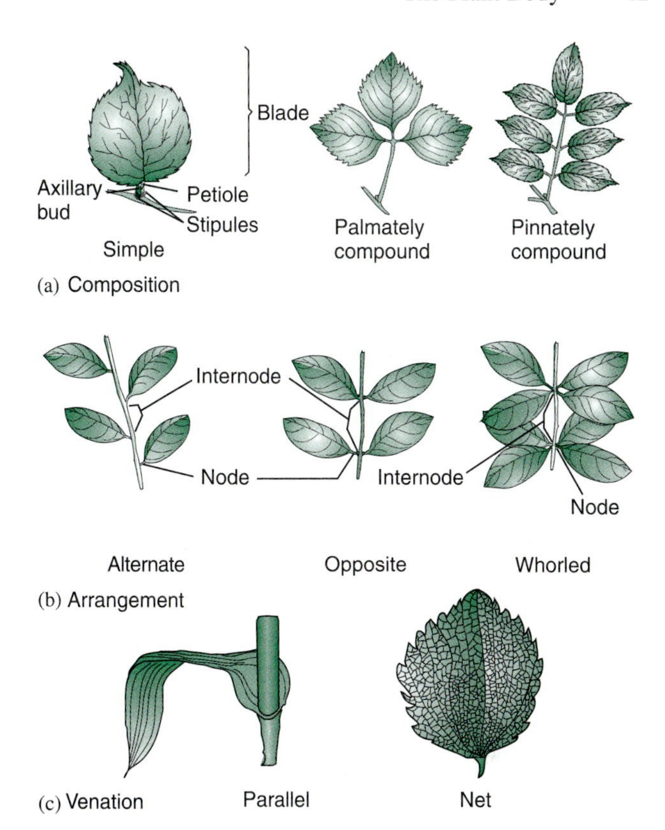

Figure 3.10 *Leaf morphology. (a) Leaf composition. Leaves may be simple, consisting of a single undivided blade or compound, in which the blade is subdivided into leaflets. (b) Leaf arrangement. Alternate, opposite, or whorled indicate the number of leaves coming off a node. (c) Leaf venation. The venation pattern is commonly parallel in monocot leaves and net in dicot leaves.*

The vascular tissues of leaves make up the **venation** patterns visible to the naked eye. Monocot leaves usually have **parallel venation** because the vascular bundles are arranged in parallel lines running the length of the blade. In contrast, dicots have **net** or **reticulate venation,** where the vascular tissue is highly branched, forming a network throughout the blade (fig. 3.10c). Perhaps you have encountered decaying leaves on the ground in late fall or winter where only the vascular tissue remains as a lacy network.

The study of tree rings is known as **dendrochronology** and is of value to fields as diverse as astronomy, ecology, and anthropology. The science began in the early twentieth century by Andrew Douglass in Arizona. Douglass, an astronomer, frequently visited logging camps to study the annual ring patterns on tree stumps. The size of a ring can indicate climatic conditions that existed when the ring was formed (box fig. 3.1). A very narrow ring may indicate a year of low rainfall or drought, while a wide ring may indicate abundant rainfall. Douglass wondered if the climatic changes brought about by the 11-year cycle of sunspots was evident in tree-ring patterns. Although Douglass did not find the answer to the sunspot question, he did see that tree-ring patterns from different areas throughout northern Arizona showed the same patterns of wide and narrow rings.

Douglass continued the study of tree rings for many years. By matching patterns from living trees, remains of fallen trees, and wood samples from Pueblo ruins, Douglass was able to date all the ancient pueblos throughout the Southwest. In 1937, Douglass founded the Laboratory of Tree Ring Research at the University of Arizona. Today, this laboratory is still a major world center for tree-ring study.

Conditions in Arizona are ideal for this type of study. Since rainfall is always limiting for tree growth, a small change in the weather will have a great effect on the width of the tree ring. Also, the arid climate prevents the decay of dead trees and wooden artifacts. In fact, in the Southwest, scientists have been able to construct a chronology of tree rings going back approximately 9,000 years.

By contrast, in other areas of the United States, and in Europe, tree-ring analysis is more difficult because more favorable growing conditions that are relatively consistent result in more uniform tree rings. Also, when trees die they decay in the moister environment. In these areas, the tree-ring chronologies only extend back 300 to 500 years.

Another aspect of tree-ring research is **dendroclimatology.** By studying the annual rings of very old trees, scientists have been able to reconstruct major climatic changes in the past. Tree-ring specialists are trying to determine if droughts occur in a cyclic pattern. Others are looking at the effects of pollution, pests, forest fires, volcanoes, or earthquakes on tree rings. Overall we know that trees are living histories; contained within the tissues of the tree is the history of the environment for the year in which a ring was formed.

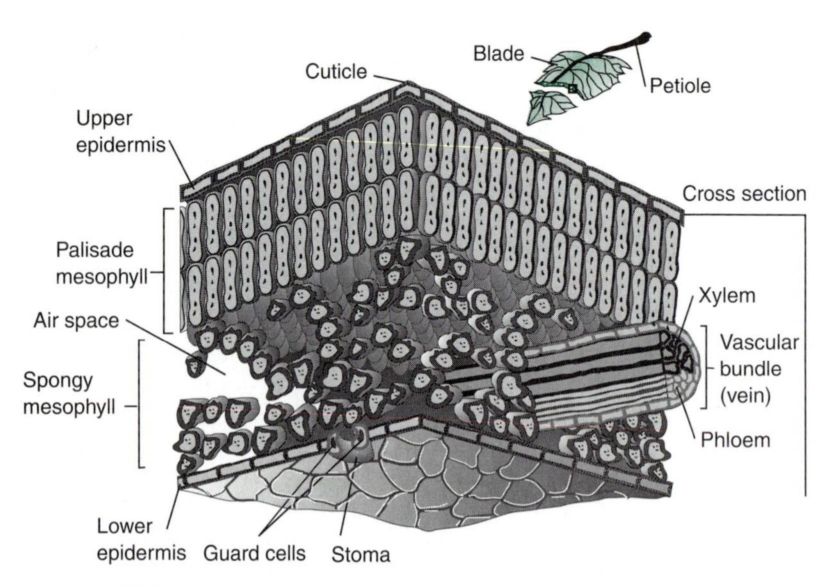

Figure 3.11 *Leaf anatomy. Diagrammatic cross section of a leaf illustrates that palisade and spongy cells make up the leaf mesophyll.*

A cross section of a blade reveals epidermis covering both the upper and lower leaf surfaces (fig. 3.11). Recall that the epidermal cells are covered by a waxy cuticle of variable thickness. Guard cells and stomata are distributed through the epidermis. Although thousands of stomata occur on both the upper and lower surfaces, their number and distribution varies considerably (table 3.2).

The middle of the leaf, or **mesophyll,** is composed mainly of photosynthetic parenchyma cells that may be of two types: **palisade** and **spongy** (fig. 3.11). The palisade parenchyma cells are tightly packed columnar cells lying just beneath the upper epidermis. Spongy parenchyma are loosely packed spherical cells with many large intercellular spaces. Scattered within the mesophyll are the vascular bundles that bring water up to the leaf and carry away the sugars produced by the mesophyll cells.

1914
When the tree was 6 years old, something pushed against it, making it lean. The rings are now wider on the lower side, as the tree builds "reaction wood" to help support it.

1924
The tree is growing straight again. But its neighbors are growing too, and their crowns and root systems take much of the water and sunshine the tree needs.

1927
The surrounding trees are harvested. The larger trees are removed and there is once again ample nourishment and sunlight. The tree can now grow rapidly again.

1930
A fire sweeps through the forest. Fortunately, the tree is only scarred, and year by year more and more of the scar is covered over by newly formed wood.

1942
These narrow rings may have been caused by a prolonged dry spell. One or 2 dry summers would not have dried the ground enough to slow the tree's growth this much.

1957
Another series of narrow rings may have been caused by an insect like the larva of the sawfly. It eats the leaves and leafbuds of many kinds of coniferous trees.

Box Figure 3.1 *The reading of annual rings correlates to events in the life of this tree.*
Source: St. Regis Paper Company, New York, NY, 1966.

Vegetables

Most of us should have more than a passing interest in plant organs because we consume many of them daily; they are the vegetables in our diet. Technically vegetables are edible parts of the vegetative plant body: stems, roots, leaves, or even flower parts. This definition excludes fruits that develop from the ovary of a flower and contain seeds. Interestingly, the legal definition of fruits and vegetables differs from these botanical definitions. In 1893, the Supreme Court ruled that tomatoes and similar fruits are legally vegetables since fruits are generally thought of as sweet while vegetables are not (*see* Chapter 6). The debate continues, but we will give the botanical definition priority. Next we will consider a few examples of some common vegetables, examining their history and folklore.

Carrots

Carrots (*Daucus carota*) are one of the most popular vegetables in the American diet; we each consume about ten pounds of this root per year. Cutting through a carrot clearly shows the organization of the root, with the stele more deeply pigmented than the surrounding cortex.

Carrots are **biennial** plants; this means that it takes two years for the plant to complete its life cycle. During the first year the plant stores a large amount of food in an enlarged tap root that we call the carrot. During the second summer the plant produces a flowering stalk. Normally we harvest carrots after the first growing season, but we can see the flowering stalk in the wild carrot, a beautiful summer weed also called Queen Anne's lace. Carrots were first introduced into North America by the early colonists, and Queen Anne's lace is the wild descendant of those first carrots (fig. 3.12).

Today we associate the color orange with carrots, but it was not always that way. Originally carrots were purple and branched. It was not until the sixteenth century that a pale yellow variety appeared in western Europe. In the seventeenth century, Dutch plant breeders developed a deep orange carrot that is the ancestor of all orange carrots grown today.

A Closer Look 3.2

Plants that Trap Animals

Usually the natural order is for animals to eat plants, but with some plants the tables are turned. These are the **carnivorous plants.** Although Audrey II from the "Little Shop of Horrors" had a taste for human flesh, most carnivorous plants "snack" on insects. Out of more than 250,000 known species of angiosperms, 400 have developed the carnivorous habit. Most of these plants are found in nutrient-poor soils such as acid freshwater bogs, and it is believed that this carnivorous trait supplies nutrients lacking in the soil. After a "meal" the plants have been reported to grow with renewed vigor.

Since plants are stationary and animals are motile, the plants have evolved elaborate traps to lure their prey. These traps are all modified leaves with various incentives such as nectar or color to attract insects. Once the insects are ensnared, digestive enzymes are released, and soon only the empty shell of the insect remains. There are various types of traps; we will consider three of the best known.

VENUS'-FLYTRAP A native to the North Carolina coastal region, *Dionaea muscipula* has a trap that imprisons its victims. Each leaf is a two-sided trap with trigger hairs on each side. When the trigger hairs are touched, the trap snaps shut tightly around the insect. Once the trap has closed, digestive enzymes begin their job. After a few days the trap opens, ready for another unsuspecting insect (box fig. 3.2a).

SUNDEW Sundews, *Drosera* spp., are small plants that use flypaperlike leaves to trap insects. Glandular hairs on the leaf surface produce an adhesive that is the "superglue of the plant kingdom." An insect that has been lured to the plant for its nectar or by its coloration sticks tight and is soon digested away (box fig. 3.2b).

PITCHER PLANTS In pitcher plants, *Sarracenia* spp. and other genera, the leaf has evolved into a vase or pitcher shape that acts like a pitfall type of trap. Once

(a)

(b)

Box Figure 3.2 *Carnivorous plants. (a) Venus'-flytrap; (b) Glandular hairs of the sundew have trapped an insect. (c) The yellow pitcher plant grows in bogs in the southeastern United States.*

(c)

lured to the pitcher, the insects slip into the pool of rainwater that has collected at the base. The pool also contains digestive juices that eat away the soft parts of the insect. At the end of a season, a pitcher may be filled with the indigestible shells of its many victims (box fig. 3.2c).

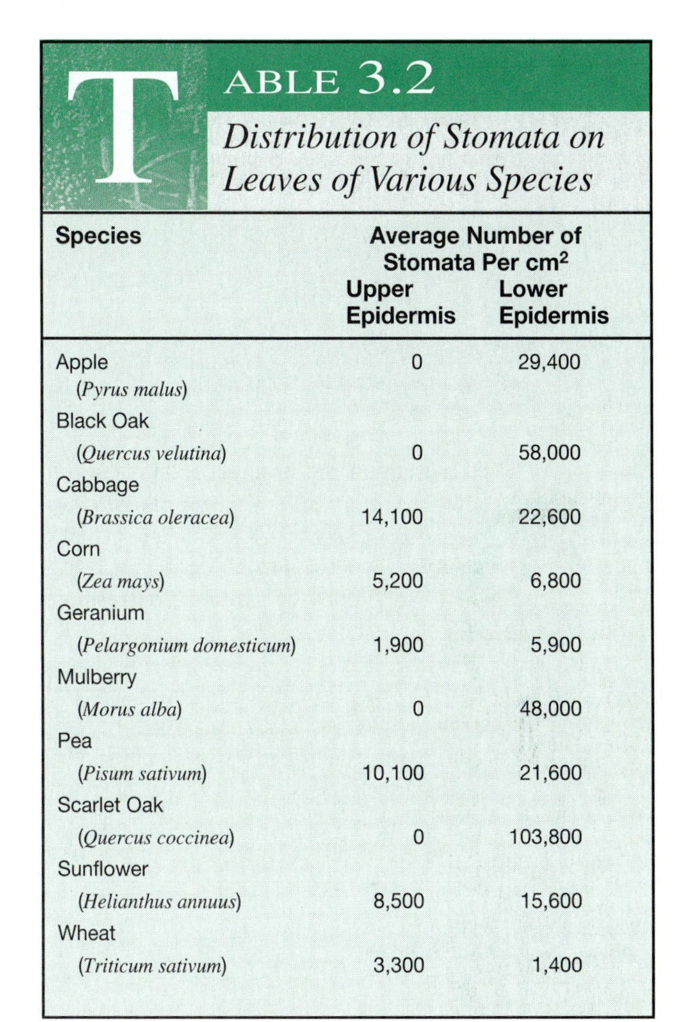

TABLE 3.2

Distribution of Stomata on Leaves of Various Species

Species	Average Number of Stomata Per cm² Upper Epidermis	Average Number of Stomata Per cm² Lower Epidermis
Apple (*Pyrus malus*)	0	29,400
Black Oak (*Quercus velutina*)	0	58,000
Cabbage (*Brassica oleracea*)	14,100	22,600
Corn (*Zea mays*)	5,200	6,800
Geranium (*Pelargonium domesticum*)	1,900	5,900
Mulberry (*Morus alba*)	0	48,000
Pea (*Pisum sativum*)	10,100	21,600
Scarlet Oak (*Quercus coccinea*)	0	103,800
Sunflower (*Helianthus annuus*)	8,500	15,600
Wheat (*Triticum sativum*)	3,300	1,400

Figure 3.12 *Queen Anne's lace, the flowers of wild carrot.*

A Closer Look 3.3

Supermarket Botany

A trip through the produce section in your neighborhood market is a chance to test your knowledge of plant structure. To play "supermarket botany," determine what plant organs are represented by the following vegetables:

1. Beet ______
2. Celery ______
3. Cabbage ______
4. Potato ______
5. Asparagus ______
6. Onion ______
7. Rhubarb ______
8. Pumpkin ______
9. Rutabaga ______
10. Brussel sprouts ______

Answers:

1. A taproot; almost all root vegetables are tap roots.
2. A petiole, although the petiole is greatly enlarged and you can see the remains of the blade.
3. Leaves, of course, but actually a whole stem with shortened internodes and tightly packed leaves.
4. An underground stem actually; even buds are present as the "eyes."
5. A stem again, one of the few aerial stems that we eat.
6. Underground leaves and stem, but the stem is usually too tough to eat.
7. A petiole again; in this case knowledge of plant anatomy may save your life since all other parts of the plant are poisonous.
8. Botanically it is not a vegetable at all, but a fruit; *see* Chapter 6.
9. A taproot again; the result of a medieval cross between cabbage and turnip.
10. These are actually leaf buds that look like miniature cabbages.

Box Figure 3.3 *Produce section in a supermarket offers a chance to study plant anatomy close up.*

The orange color is due to the pigment beta-carotene, an important dietary nutrient. When we eat carrots, the pigment is converted into vitamin A, a vitamin with many functions in the body. One important role of this vitamin is in night vision. The old wives' tale may be true: eating carrots may help you see better in the dark. Although vitamin A is necessary for healthy skin, eating too many carrots can turn the skin yellow. This condition is known as carotenemia, a harmless condition that will disappear a few weeks after the person stops eating carrots. Recently, scientific research suggested an additional benefit to eating foods rich in vitamin A; they may lower the risk of developing cancers of the larynx, esophagus, and lung. Current research is also investigating the value of vitamin A as an antioxidant (*see* Chapter 10).

Carrots also contain sizeable amounts of potassium, calcium, phosphorus, and sugar. Their high sugar content has made them the key ingredient in many desserts, such as the familiar carrot cake.

Lettuce

Contemporary Americans seem to have a love affair with lettuce (*Lactuca sativa*). It is the national favorite cold vegetable. This passion is something we share with the ancient Romans who traditionally began their feasts with a salad of lettuce. The Romans also believed that lettuce had medicinal values as a soporific (sleep inducer). Lettuce juice and lettuce teas were used for their sedative effects in colonial America; in fact, extracts from wild lettuce were used for this purpose until World War II.

Lettuce was first introduced into the New World by Columbus, who planted some in the West Indies in 1493. By the 1880s there were over 100 cultivars available in the United States and many of our modern varieties can be found in seed catalogs from that time.

Nutritionally, the lettuce leaf is mainly water with some vitamin A, calcium, and vitamin C. There are three basic types of this leafy vegetable: head lettuce, looseleaf, and cos. Head lettuce forms a dense, tightly packed head of leaves such as the familiar Iceberg lettuce, which was introduced in 1894. Iceberg lettuce dominates the United States market because of its ease in transport and storage, even though it is the least tasty and least nutritious variety. Other popular head lettuces are Boston and Bibb. Many loose-leaf varieties of lettuce can be found in today's supermarket. Among the most popular are redleaf, oakleaf, and green leaf. These lettuces are nonheading, with their ruffled leaves forming loose clusters. Cos varieties form an upright cylindrical head composed of long leaves. The heads of cos varieties are not compact; romaine lettuce is the most popular. Both the names cos and romaine reflect the ancestry of this type. Romans originally obtained the lettuce from the Greek Island of Cos.

Radishes

Today the radish, *Raphanus sativus,* is considered just a peppery garnish for salads, but in the past radishes were enormous, mild-tasting vegetables that were usually cooked. These cooking or winter radishes were valued because they could be easily stored in root cellars through the winter. Our present-day radishes, known as summer radishes, were first developed in the eighteenth century. Containing only potassium and some iron, the summer radish is not particularly nutritious, but it makes an attractive and tasty addition to salad greens.

Botanically the radish is a composite vegetable consisting of both root and hypocotyl (the base of the stem). Like the carrot, the radish is a biennial that has developed this underground storage organ to fuel the second year's growth.

Although Americans favor the small red globose varieties, radishes come in all shapes, sizes, and colors. The Japanese daikon is scarcely recognizable as a radish. It is white, carrot-shaped, and approximately 46 cm (18 inches) long. This vegetable is a staple of Oriental cuisine; the Japanese have at least 100 different ways of cooking daikon.

An unusual use of radishes is seen in the festival "La Noche de Rabanos" (The Night of the Radishes), which is held each year on December 23 in Oaxaca City in southern Mexico. Radishes grown in the rocky soils of this region are often grotesquely misshapen and may resemble human or animal forms. In 1889, these bizarre vegetables were part of displays in an agricultural exhibit. This event developed into a yearly festival. Over the years, the displays became more and more elaborate. Today the radishes are carved into detailed figures and arranged into dioramas that depict historical and religious themes such as the Mexican Revolution or the Nativity.

Delving into the backgrounds of other common vegetables will reveal other interesting facts and folklore; there's more to vegetables than just good nutrition.

Chapter Summary

1. Meristems are regions of active cell division and are the source of cells for the various tissue types in the plant body. Growth at apical meristems is primary growth. Increase in the girth in woody plants is called secondary growth.
2. There are three basic tissue types in plants: dermal, ground, and vascular. Dermal tissues include the epidermis and periderm and cover the surface of a plant. Ground tissues include parenchyma, collenchyma, and sclerenchyma. Parenchyma is the most abundant and versatile plant tissue, functioning in storage and photosynthesis. Both collenchyma and sclerenchyma are tissues of support. There are two types of sclerenchyma: fibers and sclereids.
3. Vascular tissues are the conducting tissue in plants. Xylem conducts water and minerals from the soil upward, while in phloem organic solutes are moved throughout the plant body. Tracheids and vessel elements are the water-conducting cells of xylem. Sieve tube members, assisted by companion cells, conduct organic solutes through the phloem.
4. The vegetative organs of the plant body include roots, stems, and leaves. Roots anchor the plant and absorb water and minerals from the soil. Two types of root systems are found: the taproot and fibrous root. At the tip of any root there are four distinct regions or zones: root cap, the zone of cell division, zone of cell elongation, and the zone of maturation. Vascular tissue in a nonwoody root is organized into a stele; the pattern of xylem and phloem in the stele varies in monocot and dicot roots. The stele is surrounded by a cortex and an outer layer of epidermis. Extending from the epidermal cells are root hairs through which most of the water and minerals that enter a root are absorbed.
5. Stems support leaves to maximize light absorption and are part of the conduit for the transport of water, minerals, and organic solutes. Leaves are the main

photosynthetic structures in most plants. Unlike roots, the vascular tissue in both stems and leaves is organized into vascular bundles. In stems of herbaceous dicots, the vascular bundles are arranged in a ring around a pith; in monocots, the vascular bundles are scattered. In woody dicots, the discrete vascular bundles are replaced by continuous rings of xylem that correspond to the xylem produced during a single growing season.

6. A leaf may have three parts: the blade, the petiole, and a pair of stipules. If the blade is undivided, the leaf is said to be simple; if the blade is divided into separate leaflets, the leaf is compound. According to the pattern of the leaflets, compound leaves may be pinnately or palmately compound. Leaf venation patterns are either parallel (most monocots) or net (most dicots). The entire leaf surface is covered by epidermis; the epidermis secretes a waxy layer, the cuticle. Guard cells are found in the leaf epidermis. They regulate the entry and exit of gases through the stomata. The mesophyll of the leaf is composed of photosynthetic parenchyma cells, palisade, and spongy cells.
7. An examination of the produce section of a supermarket reveals that many of our common vegetables can be identified as being one of the organs of a plant. Many vegetables have a value beyond the culinary as sources of medicine and folklore.

Review Questions

1. What is the role of meristematic tissues?
2. Describe the organization of a typical herbaceous dicot or monocot stem.
3. Describe the anatomical differences between monocots and dicots.
4. What cell types and tissues are involved in support?
5. What are the functions of roots, stems, and leaves?
6. How can scientists date wood artifacts from archeological sites?
7. Describe the trapping mechanisms of some common carnivorous plants.

Further Reading

Heslop-Harrison, Yolande. 1978. Carnivorous Plants. *Scientific American*, 238 (2): 104–116.

Hirasuna, Delphine. 1985. *Vegetables,* Chronicle Books, San Francisco, CA.

Mauseth, James D. 1988. *Plant Anatomy.* The Benjamin/Cummings Publishing Co. Inc., Menlo Park, CA.

Mauseth, James D. 1991. *Botany: An Introduction to Plant Biology.* Saunders College Publishing, Philadelphia, PA.

Pietropaolo, James and Patricia Pietropaolo. 1986. *Carnivorous Plants of the World.* Timber Press, Portland, OR.

Ross, Gary N. 1986. Night of the Radishes. *Natural History,* 95 (12): 59–64.

Rupp, Rebecca. 1987. *Blue Corn and Square Tomatoes.* Garden Way Publishing, Pownal, VT.

Simpson, Beryl B. and Molly Conner-Ogorzaly. 1995. *Economic Botany: Plants in Our World., 2nd ed.* McGraw-Hill, New York, NY.

Trefil, James. 1985. Concentric Clues from Growth Rings Unlock the Past. *Smithsonian,* 16 (4): 47–54.

4

Plant Physiology

Chapter Outline

Chapter Concepts

1. The movement of water in xylem is a passive phenomenon dependent on the pull of transpiration and the cohesion of water molecules, while the translocation of sugars in the phloem is best described by the Pressure Flow Hypothesis.
2. Plants are dynamic metabolic systems with hundreds of biochemical reactions occurring each second, which enable plants to live, grow, and respond to their environment.
3. Life on Earth is dependent on the flow of energy from the sun, and photosynthesis is the process during which plants convert carbon dioxide and water into sugars using this solar energy.
4. In cellular respiration, the chemical-bond energy in sugars is converted into an energy-rich compound, ATP, which can then be used for other metabolic reactions.

Although plants lack mobility and appear static to the casual observer, they are nonetheless active organisms with many dynamic processes occurring within each part of the plant. Materials are transported through specialized conducting systems, energy is harnessed from the sun, storage products are manufactured, stored foods are broken down to yield chemical energy, and a multitude of products are synthesized. Put simply, plants are bustling with activity. This chapter will consider some of the major transport and metabolic pathways in higher plants.

Plant Transport Systems

As described in the previous chapter, there are two conducting or vascular tissues in higher plants, the xylem and phloem, each with component cell types. Water and mineral transport in the xylem will be described first. Tracheids and vessel elements, which consist of only cell walls after the cytoplasm degenerates, are the actual conducting components in xylem.

The source of water for land plants is the soil. Even when the soil appears dry, there is often abundant soil moisture below the surface. Roots of plants have ready access to this soil water; leaves, however, are far removed from this water source and are normally surrounded by the relatively drier air. The basic challenge is moving water from the soil up to the leaves across tremendous distances, sometimes up to 100 meters (300 ft). This challenge is, in fact, met when water moves through the xylem. There are three components to this movement: the uptake of water from the soil, the conduction in the xylem, and the transpiration from the leaves (fig. 4.1).

Transpiration

Transpiration, the loss of water vapor from leaves, is the force behind the movement of water in xylem. This evaporative water loss occurs mainly through the stomata (90%) and to a lesser extent through the cuticle (10%). When stomata are open, gas exchange occurs freely between the leaf and the atmosphere. Water vapor and oxygen (from photosynthesis) diffuse out of the leaf, while carbon dioxide diffuses into the leaf (fig. 4.1a). The amount of water vapor that is transpired is astounding, with estimates of 2 liters (0.5 gal) of water per day for a single corn plant, 5 liters (1.3 gal) for a sunflower, 200 liters (52 gal) for a

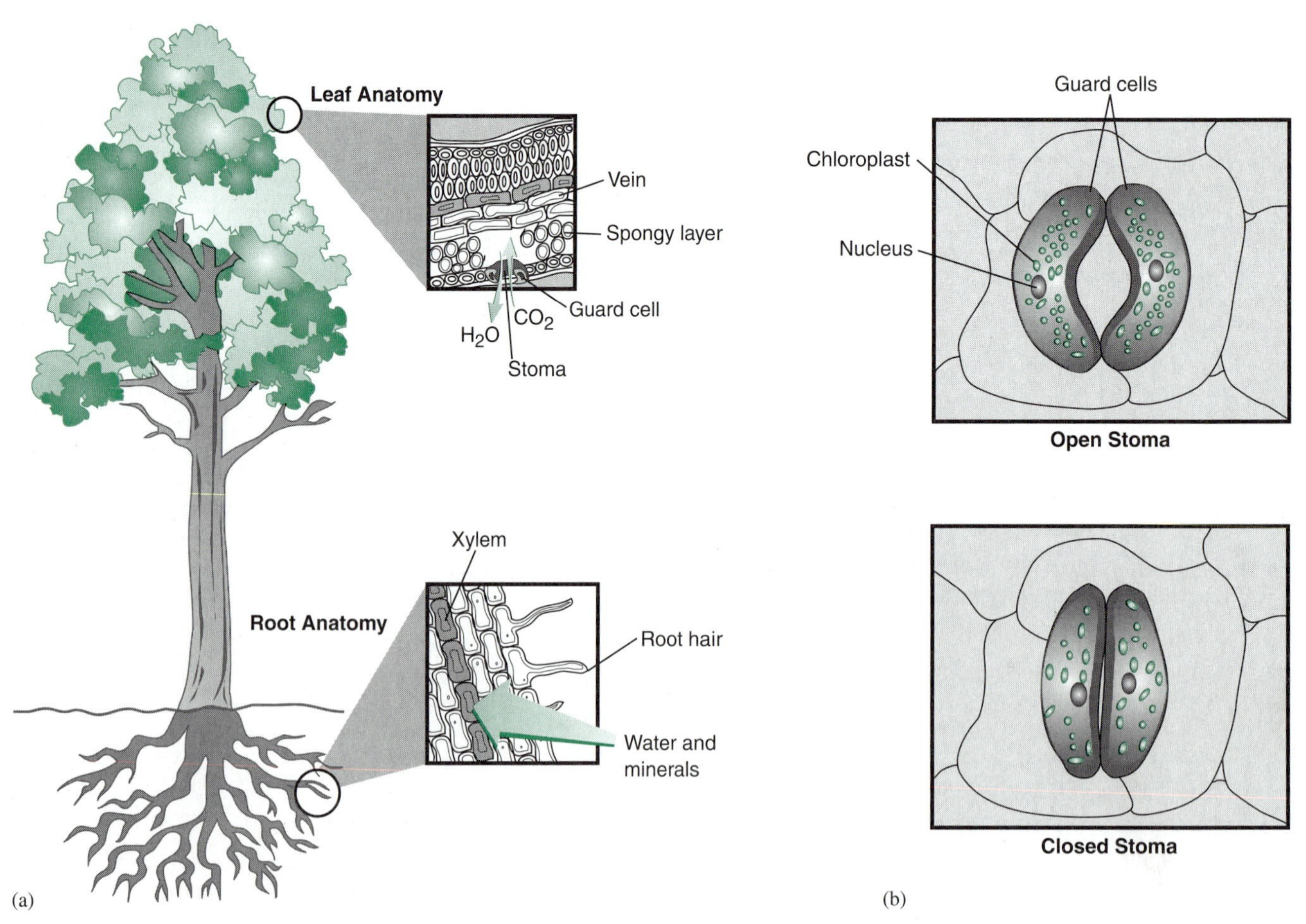

Figure 4.1 *Transpiration is the basic driving force behind water movement in the xylem. (a) When stomata are open, both transpiration and photosynthesis occur as H_20 molecules diffuse out of the leaves and CO_2 molecules diffuse in. (b) When guard cells are turgid, stoma are open and when guard cells are flaccid, stoma are closed.*

large maple tree, and 450 liters (117 gal) for a date palm. Imagine the quantities of water lost each day from the acres of corn and wheat planted in the farm belt of the United States! Clearly, transpiration by plants is a major force in the global cycling of water.

It is the action of the guard cells that regulates the rate of water lost through transpiration and, at the same time, regulates the rate of photosynthesis by controlling the CO_2 uptake. Each stoma is surrounded by a pair of guard cells, which have unevenly thickened walls. The walls of the guard cells that border the stoma are thicker than the outer walls. When guard cells become turgid they can only expand outward due to the radial orientation of cellulose fibrils; this outward expansion of the guard cells opens the stomata (fig. 4.1b). Stomata are generally open during daylight hours and closed at night. As long as the stomata are open, both transpiration and photosynthesis occur, but when water loss exceeds uptake, the guard cells lose turgor and close the stomata. On hot, dry, windy days the high rate of transpiration frequently causes the stomata to close early, resulting in a near shutdown of photosynthesis as well as transpiration. A fine balance must be struck in this photosynthesis-transpiration dilemma to allow enough CO_2 for photosynthesis, while at the same time preventing excessive water loss. Some plants have evolved an alternate pathway for CO_2 uptake at night when rates of transpiration are lower (*see* CAM pathway later in this chapter). Other plants have morphological or anatomical adaptations that reduce rates of transpiration while keeping the stomata open. These physiological and anatomical adaptations are most common in **xerophytes,** plants occurring in arid environments.

The basis of transpiration is the diffusion of water molecules from an area of high concentration within the leaf to an area of lower concentration in the atmosphere. Unless the atmospheric relative humidity is 100%, the air is relatively dry compared to the interior of a leaf where the intercellular spaces are saturated with water vapor. As long as stomata are open, a continuous stream of water vapor transpires from the leaf, creating a pull on the water column that extends from the leaf through the plant to the soil.

Absorption of Water from the Soil

Water and dissolved minerals enter a plant through the root hairs and can follow two paths, via either the **symplast** or the **apoplast.** Water molecules can diffuse through the plasma membrane into the cytoplasm of a root hair cell and continue on this intracellular movement through the cytoplasm of cells in the cortex. The cytoplasm of all cells is interconnected through plasmodesmata and is referred to as the symplast. Thus, this pathway follows the symplast from a root hair cell into the stele (fig. 4.2).

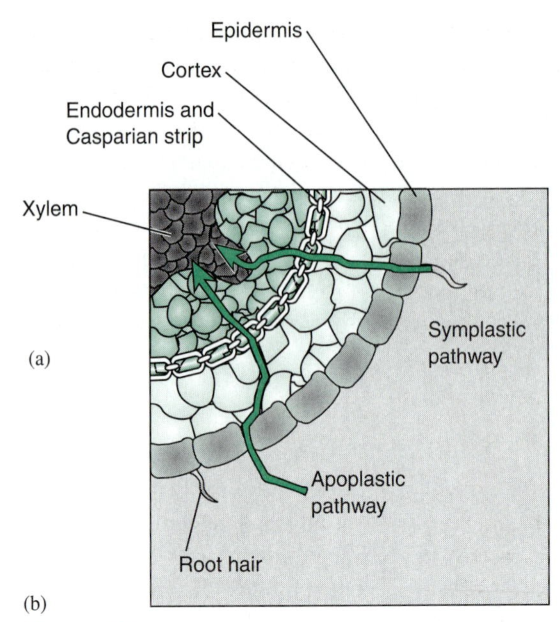

Figure 4.2 *Water and minerals can follow one of two pathways across the cortex into the vascular cylinder. (a) Apoplastic pathway in which water diffuses through the cell walls and intercellular spaces; (b) Symplastic pathway in which water diffuses into the cytoplasm of a root hair cell and continues moving through the cytoplasm from one cell to the next. In both pathways, water must move through the symplast of the endodermal cells.*

A second path is the diffusion of water through the cell walls and intercellular spaces from root hair through the cortex (fig. 4.2). The intercellular spaces and the spaces between the cellulose fibrils in the cell walls comprise the apoplast of a plant; thus the water molecules diffuse unimpeded through the apoplast until they reach the endodermis. The innermost layer of the cortex consists of a specialized cylinder of cells known as the endodermis. The presence of a **Casparian strip** on the walls of the endodermal cells regulates the movement of water and minerals into the stele. The Casparian strip is a layer of suberin (and in some instances lignin as well) on the radial and transverse walls (top, bottom, and sides) that prevents the apoplastic movement of water into the stele. The movement of water is therefore directed to the tangential walls of the endodermis and into the cytoplasm of these cells and, thus, the symplast. By forcing the water and minerals through the symplast, some control over the uptake of minerals is exerted. Some minerals are prevented from entering the stele, while others are selectively absorbed by active transport. Once inside the cytoplasm of the endodermal cells, water moves symplastically into the living cells of the pericycle, the outermost layer of the stele. The water diffuses into the conducting cells of the xylem, drawn by the pull of transpiration.

A Closer Look 4.1

Mineral Nutrition in Plants

Research has shown that certain minerals are required by plants for normal growth and development. These are included in the essential elements listed in Table 1.3. The soil is the source of these minerals, which are absorbed by the plant with the soil water. Even nitrogen, which is a gas in its elemental state, is normally absorbed from the soil in the form of nitrate (NO_3^-) ions. Some soils are notoriously deficient in micronutrients and are thereby unable to support most plant life. Serpentine soils, for example, are deficient in calcium and only plants able to tolerate the low levels of this mineral can survive. In modern agriculture, mineral depletion of soils is a major concern, since harvesting crops interrupts the natural recycling of nutrients back to the soil.

Mineral deficiencies can often be detected by specific symptoms such as chlorosis (loss of chlorophyll resulting in yellow or white leaf tissue), necrosis (isolated dead patches), anthocyanin formation (development of deep red pigmentation of leaves or stem), stunted growth, and development of woody tissue in an herbaceous plant. Soils are most commonly deficient in nitrogen and phosphorus. Nitrogen-deficient plants exhibit many of the symptoms described above. Leaves develop chlorosis; stems are short and slender; and anthocyanin discoloration occurs on stems, petioles, and lower leaf surfaces. Phosphorus-deficient plants are often stunted, with leaves turning a characteristic dark green, often with the accumulation of anthocyanin. Typically, older leaves are affected first as the phosphorus is mobilized to young growing tissue. Iron deficiency is characterized by chlorosis between veins in young leaves (box fig. 4.1a, b).

Much of the research on nutrient deficiencies is based on growing plants **hydroponically,** using soil-less nutrient solutions. This technique allows researchers to create solutions that selectively omit certain nutrients and then observe the resulting effects on the plants. Hydroponics has applications beyond basic research, since it facilitates the growing of greenhouse vegetables during winter months. Aeroponics, a technique in which plants are suspended and the roots misted with a nutrient solution, is another method for growing plants in soil-less culture.

(a)

(b)

Box Figure 4.1 *The effects of mineral deficiencies are shown in these sunflower plants grown in hydroponic culture. The plants grown in complete nutrient solution are shown on the right, while those on the left are deficient in (a) phosphorus and (b) calcium.*

While mineral deficiencies can limit the growth of plants, an overabundance of certain minerals can be toxic and also limit growth. Most research has focused on the toxic effects of heavy metals such as lead, cadmium, mercury, and aluminum; however, even copper and zinc, which are essential elements, can become toxic in high concentrations. Although most plants cannot survive in these soils, certain plants have the ability to tolerate high levels of these minerals. Saline soils, which have high concentrations of sodium chloride and other salts, also limit plant growth, and research continues to focus on developing salt-tolerant varieties of agricultural crops.

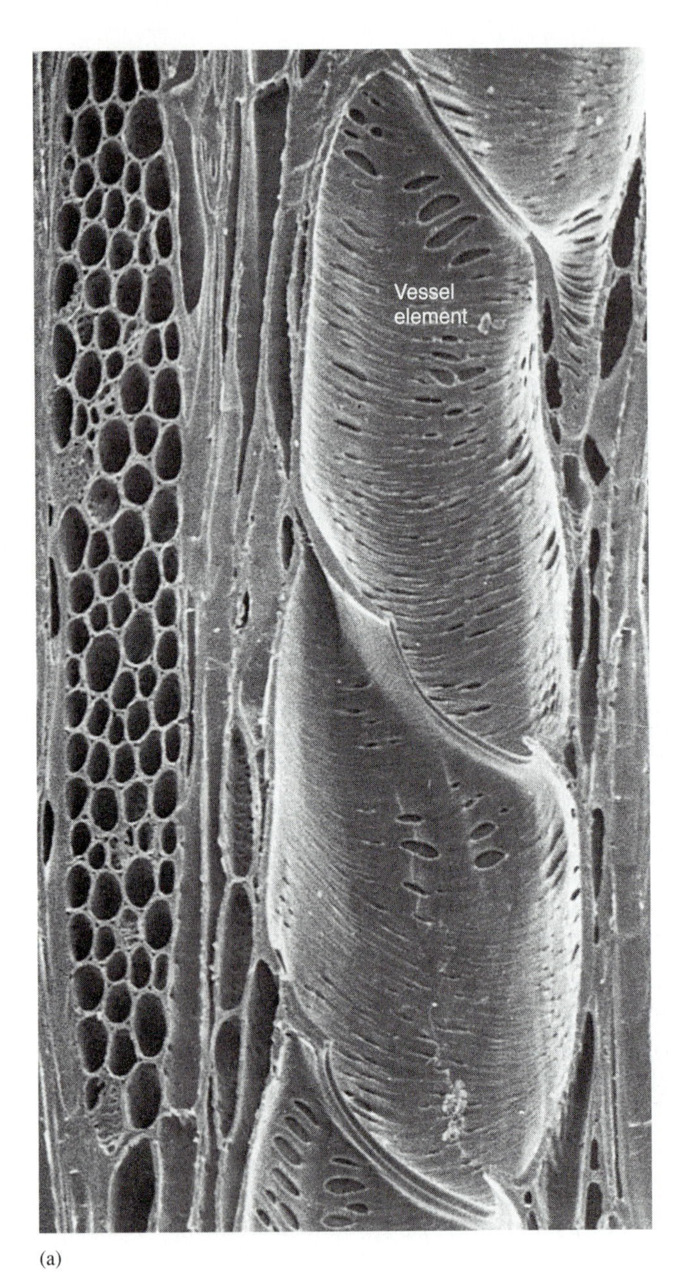

(a)

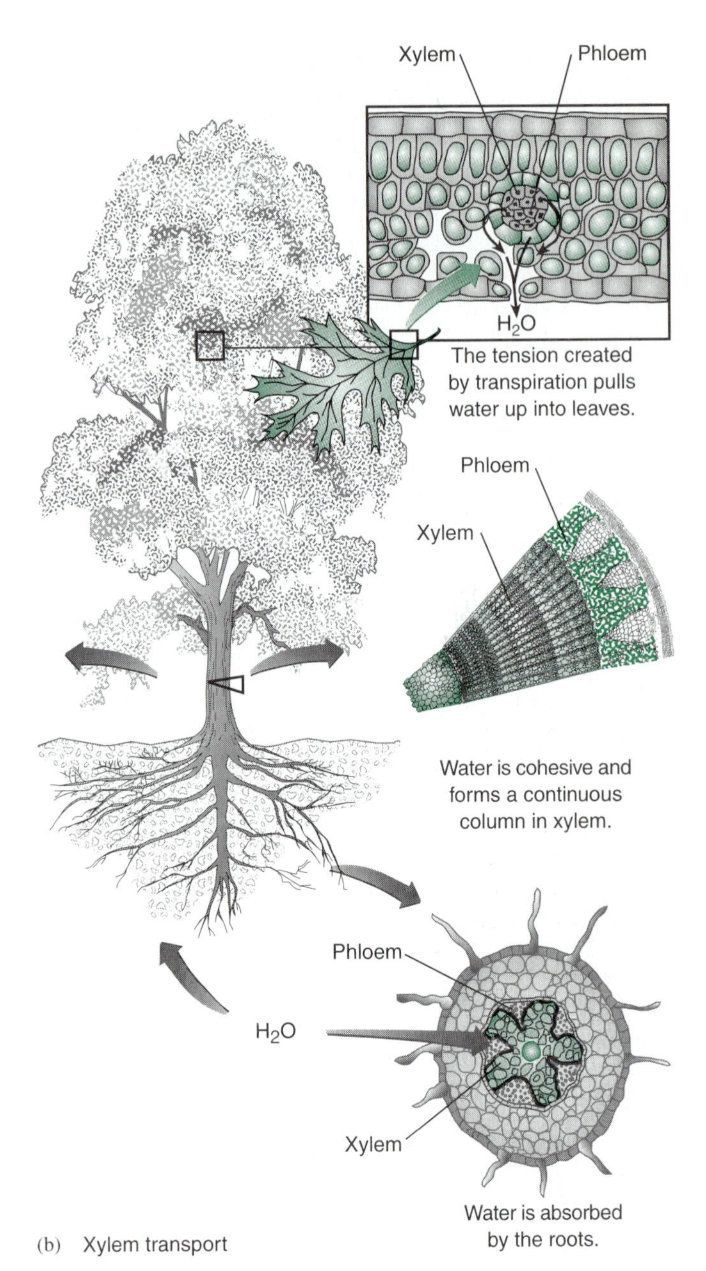

(b) Xylem transport

Figure 4.3 *Transpiration-cohesion theory of xylem transport. (a) Vessel elements join to form a long vessel that may reach from the roots to the stem tip. (b) As transpiration occurs in the leaf, it creates a cohesive pull on the whole water column downward to the roots where water is absorbed from the soil.*

Water Movement in Plants

Once water is in the xylem of the stele, its movement upward in the plant is driven by the pull of transpiration as well as certain properties of the water molecule itself, **cohesion** and **adhesion.** Recall from Chapter 1 that the polarity of water molecules creates hydrogen bonds between adjacent molecules. These hydrogen bonds may form between water molecules themselves (cohesion) or between water molecules and the molecules in the walls of vessel elements and tracheids (adhesion). The adhesion of water to the cellulose walls provides a continuous source of water that will evaporate into the intercellular spaces of the leaf and transpire through the stomata. The cohesive force is so strong that any force or pull on one water molecule acts on all of them as well. As transpiration occurs in the leaf, it creates a cohesive pull on the whole water column downward from the leaf through the xylem to the root, where water uptake occurs to replace the water lost through transpiration (fig. 4.3). This mechanism of water movement in plants is known as the **Transpiration-Cohesion Theory** and has been used to explain rates of water movement as high as 44 meters (145 ft) per hour in angiosperm trees.

Translocation of Sugar

Organic materials are translocated by the sieve tube members of the phloem. In contrast to the xylem, where the

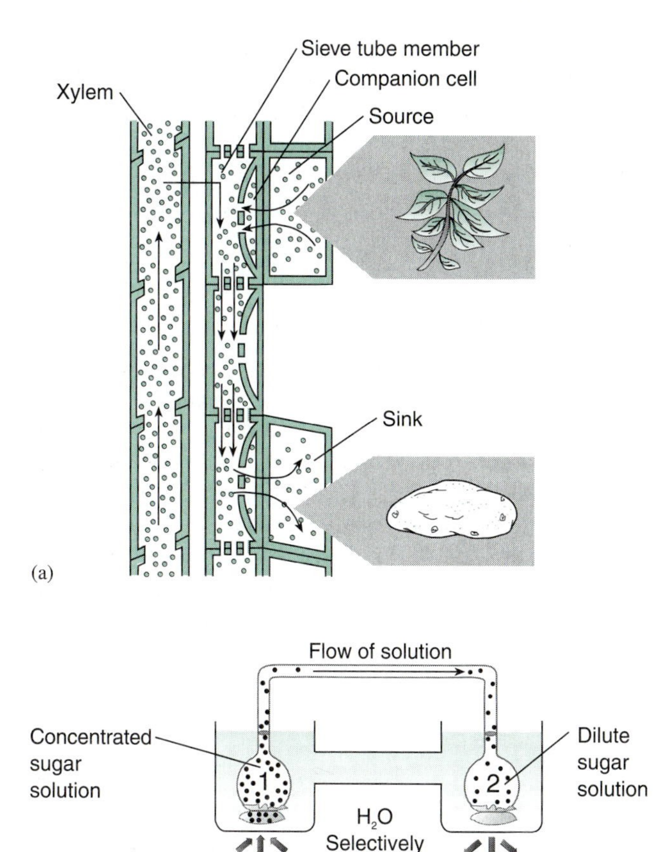

Figure 4.4 *Translocation of sugar. (a) Products of photosynthesis move from source to sink. At the source, sugar molecules are loaded into the phloem. As the sugar concentration increases, water moves in from adjacent xylem, pressure builds up, and the sugar solution is forced through the plant to the sink, where sugar is unloaded from the phloem. (b) A physical model can be used to demonstrate the Pressure-Flow Hypothesis. The concentrated sugar solution on the left is comparable to the source and the dilute solution on the right is comparable to the sink.*

conducting elements function when the cells are dead, the sieve tube members of the phloem are living but highly specialized cells (*see* Chapter 3). While water movement in the xylem is upward from the soil, phloem translocation moves in the direction from **source** to **sink.** In late winter, the source may be an underground storage organ translocating sugars to apical meristems (the sink) in the branches of a tree. In summer, the source may be photosynthetic leaves sending sugars for storage to sinks such as roots or developing fruits (fig. 4.4). In most plants, the primary material translocated in phloem is sucrose in a watery solution that also may include small amounts of amino acids, minerals, and other organic compounds. The rate of translocation in the phloem is quite rapid and has been timed at speeds averaging 1 meter (3.3 ft) per hour. The amount of material translocated is also quite impressive. In a growing pumpkin, which reaches a size of 5.5 kg (11 lbs) in 33 days, approximately 8 g (0.3 oz) of solution are translocated per hour. Each fall at state and county fairs all across the United States, prize-winning pumpkins routinely weigh well over 45.4 kg (100 lbs). All that bulk is translocated by the phloem to the growing fruit.

The hypothesis currently accepted to explain translocation in the phloem is the **Pressure Flow** (or Mass Flow) **Hypothesis.** This is a modified version of the hypothesis first proposed by Ernst Münch in 1926. According to this hypothesis, there is a bulk flow of solutes from source to sink (fig. 4.4). At the source, phloem loading takes place as sugar molecules are first actively transported into companion cells and then move symplastically into sieve tube members through plasmodesmata. This highly concentrated solution in the sieve tube members causes water to enter by osmosis from nearby xylem elements, resulting in a buildup of pressure. When pressure starts to build in these cells, the solute-rich phloem sap is pushed through the pores in the sieve plate into the adjacent sieve tube member and so on down to the sink. This movement of material *en masse* is known as mass flow. At the sink, companion cells function in active phloem unloading, which reduces the concentration of sugars and allows water to diffuse out of these cells. Sugars unloaded at the sink are picked up by nearby cells and either stored as starch or metabolized.

Metabolism

Metabolism is the sum total of all chemical reactions occurring in living organisms. Metabolic reactions that synthesize compounds are referred to as **anabolic** reactions and are generally **endergonic,** requiring an input of energy. In contrast, **catabolic** reactions, which break down compounds, are usually **exergonic** reactions that release energy. Many of these reactions also involve the conversions of energy from one form to another.

Energy

All life processes are driven by energy and, consequently, a cell or an organism deprived of an energy source will soon die. **Energy** is defined by physicists as the ability to do work and is governed by certain physical principles such as the Laws of Thermodynamics.

The First Law of Thermodynamics states that energy can neither be created nor destroyed but it can be converted from one form to another.

Among the forms of energy are radiant (light), thermal (heat), chemical, mechanical (motion), and electrical. One focus of this chapter is photosynthesis, the process that converts radiant energy from the sun into the chemical energy of a sugar molecule.

The Second Law of Thermodynamics implies that in any transfer of energy there is always a loss of useful energy to the system, usually in the form of heat.

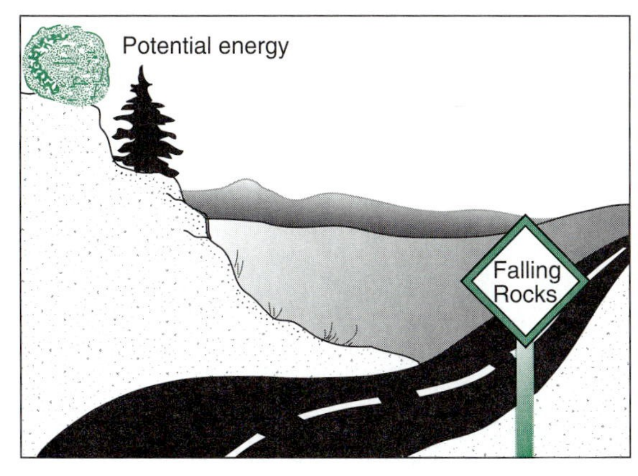

(a)

(b)

Figure 4.5 *Potential and kinetic energy. (a) The boulder at the top of a hill has a tremendous amount of potential energy. (b) If the boulder rolls down the hill, the potential energy is converted to kinetic energy. To push the boulder back up to the top would require a considerable input of energy.*

When gasoline is burned as fuel to drive an automobile engine, chemical energy is converted into mechanical energy but the conversion is not very efficient. Some of the energy is lost as heat to the surroundings.

All forms of energy can exist as either **potential energy** or **kinetic energy.** Potential energy is stored energy that has the capacity to do work, while kinetic energy actually is doing work or is energy in action. For example, a boulder at the top of a hill has a tremendous amount of potential energy. If it rolls down the hillside, the potential energy is changed into kinetic energy, an exergonic process (fig. 4.5). To push it back up to the top of the hill would be an endergonic process requiring considerable input of energy. Transformations from potential to kinetic and vice-versa occur constantly in biological systems and are part of the underlying principles of both photosynthesis and respiration.

Redox Reactions

Many energy transformations in cells involve the transfer of electrons or hydrogen atoms. When a molecule gains an electron or hydrogen atom, the molecule is said to be **reduced,** and the molecule that gives up the electron is said to be **oxidized.** A molecule that has been reduced has gained energy; likewise the oxidized molecule has lost energy. Oxidation and reduction reactions are usually coupled (sometimes called **redox reactions**); as one molecule is oxidized, the other is simultaneously reduced.

$$\underset{\text{(A-reduced)}}{AH_2} + \underset{\text{(B-oxidized)}}{B} \longrightarrow \underset{\text{(A-oxidized)}}{A} + \underset{\text{(B-reduced)}}{BH_2}$$

In many oxidation-reduction reactions, an intermediate is used to transport electrons from one reactant to another. One such electron intermediate is **NAD** (nicotinamide adenine dinucleotide), which can exist in both oxidized and reduced states **(NAD^+ = oxidized form and NADH + H^+ = reduced form).** Similarly, **NADP** and **FAD** also can exist as **$NADP^+$/NADPH + H^+** and **FAD/$FADH_2$,** respectively. NAD and FAD are common electron carriers in respiration, while NADP serves the same function in photosynthesis.

Phosphorylation

Other energy transformations involve the transfer of a phosphate group. When a phosphate group is added to a molecule, the resulting product is said to be **phosphorylated** and has a higher energy level than the original molecule. These phosphorylated compounds may also lose the high energy phosphate group and thereby release energy. The energy currency of cells, **ATP** (adenosine triphosphate), is constantly recycled in this way. When a phosphate group is removed from ATP, **ADP** (adenosine diphosphate) is formed and energy is released in this exergonic reaction.

$$ATP \longrightarrow ADP + PO_4 + \text{energy}$$

To recreate ATP, a phosphate group must be added to ADP (an endergonic reaction) with the appropriate input of energy.

$$ADP + PO_4 + \text{energy} \longrightarrow ATP$$

Enzymes

Proteins that act as catalysts for chemical reactions in living organisms are known as **enzymes. Catalysts** speed up the rate of a chemical reaction without being used up or changed during the reaction. The majority of chemical reactions that occur in living organisms require enzymes. Enzymes are highly specific for certain reactants; the compound acted upon by the enzyme is known as the **substrate.** The names of enzymes most commonly end in the suffix *-ase,* which is sometimes appended to the name of the substrate or the type of reaction. Some enzymes function properly only in the presence of **cofactors** or **coenzymes.** Cofactors are inorganic, often metallic, ions such as Mg^{++} and Mn^{++}, while organic molecules such as NAD, NADP, and some vitamins are coenzymes. Both cofactors and coenzymes are loosely associated with

A Closer Look 4.2

Sugar and Slavery

Products of photosynthesis are typically transported to growing fruits, storage organs, and other sinks throughout the plant. After being unloaded from the phloem, sugars are usually converted to starch or other storage carbohydrates. Although the disaccharide sucrose is the material translocated in the phloem of most plants, very few species store significant amounts of this sugar. Only two plants, sugarcane and sugar beet, are commercially important sources of sucrose, commonly known as table sugar.

Sugarcane is native to the islands of the South Pacific and has been grown in India since antiquity. Small amounts of sugarcane reached the ancient civilizations in the Near East and Mediterranean countries through Arab trading routes, but it was not grown in these regions until the seventh century. Even after cultivation was established, honey remained the principal sweetener in Europe until the fifteenth century. During the Middle Ages, sugar was an expensive luxury that found its greatest use in medicinal compounds to disguise the bitter taste of many herbal remedies. Early in the fifteenth century, sugar plantations were established on islands in the eastern Atlantic: on the Canary Islands by Spain, as well as Madeira and the Azores by Portugal.

Columbus introduced sugarcane to the Caribbean islands on his second voyage in 1493. By 1509, sugarcane was harvested in Santo Domingo and Hispaniola and soon spread to other islands. In fact, many Caribbean islands were eventually denuded of native forests and planted with sugar cane. The Portuguese saw the opportunities in South America and started sugar plantations in Brazil in 1521. Although late to enter the West Indies, the British established colonies in the early seventeenth century, and, by the 1640s, sugar plantations were thriving on Barbados. The first sugarcane grown in the continental United States was in the French colony of Louisiana in 1753.

The growing of sugarcane was responsible for the establishment of slavery in the Americas, and early in the sixteenth century, sugar and the slave trade became interdependent. Decimation of the native Indian populations led to the need for workers on the sugar plantations. The first suggestion to use African slaves was made in 1517. Within a short time, this was reality in both the Spanish and Portuguese colonies, with the greatest number of slaves imported into Brazil. The introduction of African slaves to these colonies was an outgrowth of the slave trade in Spain and Portugal that began in the 1440s. Initially Spain exported the slaves, but by 1530 slaves were sent directly from Africa to the Caribbean.

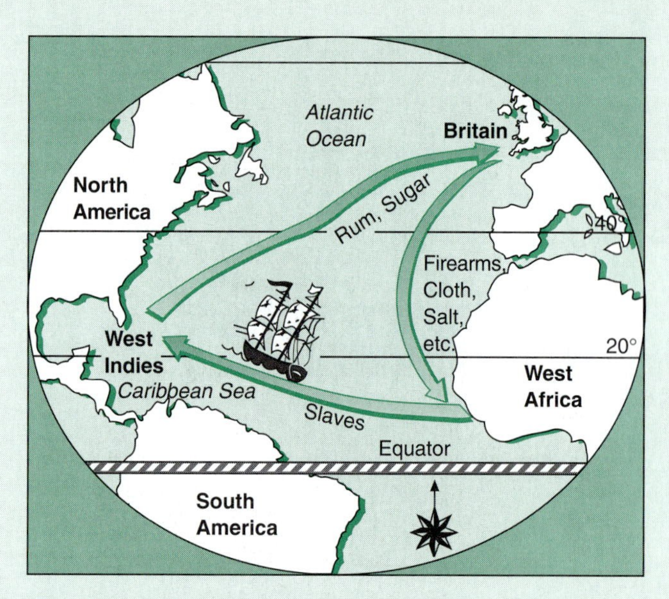

Box Figure 4.2a *The Triangular Trade.*

The sugar production in the Caribbean came at a time when supplies of honey in Europe were decreasing. The Catholic monasteries were the traditional source of honey. Beehives were kept, principally, to produce beeswax for church candles. During the Protestant Reformation, Catholic monasteries were suppressed, and the sources of honey fell short of demand. Also, in the late seventeenth century, the introduction and growing popularity of coffee, tea, and cocoa in Europe accelerated the demand for sugar, since Europeans generally disliked the naturally bitter taste of these beverages. Sugar became the most important commodity traded in the world, and eventually England became the dominant force in this enterprise. The triangular trade was the source of many fortunes. The first leg of the triangle was from England to West Africa with trinkets, cloths, firearms, salt, and other commodities that were bartered for slaves. The second leg brought slaves to the Caribbean Islands where they were sold. The final leg of the journey carried rum, molasses, and sugar back to England (box fig. 4.2a). A second triangle became important in the mideighteenth century, linking the West Indies, New England, and West Africa. The use of slaves on sugar plantations continued until the early nineteenth century,

(b)

(c)

Box Figure 4.2b and c *Major sugar crops. (b) Sugarcane and (c) Sugar beet.*

when the slave trade was abolished. During this period it has been estimated that 10–20 million African slaves had been brought to the New World.

Sugarcane, *Saccharum officinarum,* is a perennial member of the grass family, the Poaceae (box fig. 4.2b). Several species of *Saccharum* are known to exist in the tropics, and it is believed that *S. officinarum* originated as a hybrid of several species. The species owes its importance to the sucrose stored in the cells of the stem. Sugarcane, which uses the C_4 pathway for photosynthesis, is considered one of the most efficient converters of solar energy into chemical energy. Canes are often 5 to 6 meters (15 to 20 feet) tall, with individual stalks up to 10 cm (4 in) in diameter. Plants are grown from stem cuttings, with each segment containing three or four nodes and each node at least one bud. Segments are laid horizontally, bud upward, in shallow trenches. Roots soon develop from the node. Generally 12–18 months are needed before the canes can be harvested. On fertile land, subsequent crops develop from the rhizome for two or three years before replanting is necessary. Sugarcane thrives in moist lowland tropics and subtropics and today provides over 50% of the world's sugar supply.

Canes generally contain 12% to 15% sucrose. After harvesting, canes are crushed by heavy steel rollers to extract the sugary juice. The fibrous residue (bagasse) can be used to make fiberboard, paper, and other products, or used as compost. Successive boilings concentrate the sucrose; impurities are usually precipitated by adding lime water (calcium oxide solution) and removed by filtering. The solution is then evaporated to form a syrup from which the sugar is crystalized. Centrifuges separate the thick brown liquid portion from the crystals. The liquid portion is molasses, which is used in foods, or is fermented to make rum, ethyl alcohol, or vinegar. The crystallized sugar (about 96%–97% pure sucrose) is further refined to free it from any additional impurities.

Sugar beet, *Beta vulgaris,* a member of the Chenopodiaceae (Goosefoot Family), is unrelated to sugarcane (box fig. 4.2c). It is actually the same species as red beets, which are native to the Mediterranean region and have been consumed since the time of the ancient Romans. In the mideighteenth century a German chemist, Andreas Marggraf, discovered that the roots contained sugar, which was chemically identical to that from cane. The rise of the sugar beet industry can be tied to the Emperor Napoleon I. When a British naval blockade cut off the sugar imports to France, Napoleon realized the value of a domestic source of sugar and encouraged scientific research on the sugar beet. After 1815, sugarcane imports were restored, halting the developing sugar beet industry. By the early twentieth century, the industry was revived in both Europe and North America and today sugar beets provide close to 40% of the world's supply of table sugar.

Sugar beet is a biennial plant, but it is harvested at the end of the first year when the sucrose content is greatest. Selective breeding gradually raised the percent of sugar in the root from 2% to approximately 20%. After harvesting, roots are shredded, steeped in hot water, and then pressed to extract the sucrose. Further processing is similar to sugarcane and produces an identical final product.

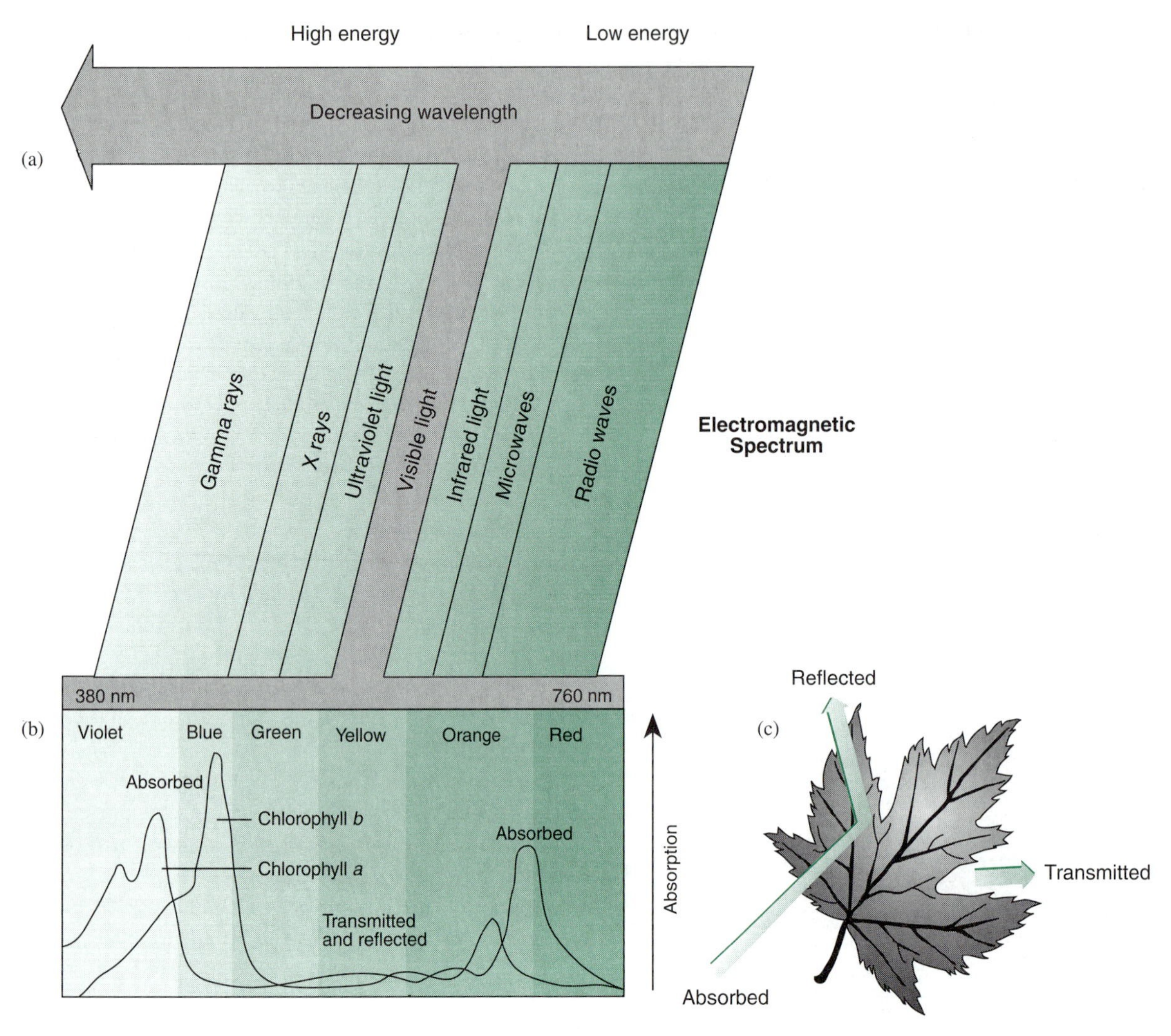

Figure 4.6 *Energy from the sun drives the process of photosynthesis. (a) Visible light is only a small portion of the electromagnetic spectrum. (b) If visible light is passed through a prism, the component colors are apparent. The chlorophyll pigments in leaves absorb the blue-violet and orange-red portions of the spectrum. (c) Leaves reflect the green and yellow portions of the spectrum.*

enzymes; however, **prosthetic groups** are nonprotein molecules that are attached to some enzymes and necessary for enzyme action.

PHOTOSYNTHESIS

Photosynthesis is the process that transforms the vast energy of the sun into chemical energy and is the basis for most food chains on Earth. The overwhelming majority of life depends on the photosynthetic ability of green plants and algae, and without these producers life, as is known today, could not survive.

Energy from the Sun

The sun is basically a thermonuclear reactor producing tremendous quantities of **electromagnetic radiation,** which bathes the Earth. Visible light is only a small portion of the electromagnetic spectrum that includes radiowaves, microwaves, infrared radiation, ultraviolet radiation, X rays, and gamma rays (fig. 4.6). This radiant energy or light has a dual nature consisting of both particles and waves. The particles are known as **photons** and have a fixed quantity of energy. It is believed that the photons travel in waves and thus display characteristic **wavelengths.** Wavelengths vary from radiowaves that may be over one kilometer in length to gamma rays that are a fraction of a nanometer (nm). The energy content also varies and is inversely proportional to the wavelength; i.e., the longer the wavelength, the lower the energy.

Approximately 40% of the radiant energy reaching the Earth is in the form of **visible light.** If visible light passes through a prism, the component colors become apparent; these range from red at one end of the visible band to blue-violet at the other end. The wavelengths of visible light range from 380 nm (violet) to 760 nm (red) and are the wavelengths most important to living organisms (fig. 4.6). In fact, the wavelengths within this range are ones absorbed by the chlorophylls and other photosynthetic pigments in green plants and algae.

Light-Absorbing Pigments

When light strikes an object, the light can pass through the object (be transmitted), be reflected from the surface, or be absorbed. For light to be absorbed, pigments must be present. Pigments absorb light selectively, with different pigments absorbing different wavelengths and reflecting others. Each pigment has a characteristic **absorption spectrum,** which depicts the absorption at each wavelength. If all visible wavelengths are absorbed, the object appears black; however, if all wavelengths are reflected, the object appears white. Green leaves appear green because these wavelengths are reflected.

In higher plants, the major organ of photosynthesis is the leaf, and the green chloroplasts within the mesophyll are the actual sites of this process. The major photosynthetic pigments are the green **chlorophylls** that are located on the **thylakoid membranes** of the chloroplasts (fig. 4.7).

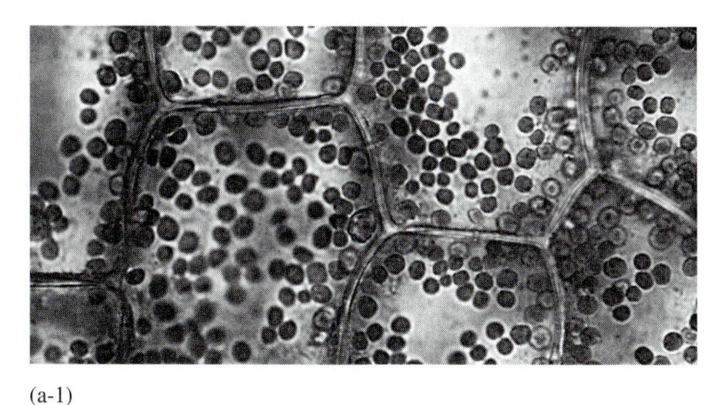

(a-1)

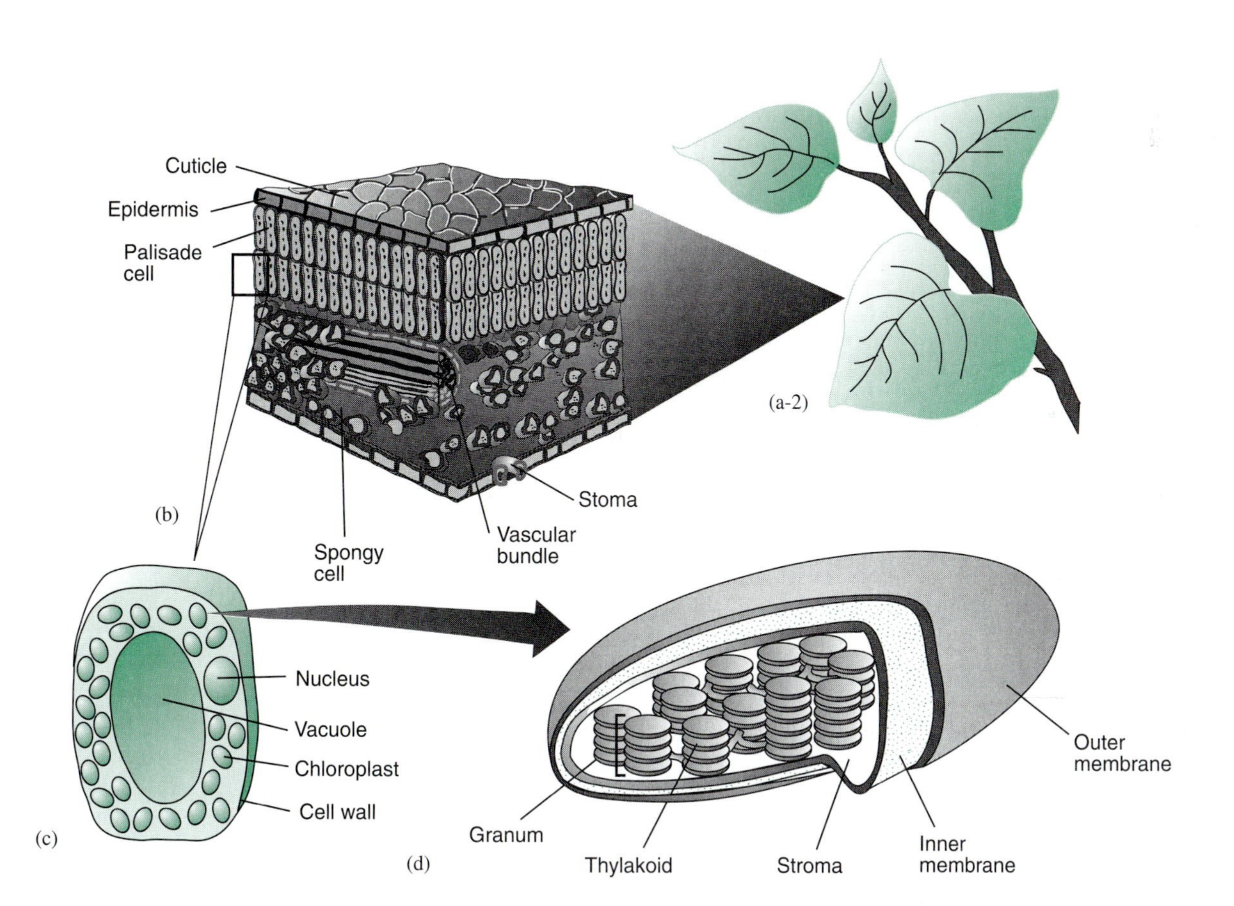

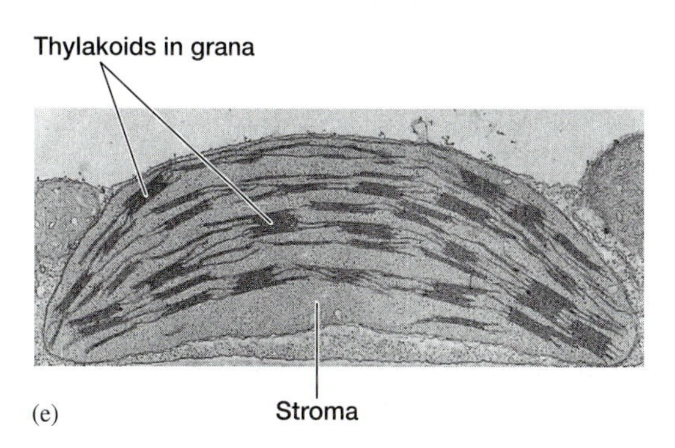

Figure 4.7 *The major organ of photosynthesis is the leaf, and the actual site of photosynthesis is the chloroplast. (a)-(c) Leaf cells with chloroplasts; (d)-(e) Internal structure of the chloroplast. Thylakoid sacs comprise the grana, sites of the light reactions. The stroma contains the enzymes that carry out the Calvin cycle.*

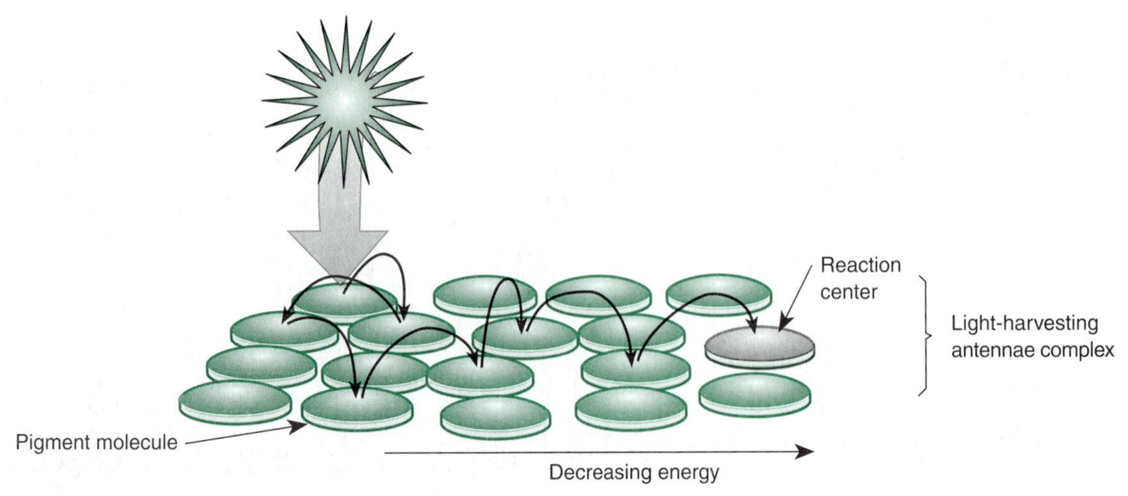

Figure 4.8 *Photosystem. Each photosystem is composed of several hundred chlorophyll and carotenoid molecules that make up light-harvest antennae. When light strikes a pigment molecule, the energy is funnelled into a reaction center.*

Thylakoids can be found in stacks known as **grana,** as well as individually in the **stroma,** the enzyme-rich ground substance of the chloroplast. Chloroplasts may have 50 to 80 grana, each with about 10 to 30 thylakoids apiece. The chlorophylls can be located on the **stroma thylakoids,** as well as in the grana, and it is the abundance of these pigments that make leaves appear green.

In green plants there are two forms of chlorophyll, *a* and *b,* which differ slightly in chemical makeup. Most chloroplasts have 3 times more chlorophyll *a* than *b.* The absorption spectra of the chlorophylls show peak absorbances in the red and blue-violet regions, with much of the yellow and green light reflected (fig. 4.6).

In addition to chlorophylls, chloroplasts also contain accessory pigments, the **carotenoids.** These include the orange **carotenes** and yellow **xanthophylls,** which absorb light in the violet, blue, and blue-green regions of the spectrum. Although present in all leaves, these pigments are normally masked by the chlorophylls. Recall that these carotenoids become apparent in autumn in temperate latitudes, when chlorophyll degrades.

The Light Reactions

Photosynthesis consists of two major phases, the **Light Reactions** and the **Calvin Cycle.** The light reactions constitute the photochemical phase of photosynthesis, during which radiant energy is converted into chemical energy. During the light reactions, water molecules are split, releasing oxygen and providing electrons for the reduction of $NADP^+$ to NADPH + H^+. The light reactions also provide the energy for the synthesis of ATP. The Calvin Cycle constitutes the biochemical phase and involves the fixation and reduction of CO_2 to form sugars using the ATP and NADPH produced in the light reactions.

The light reactions are composed of two cooperating photosystems, **Photosystem I and II,** and take place on the thylakoid membranes within the chloroplasts. Each photosystem is a complex of several hundred chlorophyll and carotenoid molecules (known as **light-harvesting antennae**) and associated membrane proteins. Countless units of these photosystems are arrayed on the thylakoid membranes throughout the chloroplast. When light strikes a pigment molecule in either photosystem, the energy is funnelled into a **reaction center,** which consists of a chlorophyll *a* molecule bound to a membrane protein (fig. 4.8). The reaction center for Photosystem I is known as $\mathbf{P_{700}}$**,** which indicates the wavelength of maximum light absorption in the red region of the spectrum, while the reaction center for Photosystem II is $\mathbf{P_{680}}$**,** again indicating the peak absorbance. Associated with the photosystems are various enzymes and coenzymes that function as electron carriers and are components of the thylakoid membranes.

When a photon of light strikes a pigment molecule in the light-harvesting antennae of Photosystem I, the energy is funnelled to P_{700} (fig. 4.9). When P_{700} absorbs this energy, an electron is excited and ejected, leaving P_{700} in an oxidized state. The ejected electron is picked up by a primary electron acceptor, which then passes the electrons on to **ferredoxin (FD),** another electron intermediate, and eventually to $NADP^+$, reducing it to NADPH + H^+, one of the products of the light reactions.

Another photon of light absorbed by a chlorophyll molecule in Photosystem II will transfer its energy to the reaction center P_{680} (fig. 4.9). When P_{680} absorbs this energy, an excited electron is ejected and passed on to another primary electron acceptor, leaving P_{680} in an oxidized state. The electron lost by P_{680} is replaced by an electron from water, in a reaction that is not fully understood, and catalyzed by

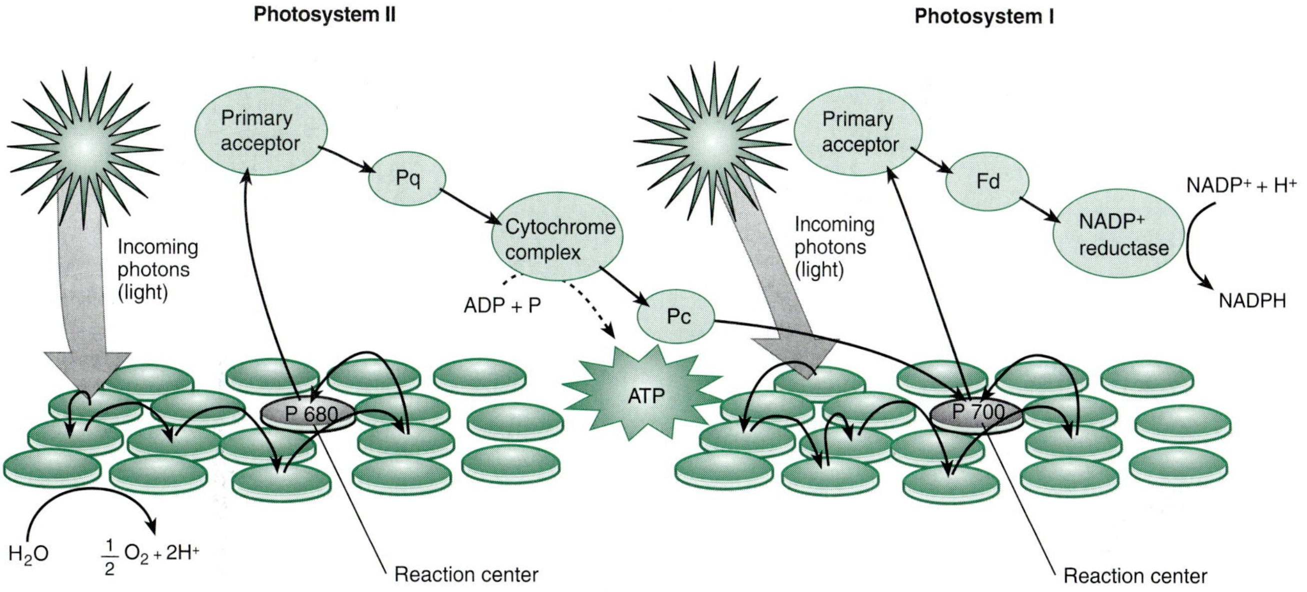

Figure 4.9 *Two cooperating photosystems work together to transfer electrons from water to NADPH. The passage of electrons from Photosystem II to Photosystem I also drives the formation of ATP, a process known as noncyclic photophosphorylation.*

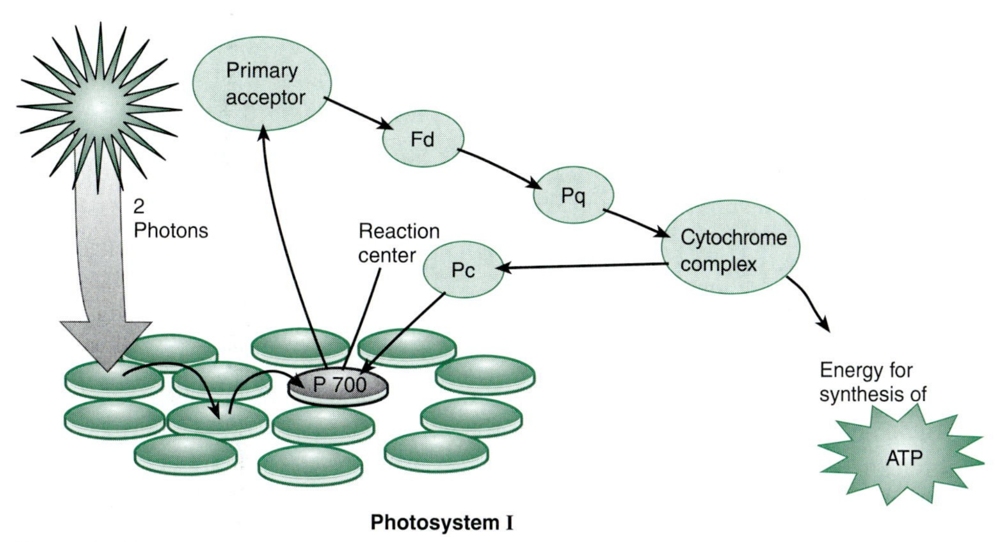

Figure 4.10 *Photosystem I can also function in a cyclic fashion. Instead of reducing NADP, electrons are passed back to P_{700}. This results in the generation of ATP by cyclic photophosphorylation.*

an enzyme on the thylakoid membrane that requires manganese atoms. In this reaction, water molecules are split into oxygen and hydrogen; the hydrogen is a source of both electrons and protons.

The primary electron acceptor passes the electron on to a series of thylakoid membrane-bound electron carriers that include **plastoquinone (PQ), cytochrome complex, plastocyanin (PC),** and others. The electron is eventually passed to the oxidized P_{700} in Photosystem I. During the transfer of electrons, ATP is synthesized as protons are passed from the thylakoid lumen into the stroma by an **ATP synthase** in the membrane. It is actually the passage of protons through this enzyme that drives the production of ATP; however, the mechanism for this reaction is still not completely understood. This synthesis of ATP is known as **photophosphorylation,** since the energy that drives the whole process is from sunlight.

In the process just described, the two photosystems are joined together by the one-way transfer of electrons from Photosystem II to Photosystem I. Water is the ultimate source of these electrons continually replenishing electrons lost from P_{680}. The photophosphorylation that occurs during this process is referred to as **noncyclic photophosphorylation,** since the electron transfer is one-way, with the reduction of NADP as the final step.

Photosystem I is also capable of functioning independently, transferring electrons in a cyclic fashion. The electrons, instead of being passed to NADP from ferredoxin, may be passed to the cytochrome complex and then back to P_{700} (fig. 4.10). ATP, but not NADPH, may be generated during this process, which is known as **cyclic photophosphorylation,** since the flow of electrons begins and ends with P_{700}.

As just described, when water is split, oxygen is released. The oxygen eventually diffuses out of the leaves into the atmosphere and is the Earth's only constant supply of this gas. The current 20% oxygen content in the atmosphere is the result of three billion years of photosynthesis. The atmosphere of early Earth did not contain this gas; it only began to accumulate after the evolution of photosynthetic organisms. Today, the vast majority of living organisms depend on oxygen for cellular respiration and, therefore, the energy that maintains life.

The overall light reactions proceed with breathtaking speed as a constant flow of electrons moves from water to NADPH, powered by the vast energy of the sun. The ATP and NADPH that result from the light reactions are needed to drive the biochemical reactions in the Calvin Cycle.

Calvin Cycle

The source of carbon used in the photosynthetic manufacture of sugars is carbon dioxide from the atmosphere. This gas makes up just a tiny fraction, approximately 0.035%, of the Earth's atmosphere and enters the leaf by diffusing through the stomata. The reactions that involve the **fixation** and reduction of CO_2 to form sugars are known as the Calvin Cycle. These reactions utilize the ATP and NADPH produced in the light reactions, but do not involve the direct participation of light, and are hence sometimes referred to as **light-independent reactions** or **dark reactions.** The Calvin Cycle takes place in the stroma of the chloroplasts, which contains the enzymes that catalyze the many reactions in the cycle. This pathway was worked out by Melvin Calvin, in association with Andrew Benson and James Bassham, during the late 1940s and early 1950s. The pathway is named in honor of Calvin, who received a Nobel Prize for his work in 1961.

The following discussion will be limited to the main events of the Calvin Cycle, which are depicted in Figure 4.11. The end product of this pathway is the synthesis of a six-carbon sugar; this requires the input of carbon dioxide. Six turns of the cycle are needed to incorporate six molecules of CO_2 into a single molecule of a six-carbon sugar. The initial event is the fixation or addition of CO_2 to **ribulose-1,5-bisphosphate (RUBP),** a five-carbon sugar with two phosphate groups. This **carboxylation** reaction is catalyzed by the enzyme **ribulose bisphosphate carboxylase (RUBISCO).** In addition to its obvious importance to photosynthesis, RUBISCO appears to be the most abundant protein on Earth, since it comprises 12.5% to 25% of total leaf protein. The product of the carboxylation is an unstable six-carbon intermediate that immediately splits into two molecules of a three-carbon compound, with one phosphate group called **phosphoglyceric acid (PGA)** or phosphoglycerate**.** In six turns of the cycle this would yield 12 molecules of PGA.

The 12 PGA molecules are converted into 12 molecules of **glyceraldehyde phosphate.** This step requires the input of 12 NADPH + H^+ and 12 ATP (both generated during the light reactions), which supply the energy for this reaction. Ten of the 12 glyceraldehyde phosphate molecules are used to regenerate the six molecules of RUBP in a complex series of interconversions that require 6 more ATP and allow the cycle to continue. Two molecules of glyceraldehyde phosphate are the net gain from 6 turns of the Calvin Cycle; these are converted into 1 molecule of fructose-1,6-bisphosphate, which is soon converted to glucose. The glucose produced is never stored as such but is converted into starch, sucrose, or a variety of other products, thus completing the conversion of solar energy into chemical energy.

The complex steps of photosynthesis can be summarized in the following simple equation, which considers only the raw materials and end products of the process:

$$6CO_2 + 12H_2O + \text{energy} \xrightarrow{\text{CHLOROPHYLL}} C_6H_{12}O_6 + 6O_2 + 6H_2O$$

Variation to Carbon Fixation

Many plants utilize a variation of carbon fixation that consists of a prefixation of CO_2 before the Calvin Cycle. There are two pathways where this prefixation occurs, the **C_4 Pathway** and the **CAM Pathway.** The C_4 pathway occurs in several thousand species of tropical and subtropical plants, including the economically important crops of corn, sugar cane, and sorghum. This pathway consists of incorporating CO_2 into organic acids, resulting in a four-carbon compound and, hence, the name of the pathway. This compound is soon broken down to release CO_2 to the Calvin Cycle. The C_4 pathway assures a more efficient delivery of CO_2 for fixation and greater photosynthetic rates under conditions of high light intensity, high temperature, and low CO_2 concentrations.

These same steps are part of the **CAM (Crassulacean Acid Metabolism)** pathway, which functions in a number of cacti and succulents, plants of desert environments. This pathway was initially described among members of the plant family Crassulaceae. CAM plants are unusual in that their stomata are closed during the daytime but open at night. Thus, they fix CO_2 during the nighttime hours, incorporating it into four-carbon organic acids. During the daylight hours these compounds are broken down to release CO_2 to continue on into the Calvin Cycle. This alternate pathway allows carbon fixation to occur at night when transpiration rates are low, an obvious advantage in hot, dry desert environments.

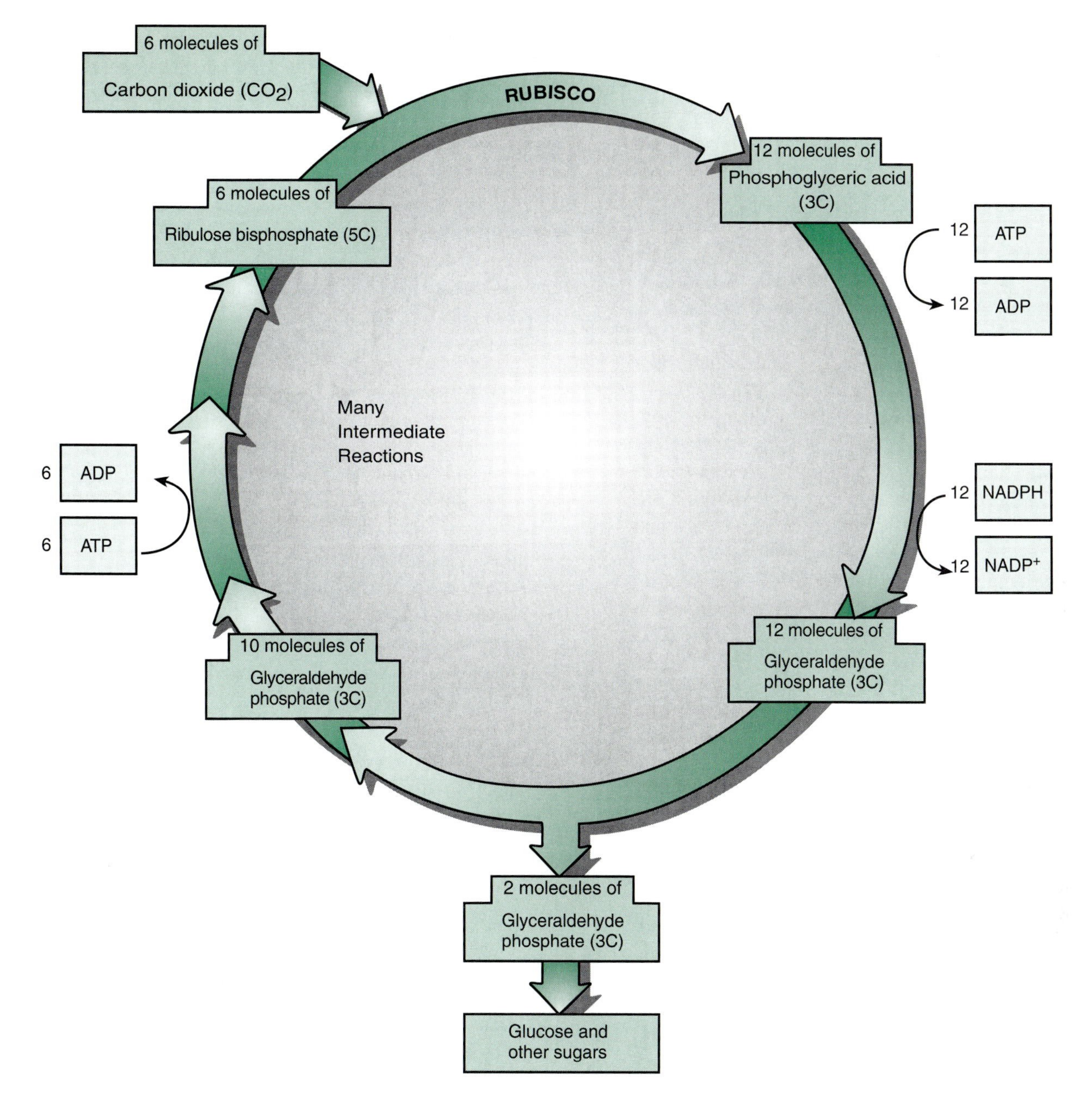

Figure 4.11 *The Calvin Cycle. For every six molecules of CO_2 that enter the cycle, one six-carbon sugar is produced. The ATP and NADPH required by this cycle are generated by the light reactions.*

Respiration

As discussed, photosynthesis converts solar energy into chemical energy, stored in a variety of organic compounds. Starch and sucrose are common storage compounds in plants and as such are the energy reserves for the plants themselves and the animals that feed on these plants. Ultimately the survival of all organisms on Earth is dependent on the release of this chemical energy through the catabolic process of cellular respiration. All living organisms require energy to maintain the processes of life. Even at the cellular level, life is a highly dynamic system, requiring continuous input of energy that is used in the processes of growth, repair, transport, synthesis, motility, cell division, and reproduction. Cellular respiration occurs continuously, every hour of every day, in all living cells; the need for energy is nonstop.

Cellular respiration is actually a step-by-step breakdown of the chemical bonds in glucose, involving many enzymatic reactions, and results in the release of useable energy in the form of ATP. The overall process is the complete oxidation of glucose resulting in CO_2 and H_2O and the formation of ATP:

$$C_6H_{12}O_6 + 6O_2 \longrightarrow 6CO_2 + 6H_2O + 36\ ATP.$$

This equation of cellular respiration is merely a summary of a complex step-by-step process that has three major stages or pathways: **Glycolysis,** the **Krebs Cycle,** and the **Electron Transport System.**

Glycolysis is a series of reactions that occur in the cytoplasm and result in the breakdown of glucose into two molecules of a three-carbon compound. Along the way, NAD is reduced and some ATPs are produced. The Krebs Cycle continues the breakdown of the three-carbon compounds in the mitochondrial matrix of the mitochondria and results in the release of CO_2. Additional ATP, NADH, and $FADH_2$ are also generated during these steps. The final stage of respiration, the Electron Transport System, occurs on the cristae, the inner membrane of the mitochondria, and consists of a series of redox reactions during which significant amounts of ATP are synthesized.

Glycolysis

The word *glycolysis* means the splitting of sugar. It starts with glucose, which arises from the breakdown of polysaccharides, most commonly either starch or glycogen (in animals and fungi), or the conversion from other substances, especially other sugars. The first few steps in glycolysis actually add energy to the molecule in the form of phosphate groups (fig. 4.12). These phosphorylations are at the expense of two molecules of ATP. In addition, the glucose molecule undergoes a rearrangement, converting it to **fructose-1,6-bisphosphate.** These steps prime the molecule for the later oxidation. The next step splits fructose-1,6-bisphosphate into **glyceraldehyde phosphate** and **dihydroxyacetone phosphate,** but the latter is converted into a second molecule of glyceraldehyde phosphate. Both glyceraldehyde phosphate molecules continue on in the glycolytic pathway so that each of the remaining steps actually occurs twice. Glyceraldehyde phosphate is phosphorylated and oxidized in the next step, which also reduces NAD^+ to $NADH + H^+$. The resulting organic acids, with two phosphate groups, give up both phosphates in the remaining steps of glycolysis, yielding two molecules of **pyruvate** (pyruvic acid), plus 4 ATP and 2 $NADH + H^+$. Note that during steps 7 and 10 a total of 4 ATP are produced; however, 2 are used during the initial steps, resulting in a net gain of only 2 ATP.

The Krebs Cycle

The remainder of cellular respiration occurs in the mitochondria of the cell (fig. 4.13). Recall from Chapter 2 that mitochondria are organelles with a double membrane. Although the outer membrane is smooth, the inner membrane is invaginated; these folds are referred to as **cristae.** The area between the outer and inner membranes is referred to as the **intermembrane space,** while the enzyme-rich area enclosed by the inner membrane is known as the **matrix.** The enzymes in the matrix catalyze each step in the Krebs Cycle.

Once inside the mitochondrial matrix, each of the two pyruvate molecules from glycolysis undergoes several changes before it enters the Krebs Cycle. The molecule is oxidized and **decarboxylated,** losing a CO_2 with the remaining two-carbon compound joining to **Coenzyme A** to form a complex known as **acetyl-CoA.** During this step, NAD^+ is also reduced to $NADH + H^+$. Acetyl-CoA enters the Krebs Cycle by combining with a four-carbon organic acid known as **oxaloacetate** (oxaloacetic acid) to form a six-carbon compound known as **citrate** (citric acid). (The Krebs Cycle, named in honor of Hans Krebs, who worked out the steps in this pathway in 1937 and later received a Nobel Prize for this work, is alternatively known as the **Citric Acid Cycle.**)

The steps in the cycle consist of a series of reactions during which two more decarboxylations (going from a six-carbon to a five-carbon, and then to a four-carbon compound) and several oxidations occur (fig. 4.14). During these steps, 3 more molecules of NAD^+ are reduced to $NADH + H^+$, a molecule of FAD is reduced to $FADH_2$, and 1 molecule of ATP is formed. At the end of these steps, the four-carbon oxaloacetate is regenerated, allowing the cycle to begin anew. For each molecule of pyruvate that entered the mitochondrion, 3 molecules of CO_2 are released and 1 ATP, 4 $NADH + H^+$ and 1 $FADH_2$ are produced. Since 2 pyruvate molecules are formed from each glucose molecule, the cycle turns twice, resulting in 6 CO_2 released and yielding 2 ATP, 8 $NADH + H^+$, and 2 $FADH_2$ as energy-rich products. At this point, the entire glucose molecule has been totally degraded; a portion of its energy has been harvested in these Krebs Cycle products as well as the 2 ATP and 2 $NADH + H^+$ produced in glycolysis.

Electron Transport System

The third and final stage of respiration, the Electron Transport System, occurs on the inner membranes of the mitochondria and involves a series of enzymes and coenzymes, including several iron-containing **cytochromes** that are embedded in this layer and function as electron

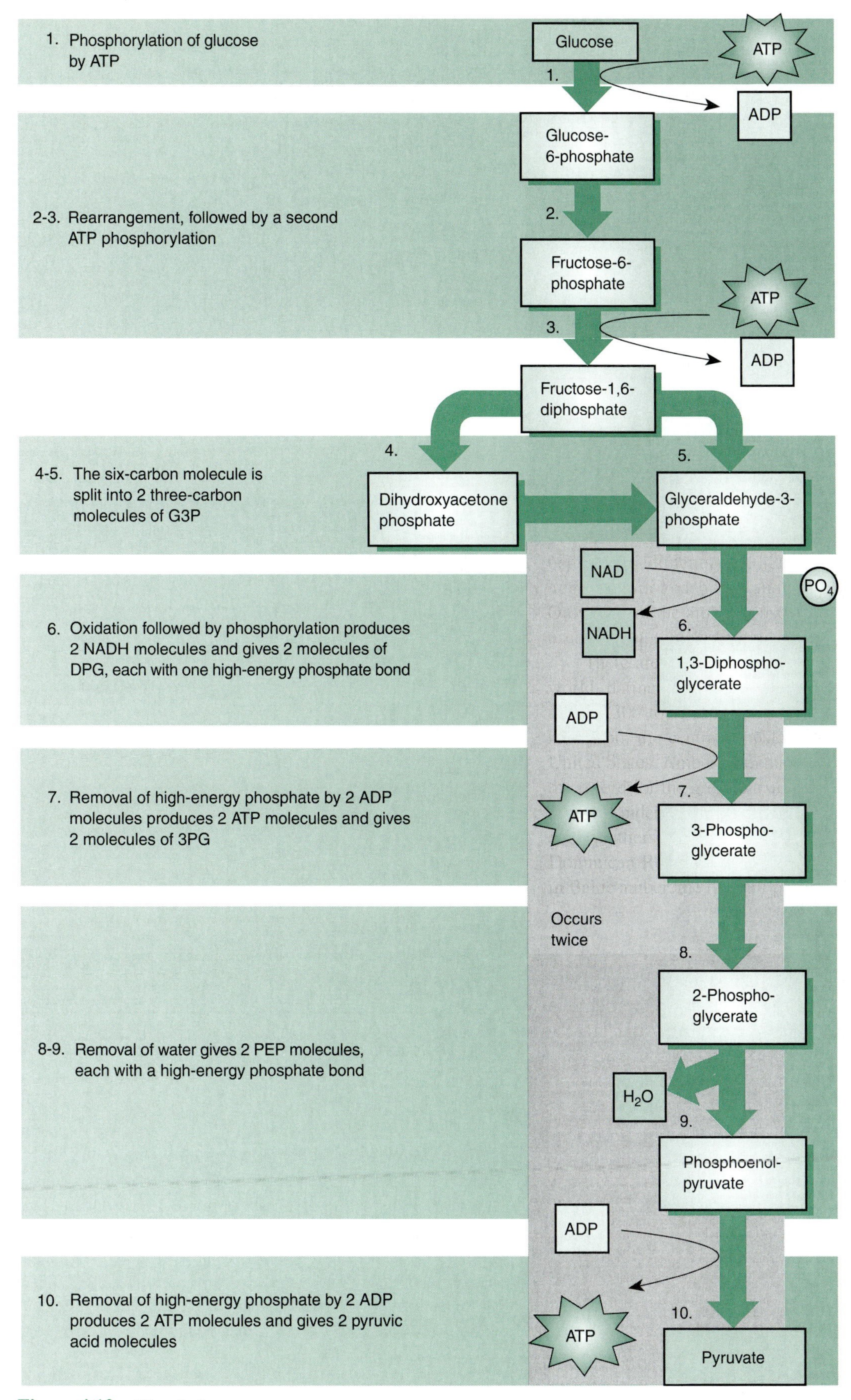

Figure 4.12 *Glycolysis.*

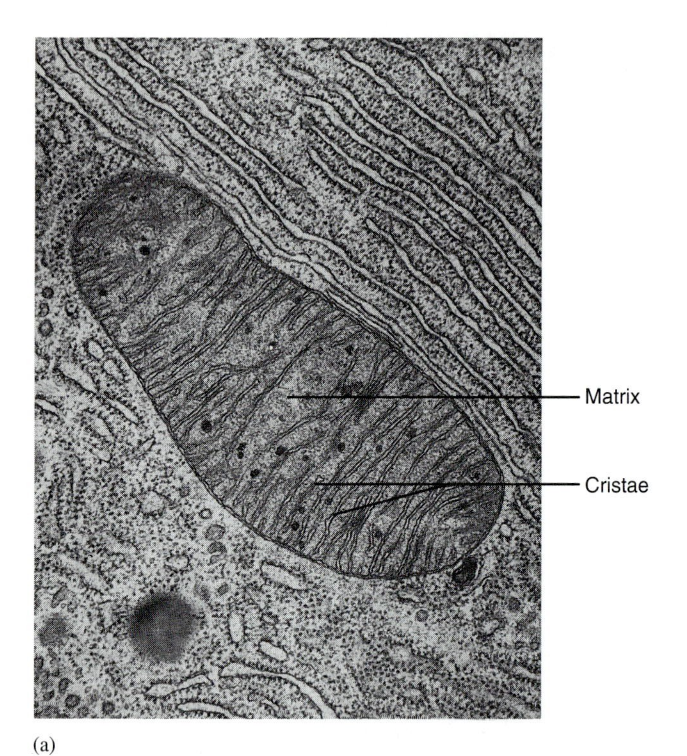

(a)

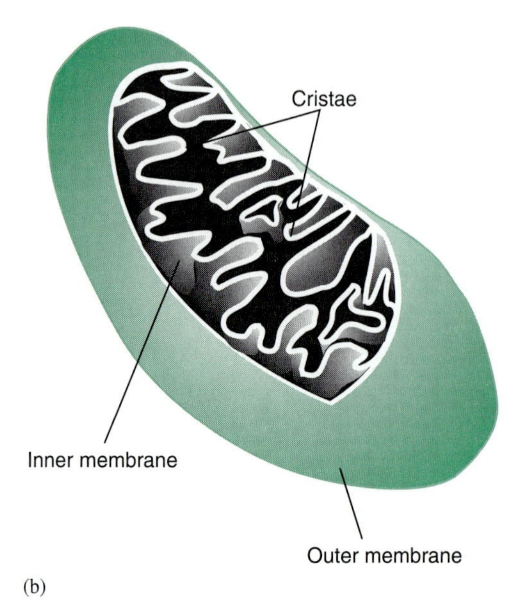

(b)

Figure 4.13 *Mitochondrial structure. The inner mitochondrial membrane has numerous infoldings known as cristae. The enzymes and coenzymes of the Electron Transport System occur on these membranes. The matrix contains the enzymes that carry out the Krebs cycle, (a) ×85,000.*

carriers. (This series of electron carriers is similar to the ones described in the light reactions of photosynthesis.) During this stage, electrons and hydrogen ions are passed from the NADH + H^+ and $FADH_2$ molecules formed in glycolysis and the Krebs Cycle, down a series of redox reactions, and are finally accepted by oxygen-forming water in the process (fig. 4.15).

The Electron Transport System is a highly exergonic process and is coupled to the formation of ATP. The ATP synthesized by this method is referred to as **oxidative phosphorylation.** When the electron flow begins from NADH produced within the mitochondria, enough energy is available to produce 3 molecules of ATP from each NADH for a total of 24 ATP from the 8 NADH. Two molecules of ATP are also synthesized during the flow of electrons from each $FADH_2$ produced in the Krebs Cycle (4 ATP) and each NADH from glycolysis (4 ATP). During the Electron Transport System then, there is a total of 32 ATP generated. This number is added to the net yield of 2 ATP from glycolysis and the 2 ATP produced in the Krebs Cycle, for a grand total of 36 ATP for each glucose molecule that completes

TABLE 4.1
Tally of ATP Produced from the Breakdown of Glucose During Cellular Respiration

Pathway	Net ATP Yield*
Glycolysis	
2 ATP	2 ATP
2 NADH	4 ATP
Acetyl-CoA Formation (2 turns)	
1 NADH x 2	6 ATP
Krebs Cycle (2 turns)	
1 ATP x 2	2 ATP
3 NADH x 2	18 ATP
1 $FADH_2$ x 2	4 ATP
TOTAL	36 ATP

*Each NADH produced in the mitochondrion yields 3 ATP, while each NADH produced during glycolysis and each $FADH_2$ yield 2 ATP.

cellular respiration (table 4.1). It should be noted, however, that this production of 36 ATP only harnesses a fraction, 39%, of the original chemical energy of the glucose molecule; the remainder is lost as heat. (Although this 39% efficiency seems low, it is actually much higher than energy conversions in mechanical systems.)

The formation of ATP during the transport of electrons is believed to occur by the same mechanism described for

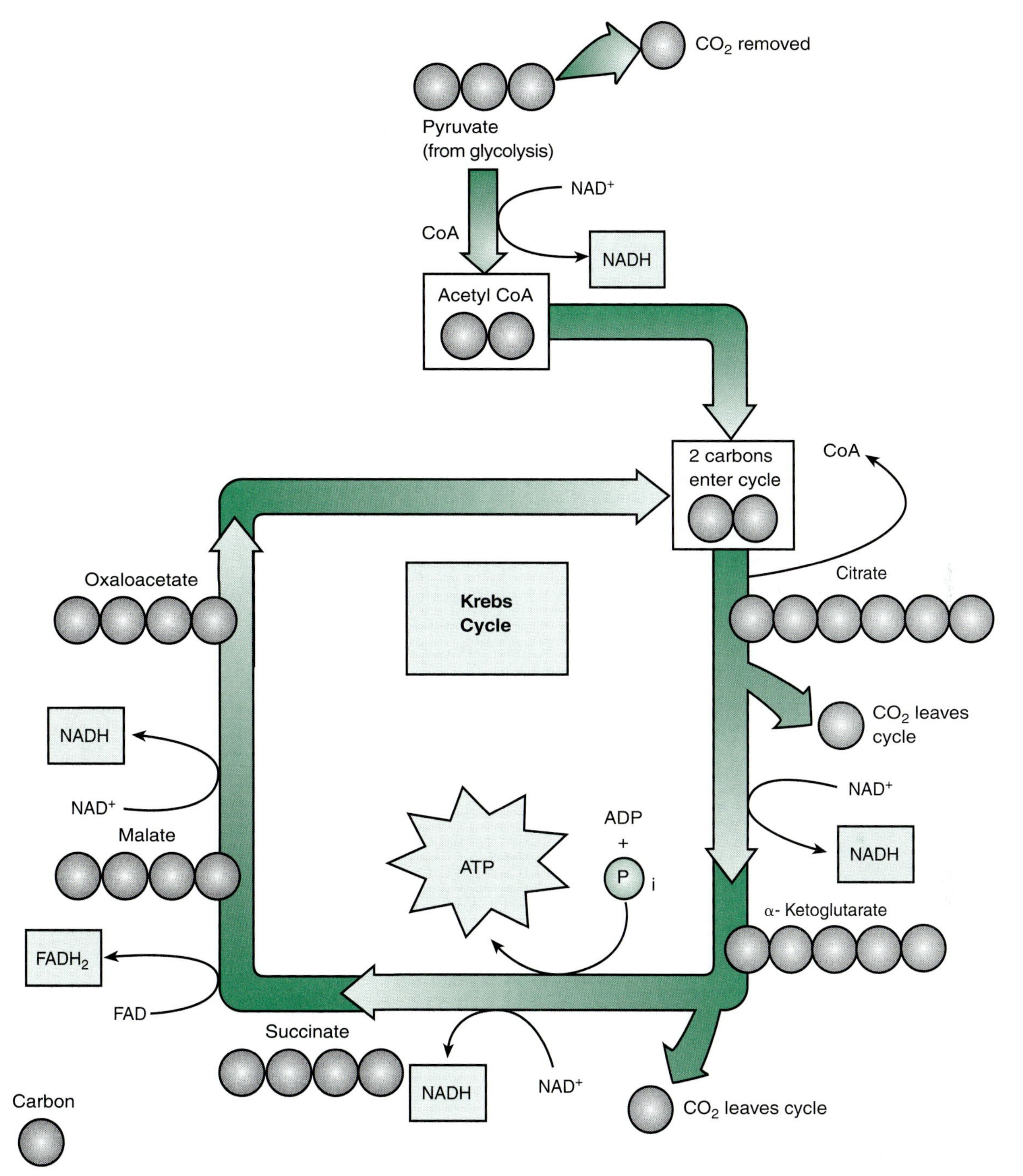

Figure 4.14 *Krebs Cycle. Before the cycle begins, pyruvate is converted to acetyl-CoA with the loss of CO_2. The two-carbon acetyl group combines with oxaloacetate to form the six-carbon citrate. Two decarboxylations and several redox reactions regenerate oxaloacetate. For every pyruvate that enters the mitochondrion, 4 NADH, 1 $FADH_2$, and 1 ATP are produced and 3 CO_2 are released.*

photophosphorylation during photosynthesis. During the transfer of electrons, protons pass from the intermembrane space into the matrix through an ATPase in the membrane (fig. 4.16). It is actually the passage of protons through this ATP synthase enzyme that somehow drives the production of ATP. This model for ATP synthesis is known as **chemiosmosis** and was first proposed by Peter Mitchell during the early 1960s. Mitchell received a Nobel Prize for this theory, which applies to ATP synthesis in both respiration and photosynthesis.

Aerobic vs. Anaerobic Respiration

The complete oxidation of glucose requires the presence of oxygen and is therefore known as **aerobic respiration.** As

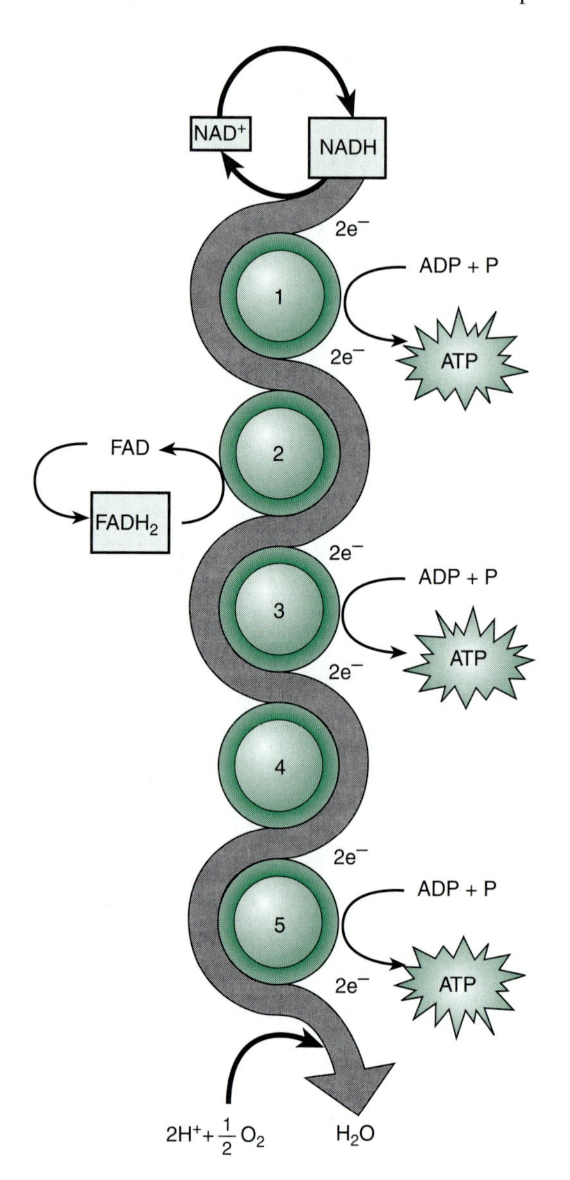

indicated, it is the last step of respiration that involves the direct participation of oxygen as the final electron acceptor. Without oxygen, this last step cannot occur, since no other compound can serve as the ultimate electron acceptor. In fact, both the Electron Transport System and the Krebs Cycle are dependent on the availability of oxygen and cannot operate in its absence.

In some organisms there are, however, metabolic pathways that allow respiration to proceed in the absence of oxygen. This type of respiration is known as **anaerobic respiration** or **fermentation** and is found in some yeast (a unicellular fungus), bacteria, and even in muscle tissue. The most familiar example is **alcoholic fermentation** found in certain types of yeast and utilized in the production of beer and wine (*see* Chapter 23). When oxygen is not available, the yeast cells can switch to a pathway that can convert pyruvic acid to ethanol and CO_2. In the process, NADH is oxidized back to NAD^+, allowing this coenzyme to recycle back to glycolysis. This allows glycolysis to continue and thus supply the energy needs of the yeast, at least in a limited way. The only energy yield from this alcoholic fermentation is the 2 ATP produced during glycolysis (compared to 36 ATP during aerobic respiration). The still energy-rich alcohol is merely a by-product of the oxidation of NADH. If oxygen becomes available, yeast can switch back to aerobic respiration with its higher energy yield. Other anaerobic pathways also exist in bacteria and muscle cells, in which the by-products are different from alcohol, but the yield of NAD^+ and ATP is similar.

Figure 4.15 *Electron Transport System. Electrons from NADH and $FADH_2$ are passed along electron-carrier molecules, including several cytochromes, and finally are accepted by oxygen. This process drives the formation of ATP by chemiosmosis.*

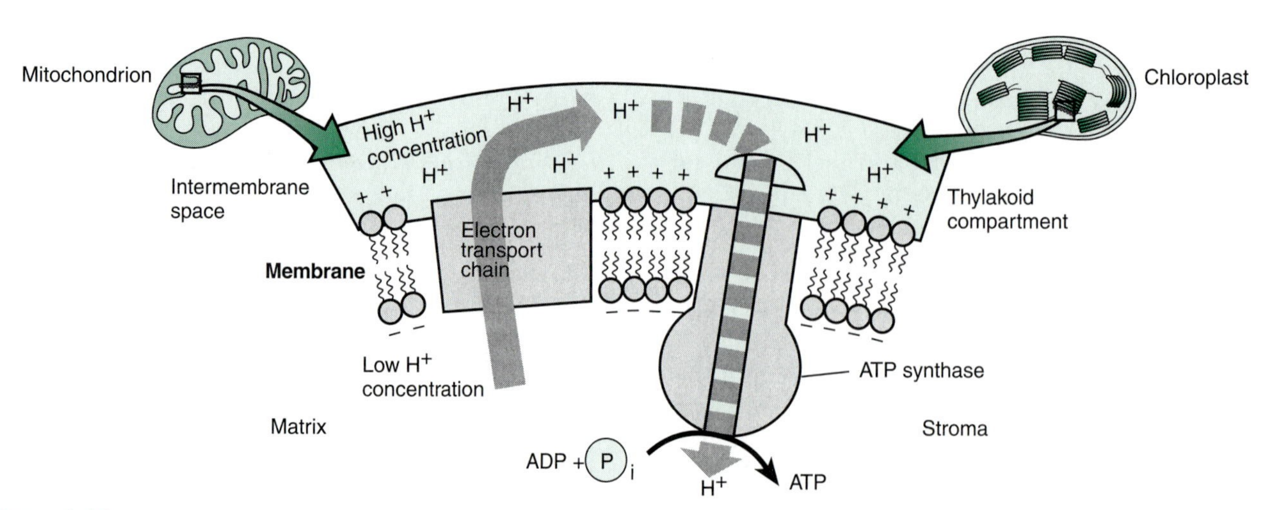

Figure 4.16 *Chemiosmosis in mitochondria and chloroplasts. In the mitochondria, protons (H^+) are translocated to the intermembrane space during the transfer of electrons down the electron transport system. The proton gradient drives ATP synthesis as protons move through the ATP synthase complex back to the matrix. In the chloroplast, protons are translocated into the thylakoid compartment. As protons move through the ATP synthase complex back to the stroma, ATP is synthesized.*

Chapter Summary

1. Plants obtain water from the soil through diffusion, moving it up within the xylem through the entire plant. This movement of water is a passive phenomenon dependent on the pull of transpiration and the cohesion of water molecules. Minerals are also obtained from the soil and transported in the xylem. The translocation of sugars occurs in the phloem, moving from source (photosynthetic leaves or storage organs) to sink (growing organs or developing storage tissue) through mass flow within sieve tube members.
2. Sucrose is the usual sugar transported in the phloem; however, very few plants actually store this economically valuable carbohydrate. Sugarcane and sugar beet are the major sucrose-supplying crops. The early development of sugarcane plantations in North America greatly influenced the course of history by introducing the slave trade to the continent.
3. Plants are dynamic metabolic systems with hundreds of reactions occurring each second to enable plants to live, grow, and respond to their environment. All life processes are driven by energy with some metabolic reactions endergonic and others exergonic. Energy transformations occur constantly in biological systems and are part of the underlying principles of both photosynthesis and respiration.
4. Photosynthesis takes place in chloroplasts of green plants and algae and results in the conversion of radiant energy into chemical energy (linking the energy of the sun with life on Earth). Using the raw materials carbon dioxide and water, along with chlorophyll and sunlight, plants are able to manufacture sugars. In the light reactions of photosynthesis, energy from the sun is harnessed, forming molecules of ATP and NADPH. During this process, water molecules are split, releasing oxygen to the atmosphere as a by-product. During the Calvin Cycle, carbon dioxide molecules are fixed and reduced to form sugars, using the energy provided by the ATP and NADPH from the light reaction.
5. Respiration is the means by which stored energy is made available for the energy requirements of the cell. Through respiration, the energy of carbohydrates is transferred to ATP molecules, which are then available for the energy needs of the cell. During aerobic respiration, each molecule of glucose is completely oxidized during the many reactions of glycolysis, the Krebs Cycle, and the Electron Transport System, resulting in the formation of 36 ATP molecules. In anaerobic respiration, only 2 molecules of ATP are formed from each glucose molecule.

Review Questions

1. Explain how water enters a root.
2. What is transpiration and how does it affect water movement in plants?
3. How are the properties of water important to the theory of water movement in plants?
4. What are the advantages and disadvantages to having stomata open during the daylight hours?
5. Explain how the pressure-flow theory accounts for translocation in the phloem.
6. How is light harnessed during the light reactions of photosynthesis and what pigments are involved?
7. How is carbon fixed during the Calvin Cycle?
8. Why is glycolysis important to living organsisms and where does it occur?
9. Describe mitochondria and the respiratory events that occur there.

Further Reading

Govindjee and William J. Coleman. 1990. How Plants Make Oxygen. *Scientific American,* 262 (2): 50–56.

Mauseth, James D. 1991. *Botany: An Introduction to Plant Biology,* Saunders College Publishing, Philadelphia, PA.

Moore, Randy and W. Dennis Clark. 1995. *Botany: Plant Form and Function,* Wm. C. Brown Publishers, Dubuque IA.

Raven, Peter, Ray F. Evert, and Susan E. Eichhorn. 1986. *Biology of Plants,* 4th Edition. Worth Publishers, Inc., NY.

Salisbury, Frank B. and Cleon W. Ross. 1992. *Plant Physiology,* 4th Edition. Wadsworth Publishing Co., Belmont, CA.

C H A P T E R

5

Plant Life Cycle: Flowers

Chapter Outline

Chapter Concepts

1. Angiosperms are unique among plants in that they have their sexual reproductive structures contained in a flower.
2. Meiosis is a type of cell division that reduces the number of chromosomes from the diploid to the haploid number and is an integral part of sexual reproduction.
3. Pollination is the transfer of pollen from the anther to the stigma and largely occurs through the action of wind or animals.
4. In angiosperms, reproduction is accomplished through the process of double fertilization.

The natural beauty of flowers has always been a source of inspiration and the appearance of the first flower of spring lightens the heart of anyone weary of winter. But what role do flowers play in the lives of plants? Their beauty notwithstanding, flowers play a pivotal role in the life cycle of angiosperms since they are the sites of sexual reproduction.

The Flower

Flowers, unique to angiosperms, are essentially modified branches bearing four sets of specialized appendages or floral organs. These appendages are grouped in whorls and consist of **sepals, petals, stamens,** and **carpels.** They are inserted into the **receptacle,** the expanded top of the **pedicel (peduncle)** or flower stalk (fig. 5.1).

Floral Organs

The outermost whorl consists of the sepals, leafy structures that cover the unopened flower bud; they are usually green and photosynthetic. The whole whorl of sepals of a single flower is called the **calyx.** The petals that make up the next whorl of flower parts are collectively called the **corolla.** Often brightly colored and conspicuous, the petals function by attracting animal pollinators. Together, the calyx and corolla comprise the **perianth.**

In the center of the flower, the male and female structures can be found. The **androecium,** the whorl of male structures, is composed of stamens, each of which consists of a pollen-producing **anther** supported on a stalk, the **filament.** Each anther houses four chambers where **pollen** develops. The pollen chambers can be seen in the cross section of the anther, Figure 5.1. The **gynoecium** is the collective term for the female structures or carpels, which are located in the middle of the flower. Flowers can have one to many carpels. A **simple pistil,** a gynoecium with just one carpel, is illustrated in Figure 5.2a. If many carpels are present, they may either be fused together to form one **compound pistil** (fig. 5.2b) or remain separate as many simple pistils. Carpels, whether individual or fused, consist of a **stigma, style,** and **ovary.** Contained within the basal ovary are one to many **ovules** (structures that will eventually become seeds); rising from the top of the ovary is a slender column called the style. The expanded tip of the style is the stigma that functions in receiving pollen. Flowers containing all four floral appendages are known as **complete** and **perfect** flowers.

Although flowers have been described in terms of only four floral appendages, some flowers may have additional floral structures called **bracts,** which are found outside the calyx. Bracts may appear leaflike or petal-like and be of various sizes. The showy red "petals" of poinsettia are actually bracts.

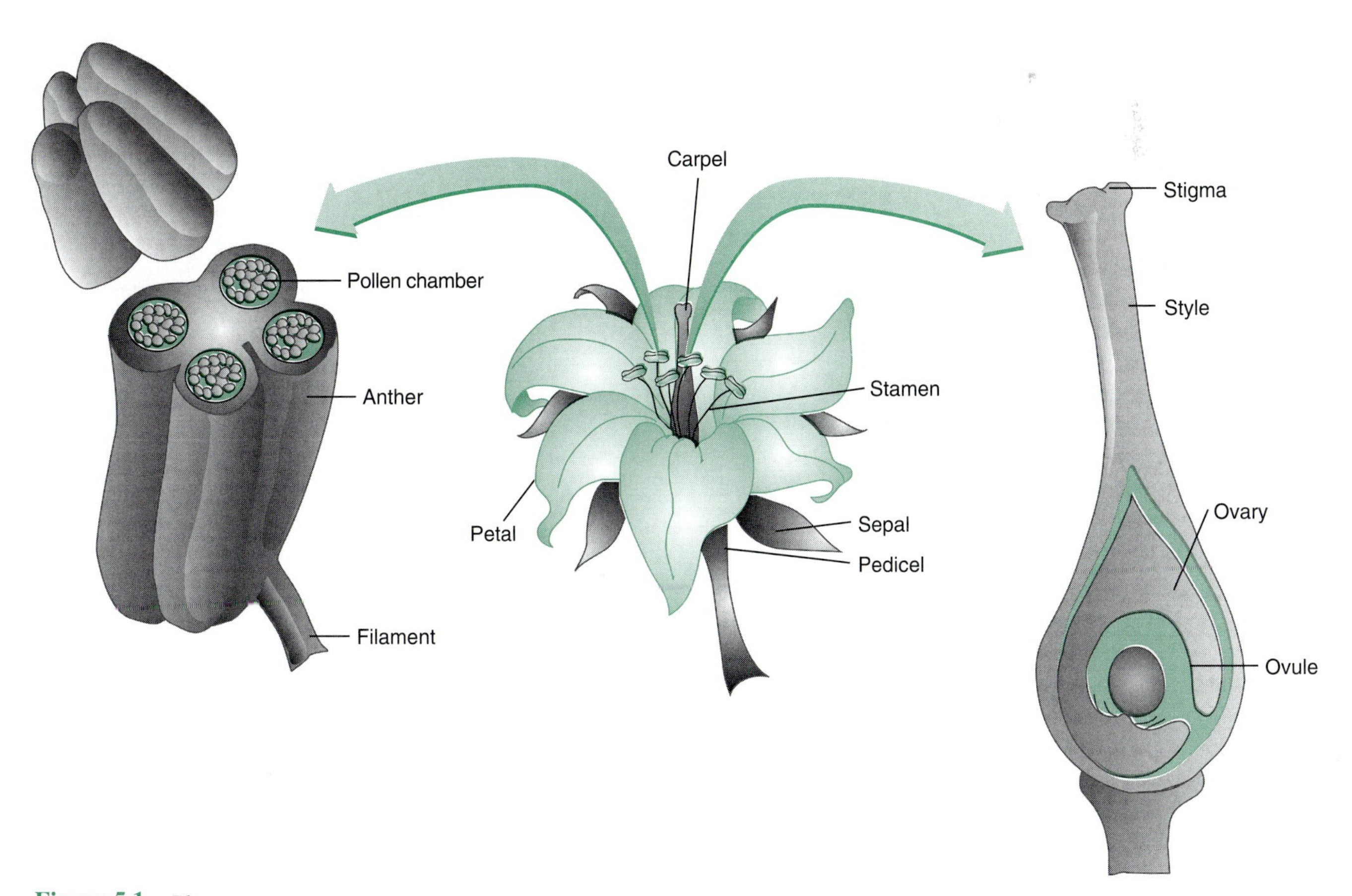

Figure 5.1 *Flower structure.*

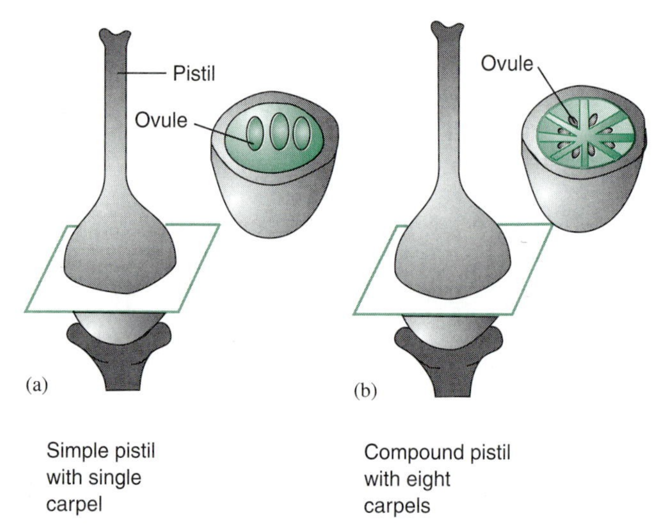

Figure 5.2 *Simple and compound pistils.*

In Chapter 3, the vegetative differences between monocots and dicots were described; there are also easily recognizable differences in the floral structures. Monocots generally have their floral parts in 3s or multiples of 3; for example, lilies have 3 sepals, 3 petals, 6 stamens, and a 3-part ovary (formed from the fusion of 3 carpels). On the other hand, dicots generally have a numerical plan of 4 or 5 or multiples; a wild geranium flower contains 5 sepals, 5 petals, 10 stamens, and 5 fused carpels with separate stigmas (fig. 5.3).

Modified Flowers

The basic pattern of flower structure is often modified. In flowers like tulips and lilies, the sepals are brightly colored and identical to the petals. In such flowers, the petals and sepals are often referred to as **tepals.** In contrast, flowers of the grasses possess neither sepals nor petals; these flowers are **incomplete** (fig. 5.4). In fact, flowers lacking any of the four floral structures are known as incomplete flowers. Some flowers, such as squash and holly, are unisexual; they are either **staminate** or **pistillate.** Incomplete flowers lacking either stamens or carpels are **imperfect.** A single plant may have both staminate and pistillate flowers; this plant is said to be **monoecious.** Alternately, **dioecious** plants have only unisexual flowers on a single individual. Corn, squash, and birches are familiar monoecious species (fig. 5.10), while some hollies are dioecious.

One feature that is important in the classification of flowers is the position of the ovary in relation to the other floral parts. If the sepals, petals, and stamens are inserted beneath the ovary, this arrangement is referred to as a **superior ovary.** The ovary is **inferior** if the sepals, petals, and stamens are inserted above it. A corresponding set of terms that refer to these arrangements are **hypogynous, epigynous,** and **perigynous.** Hypogynous (below the gynoecium) flowers have flower parts inserted beneath a superior ovary; epigynous (on the gynoecium) flowers have flower parts inserted above an inferior ovary; and perigynous (around the gynoecium) flowers have the bases of the flower parts fused into a cuplike structure surrounding a superior ovary (fig. 5.4).

(b)

Figure 5.3 *Monocot and dicot flowers. (a) Wild geranium,* Geranium *sp., illustrates a simple dicot flower with flower parts in multiples of 5. (b) Tiger lily,* Lilium canadense, *a monocot, shows flower parts in multiples of 3.*

Flowers can also be described by their pattern of symmetry. **Regular flowers (actinomorphic)** like the tulip, lily, rose, and daffodil display radial symmetry; they can be dissected into mirror-image halves along many lines. **Irregular flowers (zygomorphic)** like the orchid, iris, snapdragon, and pea display bilateral symmetry. They can only be dissected into mirror-image halves along one line (fig. 5.4).

Some flowers are borne singly on a stalk but, in many cases, flowers are grouped in clusters called **inflorescences.** Sometimes what is commonly called a single flower is actually an inflorescence, as in the case of sunflowers, daisies, and chrysanthemums. The dogwood flower is also an inflorescence, but here the pink or white "petals" are bracts surrounding a cluster of small flowers. The arrangement of flowers in the cluster determines the type of inflorescence, with many patterns possible: **spike, raceme, panicle, umbel, head,** and **catkin** (fig. 5.5). Often the type of infloresence is an important characteristic in classification.

To understand the flower's role in sexual reproduction, it is necessary to learn about a special form of cell division, **meiosis,** that occurs within stamens and carpels.

Meiosis

Sexual reproduction, whether in a plant or animal, is basically the fusion of male and female **gametes,** sperm and egg, to produce a **zygote,** which will develop into a new individual. When the egg is fertilized by the sperm, the zygote receives an equal number of chromosomes from each gamete. Gametes are different from most other cells because they are **haploid** (containing only one set of chromosomes), while other body cells are **diploid** (containing two complete sets of chromosomes). During the process of fertilization, the diploid number is restored in the zygote. When the

(a) Incomplete flower

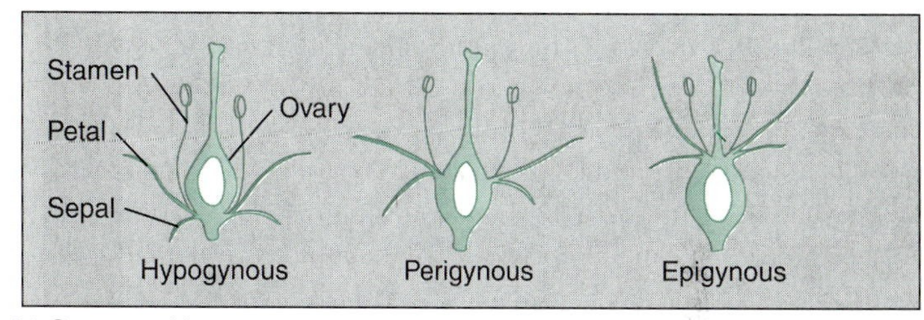

(b) Ovary position

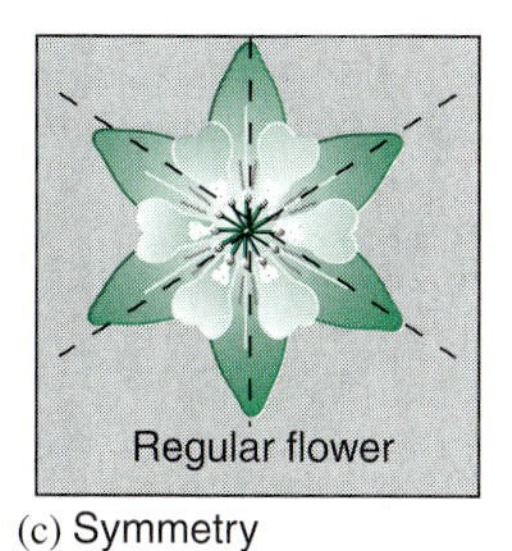

(c) Symmetry

Figure 5.4 *Modifications of the basic floral design result in diverse flower types. (a) Incomplete flowers lack one or more of the four floral organs. Grass flowers lack both sepals and petals. (b) Various positions of the floral whorls in relation to the ovary are possible. (c) Regular flowers can be bisected along many planes, while irregular flowers along only one.*

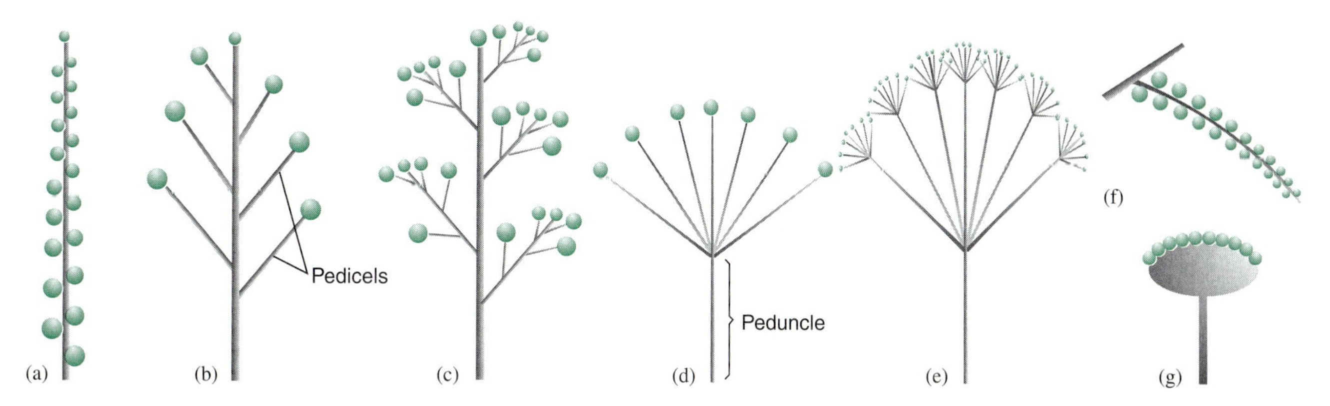

Figure 5.5 *Inflorescence types. (a) Spike. (b) Raceme. (c) Panicle. (d) Umbel. (e) Compound umbel. (f) Catkin. (g) Head.*

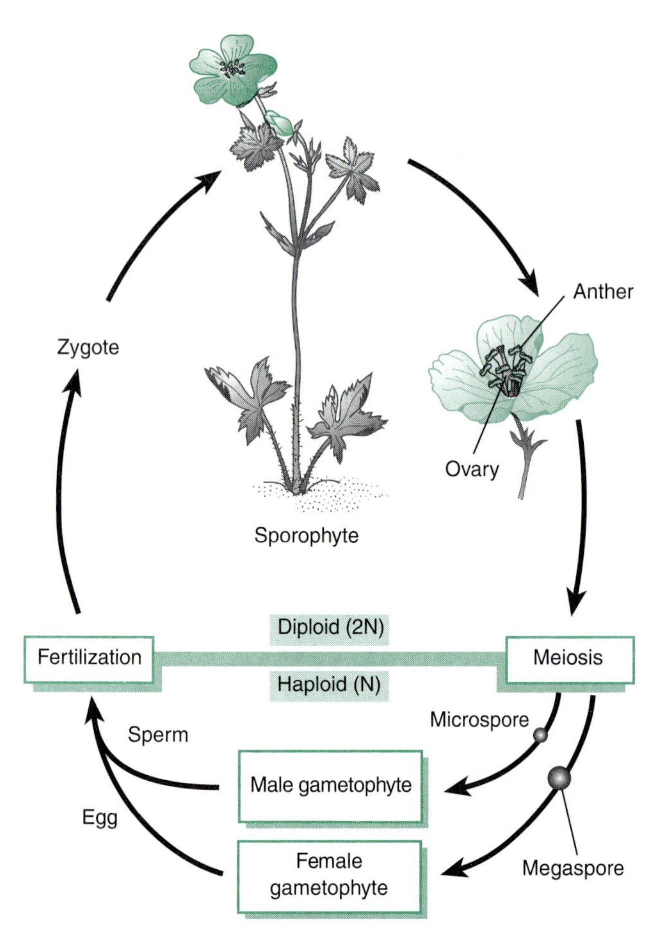

Figure 5.6 *Alternation of diploid (sporophyte) and haploid (gametophyte) generations in a flowering plant.*

chromosomes in a diploid cell are examined microscopically, it can be seen that there are two of each kind of chromosome. These pairs of chromosomes are known as **homologous chromosomes;** each member of a pair is derived from the contributing haploid gametes. Not only do the homologous chromosomes look alike, but they carry genes for the same traits.

Meiosis has a major role in all sexually reproducing organisms because it is the process that reduces the number of chromosomes from the diploid to the haploid number. This compensates for the doubling that occurs during fertilization. Without meiosis, the number of chromosomes would double with each generation.

In animals, gametes are produced directly by meiosis; however, in plants the products of meiosis are haploid **spores.** The diploid plant that undergoes meiosis to form these spores is known as a **sporophyte.** Spores develop into haploid **gametophytes** that produce gametes (fig. 5.6). (Later in this chapter the gametophytes of flowering plants will be studied.) The process of fertilization brings together egg and sperm to produce a genetically unique zygote. Sexual reproduction in this way introduces variation into a population, while offspring produced by asexual reproduction are genetic clones.

Stages of Meiosis

Meiosis is a specialized form of cell division that consists of two consecutive divisions and results in the formation of four haploid cells (fig. 5.7). Both the first and second divisions of meiosis are divided into four stages: prophase, metaphase, anaphase, and telophase. Recall that these are the same names used for the stages of mitosis.

During the first meiotic division, the chromosome number is halved; in fact, it is often called the **reduction division.** The most significant events occur during prophase I.

Prophase I

At the beginning of prophase I, the chromosomes appear threadlike. As in mitosis, the DNA is duplicated during the "S" phase of the preceding interphase, so that each chromosome actually consists of two chromatids. As the chromosomes continue to condense and coil, the homologous chromosomes pair up gene for gene in a process called **synapsis.** Since each chromosome is doubled, the synapsed homologous chromosomes actually consist of four chromatids. As the synapsed chromosomes continue to condense, breaks and exchanges of genetic material can occur between the chromatids in an event called **crossing-over.** This results in chromatids that are complete but have new genetic combinations. Soon synapsis starts to break down, and the homologous chromosomes repel each other; however, they are held together at points where crossing-over occurred. These places are referred to as **chiasmata** (singular, **chiasma**). While these chromosome events are occurring, the nucleolus and nuclear membrane break down, leaving the chromosomes free in the cytoplasm. Prophase I is the longest and most complex stage of meiosis (fig. 5.7).

Metaphase I

During the next stage, the homologous chromosome pairs line up at the equatorial plane (across the center of the cell). Spindle fibers that actually begin to appear in late prophase attach to the centromeres of each homologous pair. Two types of spindle fibers occur; those that run from pole to pole and those that run from one pole to a centromere. Recall that spindle fibers are composed of microtubules (fig. 5.7).

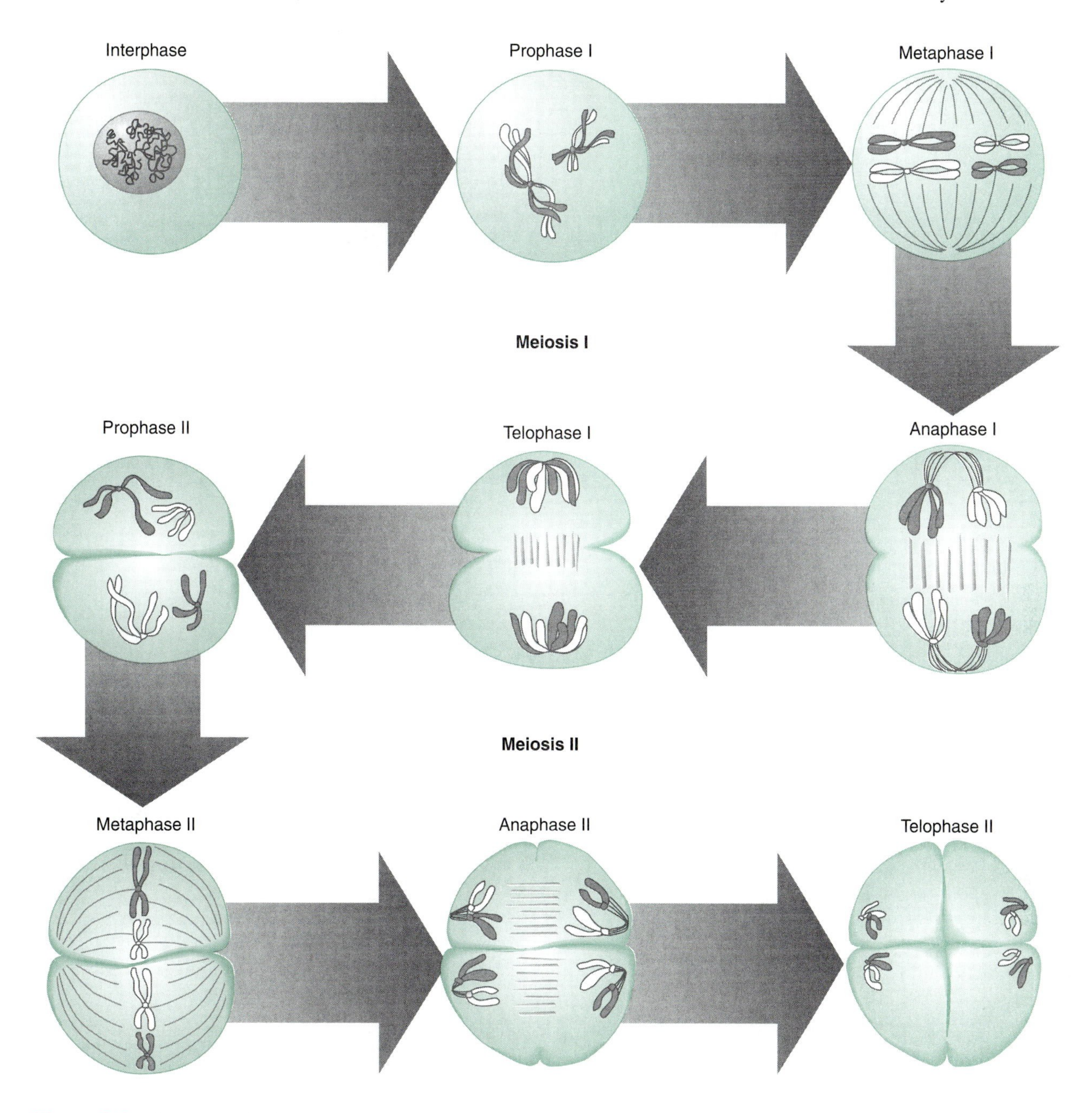

Figure 5.7 *Stages of meiosis.*

Anaphase I

Homologous chromosomes separate during this stage; they are pulled by the spindle fibers to opposite poles of the cell. In contrast to mitotic anaphase, the chromatids of each chromosome are still united; it is only the homologous pairs that are separating. By the end of this stage, the chromosome number has been halved (fig. 5.7).

Telophase I

This stage is similar to telophase of mitosis in that the spindle disappears, the chromosomes become less distinct, the nuclear membrane reforms, and the nucleoli reappear. Cytokinesis generally follows, dividing the cell into two daughter cells, each with half the number of chromosomes of the original parent cell (fig. 5.7).

A Closer Look 5.1

Pollen is More than Something to Sneeze at

The essential role that pollen plays in the life cycle of seed plants is well documented. Less well-known is the significance that palynology (the study of pollen) has had in many diverse fields: petroleum geology, anthropology, archeology, criminology, aerobiology, and the study of allergy.

When pollen is released by wind-pollinated plants, only a tiny percentage reaches the stigma. At the proper season, pollen is so abundant that clouds of it can be seen emanating from vegetation disturbed by wind or shaking (box fig. 5.1a). Most of it is carried by the wind and eventually settles back to Earth. It is this excess pollen that is the focus of study.

The distinctive ornamentation on the outer wall of a pollen grain allows for the identification of most types of pollen, sometimes even to the species level (box fig. 5.1b). Under certain conditions, pollen can be preserved, leaving a record of area vegetation. Some fossil pollen dates back over 200 million years and has revealed information about the changing vegetation patterns over evolutionary time.

Palynology is essential to the petroleum industry; an examination of fossil pollen from core samples can determine if an area is likely to be a rich source of oil. Certain fossil species in a particular region are known to be associated with oil deposits; palynologists look for the pollen of these indicator species in core samples.

(a)

Box Figure 5.1a *Clouds of pollen can be seen wafting from the pollen cones when a conifer branch is disturbed.*

Archeologists have sought the help of palynologists in determining when agriculture originated in certain areas and what plants were consumed by ancient peoples. By examining the fossil pollen, it is possible to pinpoint the shift from gathering native vegetation to the cultivation of cereal grains. Pollen residues found in storage vessels or coprolites (fossilized feces) give direct evidence of the diet of prehistoric groups; for example, the Viking recipe for mead was determined by examining pollen scrapings in drinking horns.

Pollen has proven instrumental in solving many criminal cases. The scene of a crime or the whereabouts of a suspect at the time of the crime can often be determined by analyzing pollen clinging to the victim's body, or to the shoes and clothing of a suspect.

Anthropologists have learned that pollen has symbolic meaning to several Native American tribes in the Southwest. Among the Navajos, pollen is revered as a symbol of life and fulfillment; it is used in sacred ceremonies and chants throughout the stages of life from birth to death

Second Meiotic Division

In some organisms, an interphase occurs between the two meiotic divisions; however, in other cases the cells proceed directly from telophase I to prophase II. The second meiotic division is essentially similar to mitosis; the chromatids, which are still joined together, finally separate. Prophase II is identical to mitotic prophase; in each cell the chromosomes become evident, the nuclear membrane breaks down, and the nucleoli disappear (fig. 5.7). During metaphase II, the chromosomes line up at the equatorial plane of each cell and spindles appear, with the spindle fibers stretching from pole to pole and pole to centromere. During anaphase II, the chromatids separate, pulled to the poles by the spindle fibers. Cytokinesis occurs in telophase II

(b-1)

Box Figure 5.1b *The pollen grains shown here vary in the types of exine ornamentation. (b-1) Netlike ridges; (b-2) cuboidal pollen with long projections.*

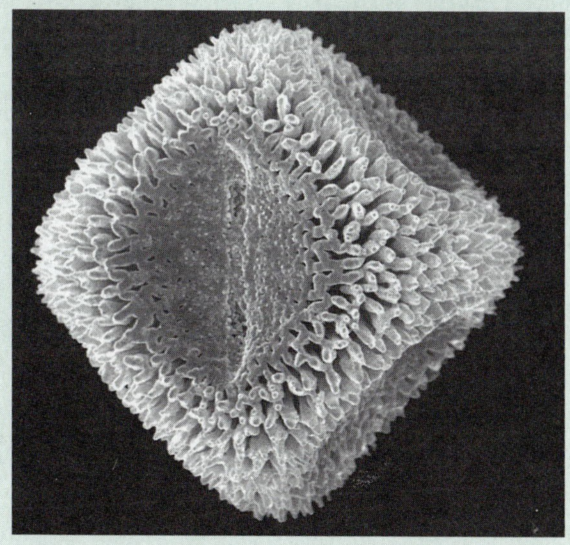

(b-2)

(box fig. 5.1c). The mystique of pollen has even been adopted by current health food fadists who claim that eating bee pollen (pollen collected by bees) is a power food that cures ailments, prevents disease, and promotes fitness. A few athletes take daily supplements to maintain a winning edge, but most nutritionists discount these claims and even express concern about allergic reactions.

Airborne pollen is well-known to trigger hay fever, asthma, and other allergic reactions in sensitized individuals. Despite its name, hay fever is not caused by hay, but is due to pollen from inconspicuous flowers of wind-pollinated trees, grasses, and weeds. This pollen is responsible for the misery of at least the 15% of the United States population identified as allergy sufferers. Aerobiologists study airborne pollen to document the species responsible and the factors influencing pollen abundance and distribution. Their findings suggest that there should be greater care in the selection of landscaping plants so that hay-fever-causing plants are avoided. More information about allergies and their causes will be discussed in Chapter 21.

(c)

Box Figure 5.1c *Pollen has symbolic meaning to several Native American tribes. This painting by Harrison Begay illustrates a Navajo woman gathering corn pollen.* ("Navajo woman and child gathering corn pollen," Harrison Begay, 0237.48 from the Collection of the Gilcrease Museum, Tulsa.)

and the nuclear membranes and nucleoli reappear as the single-stranded chromosomes become threadlike chromatin. By the end of telophase II, four haploid cells are produced. The four cells contain unique genetic combinations that differ from each other and the parent cell from which they originated. This contrasts with the process of mitosis in which the two daughter cells are genetically identical to the parent cell.

Meiosis in Flowering Plants

Within the flower, meiosis occurs during the formation of pollen in the stamen and the formation of ovules within the carpel (fig. 5.8).

Male Gametophyte Development

During the development of the stamen, certain cells in the pollen chambers of the anther become distinct as

A Closer Look 5.2

Alluring Scents

Since earliest times the fragrances of certain plants, due to their essential oils, have been valued as a source of perfumes. It is difficult to pinpoint when people first began using plant fragrances to scent their bodies, but by 5,000 years ago the Egyptians were skilled perfumers, producing fragrant oils that were used by both men and women to anoint their hair and bodies. Fragrances were also used as incense to fumigate homes and temples in the belief that these aromas could ward off evil and disease. In fact, our very word *perfume* comes from the Latin *per* meaning through and *fumus* meaning smoke, possibly referring to an early use of perfumes as incense.

Today most perfumes are a mixture of several hundred different scents that are carefully blended, using formulas that are highly guarded secrets. Many of these scents are now synthetics that resemble the natural essences from plants, but many costly perfumes still rely on the natural essential oils extracted directly from plants.

Various methods are used to extract the essential oils from plant organs including distillation, solvent extraction, expression, and enfleurage. The method used depends to a large extent on the location and chemical properties of the essential oil. Distillation, one of the most common methods used, employs exposing the tissue to boiling water or steam, thereby volatilizing the essential oil, which can then be separated from the condensate (box fig. 5.2). Since this method employs heat, only the most stable essential oils can be extracted by distillation. For solvent extraction, plant material is immersed in an organic solvent at room temperature; later the essential oil is recovered from the solution. At no time is heat used, thus avoiding damage to temperature-sensitive essential oils. Expression is the simplest method and is mainly used to express the oil from citrus rinds with mechanical pressure. In enfleurage, flower petals are layered on trays containing cold fat that absorbs the essential oils

Box Figure 5.2 *Rose petals undergo distillation to extract rose oil, one of the perfume industry's most valued scents.*

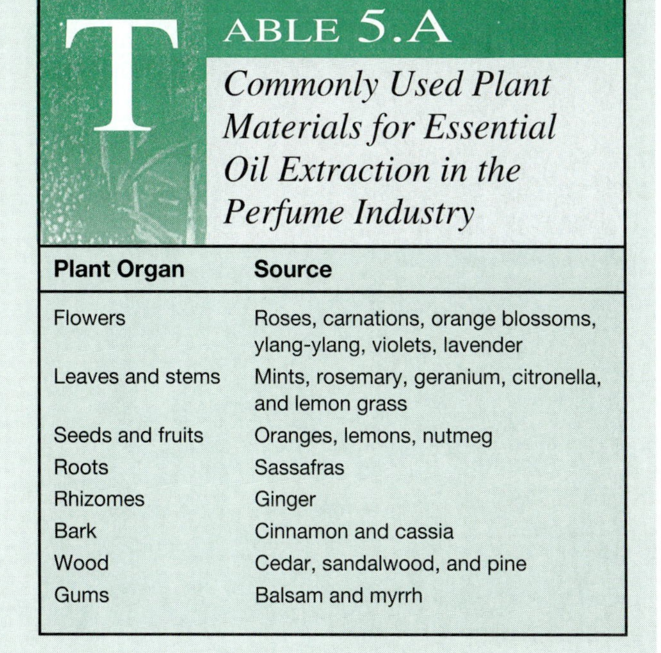

TABLE 5.A
Commonly Used Plant Materials for Essential Oil Extraction in the Perfume Industry

Plant Organ	Source
Flowers	Roses, carnations, orange blossoms, ylang-ylang, violets, lavender
Leaves and stems	Mints, rosemary, geranium, citronella, and lemon grass
Seeds and fruits	Oranges, lemons, nutmeg
Roots	Sassafras
Rhizomes	Ginger
Bark	Cinnamon and cassia
Wood	Cedar, sandalwood, and pine
Gums	Balsam and myrrh

from the blossoms. The petals are replaced until the fat is saturated with the floral essence. The essential oil is then extracted from the fat with alcohol. Enfleurage is slow and labor-intensive and is only used to extract those delicate essential oils that would be destroyed by other methods. Regardless of the method used, tremendous quantities of plant material are needed to produce even small quantities of the pure essential oil; for example 60,000 roses are needed for one ounce (28 grams) of rose oil.

Once the essential oils are extracted and blended for the characteristic fragrance of a particular perfume, fixatives are added to retard the evaporation of highly volatile essential oils. The fixatives may be plant or animal oils, such as musk oil from the musk deer. However, today most animal oils have been replaced by synthetics.

The final perfume concentrate is then diluted with alcohol and a small amount of distilled water: perfumes generally contain 18%-25% concentrate; eau de parfum contains 10%-15%; eau de cologne, 5%-8%; and eau de toilette, 2%-4%.

microspore mother cells; in fact, the pollen chambers are technically referred to as **microsporangia** (singular, **microsporangium**). Each microspore mother cell undergoes meiosis to produce four **microspores** (male spores). Initially, the four spores stay together as a **tetrad,** but eventually they separate and each will develop into a pollen grain, an immature male gametophyte. In the development of the pollen grain, the microspore undergoes a mitotic division to produce a **generative nucleus** and a **tube nucleus.** Also, the wall of the microspore becomes chemically and structurally modified into the pollen wall. The pollen wall consists of an inner layer, the **intine,** and an outer layer, the **exine,** which may be ornamented with spines, ridges, or pores. When the pollen grains are fully developed, they are released as the anthers open or dehisce. (*See* A Closer Look: Pollen is More than Something to Sneeze at for a further discussion on pollen.)

Female Gametophyte Development

Within the ovary one or more ovules develop; an ovule can be considered a **megasporangium** enveloped by two layers of tissue called **integuments.** The integuments completely surround the megasporangium except for an opening called the **micropyle.** The ovule first appears as a bulge in the ovary wall. During the development of the ovule, one cell becomes distinct as a **megaspore mother cell;** it is surrounded by tissue called the **nucellus.** The megaspore mother cell undergoes meiosis to produce four **megaspores;** three of these degenerate, leaving one surviving megaspore. This megaspore then undergoes a series of mitotic divisions, eventually producing a mature female gametophyte, which is often called the **embryo sac.** In the typical pattern of development, a series of three mitotic divisions produces eight nuclei within the greatly enlarged megaspore. These eight nuclei are distributed with three (the **egg apparatus**) near the micropyle end of the ovule, three **antipodals** at the opposite or **chalazal** end, and two **polar nuclei** in the center. The egg apparatus consists of two **synergids** and one **egg.** Cell walls soon develop around the egg, synergids, and antipodals; at this stage the female gametophyte is mature (fig. 5.8).

POLLINATION AND FERTILIZATION

Pollination involves the transfer of pollen from the anther to the stigma. If the pollen transfer is within the same flower, it is known as **self-pollination. Cross-pollination**

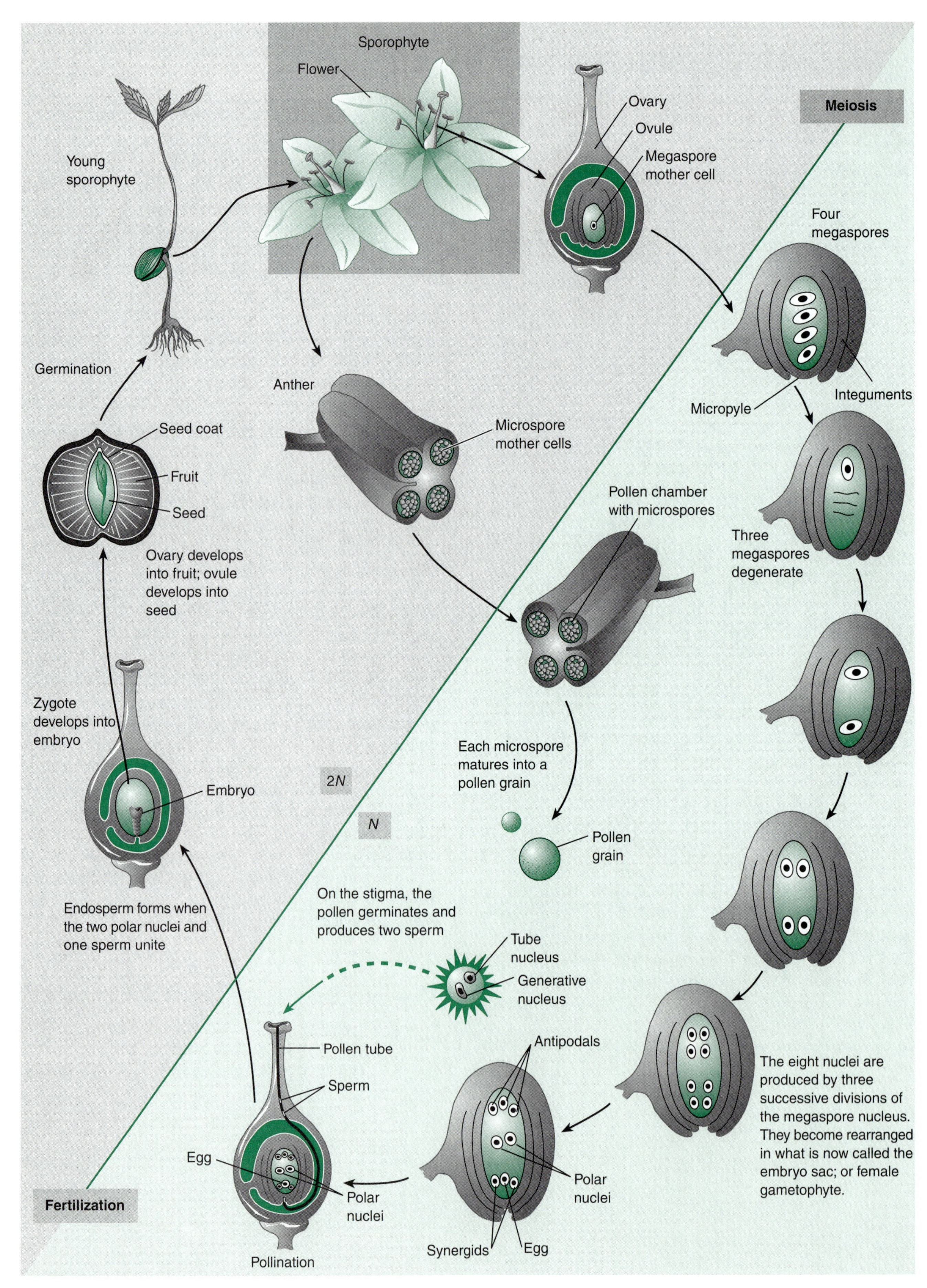

Figure 5.8 *Reproductive cycle of an angiosperm.*

(a)

(b)

(c)

Figure 5.9 *Flowers and their animal pollinators. (a) Bee-pollinated flowers often have the stamens brush up against the bee's body. (b) Butterfly-pollinated flowers have a broad expanse for the butterfly to land. (c) Hummingbird-pollinated flowers are curved back, allowing the bird to insert its beak to reach the nectar.*

involves the transfer of pollen from one flower to another. Pollination can be accomplished by various methods. Large showy flowers usually attract animal pollinators, while small inconspicuous flowers are often wind pollinated.

Animal Pollination

Although bees are the most familiar animal pollinators, a host of different species are involved in the transfer of pollen (fig. 5.9). Other insects such as wasps, flies, ants, butterflies, and moths are equally important pollinators for many flowers. Even larger animals, such as birds and bats, are efficient pollinators for some species. It should be noted that pollination is accomplished inadvertently when the animal visitor, dusted with pollen from one flower, visits a second flower of the same species.

Color and scent are what attract animals to flowers. Certain colors are associated with specific pollinators; for example, bee-pollinated flowers are often yellow, blue, or purple, or some combination of these colors. Many bird-pollinated flowers, such as columbine and trumpet creeper, are red. In addition, many white or light-colored flowers are pollinated in the evening by night-flying visitors. Various contrasting color patterns **(nectar guides)** seen on petals serve to direct insects toward the nectar. Often nectar guides cannot be seen by the human eye but are visible in ultraviolet light, which can be perceived by certain insects. To the insect eye, the nectar guides seem like airport lights lining a runaway. **Essential oils,** volatile oils that impart a fragrance, attract pollinators by scent. The essential oils of flowers like the rose, orange, and jasmine have been used in perfumes for hundreds of years (*see* A Closer Look: Alluring Scents on page 78). Not all scents are appealing to humans; for example, the carrion flower (*Stapelia* sp.), which is fly-pollinated, gives off an aroma of rotting meat.

In most cases the flower provides a reward of **nectar,** pollen, or both to the animal. Nectar is a sugary liquid produced in glands called **nectaries** found in the epidermis of a floral organ. Many children have tasted its sweetness when they sip the nectar of honeysuckle (*Lonicera*

Figure 5.10 *Wind-pollinated flowers of birch,* Betula, *are grouped into staminate and pistillate catkins.*

japonica) blossoms. The amount of nectar produced by flowers varies greatly; flowers pollinated by birds generally produce copious amounts of nectar.

Wind Pollination

As described, animal-pollinated flowers have a variety of mechanisms used to attract the pollinator; wind-pollinated flowers, on the other hand, have a much simpler structure. Color, nectar, and fragrance that play an integral role in animal pollination are usually not prominent in the wind-pollinated flower.

Wind-pollinated flowers are often small and inconspicuous, usually lacking petals and sometimes even sepals; these drab flowers are frequently arranged in inflorescences such as the catkins of oak, birch, and willow, and the panicles, racemes, and spikes of the grasses (fig. 5.10). Although most grasses have perfect flowers, many other wind-pollinated species are imperfect. Stamens and stigmas are also modified for this method of pollination. The filaments are usually long, allowing the anthers to hang free away from the rest of the flower, thereby enabling the pollen to be caught by the wind. Stigmas are often feathery, increasing the surface area for trapping pollen.

Although individual flowers are small, this is offset by the large number of flowers formed and by the production of copious amounts of dry, lightweight pollen. A single stamen of corn contains between 2,000 and 2,500 pollen grains, with the whole plant producing about 14 million pollen grains. One wind-pollinated plant whose pollen causes misery to hay fever sufferers is ragweed; one healthy ragweed plant can release one billion pollen grains. It is estimated that one million tons of ragweed pollen are produced in the United States each year!

Double Fertilization

Once pollination is accomplished, the stage is set for fertilization. Recall that the pollen grain at the time of pollination contains a tube cell and a generative cell. On a compatible stigma, the pollen grain germinates; a **pollen tube** begins growing down into the style toward the ovary. The **tube nucleus** is generally found at the growing end of the pollen tube, while behind it the **generative nucleus** divides mitotically, producing two **nonmotile sperm.** The pollen tube continues to grow until it reaches and grows into the micropyle of an ovule, penetrating the ovule at one synergid. The tube nucleus, synergids, and antipodals usually degenerate during the fertilization process, leaving the two sperm, egg, and polar nuclei as the remaining participants (fig. 5.8).

Both sperm are involved in fertilization. One sperm fertilizes the egg to produce a zygote that will develop into an embryo. The zygote produced from the fusion of haploid egg and sperm is diploid; this restores the chromosome number for the sporophyte generation. The second sperm fertilizes or fuses with the two polar nuclei, producing the **primary endosperm nucleus,** which develops into **endosperm,** a nutritive tissue for the developing embryo. The fusion of the haploid sperm and polar nuclei normally produces **triploid** endosperm. This **double fertilization** is a distinctive feature of angiosperm reproduction.

The value of endosperm as a food source for the human population cannot be overemphasized. The nutritive value of wheat, rice, and corn, the world's major crops, is due to the large endosperm reserves in these grains. The development of early civilizations in various parts of the world is linked to the cultivation of these grains providing stable food sources.

Following fertilization, changes begin to occur within the whole flower. Sepals, petals, and stamens often wither and drop off as the ovary greatly expands, becoming a **fruit.** Within the ovary each fertilized ovule becomes a **seed;** the integuments of the ovule develop into the **seed coat,** the outer covering of the seed. The discussion of fruits and seeds continues in Chapter 6.

Chapter Summary

1. Flowers, the characteristic reproductive structures of angiosperms, are composed of sepals, petals, stamens, and carpels. Modifications of the basic floral organs are common, often resulting in incomplete and imperfect flowers.
2. Meiosis is a form of cell division that reduces the number of chromosomes from diploid to haploid. The process consists of two consecutive divisions, with the reduction in chromosome number occurring in the first division. The most significant events of meiosis occur in prophase I when synapsis occurs, and anaphase I when the homologous chromosome pairs separate.
3. In angiosperms, meiosis occurs prior to the formation of male and female gametophytes, which are small and relatively short-lived. Ovules, which include the female gametophytes, develop within the carpels, while the pollen grains, or male gametophytes, develop in the stamen.
4. Pollen is transferred passively by animals or wind from stamen to stigma. Insect-pollinated flowers typically have bright, showy petals, fragrant aromas, and are rich in nectar. Pollen in these flowers is often sticky, adhering to the insect body. Wind-pollinated flowers are usually small and inconspicuous but produce copious amounts of dry, lightweight pollen. Only a small amount of pollen from wind-pollinated plants reaches the female organ. Most pollen grains settle to Earth where they can leave a lasting record in the sediment.
5. Prior to fertilization, the pollen tube grows down the style into the ovary and ovule. The generative nucleus gives rise to two sperm. Within the ovule, double fertilization occurs as one sperm fertilizes the egg producing the zygote, while the second sperm fuses with the polar nuclei, giving rise to the primary endosperm nucleus. Following fertilization, the ovary becomes a fruit and each ovule becomes a seed.

Review Questions

1. Describe the parts of a flower and indicate some common modifications.
2. Detail the events of meiosis. Why is prophase I of meiosis an important stage?
3. Describe the male and female gametophytes.
4. What is the relationship between animal pollinators and floral morphology?
5. What is meant by double fertilization?
6. What fields depend on the study of pollen as a useful tool? What types of information have been gained from this approach?

Further Reading

Berger, Terry. 1977. "Tulipomania" Was No Dutch Treat to Gambling Burghers. *Smithsonian,* 8 (1): 70–77.

Green, Timothy. 1991. Making Scents is More Complicated Than You'd Think. *Smithsonian,* 22 (5): 52–61.

Meeuse, Bastian and Sean Morris. 1984. *The Sex Life of Flowers,* Rainbird Publishing, London.

Moize, Elizabeth A. May, 1978. Tulips: Holland's Beautiful Business. *National Geographic,* 153 (5): 712–728.

Newman, Cathy. 1984. Pollen: Breath of Life and Sneezes. *National Geographic,* 166 (4): 496–521.

Raven, Peter H. and George B. Johnson. 1992. *Biology,* 3rd Edition. Mosby, St. Louis, MO.

Tanner, Ogden. 1985. The Flowers That Afflict Us With "A Sort of Madness." *Smithsonian,* 16 (8): 168–178.

CHAPTER

6

Plant Life Cycle: Fruits and Seeds

Chapter Outline

Chapter Concepts

1. Fruits are ripened ovaries that are the end products of sexual reproduction in angiosperms and are a major vehicle for the dispersal of their enclosed seeds.
2. Seeds contain an embryonic plant plus some nutritive tissue and are the starting point for the next generation.
3. Edible fruits of various types play a major role in the human diet.

Fruits, as with flowers, are unique aspects of sexual reproduction in angiosperms; they protect the enclosed seeds and aid in their dispersal. Not only are fruits essential in the angiosperm life cycle, but they are widely utilized as a significant food source.

Fruit Types

The fruit wall that develops from the ovary wall is known as the **pericarp** and is composed of three layers: the outer **exocarp,** the middle **mesocarp,** and the inner **endocarp.** The thickness and distinctiveness of these three layers vary among different fruit types.

Simple Fleshy Fruits

Simple fruits are derived from the ovary of a single carpel or several fused carpels and are described as **fleshy** or **dry.** When ripe, the pericarp of fleshy fruits is often soft and juicy. Seed dispersal in the fleshy fruits is accomplished when animals eat the fruits. The following describes the most common types of fleshy fruits (fig. 6.1):

A **berry** has a thin exocarp and a soft fleshy mesocarp and endocarp with one to many seeds. Tomatoes, grapes, and blueberries are familiar berries.

A **hesperidium** is a berry with a tough leathery rind such as oranges, lemons, and other citrus fruits.

A **pepo** is a fleshy fruit with a tough outer rind (consisting of both receptacle tissue and exocarp); the mesocarp and endocarp are fleshy. All members of the squash family, including pumpkins, melons, and cucumbers, form pepos.

A **drupe** has a thin exocarp, a fleshy mesocarp, and a hard stony endocarp that encases the seed; cherries, peaches, and plums are examples.

Apples and pears are **pomes;** most of the fleshy part of pomes develops from the enlarged base of the perianth that has fused to the ovary wall.

As described in both the pepo and pome, some fruits develop from flower parts other than the ovary; fruits of these types are termed **accessory fruits.**

Dry Dehiscent Fruits

In dry fruits, the pericarp may be tough and woody or thin and papery; dry fruits fall into two categories, **dehiscent** and **indehiscent.** Dehiscent fruits split open at maturity and so release their seeds. Wind often aids the dispersal of seeds from dehiscent fruits. Three common types of dehiscent fruits, **follicles, legumes,** and **capsules,** are characterized by the way in which they open. Follicles, as found in magnolia and milkweed, split open along one seam, while legumes like bean pods and pea pods split along two seams (fig. 6.1). The most common dehiscent fruit is a capsule that may open along many pores or slits; cotton and poppy are representative capsules.

Dry Indehiscent Fruits

Indehiscent fruits do not split open. Instead, they use other means of dispersing the seeds. **Achenes, samaras, grains,** and **nuts** are examples of indehiscent fruits. Sunflower "seeds" are, in fact, achenes, one-seeded fruits in which the pericarp is free from the seed (fig. 6.1). Carried by the wind, the winged fruits of maple and ash trees are familiar types of samaras. Samaras are usually described as modified achenes. The fruits of all our cereal grasses are grains, single-seeded fruits in which the pericarp is fused to the seed coat. Also called a **caryopsis,** this type of fruit is found in wheat, rice, corn, and barley. Botanically, nuts are one-seeded fruits with hard stony pericarps such as hazelnuts, chestnuts, and acorns. In common usage, however, the term nut has also been applied to seeds of other plants; peanuts, cashews, and almonds are actually seeds, not nuts.

Aggregate and Multiple Fruits

Aggregate fruits develop from a single flower with many separate pistils (carpels), all of which ripen at the same time as in raspberries and blackberries. Strawberries, another aggregate fruit, also contain accessory tissue. The brownish-yellow spots on the surface are actually achenes inserted on the enlarged, fleshy, red receptacle (fig. 6.1).

Multiple fruits result from the fusion of ovaries from many separate flowers on an inflorescence. Figs and pineapples are examples of multiple fruits (fig. 6.1).

Seed Structure and Germination

A seed contains the next generation and so completes the life cycle of a flowering plant. The seed develops from the fertilized ovule and includes an embryonic plant and some form of nutritive tissue within a seed coat. Differences between dicots and monocots are apparent within seeds. The very names, dicot and monocot, refer to the number of seed leaves, or **cotyledons,** present in the seed. Dicot seeds have two cotyledons that are attached to and enclose the embryonic plant. The cotyledons, which are often large and fleshy, occupy the greatest part of the dicot seed and have absorbed the nutrients from the endosperm. Thus, the endosperm in many dicot seeds is either lacking entirely or very much reduced. Monocots have a single small cotyledon that functions to transfer food from the endosperm to the embryo. In several monocot families, large amounts of endosperm are apparent. Because of these stored nutrient reserves (either in the cotyledons or the endosperm), many seeds, like many fruits, are valuable foods for humans and other animals.

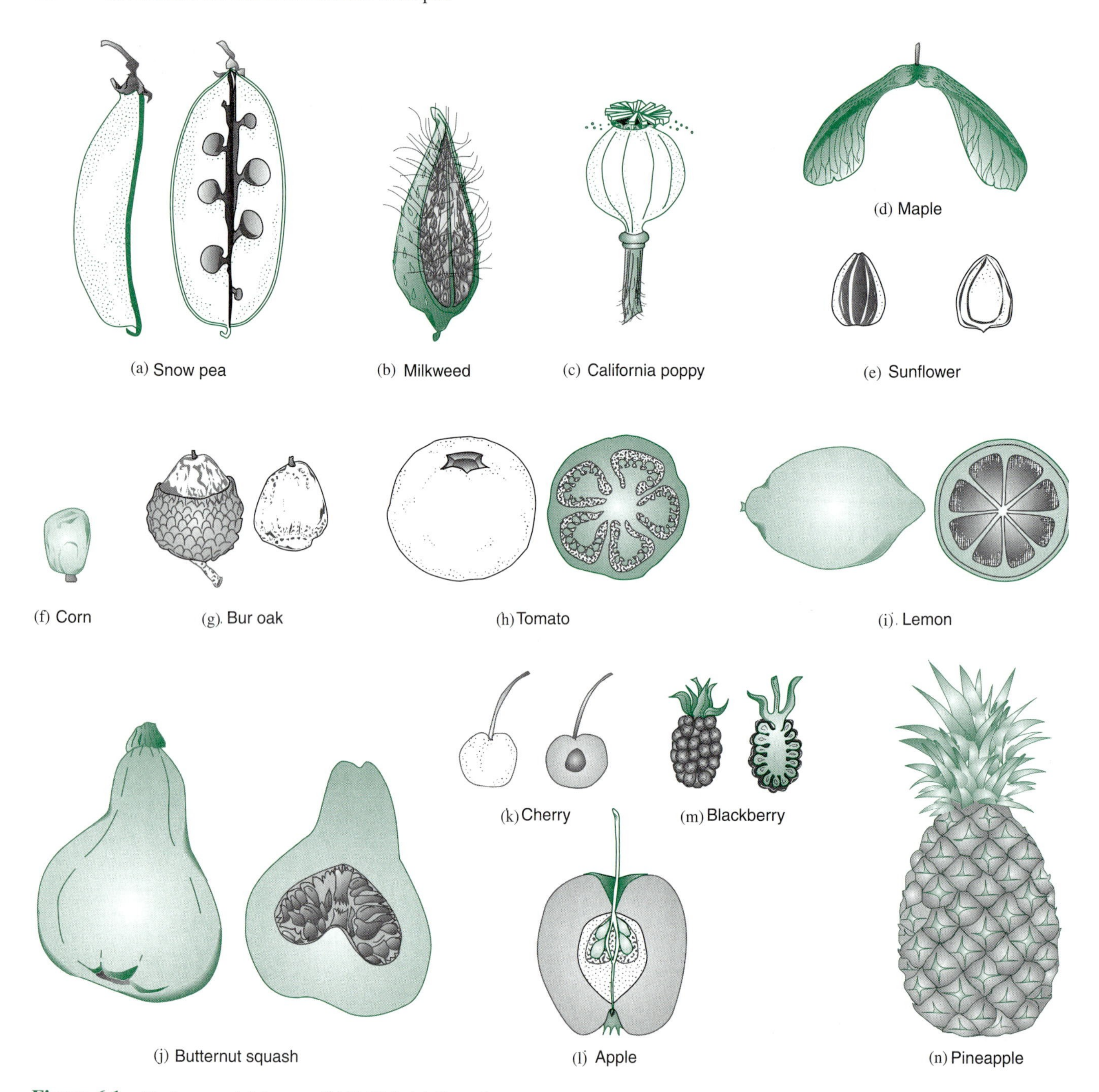

Figure 6.1 *Fruit types. (a) Legume (b) Follicle (c) Capsule (d) Samara (e) Achene (f) Grain (g) Nut (h) Berry (i) Hesperidium (j) Pepo (k) Drupe (l) Pome (m) Aggregate fruit (n) Multiple fruit*

Dicot Seeds

The lima bean, because of its large size, is a good example of a dicot seed. A thin membranous seed coat, also known as the **testa,** encloses the seed. A **hilum** and **micropyle** are visible on the surface of the testa (fig. 6.2a). The hilum is a scar that results from the separation of the seed from the ovary wall. Recall that the micropyle, seen as a small pore, is the opening in the integument through which the pollen tube enters the ovule. If the testa is removed, the two large food-storing cotyledons are easily seen and separated. Sheltered between the cotyledons is the embryo axis consisting of the **epicotyl,** the **hypocotyl,** and the **radicle.** The epicotyl develops into the shoot (stems and leaves) of the seedling and typically bears embryonic leaves within the seed. The hypocotyl is the portion of the embryo axis between the cotyledon attachment and the radicle, the embryonic root.

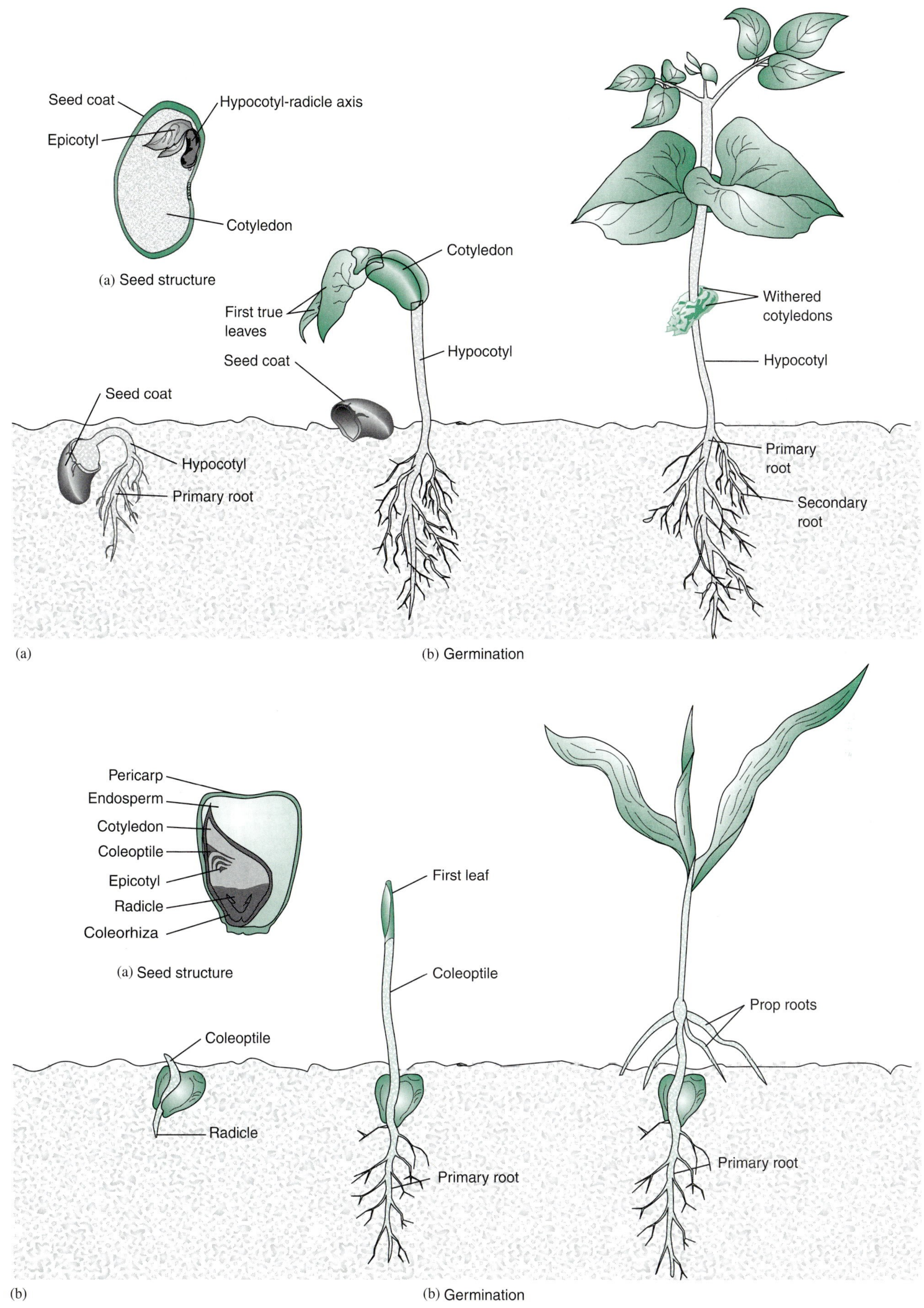

Figure 6.2 *Seed structure and germination of (a) a dicot, the garden bean; (b) a monocot, corn.*

Monocot Seeds

The corn kernel is a familiar grain that can be used to illustrate the composition of a monocot seed (fig. 6.2b). It is important to remember that a grain is a fruit in which the testa of the single seed is fused to the pericarp. One major difference from the lima bean is the presence of extensive endosperm that occupies much of the volume of the seed. The small embryo has only a single cotyledon called a **scutellum.** Other differences include the presence of a **coleoptile** (a protective sheath that surrounds the epicotyl) and the **coleorhiza** (a protective covering around the radicle).

Seed Germination and Development

With appropriate environmental conditions (adequate moisture, oxygen, and appropriate temperature) seeds germinate (fig. 6.2). The first structure to emerge from the seed is the radicle, which continues to grow and produces the primary root. In monocots, the radicle first breaks through the coleorhiza. This early establishment of the root system enables the developing seedling to absorb water for continued growth. Next the shoot emerges. In a dicot the hypocotyl elongates and breaks through the soil in a characteristic arch that protects the epicotyl tip with its embryonic leaves. In most dicots the cotyledons are carried aboveground with the expanding hypocotyl, while in others the cotyledons remain belowground. Soon after the tissues of the seedling emerge from undergound and are exposed to sunlight, they develop chlorophyll and begin to photosynthesize. The exposure to sunlight also triggers the hypocotyl to straighten into an erect position. In monocots, the coleoptile emerges from the soil; the epicotyl soon breaks through the coleoptile and the embryonic leaves begin expanding. Establishment of the seedling is the most critical phase in the life of a plant and high mortality is common. Seedlings are sensitive to environmental stress and vulnerable to attack by pathogens and predators, while established plants have a greater array of defenses.

Representative Edible Fruits

Of the more than 250,000 known angiosperms, only a small percentage produce fruits that have been utilized by humans; however, these have made a significant impact on our diet and economics (*see* A Closer Look: Exotic Fruits). Fruits are packed with nutrients and are particularly excellent sources of vitamin C, potassium, and fiber. As this chapter describes, a fruit is a mature or ripened ovary, but this botanical definition has been ignored in the marketplace. Even the United States Supreme Court has debated the question "What is a Fruit?"

It all started in the late nineteenth century when an enterprising New Jersey importer, John Nix, refused to pay the import tariff on a shipment of tomatoes from the West Indies. He argued that the 10% duty placed on vegetables by the Tariff Act of 1883 was not applicable to tomatoes since, botanically, they are fruits not vegetables. This fruit-vegetable debate eventually reached the Supreme Court in 1893 (fig. 6.3). Justice Horace Gray wrote the decision stating

> "Botanically speaking, tomatoes are the fruits of the vine, just as are cucumbers, squashes, beans and peas. But in the common language of the people, whether sellers or consumers of provisions, all these are vegetables which are grown in kitchen garden, and which, whether eaten cooked or raw, are, like potatoes, carrots, parsnips, turnips, beets, cauliflower, cabbage, celery, and lettuce, usually served at dinner, in, with, or after the soup, fish, or meats, which constitute the principal part of the repast, and not, like fruits, generally as dessert."

Tomatoes were legally declared vegetables and Nix paid the tariff. Despite the legal definition, botanists still consider tomatoes to be berries.

Figure 6.3 *An 1893 decision of the U.S. Supreme Court decided that the tomato was legally a vegetable.*
Jack E. Davis © 1990.

Tomatoes

Tomatoes, *Lycopersicon esculentum,* are native to South America, the Andes region of Chile, Colombia, Ecuador, Bolivia, and Peru and are believed to have been first domesticated in Mexico. The Spanish conquistadors introduced the tomato to Europe, where it was first known as the "Apple of Peru," the first of many names for this fruit. Later it was known as *pomo doro,* golden apple (early varieties were yellow) in Italy, and pomme d'amour, love apple (it was believed by many to be an aphrodisiac) in France. The common name for the fruit comes from the Mexican Indian (Nahua dialect) word for it, *tomatl.* The scientific name also reflects another early myth about the tomato; literally translated *Lycopersicon* means "wolf peach," a reference to its poisonous relatives including the deadly nightshade and henbane, while *esculentum* refers to its edibility. Despite its described edibility, it took years for the tomato to live down its poisonous reputation. One bizarre demonstration of its lack of toxicity took place in Salem, New Jersey in 1820 when Colonel Robert Gibbon Johnson ate a bushel of tomatoes in front of a crowd gathered to witness his certain demise. His obvious survival, without ill effects, finally settled the issue of the tomato's edibility.

Over the years since Johnson's demonstration, the tomato, due to its attractiveness, taste, and versatility, has become one of the most commonly eaten "vegetables" in the United States; we each consume approximately 36 kilograms (80 lbs) every year. Individually, the tomato is largely water with only small amounts of vitamins and minerals, but because of the large volume consumed, the tomato leads all fruits and vegetables in supplying these dietary requirements. Its versatility is almost unequaled; what would pizza, ketchup, bloody Marys, salads, and lasagna be without tomatoes?

Almost all the **cultivars,** cultivated varieties, (over 500 exist) belong to the species *Lycopersicon esculentum;* however, there are eight other recognized species in this genus. The most familiar varieties have large fruits such as Big Boy, Mammoth Wonder, and Beefsteak. At the other extreme are the cherry tomatoes that scientists believe are most similar to the ancestral wild type. Today's large scale commercial production yields a tomato very different from the home garden varieties. Commercial tomatoes have been bred for efficient mechanical harvesting, transportability, and long shelf life, not necessarily for taste. Another characteristic that has improved the commercial varieties is the determinate habit. This trait originally appeared as a spontaneous mutation in Florida in 1914. Determinate plants are shorter, bushier, and more compact than the indeterminate habit, which has a more sprawling pattern of growth that requires extensive staking or trellising. While no other species are widely cultivated, scientists are interested in the salt tolerance of *L. cheesmanii,* which is native to the Galápagos Islands. This species, unlike *L. esculentum,* is able to survive in seawater with its high salt concentration. Developing salt tolerance in crop plants is one of the aims of plant-breeding programs, since salt-tolerant crops will allow agriculture to expand into areas with saline soils.

Tomatoes are one of the most common plants grown in the home garden. Even in large metropolitan areas, many apartment dwellers will grow patio varieties in containers. Home gardeners have all watched the stages of growth, flowering, and fruit ripening as they wait for the first harvest. Tomato fruits require 40 to 60 days from flowering to reach maturity. Once fertilization has occurred, the fruit rapidly increases in size, reaching its mature size in 20 to 30 days. In the latter half of fruit development, color changes reflect internal changes in the acidity, sweetness, and vitamin C content of the fruit. The first hint of ripening is seen when the green fruit lightens as a result of chlorophyll breakdown. As the chlorophyll content continues to decrease, additional carotenoids are synthesized. The carotenoids, β-carotene (orange) and especially **lycopene** (red), give the mature fruit its characteristic color. As the red color deepens, the acidity decreases, the sugars and vitamin C increase, the flavor develops, and the fruit softens. These changes coincide with increases in **ethylene** and a sudden peak in respiration in the fruit. Ethylene is a gaseous plant hormone that is involved in several developmental stages but is best known for its involvement with fruit ripening. When tomatoes are picked green and ripened in storage, they have lower vitamin C and sugar content and poorer flavor; this explains why most people prefer a vine-ripened tomato.

Many home gardeners have noticed the role that temperature plays in tomato development and ripening. Most varieties do best when air temperatures are between 18°–27° C (65°–80° F). When temperatures are either too hot or too cold, fruit set is inhibited. Also, temperatures above 29° C (85° F) inhibit the development of the red pigments.

During the spring of 1990, millions of elementary school children and high school and college students in the United States and 30 foreign countries were given the opportunity to participate in the SEEDS project. The project was a cooperative venture between NASA and the George W. Park Seed Company to contrast the growth of space-exposed seeds to the responses of control seeds stored on Earth. The space-bound seeds were launched into orbit aboard the LDEF satellite on April 7, 1984 by the Space Shuttle *Challenger.* The LDEF (Long Duration Exposure Facility) satellite is a cylindrical structure approximately 9 meters long and 4 meters in diameter and built to house many separate experiments that were designed to test the effects of space on various systems (fig. 6.4). Twelve and one-half million Rutgers California Supreme tomato seeds were aboard LDEF for almost six years, experiencing weightlessness and exposure to cosmic radiation for longer than any previous NASA experiment involving biological

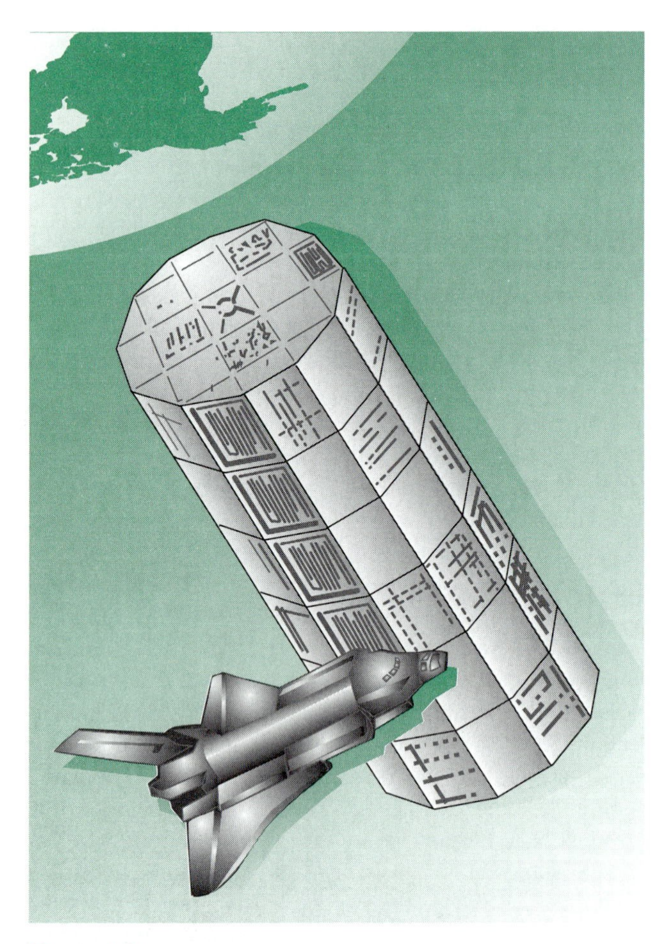

Figure 6.4 *The LDEF remained in space for almost six years to test the effects of weightlessness and exposure to cosmic radiation on tomato seeds.*

tissue. The satellite was recovered by the Space Shuttle *Columbia* and returned to Earth on January 20, 1990. The first germination test began in late February and continued throughout the spring as seed packages were mailed to teachers and students. The first conclusion reached was that after nearly six years in space, the tomato seeds were still viable. In fact, they consistently exhibited an 18%–30% faster germination rate than their earth-based counterparts. Growth rates of the space seedlings were also accelerated for the first three or four weeks; after that the Earth-based seedlings caught up and no significant differences between earth and space tomato plants or their fruits were observed (table 6.1). Researchers also observed a greater number of mutant individuals such as albino plants, stunted individuals, and one plant with no leaves at all, in the space-exposed group. Interestingly, plants of space-exposed seeds had greater levels of chlorophylls and carotenes than in the earth group. Perhaps, the most significant conclusion of the SEEDS project is the proof that seeds can survive relatively undamaged in space for long periods of time.

TABLE 6.1

Comparisons of Space-Exposed and Earth-Based Tomato Seeds, SEEDS Project

	Space-Exposed	Earth-Based
Percent germination	66.3	64.6
Average height (cm) at 56 days	21.2	20.9
Percent flowering	73.4	72.3
Percent fruiting	74.6	76.1

Apples

Apples, *Malus pumila,* have a long history of human use; they were among the first tree fruits to be domesticated in temperate regions. Most of today's cultivated varieties are descendants of apples native to the Caucasus Mountains of western Asia where the apple has been domesticated for thousands of years. Because of its long history, the apple also has a place in the imagery and folklore of many cultures. The following expressions reveal the apple as more than a tasty fruit in American culture: the "Big Apple," "American as apple pie," "apple of your eye," "apple polishing," "apple pie order," and "an apple a day keeps the doctor away."

Most are familiar with the biblical story of Adam and Eve and the presumed role of the "apple" in the downfall of mankind. In reality, the apple's only involvement was due to a faulty translation from the Latin version of the Old Testament. The confusion starts with the Latin word *mali,* which could refer to *malum* meaning evil or *malus* meaning apple tree. In Genesis, Adam and Eve were told not to partake of the fruit of the tree of the knowledge of good and evil. The erroneous association with apple trees began with thirteenth and fourteenth century artists, whose Latin was poor; they incorrectly translated the *mali* to mean apple tree.

The legend of Johnny Appleseed is another story familiar to most Americans. There really was a "Johnny Appleseed"; his name was John Chapman. He was born in Massachusetts in 1774, but we know nothing of his early life. He appeared in 1797 sowing apple seeds in what was the Northwest Territory in frontier America: western Pennsylvania, Ohio, and eastern Indiana. Chapman probably saw more of America than most of his contemporaries; he traveled hundreds of miles by foot, horseback, and canoe. Chapman was an itinerant orchardist who gave away or sold, and even planted with his own hands, apple seeds and

Figure 6.5 *John Chapman, known better as "Johnny Appleseed," sowed apple seeds on the American frontier from the 1790s to 1845.*

(a)

(b)

Figure 6.6 *(a) These apple blossoms are being pollinated by bees. (b) Scions from a Granny Smith tree have been grafted onto a rootstock.*

seedlings that gave rise to acres of apple trees throughout the region (fig. 6.5). Some of the orchards are still in existence today. He continued his mission until his death in 1845 in Fort Wayne, Indiana, a city that still honors his memory every summer with a Johnny Appleseed Festival.

Apple trees are medium-size trees with a broad, rounded crown and short trunk. They are generally spreading, long-lived trees that can bear fruit for up to a century. In modern commercial orchards, however, dwarf trees have become the norm. Often about six feet tall, these trees are easily pruned and mechanically harvested.

The apple blossoms appear in profusion early in the spring before the leaves develop. The fragrant pinkish-white blossoms are 5-merous flowers (containing five sepals, five petals, numerous stamens, and a five-carpeled ovary) that are usually pollinated by bees (fig. 6.6). The apple tree requires cold winter temperatures in order to flower, and therefore cannot be grown in tropical and subtropical climates. The leading apple-growing areas in the United States are Washington, Oregon, and northern California in the West and Michigan, New York, and Virginia in the East.

Although Johnny Appleseed distributed thousands of apple seeds, today's modern orchardist does not grow the trees from seeds. Each seed is a unique combination of traits that are not identical to either parent; in a sense, planting a seed is a genetic experiment. Although some seeds may develop into valuable varieties, most will not. (Interestingly, most of our familiar cultivars did develop as chance or volunteer seedlings from naturally produced seeds.) Also, planting seeds is a long-term experiment, and it would take a number of years to know the results. Today's apple growers need to insure uniformity in their orchards, not only to produce apples with the desired flavor and taste, but also to maximize efficiency at harvest time. As a result, most apple trees are produced by **grafting** (a form of asexual reproduction or cloning) in which stem

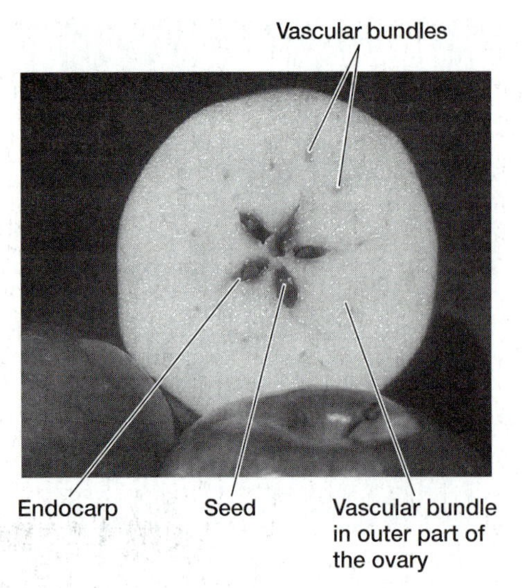

Figure 6.7 *In this cross section of the apple fruit, the vascular bundles delineate the outer edge of the true ovary wall from the accessory tissue. The endocarp is the brown, papery material surrounding each seed.*

cuttings or buds from a desirable cultivar are joined to the base of a second tree. The cutting or bud, called the **scion,** will become the upper or top portion of the new tree, while the **rootstock,** or simply **stock,** of the second tree is the root system of the graft combination (fig. 6.6). Grafting can create thousands of identical copies of a variety that will continue to produce apples with the desired characteristics.

As described earlier, the apple is a pome, a simple fleshy fruit with accessory tissue. The core of the apple is a five-carpeled ovary with seeds; the ovary wall is visible as a fine brownish line and the endocarp is prominent as the parchment-like material around the seeds (fig. 6.7). The skin and flesh of the apple develop from the receptacle and base of the perianth. A ripe apple, although mostly water, contains about 12% sugar, 1% fiber, and negligible amounts of fat and protein.

Although there are thousands of varieties of apples, only a few can be found in the modern supermarket. In the past many more varieties were available to consumers; virtually gone from the market are old staples such as Baldwin, Early Harvest, Fall Pippin, and Gravenstein. Today, Red Delicious, Golden Delicious, and Granny Smith are the most common apples seen. The red delicious apple was first discovered in Iowa in the 1870s but its popularity can be traced to the growth of the large supermarket chains during the 1950s and 1960s. It became the generic "red apple" to the American consumer. Other commercially important red apples include the Rome, Jonathan, and McIntosh. The Golden Delicious, the leading yellow apple, arose by chance on a West Virginia farm in 1910. The Granny Smith is a tart green apple that originated in New Zealand and is rising in popularity. Growers are constantly developing new varieties and one of the newest is Gala, a red and yellow apple also developed in New Zealand. Approximately 50% of the apples are consumed as fresh fruit, with the remainder being processed into applesauce, apple butter, apple cider, apple juice, cider vinegar, dried apples, and canned apples.

Oranges

The citrus family (Rutaceae) is the source of many edible fruits; sweet oranges (*Citrus sinensis*), grapefruits (*C. paradisi*), tangerines (*C. reticulata*), lemons (*C. limon*), and limes (*C. aurantifolia*) come to mind most immediately. But there are also pummelos (*C. grandis*), citrons and etrogs (*C. medica*), bergamots (*C. bergamia*), sour oranges (*C. aurantium*), and kumquats (*Fortunella japonica*). Most of these species are native to southeastern Asia, where they were undoubtedly cultivated by native peoples. Brought by caravan from the East, the citron was the first citrus fruit known to western civilization; it was widely grown throughout the Mediterranean area during Greek and Roman times, being spread by the Jewish people for whom it had religious significance. Centuries later Europeans became acquainted with sour oranges, lemons, and limes when Arabic traders during the Middle Ages introduced them from the East. It was not until the sixteenth century, however, that the sweet orange, the most familiar type today, was brought to Europe by Portuguese traders. It is believed that these first sweet oranges came from India, but later oranges were brought from China, which is believed to be the country of origin; this is reflected in its scientific name, *Citrus sinensis.*

The Spanish and Portuguese explorers introduced citrus to the New World; by 1565 sour orange trees were growing in Florida. After Florida became a state in 1821, the wild groves of sour oranges became the rootstocks for the sweet orange industry. Florida soon became, and remains, the leading orange-producing state. Valencia oranges are the main variety cultivated in Florida; this is a thin-skinned, seeded orange grown primarily for juice. Citrus was brought to California by the Spanish missionaries in the eighteenth century. Today we associate the navel orange with California, but these oranges have their origins in Bahia, Brazil. In 1870, an American missionary stationed in Bahia was impressed by the appearance and flavor of a local variety and sent 12 saplings to the USDA in Washington, D.C. The USDA propagated the trees and offered free plants to anyone wishing to grow them. In 1873, a Riverside, California resident, Mrs. Luther Tibbets, received two trees that were such a success that they were soon widely propagated. In fact, it is believed that all navel oranges today are descendants of Mrs. Tibbets' trees.

Botanically, a citrus fruit is a hesperidium. The thick rind is impregnated with oil glands; fragrant essential oils are characteristic of the family and are important commercially for perfumes and cosmetics. Within the fruit, the individual carpels are filled with many one-celled juice sacs. Although the carpels are always distinct, their ease in separation varies with the type and variety of fruit. Navel oranges are noted for their seedless condition and also for the navel, which is actually a small aborted ovary near the top of the fruit. Although widespread cultivation of navel oranges is a fairly recent phenomenon, Europeans were aware of navel oranges in the seventeenth century. Navel oranges have probably appeared spontaneously many times throughout the history of orange growing.

The color development in oranges is not related to ripening. The orange color associated with the fruit develops only under cool nighttime temperatures; in tropical climates the fruits stay green. In most areas, a deep orange color is needed for successful marketing and growers use various methods to achieve the desired color. The most widely used method involves exposing the ripened fruit to ethylene, which promotes the loss of chlorophyll, thereby making the orange carotenes visible. As a group, the citrus fruits are high in vitamin C. Those that are orange-colored also provide some beta-carotene (precursor to vitamin A) and calcium.

Chestnuts

The images of the winter holidays are often filled with the aroma of "chestnuts roasting on an open fire." These flavorful nuts are products of *Castanea* spp., a genus native to temperate regions of eastern North America, southern Europe, northern Africa, and Asia. In North America, *Castanea dentata,* the American chestnut tree, was one of the most useful trees to the native peoples and settlers. The trees, once some of the most abundant in the Eastern forests, were towering specimens, often reaching heights of 36 meters (120 feet) and diameters of 2 meters (7 feet). The wood was highly prized for furniture, fence posts, telegraph poles, shingles, ship masts, and railroad ties. The high **tannin** content of the wood made the lumber resistant to decay and was a source of tannins for the tanning of leather.

The nuts have long been consumed by humans and animals alike; they can be eaten raw, but are more commonly boiled or roasted. The chestnuts can also be made into a rich confectionery paste known as *creme de marrons* (chestnut spread) used in French desserts. Nutritionally, chestnuts are high in starch (approximately 78%) and unusually low in fat (4% to 5%) for a nut.

Usually three nuts (one-seeded fruits with a stony pericarp) are borne in a spiny bur or husk that splits open at maturity (fig. 6.8a). Each nut is produced by a single female flower, which is borne in a cluster of three; the cluster is subtended by an **involucre,** a collection of bracts that develop into the spiny bur as the fruits mature. The trees are monoecious, with the staminate flowers borne in long slender catkins on the same individual.

The reign of the chestnut trees in American forests began to decline early in the twentieth century due to **chestnut blight,** a disease caused by the fungus *Cryphonectria parasitica* (formerly known as *Endothia parasitica*). The fungus is believed to have been introduced in 1890 from some Asian chestnut trees brought to New York. The first reported case of the disease was in 1904 at the Bronx Zoological Park. **Cankers,** localized areas of dead tissue, were noted on several of the trees in the park. Although attempts were made to stop the disease by pruning away diseased branches, the fungus soon spread to all the chestnut trees in the park. By 1950, chestnut blight had spread throughout the natural range of *Castanea dentata* from Maine to Alabama and west to the Mississippi River. Although the trunk of an infected tree dies, the roots are usually not infected. The chestnut's ability to resprout from the roots has saved the species from extinction (fig. 6.8b). These young saplings can reach heights of 4 to 6 meters (12–20 ft) before they succumb to the blight. Today, intense research efforts are focusing on various techniques to restore the American chestnut to its former glory. Researchers at the University of Florida are field testing a blight-resistant variety of American chestnut in hopes of restoring this species.

(a)

(b)

Figure 6.8 *(a) The three nuts of a single chestnut fruit are borne in a spiny bur that splits open at maturity. (b) Chestnut sprouts from the root continue to appear long after the tree has succumbed to chestnut blight.*

A Closer Look 6.1

Exotic Fruits

Many of the dessert and snack fruits that we commonly enjoy are not native to North America but had their origins in exotic lands and faraway places. Consider such familiar fruits and their origins: apples (central Asia), oranges (southeastern Asia), peaches (China), bananas (southeastern Asia), watermelons (Africa), and pineapples (Latin America). Today this trend is accelerating and every year new fruits are introduced to the North American public. What follows is just a sampling of the many exotics that may become as popular as apples and oranges to future generations.

Around 1980 the kiwifruit, *Actinidia chinensis,* began appearing in supermarkets throughout the United States, and soon after many were enjoying this fuzzy brown egg-shaped fruit (box fig. 6.1a). The distinctive flavor of its emerald green flesh is reminiscent of a strawberry-banana-pineapple combination. Its current popularity is the result of successful marketing that began in New Zealand. Originally native to China, the plants were introduced into New Zealand in 1904, where it was known as the Chinese gooseberry and often grown as an ornamental vine. Commercial farming began in the 1930s and the first exports were delivered to England in 1952. Around this time, marketing strategists renamed the Chinese gooseberry; the new name, kiwifruit, fit because its fuzzy rind resembles New Zealand's flightless bird, the kiwi. The name change paid off—sales and exports increased steadily and by the late 1980s it was an American produce staple. In the United States, cultivation of this berry is found mainly in northern California, with a large percentage of the crop exported to Europe, Japan, and Canada.

One exotic fruit that may be seen more frequently in the near future is the cherimoya, *Annona cherimola.* This tree fruit, native to the uplands of Peru and Ecuador, was first cultivated by the ancient Incas. In appearance, the aggregate fruit looks like a leathery green pine cone. Its custardlike flesh can be scooped out and is delicious with cream or orange juice. Today cherimoya, described by some as the aristocrat of fruits, is widely grown in Chile, Spain, and Israel and is making inroads in North America through California growers. Atemoya, a new hybrid resulting from a cross between the cherimoya and the closely related sugar apple (*Annona squamosa*), has the advantage of being more tolerant of environmental conditions than either parent and can, therefore, be grown in a wide variety of climates. The sweet taste of atemoya makes this a superb fruit for fresh consumption as well as for frozen desserts.

Box Figure 6.1a *Kiwifruit is the fruit of the woody vine* Actinidia chinensis.

Box Figure 6.1b *Carambola is a fluted fruit from Malaysia. When cut crosswise, yellow stars are produced.*

Carambola, *Averrhoa carambola,* is another ancient fruit that has recently been introduced to North American markets (fig. 6.1b). Native to Malaysia, the carambola is also known as star fruit because a series of five-pointed yellow stars results from slicing the elongate fluted fruit. This tart fruit adds an appealing shape that brightens up seafoods, salads, desserts, and fruit punches. The fruit can also be squeezed to make a refreshing juice or be picked green, cooked, and eaten as a vegetable. Presently star fruits are grown throughout the tropics, with the U.S. cultivation centered in Florida.

Chapter Summary

1. Fruits are unique to the sexual reproduction of angiosperms. They protect the enclosed seeds and aid in seed dissemination. Botanically, a fruit is a ripened ovary, although in the United States, the legal definition of a fruit is something that tastes sweet and is eaten as dessert.
2. Fruits can be classified according to the characteristics of the fruit wall or pericarp. In fleshy fruits, the pericarp is soft and juicy; berry, hesperidium, pepo, drupe, and pome are all examples of fleshy fruits. In dry fruits, the pericarp is often tough or papery. Dry fruits can also be dehiscent, splitting open along one or more seams to release their seeds. Follicles, legumes, and capsules are examples of dehiscent fruits. Dry fruits that do not split open are indehiscent; examples of this fruit type are achenes, samaras, grains, and nuts. Simple fruits are derived from a single ovary. Aggregate fruits develop from the separate ovaries within a single flower, while multiple fruits result from the fusion of ovaries from separate flowers in an inflorescence.
3. Seeds are the end products of sexual reproduction in flowering plants. Each seed contains an embryonic plant, nutrient tissue to nourish the embryo, and a tough outer seed coat. Differences exist between monocot and dicot seeds. Monocotyledonous seeds have a single small cotyledon, while dicotyledonous seeds typically possess two large cotyledons.
4. Edible fruits have played an important role not only as a significant contribution to the human diet but also in scientific studies and folklore. Once-exotic fruits are becoming commonplace as they become incorporated into the world's marketplace.

Review Questions

1. What is the function of the fruit in the life cycle of an angiosperm?
2. Give the botanical meaning of the following: berry, nut, legume, grain.
3. What is a seed?
4. Compare and contrast monocot and dicot seeds in structure and germination.
5. How did the tomato, an Aztec fruit, become the staple of Italian cookery?
6. What factors contributed to the successful introduction to the kiwifruit into North American markets?
7. Although "chestnuts roasting on an open fire" is an American image, most chestnuts eaten in the United States are from Italy. Why?

Further Reading

Chasan, Daniel J. 1986. Varieties Come and Go But Apples Remain a Staple. *Smithsonian,* 17(6): 123–133.

Conniff, Richard. 1987. How the World Puts Gourds to Work. *International Wildlife,* 17(3): 18–24.

McPhee, John. 1967. *Oranges.* Farrar, Straus, and Giroux, New York.

Newhouse, Joseph R. 1990. Chestnut Blight. *Scientific American,* 263 (1): 106–111.

Rick, Charles M. 1978. The Tomato. *Scientific American,* 239 (2): 76–87.

Rosengarten, Jr., Frederic. 1984. *The Book of Edible Nuts.* Walker and Co., NY.

Rupp, Rebbecca. 1987. *Blue Corn and Square Tomatoes.* Garden Way Publishing, Pownal, VT.

Vietmeyer, Noel. 1985. Exotic Edibles Are Altering America's Diet and Agriculture. *Smithsonian,* 16 (9): 34–43.

Vietmeyer, Noel. 1987. The Captivating Kiwifruit. *National Geographic,* 171 (5): 682–688.

Wolf, Thomas H. 1990. How the Lowly Love Apple Rose in the World. *Smithsonian,* 21(5): 110–117.

CHAPTER

7

GENETICS

Chapter Outline

Chapter Concepts

1. Gregor Mendel, an Austrian monk, discovered the basic principles of inheritance in the 1860s.
2. Mendelian genetics explains the inheritance of genes that are located on separate chromosomes, but other patterns of inheritance occur that are not governed by Mendel's principles.
3. The chemical basis of inheritance is the gene, a segment of the DNA molecule that codes for the formation of one polypeptide.

Genetics is the study of **inheritance,** the transmission of traits from parent to offspring. From earliest times, people have realized that certain traits in both plants and animals are passed on from parents to offspring. Artificial selection was practiced by farmers both consciously and unconsciously in establishing many domesticated plants and animals. It has only been in the twentieth century that science has provided a clear understanding of the nature of **genes** and the scientific basis for selective breeding. The hereditary material, DNA, found in chromosomes is organized into units called genes. Each gene codes for the formation of one polypeptide or one protein. Genes control all phases in the life of an organism, including its metabolism, size, color, development, and reproduction.

A gene is a segment of the DNA molecule but it can also be considered as simply a **locus,** or site, on a chromosome. The discussion of meiosis in Chapter 5 indicated that each gamete contains only one of each homologous chromosome; so diploid offspring receive one set of chromosomes (a haploid set) from each parent. For any genetic character, the offspring, therefore, will have two genes. It will receive one gene on a specific chromosome from one parent and another gene for that character on the homologous chromosome from the other parent. Genes often exist in at least two alternate forms, known as **alleles.** For example, a gene that controls flower color may have alleles that specify purple or white flowers. An offspring can receive two identical alleles for a specific character, a condition known as **homozygous,** or two different alleles for that character, a condition called **heterozygous.**

Today, these concepts constitute a clear foundation for understanding the principles of genetics. When Gregor Mendel, considered the founder of genetics, began his studies of inheritance in the middle of the nineteenth century, none of these concepts was known. The cell theory was generally accepted among scientists of the time, including the view that cells arose only from preexisting cells. In 1831, Robert Brown had described the nucleus as a fundamental and constant component of the cell, but its importance was not recognized. Nucleic acids and chromosomes were not known and neither mitosis nor meiosis had been described. This makes Mendel's discoveries all the more spectacular.

Mendelian Genetics

The science of genetics can be traced back to the work of Gregor Mendel (1822–1884), an Augustinian monk, who lived and worked in an Austrian monastery in the mid-nineteenth century (fig. 7.1). The monastery at Brunn (now Brno in the Czech Republic) was a center of enlightenment and scientific thought. It had an excellent library, botanical garden, and herbarium. Much of this was due to Abbot Cyril Napp, who was a skilled plant breeder as well as a leader in church matters. From 1851 to 1853, Napp sent Mendel to study biology, physics, and math at the University of Vienna, where he was influenced by some of the leading scientists of the day. Here, Mendel learned to apply a quantitative experimental approach to the study of natural phenomena.

Figure 7.1 *Gregor Mendel (1822–1884) is considered the father of genetics.*

Mendel was interested in the inheritance of traits and when he returned to the monastery, he began a series of experiments that eventually demonstrated the nature of heredity. In 1865, Mendel summarized the results of his research in a paper he read to the Brunn Society for the Study of Natural Science and published a detailed written account the following year. Although Mendel sent copies of his paper to leading biologists of the day, his work was ignored for over 30 years. In 1900, 16 years after Mendel's death, three scientists independently carried out genetic studies of their own. They reached the same conclusions as Mendel and cited his pioneering work.

Gregor Mendel and the Garden Pea

Mendel chose the garden pea for his experiments, possibly because they were available in many easily distinguishable varieties and were easy to cultivate. Mendel obtained 34 distinct strains of peas from local farmers and raised generations of the plants for two years to determine which plants bred true. (Today, of course, true breeding plants are called homozygous.) In plants that breed true, the offspring are

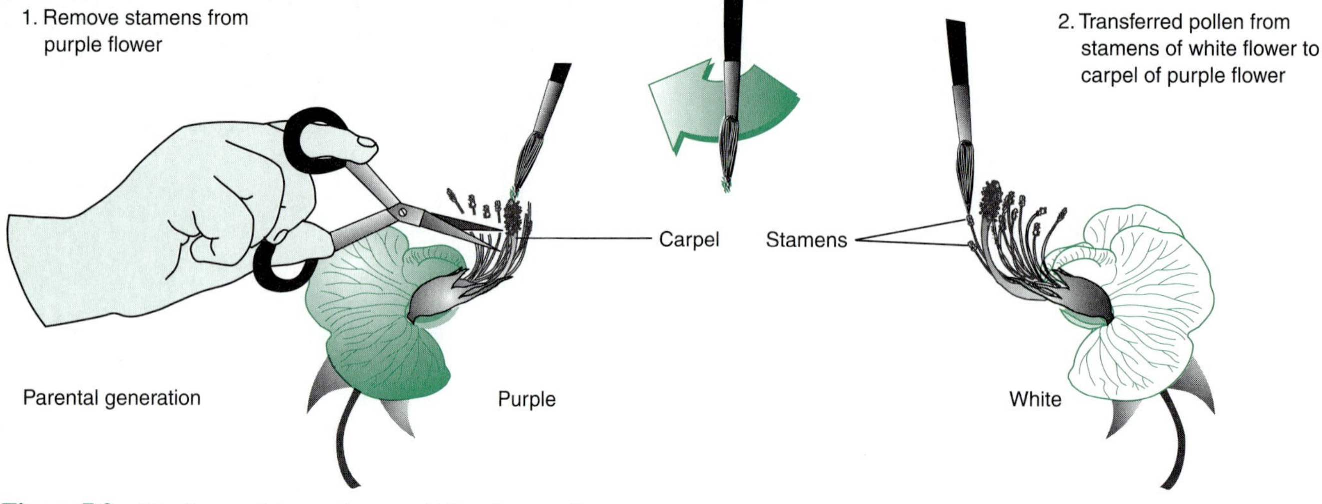

Figure 7.2 *The flower of the garden pea. Although normally self-pollinated, experimental crosses can be made by opening the flower and removing the anthers before they mature. Pollen from another pea flower is then transferred.*

Trait	Stem length	Pod shape	Seed shape	Seed color	Flower position	Flower color	Pod color
Characteristics	Tall	Inflated	Smooth	Yellow	Lateral	Purple	Green
	Dwarf	Constricted	Wrinkled	Green	Terminal	White	Yellow

Figure 7.3 *The seven pairs of characteristics studied by Mendel.*

identical to the parents for the trait in question. In his initial crosses, Mendel only used those plants that bred true. It was possible to maintain these strains because garden peas, unlike many plants, are selfpollinating. The flower does not open fully, so pollen from the anther lands on the stigma within the same flower and fertilizes it. Although this is the normal method of sexual reproduction, experimental crosses can also be made by carefully opening the flower and removing the anthers before the pollen is mature. Pollen from the flower of another pea plant can be transferred to the stigma of the first plant (fig. 7.2). Mendel made many of these crosses. As added insurance, Mendel covered these flowers with small cloth bags to prevent any pollinating insects from visiting the flower and introducing unwanted pollen.

For his experiments Mendel eventually selected strains with seven clearly contrasting pairs of traits: **Tall** versus **dwarf** plants, **yellow** versus **green seeds, smooth** versus **wrinkled seeds, green** versus **yellow pods, inflated** versus **constricted pods, purple** versus **white flowers,** and **terminal** versus **lateral flowers** (fig. 7.3). Mendel studied only one or two of these pairs of traits at a time. He traced and recorded the type and number of all offspring produced from each pair of plants that he cross-fertilized. Mendel also followed the results of each cross for at least two generations.

Monohybrid Cross

Mendel carried out a series of **monohybrid** crosses mating individuals that differed in only one trait. The members of the first generation of offspring all looked alike and resembled one of the two parents. In one set of experiments, Mendel crossed a pea plant that produced yellow seeds with a strain that produced green seeds. These were called the **P,** or **parental, generation.** The offspring from this cross called the **F_1** generation, or **first filial generation,** produced only yellow seeds. Mendel then allowed the F_1 generation to

ABLE 7.1

Mendel's Results from Monohybrid Crosses

Parents		F_1	F_2		Ratio
Dominant	Recessive	Generation	Generation		
Yellow seeds	Green seeds	All yellow seeds	6,022 yellow seeds	2,001 green seeds	3.01:1
Purple flowers	White flowers	All purple flowers	705 purple flowers	224 white flowers	3.15:1
Smooth seeds	Wrinkled seeds	All smooth seeds	5,474 smooth seeds	1,850 wrinkled seeds	2.96:1
Inflated pods	Constricted pods	All inflated pods	822 inflated pods	299 constricted pods	2.95:1
Green pods	Yellow pods	All green pods	428 green pods	152 yellow pods	2.82:1
Tall plants	Dwarf plants	All tall plants	787 tall plants	277 dwarf plants	2.84:1
Lateral flowers	Terminal flowers	All lateral flowers	651 lateral flowers	207 terminal flowers	3.14:1

self-pollinate and produce the **F_2 generation,** or **second filial generation,** which developed 8,023 seeds. Out of the total, 6,022 were yellow and 2,001 were green, or approximately three-fourths yellow and one-fourth green. This represented a 3:1 ratio. The green trait reappeared in the F_2 generation, having been masked in the F_1 generation. The yellow trait had dominated over the green trait in the F_1 generation.

Mendel got similar results from other crosses (table 7.1). From these results he proposed that each kind of inherited character is controlled by two hereditary factors. (Today we call these hereditary factors genes and know that they are located on homologous chromosomes.) For each pair of traits he studied, one allele masked, or was dominant over, the other allele in the F_1 generation. This concept is usually referred to as Mendel's **principle of dominance.** Mendel called the nondominant allele **recessive.** Each recessive trait reappeared in the F_2 generation in the ratio of three dominant to one recessive.

Although nothing was known about meiosis at this time, Mendel proposed that when gametes are formed, the pairs of hereditary factors (gene pairs) become separated so that each sex cell (egg or sperm) only receives one of each kind of factor or gene. This concept is known as the **principle of segregation.**

The actual appearance of a trait is known as the **phenotype.** In the example above, the phenotypes of the true-breeding parents were yellow seeds for one parent and green seeds for the other. If the symbol *Y* is used to designate yellow and *y* is used to designate green, then the genetic makeup (known as the **genotype**) of one parent is *YY* and the genotype of the other parent is *yy.* One parent was homozygous for yellow seeds and the other parent homozygous for green seeds. The gametes produced by the yellow-seeded parent all contain the *Y* allele and those produced by the green-seeded parent all contained the *y* allele. The phenotype of all the F_1 generation was yellow seeds, but the genotype was *Yy.* These plants were heterozygous, but the dominant allele masked the expression of the recessive trait.

Parent with Yellow Seeds — X (crosses with) — Parent with Green Seeds
Genotype: YY x yy
Possible gametes: Y, Y, y, y
F_1: Yy, Yy, Yy, Yy
Phenotype ....All Yellow Seeds.............................

When the F_1 plants were allowed to self-pollinate and produce the F_2 generation, gametes containing either *Y* or *y* were formed. When fertilization took place, a random combination of the gametes produced the following results:

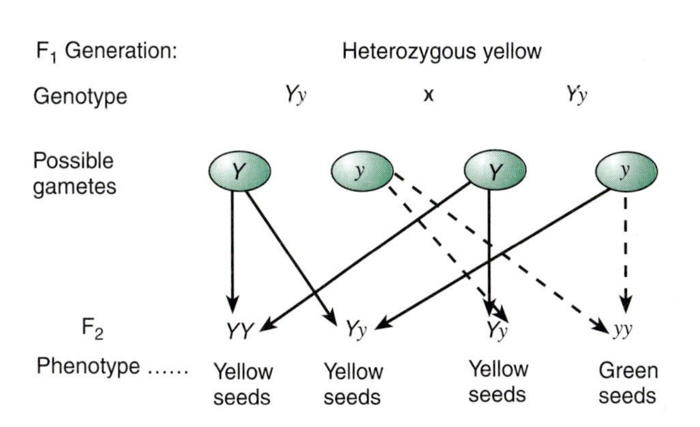

Although three-fourths of the F_2 generation contained the recessive allele (*y*), only one-fourth expressed the recessive trait, green seeds.

Another way of displaying the probable recombinants that can occur is the use of a Punnett square. The alleles representing the gametes from one parent are on one side of the square, and those representing the other parent are along the other side. Combining gametes with the indicated alleles fills in all the boxes of the square, thereby displaying all possible combinations of offspring. Using a Punnett square to show the self-pollination of the F_1 generation gives:

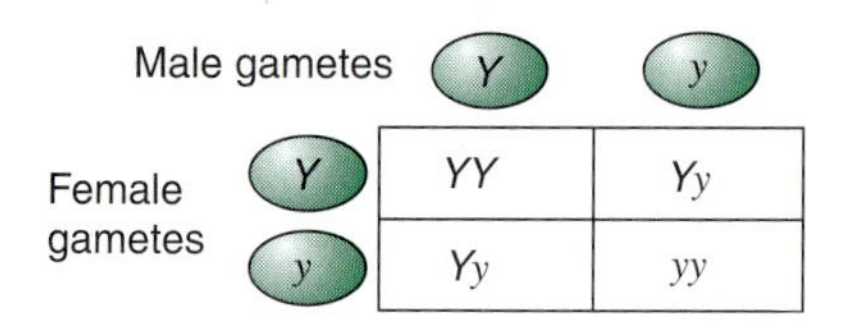

	Male gametes *Y*	Male gametes *y*
Female gametes *Y*	*YY*	*Yy*
Female gametes *y*	*Yy*	*yy*

To test his theory, Mendel allowed the F_2 plants to grow to maturity and self-pollinate and thus produce the F_3 generation. Green-seeded plants only gave rise to green seeds in the third generation. One-third of the yellow-seeded plants (those that were homozygous *YY*) gave rise only to yellow seeds, while two-thirds of the yellow-seeded plants (which were heterozygous *Yy*) gave rise to both yellow and green in the same 3:1 ratio.

Mendel also used testcrosses to support his hypotheses. A testcross involves mating an individual with an unknown genotype to a homozygous recessive individual. For example, a pea plant with yellow seeds can be *Yy* or *YY*. If the unknown were homozygous *YY*, then all of the offspring would be yellow-seeded when the unknown was crossed with a green-seeded plant. If the unknown were heterozygous *Yy*, then half the offspring would have yellow seeds (*Yy*), and half would have green seeds (*yy*).

Testcrosses are used today to determine unknown genotypes, especially in commercial seed production when the breeder is trying to determine if a strain will breed true for a certain trait.

Dihybrid Cross

Mendel also analyzed a series of **dihybrid crosses,** matings that involved parents that differed in two independent traits. For the parent, Mendel crossed true-breeding tall plants with purple flowers (*TTPP*)and true-breeding dwarf plants with white flowers (*ttpp*). The entire F_1 generation was tall with purple flowers, heterozygous for both characters (*TtPp*). The results were predictable in keeping with the dominance of both the tall and purple alleles.

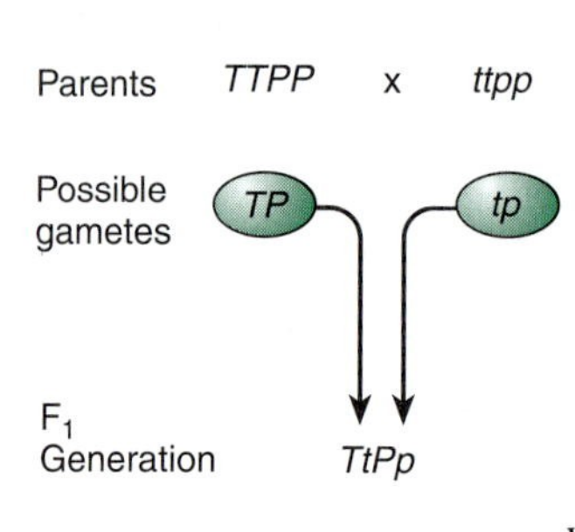

Mendel allowed the F_1 generation to self-pollinate and observed the phenotypes in the F_2 generation. Four phenotypes appeared in the F_2 generation, including two phenotypic combinations not seen in either parent (tall plants with white flowers and dwarf plants with purple flowers). In order for this to happen, the alleles for the two genes (for height and flower color) must have segregated independently from each other during gamete formation. Instead of gametes having only *TP* and *tp* combinations, four kinds of gametes would be produced in equal numbers: *TP, Tp, tP, tp*. When these gametes

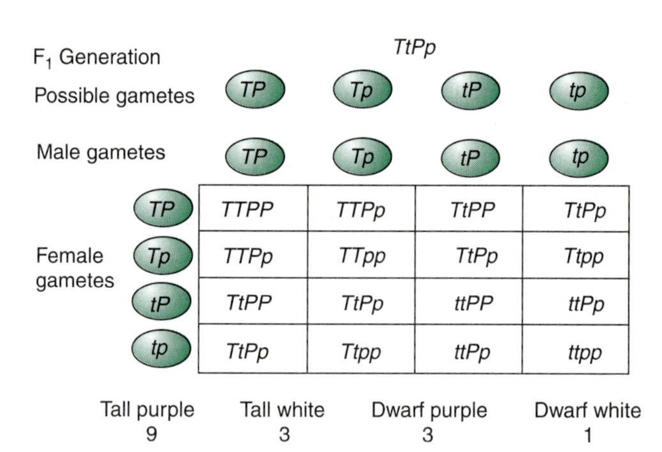

	Male gametes *TP*	*Tp*	*tP*	*tp*
Female gametes *TP*	*TTPP*	*TTPp*	*TtPP*	*TtPp*
Tp	*TTPp*	*TTpp*	*TtPp*	*Ttpp*
tP	*TtPP*	*TtPp*	*ttPP*	*ttPp*
tp	*TtPp*	*Ttpp*	*ttPp*	*ttpp*

combined randomly, they would produce nine different genotypes and four phenotypes (tall purple, tall white, dwarf purple, and dwarf white) in the F_2 generation. The four phenotypes occurred in a 9:3:3:1 ratio. Not only were the original parental types represented in the F_2 generation, but two new recombinant types were also present.

Mendel's actual results were close to the 9:3:3:1 ratio, not only for this cross but also with several

Test Cross

Y? Yellow-seeded plant of unknown genotype x *yy* Green-seeded plant

If the plant being tested is homozygous:

	Male gametes *Y*	Male gametes *Y*
Female gametes *y*	*Yy*	*Yy*
Female gametes *y*	*Yy*	*Yy*

All the seeds are yellow.

If the plant being tested is heterozygous:

	Male gametes *Y*	Male gametes *y*
Female gametes *y*	*Yy*	*yy*
Female gametes *y*	*Yy*	*yy*

Half the seeds are yellow and half are green.

other dihybrid crosses. On the basis of these results, Mendel postulated the **principle of independent assortment,** which states that members of one gene pair segregate independently from other gene pairs during gamete formation. In other words, different characters are inherited independently. Each gamete receives one gene for each character, but the alleles of different genes are assorted at random with respect to each other when forming gametes. Today it is known that this principle applies only to genes that occur on separate chromosomes, not to those that lie on the same chromosome.

Beyond Mendelian Genetics

The patterns of heredity explained by Mendel's principles, or laws of inheritance, are often called Mendelian genetics. The characters and traits that Mendel chose to study were lucky choices because they showed up very clearly. For each of the characters, one allele was dominant over the other, and both phenotypes were easy to recognize. Sometimes, however, the outcome is not that obvious because of variations in the usual dominant-recessive relationship.

Incomplete Dominance

When homozygous red snapdragons (*RR*) are crossed with homozygous white snapdragons (*rr*), the flowers produced by the F_1 generation are pink (*Rr*). When the F_1 generation is self-pollinated, the resulting F_2 plants are produced in a ratio of 1 red: 2 pink: 1 white. When a pink snapdragon is crossed with a white one, one-half the offspring are pink and one-half white. The pink flowered plants are heterozygous, but neither the red allele nor the white allele is completely dominant (fig. 7.4).

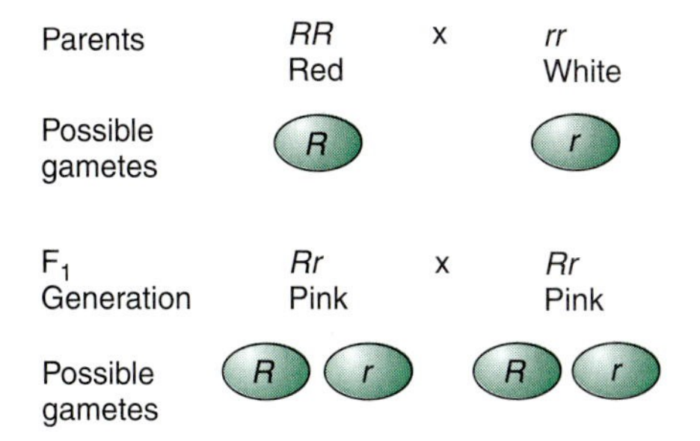

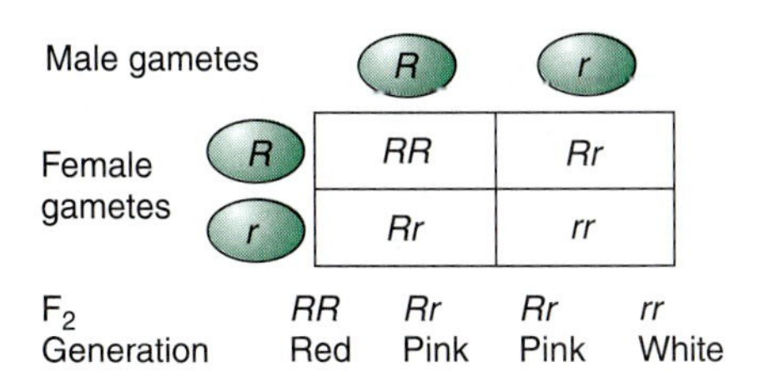

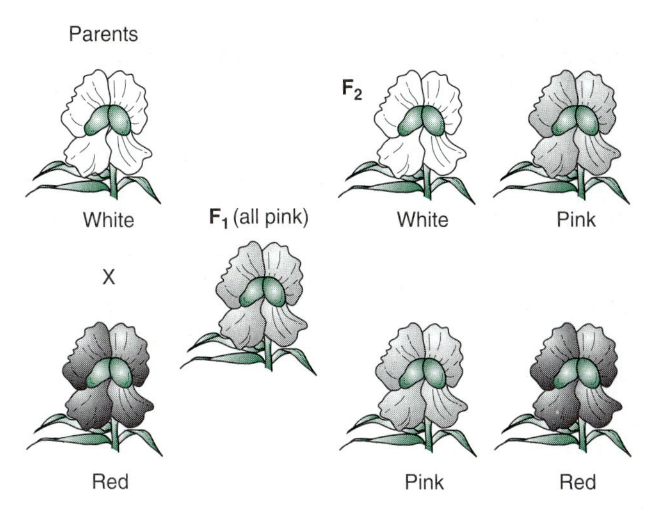

Figure 7.4 *Incomplete dominance. In a cross between a red and white snapdragon, all the offspring in the F_1 generation have pink flowers. In the F_2 generation, there is a 1:2:1 ratio of red to pink to white flowers.*

The *R* allele codes for a protein that is involved in the synthesis of red pigment, while the *r* allele codes for an altered protein that does not lead to pigment production. Plants that are homozygous *rr* cannot synthesize pigment at all and the flowers are white. Heterozygous plants with one *R* gene produce just enough red pigment so that the flowers are pink, while homozygous *RR* plants produce more red pigment. Whenever the heterozygous phenotype is intermediate, the genes are said to show **incomplete dominance.** This is not unique to snapdragons; in fact, it is relatively common.

Sometimes when neither allele is dominant, both alleles are expressed independently in the heterozygote. This condition is known as **codominance.** (In incomplete dominance, the heterozygote shows an intermediate phenotype, but in codominance both phenotypes are expressed.) The classical example of codominance is seen in the human AB blood type. Some individuals have type A blood and others have B; the alleles refer to the presence of A or B antigens on the surface of a red blood cell. Individuals with type AB blood have both types of antigens present on their red blood cells.

Multiple Alleles

Sometimes more than two alleles exist for a given character. Of course, any single individual only has two alleles (on homologous chromosomes). Human blood types also serve as an excellent example of this non-Mendelian feature known as **multiple alleles.** Three alleles exist for human blood types I^A, I^B, and *i*. Both I^A and I^B code for the formation of surface antigens, while *i* does

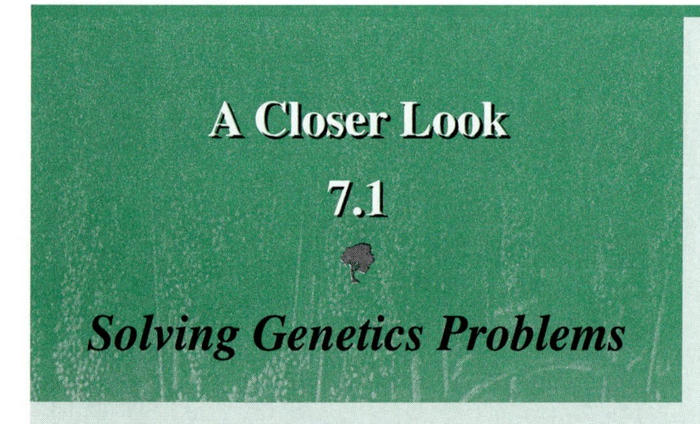

A Closer Look 7.1

Solving Genetics Problems

Solving genetics problems is like solving a puzzle. They can be challenging and perplexing, but rewarding when finally solved. By using the following steps, the problems will be easier to unravel.

Step 1. Establish a key to the alleles for each trait described. Use an uppercase letter to designate the dominant allele and the same letter, but lower case for the recessive allele. If the problem does not indicate which trait is dominant, the phenotype of the F_1 generation is often a good clue.

Step 2. Determine the genotypes of the parents. If the problem states the parents are true breeding, they should be considered homozygous. If one parent expresses the recessive phenotype, the genotype can be readily deduced. The phenotype of the offspring can also provide evidence of the parental genotypes.

Step 3. Determine the possible types of gametes formed by each parent. If the problem is a monohybrid cross, the principle of segregation must apply. For example, if the parent is heterozygous *Tt*, two kinds of gametes are produced: *T* and *t*. A homozygous parent only forms one kind of gamete: *TT* can only produce gametes with *T*, and *tt* can only produce gametes with *t*. If the problem involves a dihybrid cross, the principle of segregation and the principle of independent assortment both apply. If a parent is heterozygous for two traits, *TtSs*, and the genes for the two traits occur on separate chromosomes, then four possible types of gametes are formed: *TS, Ts, tS,* and *ts*. Often the most difficult part of the problem is solved by determining the gametes.

Step 4. Set up a Punnett square, placing the gametes from one parent across the top, and the gametes from the other parent down the left-hand side. Fill in the Punnett square. If a dominant allele is present, always place it first in each pair of genes (i.e., always write the dihybrid as *TtSs* and never *tTsS*). Read off the phenotypes (and genotypes, if requested in the problem) and calculate the phenotypic ratios.

Try the following problems:

Problem 1. In squash, an allele for white color *(W)* is dominant to the allele for yellow color *(w)*. Give the phenotypic ratios for each of the following crosses:

a. $WW \times ww$
b. $Ww \times ww$
c. $Ww \times Ww$

Answers: Solving these crosses step by step as suggested above, reveals that Steps 1 and 2 are given in the problem. The key to the alleles is provided (White-*W*-is dominant to yellow-*w*) and the genotypes of the parents are provided for each cross. Starting with Step 3

a. Both parents, *WW* and *ww*, are homozygous; therefore, each can form only one kind of gamete. Parent *WW* can only produce gametes with the gene for white color (*W*) and parent *ww* can only form gametes with the gene for yellow color (*w*). Step 4 involves setting up a Punnett square. For this simple monohybrid cross, it may not be necessary since each parent can produce only one kind of gamete:

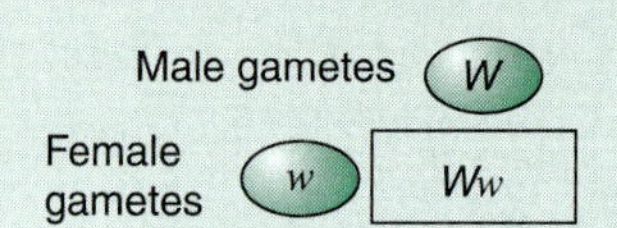

Answer to a. Since all the offspring will have the genotype *Ww*, all will show the white phenotype.

b. One parent is heterozygous and one is homozygous. The heterozygous parent *Ww* can produce two kinds of gametes, either (*W*) or (*w*), while the homozygous parent *ww* can only form gametes with the *w* allele. The Punnett square should show the following:

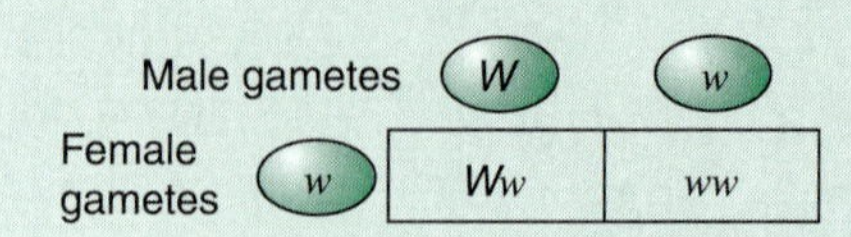

Since one-half of the offspring are heterozygous, they will show the dominant phenotype, white. The other half are homozygous recessive and will be yellow.

Answer to b. There is a 1:1 ratio of white to yellow.

c. Both parents are heterozygous and can therefore form two kinds of gametes, either (*W*) or (*w*). The Punnett square should show the following:

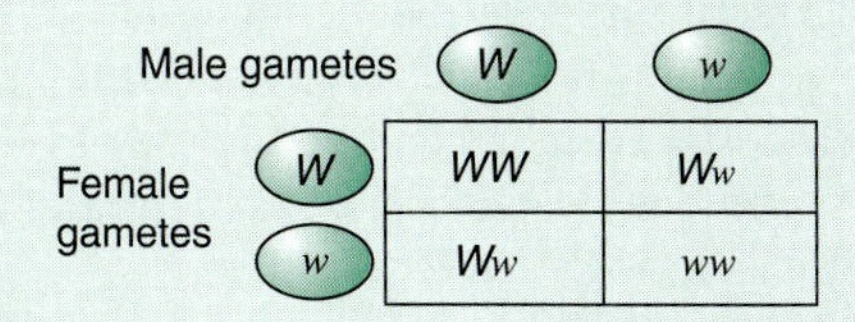

Three-fourths of the offspring (homozygous *WW* and the heterozygous *Ww*) will show the dominant phenotype, white. One-fourth of the offspring (homozygous *ww*) will show the recessive phenotype, yellow.

Answer to c. There is a 3:1 ratio of white to yellow.

Problem 2. In garden peas, an allele for tall plants (*T*) is dominant over the allele for dwarf plants (*t*) and the allele for smooth peas (*S*) is dominant over the allele for wrinkled peas (*s*). Determine the phenotypic ratios for each of the following crosses:

a. *TtSs* × *ttss*
b. *TtSs* × *TtSs*

Answers: Again Steps 1 and 2 are provided in the problem and the solutions can begin with Step 3, determining the gametes.

a. One parent is heterozygous for both traits, *TtSs*. Remember that the principles of segregation and independent assortment both apply. This will result in the formation of four possible types of gametes: (*TS*), (*Ts*), (*tS*), and (*ts*). The other parent is homozygous for both traits, *ttss*, so only one type of gamete can form, (*ts*). The Punnett square should show the following:

Female gametes \ Male gametes	TS	Ts	tS	ts
ts	TtSs	Ttss	ttSs	ttss

One-fourth of the offspring will show both dominant characters, tall with smooth peas; one-fourth will be tall with wrinkled peas; one-fourth will be dwarf with smooth seeds; and one-fourth will show both recessive characters, dwarf with wrinkled seeds.

Answer to a. There will be a 1:1:1:1 ratio of tall smooth to tall wrinkled to dwarf smooth to dwarf wrinkled.

b. Both parents are heterozygous for both traits; therefore, the Punnett Square should show the following:

Female gametes \ Male gametes	TS	Ts	tS	ts
TS	TTSS	TTSs	TtSS	TtSs
Ts	TTSs	TTss	TtSs	Ttss
tS	TtSS	TtSs	ttSS	ttSs
ts	TtSs	Ttss	ttSs	ttss

Reading off the phenotypes from the Punnett square will show that nine-sixteenths of the offspring will show both dominant characters, tall plants with smooth seeds, three-sixteenths will be tall with wrinkled seeds, three-sixteenths will be dwarf plants with smooth peas, and one-sixteenth of the offspring will show both recessive characters, dwarf plants with wrinkled peas.

Answer to b. There will be a 9:3:3:1 ratio of tall smooth to tall wrinkled to dwarf smooth to dwarf wrinkled.

Problem 3. In watermelons, the genes for green color and for short shape are dominant over their alleles for striped fruit and long shape. Suppose a plant that has genes for long striped fruit is crossed with a plant heterozygous for both these traits. What would be the phenotypic ratio of offspring from this cross? **Answer:** To solve this problem, begin at Step 1. Assign letters to represent the dominant and recessive alleles for each trait. Green fruit (*G*) is dominant to striped fruit (*g*), and short shape (*S*) is dominant to long shape (*s*).

Next, determine the genotypes of the parents. The plant with long striped fruit will be *ssgg,* while the heterozygous parent will be *SsGg*.

The homozygous recessive parent *ssgg* can only form one type of gamete, (*sg*). The heterozygous parent can form four possible types of gametes (*SG*), (*Sg*), (*sG*), and (*sg*). The Punnett square should show the following:

Female gametes \ Male gametes	SG	Sg	sG	sg
sg	SsGg	Ssgg	ssGg	ssgg

Among the offspring, there will be a 1:1:1:1 ratio of short green to short striped to long green to long striped watermelon.

not result in antigen production. Different combinations of these alleles result in four different phenotypes (blood types): A, B, AB, and O. Although I^A and I^B are codominant with respect to each other, both are dominant to *i* and will produce antigens if one allele is present. The homozygous recessive condition, *ii,* produces blood type O with no surface antigens.

Other well-studied examples of multiple alleles are found in fruit flies and clover leaves. In the fruit fly, a large number of alleles affect eye color by determining the amount of pigment produced. The final eye color depends on which two alleles are inherited. Sometimes leaves of clover are solid green, while other clover plants have patterns (stripes, chevrons, or triangles) of white present on the leaves. Seven different alleles exist that influence this trait. The resulting coloration and pattern depends on which two alleles are inherited.

Polygenic Inheritance

Often a character is controlled by more than one pair of genes, and each allele has an additive effect on the same character. Examples of **polygenic inheritance** are relatively common, especially for human characteristics such as skin color and height. The inheritance of seed color in wheat, controlled by three sets of alleles, is also an example of polygenic inheritance. When plants producing white seeds are crossed with plants producing dark red seeds, the F_1 generation is intermediate in color. However, if the F_1 are allowed to self-pollinate, the second generation shows seven phenotypes ranging from white to dark red. Each dominant allele has a small but additive effect on the pigmentation of the seeds. If *A,B,* and *C* represent the dominant alleles that code for pigment production and *a,b,* and *c* represent the recessive alleles that lead to no pigment production, the three generations can be described as follows:

Parents: *AABBCC* x *aabbcc*
Dark red seeds White seeds

F_1 Generation: *AaBbCc*
Medium red seeds

F_2 Generation:
All dominant alleles *(AABBCC)* - Dark red seeds
Any five dominant alleles - Deeper red seeds
Any four dominant alleles - Deep red seeds
Any three dominant alleles - Medium red seeds
Any two dominant alleles - Light red seeds
Any one dominant allele - Pale red seeds
All recessive alleles *(aabbcc)* - White seeds

Most of the F_2 generation show intermediate phenotypes. On average, only 1 out of 64 would show the dark red seeds of one parental type and 1 out of 64 would show the white seeds of the other parental type (fig. 7.5).

Linkage

When Mendel conducted his experiments and developed his principles of inheritance, the existence of chromosomes was not known. Mendel considered his "hereditary factors" as separate particles, and this idea was integrated into his principle for independent assortment. However, chromosomes are inherited as units, so genes that occur on one chromosome tend to be inherited together, a condition known as **linkage.** Because the genes are on the same chromosome, they move together through meiosis and fertilization.

Linkage was first described by British geneticists W. Bateson and R. C. Punnett in 1905 while working with sweet peas. In the years from Mendel's work until 1905, much progress had been made studying cells and reproduction. Chromosomes had been described, as well as the events of mitosis and meiosis. Bateson and Punnett knew from previous experiments that purple flowers were dominant to red flowers and the oval (long) pollen shape was dominant to round. When they crossed a homozygous purple-flowered plant with long pollen (*PPLL*) to a homozygous red-flowered plant with round pollen (*ppll*), the F_1 generation plants (*PpLl*) all resembled the dominant parent with purple flowers and long pollen. However, when they allowed the F_1 to self-pollinate, the F_2 generation did not occur in a 9:3:3:1 ratio that was expected based on Mendel's experiments. Instead, there were 284 purple-flowered plants with long pollen, 21 plants with purple flowers and round pollen, 21 plants with red flowers and long pollen, and 55 plants with red flowers and round pollen. Bateson and Punnett reasoned that genes for both flower color and pollen shape occurred on the same chromosome. The linkage of these genes should yield a 3:1 ratio of purple flowers with long pollen to red flowers with round pollen. The linking of these genes is indicated by:

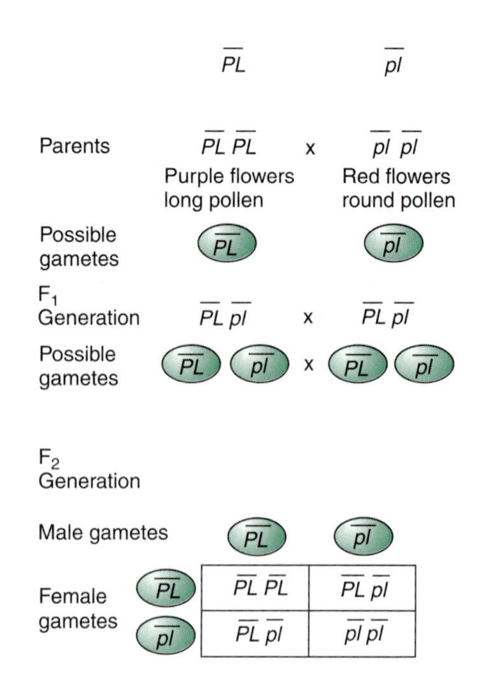

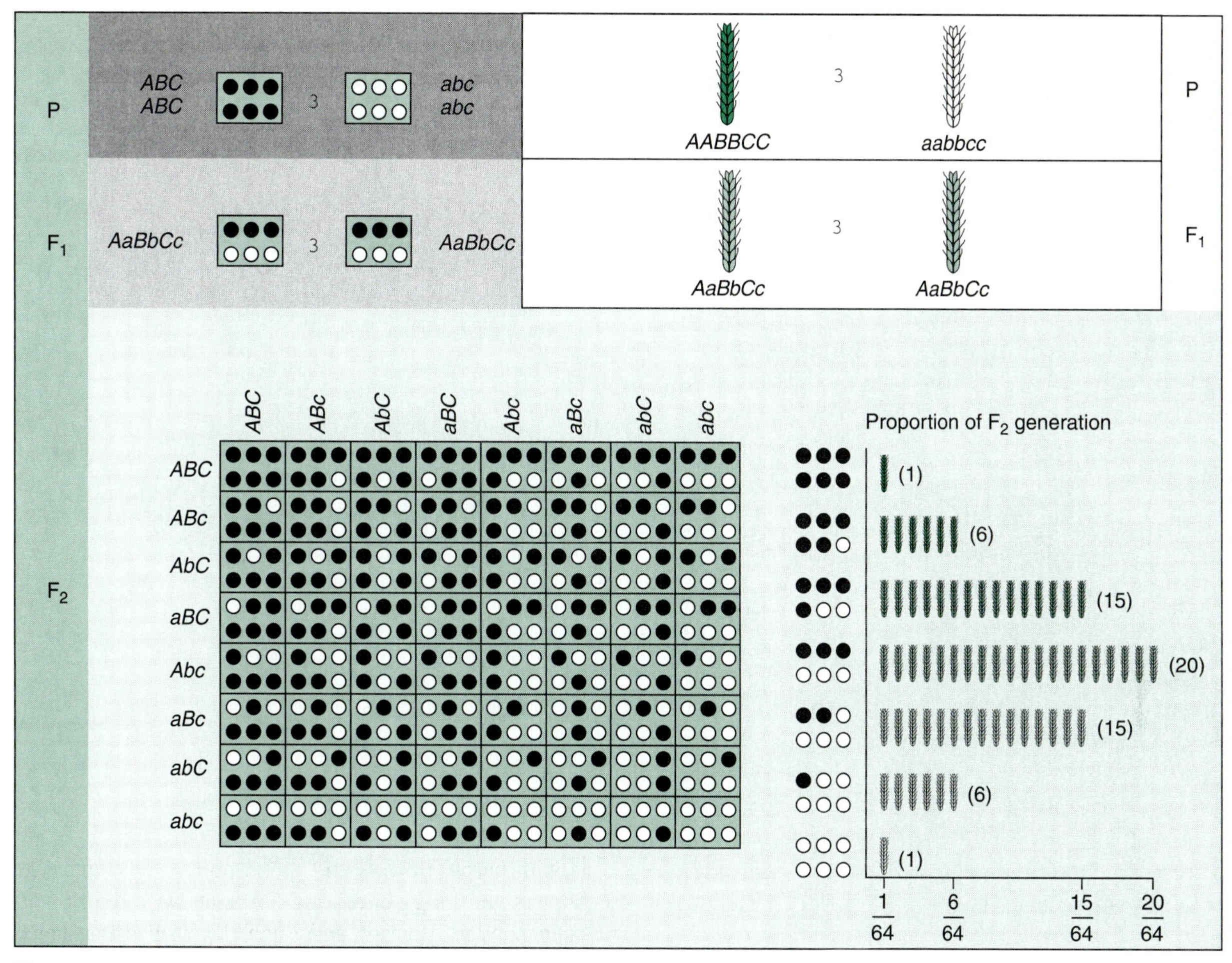

Figure 7.5 *The inheritance of seed color in wheat is controlled by three sets of alleles, each of which have an additive effect on the phenotype.*

Although Bateson and Punnett correctly identified the linkage of these genes, they could not explain the 21 purple-flowered plants with round pollen and the 21 red-flowered plants with long pollen that occurred in the F_2 generation. It was the work of Thomas Hunt Morgan that explained the phenotype of these plants. In the early 1900s, Morgan was conducting genetics studies on the fruitfly, *Drosophila melanogaster* at Columbia University. Morgan's work identified linked genes and also showed that a small number of offspring received new combinations of alleles. Morgan proposed that an occasional exchange of segments between homologous chromosomes takes place during meiosis, thereby breaking the linkage between genes. Recall from Chapter 5 that this exchange of segments, called **crossing-over,** occurs during prophase I of meiosis. Nonsister chromatids break at corresponding places, but the fragments reunite with the opposite chromatid. Thus, crossing-over accounts for the recombination of linked genes (fig. 7.6).

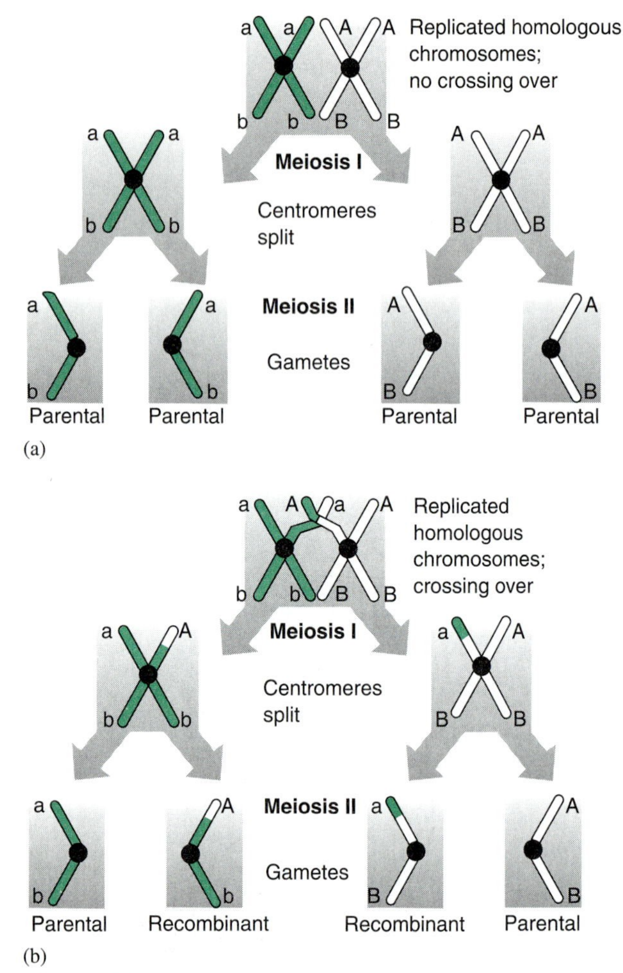

Figure 7.6 *Crossing-over during prophase I of meiosis can result in recombination of linked genes. In (a) no crossing-over has occurred so only two types of gametes are formed. In (b) crossing-over occurred, resulting in four possible types of gametes that include two recombinant types.*

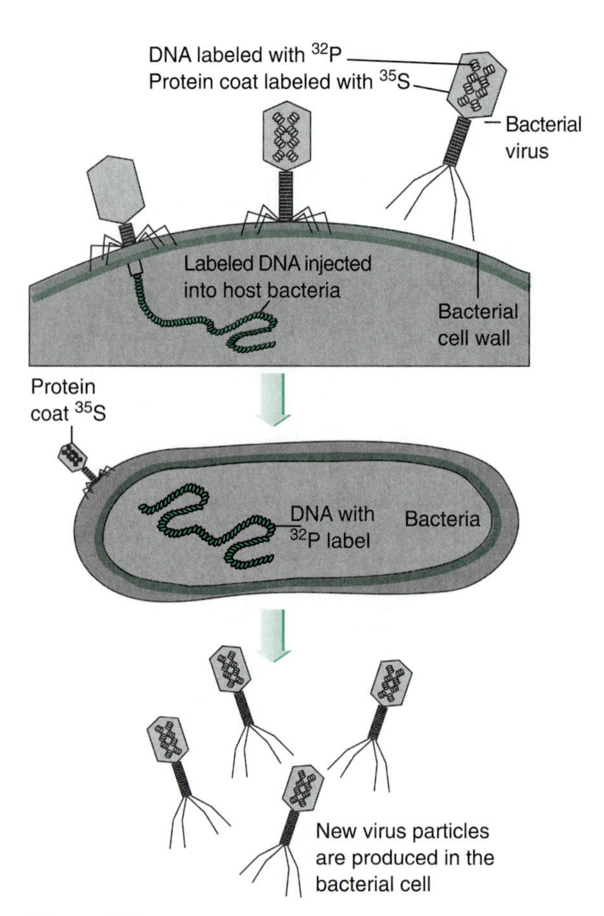

Figure 7.7 *Hershey and Chase experiment that proved conclusively that DNA is the genetic material. They used bacterial viruses with radioactively labeled protein and DNA. They found that only the viral DNA entered the bacterial cell and provided the information to produce new viral particles.*

Modern Genetics

Probably the greatest achievement in biology in the twentieth century has been the development of the field of molecular genetics. Because of the advances in this field, scientists now have a precise understanding of the chemical nature of the gene, how it functions, how mutations can occur, and how to transfer genes from one organism to another.

DNA—The Genetic Material

While genetics studies continued through the first four decades of the twentieth century, the exact nature of the gene remained unknown. Scientists sought to learn what genes were made of and how they worked. Although it was known that genes occurred on chromosomes and that chromosomes were composed of DNA and protein, it was widely assumed that proteins were the genetic material. The proof that DNA is the genetic material was established through the efforts of several researchers in the 1940s and early 1950s. The final proof was provided by Alfred Hershey and Martha Chase, who showed that when bacterial cells are infected with viruses, only the viral DNA enters the cell. The protein coat remains on the outside, while the viral DNA provides all the instructions for the manufacture of new virus particles (fig. 7.7).

In 1952, about the same time that Hershey and Chase were proving that DNA was the genetic material, James Watson and Francis Crick described the structure of the DNA molecule. Recall from Chapter 1 that DNA is a double helix with alternating sugar molecules **(deoxyribose)** and phosphate groups making up sides of the helix and nitrogenous bases forming the cross-rungs of the helix. Watson and Crick suggested that the nitrogenous bases always pair up in a specific pattern with one **purine** base (**adenine** or **guanine**) hydrogen bonding to one **pyrimidine** base (**thymine** or **cytosine**). Adenine always pairs with thymine, and guanine with cytosine; this is referred to as complementary base pairing. Watson and Crick also proposed that complementary base pairing provided a mechanism for the replication of the hereditary material. The model suggested that the double helix unwound and each strand served as a template for manufacturing a complementary strand (fig. 7.8). This is known as semiconservative replication; each new molecule of DNA contains one old strand and one new strand. Today it is known that the process is more complicated than originally proposed, but the basics of the process are unchanged.

Genes Control Proteins

As early as 1908, Archibald Garrod, a British physician, suggested that genes and enzymes were somehow related. In studying the inheritance of genetic diseases, he suggested that the absence of a certain enzyme was associated with a specific gene. Little work was done in this area until early 1940 when George Beadle and Edward Tatum found mutants of the fungus *Neurospora crassa* that interfered with known metabolic pathways. Each mutant strain was shown to have a mutation in only one gene. Beadle and Tatum found that for each individual gene identified, only one enzyme was affected. This was summarized into a *one gene-one enzyme hypothesis.* Beadle and Tatum received the 1958 Nobel Prize in Physiology and Medicine for their findings. It was soon realized, however, that many genes code for proteins other than enzymes, and also that many proteins contain more than one polypeptide chain. The Beadle and Tatum gene hypothesis was later modified to state that *each gene codes for one polypeptide chain.* Nevertheless, it was another 20 years before scientists understood how cells were able to convert the DNA information into the polypeptide chains.

Transcription and the Genetic Code

The sequence of bases in the DNA determines the sequence of amino acids in proteins, but the information in the DNA is not used directly. A molecule of **messenger RNA (mRNA)** is made as a complementary copy of a gene, a portion of one strand of the double helix. The process of RNA synthesis from DNA is known as **transcription.** Messenger RNA forms as a single strand by a mechanism that is similar to that of DNA replication. Enzymes cause a portion of the DNA molecule to unwind,

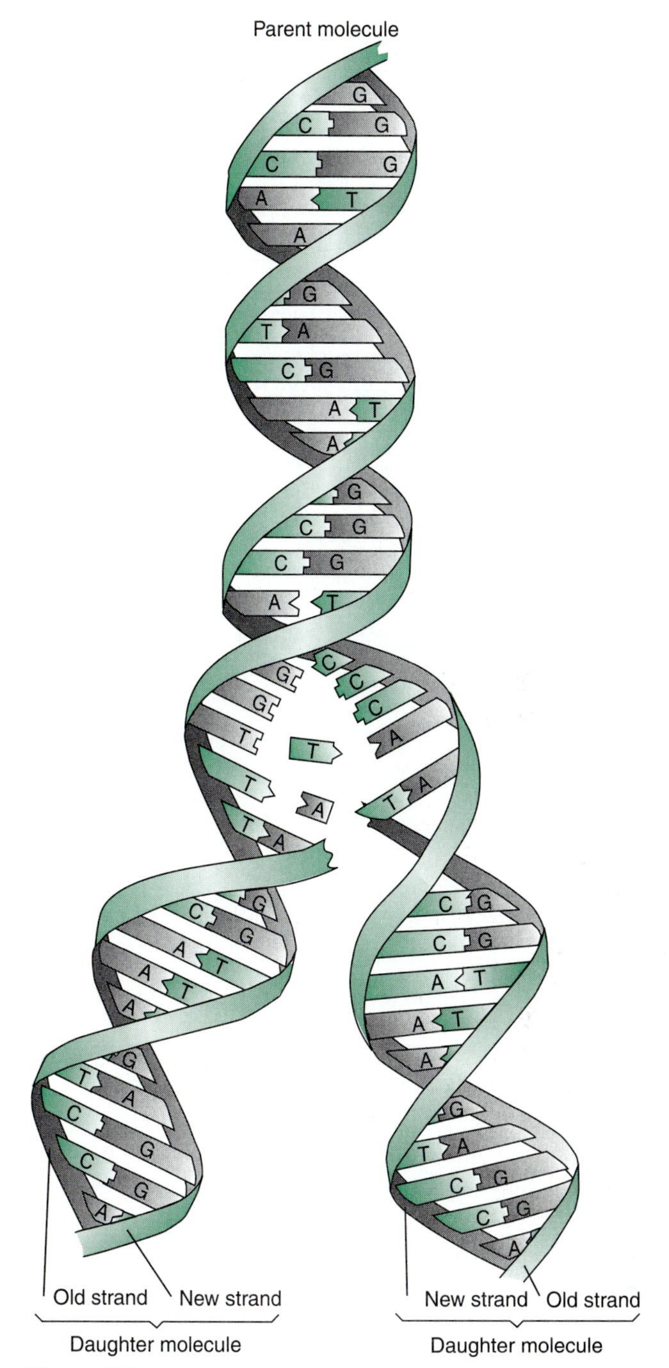

Figure 7.8 *Semiconservative replication of DNA. Each strand serves as a template for the formation of a new strand of DNA. Each of the daughter molecules produced is composed of one old strand and one new strand.*

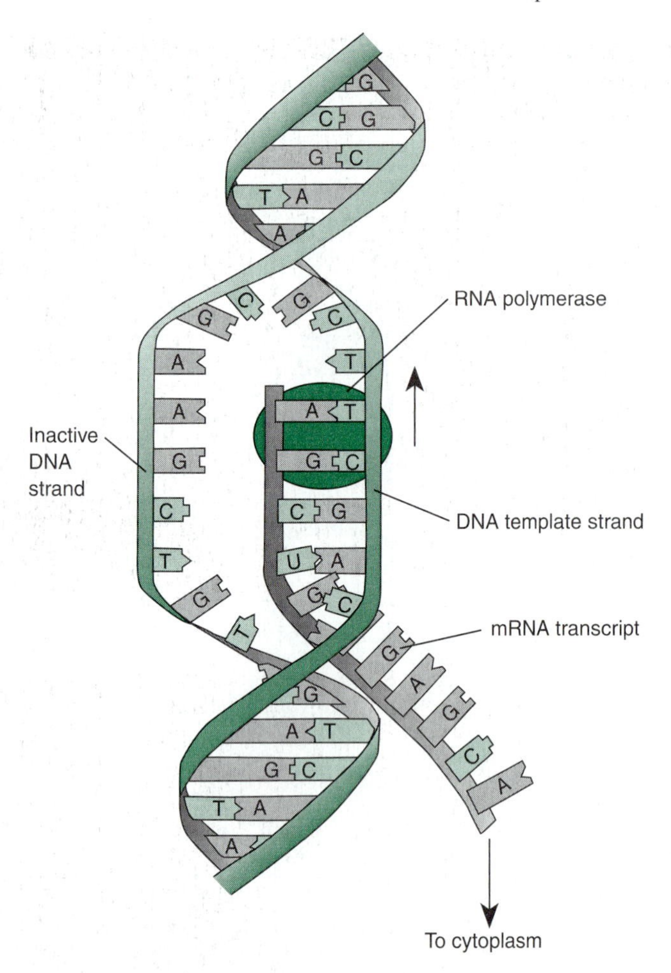

Figure 7.9 *Transcription. During transcription, a molecule of messenger RNA is formed as a complementary copy of a region on one strand of the DNA molecule.*

and bases of RNA nucleotides bond with the exposed bases on one strand of the DNA. The enzyme **RNA polymerase** is responsible for attaching nucleotides together in the sequence specified by the DNA (fig. 7.9). For example, if the sequence of bases on one strand of the DNA molecule is ATTAGCAT, the synthesized sequence on the RNA stand is UAAUCGUA. (Remember that the pyrimidine base uracil replaces thymine in RNA.) This molecule of mRNA represents a gene, and each gene in an organism is represented by a different mRNA molecule. Each mRNA contains in its sequence of bases information that will be **translated** into the sequence of amino acids that constitute a specific protein. Once the mRNA has been formed, enzymes within the nucleus process the molecule, modifying and removing segments called **introns,** that are not specifically involved in the coding for the polypeptide. The remaining segments, **exons** (or expressed segments), are spliced together to form the final mRNA molecule that leaves the nucleus and goes into the cytoplasm.

Each of three consecutive bases on the mRNA molecule constitutes a code word, or **codon,** that specifies a particular amino acid. There are a total of 64 codons, with 61 of these coding for amino acids (table 7.2). Three codons do not code for amino acids but act to signal termination (UAA, UAG, and UGA). One codon, AUG, which codes for the amino acid methionine, acts as a signal to start translation. The genetic code is redundant because two or more codons that differ in the third base code for the same amino acid (table 7.2). With very few exceptions, the genetic code is a universal code among living organisms. Codons specify the same amino acid in all organisms. This means that bacteria can translate genetic information from potato cells or human cells, or vice versa, the eukaryotic cells can translate the genetic information of the bacterial cell (*see* Chapter 15).

Translation

The translation of the mRNA codons into an amino acid sequence occurs on ribosomes in the cytoplasm of the cell. In addition to mRNA, two other types of RNA function in translation. **Ribosomal RNA (rRNA)** joins with a number of proteins to form ribosomes, the sites of protein synthesis. A ribosome consists of two subunits (small and large), each with rRNA and protein. Subunits are packaged in the nucleolus and float free in the cytoplasm. Two subunits come together when they attach to the end of an mRNA molecule.

Transfer RNA (tRNA) molecules are transport molecules that carry specific amino acids to a ribosome and align the amino acids to form a polypeptide chain. There are binding sites for two tRNA molecules on the ribosome where each tRNA recognizes the correct codon on the mRNA molecule.

There are at least 20 different types of tRNA molecules, one specific for each amino acid. They are all relatively small molecules containing between 70 and 80 nucleotides. Although single-stranded, some of the bases hydrogen bond with each other, folding the molecule into a specific shape. The specified amino acid attaches to one end of the tRNA molecule, while the bottom of the molecule attaches to an mRNA codon by base pairing with the anticodon. The **anticodon** is a series of three bases on the tRNA molecule that pairs with the complementary codon on mRNA. For example, the mRNA codon UUU specifies the amino acid phenylalanine. The tRNA molecule that brings phenylalanine to the ribosome has the anticodon AAA and will, thereby, base pair with the codon, inserting

TABLE 7.2

The Genetic Code

First Base	Second Base: U		Second Base: C		Second Base: A		Second Base: G		Third Base
	U		**C**		**A**		**G**		
U	UUU	Phe	UCU	Ser	UAU	Tyr	UGU	Cys	U
	UUC		UCC		UAC		UGC		C
	UUA	Leu	UCA		UAA-Stop		UGA-Stop		A
	UUG		UCG		UAG-Stop		UGG-Trp		G
C	CUU	Leu	CCU	Pro	CAU	His	CGU	Arg	U
	CUC		CCC		CAC		CGC		C
	CUA		CCA		CAA	Gln	CGA		A
	CUG		CCG		CAG		CGG		G
A	AUU	Ile	ACU	Thr	AAU	Asn	AGU	Ser	U
	AUC		ACC		AAC		AGC		C
	AUA		ACA		AAA	Lys	AGA	Arg	A
	AUG-Met or Start		ACG		AAG		AGG		G
G	GUU	Val	GCU	Ala	GAU	Asp	GGU	Gly	U
	GUC		GCC		GAC		GGC		C
	GUA		GCA		GAA	Glu	GGA		A
	GUG		GCG		GAG		GGG		G

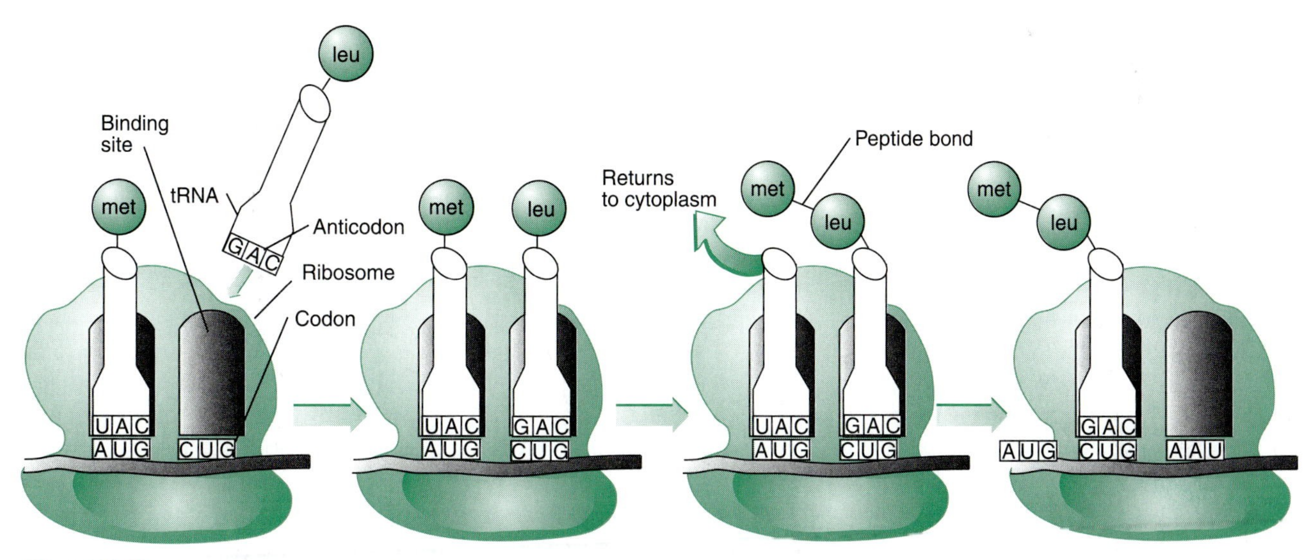

Figure 7.10 *Translation. Transfer RNA molecules arrive at the ribosome bringing amino acids. Codon-anticodon base pairing insures that the amino acids will be incorporated in the sequence specified on the mRNA.*

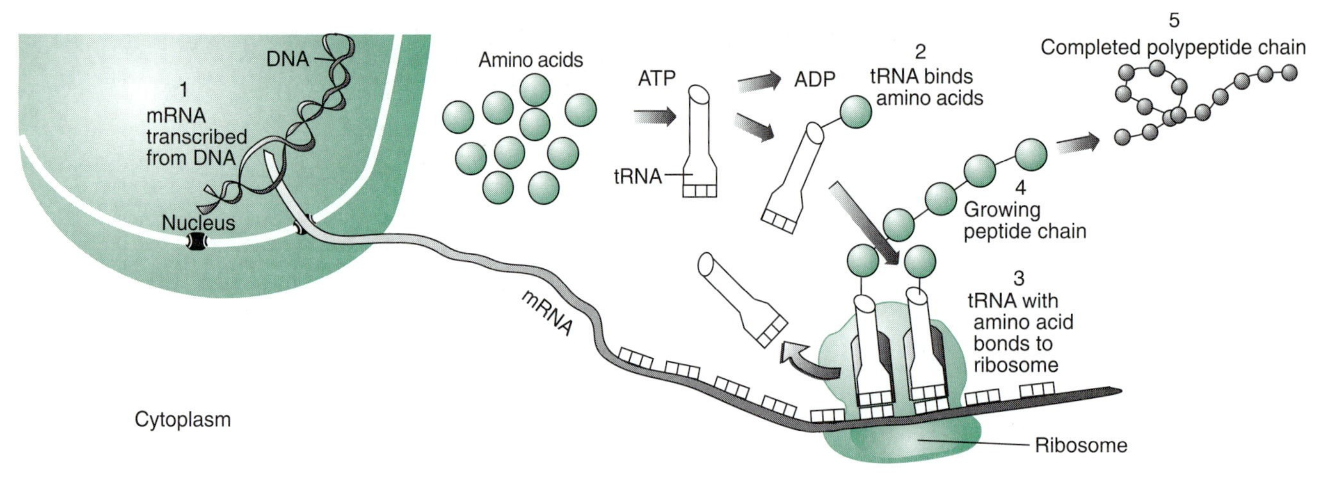

Figure 7.11 *Summary of protein synthesis.*

phenylalanine in the correct place in the growing polypeptide chain. Once the amino acid becomes part of the polypeptide, the bond with the tRNA is broken, allowing the tRNA to cycle back to the cytoplasm and bind another amino acid molecule (fig. 7.10). The ribosome now moves along the mRNA molecule reading a new codon, and a tRNA brings in the specified amino acid. Translation continues with the ribosome sliding over the mRNA, codon by codon, translating the sequence of bases until a termination codon is reached, signaling the end of the process. A summary of the events involved in gene expression, from transcription through translation, is presented in Figure 7.11.

Mutations

Mutations are changes in the DNA. Once the DNA sequence is changed, DNA replication copies the altered sequence and passes it along to future generations of that cell line. The smallest mutations are called **point mutations,** caused by a base substitution in which one base is substituted for another. When the gene is transcribed, the altered mRNA codon may insert a different amino acid and possibly alter the shape and function of the final protein. A single base substitution is responsible for the altered protein that causes the inherited disease, sickle cell anemia. If a small segment of the DNA is lost, the mutation is known as a **deletion,** and if a segment is added it is called an **insertion.** The segment lost or added may be as small as a single base. This may result in a modification of the mRNA reading frame known as a **frame-shift mutation,** and all codons downstream of this site will specify different amino acids. An analogy to a frame-shift mutation can be seen in the following sentence, THE RAT SAW THE CAT AND RAN. The deletion of the letter "R" in RAT can create a shift of the reading frame. The resulting sentence, THE ATS AWT HEC ATA NDR AN, is meaningless. Similarly, the deletion of a nucleotide can create a meaningless mRNA message.

Mutations can occur in any cell. If they occur in cells that do not lead to gametes, they are called **somatic mutations.** Somatic mutations can occur in cells of leaves, stems, or roots and are not usually passed on to offspring. However, some plants undergo extensive asexual reproduction. If the organ affected is involved in this process, the resulting clone will also incorporate the somatic mutation. If the mutation occurs during the formation of gametes, it will be passed on to the offspring as well, possibly creating a new allele for a genetic character. This is one small way in which genetic change can lead to the evolution of new species (*see* Chapter 8).

Recombinant DNA

By the 1970s, scientists had an understanding of the chemical nature of the gene, how it functioned, and how mutations occurred. In the past two decades this knowledge has been applied to practical uses in a number of areas that affect society such as health care, agriculture, plant breeding, and food processing. New applications are occurring so rapidly that it is difficult to predict future directions.

Many of these applications involve **recombinant DNA** technology, which entails the introduction of genes from one organism into the DNA of a second organism. This technique has allowed bacteria to produce human insulin and allowed plants to express a gene for herbicide-resistance normally found in bacteria (*see* Chapter 15). The formation of recombinant DNA makes use of proteins called **restriction enzymes** to cut a gene from its normal location. Restriction enzymes, which normally occur in bacterial cells, are able to cut DNA at specific base sequences.

Transferring the isolated gene to another species requires the use of a vector, usually a **plasmid,** which is a small circular strand of DNA that also occurs in bacterial cells. The plasmid is cut with the same restriction enzyme, then mixed with the isolated gene. The ends of the plasmid join to the ends of the gene, with the result being a **recombinant DNA molecule** that is transferred to a cell in another organism (fig. 7.12).

In much of the early work on recombinant DNA, the plasmids were transferred to bacterial cells. When a recombinant bacterial cell divides, so does the plasmid, and within hours a whole colony of cells are produced, all with the transferred, or foreign, gene. Because the genetic code is a universal code, the bacterial cell is able to transcribe and translate the information on the plasmid producing the desired protein. This technique has been used to manufacture high quality human insulin, human growth factor, and many other hormones. Recombinant DNA technology is now applied to higher organisms and both animals and plants have been produced expressing genes from other organisms (*see* Chapter 15).

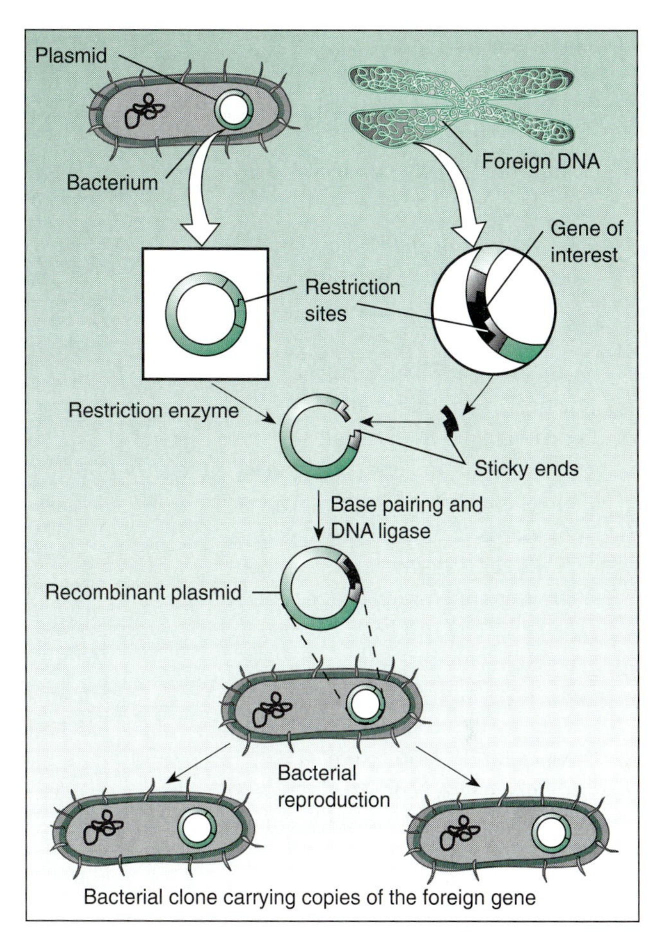

Figure 7.12 *Recombinant DNA technology. A gene of interest from one organism is cut from the DNA molecule using a restriction enzyme. The same enzyme is used to cut a bacterial plasmid. The two DNA samples are mixed together and the DNA sealed with the enzyme DNA ligase. The recombinant plasmid is taken up by the bacterial cell, reproducing when the cell divides and expressing the foreign gene.*

Chapter Summary

1. Genes control all phases of life in an organism, including its metabolism, size, color, development, and reproduction. Genes occur in pairs and are located on homologous chromosomes, which are inherited from parents. Through the use of monohybrid and dihybrid crosses with the garden pea, Gregor Mendel discovered the basic principles that governed the inheritance of genes.
2. Mendelian genetics explains the inheritance of genes that are located on separate chromosomes and have two allelic forms, but other patterns of inheritance occur that are not governed by Mendel's principles. Incomplete dominance and codominance occur when neither allele is completely dominant. Multiple phenotypes can occur when more than two alleles exist for a given trait or when that trait is controlled by

more than one pair of genes. Genes that occur on one chromosome are inherited together or linked. Crossing-over can result in the recombination of linked genes.

3. James Watson and Francis Crick described the structure of the DNA molecule as a double helix of nucleotides. A gene is a segment of the DNA molecule that codes for the formation of one polypeptide. Through transcription and translation, the information coded in the sequence of bases is expressed as a specific sequence of amino acids. Some mutations arise from changes in the DNA that results in altered proteins with different characteristics. Since the genetic code is a universal code, the genetic information from organisms can be translated by another organism. Recombinant DNA technology is now being used for practical applications.

Review Questions

1. What can the phenotype tell you about the genotype of an individual? Distinguish between phenotype and genotype.
2. In the commercial development of seeds, how would you determine if a newly established strain will breed true?
3. What do we mean when we say that genes are linked? Why does this alter the normal ratio of phenotypes?
4. How are the various types of RNA involved in the process of protein synthesis?
5. What is a gene?
6. What is the relationship between inherited traits and proteins?
7. What is a codon? How many exist? What does it represent?

Additional Genetics Problems

1. In the garden peas, yellow seeds are dominant to green. What are the phenotypic ratios of offspring in the following crosses?
   a. homozygous yellow x heterozygous yellow
   b. heterozygous yellow x homozygous green
   c. heterozygous yellow x heterozygous yellow
2. Also in garden peas, the tall allele (*T*) is dominant to the dwarf allele (*t*). What are the phenotypic ratios of offspring in the following crosses?
   a. $TT \times Tt$
   b. $tt \times TT$
   c. $Tt \times Tt$
3. Using these same alleles, when tall green garden peas are crossed with dwarf yellow peas, what would be the appearance of the F_1 and F_2 generations?
4. In tomatoes, red fruit is dominant to yellow fruit and tall plants are dominant to short plants. A short plant with yellow fruit is crossed with a plant that is heterozygous for both traits. What would be the phenotypic ratio of the offspring from such a cross?

Further Reading

Campbell, N. A. 1993. *Biology,* 3rd Edition. Benjamin/Cummings Publishing Co., Redwood City, CA.

Darnell, J. E. October, 1985. RNA. *Scientific American,* 253 (4): 68–78.

Doolittle, R. F. October, 1985. Proteins. *Scientific American,* 253 (4): 88–99.

Felsenfeld, G. October, 1985. DNA. *Scientific American,* 253 (4): 58–67.

Mendel, Gregor. 1967. *Experiments in Plant Hybridization.* Harvard University Press, Cambridge, MA.

Rennie, J. March, 1993. DNA's New Twists. *Scientific American,* 266 (3): 122–132.

Sayre, Anne. 1975. *Rosalind Franklin & DNA,* W. W. Norton, New York.

Watson, J. D. 1968. *The Double Helix,* Antheneum, NY.

C H A P T E R

8

Plant Systematics

Chapter Outline

Chapter Concepts

1. Scientific names are two-word names called binomials that are internationally recognized by the scientific community.
2. Carolus Linnaeus, an eighteenth century Swedish botanist, started the binomial system and is therefore known as the Father of Taxonomy.
3. Natural selection favors the survival and reproduction of those individuals in a species that possess traits that better adapt them to a particular environment.

Plant systematics is the branch of botany that is concerned with the naming, identification, evolution, and classification (arrangement into groups with common characteristics) of plants. In a strict sense, plant taxonomy is the science of naming and classifying plants; however, in this book the terms *taxonomy* and *systematics* are used interchangeably. The simplest form of classification is a system based on need and use; early humans undoubtedly classified plants into edible, poisonous, medicinal, and hallucinogenic categories.

Early History of Classification

The earliest known formal classification was proposed by the Greek naturalist Theophrastus (370–285 B.C.), who was a student of Aristotle. In his botanical writings (*Enquiries into Plants* and *The Causes of Plants*), he described and classified approximately 500 species of plants into herbs, undershrubs, shrubs, and trees. Because his influence extended through the Middle Ages, he is regarded as the Father of Botany.

Two Roman naturalists who also had long-lasting impacts on plant taxonomy were Pliny the Elder (A.D. 23–79) and Dioscorides (first century A.D.). Both described medicinal plants in their writings, and Dioscorides' *Materia Medica* remained the standard medical reference for 1500 years. From this period through the Middle Ages, little new botanical knowledge was added. Blind adherence to the Greek and Roman classics prevailed, using manuscripts painstakingly copied by hand in monasteries throughout Europe.

The revival of botany after its stagnation in the Middle Ages began early in the Renaissance with the renewed interest in science and other fields of study. The invention of the printing press in the middle of the fifteenth century allowed botanical works to be more easily produced than ever before. These richly illustrated books, known as **herbals,** largely dealt with the medicinal plants and their identification, collection, and preparation. The renewal of interest in taxonomy can be traced to the work of several herbalists; in fact, this period of botanical history from the fifteenth through the seventeenth centuries is known as the Age of Herbals. Another factor in the revival of taxonomy was the global explorations by the Europeans during this period, which led to the discoveries of thousands of new plant species. In less than 100 years more plants were introduced to Europe than in the previous 2,000 years.

Carolus Linnaeus

By the beginning of the eighteenth century, it was common to name plants using a **polynomial,** which included a single word name for the plant (today called the genus name), followed by a lengthy list of descriptive terms, all in Latin. This system had flaws. It was not standardized, different polynomials existed for the same plant, and it was cumbersome to remember some of the longer polynomials that could be a paragraph in length. This was the state of taxonomy during the time of Linnaeus.

Figure 8.1 *Statue of the young Carolus Linnaeus (1707–1778) in the Linnean Gardens at the University of Uppsala, Sweden.*

Carolus Linnaeus (fig. 8.1) was born in May, 1707, in southern Sweden, the son of a clergyman. He became interested in botany at a very young age through the influence of his father, who was an avid gardener and amateur botanist. It was expected that Linnaeus would also become a clergyman, but in school he did not do well in theological subjects. He did, however, excel in the natural sciences and entered the University of Lund in 1727 to pursue studies in natural science and medicine. (At this time medical schools were the centers of botanical study because physicians were expected to know the plant sources of medicines in use.) After one year he transferred to the University of Uppsala, the most prestigious university in Sweden. It was here that he published his first botanical papers, which laid the foundations for his later works in classification and plant sexuality. In 1732, he undertook a solo expedition to Lapland to catalog the natural history of this relatively unknown area. He later published *Flora Lapponica,* a detailed description of the plants of this area.

SYSTEMA
NATURÆ
REGNA TRIA NATURÆ,
CLASSES ET ORDINES
GENERA ET SPECIES

Figure 8.2 *Frontispiece of* Systema Naturae, *one of the writings of Linnaeus, in which he expounded on his ideas of classification.*

(a)

(b)

Figure 8.3 *(a) The cottage behind Linnaeus's country home at Uppsala, where he often brought students. (b) The wallpaper inside Linnaeus's country house in Uppsala, Sweden.*

Linnaeus received his medical degree in 1735 from the University of Harderwijk in The Netherlands. Soon he came under the patronage of George Clifford, a director of the Dutch East India Company and one of the wealthiest men in Europe. He served as Clifford's personal physician, as well as curator of his magnificent gardens, which housed specimens from around the world. The three years he spent in The Netherlands were the most productive period in his life, completing several books and papers including *Systema Naturae, Fundamenta Botanica,* and *Genera Plantarum,* which expanded on his ideas of classification (fig. 8.2).

He returned to Sweden in 1738 and soon married Sara Elisabeth Moraea. After setting up a medical practice in Stockholm, he was appointed physician to the Swedish Admiralty, specializing in the treatment of venereal diseases. In 1741, he returned to the University of Uppsala as professor of medicine and botany, a position he retained until retirement in 1775 (fig. 8.3). Linnaeus was a popular teacher who attracted students from all over Europe. Many of his students became famous professors in their own right, while others travelled to distant lands collecting unknown specimens for Linnaeus to classify. After suffering several strokes, he died in January, 1778.

One of Linnaeus's achievements was his sexual system of plant classification, which did much to popularize the study of botany. This system was based on the number, arrangement, and length of stamens and thus divided flowering plants into 24 classes. Using this system it was possible to identify and name unknown plants. At the time, his language was risque because he compared floral parts to human sexuality, with stamens referred to as husbands and pistils as wives; i.e., "husband and wife have the same bed" meant stamens and pistils in the same flower (fig. 8.4). Dr. Johann Siegesbeck, a contemporary of Linnaeus and director of the botanical garden in St. Petersburg, was shocked at the analogies and said such

> "loathsome harlotry as several males to one female would not be permitted in the vegetable kingdom by the Creator . . . Who would have thought that bluebells, lilies, and onions could be up to such immorality?"

Despite the opposition of some, the Linnaean sexual method was easy to understand and simple for even the amateur botanist to use. This method, however, was an artificial system grouping together clearly unrelated plants (in his system cherries and cacti were grouped together); by the early nineteenth century it was abandoned in favor of systems that reflected natural relationships among plants.

Linnaeus's greatest accomplishment was his adoption and popularization of a **binomial system of nomenclature.** When he described new plants, he conformed to the current practice of using a polynomial. For convenience, however, he began to add in the margin a single descriptive adjective

VEGETABLE KINGDOM
KEY OF THE SEXUAL SYSTEM

MARRIAGES OF PLANTS.
Florescence.
PUBLIC MARRIAGES.
Flowers visible to every one.
IN ONE BED.
Husband and wife have the same bed.
All the flowers hermaphrodite: stamens and pistils in the same flower.
WITHOUT AFFINITY.
Husbands not related to each other.
Stamens not joined together in any part.
WITH EQUALITY.
All the males of equal rank.
Stamens have no determinate proportion of length.

1. ONE MALE.
2. TWO MALES.
3. THREE MALES.
4. FOUR MALES.
5. FIVE MALES.
6. SIX MALES.
7. SEVEN MALES.
8. EIGHT MALES.
9. NINE MALES.
10. TEN MALES.
11. TWELVE MALES.
12. TWENTY MALES.
13. MANY MALES.

WITH SUBORDINATION
Some males above others.
Two stamens are always lower than the others.

14. TWO POWERS.
15. FOUR POWERS.

WITH AFFINITY
Husbands related to each other.
Stamens cohere with each other, or with the pistil.

16. ONE BROTHERHOOD.
17. TWO BROTHERHOODS.
18. MANY BROTHERHOODS.
19. CONFEDERATE MALES.
20. FEMININE MALES.

IN TWO BEDS.
Husband and wife have separate beds.
Male flowers and female flowers in the same species.

21. ONE HOUSE.
22. TWO HOUSES.
23. POLYGAMIES.

CLANDESTINE MARRIAGES.
Flowers scarce visible to the naked eye.

24. CLANDESTINE MARRIAGES.

Figure 8.4 *Linnaeus's sexual system related floral parts to human sexuality.*

W.T. Stern, *Linnean Classification,* © 1957 and 1971.

that would identify unequivocally a particular species (fig. 8.5). He called this adjective the trivial name. This combination later developed into the two-word scientific name or binomial, described in the next section. Linnaeus used this system consistently in *Species Plantarum* published in 1753. This work contains descriptions and names of 5,900 plants, all the plants known to Linnaeus. The binomial system simplified scientific names and was soon in wide use. In 1867, a group of botanists at the International Botanical Congress in Paris established rules governing plant nomenclature and classification. They established *Species Plantarum* as the starting point for scientific names. Although the rules (formalized in the *International Code of Botanical Nomenclature*) have been modified over the years, the 1753 date is still valid, and many names first proposed by Linnaeus are still in use today.

Linnaeus's contributions were not limited to botany, since the binomial system is used for all known organisms. During his life, he is credited with naming approximately 12,000 plants and animals; for all his contributions to the field of taxonomy, he is known as the Father of Taxonomy.

How Plants Are Named

Names are useful because they impart some information about a plant; it may relate to flower color, leaf shape, flavor, medicinal value, season of blooming, or location. Names are necessary for communication;

> "if you know not the name, knowledge of things is wasted."

1146 SYNGENESIA: POLYGAMIA ÆQUALIS.

Cirſium inerme, caulibus adſcendentibus, foliis linearibus infra cinereis. *Gmel. ſib.* 2. *p.* 71. *t.* 28.? *ſed flores majores.*
Habitat in Sibiria. *D. Gmelin.*
Caulis *angulatus, corymboſus, ramis itidem corymboſis, ut terminetur denſiſſima ſylva florum, fere infinitorum.* Folia *ſaligna, ſubtus albo villo veſtita.* Calyces *cylindrici Squamis glabris, acutis, purpuraſcentibus. Similis præcedenti, ſed folia baſi parum decurrentia, ſubtus villoſa,* & Calyces *copioſiores, argutiores, glabri magis* & *lætius colorati.*

noveboracenſis. 6. SERRATULA foliis lanceolato-oblongis ſerratis pendulis. *Hort. cliff.* 392. *Roy. lugdb.* 143.
Serratula noveboracenſis maxima, foliis longis ſerratis, *Dill. elth.* 255. *t.* 263. *f.* 342.
Serratula noveboracenſis altiſſima, foliis doriæ mollibus ſubincanis. *Moriſ. hiſt.* 3. *p.* 133. *Raj. ſuppl.* 208.
Centaurium medium noveboracenſe luteum, ſolidaginis folio integro tenuiter crenato. *Pluk. alm.* 93. *t.* 109. *f.* 3.
Habitat in Noveboraco, Virginia, Carolina, Canada, Kamtſchatca. ♃

præalta 7. SERRATULA foliis lanceolato oblongis ſerratis patentibus ſubtus hirſutis. *Mill. dict. t.* 234.
Serratula virginiana, perſiæ folio ſubtus incano. *Dill. elth.* 356. *t.* 264. *f.* 343.
Serratula præalta, anguſto plantaginis aut perſicæ folio. *Bocc. muſ.* 2. *p.* 45. *t.* 32.
Eupatoria virginiana, ſerratulæ noveboracenſis latioribus foliis. *Pluk. alm.* 141. *t.* 280. *f.* 6.
Habitat in Carolina, Virginia, Penſylvania.
Receptaculum nudum, nec villoſum. Tozzet. app. 166.

glauca. 8. SERRATULA foliis ovato oblongis acuminatis ſerratis, floribus corymboſis, calycibus ſubrotundis. *Gron. virg.* 116.
Serratula marilandica, foliis glaucis cirſii inſtar denticulatis. *Dill. elth.* 354. *t.* 262. *f.* 341.
Centaurium medium marianum folio integro cirſii noſtratis more ſpinulis fimbriato. *Pluk. mant.* 40.
Habitat in Marilandia, Virginia, Carolina ♃

ſquarroſa. 9. SERRATULA foliis linearibus, calycibus ſquarroſis

Figure 8.5 *A photograph from* Species Plantarum *illustrates the beginning of the binomial system. Note the trivial names in the margin next to the polynomial description for each species. The trivial name was later designated as the species epithet that, together with the generic, forms the binomial.*

This discussion begins with a look at names, common names or what plants are called locally, and internationally recognized scientific names.

Common Names

A closer look at common names often reveals a keen sense of observation, a fanciful imagination, or even a sense of humor: trout lily, milkweed, Dutchman's pipe, Texas bluebonnet, ragged sailor, and old maid's nightcap (table 8.1).

Names have evolved over centuries, but are sometimes only used in a limited geographical area. Even short distances away, other common names may be used for the same plant. Consider, for example, the many names for the tree that many people call osage orange (fig. 8.6a): bodeck, bodoch, bois d'arc, bow-wood, osage apple tree, hedge, hedge apple, hedge osage, hedge-plant osage, horse apple, mock orange, orange-like maclura, osage apple, and wild orange.

On the other hand, different plants may share the same common name. Although mock orange is one of the common names for osage orange, the name mock orange

TABLE 8.1
Some Common Names and their Meanings

Names in Commemoration	
Douglas Fir	David Douglas, plant collector (1798–1834)
Camellia	Georg Josef Kamel, pharmacist (1661–1706)
Gerber Daisy	Traugott Gerber, German explorer (?–1743)
Freesia	Friedrich H.T. Freese, German physician (?–1876)
Names that Describe Physical Qualities	
Dusty miller	Refers to the white woolly leaves
Dutchman's Breeches	Shape of flower
Goldenrod	Shape and color of inflorescence
Indian pipe	Refers to shape of flower with stem
Cattail	Refers to inflorescence of pistillate flowers
Lady's Slipper	Shape of this orchid's flower
Milkweed	Milky juice when plant is cut
Skunk cabbage	Fetid odor of inflorescence
Cheeses	Fruit resembles a round head of cheese
Smoke tree	Plumelike pedicels
Shagbark hickory	Refers to the shedding bark
Redbud	Color of flower buds
Quaking aspen	Refers to the rustling leaves
Bluebell	Color and shape of flower
Crape myrtle	Wavy edges of petals
Scientific Names that Have Become Common Names	
Hydrangea	Abelia
Vanilla	Narcissus
Coreopsis	Gladiolus
Names that Indicate Use	
Daisy fleabane	Gets rid of fleas
Boneset	A tonic from this plant can heal bones
Feverwort	Medicinal property to reduce fever
Kentucky coffeetree	Seeds roasted for coffee substitute
Belladonna	Juice of this plant used to beautify by producing pallid skin and dilated, mysterious eyes
Names that Indicate Origin, Location, or Season	
Pacific Yew	Grows along northern Pacific coast
Spring Beauty	One of the first flowers of spring
Marshmallow	Found in wet, marshy habitat
Daylily	Flowers last only a day
Four-o'clock	Flowers in late afternoon
Japanese honeysuckle	Country of origin is Japan

(a)

(b)

Figure 8.6 *(a)* Maclura pomifera *and (b)* Philadelphus sp. *two entirely different species of plants that share the same common name of mock orange.*

TABLE 8.2

Genus Names and their Meanings

Genus	Meaning
Names in Commemoration	
Begonia	Michel Begon, patron of botany (1638–1710)
Forsythia	William Forsyth, gardener at Kensington Palace (1737–1804)
Bougainvillea	Louis Antoine de Bougainville, explorer and scientist (1729–1811)
Fuchsia	Leonhard Fuchs, German physician and herbalist (1501–1566)
Zinnia	Johann Gottfried Zinn (1727–1759)
Wisteria	Caspar Wistar, American professor of plant anatomy (1761–1818)
Names that Describe Physical Qualities	
Myriophyllum	Finely divided leaves
Chlorophytum	Green plant
Lunaria	Moon, refers to appearance of pods
Helianthus	Sun flower
Zebrina	Zebra, refers to striped leaves
Trillium	Floral parts in three's
Tetrastigma	Four-lobed stigma
Ribes	Acid tasting; refers to fruit
Polygonum	Many knees; refers to jointed stems
Zanthoxylum	Yellow wood
Sagittaria	Arrow; refers to arrowhead leaves
Names from Aboriginal or Classic Origins	
Avena	Oats (Latin)
Triticum	Wheat (Latin)
Allium	Garlic (Greek)
Catalpa	Catalpa (North American Indian)
Vitis	Grape vine (Latin)
Ulmus	Elm (Latin)
Pinus	Pine (Latin)
Names that Indicate Use	
Solidago	Make whole or strengthen
Angelica	Angelic medicinal properties
Cimcifuga	Repel bugs
Saponaria	Soap; refers to soap that can be made from that plant
Pulmonaria	Lung; used to treat infections of the lung
Potentilla	Powerful; refers to its potent medicinal properties
Names that Indicate Location	
Elodea	Grows in marshes
Petrocoptis	Break rock; refers to habit of growing in rock crevices

is usually associated with a completely unrelated group of flowering shrubs (fig. 8.6b). This points out the difficulties with common names; one plant may be known by several different names and the same name may apply to several different plants. The need to have one universally accepted name is fulfilled with scientific names.

Scientific Names

Each kind of organism is known as a **species,** and similar species form a group called a **genus** (plural **genera**). Each species has a scientific name in Latin that consists of two elements; the first is the genus and the second is the species epithet (or specific epithet). Such a name is a binomial, literally two names and is always italicized or underlined; for example, *Maclura pomifera* is the scientific name for osage orange. A rough analogy of the binomial concept can be seen in a list of names in a telephone directory, where the surname "Smith" (listed first) represents the genus and the first names (John, Frank, and Mary) define particular species within the genus.

TABLE 8.3 Common Scientific Epithets and their Meanings

acidosus, -a, -um	Sour
aestivus, -a, -um	Developing in summer
albus, -a, -um	White
alpinus, -a, -um	Alpine
annus, -a, -um	Annual
arabicus, -a, -um	Of Arabia
arborues, -a, -um	Treelike
arvensis, -a, -um	Of the field
biennis, -a, -um	Biennial
campester, -tris, -tre	Of the pasture
canadensis, -is, -e	From Canada
carolinianus, -a, -um	From the Carolinas
chinensis, -is, -e	From China
coccineus, -a, -um	Scarlet
deliciosus, -a, -um	Delicious
dentatus, -a, -um	Having teeth
domesticus, -a, -um	Domesticated
edulis, -is, -e	Edible
esculentus, -a, -um	Tasty
europaeus, -a, -um	From Europe
fetidus, -a, -um	Bad smelling
floridus, -a, -um	Flowery
foliatus, -a, -um	Leafy
hirsutus, -a, -um	Hairy
japonicus, -a, -um	From Japan
lacteus, -a, -um	Milky white
littoralis, -is, -e	Growing by the shore
luteus, -a, -um	Yellow
mellitus, -a, -um	Honey-sweet
niger, -ra, -rum	Black
occidentalis, -is, -e	Western
odoratus, -a, -um	Fragrant
officinalis, -is, -e	Used medicinally
robustus, -a, -um	Hardy
ruber, -ra, -rum	Red
saccharinus, -a, -um	Sugary
silvaticus, -a, -um	Of the woods
sinensis, -is, -e	Chinese
speciosus, -a, -um	Showy
tinctorius, -a, -um	Used for dyeing
utilis, -is, -e	Useful
vernalis, -is, -e	Spring flowering
virginianus, -a, -um	From Virginia
vulgaris, -is, -e	Common

In the binomial, the first name is a noun and is capitalized, while the second written in lower case is usually an adjective. After the first mention of a binomial, the genus name can be abbreviated to its first letter, as in *M. pomifera,* but the species epithet can never be used alone. The genus name, however, can be used alone, especially when referring to several different species within a genus; for example, *Philadelphus* refers to over 50 different species of mock orange. The specific epithet can be replaced by an abbreviation for species, "sp." (or "spp." plural) when the name of the species is unknown or unnecessary for the discussion. Using the previous example, *Philadelphus* sp. refers to one species of mock orange, whereas *Philadelphus* spp. refers to more than one species.

Scientific names may be just as descriptive as common names and translation of the Latin (or latinized Greek) is informative (table 8.2). Sometimes either the genus name or the species epithet is commemorative, derived from the name of a botanist or other scientist. Some specific epithets are frequently used with more than one genus and knowledge of their meanings will provide some insight into scientific names encountered later in this text (table 8.3).

A complete scientific name also includes the name(s) of the author or authors (often abbreviated) who first described the species or placed it in a particular genus. For example, the complete scientific name for corn is *Zea mays* L.; the "L" indicates that Linnaeus named this species. On the other hand, the complete name for osage orange is *Maclura pomifera* (Raf.) Schneid. This author citation indicates that Rafinesque-Schmaltz first described the species giving it the specific epithet *pomifera,* but Schneider later put it in the genus *Maclura.* In this text, the author citations are omitted for simplicity.

A scientific name is unique, referring to only one species and universally accepted among scientists. It is the key to unlocking the door to the accumulated knowledge about a plant. Imagine the confusion if only common names were used and a reference was made to mock orange? Would this allude to *Maclura pomifera* or a species of *Philadelphus?*

TAXONOMIC HIERARCHY

In addition to genus and species, other taxonomic categories exist to conveniently group related organisms. As pointed out, Linnaeus used an artificial system; however, today scientists use a **phylogenetic** system to group plants. In a phylogenetic system, information is gathered from morphology, anatomy, cell structure, biochemistry, genetics, and the fossil record to determine evolutionary relationships and, therefore, natural groupings among plants.

ABLE 8.4

Economically Important Angiosperm Families

Scientific Family Name	Common Family Name	Economic Importance
Aceraceae	Maple	Lumber (ash, maple), maple sugar
Apiaceae	Carrot	Edibles (carrot, celery), herbs (dill), poisonous (poison hemlock)
Arecaceae	Palm	Edibles (coconut), fiber oils and waxes, furniture (rattan)
Asteraceae	Sunflower	Edibles (lettuce), oils (sunflower oil), ornamentals (daisy)
Brassicaceae	Mustard	Edibles (cabbage, broccoli)
Cactaceae	Cactus	Ornamentals, psychoactive plants (peyote)
Cannabaceae	Hemp	Psychoactive (marijuana), fiber plants
Chenopodiaceae	Goosefoot	Edibles (spinach), beet sugar
Cucurbitaceae	Gourd	Edibles (melons, squashes)
Euphorbiaceae	Spurge	Rubber, medicinals (castor oil), edibles (cassava), ornamentals (poinsettia)
Fabaceae	Bean	Edibles (beans, peas), oil, dyes, forage, ornamentals
Fagaceae	Beech	Lumber (oak), dyes (tannins), ornamentals
Iridaceae	Iris	Ornamentals
Juglandaceae	Walnut	Lumber, edibles (walnut, pecan)
Lamiaceae	Mint	Aromatic herbs (sage, basil)
Lauraceae	Laurel	Aromatic oils (bay leaves), lumber
Liliaceae	Lily	Ornamentals, edibles (onion), poisonous
Magnoliaceae	Magnolia	Ornamentals, lumber
Malvaceae	Mallow	Fiber (cotton), seed oil, edibles (okra), ornamentals
Musaceae	Banana	Edibles (bananas), fibers
Myrticaceae	Myrtle	Timber, medicinals (eucalyptus), spices (cloves)
Oleaceae	Olive	Lumber (ash), edible oil and fruits (olive)
Orchidaceae	Orchid	Ornamentals, spice (vanilla)
Papaveraceae	Poppy	Medicinal and psychoactive plants (opium poppy)
Piperaceae	Pepper	Black pepper, houseplants
Poaceae	Grass	Cereals, forage, ornamentals
Ranunculaceae	Buttercup	Ornamentals, medicinal and poisonous plants
Rosaceae	Rose	Fruits (apple, cherry), ornamentals (roses)
Rubiaceae	Coffee	Beverage (coffee), medicinals (quinine)
Rutaceae	Citrus	Edible fruits (orange, lemon)
Salicaceae	Willow	Ornamentals, furniture (wicker), medicines (aspirin)
Solanaceae	Nightshade	Edible (tomato, potato), psychoactive, poisonous (tobacco, mandrake)
Theaceae	Tea	Beverage (tea)
Vitaceae	Grape	Fruits (grapes), wine

Higher Taxa

Species that have many characteristics in common are grouped into a genus, one of the oldest concepts in taxonomy (fig. 8.7). In almost every society, the concept of genus has developed in colloquial language; in English the words oak, maple, pine, lily, and rose represent distinct genera. These intuitive groupings reflect natural relationships based on shared vegetative and reproductive characteristics. Many of the scientific names of genera are directly taken from the ancient Greek and Roman common names for these genera (*Quercus,* old Latin word for oak).

The next higher category or **taxon** (plural **taxa**) above the rank of genus is the **family.** Families are composed of related genera that again (as in a genus) share combinations of morphological traits. In the angiosperms, floral and fruit features are often used to characterize a family. Ideally, the family represents a natural group with a common evolutionary lineage; some families may be very small, while others very large, but still cohesive, groups. A few common angiosperm families that have special economic importance are listed in Table 8.4. According to the *International Code of Botanical Nomenclature,* each family is assigned one name, which is always capitalized and ends in the suffix *-aceae.* The old established names of several well-known families present exceptions to this rule. Both the traditional and standardized names are used

Figure 8.7 *A genus is a group of species that share many characteristics in common. Although willow oak* (Quercus phellos) *and red oak* (Quercus rubra) *are clearly distinct species, they are both recognizable as belonging to the oak* (Quercus) *genus.*

for these families (table 8.5). The taxa above the rank of family and their appropriate endings are presented in Table 8.6. The higher the taxonomic category, the more inclusive the grouping (fig. 8.8). Families are grouped into **orders,** orders into **classes,** classes into **divisions,** and finally, divisions into **kingdoms.** The complete classification of a familiar species is also illustrated in Table 8.6. In addition to the categories already described, biologists also recognize intermediate categories with the "sub" for any rank; for example, divisions may be divided into subdivisions and species may be divided into subspecies (varieties and forms are also categories below the rank of species).

Although the *International Code of Botanical Nomenclature* has rules that govern the assignment of names and defines the taxonomic hierarchy, it does not set forth any particular classification system. As a result, there are several different organizational schemes that have supporters. These systems differ in the numbers of classes, divisions, and even kingdoms and how they are related to one another. Presently most biologists use a five-kingdom system, which will be described fully in the next chapter. While there is general agreement about the use of a five-kingdom system, biologists still debate the definition of a species.

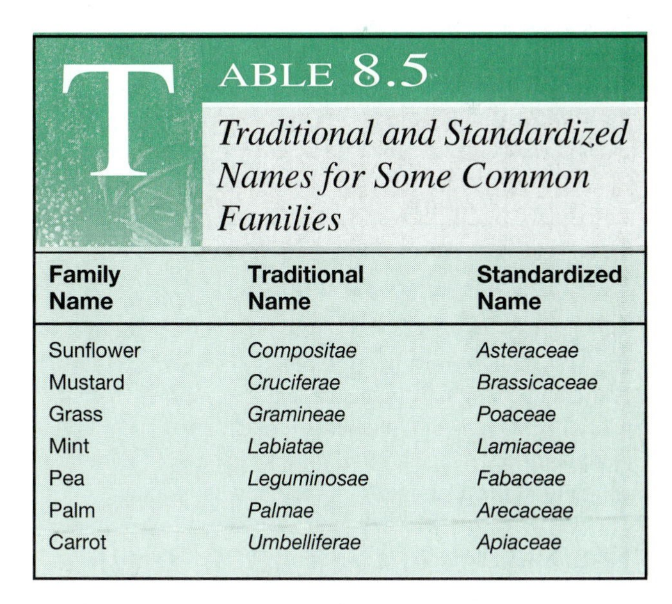
TABLE 8.5

Traditional and Standardized Names for Some Common Families

Family Name	Traditional Name	Standardized Name
Sunflower	*Compositae*	*Asteraceae*
Mustard	*Cruciferae*	*Brassicaceae*
Grass	*Gramineae*	*Poaceae*
Mint	*Labiatae*	*Lamiaceae*
Pea	*Leguminosae*	*Fabaceae*
Palm	*Palmae*	*Arecaceae*
Carrot	*Umbelliferae*	*Apiaceae*

What is a Species?

As indicated previously, each kind of organism is known as a species. While this intuitive definition, based on morphological similarities, works fairly well in many circumstances, it is limited; scientists have given much thought to the biological basis of a species. Many accept

TABLE 8.6

The Taxonomic Hierarchy and Standard Endings

Rank	Standard Ending	Example
Division	*-phyta*	Anthophyta
Class	*-opsida*	Liliopsida
Order	*-ales*	Lilales
Family	*-aceae*	Liliaceae
Genus		*Lilium*
Species		*Lilium superbum* L.

the **biological species concept** first proposed by Ernst Mayr in 1942 that defines a species as

> "a group of interbreeding populations reproductively isolated from any other such group of populations."

This definition presents problems when defining plant species. Many closely related plant species that are distinct morphologically are, in fact, able to interbreed; this is true for many species of oaks and sycamores. By contrast, a single plant species may have diploid and **polyploid** (more than the diploid number of chromosomes) individuals that may be reproductively isolated from each other. It is estimated that as many as 40% of flowering plants may be polyploids, with the evening primrose group a thoroughly studied example; an even higher percentage of polyploid species occurs in ferns. Despite the lack of an all-inclusive botanical definition, the concept of "species" facilitates the naming, describing, and classifying of plants in a uniform manner.

The Influence of Darwin's Theory of Evolution

The theory of evolution by means of natural selection was to irrevocably change the way biologists viewed species. Instead of unchanging organisms and generations created all alike, it was realized that species are dynamic and variable, continually evolving through the mechanism of natural selection to refine adaptions to a changing environment.

The Voyage of the H.M.S. *Beagle*

Charles Robert Darwin (fig. 8.9) was born in England in 1809 to a family of distinguished naturalists and physicians. His grandfather was Erasmus Darwin, a well-known poet and physician, and his father Robert Darwin was a successful country doctor. At 15 years of age, Charles was sent to the University of Edinburgh Medical School to study medicine. Not finding it to his liking, he transferred after two years to Cambridge University to study theology. While at Cambridge, he spent much of his free time with the students and professors of natural history. This association later proved invaluable. In 1831, at the age of 22, Darwin graduated from Cambridge with a degree in theology. Shortly thereafter he was recommended as ship naturalist by John Henslow, one of the natural history professors at Cambridge. The ship in question was the H.M.S. *Beagle,* commissioned by King William IV to undertake a voyage around the world for the purpose of charting coastlines, particularly that of South America, for the British navy. The voyage of the *Beagle* began on December 27, 1831 and was to last 5 years (fig. 8.10). During his time on the *Beagle,* Darwin collected thousands of plants and specimens from South America, the Galápagos Islands (off the coast of Ecuador), Australia, and New Zealand. He studied geological formations and noted fossil forms of extinct species. He found that some fossils of extinct species bore a striking resemblance to extant species, as though the former had given rise to the latter. Darwin spent some time studying the species found on the Galápagos Islands off the coast of South America. He noted that animals and plants found in the Galápagos were obviously similar to species found in South America, but there were distinct differences. These observations led Darwin to question the fixity of species concept. According to this concept, widely held at the time of Darwin, species were acts of Divine Creation, unchanging over time.

When the *Beagle* returned to England in 1836, Darwin married his cousin Emma Wedgwood (of the famous Wedgwood china family) and settled, at age 27, in the English countryside. He continued his work in natural history with experiments, writing papers, and corresponding with other naturalists. Among his works was a four-volume treatise on the classification and natural history of barnacles.

In 1842, he began putting his thoughts together on what was to become his theory of evolution by natural

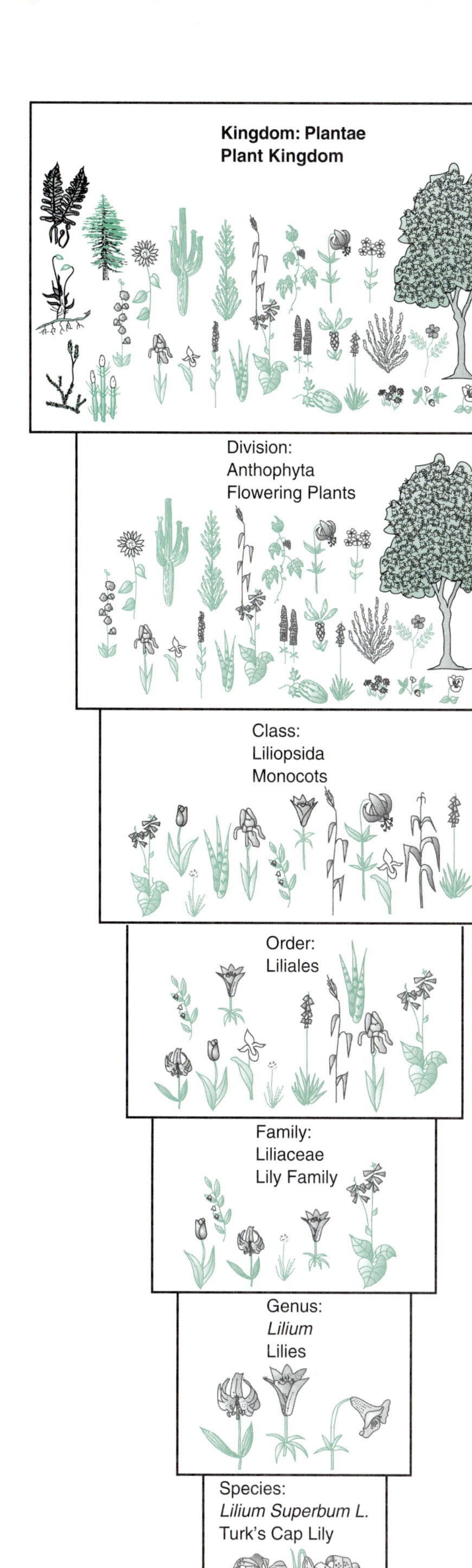

Figure 8.9 *Charles Darwin (1809–1882) at age 29, a few years after his return from the voyage of the H.M.S.* Beagle.

selection. Darwin continued to expand and fine-tune his thoughts over the next 16 years. In June of 1858, he received a manuscript from Alfred Russel Wallace (1823–1913), a young British naturalist working in Malaysia. Wallace's work was entitled *On the Tendency of Varieties to Depart Indefinitely From the Original Type* and had independently arrived at the concept of natural selection. Wallace and Darwin jointly presented their ideas on July 1, 1858 at a meeting of the Linnean Society in London. During the next few months, Darwin completed writing what was to become one of the most influential texts of all time. With the publication on November 24, 1859 of *On the Origin of Species by Means of Natural Selection, or the Preservation of Favoured Races in the Struggle for Life* by Charles Darwin, biological thought was changed forever.

Figure 8.8 *Major ranks in the taxonomic hierarchy. Note that the higher the ranking, the broader the defining characteristics and the more inclusive the group.*

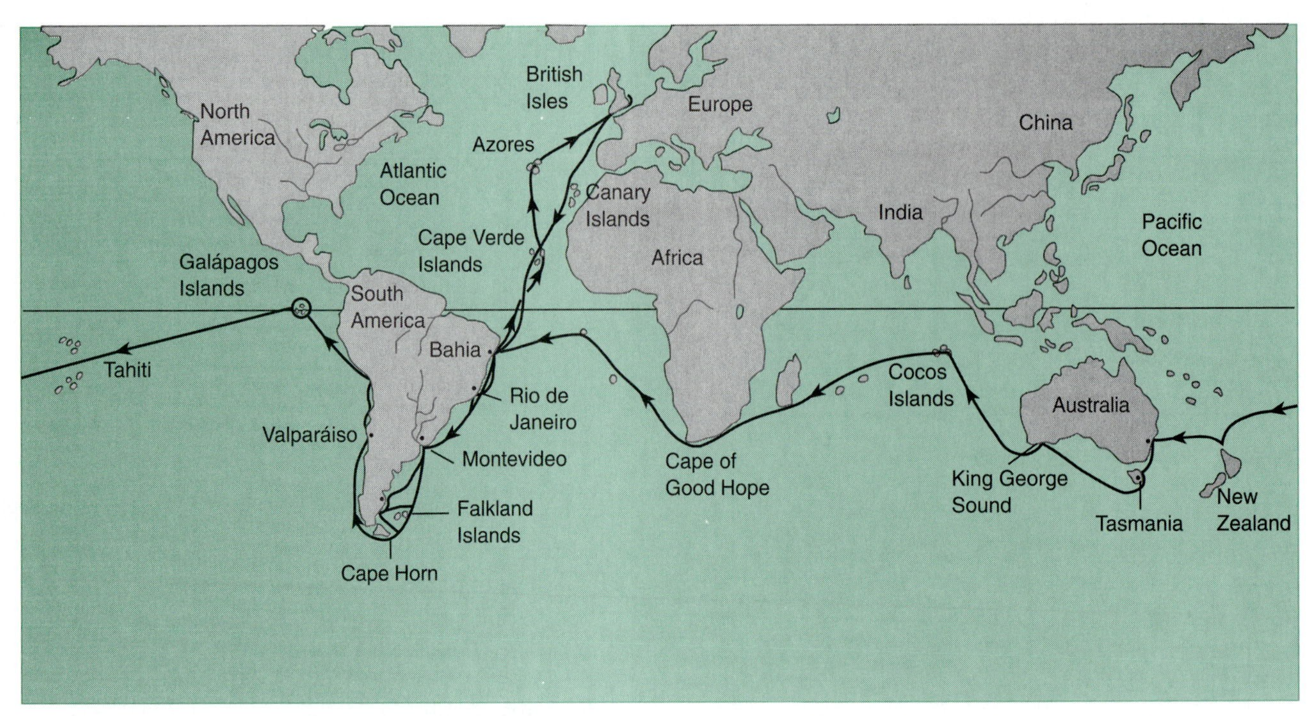

Figure 8.10 *The 5-year voyage of the H.M.S.* Beagle. *Darwin's observations on the geology and distributions of plants and animals in South America and the Galápagos Islands were the groundwork for the development of the theory of evolution by means of natural selection.*

From Peter H. Raven and George B. Johnson, *Biology*, Updated version, 3d ed., © 1995 Wm. C. Brown Communications, Inc. Reprinted by permission of Times Mirror Higher Education Group, Inc., Dubuque, Iowa. All Rights Reserved.

Natural Selection

There are four underlying premises to Darwin's theory of evolution by natural selection:

1) *variation:* members within a species exhibit individual differences and these differences are inheritable
2) *overproduction:* natural populations increase geometrically, producing more offspring than will survive
3) *competition:* individuals compete for limited resources, what Darwin called "a struggle for existence"
4) *survival to reproduce:* only those individuals that are better suited to the environment survive and reproduce (survival of the fittest), passing on to a proportion of their offspring the advantageous characteristics.

Offspring that inherit the advantageous traits are selected for; their chances of survival are enhanced and many live to reproductive age, passing on the desirable attributes. Those that do not inherit these traits are not likely to survive nor reproduce. Gradually, the species evolves, or changes, as more and more individuals carry these traits. Darwin gave this example

> "If the number of individuals of a species with plumed seeds could be increased by greater powers of dissemination within its own area (that is, if the checks to increase fell chiefly on the seeds), those seeds which were provided with ever so little more down, would in the long run be most disseminated; hence a greater number of seeds thus formed would germinate, and would tend to produce plants inheriting the slightly better-adapted down."

The most serious flaw in Darwin's Theory of Evolution was the mechanism of heredity. Darwin had not worked out the source of variation in species, nor did he understand the means by which traits are passed down from generation to generation. It would take an Austrian monk, Gregor Mendel (*see* Chapter 7), working in relative obscurity with pea plants to come up with the answers to Darwin's questions about inheritance.

A well-known example of natural selection is the case of heavy metal tolerance in bent grass, *Agrostis tenuis.* Certain populations of bent grass were found growing near the tailings or soil heaps excavated from lead mines in Wales, despite the fact that mine soils had high concentrations of lead and other heavy metals (copper, zinc, and nickel). When mine plants were transplanted into uncontaminated pasture soil, all survived but were small and slow-growing. A nearby population of bent grass from uncontaminated pasture soil exhibited no such tolerance when transplanted into mine soil; in fact, most (57 out of 60) of the pasture plants died in the lead-contaminated soil. The survival of the three pasture

A Closer Look 8.1

The Language of Flowers

Through traditions, some flowers became symbolic of certain emotions and feelings. This was sometimes even reflected in their common names; two straightforward examples are forget-me-not flowers, which conveyed the sentiment "remember me" and bachelor's button, which indicated the single status of the wearer. This symbolism reached its peak during the Victorian Era when almost every flower and plant had a special meaning. In Victorian times, it was possible to construct a bouquet of flowers that imparted a whole message. A Victorian suitor might send a bouquet of jonquils, white roses, and ferns to his intended, which indicated that he desired a return of affection, he was worthy of her love, and was fascinated by her. This "language" became so popular that dictionaries were printed to interpret floral meanings. One of the most popular dictionaries was the *Language of Flowers* (1884) by Kate Greenaway, a well-known illustrator of children's books. Following is a small sampling of some common flowers and plants, and what they symbolized:

Amaryllis . . . pride

Apple blossom . . . preference

Bachelor's buttons . . . celibacy

Bluebell . . . constancy

Buttercup . . . ingratitude

Yellow chrysanthemum . . . slighted love

Daffodil . . . regard

Daisy . . . innocence

Dogwood . . . durability

Elm . . . dignity

Goldenrod . . . precaution

Holly . . . foresight

Honeysuckle . . . generous and devoted affection

Ivy . . . fidelity

Lavender . . . distrust

Lichen . . . dejection

Lily of the valley . . . return of happiness

Live oak . . . liberty

Box Figure 8.1 *Flowers convey a message all their own.*

Magnolia . . . love of nature

Marigold . . . grief

Mock Orange . . . counterfeit

Oak leaves . . . bravery

Palm . . . victory

Pansy . . . thoughts

Spring crocus . . . youthful gladness

Dwarf sunflower . . . adoration

Tall sunflower . . . haughtiness

Yellow tulip . . . hopeless love

Blue violet . . . faithfulness

Wild grape . . . charity

Zinnia . . . thought of absent friends

Even today, several plants have well-known symbolic meanings: red roses convey passionate love, four-leaf clover means luck, orange blossoms symbolize weddings, and an olive branch indicates peace. Floral colors can also communicate feelings with red indicating passion, blue-security, yellow-cheer, white-sympathy, and orange-friendship. With some thought, it is possible to find the right flower and color to express the exact message.

plants in mine soil is significant; undoubtedly these three possessed an advantageous trait, the ability to tolerate heavy metal soil. A trait that promotes the survival and reproductive success of an organism in a particular environment is an **adaptation.** The mine plants had descended from bent grass plants that possessed the adaptation that conferred tolerance to the mine soil; over time (less than 100 years in this case) populations of *Agrostis* tolerant to heavy metal evolved from those few tolerant individuals.

Although Darwin's theory of natural selection is the foundation of modern evolutionary concepts, biologists today are still learning about the forces that shape evolution.

Chapter Summary

1. Plant systematics has its origins in the classical works of Theophrastus of ancient Greece, who is generally regarded as the Father of Botany. The study of plants, as with many other intellectual endeavors, went into a decline during the Dark Ages of Europe, but was later revived due to renewed interest in herbalism during the fifteenth to seventeenth centuries.
2. Linnaeus, a Swedish botanist of the eighteenth century, is credited with the creation of the binomial or scientific name. Although common names are often informative and readily accessible, scientific names have the advantage of being recognized the world over and unique to a single species.
3. The taxonomic hierarchy includes the major ranks: Kingdom, Division, Class, Order, Family, Genus, and Species.
4. Biologists have wrestled with the concept of the species; the biological concept describes a species as a group of interbreeding populations, reproductively isolated from other populations.
5. Charles Darwin and his theory of evolution by natural selection irrevocably changed the way biologists viewed species. Natural selection favors those individuals that possess traits that better enable them to survive in the environment. These individuals survive to reproduce, and many of their offspring will tend to have these adaptations and pass them on to future generations. In this way, populations change over time. The four underlying conditions of Darwin's theory of evolution by natural selection are variation, overproduction of offspring, competition, and survival to reproduce.

Review Questions

1. List the common names of some of the wild flowers in your area. Determine the type of information each name imparts.
2. Using a plant dictionary (see Further Reading) look up the scientific names and their meanings for common houseplants and landscape plants in your area.
3. Briefly describe the concept of evolution by natural selection.
4. Why are only inherited traits important in the evolutionary process?
5. How do mutations (Chapter 7) lead to the evolution of new species?
6. What was the lasting contribution of Linnaeus? How was the binomial system an improvement over polynomials?

Further Reading

Blunt, Wilfrid. 1971. *The Compleat Naturalist: A life of Linnaeus.* The Viking Press, NY.

Briggs, D. and S. M. Walters. 1984. *Plant Variation and Evolution,* 2nd Edition, Cambridge University Press, Cambridge, MA.

Coombes, Allen J. 1985. *Dictionary of Plant Names.* Timber Press, Beaverton, OR.

Darwin, Charles. 1975. *On the Origin of Species by Means of Natural Selection or the Preservation of Favored Races in the Struggle for Life.* Cambridge University Press, New York.

Gilbert, Bil. 1984. The Obscure Fame of Carl Linnaeus. *Audubon,* 86: 102–115.

Greenaway, Kate. 1978. *The Illuminated Language of Flowers.* Holt, Rinehart and Winston, New York.

Irvine, William. 1964. *Apes, Angels, and Victorians: Darwin, Huxley and Evolution.* Meridian Books, New York.

Jones, Samuel B. and Arlene E. Luchsinger. 1986. *Plant Systematics.* McGraw-Hill Book Company, New York.

CHAPTER

9

Diversity of Plant Life

Chapter Outline

Chapter Concepts

1. Living organisms are classified into five kingdoms: Monera, Protista, Animalia, Plantae, and Fungi.
2. Organisms traditionally known as plants are found in four of the five kingdoms.
3. This diverse group of organisms impacts our lives in many economically and ecologically important ways.

With the overwhelming diversity of life on Earth, scientists have long sought to categorize these organisms into a meaningful system. For many years organisms were classified as either plants or animals. As our knowledge increased, it became evident that many organisms could not be conveniently classified as either. To bypass the problems with this two-kingdom system, other kingdoms have been suggested. Currently, most biologists accept the Five-Kingdom System first proposed by Robert H. Whitaker in 1969 (fig. 9.1).

The Five-Kingdom System

The **Kingdom Monera** contains the organisms commonly known as bacteria and cyanobacteria. It is the only kingdom in which the cells are **prokaryotic;** in the other four kingdoms the cells are **eukaryotic.** Recall that a prokaryotic cell lacks a nucleus and membrane-bound organelles that occur in eukaryotes. The **Kingdom Protista** consists of unicellular and simple multicellular organisms that can be plantlike, fungal-like, or animal-like. The remaining three kingdoms are all multicellular in organization and can be distinguished by their modes of nutrition. The **Kingdom Plantae** contains land plants that are **autotrophic,** capable of manufacturing their own food through photosynthesis. The organisms in the **Kingdoms Animalia** and **Fungi** cannot make their own food and rely on external sources of nutrition. They are, therefore, considered **heterotrophic.** Animals, from primitive sponges to highly evolved mammals, are **ingestive heterotrophs,** engulfing their food and digesting it internally. The fungi, from molds to mushrooms, are **absorptive heterotrophs,** secreting enzymes into their surroundings that break down food, which is then absorbed. Although historically fungi were considered members of the plant kingdom, recent molecular evidence suggests a closer evolutionary relationship between fungi and animals.

Survey of the Kingdoms

Organisms that were once regarded as plants in the old two-kingdom system now are included in four of the five kingdoms just described. Let us take a kingdom-by-kingdom survey of those organisms that have traditionally been called plants. In the last chapter it was noted that within each kingdom large groupings of similar organisms are called divisions.

Cyanobacteria

Within the Kingdom Monera, the organisms formerly called blue-green algae make up the division **Cyanobacteria.** This group of photosynthetic prokaryotes includes microscopic unicells, filaments, and colonial forms that are usually bluish-green in color. These primitive organisms first appeared in the fossil record approximately 3.5 billion years ago and are still common in oceans, fresh water, and even terrestrial environments. They often form a slimy scum on greenhouse floors, clay flower pots, and pool decks. Even though the algae on the greenhouse floor may look unappetizing, some societies use a few types of cyanobacteria as a food source.

The Algae and Seaweeds

In the Kingdom Protista, we will consider six divisions of algae: the **Pyrrophyta,** the **Chrysophyta,** the **Euglenophyta,** the **Chlorophyta,** the **Rhodophyta,** and the **Phaeophyta.** The Pyrrophyta, more commonly known as dinoflagellates, are a group of bizarre unicells. Many dinoflagellates are covered with cellulose plates, giving them an armored appearance (fig. 9.2). Dinoflagellates typically possess two flagella, one trailing and one wrapped around the middle, that enable them to roll through the water like a top. (A **flagellum** is a whip-like organelle composed of microtubules that imparts motility.) These organisms have gained notoriety because some marine species are responsible for the phenomenon known as Red Tide. Under certain environmental conditions these dinoflagellates undergo a population explosion, becoming so abundant and containing up to 20 million organisms per liter (77 million/gal), that they color the water red, orange, yellow, brown, or any hue in between. These organisms produce a powerful toxin that can cause massive fish kills. Also, shellfish can accumulate, but not be affected by, these toxins; this makes the shellfish poisonous to humans, other mammals, and birds. Red tides have occurred in coastal areas of the United States as well as other countries, resulting in tremendous economic losses to the shellfish industry, tourism, and fishing.

A recently recognized dinoflagellate phenomenon is ciguatera poisoning, the leading cause of nonbacterial food poisoning in the United States, Canada, and Europe. Many species of tropical fish accumulate ciguatoxin, a dinoflagellate toxin, in their fatty tissues. Although the fish themselves are unaffected, people eating the tainted fish experience nausea, vomiting, and diarrhea, and develop bizarre neurological symptoms that can be long-lasting. People have reported reverse sensations; for example, hot objects feel cold to the touch.

Other dinoflagellate species are known for their ability to luminesce (emit light). You may have seen tiny specks of light at night, especially as ocean waves break on the shore; these were caused by bioluminescent dinoflagellates.

In the Division Chrysophyta there are several different types of algae, but the most abundant and the most important economically are the diatoms. Diatoms are mainly unicells surrounded by a silicon-based wall composed of two halves that fit together like the top and bottom of a box. In surface view the diatoms are usually circular or rectangular in outline. The glasslike walls are often intricately marked with pits, grooves, and ridges, giving these microscopic forms the appearance of cut crystal (fig. 9.3). It is the walls of the diatoms that are used commercially. When diatoms die, their virtually indestructible walls accumulate on the ocean floor. Deposits from past geological ages are known

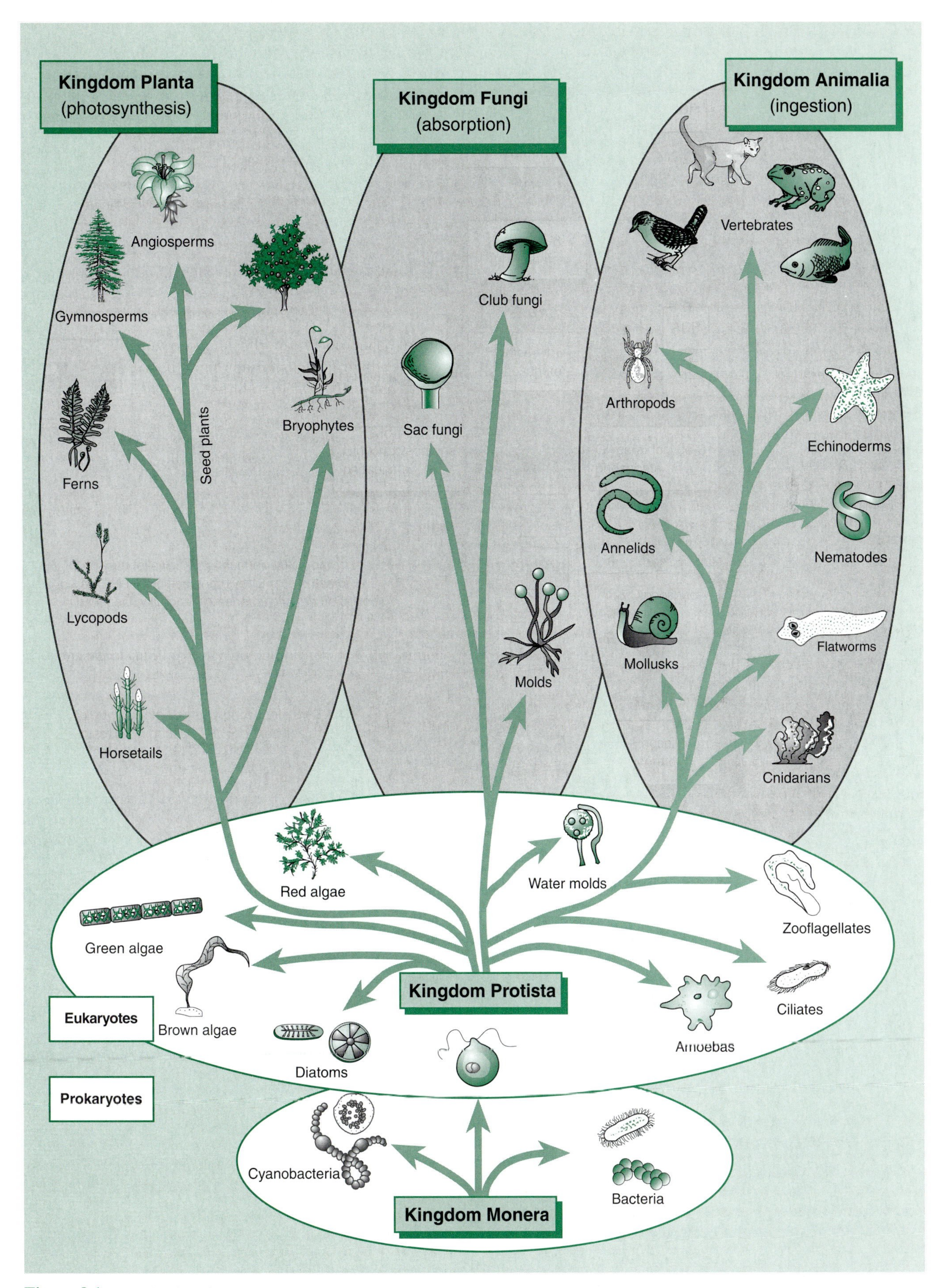

Figure 9.1 *The five kingdoms.*

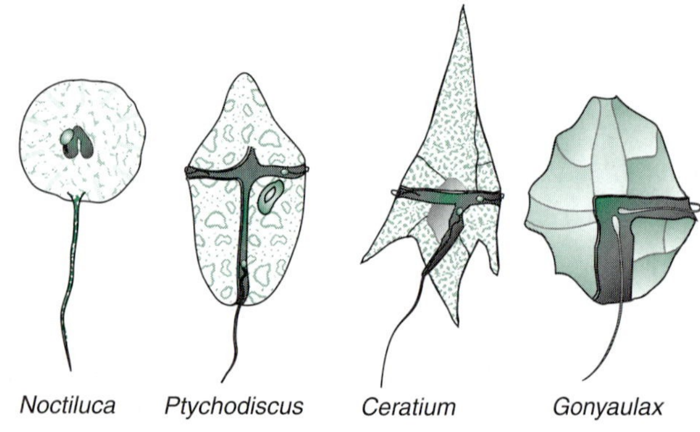

Figure 9.2 *Dinoflagellates. The four genera illustrated here include* Noctiluca, *which is bioluminescent, and* Ptychodicus, *a dinoflagellate responsible for red tides off the coast of Florida.*

as diatomaceous earth and can be mined (fig. 9.3). Diatomaceous earth is useful as a polishing agent in silver polish, as a filter in wine and petroleum industries, and as the reflective material in highway paint. A recent use of diatomaceous earth is as a soil additive; it acts like a field of broken glass to discourage some microscopic garden pests. Ecologically the diatoms and the dinoflagellates are vitally important producers for both marine and freshwater food chains as the food for other organisms.

The Euglenophyta are a small group of mainly freshwater unicells that are grass green in color. Euglenoids can be extremely abundant in nutrient-enriched or polluted waters such as farm ponds and ditches. Instead of a rigid cell wall, they have a proteinaceous covering called a pellicle that allows for flexibility and shape change during swimming. In addition to their ability to change shape, euglenoids have several other distinctive features,

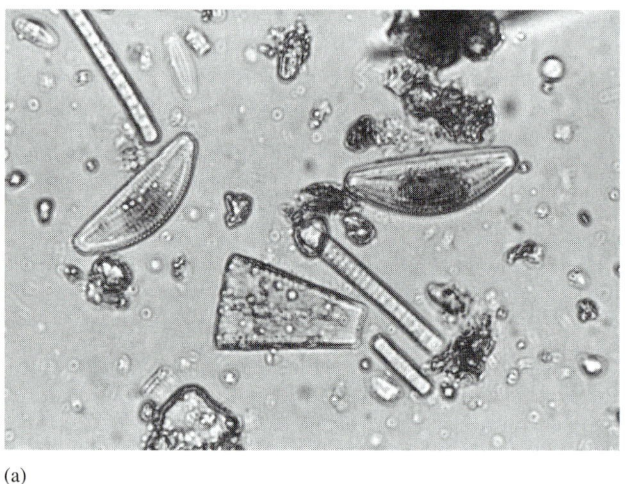

(a)

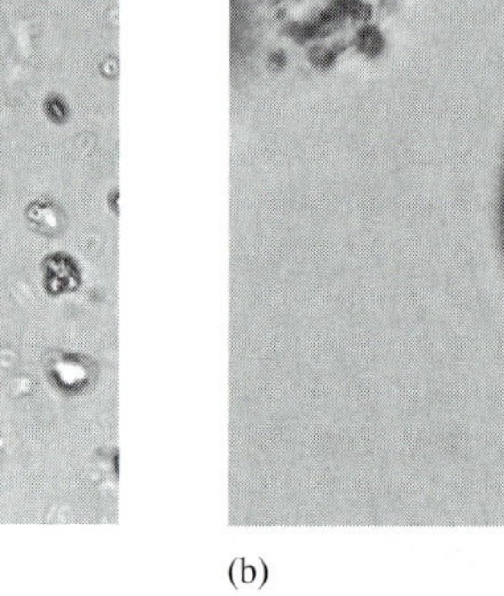

(b)

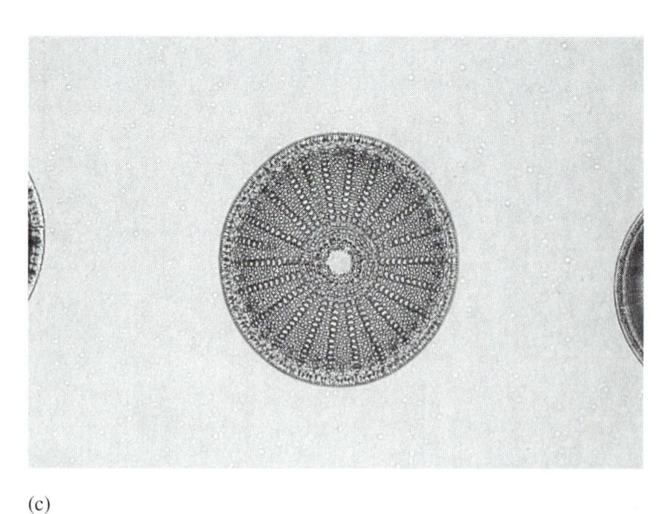

(c)

(d)

Figure 9.3 *Diatoms. (a) and (b) Diatoms are abundant in both freshwater and marine habitats, where they make up the base of the food chain, ×700.(c) The ornamentation on the diatom walls has the appearance of cut crystal when viewed microscopically, ×250. (d) Mining diatomaceous earth in northern Mexico.*

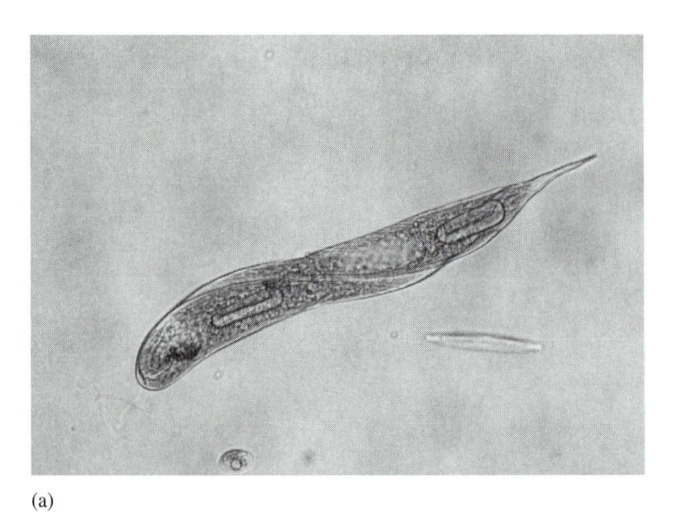

(a)

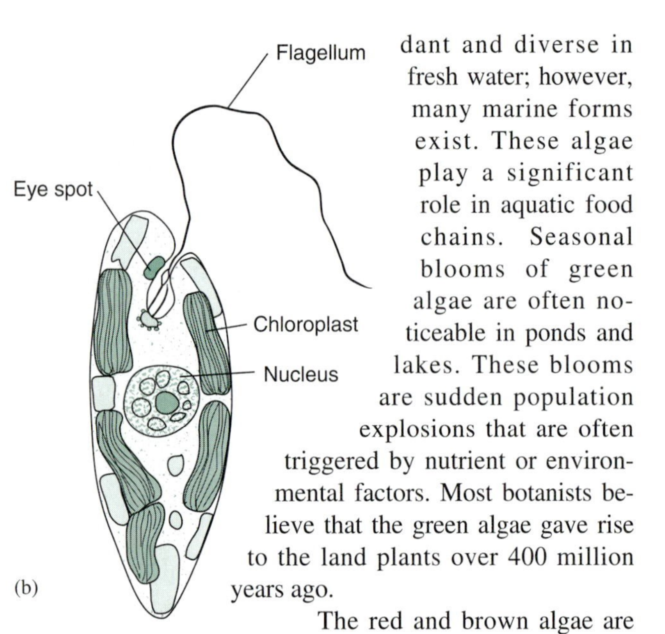

(b)

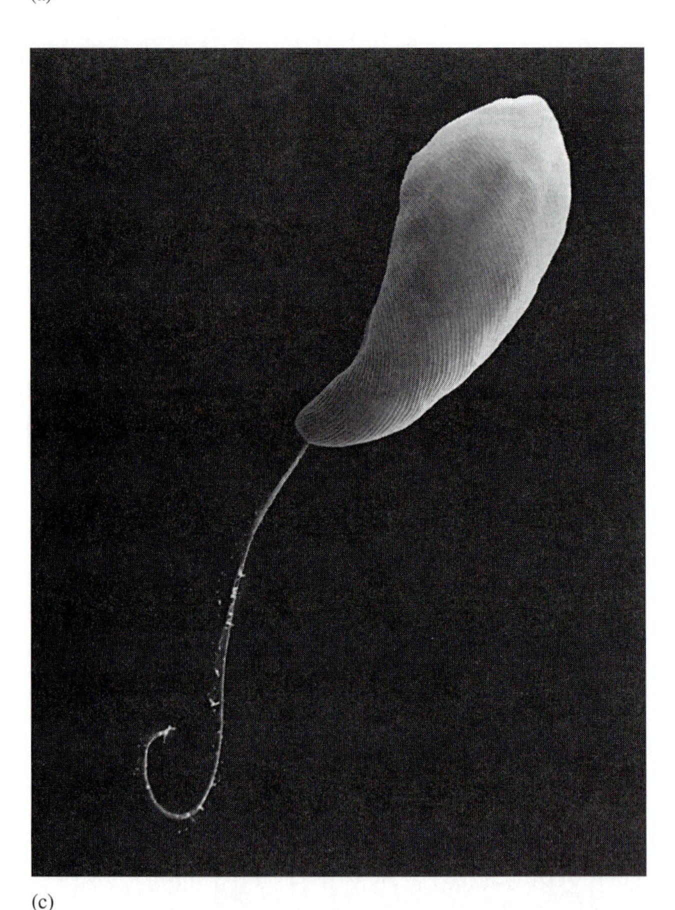

(c)

Figure 9.4 Euglena *shown in a light micrograph ×1400 (a) and (b) drawing, and scanning electron micrograph ×1200 (c).*

one of which is a large red eyespot involved in sensing the direction of light (fig. 9.4). Euglena have two flagella, one short and apparently nonfunctional and the second long and whip-like.

The **Division Chlorophyta** (the green algae) contains a great many morphological types: motile and nonmotile unicells, colonies, branched and unbranched filaments, tubular forms, and sheets (fig. 9.5). The green algae are most abundant and diverse in fresh water; however, many marine forms exist. These algae play a significant role in aquatic food chains. Seasonal blooms of green algae are often noticeable in ponds and lakes. These blooms are sudden population explosions that are often triggered by nutrient or environmental factors. Most botanists believe that the green algae gave rise to the land plants over 400 million years ago.

The red and brown algae are commonly known as seaweeds. They are mainly large multicellular marine forms occurring in coastal waters and often attached to rocks. The red algae, **Division Rhodophyta,** are generally reddish in color and many consist of highly branched filaments, giving them a feathery appearance (fig. 9.6). Other forms are sheetlike. Two economically useful products, **carrageenan** and **agar,** obtained from this group of algae are used as stabilizing agents. Carrageenan is a common ingredient in ice cream, pudding, cottage cheese, weight-loss products, toothpaste, lotions, and paints; this natural additive imparts a creamy texture. Although agar is also used in a variety of commercial products, its most important use is as a solidifying agent in culture media used in laboratory research on bacteria, fungi, and plant tissue culture (*see* Chapter 15).

The **Division Phaeophyta,** the brown algae, includes the huge kelps that form extensive underwater "forests" off the California coast, and the rockweeds commonly found in the intertidal zone in coastal areas. Some species of kelps are among the largest plants on Earth, reaching 100 meters (over 300 feet) in length. The brown algae are the most complex, structurally, of all the algal divisions, almost resembling a land plant in having a rootlike **holdfast,** stemlike **stipe** and leaflike **blade** (fig. 9.6). Several species of brown algae are the source of **alginic acid,** which has broad commercial use in the treatment of latex during tire manufacturing, as a binding agent for charcoal briquettes, and in confections, ice cream, and other products where it acts similarly to carrageenan.

Mosses

The Kingdom Plantae includes a diverse group of organisms ranging from mosses to flowering plants. The **Division Bryophyta** is the first group of land plants to be considered and includes primarily the **mosses** and **liverworts.** These are small plants restricted to moist environments. The

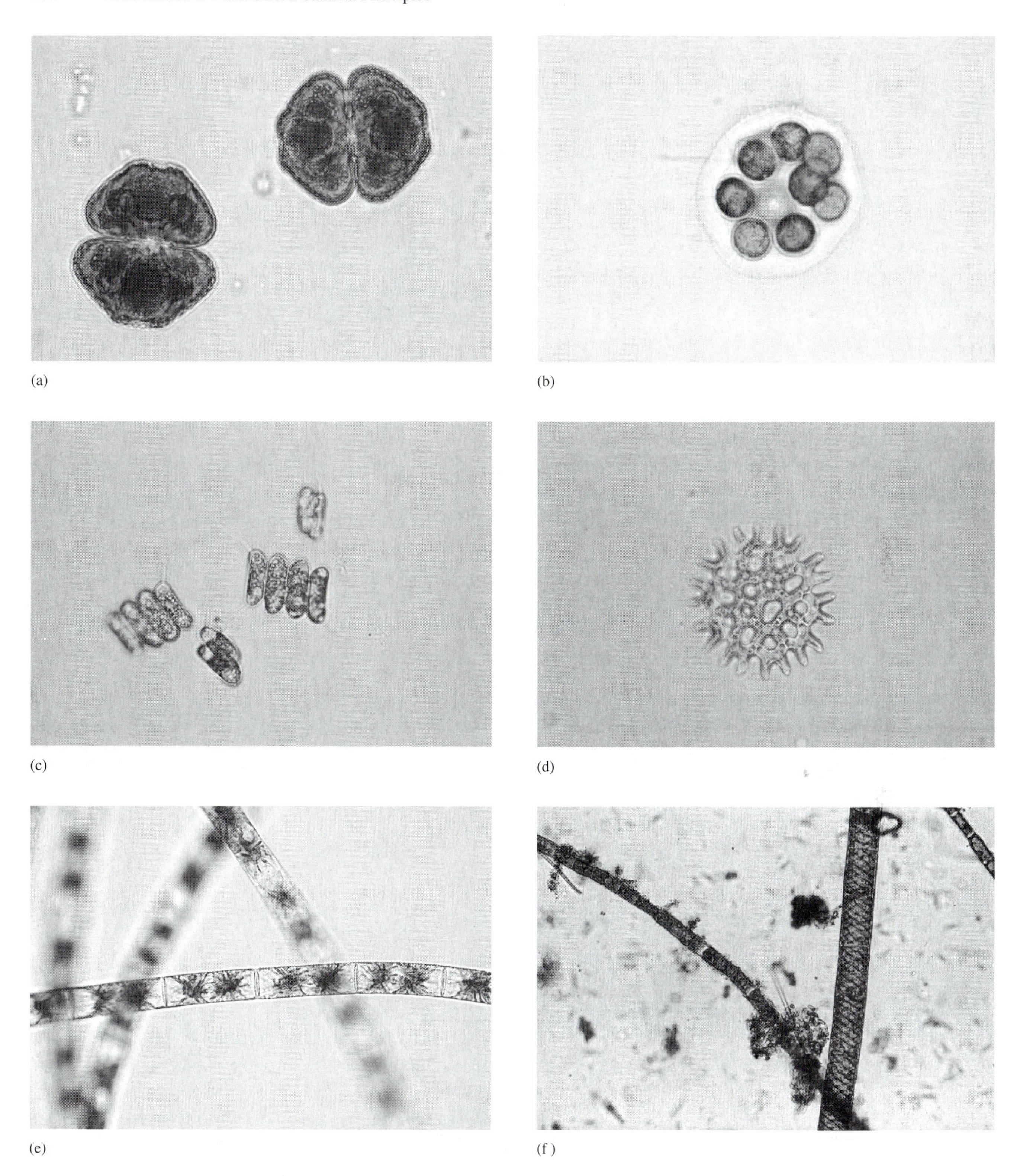

Figure 9.5 *Green algae occur in various growth forms. (a)* Cosmarium, *a unicell, ×600; (b)* Pandorina, *×850; (c)* Scenedesmus, *×700; and (d)* Pediastrum *are all colonial forms, ×750; (e)* Zygnema, *×250 and (f)* Spirogyra *are filamentous genera, ×60.*

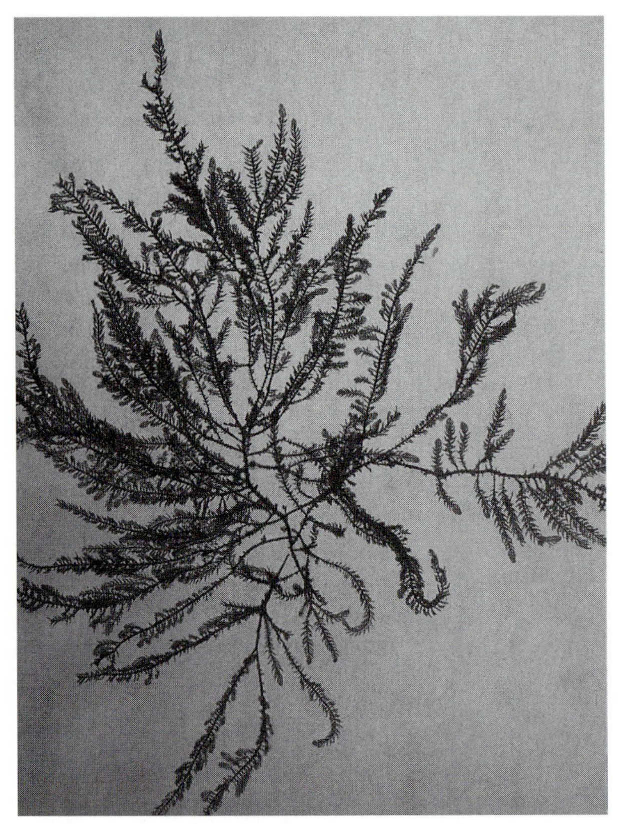

(a)

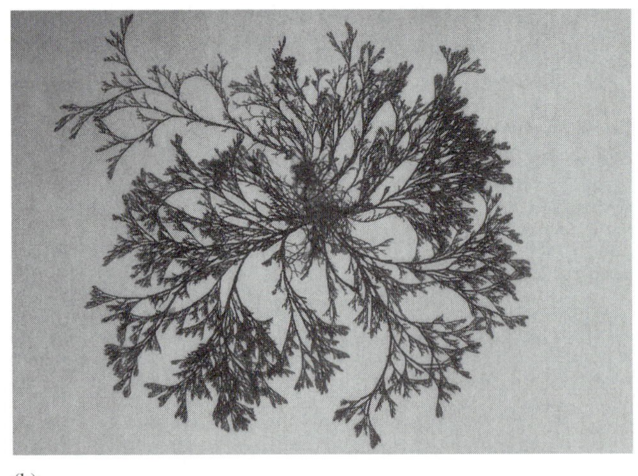

(b)

(d)

(c)

(e)

Figure 9.6 *Seaweeds—red and brown algae. Two red algae common along the northern California coast (a)* Plumaria *and (b)* Microcladia; (c) *Rocky shoreline in New England with numerous seaweeds attached;* (d) Fucus, *or rock weed, is a common brown alga along the New England coast; (e)* Postelsia, *another brown from the Pacific coast.*

A Closer Look 9.1

Seaweed for Supper

In addition to the use of algal extracts in food products, algae have played an important role in the diet of many cultures. It has even been speculated that the "manna" the Israelites consumed during their forty-year sojourn in the desert might have been the cyanobacterium or blue-green alga *Nostoc*. Another example of an edible blue-green alga is *Spirulina,* a corkscrew-shaped filament that has significant quantities of protein, vitamins, and minerals. The ancient Aztecs cultivated mats of *Spirulina* in shallow ponds and lakes; once dried and cooked, it was eaten like a vegetable. This alga has also been a traditional part of meals in Chad, where it is made into a sauce for beans, meat, or fish. Today, *Spirulina* is used to prepare a high-protein food additive that has been considered one solution to the malnourishment problems in Third World countries (box fig. 9.1a). Some scientists are concerned about this current use of *Spirulina* because of the danger of contamination with toxin-producing cyanobacteria.

In coastal areas, the red and brown seaweeds have long been used as a source of food. In Oriental countries, this use dates back to prehistoric times. It has been estimated that over 100 species of marine algae are eaten in one form or another. Among the brown algae, the most popular is kombu (*Laminaria*). Some favorite red seaweeds are dulse (*Rhodymenia*) and nori (*Porphyra*). These can be eaten many different ways: as a confection, as a relish, with rice, or in soups and salads (box fig. 9.1b, c). The nutritional value of seaweeds lies in their high protein content as well as their essential vitamin and mineral content, particularly iodine and potassium. Many of these are available in the United States in health food stores (or at the ocean!) (box fig. 9.1d).

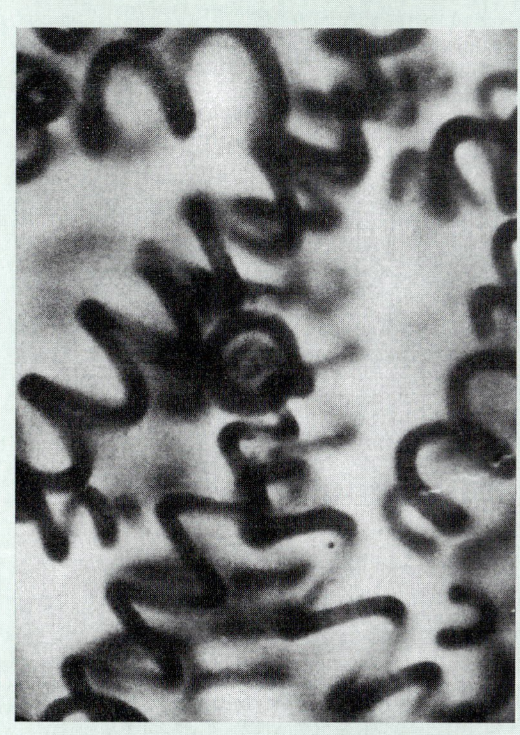

(a)

Box Figure 9.1 *Cyanobacteria have been used as a source of food by many people. (a) Individual filaments of* Spirulina; *(b) A mat of* Spirulina *on a conveyor belt being processed for food in Mexico; (c)* Nostoc *colonies develop in a gelatinous sheath; (d) Edible seaweed.*

mosses have small appendages appearing somewhat similar to leaves, while the liverworts are either leafy or flat and ribbonlike (fig. 9.7). The well-known peat moss, *Sphagnum,* is unusual because it grows in acid water, making the water even more acidic and often converting the pond into a bog. The acidity restricts the growth of other plants and prevents the growth of microorganisms that cause decay. This creates a perfect environment for preservation. Important archaeological finds have been made in peat bogs; several 2,000-year-old fully-clothed bodies have been found in bogs in Denmark. In addition, the 7,000-year-old skeletal remains of American Indians were recently found in Florida. Although most of the flesh was decayed, many of the brains were preserved. Scientists are hoping to examine the DNA from this tissue to see if major genetic changes have taken place over the past 7,000 years.

There are several economic uses for *Sphagnum* as well. Dried peat moss is commonly used as a potting and bedding material by gardeners, since it has a great water-holding capacity. For centuries, the peat bogs of northern Europe, especially those in Ireland, have been the source of a home heating fuel. The peat from these bogs, which is partially decomposed *Sphagnum,* is harvested, dried, (fig. 9.7) and burned in large quantities. As late as World

(b)

(c)

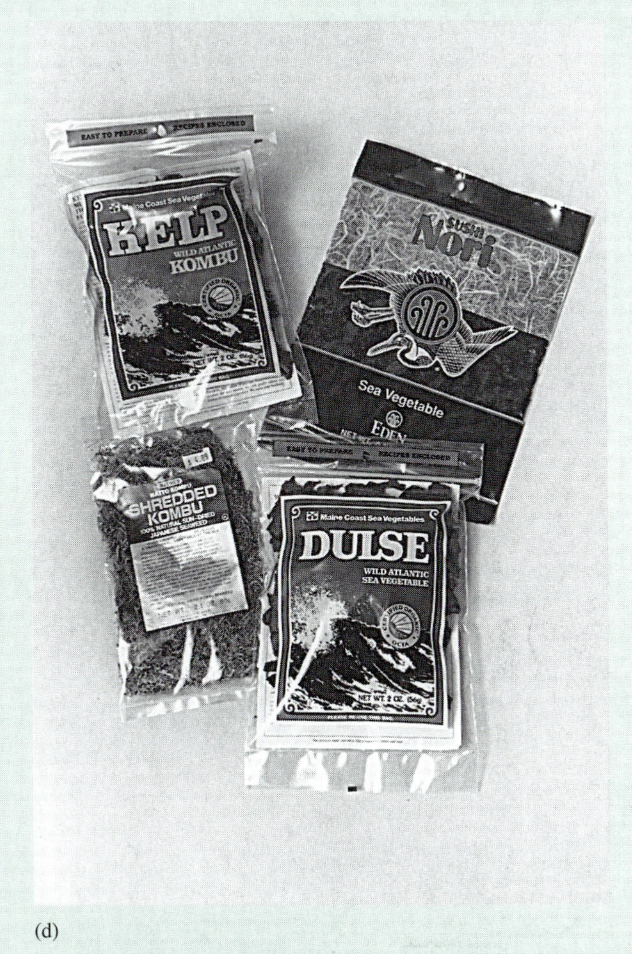

(d)

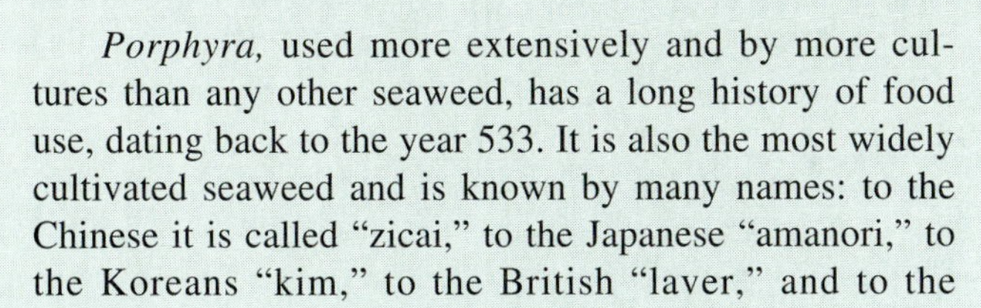

Porphyra, used more extensively and by more cultures than any other seaweed, has a long history of food use, dating back to the year 533. It is also the most widely cultivated seaweed and is known by many names: to the Chinese it is called "zicai," to the Japanese "amanori," to the Koreans "kim," to the British "laver," and to the Maoris of New Zealand it is "karengo." This red alga not only tastes good but is nutritious and enhances the flavor of other foods.

With current interest in developing new foods, perhaps the algae will play a bigger role in the human diet in the future.

War I, *Sphagnum* also was used as a dressing for wounds because of its antiseptic properties. In contrast, most other bryophytes are of limited economic importance.

Fern-Allies and Ferns

The remaining divisions in the Plant Kingdom are all **vascular plants,** i.e., they all contain **vascular tissues** that conduct both water and food throughout the plant body. The evolution of vascular tissue allowed the establishment of land plants in areas where free-standing water was limited. The first four divisions of vascular plants are nonseed plants, reproducing only by **spores,** and include the **ferns** and **fern-allies.** The most primitive vascular plants belong to the Division **Psilophyta** or **whisk ferns.** This is a very small group of simple plants; *Psilotum* consists of green branching stems with tiny scalelike appendages and globose yellow **sporangia** (singular, **sporangium**), structures in which spores develop (fig. 9.8). Scientists have an interest in whisk ferns because they are morphologically similar to the fossils of the earliest vascular land plants.

The **club mosses** make up the Division **Lycophyta.** Today these plants range from small prostrate forms commonly found on the forest floor to larger epiphytic forms in

(a)

(b)

(c)

Figure 9.7 *Bryophytes include mosses and liverworts; (a) Hairy cap moss,* Polytrichum, *and (b)* Marchantia. *Gametes are produced in the umbrellalike structures shown arising from the flattened liverwort. (c) A peat bog being drained.*

Figure 9.8 *Branching stem of* Psilotum.

the tropics. (**Epiphytes** are plants hanging from other larger forms of vegetation.) Superficially these plants resemble mosses because their stems are covered with small overlapping scalelike leaves; however, they are not mosses, but vascular plants with conducting tissue occurring in the roots, stems, and leaves. Two to three hundred million years ago there were treelike species of lycopods that formed extensive forests. These prehistoric forests are the basis for many of our coal deposits today (fig. 9.9).

In many species of *Lycopodium,* a member of the Division Lycophyta, the sporangia are arranged in compact clusters at the ends of erect stems. The clusters of sporangia often have a clublike appearance (fig. 9.9). Creeping jenny and ground pine are two well-known club mosses often used in Christmas wreaths. In the early days of photography, *Lycopodium* spores were the original flash powder. *Lycopodium* powder was also used in the past as a type of talcum powder, as a coating for pills, to stop bleeding from wounds and for other medicinal proposes. Another common plant in this division is the resurrection plant, *Selaginella lepidophylla,* a native of arid regions, which appears as a dried brown clump when water is scarce yet quickly becomes green and photosynthetic when water is available. This plant is often sold as a novelty in shops.

Today the Division **Sphenophyta** contains only one genus, *Equisetum,* the **horsetails.** These plants have ribbed, jointed, photosynthetic stems with whorls of tiny leaves that soon become brown and nonphotosynthetic. Sporangia are grouped into conelike structures at the tips of some stems. Although most horsetails are small, some species may be tall, 2 meters (6 ft) or more, but have very slender stems (fig. 9.10). Like the Division Lycophyta, tree-sized

Figure 9.9 *Reconstruction of a Carboniferous forest from 300 million years ago. The large trees on the left and lying in the foreground are tree-sized lycopods. Those on the right are sphenopsids related to modern horsetails.*

(a)

(b)

Figure 9.10 *(a)* Lycopodium, *a club moss; (b)* Equisetum, *the only living genus in the Sphenophyta.*

A Closer Look 9.2

Alternation of Generations

One characteristic of plants, including those organisms in the kingdoms fungi and protista that were traditionally classified as plants, is a life cycle with an alternation of **haploid** and **diploid** generations. Each cell in the diploid generation has two sets of chromosomes, while the cells in the haploid generation have only one set. The haploid generation is called the gametophyte because it gives rise to gametes; the sporophyte, which produces spores, is the diploid generation. The links between the two generations are the processes of fertilization and meiosis. During fertilization, haploid gametes fuse to form a diploid zygote, which develops into the sporophyte. In time the diploid sporophyte undergoes meiosis to produce haploid spores, which begin the gametophyte generation (box fig. 9.2a). The prominence of each generation is variable among the different divisions. For example, the gametophyte is highly reduced and dependent on the sporophyte in angiosperms, but in some nonflowering plants the gametophyte generation may be prominent and often independent.

One division in which the gametophyte is conspicuous is the Bryophyta, the mosses and liverworts. In fact, in this group the gametophyte is the dominant form; the mossy carpet often seen growing on a rock or tree trunk consists of gametophytes. At certain times of the year the moss undergoes sexual reproduction, resulting in a sporophyte that is attached to and dependent on the gametophyte. In mosses the gametophytes are either male or female. The male gametophyte develops spherical structures known as **antheridia** (singular, **antheridium**) where the sperm form. The female gametophyte produces flask-shaped **archegonia** (singular, **archegonium**); each contains a single egg cell. When mature, the flagellated sperm will leave the antheridium and, if sufficient moisture is present, will swim to the archegonium of a female gametophyte. The sperm will fertilize the egg, forming a zygote, which establishes the diploid sporophyte. The sporophyte consists of a stalk and **capsule;** the stalk is embedded in the female gametophyte and receives both water and nutrients from it. Within the capsule meiosis occurs, generating haploid spores. If the spores released from the capsule land on a suitable environment, each will germinate into a threadlike structure called a protonema. From the protonema the mature moss gametophyte develops. The moss life cycle is shown in Box Figure 9.2b.1; similar life cycles exist for liverworts.

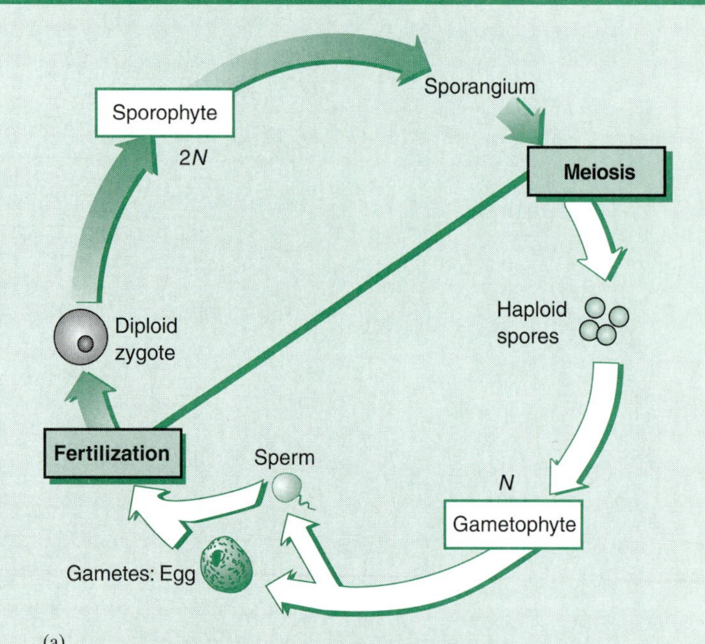

(a)

Box Figure 9.2 *(a) Alternation of generations. Plants display an alternation of haploid and diploid phases in their life cycles; (b.1) Moss life cycle; (b.2) Moss sporophytes, (c) Fern life cycle; (d) Pine life cycle.*

The lower vascular plants in the divisions Psilophyta, Lycopohyta, Sphenophyta, and Pterophyta all share a similar life cycle with a dominant sporophyte and a small but free-living gametophyte. The ferns will be used as a model for this group (box fig. 9.2c). Unlike the mosses, it is the sporophyte stage of ferns that is the dominant form. In the fern sporophyte, sporangia generally form on the underside of the leafy frond. Sporangia are often clustered into **sori** (singular, **sorus**), which are visible with the naked eye; meiosis occurs in these sporangia, producing haploid spores. In a hospitable environment, spores give rise to small, flat, often heart-shaped gametophytes that bear both archegonia and antheridia. Flagellated sperm from the antheridia swim to the archegonia, fertilizing the eggs; however, only one sporophyte will emerge from each gametophyte. Although initially attached to the gametophyte, the sporophyte soon grows and becomes independent while the gametophyte dies.

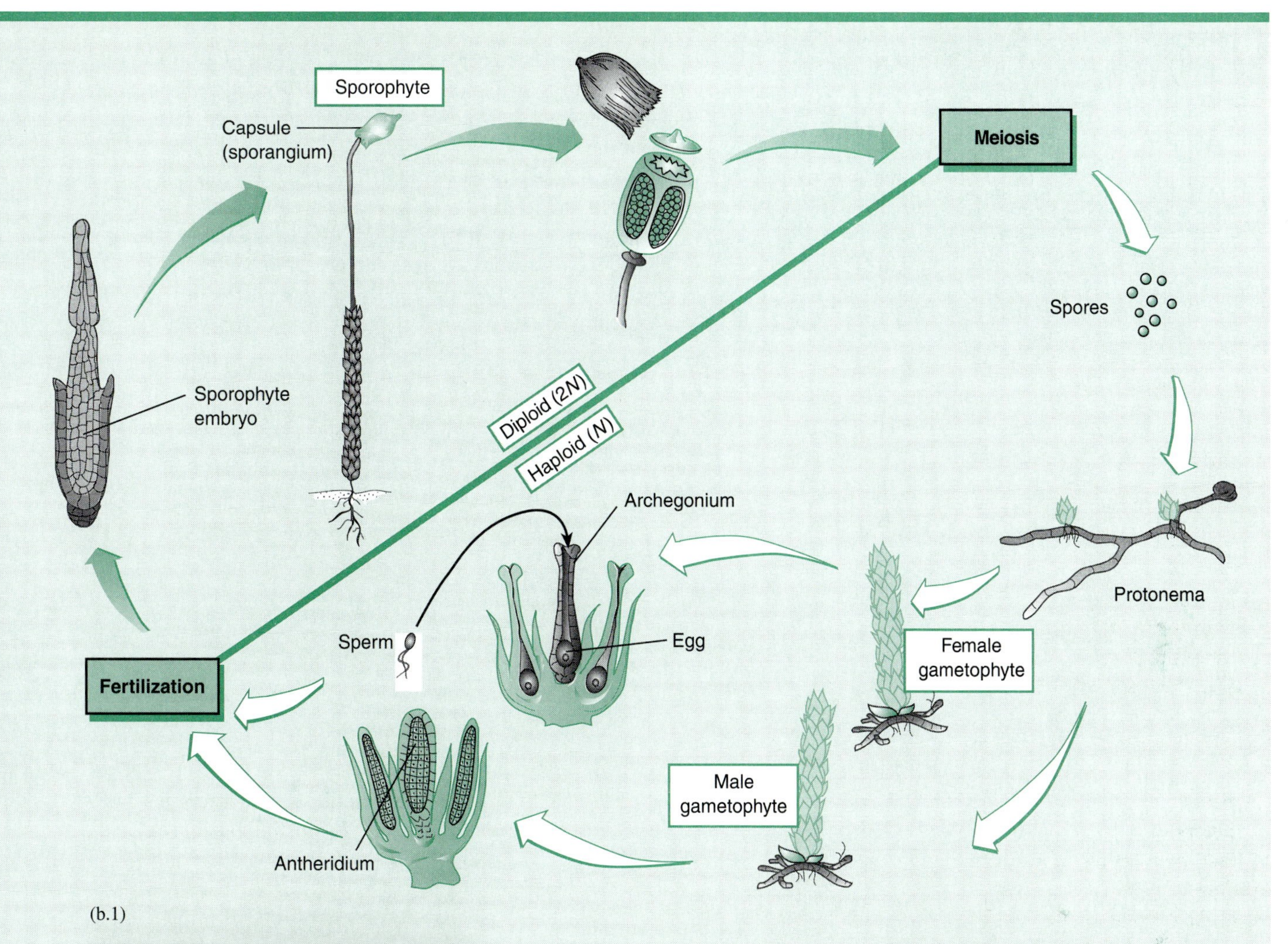

(b.1)

(b.2)

Like the angiosperms (*see* Chapter 5), the gymnosperms have an extremely reduced, dependent gametophyte generation, with the sporophyte generation being the dominant and familiar form. A pine in the Division Coniferophyta will be used as a example of the gymnosperm life cycle (box fig. 9.2d). The characteristic pollen and seed cones are the reproductive structures of a pine. The pollen cone consists of sporophylls, which are modified leaves that bear **microsporangia;** two occur on each sporophyll. Microspores are produced in these microsporangia. The microspore develops into the pollen grain that constitutes the male gametophyte generation. At the time the pollen grain is released from the cone, it consists of four cells: two body cells, one generative cell, and one tube cell. Pines are wind-pollinated; air bladders in the walls of the pollen grain contribute to their buoyancy. The pollen grains are carried to the seed cone, where they are trapped by a sticky fluid. The pollen grain germinates with the tube cell producing a pollen tube that will eventually carry nonflagellated sperm to the egg.

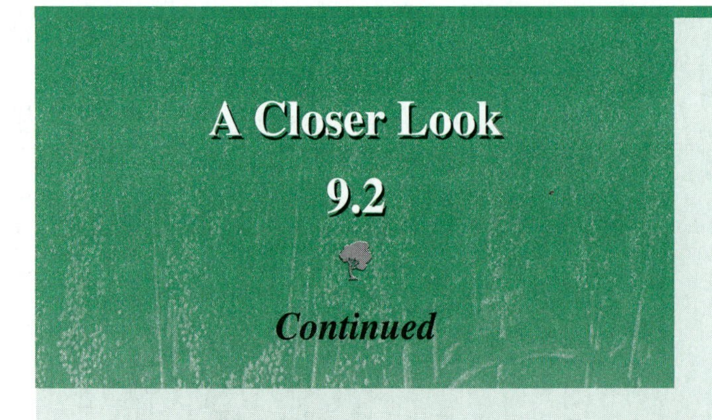

The seed cone is larger and more complex structurally than the pollen cone. Ovules develop on the upper surface of cone scales, which are arranged around a central axis. Normally two ovules are borne on each cone scale. Recall from the life cycle of angiosperms that the ovule is a **megasporangium** surrounded by integuments with an opening, the micropyle, facing the central axis. Meiosis produces four megaspores within each ovule, but three degenerate. The one surviving megaspore develops into the female gametophyte. At maturity the female gametophyte

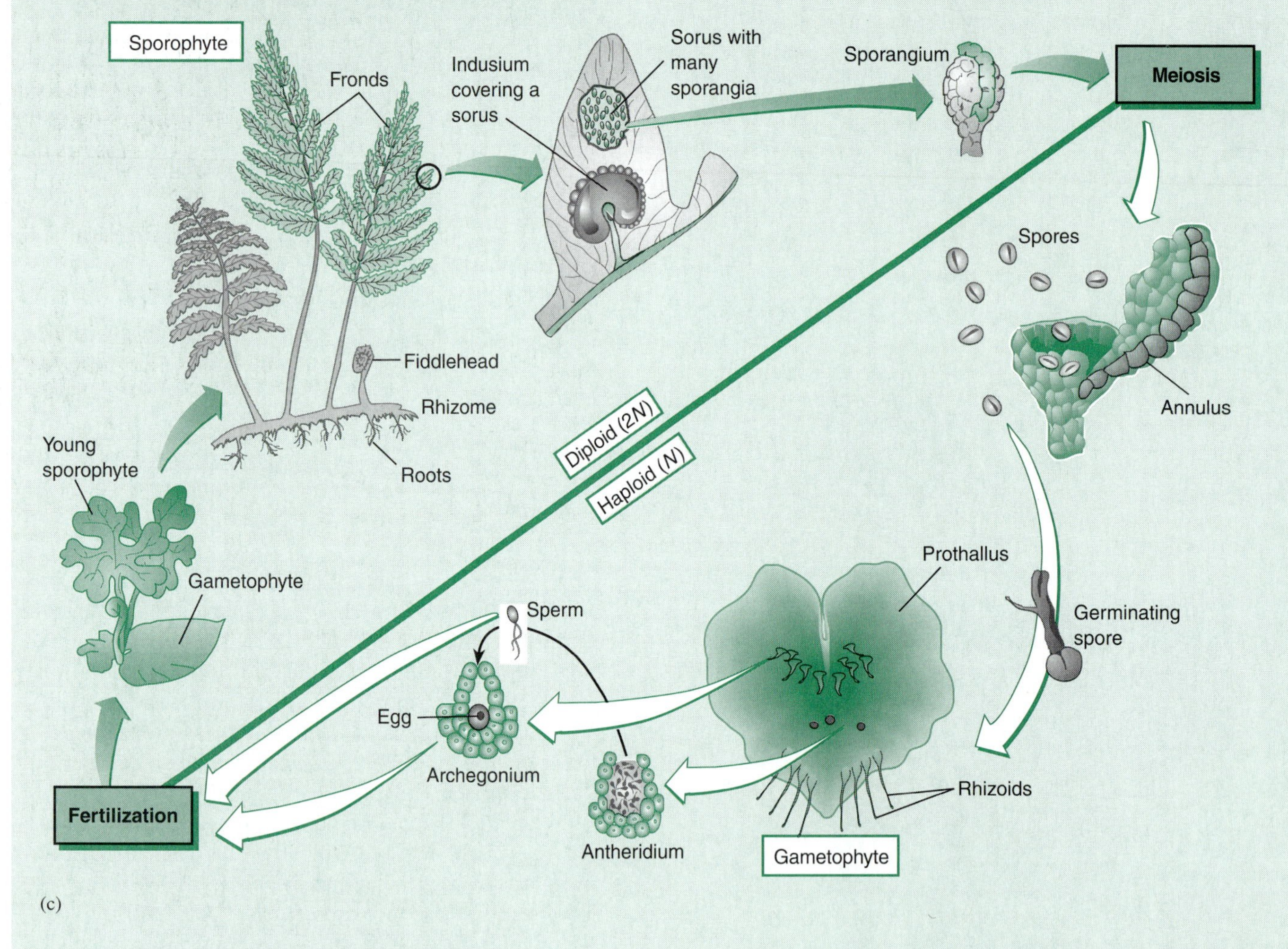

(c)

species existed 2 to 3 hundred million years ago and also contributed to coal deposits that are mined today. One interesting feature of the horsetails is the presence of silica in their cell walls. The silica in the walls makes the stems abrasive and, for this reason, pioneers in North America used these plants as a primitive scouring pad to clean pots and pans. People in developing nations still use them for this purpose. This explains another common name for these organisms—the scouring rushes. Musicians also use the scouring rushes to sand the reeds on wind instruments.

The ferns, Division **Pterophyta,** are the largest group of nonseed vascular plants. Ferns range from small aquatic types less than 1.25 cm (0.5 in) in size to large tropical tree ferns that grow to 20 meters (60 ft) in height. The ferns that we are most familiar with, from temperate forests and as houseplants, have large divided leaves

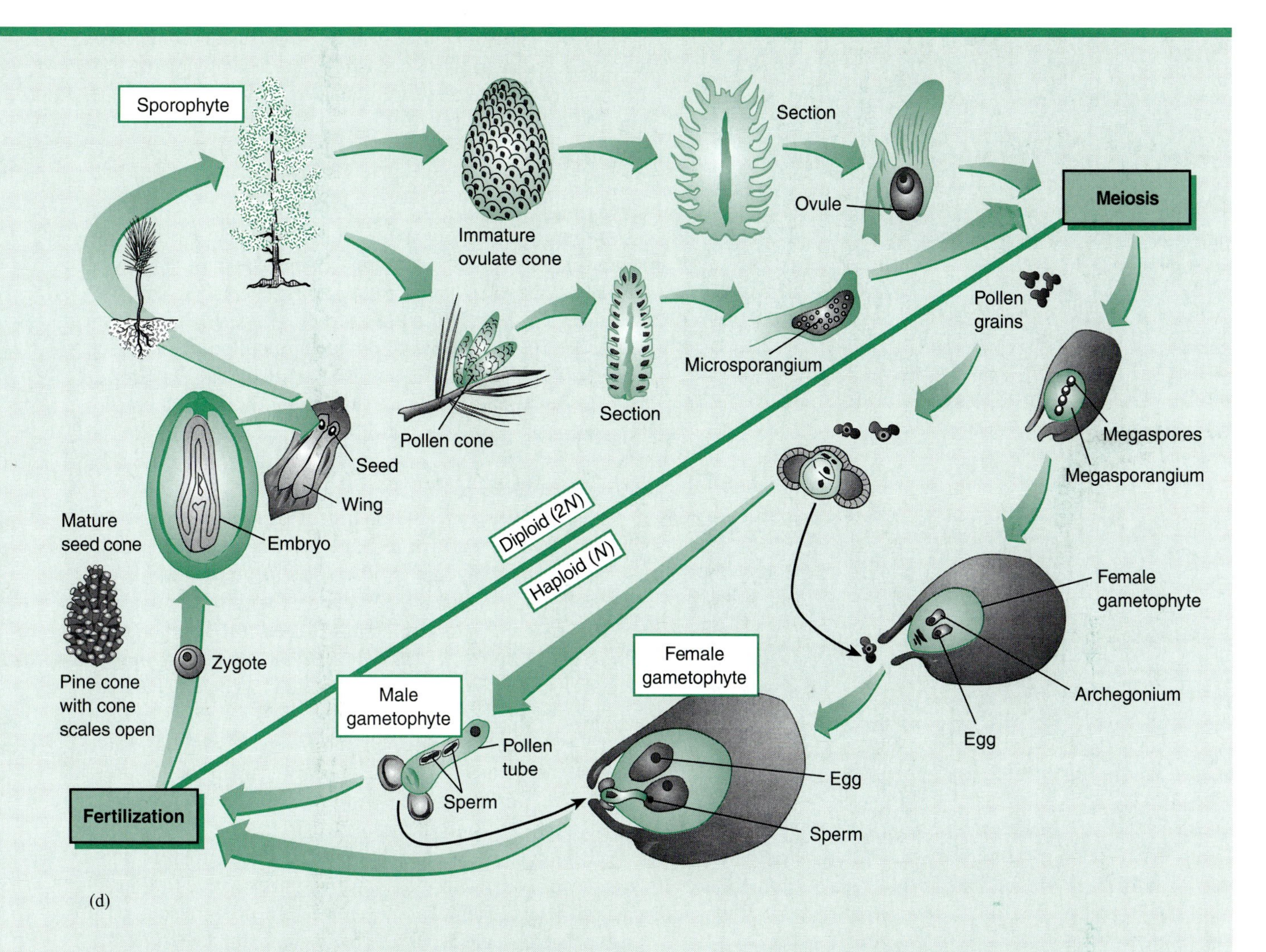

(d)

contains two archegonia, each with a single egg. While the female gametophyte is developing, the pollen tube is slowly growing through the surrounding tissues. During this time the generative cell within the pollen tube divides to form two sperm. Although both nonmotile sperm are carried to the archegonium, only one sperm fertilizes the egg, giving rise to the new sporophyte generation. Within the ovule only one embryo is produced, with the female gametophyte as its nutrient tissue. As the fertilized ovule develops into the seed, the integuments develop into the seed coat. When the seeds are mature, the seed cone opens and sheds the seeds, which are wind dispersed.

The three life cycles described here illustrate the variation of the basic alternation of a haploid generation with a diploid generation found in the plant kingdom.

called fronds that arise from a horizontal underground stem, a rhizome, which also gives rise to (fig. 9.11) roots. The sporangia of some ferns are borne on the underside of the fronds; the distribution of sporangia on the frond is a characteristic that can be used as an aid in identification. Young fern leaves first appear as tightly coiled fiddleheads, which gradually unroll as they grow. The fiddleheads of some ferns are considered a gourmet delicacy. Again, ferns were more abundant during past geological times and contributed to our coal deposits.

Gymnosperms

The remaining five divisions in the plant kingdom are all **seed plants.** The seed plants are the dominant vegetation in the world. Recall that a seed contains a small embryonic plant with a food supply encased in a tough protective coat.

(a)

(b)

(c)

Figure 9.11 *Ferns. (a) Fiddleheads unrolling in springtime; (b)* Osmunda cinnamomea, *the cinnamon fern and (c)* Osmunda claytoniana, *the interrupted fern.*

Seeds are found in either cones or fruits. In cones, the seeds are exposed at maturity and said to be "naked." Plants with naked seeds are called **gymnosperms,** while plants with seeds enclosed or hidden within fruits are called **angiosperms.**

There are four divisions of gymnosperms: **Coniferophyta, Cycadophyta, Ginkgophyta,** and **Gnetophyta. Conifers,** members of the Coniferophyta, are the most familiar gymnosperms and include pines, spruces, yews, firs, cedars, redwoods, and larches. The largest, tallest, and oldest trees in the world are all conifers (fig. 9.12). Conifers are cone bearers, typically producing both pollen and seed cones. The pollen cones are usually small and inconspicuous and produce pollen while the seed cones, which produce seeds, are larger and often take several years to mature (fig. 9.13). The pollen and seed cones may be on the same or different trees.

The conifers are the dominant trees in the northernmost forests of the world. They are important sources of lumber, paper pulp, and products such as turpentine, rosin, and pitch (*see* Chapter 18). Although conifers are not usually considered as food, pine nuts, seeds of the pinyon pine, are enjoyed by many. Also, some types of cedars, *Juniperus* spp., are essential in the production of gin, a popular alcoholic beverage. The seed cones of these cedars are unusual because they are small and fleshy, resembling waxy blue berries (fig. 9.13). These "berries" impart the flavoring to gin. Next time you see a *Juniperus,* pick a few "berries," crush them, and sniff the aroma of gin.

The **cycads** (Cycadophyta) are a small group of short shrubs to moderate-sized, long-lived trees native to tropical and subtropical regions. These trees are often mistaken for palms because the leaves are large and palmlike (fig. 9.14). Like the conifers, most cycads bear cones. The seed cones of certain species can be quite large; some weigh up to 100 pounds! The sago palm, a cycad commonly used for landscaping in subtropical to tropical climates, is also a popular houseplant.

Ginkgo biloba, the maidenhair tree, is the only living species of the Ginkgophyta. It is a moderate-sized tree with unique fan-shaped leaves that may or may not be notched in the middle (fig. 9.14). Although rarely, if ever, found in the wild, ginkgo has been cultivated for centuries in the Orient. Today ginkgo is a popular tree in city landscaping because it is air-pollution-tolerant and hardy. Ginkgo seeds are partially surrounded by a fleshy coat that smells like rancid butter when the seed is mature. For this reason efforts are made to screen for pollen-bearing trees before planting.

The Gnetophyta is a very small group of gymnosperms with unusual morphological and anatomical features that have intrigued botanists for years. *Ephedra* is the only economically important member of this group. This desert shrub produces the alkaloid ephedrine, which is useful in the treatment of bronchial asthma, sinusitis, the common cold, and hay fever (fig. 9.14).

(a)

(b)

Figure 9.12 *(a) Bristlecone pine,* Pinus longaeva. *Individual trees of this species may live 3,000 to 5,000 years, making them the oldest living trees. (b) Coast redwood,* Sequoia sempervirens; *as a group, the coast redwoods are the world's tallest trees, often reaching heights over 110 meters (350 feet).*

(a)

(b)

Figure 9.13 *(a) Seed cone of* Pinus longaeva, *bristlecone pine. (b) Seed cones of* Juniperus asheii, *mountain cedar. The small berry-like cones of* Juniperus *are a sharp contrast to the seed cones of* Pinus.

(a)

(b)

(c)

Figure 9.14 *(a)* Dioon edule, *a cycad growing in the Fairchild Tropical Gardens in Florida. (b)* Ginkgo biloba, *the maidenhair tree, branch with leaves and seeds. (c)* Ephedra viridis, *a desert shrub.*

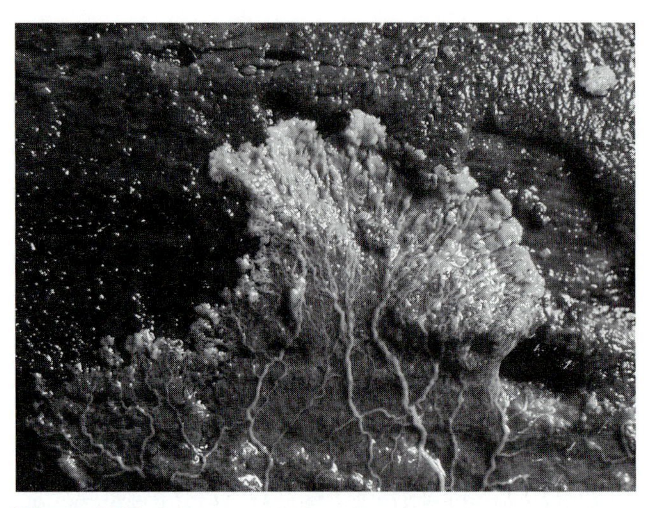

Figure 9.15 *Slime molds. Vegetative stage of a slime mold is an amoeboid plasmodium that moves along the forest floor feeding on organic matter or bacteria.*

Angiosperms

The last remaining division in the Plant Kingdom is the **Anthophyta,** the flowering plants or angiosperms. The angiosperms constitute the most widespread vegetation on Earth today, ranging in form from small herbaceous plants to large trees. This division includes most of our familiar plants such as lawn grasses, crop plants, vegetables, weeds, and houseplants, as well as oaks, elms, maples, and many other trees. Over 250,000 species of angiosperms have been described, and it has been suggested that as many as 1,000,000 undescribed species may exist in tropical forests. In the previous eight chapters we have focused on the anatomy, reproduction, and physiology of angiosperms. In the chapters that follow we shall see that most of our economically useful plants are, in fact, angiosperms.

Fungi

Mushrooms, molds, yeasts, and mildews are all fungi. Organisms traditionally called fungi are now placed in two kingdoms, Kingdom Protista and Kingdom Fungi. The Protista includes several divisions of fungi that evolved along separate evolutionary paths; most fungi are classified in the Kingdom Fungi. The Myxomycota, or **slime molds,** are an obscure but biologically interesting group of protists since they have characteristics that straddle the plant and animal worlds. The vegetative body is an animal-like amoeboid form but the reproduction is by spores (fig. 9.15).

The Kingdom Fungi is a large and diverse group consisting of several different types of organisms. They are generally characterized by threadlike structures called **hyphae.** All the hyphae are collectively called the **mycelium.** The mycelium grows in the substrate and is generally hidden (fig. 9.16). Perhaps you have encountered the mycelium as fine white threads when digging in moist soil. The mycelium can also be found in decaying plants and animals, as well as living hosts. Fungi reproduce by spores

A Closer Look 9.3

Amber: A Glimpse into the Past

Many economically useful products are derived from conifers; much of the lumber and paper industry (*see* Chapter 18) is based on pine, spruce, and other coniferous species. Resin is another product that is usually obtained from conifers; it is the material often seen oozing from the trunk of a pine tree. Resins are sticky secretions that harden as they dry. Wounding a plant initiates the release of resins that promote healing by sealing off a wound against water loss, pathogens, and insects. Resins are best known as the source of turpentine and rosin (*see* Chapter 18).

If conditions are right and the resin is buried before it can oxidize, it fossilizes. Amber is fossilized tree resin. Translucent and typically golden-brown in color, amber has been valued as a semiprecious stone for millennia. Stone Age figurines of amber have been found in the Baltic area; they may have been talismans. Roman soldiers wore amber-studded armor for good luck, and the emperor Nero established trade routes with the Germanic tribes of the North to obtain it. The pinnacle of amber artistry was the famed Amber Room in the Catherine Palace of Russia. A gift from the Prussian King Frederick Wilhelm I to the Russian Czar Peter the Great, the room was lined with intricately carved pieces of amber fitted together on wooden wall panels. The panels were disassembled and stolen by the Nazis during World War II. The reassembled panels were later put on display in Königsberg, Germany but disappeared during the last days of the War. The fate of the panels remains a mystery.

Amber has been gathered for centuries on the shores of the Baltic Sea, an area well-known for its rich deposits. As waves erode away the shoreline, amber is washed free and can be found floating in the water. The largest amber-producing mines are located in the town of Palmnicken on the Baltic Sea. Approximately 30 million years ago, this area was subtropical in climate with forests of conifers. The resins of these conifers are the source for the 700 tons of amber mined each year. Amber-rich beds are found 40 m (120 ft) below the surface; as much as 4.05 kg (7.7 lbs/cubic yd) can be found per cubic meter. The Earth is strip-mined to the amber beds; the sediments are then washed with water from huge pressurized hoses and the amber separated out. Only 13% of the amber is useable for jewelry; the rest is used in paint thinners, varnishes, and polishes.

There are other amber deposits in many parts of the world, dating from the late Carboniferous to the Pleistocene (300 to 1.5 million years ago). The oldest amber hails from the central Appalachian region in the eastern United States. Amber from the Dominican Republic holds the record for the greatest number of inclusions, encasing insects, spiders, bits of wood, flowers, seeds and leaves, even feathers. For every 100 pieces of amber from the Dominican Republic, one piece contains an insect fossil; in Baltic amber, the ratio drops to a mere one in 1,000.

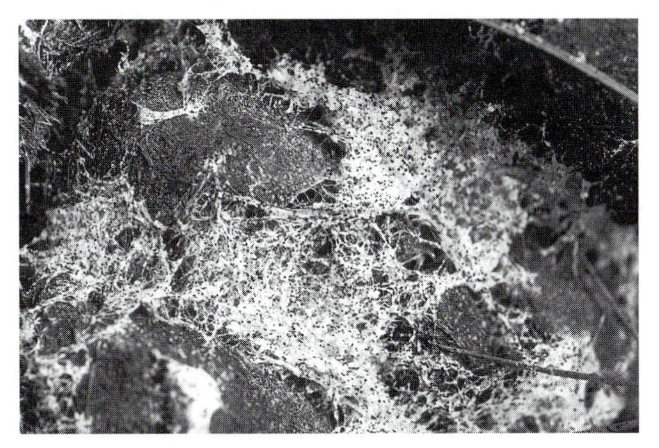

Figure 9.16 *True fungi are characterized by threadlike hyphae, which make up the mycelium. The underground mycelium of a mushroom is visible in the soil.*

and when conditions are right, the mycelium gives rise to spore-forming structures. Sometimes the spore-forming structures are large and conspicuous such as mushrooms, puffballs, and bracket fungi. The mycelium you encounter in your garden soil may give rise to the mushrooms that appear after sufficient rainfall.

Other than the angiosperms, the fungi are the most economically important group in both beneficial and deleterious ways. In addition to edible mushrooms, fungi are used in the production of bread, alcoholic beverages, many cheeses, and many antibiotics. On the other hand, pathogenic fungi are responsible for devastating crop losses that have impacted human history, and debilitating diseases of humans and domestic animals. Because of their economic impact, the fungi will be considered in detail in later chapters (*see* Unit VI).

A Closer Look 9.3

Continued

Recently, amber and the biological specimens trapped within, have been the focus of much interest because of the possibility of extracting and deciphering the entrapped fossil DNA. Amber is especially good at preserving ancient DNA because the resin contains compounds that dry and fix living tissue and, at the same time, inhibit bacterial decomposition. Forty-million-year-old fungus gnats were so perfectly preserved in amber that organelles such as mitochondria, ribosomes, and cell membranes could be clearly seen in the abdominal cells. The first extraction of amber DNA was from a species of extinct bee. Since that first extraction, the genetic material of a 120-million-year-old weevil and that of a 30-million-year-old extinct species of termite have also been isolated.

One problem with the technique is contamination with extant DNA. One research team had thought they isolated fossil DNA only to find out it was contaminated with the DNA from modern species. The method of extraction presents another problem. The amber is cracked and destroyed in order to scrape out the specimen's DNA. Less destructive methods are being investigated such as drilling a tiny hole in which a small needle is inserted so that only a portion of the insect is extracted and the amber is saved.

Box Figure 9.3 *Fossilized in amber, a 100-million-year-old ant reveals new information about the evolutionary lineage of ants.*

Once the DNA has been isolated, it can be sequenced to determine the base code. This information will be invaluable in tracing evolutionary lineages (box fig. 9.3). It may even be possible to some day reconstruct long extinct species from a deciphered genetic blueprint. The plot of the science fiction book and movie *Jurassic Park* is based on this possibility. Right now reconstructing dinosaur DNA is remote, since the most successful isolation of amber-fossilized DNA has only sequenced 200 base pairs. Even a bacterium contains nearly 10 million base pairs and humans have close to 6 billion. Nevertheless, amber has opened the gateway to looking at fossil DNA and examining the evolutionary past of ancestral and fossil organisms.

Chapter Summary

1. Currently most biologists accept the Five-Kingdom System of classification first proposed by Robert H. Whitaker in 1969. In this system, living organisms are classified as Monera, Protista, Animalia, Plantae, and Fungi. Organisms traditionally known as plants are found in four of the five kingdoms.
2. The Kingdom Monera consists of organisms with prokaryotic cells, including cyanobacteria, also known as blue-green algae. These primitive organisms are common in oceans, freshwater, and even terrestrial environments. Some cyanobacteria have proven to be useful as food or food supplements.
3. Several divisions of algae and simple fungal-like organisms are included in the Kingdom Protista. Among the algae are the diatoms and the dinoflagellates, which serve a vital role in both marine and freshwater food chains as the food for other organisms. Some dinoflagellates have gained notoriety as the cause of Red Tides and other problems due to toxin production. The greatest diversity of form is found among the green algae, which also play a major role in aquatic food chains. The red and brown algae, commonly called seaweeds, include large multicellular marine organisms that provide products for both food and industrial applications. Some seaweeds are

widely used directly as food, while others are used as sources of alginic acid, carrageenan, and agar. Fungal-like organisms in this kingdom include slime molds, organisms with a vegetative body having an animal-like amoeboid form. Like fungi, however, reproduction in the slime molds is by spores.

4. The Kingdom Plantae contains land plants that manufacture their own food through photosynthesis. All plants have an alternation of generations. Some plants, such as mosses, have a dominant gametophyte, while all vascular plants have a dominant sporophyte. Although the fern allies are small and often inconspicuous plants today, in past geological ages they dominated the landscape. The seed plants are currently the dominant vegetation in the world. Four divisions of seed plants are gymnosperms. Conifers are the most familiar gymnosperms and include the largest, tallest, and oldest trees in the world. Conifers are of great economic importance as the source of wood, pulp, and chemicals. The angiosperms constitute the most widespread and diverse vegetation today. They provide most of the food for life on the planet, as well as economically useful products.
5. The organisms in the Kingdom Fungi cannot make their own food and rely on external sources of nutrition. The fungi, from molds to mushrooms, are absorptive heterotrophs, secreting enzymes into their surroundings that break down food, which is then absorbed. Fungi generally have a threadlike body and reproduce by spores. Next to angiosperms, fungi are the most economically important group in both beneficial and deleterious ways.

Review Questions

1. How is the Five-Kingdom System a better representation of the diversity of life than a two-kingdom system?
2. Organisms traditionally called plants are now assigned to four kingdoms. Describe which groups have been reassigned and why.
3. Describe the divisions of seed plants.
4. What are the basic features of the ferns and fern-allies? What is the economic impact of these groups, both living and fossil?
5. Describe the economic roles of the various groups of algae.
6. Describe the alternation of generations in vascular plants.
7. Define sporangium, archegonium, antheridium, sporophyte, and gametophyte.
8. Discuss the economic and ecological value of the gymnosperms.

Further Reading

Anderson, Donald M. 1994. Red Tides. *Scientific American,* 271 (2): 62–68.

Bold, H. C., C. J. Alexopoulos, and T. Delevoryas. 1987. *Morphology of Plants and Fungi,* 5th Edition, Harper & Row, NY.

Carmichael, Wayne W. 1994. The Toxins of Cyanobacteria. *Scientific American,* 271 (1): 78–86.

Corner, E. J. H. 1964. *The Life of Plants.* University of Chicago Press, Chicago, IL.

Gifford, E. M. and A. S. Foster. 1989. *Morphology and Evolution of Vascular Plants,* 3rd Edition, W. H. Freeman & Co., New York.

Langenheim, Jean H. 1990. Plant Resins. *American Scientist,* 78: 16–24.

Pritchard, Hayden N. and Patricia Bradt. 1984. *Biology of Nonvascular Plants.* Times Mirror/Mosby, St. Louis, MO.

Ross, John. F. 1993. Treasured in Its Own Right, Amber Is a Golden Window on the Long Ago. *Smithsonian,* 23 (10): 30–41.

Scagel, R. F., R. J. Bandoni, J. R. Maze, G. E. Rouse, W. B. Schofield, and J. R. Stein. 1984. *Plants: An Evolutionary Survey.* Wadsworth Publishing Co, Belmont, CA.

Sze, Phillip. 1993. *A Biology of the Algae,* 2nd Edition, Wm C. Brown, Dubuque, IA.

UNIT

III

Plants as a Source of Food

Plant crops, like this field of soybeans, are invaluable food sources to the world's peoples.

10

Human Nutrition

Chapter Outline

Chapter Concepts

1. Human nutritional needs are supplied by macronutrients (carbohydrates, proteins, and fats) and micronutrients (vitamins and minerals).
2. If nutritional requirements are not satisfied, deficiency diseases can result that have widespread effects in the bodily systems.
3. Plants can supply the majority of human nutritional requirements and there is evidence that increasing the proportion of plant foods in the diet can have positive health benefits.

New ideas in nutrition are quickly incorporated by the health conscious segment of society and those advertisers looking for a new marketing gimmick. In the past decade, the benefits of several nutritional concepts such as fiber, oat bran, and low-fat diets have made headlines and influenced life-style changes. All of these promise better health and are, in fact, dependent on a greater consumption of plants in the human diet. This chapter will examine human nutritional needs and how plants can satisfy these needs.

Macronutrients

The basic nutritional needs of humans are to supply energy and raw materials for all the various activities and processes that occur in the body. In addition to the need for water, humans require five types of nutrients from their food supply; three of these are required in relatively large amounts and are called **macronutrients,** consisting of **carbohydrates, proteins,** and **fats.** The other two types of nutrients, **vitamins** and **minerals,** are required in small amounts and are known as **micronutrients.** If water is removed, the macronutrients would make up almost all the dry weight of foods.

Human energy requirements vary with the age, sex, and activity level of the individual within a wide range of 1,200 to 3,200 Calories per day. (A **calorie** is a measure of energy, technically, the amount of energy needed to raise the temperature of one gram of water one degree Celsius.) Food energy is normally measured in **kilocalories,** 1,000 calories = 1 kilocalorie, which can be abbreviated as **kcal** or **Calories** with a capital "C." Most dietary guides simply use the term calories, but this book will use the more accurate Calories or kcal. Each gram of carbohydrate or protein can supply 4 Calories, while, for each gram of fat consumed, the amount of energy supplied is more than double at 9 Calories. Although all the macronutrients can be used as a source of energy, normally only carbohydrates and fats do so, while proteins provide the raw materials, or building blocks, required for the synthesis of essential metabolites, growth, and tissue maintenance.

Sugars and Complex Carbohydrates

Although carbohydrates are commonly grouped into sugars and starches, recall (*see* Chapter 1) that these compounds can be chemically classified into **monosaccharides, disaccharides,** and **polysaccharides,** based on the number of sugar units in the molecule.

Monosaccharides

Monosaccharides are the basic building block of all carbohydrates, and **glucose** is the most abundant of these sugars. During the process of digestion, many carbohydrates are broken down or converted into glucose, which is then transported by the blood to all the cells in the body. Within cells, the process of cellular respiration metabolizes glucose to produce the energy necessary to sustain life. Other common monosaccharides are **fructose** and **galactose,** which have the same chemical makeup as glucose, $C_6H_{12}O_6$, differing only in the arrangement of the atoms within the molecules. In the body, most fructose and galactose are converted into glucose and metabolized as such. In the United States, a common sweetener for some types of processed foods is high fructose corn syrup, often preferred because fructose is sweeter tasting than table sugar. Fructose is commonly found in many fruits, unlike galactose, which does not normally occur free in nature.

Disaccharides

Disaccharides are composed of two monosaccharides chemically joined together. The most common disaccharide is **sucrose,** or table sugar, formed from a molecule of glucose and a molecule of fructose. Other disaccharides are the milk sugar **lactose** (a combination of glucose and galactose) and **maltose** (formed by two glucose molecules), which is largely found in germinating grains. Table sugar, which primarily comes from sugar cane and sugar beet, is at least 97% pure sucrose with no other nutritional value, thereby supplying only Calories. During digestion, these disaccharides are broken down to yield their component monosaccharides.

Polysaccharides

Polysaccharides, also known as complex carbohydrates, contain hundreds to thousands of individual sugar units, and, for the most part, glucose is the only monosaccharide present. The different polysaccharides are distinguished by the way in which the glucose units are joined together, their arrangement, and their number. **Starch** is the storage form of glucose found in plants occurring abundantly in seeds, some fruits, tubers, and taproots. The presence of starch in foods can be traced directly to its plant origin; the starch in white bread or pasta was originally stored in the grain of a wheat plant. The major grain crops (wheat, rice, and corn), the major underground crops (potato, sweet potato, and cassava), and the major legumes (beans and peas) supply the majority of starch in the human diet. In the body, starch is broken down into glucose by enzymes in saliva and the small intestine and transported as described earlier.

Glycogen is the body's storage form of glucose found in the liver and skeletal muscles. When the levels of glucose in the blood are higher than the demands of the cells, the excess is used for the synthesis of glycogen in liver and muscle cells. Only a limited amount of glycogen can be stored as a reserve, no more than a day's worth of energy needs. Excess glucose beyond this is generally converted to fat. During strenuous exercise the body's glycogen reserves are called upon; therefore, athletes training for a competition practice a regime of carbohydrate loading by eating lots of starchy foods to build up muscle glycogen reserves.

A Closer Look 10.1

The Faces of Famine

In 1994, estimates of the world population size were placed at 5.6 billion, with projections of over 6 billion by the year 2000 and 8 billion by 2017 (box fig. 10.1a). This unprecedented population growth presents many problems related to the production and distribution of sufficient food to meet human nutritional needs. Although global food production has increased during recent decades, chronic hunger and malnutrition are ever-present problems in many developing nations, especially in subSaharan Africa. Estimates of hunger vary greatly, ranging from 17%–40% of the world's population who suffer from undernutrition or malnutrition, with approximately 20 million deaths each year attributed to these conditions. Undernutrition is defined as an insufficient number of Calories to maintain daily energy requirements, while malnutrition is a quality deficiency in which one or more essential nutrients is lacking, even though caloric intake may be sufficient. The majority of starvation-related deaths are children, with about half of those dying under the age of five. Starvation is usually the underlying cause of death, but most die of diseases such as diarrhea or measles, which would not be fatal in a properly nourished individual.

Two devastating conditions specifically related to undernourishment and malnourishment are **kwashiorkor** and **marasmus** (box fig. 10.1b). Kwashiorkor occurs when the diet is deficient in protein but has sufficient Calories. It is particularly prevalent after weaning, when a child no longer receives the protein-rich breast milk and is switched to a starchy diet low in protein content

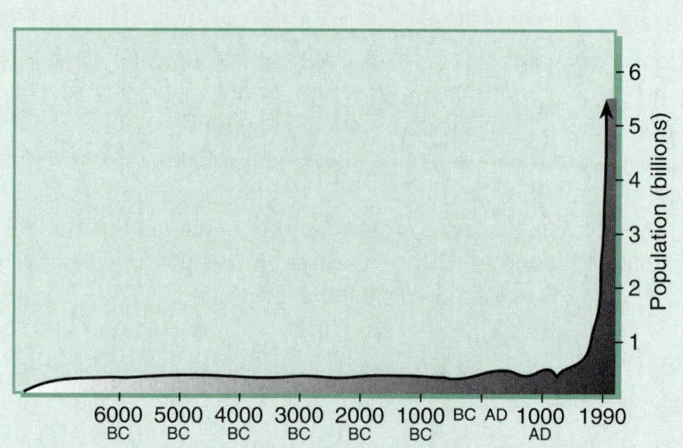

Box Figure 10.1 (*a*) *The human population has continued to grow rapidly, up to 5.6 billion in 1994, as indicated by the J-shaped curve.*

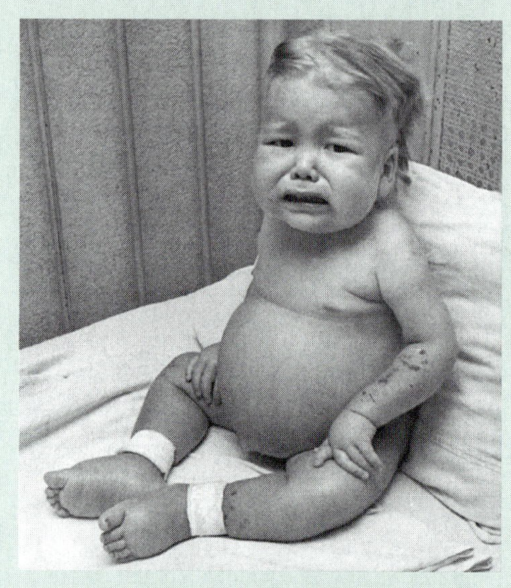

(b.1). The child is suffering from kwashiorkor, a protein-deficiency disease. (b.2) In marasmus, victims have a skeletal appearance as the body wastes away due to starvation.

Fiber in the Diet

Another important dietary component is **fiber,** which is derived from plant sources. Although not digestible, it does provide bulk and other benefits. There are many different types of dietary fiber: cellulose, lignin, hemicellulose, pectin, gums, mucilages, and others. Cellulose, a principal component of plant cell walls, is another polysaccharide composed of glucose; unlike starch and glycogen, however, humans do not have the enzymatic ability to break the bonds connecting the glucose molecules and thus cellulose passes through the digestive tract as roughage, largely unaltered. Other cell-wall components considered dietary fiber include lignin, pectins, and hemicelluloses. Lignin, a cell-wall component in plant cells that have secondary walls, is not a

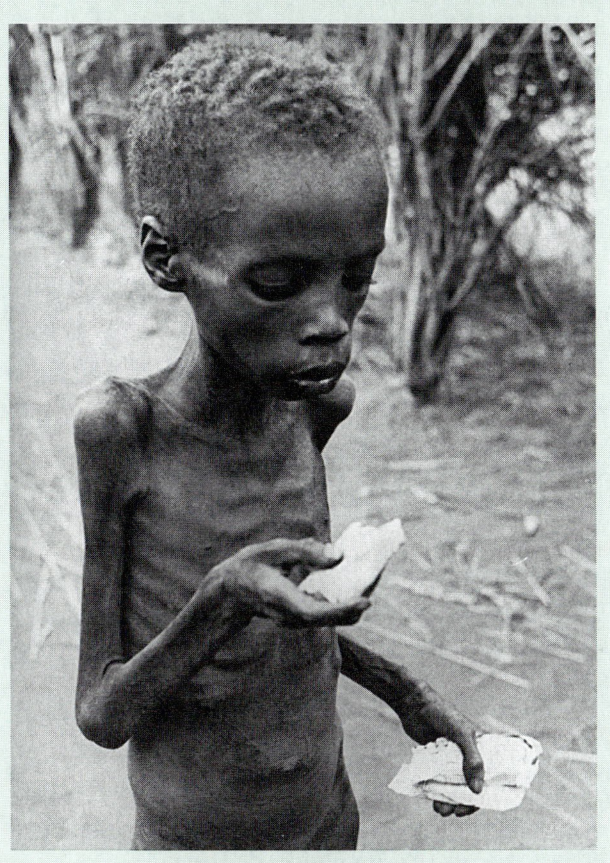

(b.2)

and/or quality. Symptoms of kwashiorkor include puffy skin and swollen belly due to edema, a fatty liver, a reddish-orange cast to the hair, dermatitis, and listlessness.

Marasmus results from starvation when the diet is low in both Calories and protein; other nutrients are probably deficient as well. Sufferers from marasmus are extremely thin and shriveled—literally skin and bones as the muscles of the body, even the heart muscle, are wasted away because muscle protein is digested to supply energy needs. The overt symptoms of both marasmus and kwashiorkor can be reversed if treated in time, but especially in infants and young children, mild mental retardation and stunted growth may be permanent results.

polysaccharide but a complex polymer. Pectins and hemicelluloses, which are also cell-wall polysaccharides, form the matrix in which cellulose fibrils are embedded. Pectins also occur in the middle lamella between adjacent cells. Gums and mucilages are exudates from various plants that are used commercially as thickening agents in prepared foods. Cell-wall polysaccharides, refined from some species of red and brown algae, can also be considered dietary fiber. Although not digestible by human enzymes, some fiber, especially some hemicelluloses, can be broken down by intestinal bacteria and the nutrients made available to the body.

Dietary fiber can be conveniently grouped into two types, **soluble** and **insoluble,** relating to their solubility in water. Insoluble fiber includes cellulose, lignin, and some hemicelluloses, while soluble fiber includes other hemicelluloses, pectins, gums, mucilages, and the algal polysaccharides. Fruits, vegetables, seeds, and whole grains supply most of the fiber in the human diet. Some plants are higher in one or more of these types of fiber and the beneficial effects of high-fiber foods differ, depending on which fiber is abundant. For example, the soluble fiber present as gum in oat bran and as pectin in apples is believed to lower cholesterol levels in the blood. Wheat bran, which is largely cellulose, an insoluble fiber, has no particular cholesterol-lowering ability, but seems to be most effective in speeding passage through the colon, which may reduce the risk of colon cancer.

Proteins and Essential Amino Acids

Proteins are a group of large complex molecules that serve as structural components as well as regulating a large variety of bodily functions (table 10.1). Recall that the constituents of proteins are **amino acids;** there are 20 naturally occurring amino acids that can be assembled in various combinations and numbers to make the thousands of different types of proteins. During digestion, proteins may be broken down into their component amino acids by enzymes in the digestive tract and transported in the bloodstream to the liver and body tissues.

Essential Amino Acids

The necessary role of dietary proteins is to supply amino acids so that the body can construct human proteins. All 20 amino acids are necessary for protein synthesis, and cells in the human body have the ability to synthesize 11 amino acids from raw materials; the other nine cannot be made by the body. These nine are called the **essential amino acids** (table 10.2) and must come from the diet. It is important to note that these essential amino acids cannot be stored by the body, and they must be present simultaneously in the diet. For this reason it is critical that the body receives all of the essential amino acids in a single meal. Persistent lack of these essential amino acids prevents synthesis of necessary proteins and results in protein deficiency diseases.

Complete Proteins

Complete proteins contain all the essential amino acids and in the right proportions. Almost all proteins derived from animals are complete proteins, whereas proteins derived from plants are usually **incomplete,** deficient in one or more essential amino acids. Although plant proteins are

TABLE 10.1

Functions of Proteins

Type of Protein	Function	Examples
Structural	Support	Collagen and keratin
Enzymes	Catalysts	Digestive enzymes
Hormones	Regulation	Insulin
Transport	Transport substance	Hemoglobin
Storage	Storage of amino acids	Ovalbumin in egg white; casein in milk
Contractile	Movement	Actin and myosin in muscles
Defensive	Protection	Antibodies (immunoglobins)

TABLE 10.2

Essential and Nonessential Amino Acids

Essential	Nonessential
Histidine	Alanine
Isoleucine	Asparagine
Leucine	Aspartic acid
Lysine	Arginine
Methionine	Cysteine
Phenylalanine	Glutamic acid
Threonine	Glutamine
Tryptophan	Glycine
Valine	Proline
	Serine
	Tyrosine

incomplete, by combining complementary plant proteins the essential amino acid requirements can be met. For example, the traditional diet of the Mexican Indians, beans and corn, contains complementary protein sources. The beans are low in methionine but adequate in tryptophan and lysine, while corn, which is poor in tryptophan and lysine, contains adequate amounts of methionine.

Proteins can be assigned a numerical value that reflects how well they supply the essential amino acids. The protein in eggs has been assigned a biological value of 100 and all other foods are given values using egg protein as the reference standard. Another factor that needs to be considered is the digestibility of a particular protein. Some proteins cannot be broken down completely; i.e., the amino acids are not fully released during digestion. This reduces the dietary value of the protein. For example, when digestibility is taken into account even egg protein, considered the perfect protein source, drops to a value of 94. High-quality proteins contain all the essential amino acids in the right proportions and are fully digestible, freeing their amino acids that are then absorbed into the blood and transported to the body's cells.

Fats and Cholesterol

Fats are usually considered the culprits in the diet, leading to obesity and cardiovascular disease, but some fat is necessary since it serves several vital functions. Fats and related compounds belong to a larger category of organic molecules called **lipids.** Although a very diverse group of compounds, all lipids share the common characteristic of insolubility in water (table 10.3).

Triglycerides

Ninety-five percent of the lipids in foods are fats and oils; both of these compounds are chemically classified as **triglycerides,** which are formed from glycerol and three **fatty acids** (*see* Chapter 1). Fatty acids themselves are the simplest type of lipid and serve as building blocks for triglycerides and **phospholipids.** A glycerol backbone is common to all triglycerides, but many types of fatty acids can occur. It is the nature of the fatty acids that determines the chemical and physical properties of the triglyceride. Each fatty acid contains a carbon chain with hydrogen attached to the carbon atoms; different fatty acids vary in the number of carbon and hydrogen atoms.

During digestion in the small intestine, triglycerides are first acted on by bile made by the liver, stored and released by the gallbladder. Bile contains a complex mixture of lipids, bile salts, and pigments; prominent among the lipid components are cholesterol and lecithin. Bile acts as an emulsifier, breaking up the triglycerides into smaller droplets that can be acted on by enzymes. Enzymes from the pancreas and intestinal cells split these smaller droplets of triglycerides into monoglycerides and two fatty acids or into glycerol and three fatty acids. These end products are absorbed into the intestinal cells, where they are resynthesized into new triglycerides that enter the lymphatic system and eventually the bloodstream.

Essential Fatty Acids

The body is capable of synthesizing most fatty acids, but three must be supplied in the diet. Linoleic, linolenic, and arachidonic acids are designated essential fatty acids, but

Table 10.3 Functions of Lipids

Type of Lipid	Function	Examples
Triglyceride	Energy; storage	Animal fat; vegetable oils
	Insulation	Subcutaneous fat
Steroid	Structural	Cholesterol in membranes
	Hormonal regulation	Cortisol; estrogen; testosterone
Phospholipid	Structural	Phosphatidylcholine in cell membranes

few adults suffer deficiency symptoms since they are widely found in foods, especially vegetable oils. Even if an adult were consuming a totally fat-free diet, as little as one teaspoon of corn oil, as an ingredient in foods, would supply the essential fatty acids. Deficiency symptoms, such as poor growth and skin irritation, have been seen in infants fed a formula lacking these essential nutrients.

Saturated and Unsaturated Fats

Fatty acids can be separated into two types, **saturated** and **unsaturated.** Saturated fatty acids contain all single bonds between the carbon atoms and have the maximum number of hydrogen atoms (it is said to be saturated with hydrogen). Unsaturated fatty acids have one or more double bonds between carbon atoms and consequently fewer hydrogen atoms. Each carbon atom can only form four bonds, so if a double bond occurs between two carbons, then less than the full complement of hydrogen atoms can be attached. A fatty acid with one double bond is called **monounsaturated** and lacks two hydrogen atoms; a **polyunsaturated** fatty acid has two or more double bonds and lacks four or more hydrogen atoms (fig. 10.1).

All food fats contain a mixture of both saturated and unsaturated fatty acids. Saturated fats contain mostly saturated fatty acids and are solid at room temperature; animal fats such as lard, butter, and beef fat are familiar examples. Vegetable oils are generally composed of unsaturated fatty acids and are liquid at room temperature. Oils containing mostly monounsaturated fatty acids are olive oil, peanut oil, and canola oil, while other vegetable oils like corn oil, soybean oil, and safflower oil contain mostly polyunsaturated fatty acids. Coconut oil, palm oil, and cocoa butter are exceptions to the rule. Although they are of plant origin, they consist mostly of saturated fatty acids. On the other hand, certain fish oils are actually unsaturated. In many prepared foods containing vegetable oils, the fatty acids on the unsaturated oil have been chemically modified by **hydrogenation** (adding hydrogen to make the unsaturated fat a saturated one). These modified fatty acids are called trans fatty acids. This artificially saturated vegetable fat has all the properties of a naturally occurring animal fat and all the negative health effects.

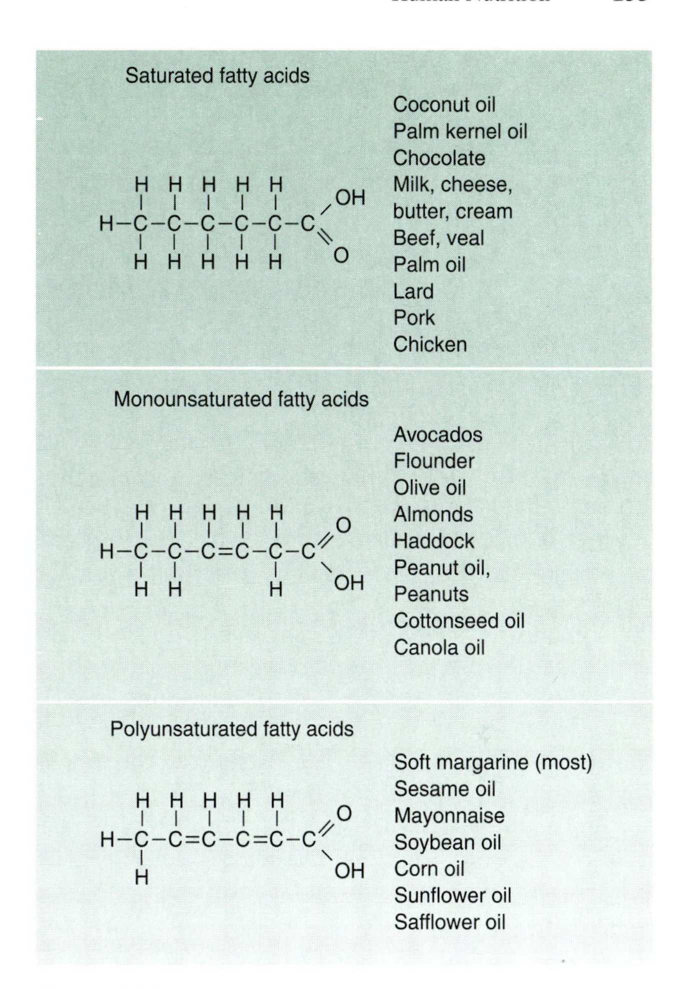

Figure 10.1 *Structure of saturated, monounsaturated, and polyunsaturated fatty acids.*

The health implications of saturated vs. unsaturated fats have been intensely studied by the scientific community in recent years. Diets high in saturated fats may lead to obesity, which has many inherent health risks . These diets have also been implicated in colon, breast, and prostate cancers. Because saturated fats increase blood cholesterol levels, they are also linked to cardiovascular diseases as well. Unsaturated fats, on the other hand, lower the risk of cardiovascular disease by lowering blood cholesterol levels.

Cholesterol

Cholesterol belongs to a subcategory of lipids known as steroids, which are compounds containing four carbon rings (*see* Chapter 1). Several steroids, including cholesterol, also have a hydrocarbon tail and an –OH group, making them sterols. Cholesterol is a vital constituent of cells, since it is part of the lipid component of cell membranes and is also used in the synthesis of sex hormones and several other hormones.

Cholesterol is synthesized in the liver from saturated fatty acids and is also absorbed by intestinal cells from animal foods, especially eggs, butter, cheese, and meat. If the diet is high in saturated fats, even if it is low in cholesterol, the liver responds by increased cholesterol synthesis. Since cholesterol, like all lipids, is insoluble in the watery medium of the blood, it is transported by a special complex that consists of a cholesterol center with a coating of lipids and water-soluble proteins. These transport molecules are known as lipoproteins and exist in several forms. Two of the most significant are the **low density lipoproteins (LDL)** and **high density lipoproteins (HDL).** LDLs transport cholesterol to all the body cells, while HDLs remove excess cholesterol from the body's tissues and carry it to the liver for degradation and elimination. In the popular press the LDLs are considered the "bad" cholesterol because they can be taken up by the cells lining the arteries. The resulting deposition of cholesterol blocks the arteries, restricting the blood flow, and is known as atherosclerosis. This condition can lead to heart attacks if the coronary arteries are blocked and strokes if the arteries delivering blood to the brain are blocked. HDLs are considered "good" cholesterol because they can prevent atherosclerosis by preventing the buildup of cholesterol deposits on the lining of the arteries.

Diets high in cholesterol and/or saturated fat contribute to high blood cholesterol levels, especially in the form of LDLs. Plant sources do not contribute dietary cholesterol directly although, as pointed out previously, some may be high in saturated fats. On the other hand, plant oils are generally rich in unsaturated fats, which are known to lower blood cholesterol levels. However, monounsaturated and polyunsaturated fats act differently. Polyunsaturated fats tend to lower all cholesterol levels, including the protective HDLs, whereas monounsaturated fats preserve the HDLs while lowering total and LDL levels. Although hereditary factors, cigarette smoking, and exercise play a role in the LDL/HDL balance, for the majority of people diet is the single most important factor in controlling cholesterol levels and the inherent disease risks.

Micronutrients

Like macronutrients, the micronutrients are essential for proper nutrition but are required in much smaller amounts. While macronutrients make up the bulk dry weight of food, micronutrients comprise only 1%–2% of the dry weight. There are two categories of micronutrients, the organic compounds known as vitamins and the inorganic compounds, the minerals.

TABLE 10.4

Fat-Soluble Vitamins

Vitamin	Dietary Source	Results of Deficiency
A	Yellow/orange, dark green vegetables and fruits; dairy products	Night blindness; xeropthalmia
D	Eggs and enriched dairy products	Rickets
E	Seeds; leafy green vegetables	Unknown
K	Leafy green vegetables	Poor blood clotting

Vitamins

Many vitamins play roles as **coenzymes** (molecules that are required for the proper functioning of certain enzymes) in many metabolic pathways in the body, while others are directly involved in the synthesis of indispensable compounds. Vitamins are classified according to their solubility, with four **fat-soluble vitamins (A, D, E, and K)** and nine **water-soluble vitamins (eight B-complex vitamins and C).** The water-soluble vitamins are not readily stored in the body, and any excess is eliminated in the urine; therefore, these are unlikely to become toxic. On the other hand, the fat-soluble vitamins are easily stored in the fatty tissues of the body, and excessive intake can lead to toxicity symptoms. The dietary sources and deficiencies of these are described in Tables 10.4 and 10.5. The following discussion is limited to vitamins A, D, C, and the B vitamins, thiamine, niacin, and B_{12}.

Vitamin A

Vitamin A has many roles in the body. One of the best known involves the formation of visual pigments (rhodopsin and others) present in the retina of the eye. Each pigment is composed of a molecule of retinal (a form of vitamin A) and a protein molecule, called an opsin, that differs from pigment to pigment. The pigments are contained in two types of photoreceptor cells, rods and cones, located deep in the retina of the eye. All light stimulates the rods, providing only black and white vision, while the cones are selectively stimulated by different colors, providing the full spectrum of color vision. In dim light or at night, only the rods are used for vision; therefore, a shortage of retinal is especially pronounced, leading to one of the earliest signs of vitamin A deficiency, **night blindness.**

Vitamin A is also necessary for the maintenance of epithelial tissues that line both internal and external body

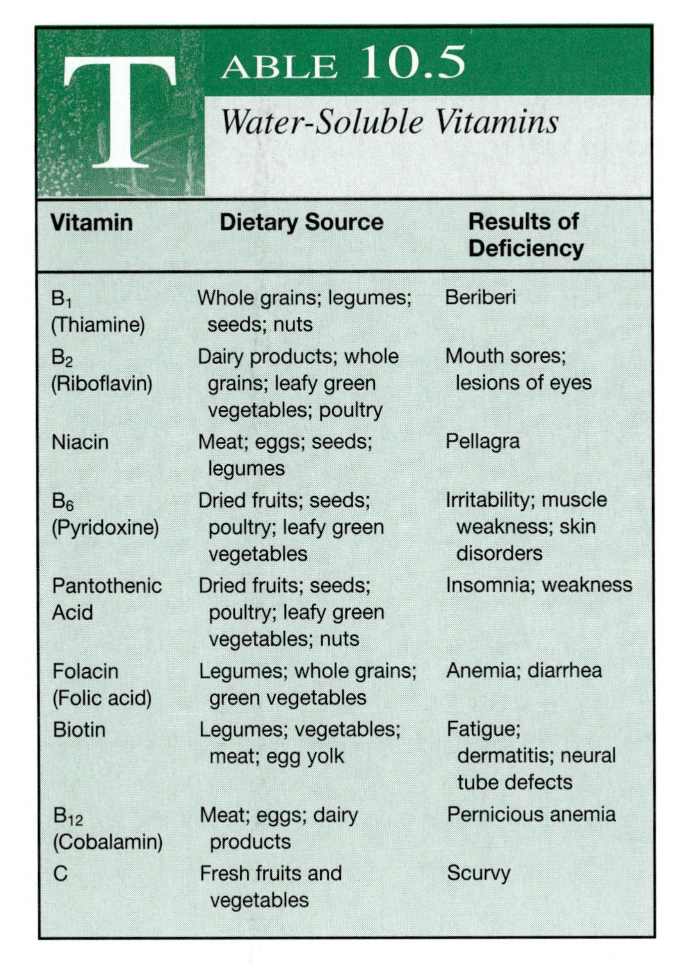

TABLE 10.5

Water-Soluble Vitamins

Vitamin	Dietary Source	Results of Deficiency
B_1 (Thiamine)	Whole grains; legumes; seeds; nuts	Beriberi
B_2 (Riboflavin)	Dairy products; whole grains; leafy green vegetables; poultry	Mouth sores; lesions of eyes
Niacin	Meat; eggs; seeds; legumes	Pellagra
B_6 (Pyridoxine)	Dried fruits; seeds; poultry; leafy green vegetables	Irritability; muscle weakness; skin disorders
Pantothenic Acid	Dried fruits; seeds; poultry; leafy green vegetables; nuts	Insomnia; weakness
Folacin (Folic acid)	Legumes; whole grains; green vegetables	Anemia; diarrhea
Biotin	Legumes; vegetables; meat; egg yolk	Fatigue; dermatitis; neural tube defects
B_{12} (Cobalamin)	Meat; eggs; dairy products	Pernicious anemia
C	Fresh fruits and vegetables	Scurvy

surfaces, an area roughly equivalent to one-fourth of a football field. When vitamin A is lacking, these tissues fail to secrete their protective mucous covering, producing instead a protein called keratin (normally found in hair and nails), which results in the tissue becoming dried and hardened. This aspect of vitamin A deficiency can therefore affect many different areas in the body. One of the most tragic consequences is a type of blindness known as **xerophthalmia,** in which severe vitamin A deficiency results in irreversible drying and degeneration of the cornea. This permanent blindness, easily preventable with proper nutrition, is found most frequently among malnourished children in developing nations (fig. 10.2). In the skin, keratinization results in rough, dry, scaly, and cracked skin, often with an accumulation of hard material around hair follicles that looks like a permanent goose bump. Other epithelial tissues such as those in the mouth, gastrointestinal tract, and respiratory system are also affected by the decrease in mucous production, becoming progressively drier and subject to infection. Vitamin A is additionally involved in dozens of other roles in the body, including normal bone and tooth development and hormone production in the adrenal and thyroid glands.

In food derived from animal sources, especially liver, the vitamin A occurs primarily as retinol, which is readily absorbed by the body. In plant sources, no retinol is present, but a vitamin A precursor, **beta-carotene** (first isolated from carrots) occurs abundantly in many yellow, orange, and dark green fruits and vegetables (*see* Chapter 3) and can be split into two molecules of retinol in the body. Recently the importance of beta carotene as an antioxidant has been investigated. Antioxidants may protect the body against the destructive action of reactive ions called free radicals and thus reduce the risk of cancer and heart disease and even delay aging.

Vitamin D

The primary function of vitamin D is the regulation of calcium and phosphorus levels, especially for normal bone development. The action of vitamin D in controlling the blood levels of these minerals takes place in three ways:

(1) the absorption of calcium and phosphorus from food in the gastrointestinal tract,
(2) the removal of these minerals from bones to maintain the concentration in the blood, and
(3) the retention of calcium by the kidneys.

Vitamin D is unique in that it can be synthesized by the human body on exposure to sunlight; in fact, it has long been called the "sunshine vitamin." The precursor (a cholesterol derivative) is manufactured by the liver and transported to the skin where exposure to the sun's ultraviolet rays converts it to provitamin D; the final steps in the manufacture of active vitamin D occur in the liver and kidney. The amount of pigmentation in the skin affects the synthesis of vitamin D because pigment blocks ultraviolet absorption. Darker skin requires longer exposure to sunlight to produce adequate amounts of vitamin D. Thirty minutes of sunlight is adequate for light skin, while darker skin may require up to three hours. This exposure to sunlight must be achieved with caution since overexposure to the ultraviolet rays in sunlight is linked to higher risk of skin cancer. It has been estimated by the Environmental Protection Agency that the average American today is indoors 93% of the time, in transit 5% (cars, buses, trains), and is outdoors only 2% of the time daily. With this limited exposure to sunlight, it is essential that the diet contains vitamin D to avoid deficiency symptoms. Vitamin D is not naturally abundant in any food. None occurs in plant sources, but vitamin D does occur in limited amounts in animal sources such as egg yolks, liver, cream, some fish, and butter. Since there is concern about meeting the nutritional needs for vitamin D, especially for children, and since milk does not contain adequate vitamin D, it is routinely fortified with this vitamin.

Because of the role of vitamin D in calcium regulation, the deficiency symptoms are most evident in bone formation. The effects of vitamin D deficiency are most pronounced in children and result in a characteristic malformation of the skeleton known as **rickets.** The bowing of the legs so commonly associated with the condition is just one of many abnormalities of the skeletal system

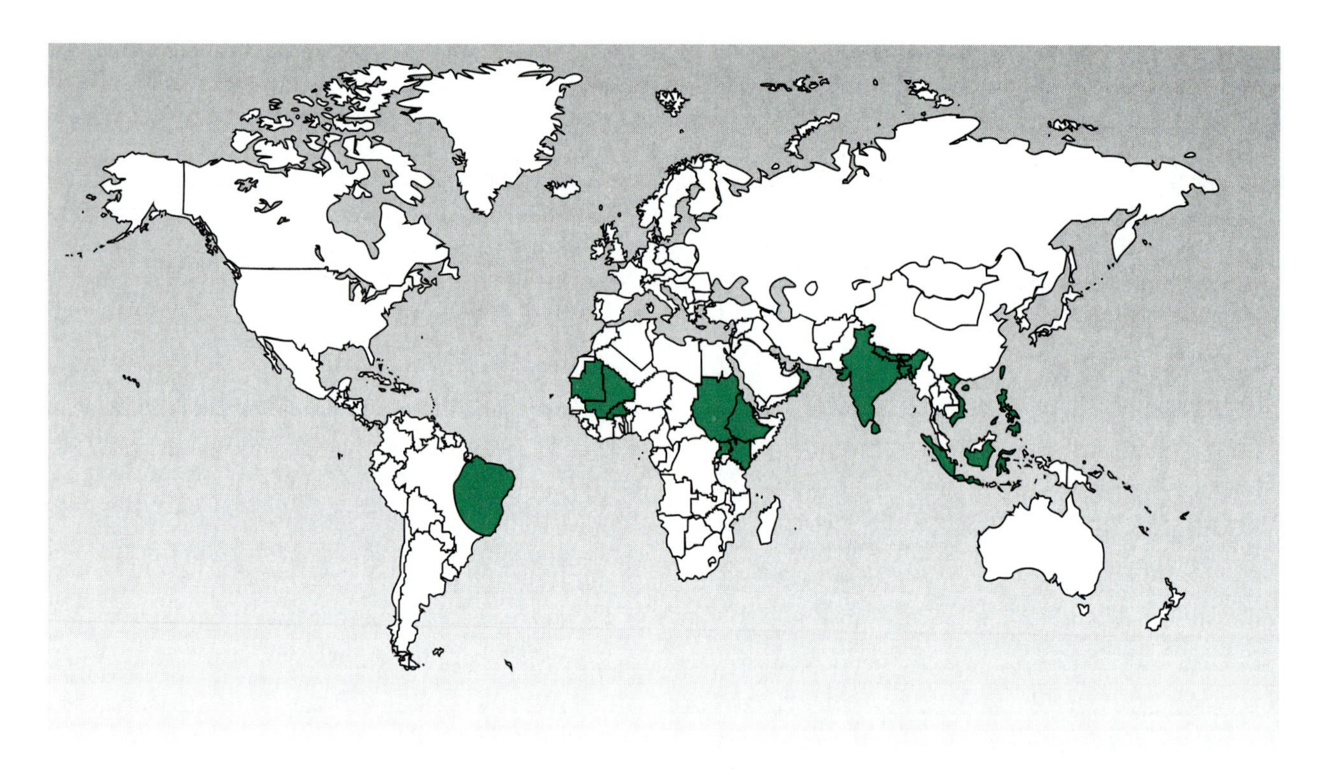

■ = Countries or regions where xerophthalmia is a significant public health problem according to World Health Organization criteria

Figure 10.2 *World map indicating areas of blindness (xeropthalmia) caused by vitamin A deficiency.*

(fig. 10.3a). Children with dark skin living in northern, smoggy cities are especially vulnerable to developing this deficiency. Adult rickets, also known as **osteomalacia** is rare but does occur in women who have undergone repeated pregnancies, have low calcium intake, and have inadequate exposure to sunlight.

Excess vitamin D causes abnormally high levels of calcium in the blood; this often leads to calcium deposits in soft tissues such as kidneys and blood vessels. If not caught in time, this may result in irreversible damage to the cardiovascular system, kidney failure, and even death. In excess, vitamin D is the most toxic of all the vitamins. However, this toxicity cannot occur from sunlight or food, but only results when megadoses of the vitamin are administered without medical supervision.

Vitamin C

Fresh fruits and vegetables are the richest sources of vitamin C, **ascorbic acid,** with organ meats the only significant animal sources. The most important role of vitamin C in the body is in the synthesis of collagen, a connective tissue protein that serves as a "cellular cement" holding cells and tissues together. Collagen, the most abundant protein in the body, is found in the matrix of bones, teeth, and cartilage and provides the elasticity of blood vessels and skin. Vitamin C also functions as an antioxidant in the body, preventing other molecules from being oxidized (losing electrons). Because vitamin C prevents oxidation, it is sometimes added to packaged foods to extend the shelf life; in a similar manner, orange juice or lemon juice, with its high vitamin C, prevents the oxidation (browning) of sliced apples or bananas. Vitamin C is additionally involved in promoting iron absorption through the intestines. When foods containing vitamin C are consumed with foods containing iron, absorption is enhanced. Finally, vitamin C is involved in a number of other metabolic reactions including the production of various hormones.

For centuries, sailors on long ocean voyages faced the possibility of **scurvy,** a disease that could cause bleeding of the gums, pinpoint hemorrhages under the skin, severe fatigue, wounds failing to heal, brittle bones, and even

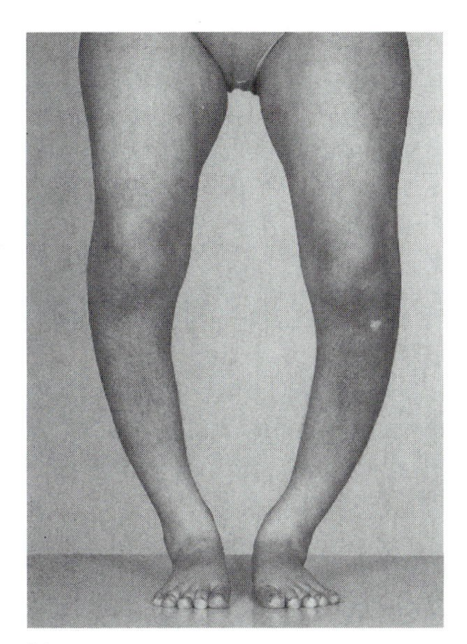

(a)

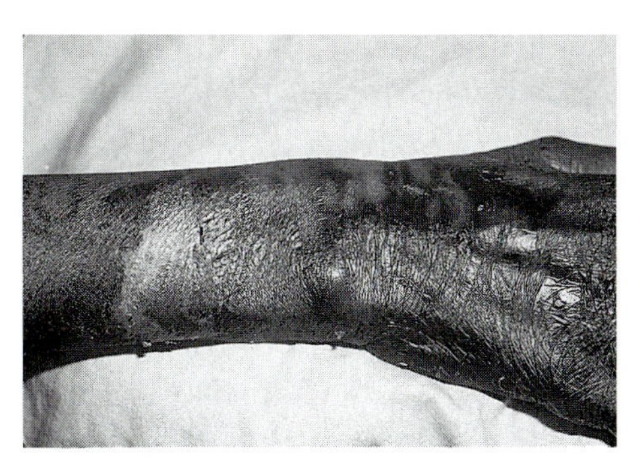

(b)

Figure 10.3 *Vitamin deficiency diseases. (a) Characteristic bowing of the legs in rickets, a disease due to insufficient vitamin D. (b) Dermatitis is one of the symptoms of pellagra, caused by a lack of niacin in the diet.*

sudden death due to massive internal bleeding. It was not uncommon for half to two-thirds of the ship's company to die of scurvy on a long voyage. The first cure for this disease was identified in 1747 by Dr. James Lind, who experimented with 12 seamen afflicted with scurvy. Lind tried various dietary supplements and found that those sailors given either oranges or lemons for 6 days improved rapidly. Interestingly enough, it was almost 50 years after these findings that the British admiralty took measures to prevent scurvy by dictating that all sailors receive lemon or lime juice daily. (British sailors were soon nicknamed "limeys" because of this practice, a name that is still heard today.) Now it is well-known that the vitamin C in the citrus juice prevented scurvy. The typical symptoms of scurvy can be traced directly to the inability of the body to make collagen. The scurvy-vitamin C connection is reflected in the name ascorbic acid, which literally means "without scurvy."

Since the 1970s, the daily intake of vitamin C has been the focal point of a heated debate after the publication of *Vitamin C and the Common Cold* by Linus Pauling, a Nobel-Prize-winning chemist. The RDA (Recommended Daily Allowance) for vitamin C is 60 mg (approximately 10 mg daily can prevent scurvy), but Pauling recommended megadoses as high as 2,000 to 10,000 mg for optimum health. Pauling maintained that large doses can prevent colds and other viral infections, is bactericidal, and even cures cancer. Recently some of these claims have been substantiated by other research groups. However, other authorities believe that tissues become saturated with vitamin C at levels of 80 to 100 mg per day and intakes above this level are generally excreted in the urine. Controversy also exists regarding toxicity of vitamin C. Some scientists maintain that high doses are nontoxic, while others feel that toxicity can occur. The most common toxicity symptoms are nausea, abdominal cramps, and diarrhea. Another possible complication of megadosing is rebound scurvy, a condition that may occur after an abrupt cessation of these high doses. Symptoms mimic scurvy, even though vitamin C intake is not deficient; these symptoms do not occur if the megadoses are decreased gradually.

Vitamin B Complex

The vitamin B complex includes a group of eight vitamins that are often found in foods together and have similar roles in the body; i.e., they function as coenzymes, involved in thousands of metabolic reactions. They are found in each cell of the body and must be present for normal cell functioning. For each of these vitamins, specific deficiency symptoms occur when the vitamin is lacking in the diet; however, in general no toxicity symptoms have been reported since excess is excreted in the urine. Like vitamin C, the B vitamins are water-soluble and can be leached out during food preparation when excess water is used and discarded. In addition, some of the B vitamins may be destroyed by high temperatures during cooking. **Thiamine, niacin,** and **B_{12}** will be considered, since these may be deficient in plant sources.

Thiamine Thiamine, also known as **vitamin B_1,** is part of the coenzyme thiamine pyrophosphate, which is involved in the metabolic breakdown of carbohydrates just prior to the Krebs cycle. Because of its central role in metabolism, the symptoms of thiamine deficiency are profound: fatigue, depression, mental confusion, cramping, burning, and numbness in the legs, edema, enlarged heart, and eventually death from cardiac failure. This thiamine deficiency is known as **beriberi** and was found mainly in the Orient where diets were based mainly on polished or white rice rather than the whole-grain brown rice that still has the outer bran (husk) intact. The thiamine that occurs in the outer layer of the rice is removed during the polishing process. Beriberi became more prevalent when improved techniques for polishing rice were developed that removed more of the bran and, inadvertently, more of the thiamine. In the 1880s, the thiamine deficiency in the Japanese navy was particularly widespread, with 25% to 40% of the sailors developing beriberi. A Japanese physician, Dr. K. Takaki, observed that few sailors developed the disease when milk, meat, and eggs were added to the normal staple of white rice. Unfortunately, he did not realize that the enriched diet was supplying a nutrient missing in the white rice diet. A few years later a Dutch physician, Dr. Christian

Eijkman, studied beriberi in the East Indies; he showed that, in diets based almost exclusively on rice, the consumption of brown rice instead of white rice prevented the appearance of beriberi. It wasn't until the twentieth century that thiamine deficiency was actually identified as the cause of the disease. Good dietary sources of thiamine included meat, especially pork and liver, whole grains, seeds and nuts, and legumes.

Niacin Niacin is the collective term for two compounds, **nicotinic acid** and **nicotinamide,** either one of which is used to form the coenzymes NAD^+ and $NADP^+$. Recall the importance of these coenzymes for oxidation-reduction reactions in many energy yielding metabolic pathways (*see* Chapter 4). Without these coenzymes, the release of energy from the breakdown of foods cannot occur and cellular death results. Interestingly enough, niacin can be supplied directly through foods rich in niacin itself, or foods rich in the essential amino acid tryptophan, because the body can synthesize niacin from the amino acid. Niacin deficiency, therefore, is coupled with a low protein diet.

Without niacin, every organ of the body is severely impacted, and a severe deficiency disease, **pellagra,** develops. The symptoms of pellagra are referred to as the 4 Ds: dermatitis (skin disorders), dementia (mental confusion), diarrhea, and eventually death if niacin is not supplied. The dermatitis is characterized by rough, reddened skin with lesions developing in exposed areas; in fact, *pellagra* means rough skin (fig. 10.3b). The central nervous system is affected, and confusion, memory loss, dizziness, and hallucinations occur. The gastrointestinal system is also involved and diarrhea, along with abdominal discomfort, nausea, and vomiting, is common. Another characteristic of the condition is a bright red or strawberry tongue.

Pellagra is especially common in areas where corn is the dietary staple, and outbreaks have occurred in both southern Europe, particularly Italy, parts of southern India, and the rural South in the United States, where the disease became epidemic early in the twentieth century. It was estimated that 10,000 deaths occurred and another 200,000 were afflicted each year. During this period, about half of the patients in mental hospitals in the South were suffering from the dementia caused by pellagra.

Although corn does contain some niacin, it is in a form that makes it unavailable; furthermore, corn is also deficient in tryptophan. However, pellagra was not a problem in the traditional diet of the Indians of Mexico, Central America, and parts of South America for two reasons. Lime (calcium oxide from wood ash or shells), used in the preparation of the corn meal, is able to release the bound niacin. Additionally the beans, squash, tomatoes, and peppers commonly eaten with the corn also supplied niacin.

The addition of milk and meat to the diet was recommended to prevent pellagra long before the vitamin was identified. Although meat contains niacin, milk is low in the vitamin but does supply tryptophan. Sources rich in niacin include meat, poultry, fish, eggs, nuts, seeds, and legumes. One way the pellagra in the South could have been prevented was by the consumption of a handful of peanuts every other day, since peanuts are an excellent source of niacin.

In recent years there has been an increased interest in the therapeutic value of megadoses of niacin (the nicotinic acid form) for reducing blood cholesterol levels. Unfortunately the megadoses can cause some toxicity symptoms, the most common of which is a niacin flush. It produces a temporary warm flush of the skin, with a tingling or stinging sensation. Intestinal irritation and liver damage have also been reported. The other form of niacin, nicotinamide, does not produce these toxicity symptoms, but is not at all effective in lowering blood cholesterol levels.

Vitamin B_{12} Vitamin B_{12} **(cobalamin)** is unique in that it does not occur naturally in any foods of plant origin but only occurs in animal sources, where it is widely available. Those who completely eliminate meat, dairy products, and eggs from their diets are at risk of developing a B_{12} deficiency unless vitamin supplements are taken or fortified foods are eaten. Soy milk, breakfast cereals, and meat substitutes are often fortified with B_{12}.

The absorption of vitamin B_{12} in the small intestines requires the presence of a substance secreted by the stomach called an intrinsic factor. Poor absorption of B_{12} has been reported in people who, because of a genetic defect, do not produce the intrinsic factor. This defect most often shows up after the age of 60, when intrinsic factor production becomes impaired. In this case, B_{12} deficiency symptoms show up even though there are sufficient quantities in the diet, and the vitamin must be received by injection.

The most common result of B_{12} deficiency is **pernicious anemia** characterized by the production of improperly formed red blood cells. The associated symptoms include fatigue and weakness, because the delivery of oxygen to the body's tissues is impaired. A more serious consequence of B_{12} deficiency is nerve damage that begins as a creeping numbness of the lower extremities.

In general, vitamin B_{12} is involved in nucleic acid synthesis and interacts with **folacin** (another B vitamin) in this function. Because blood cells are constantly being formed in the bone marrow, this site of rapid cell division is one of the first affected by impaired synthesis of DNA due to a deficiency of either vitamin. These vitamins are, therefore, involved in the normal development of red blood cells, and a deficiency of either vitamin causes anemia. The anemia can be treated by either B_{12} or folacin supplements. However, folacin supplements have no effect on the nerve damage caused by a B_{12} deficiency, and the administration of folacin for the anemia can mask a true B_{12} deficiency and result in permanent neurological degeneration. In this regard, vitamin B_{12} functions in maintaining the sheath surrounding nerve fibers, which is necessary for the transmission of nerve impulses.

TABLE 10.6
Dietary Mineral Requirements

Mineral	Function
Major Minerals	
Calcium	Bone and tooth formation; blood clotting; nerve transmission; muscle contraction
Phosphorus	Nucleic acids; bone and tooth formation; cell membranes, ATP formation
Sulfur	Protein formation
Potassium	Muscle contraction; nerve transmission; electrolyte balance
Chlorine	Gastric juice
Sodium	Nerve transmission; body water balance
Magnesium	Protein formation; enzyme cofactor
Trace Minerals	
Iron	Hemoglobin
Zinc	Component of many enzymes and insulin; wound healing
Iodine	Component of thyroid hormones
Fluorine	Bone and tooth formation
Copper	Enzyme component; red blood cell formation
Selenium	Antioxidant
Cobalt	Component of vitamin B_{12}
Chromium	Normal glucose metabolism
Manganese	Enzyme cofactor
Molybdenum	Enzyme cofactor

Minerals

Minerals are inorganic compounds that exist in the body as ions (charged atoms) or as part of complex molecules. At least 17 minerals are required for normal metabolic activities (table 10.6) and some, known to be essential for other animals, may be added to the list as research continues. Minerals are subdivided into two categories, the **major minerals,** needed in amounts greater than 100 mg/day and the **trace minerals,** needed in amounts no more than a few mg/day. The following discussion will be limited to **calcium,** a major mineral whose RDA has been recently revised, and two trace minerals that have been extensively studied, **iron** and **iodine.**

Calcium

Calcium is the most abundant mineral in the body, with the average adult containing 800 to 1,300 g of the element. Ninety-nine percent of the body's calcium is found in the bones and teeth, while the other 1% is in the blood and tissues. The concentration of calcium is under the control of several hormones and vitamin D. If the amount in the blood gets too low, calcium reserves in the bone are drawn upon to restore levels to the normal range. If the amount of calcium in the blood is too high, more calcium is deposited in the bone and more is excreted by the kidneys. Excess calcium intakes (12,000 mg per day and above are considered toxic) have been associated with increased risk of kidney stone formation. In addition to forming the matrix of bones and teeth, calcium in the body fluids is involved in many important functions: nerve transmission, muscle action including heartbeat, blood clotting, cell membrane integrity, intracellular communication, and as a **cofactor** for enzymes (cofactors are mineral ions and, like coenzymes, are necessary for the proper functioning of certain enzymes).

Calcium deficiency may lead to **osteoporosis,** a degenerative bone disease that may strike older individuals without warning. In osteoporosis, the bone density is greatly reduced (osteoporosis literally means *porous bone*), resulting in bones that fracture readily. This condition can result from years of low dietary calcium intake or poor absorption of calcium from the intestines (caused by lack of vitamin D or other factors). To maintain blood calcium levels, the reserves in the bone are dangerously depleted. Postmenopausal women are particularly at risk for developing osteoporosis since bone loss is accelerated at this time. Estrogen replacement therapy appears to retard this bone loss. Adequate dietary calcium and regular exercise also prevent osteoporosis; however, there are many interacting factors (both genetic and environmental) and much more research is needed in this area.

Milk and milk products are among the best sources of calcium, but the element is also present in dark-green leafy vegetables, many seeds, and other foods. Unfortunately, in some vegetables, the presence of oxalic acid inhibits the absorption of calcium. Recently the RDA for dietary calcium for adults has been increased to 1,200 mg to prevent the development of osteoporosis. New evidence has also shown that calcium, together with vitamin D, may also provide protection against colon cancer; the amount of calcium required for this beneficial action is 1,500 mg per day.

Iron

Although most of the trace minerals are usually found in adequate amounts in a well-balanced diet, iron and iodine present special problems. Iron deficiency is common in women and children and care must be taken to insure that the diet supplies sufficient quantities of the element. Meat, especially liver and other organ meats, shellfish, fish, and poultry are excellent iron sources. Many foods from plants are also rich in iron, including dark-green leafy vegetables, dried fruits, legumes, whole grains, and enriched breads and cereal products. Overall, only about 10% of the dietary iron is actually absorbed by the body, with the absorption dependent on the type of iron compound present; the rest is eliminated in the feces. The iron from animal sources may be present as **heme iron** (40%) or **nonheme iron** (60%), while the iron in plant sources is nonheme. Heme iron is more readily absorbed by the body, but the absorption of nonheme iron can be improved by the presence of vitamin C.

The most important role of iron is as a component of hemoglobin, the molecule that carries oxygen in red blood cells; in fact, it is the iron that imparts the red color to these cells. Additionally, iron occurs in myoglobin, a molecule similar to hemoglobin, the oxygen carrier in muscle cells, several storage proteins in the liver, bone marrow, and spleen, and also in enzymes present in each cell.

Since the majority of iron is found in hemoglobin, iron deficiency has its greatest impact on red blood cells. When iron reserves in the body are low, not enough hemoglobin can be synthesized for newly formed red blood cells. These cells are smaller, pale in color, and less efficient in oxygen transport and are characteristic of **iron deficiency anemia,** the most common dietary deficiency disease in the world. The symptoms of iron deficiency anemia include fatigue, inability to concentrate, pale coloration, weakness, and listlessness.

The greatest risk of iron toxicity comes from overdosing on iron supplements, which can result in damage to the liver and pancreas, and even sudden death in young children.

Iodine

The presence of iodine in food is dependent on the availability of iodine in the natural environment where the plant or animal developed. Foods from the ocean are reliable sources of iodine since this element is plentiful in seawater. Generally, inland areas, especially mountainous regions, are more likely to have iodine deficient soils. It is in these areas where people may develop the iodine deficiency disease, **endemic** or **simple goiter.** In the United States, the area around the Great Lakes was formerly known as the "goiter belt" because of the high incidence of goiter. The most obvious symptom of goiter is a swelling of the neck caused by an enlargement of the thyroid gland that straddles the trachea (fig. 10.4).

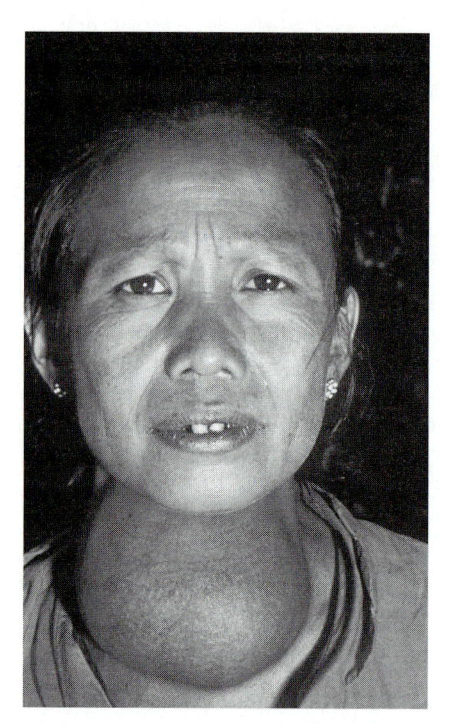

Figure 10.4 *This woman is suffering from goiter, a swelling of the thyroid gland, caused by an iodine deficiency in the diet.*

Iodine is required for the formation of thyroid hormones that regulate metabolism in all cells of the body, control body temperature, growth, development, and reproduction. When the amount of iodine is low, hormone production is impaired. The thyroid enlarges in an attempt to produce more of the needed hormones; this attempt is futile without the necessary iodine. A person suffering from simple goiter exhibits a lack of energy, decreased blood pressure, sensitivity to cold temperatures, and weight gain.

Goiter has been a recognized ailment since ancient times, and various treatments have been suggested as a cure. The earliest known treatment is recorded in a Chinese source from 5,000 years ago that recommended eating seaweed and burned marine sponge. Today it is known, of course, that organisms from the ocean are naturally rich in iodine. The relationship between iodine and goiter was confirmed in 1820 by the French physician Coindet who reported the treatment of goiter using doses of iodine salts. Today goiter is rare in the United States and Europe because of iodized table salt first introduced in 1924. However, in other areas of the world, almost 200 million people still suffer from goiter, an easily preventable disorder.

More than half of the salt sold in the United States is iodized and a single teaspoon of this salt supplies almost twice the RDA of 0.15 mg. However, overconsumption of iodine-containing substances can also be a problem because iodine can be toxic. As little as 2.0 mg per day is considered toxic, resulting also in an enlargement of the thyroid gland.

In addition to a lack (or even an excess) of dietary iodine, goiter can result from the overconsumption of goitrogenic compounds. Certain medications, including some of the sulfa drugs, and vegetables in the cabbage family, are known to contain compounds that block the utilization of iodine in the thyroid. In a varied diet, these compounds are harmless, but they may be a problem in diets restricted solely to these vegetables.

Dietary Guidelines

Research has shown that many significant diseases are influenced by nutrition, and beneficial changes in diet can, therefore, reduce the risk of developing these conditions. Diseases linked to nutrition are some of the major causes of death in the United States: cardiovascular diseases, hypertension, some forms of cancer, and noninsulin dependent diabetes. These diseases may arise in part from

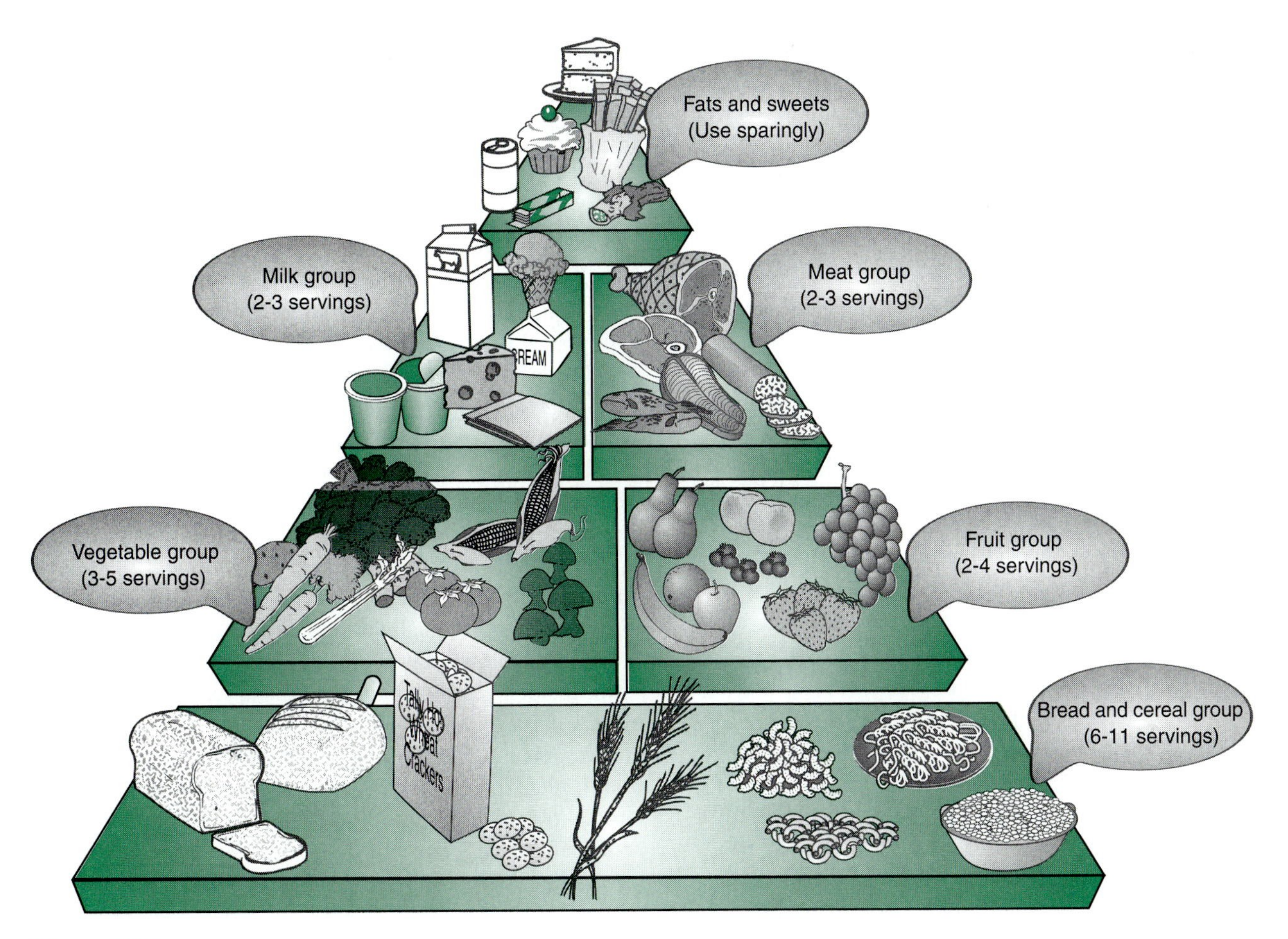

Figure 10.5 *The food pyramid illustrates the recommended dietary guidelines set forth by the U.S. Department of Agriculture.*

excess consumption of fat, especially saturated fat, cholesterol, refined sugar, and salt. In light of these findings, governmental agencies and health professionals have recommended dietary guidelines for better health and the prevention of disease.

Balancing Nutritional Requirements

The U.S. Senate Select Committee on Nutrition and Human Needs issued the *Dietary Goals for the United States* in 1977 that defined seven nutritional recommendations for the American people (fig. 10.5).

1. To avoid becoming overweight, consume only as much energy (kcal) as is expended; if overweight, decrease energy intake and increase energy expenditure.
2. Increase the consumption of complex carbohydrates and "naturally occurring" sugars from about 28% of energy intake to about 48% of energy intake.
3. Reduce the consumption of refined and other processed sugars by about 45% to account for about 10% of total energy intake.
4. Reduce overall fat consumption from approximately 40% to about 30% of energy intake.
5. Reduce saturated fat consumption to account for about 10% of total energy intake; and balance that with polyunsaturated and monounsaturated fats, which should account for about 10% of energy intake each.
6. Reduce cholesterol consumption to about 300 mg a day.
7. Limit the intake of sodium by reducing the intake of salt to about 5 g a day.

Since these recommendations were issued, there have been some healthy trends developing among the American people. The consumption of complex carbohydrates has increased, adding both starch and fiber to the diet. The starch increase has largely resulted from the rising popularity of pasta, noodles, and baked potatoes. Fresh fruits, vegetables, and whole grain products have undergone a resurgence in popularity due to expanded selections available in the supermarkets and increased nutritional awareness of the value of fiber.

Although the goals call for a reduction in the consumption of refined and processed sugars, the consumption of these sugars has continued to rise (a trend that began early in the twentieth century). At least part of the recent rise is attributable to the increased consumption of soft drinks, which contain high fructose corn syrup. One positive sign, however, is the continuing switch to diet beverages, which now account for 40% of the market.

Americans are eating less red meat, eggs, and whole dairy products than ever before, which is a good trend for reducing saturated fats and cholesterol. Consumption of poultry and fish has increased as beef has declined, and low fat dairy products have become the preferred choice for many. Although the intake of saturated fat has decreased (dramatically for butter and lard), overall fat consumption is still high, with plant oils the main source. This has had a significant agricultural impact, as farmers have increased the acreage devoted to growing oil crops such as sunflower, safflower, rape seed (canola oil), corn, and soybean. Margarine, shortening, salad oils, and cooking oils account for the expanded use of plant oils, but remember, if the plant oils are hydrogenated, they have reduced health value.

Since the early 1980s there has been a heightened awareness about the dangers of high blood cholesterol levels and Americans have responded with changes in diet and exercise. A report on several thousand men and women showed a decrease in cholesterol levels: for men, the 1980–82 average was 205 mg lowered to 200 mg in the 1985–87 study; likewise, women showed a drop from 201 mg to 195 mg in the same period. This trend indicates that many Americans are attempting to keep blood cholesterol at or below the recommended level of 200 mg.

Although sodium is one of the major minerals, most Americans ingest more sodium (in the form of sodium chloride—table salt) than required; high sodium intake is related to hypertension (high blood pressure). Excessive salt ions in the bloodstream draw water from the tissues, thereby raising the fluid pressure in the blood vessels. A low salt diet is often effective in lowering blood pressure. Some foods naturally contain sodium, but much of the sodium in our diet generally comes from the table salt added during cooking, during the meal, or in prepared foods. In fact, two-thirds of dietary salt is actually "hidden" in commercially prepared food and beverages.

Meatless Alternatives

With the awareness about the dangers of saturated fat and cholesterol inherent in animal products, many Americans are incorporating a greater percentage of vegetables, fruits, grains, and legumes in their diet. Some even choose a totally **vegetarian** life-style. There are many different forms of vegetarianisms: **lacto vegetarians, lacto-ovo vegetarians,** and **vegans.** Some vegetarians, like the lacto and the lacto-ovo, consume dairy products or dairy products and eggs, but do not consume animal flesh. Vegans are pure vegetarians, consuming no animal products at all. Some other vegetarians stretch the concept by consuming fish and poultry, only avoiding red meat.

There are several health benefits to increasing the consumption of plant products while decreasing the consumption of animal products. Vegetarians are less likely to suffer from chronic diseases that afflict many Americans whose diets are high in animal products. Blood cholesterol and triglyceride levels usually reflect the amount of animal fat in the diet and are lowest in vegans who, consequently, have a lower incidence of heart disease. Those cancers linked to fat consumption (colon, breast, and prostate) are less common in vegetarians. High fiber in a vegetarian diet also plays a role in reducing risks of colon cancer and lowering cholesterol levels. A further benefit of high fiber diets is in weight control; the filling effect of fiber suppresses overeating.

Vegetarians, especially vegans, must be knowledgeable about all the nutritional requirements and use care in selecting those plant sources that meet these requirements. Special attention must be given to insure that iron, the B vitamins (especially B_{12}), calcium, vitamin D, and the essential amino acids are supplied. Complementing the essential amino acid is of prime importance, since all plant proteins are incomplete. This can be easily accomplished by eating legumes and grains that, in combination, provide an excellent source of protein for the diet. In fact, a recent study indicates that proteins from plant foods may be preferable to animal protein since consumption of excessive animal protein itself, not just animal fat, may be linked to heart disease and cancer. The suggestion is that meat should no longer be the centerpiece of a meal, but instead relegated to a side dish for a vegetarian entree. Perhaps vegetarians have the right idea.

1. The nutritional needs of the human diet can be categorized into the macronutrients (carbohydrates, proteins, lipids) and micronutrients (vitamins and minerals).
2. Carbohydrates include simple sugars like the monosaccharides, fructose and glucose, and the disaccharides, sucrose and lactose. Complex carbohydrates or polysaccharides include starch and glycogen and fulfill the role as the body's fuel. Dietary fiber includes both soluble and insoluble forms that provide beneficial health effects acting as roughage, promoting regularity and lowering blood cholesterol levels.
3. All plant proteins are incomplete, lacking one or more essential amino acids. But complementing incomplete plant proteins such as combining legumes and cereals in a single meal can overcome this deficiency.

4. Lipids includes fats and oils, and most of the dietary lipids are classified as triglycerides. A diet high in saturated fats raises the risk of cardiovascular disease and certain cancers. Unsaturated fats, whether monounsaturated or polyunsaturated, lower the risk of cardiovascular disease by lowering blood cholesterol levels. Foods of animal origin generally are high in saturated fats and cholesterol, but food of plant origin lack cholesterol and contain mainly unsaturated fats.
5. Deficiency diseases result if a diet lacks any of the essential vitamins and minerals. Fat-soluble vitamins (A, D, E, and K) can be stored by the body but can build up to toxic levels if excessive amounts are taken. Water-soluble vitamins (C and B complex) cannot be stored in appreciable quantities by the body. Most vitamins and minerals (the exception is vitamin B_{12}) can be found in foods of plant origin. The category of minerals (major or trace) is defined by the quantity needed in the body.
6. Americans should modify their diet for a healthier life-style. Dietary guidelines suggest that carbohydrates should make up about 58% of the daily caloric intake, followed by no more than 30% total fats (10% each saturated, monounsaturated, and polyunsaturated). Increasing the percentage of plant-based foods in the diet, while at the same time decreasing the amount of animal-based foods, can lower the risks of cardiovascular disease and certain cancers.

Review Questions

1. What is the importance of fiber to the diet?
2. What are the essential amino acids? Why is it important that they be consumed?
3. What is the role of lipids in the body? What is the dietary significance of saturated and unsaturated fats?
4. Describe the following deficiency diseases: scurvy, rickets, marasmus, beriberi, osteoporosis.
5. What are the latest dietary recommendations?
6. What can cause anemia?
7. What differentiates simple carbohydrates from complex carbohydrates?
8. What dietary factors influence the development of cardiovascular diseases or cancer?

Further Reading

Brody, Jane E. 1985. *Jane Brody's Good Food Book: Living the High-Carbohydrate Way.* Bantam Books, Toronto.

Hoff, Johan E. and Jules Janick. 1973. *Food: Readings from Scientific American.* W. H. Freeman and Co., San Francisco, CA.

Kretchmer, Norman and William van B. Robertson. 1978. *Human Nutrition.* W. H. Freeman and Co., San Francisco, CA.

Lappe, France Moore, 1982. *Diet for a Small Planet.* Ballantine Books, New York.

Nieman, David C., Diane E. Butterworth, and Catherine N. Nieman. 1990. *Nutrition.* Wm. C. Brown, Dubuque, IA.

Whitney, Eleanor N. and Eva N. M. Hamilton. 1984. *Understanding Nutrition.* West Publishing Co., St. Paul, MN.

11

Origins of Agriculture

Chapter Outline

Chapter Concepts

1. Early human societies were based on a foraging lifestyle in which wild plants were gathered and animals hunted.
2. Agriculture evolved independently in several areas of the world, most likely as a natural consequence of intensified foraging.
3. The earliest evidence of agriculture dates back approximately 10,000 years in the Near East, with dates more recent for the Far East and Mesoamerica.
4. Domesticated plants are genetically different from their wild counterparts, and many can no longer survive without human intervention.

The human species known as *Homo sapiens* has existed for possibly as long as 400,000 years. For most of that time, humans survived as **foragers** or **hunter-gatherers,** gathering wild plants and hunting animals in their natural environment. Around 10,000 years ago in many areas of the world, there was a shift in human endeavor from foraging to farming. Most authorities agree that agriculture arose independently in different areas over several thousand years. Why this shift occurred can only be theorized, but the development of agriculture formed the basis of advanced civilization in both the Old and the New Worlds. Over the centuries, agricultural societies spread into those environments that could be easily adapted to agriculture, and foragers gradually became restricted to marginal areas. By the late twentieth century, foraging societies have largely disappeared, comprising only a tiny percentage of the human population and limited to a few tropical rain forests, deserts, savannahs, tundras, and boreal forests (fig. 11.1).

Foraging Societies and Their Diets

Foraging societies are by no means all alike in the types of food they eat. Some groups, such as the Arctic Eskimo, subsist almost entirely on meat, while at the other extreme the Hadza of Tanzania are largely vegetarian, rarely hunting for meat. Other groups such as the !Kung (the exclamation point is pronounced as a click) of southern Africa, however, have had a more varied diet, largely plant-based but supplemented by eggs, insects, fish, small animals, and meat from the hunt.

Many hunter-gatherers have utilized plants by gathering seeds, flowers, roots, and tubers and have a fairly thorough knowledge of the botany in the area. From experience they know which plants are edible, which are poisonous, which have medicinal properties, which are sources of dyes, which could be used for weaving or building materials, and even which have psychoactive properties. By looking for certain visible clues such as flowering on so-called "calendar plants," foragers know if tubers are ready to be dug up or if turtles are laying their eggs. They have developed remarkable methods to prepare edible foods, even from plants with toxins such as cassava, which contains poisonous hydrocyanic acid.

Early Foragers

Archaeological investigations have supplied knowledge about the diet of early humans from many sources and radiocarbon dating of artifacts can provide an estimated time frame. Fossilized remains of both plants and animals have been found in early settlements (fig. 11.2). Plants in the diet have been identified from charred seeds and preserved fruits or other plant parts, while bones, teeth, feathers,

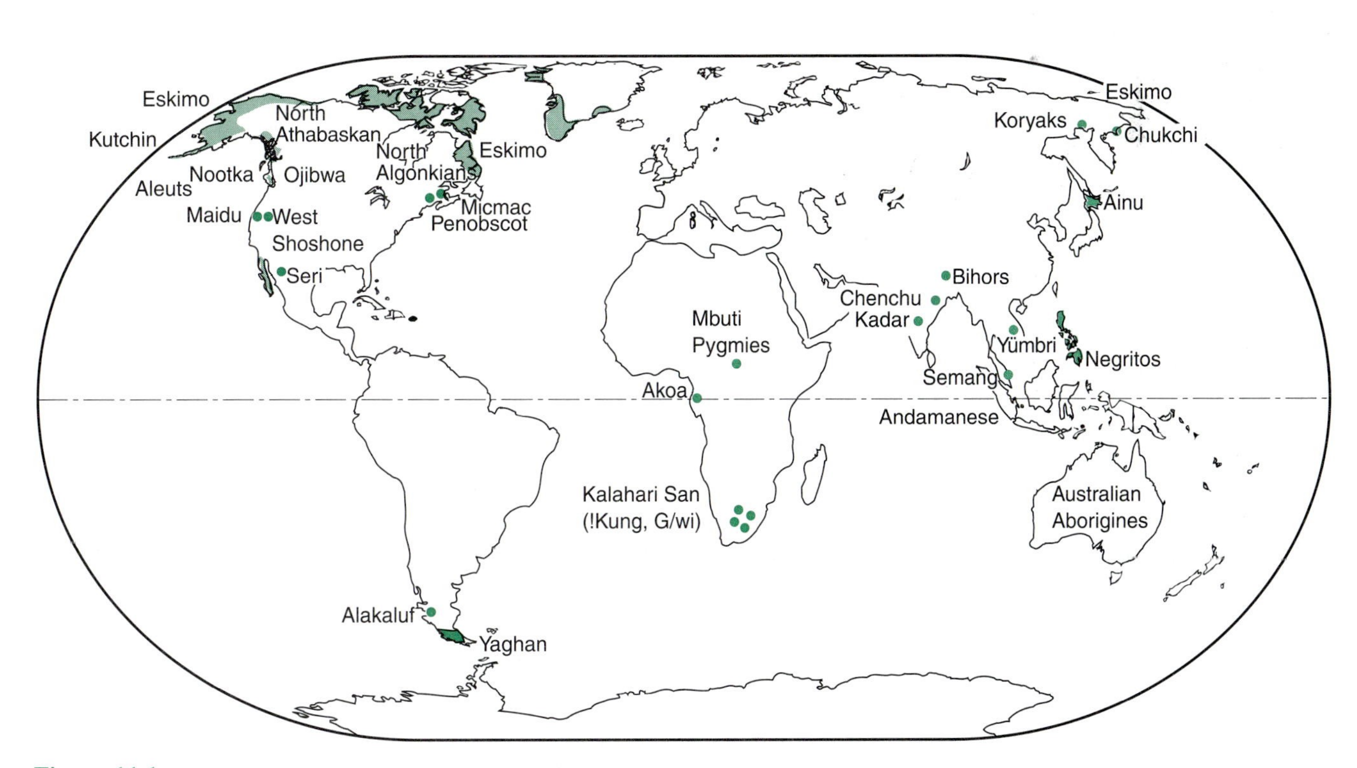

Figure 11.1 *Few foraging societies exist today. The location of these groups is indicated on this world map.*

Richard E. Leakey and Roger Lewin, *Origins,* 1977 E.P. Dutton, New York.

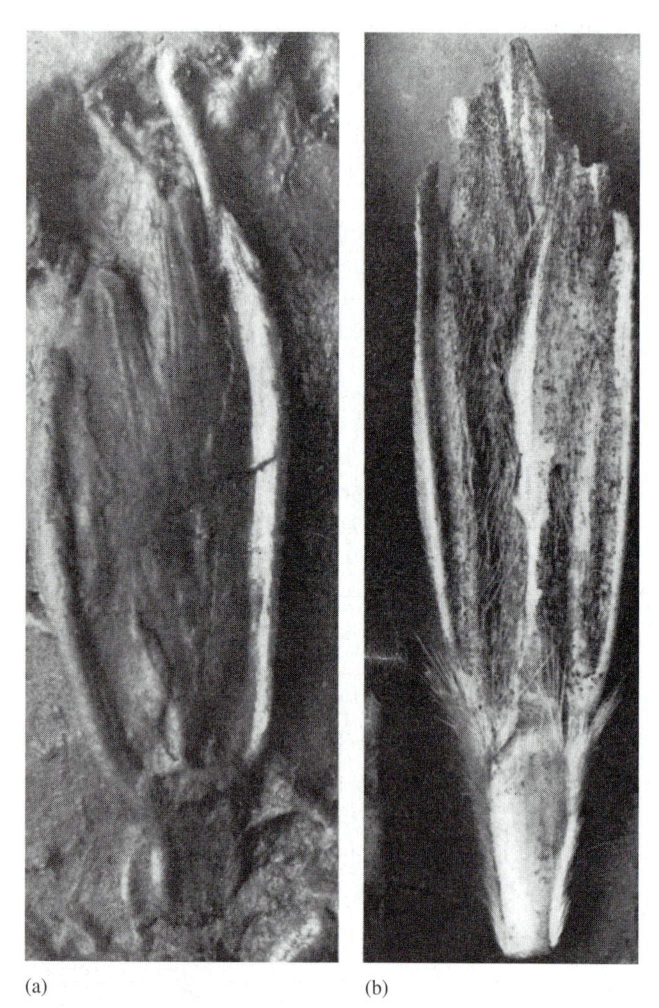

(a) (b)

Figure 11.2 *(a) A clay impression fossil of wheat as compared to (b) a present day species of wild wheat.*

scales, fur, and shells indicate the animals in the diet. Microscopic remains include plant fibers, plant crystals, and pollen, with the latter especially useful in identification. **Coprolites,** fossilized fecal materials, provide direct evidence of the diet since some plant materials, especially seeds and pollen, can pass through the digestive tract largely intact. Often middens, or dump sites, from a human encampment provide a concentrated source of plant and animal remains.

Grinding stones, sickles, and digging implements provide information on the food in the diet, but do not indicate if the plants harvested were wild or domesticated. Tools of the hunt also provide some information on the size of the animals hunted and the method of the kill. Depictions in cave paintings (of animals as well as hunting and gathering activities), pottery fragments, and clay figurines are other windows to the past.

In some excavations at Wadi Kubbaniya in the Nile Valley of Upper Egypt, archaeologists have dated plant remains of a hunter-gatherer settlement from 17,000 to 18,000 years ago. The charred remains of fruits, seeds, and tubers from 25 different plant species have been found; interestingly, contemporary foraging societies have utilized these same plants or closely related species. The most abundant plant remains found were the tubers of wild nut grass, a type of sedge. Analysis of modern samples of these tubers indicates that they are high in carbohydrates and fiber but low in protein; however, the protein is of good quality since it is high in lysine, one of the essential amino acids. Mature nut grass tubers also contain toxins, but these could be easily removed by various methods still in use today. It is believed that this tuber served as one of the dietary staples, along with acacia seeds, cattail rhizomes, and palm fruits. Evidence from this and other **Paleolithic** (Old Stone Age or preagricultural societies) sites indicate that early foraging groups had a remarkably varied plant diet.

Modern Foragers

Much of the knowledge about the foraging way of life has been documented by studying modern foraging societies such as the !Kung San of the Kalahari Desert of southern Africa. These people live in the tropical savannahs that border the Kalahari in what is now southeast Angola, northeast Namibia, and northwest Botswana. The !Kung have foraged in this area for at least 10,000 years, continuing the hunter-gatherer way of life until recent times. Extensive studies of the !Kung during the 1960s revealed that they utilized over 100 species of plants and 50 animal species, with two-thirds of their diet plant-based. The plants included a mixture of fruits, nuts, berries, melons, roots, and greenery (fig. 11.3); one particular nut, the mongongo nut, was a very important high protein component of their diet. The !Kung diet was very nutritious; they consumed an average of 2,355 Calories per person per day, with 96 grams of protein and adequate vitamins and minerals. This more than meets the nutritional requirements established for people of the !Kung stature and physical activity. As with most foraging societies, the division of labor was along gender lines, with women doing most of the gathering and men the hunting. Surprisingly, the amount of time spent on foraging activities averaged only about 2.5 days per week, which left plenty of time for leisure and socializing.

Agriculture: Revolution or Evolution?

Archaeological evidence indicates that about 10,000 years ago human cultures began the practice of agriculture in several different areas of the world. Over the next few thousand years in the Near East, the Far East, and Mesoamerica, agriculture flourished. The question that has puzzled archaeologists and other scientists is "Why, after thousands of years of foraging, did hunter-gatherers switch to agriculture?"

Many theories have been proposed to answer this question. Some state that agriculture was the discovery of a brilliant sage who, with a flash of insight, realized that if you

(a)

(b)

Figure 11.3 *(a) The !Kung are a modern day example of a hunter-gatherer society. A guinea fowl is the trophy of this hunt. (b) The diet also includes a wide variety of plants.*

sow seeds, the crop will grow. There were many variations of this theme. Some held that the brilliant sage realized that plants growing at middens, or dumpsites, were growing from discarded seeds. Others held that the sage made his observations from seeds (buried with the dead as food in the afterlife), which gave rise to plants at grave sites. These theories viewed agriculture as a revolution; in fact, the term "Agricultural Revolution" was used to describe the transition from foraging to agriculture. It was suggested that this revolution spread quickly because agriculture was thought to be an improvement over the hunter-gatherer life-style in that a dependable food source could be easily grown rather than collected from the wild.

Beginning in 1960, archaeologists questioned this view, suggesting instead that the origin of agriculture was not a revolution but the result of a gradual cultural evolution. They reasoned that hunter-gatherers knew the wild plants, knew how they grew, and would incorporate farming along with foraging as part of an overall food collection strategy when necessary. For example, certain early Indian groups in coastal Peru abandoned their farming practices whenever fish became plentiful.

Many archaeologists believe that there was a transitional stage between simple foraging, where small nomadic bands followed the wild plants and animals, and agricultural societies with their sedentary life-style. During this transitional stage, foraging groups formed settlements but sent out members to hunt and gather. This more complex strategy resulted in changes in the social organization of the groups and permitted populations to increase. This transitional stage lasted for several thousand years in some locations, until resource stress or environmental change led to the switch to agriculture. In the Near East, for example, archaeologists believe that the climatic dry period around 11,000 years ago brought about a change in the distribution of cereal grains (wheat and barley). Applying their botanical knowledge, these foragers gradually changed from collecting these wild cereals to cultivating them.

A Closer Look 11.1

Forensic Botany

Archaeologists have made extensive use of plant remains in reconstructing the lifestyle of ancient foraging, as well as early agricultural, peoples. The recovering and processing of plant remains begin the painstaking task of botanical detection. Not all parts of a plant are equally well preserved; seeds, wood, pollen, phytoliths, and fibers are among the most informative recoverable remains. The lignified walls of wood and fibers are more resistant to decay than purely herbaceous tissue, since lignin can only be degraded by a few types of fungi. Likewise the lignin, in the sclerified cells of the seed coat or testa, makes seeds among the most common dietary material recovered from archaeological sites. Buried pollen grains are virtually indestructible due to the chemical properties of the exine (outer pollen wall). Phytoliths are plant crystals that are formed in the vacuoles of epidermal cells in the stems and leaves of some herbaceous plants (box fig. 11.1a). The crystals are usually composed of either calcium salts or silica and can remain intact for thousands of years under certain conditions. The distinctive structure of many plant remains permits identification at least to the family level and usually to the genus level (box fig. 11.1b).

These same principles of plant detection are applicable to forensic science, where the identification of plants can be used as incriminating evidence. The first criminal case that used botanical information was the famous 1935 trial of Bruno Hauptmann, who was accused and convicted of kidnapping and murdering the young son of Charles and Anne Morrow Lindbergh. The botanical evidence centered on a homemade wooden ladder used during the kidnapping and left at the scene of the crime. After extensive investigation, plant anatomist Arthur Koehler showed that parts of the ladder were made from wooden planks from Hauptmann's attic floor.

In forensics, even herbaceous plant parts can be useful for identification. In one case, botanical evidence disputed the testimony of an accused rapist. Fragments of tree leaves and bark in his pants cuff indicated that the accused climbed a tree to get into a window of the victim's home, rather than being admitted through the front door as he claimed. In cases of suspected plant poisonings, identification can be made from leaves or fruits of intact plants or even from analysis of stomach contents. In this type of situation, proper medical treatment depends on accurate identification of the plant or mushroom. Trained botanists and mycologists (scientists who study fungi) are routinely called to hospital emergency rooms for this purpose. Botanical analyses of stomach contents have also played roles in other types of cases; a hunting guide killed a grizzly bear he claimed was eating his supply of alfalfa hay. Botanical evidence showed no alfalfa in the stomach, only native vegetation. He was subsequently fined and imprisoned for killing an endangered species.

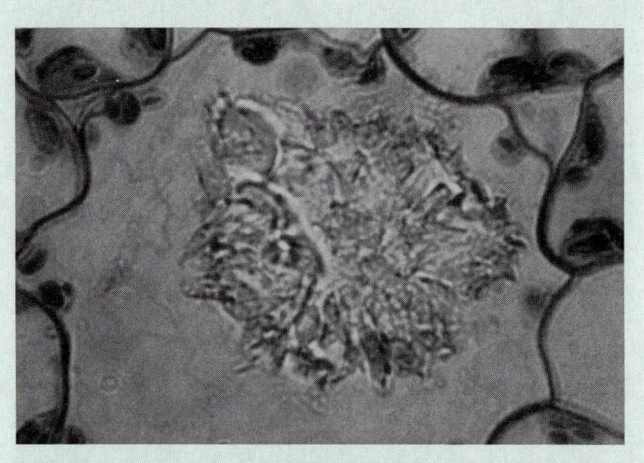

Box Figure 11.1a *Calcium oxalate crystal in the vacuole of a leaf cell, ×270.*

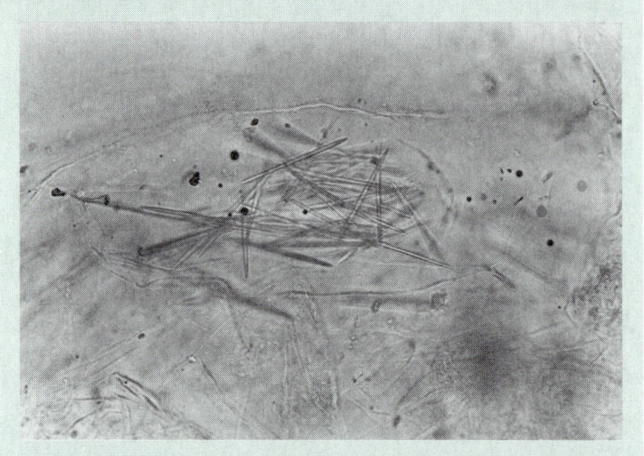

Box Figure 11.1b *Needlelike crystals are still recognizable in chewed pineapple pulp, ×300.*

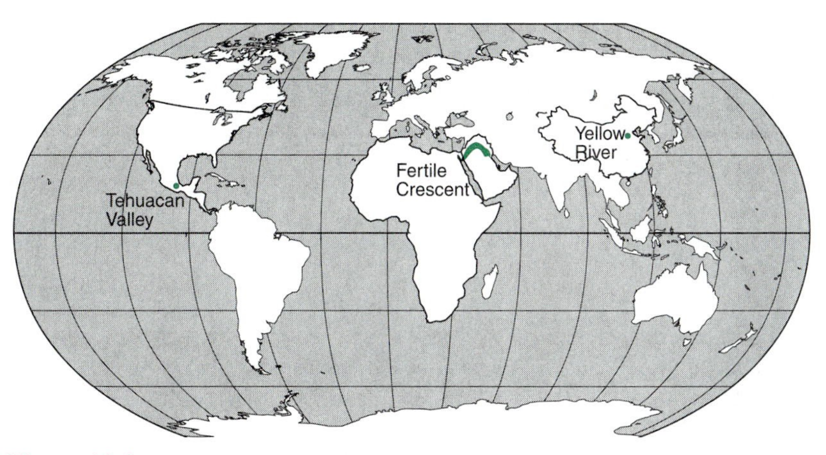

Figure 11.4 *Evidence of the beginnings of agriculture have been found in the Fertile Crescent of the Near East, the Yellow River Valley of China (Far East), and the Tehuacan Valley in Mexico (Mesoamerica).*

Early Sites of Agriculture

Archaeological excavations have documented many sites of early agriculture in both the Old and New Worlds. The evidence indicates that the earliest sites were in the Near East, dating back approximately 10,000 years.

The Near East

Some of the oldest sites of agriculture are in southwestern Asia, in the foothills around the area known as the **Fertile Crescent,** which today includes parts of Iran, Iraq, Turkey, Syria, Lebanon, and Israel (fig. 11.4). From sites such as Jarmo, Jericho, Ali Kosh, Cayonu, and others, remains of both plants and animals date back 9,000 to 14,000 years ago. Early plant domesticates include einkorn wheat, emmer wheat, barley, pea, lentil, and vetch, while dogs, goats, and sheep were among the domesticated animals. Research suggests that the animals were domesticated several thousand years before the plants.

One of the sites that had been particularly well studied was Jarmo in northeast Iraq, in the foothills of the Zagros Mountains. This area was inhabited approximately 9,000 years ago; it was a permanent farming village with about 24 mud-walled houses and a population of about 150 people. The charred grains of wheat and barley found here were a domesticated type already changed from the wild type; for this genetic change to have taken place, initial cultivation must have predated the age of the remains. In addition to the domesticated plants and animals, the inhabitants of Jarmo also continued foraging, as evidenced by the bones of wild animals, snail shells, acorns, and pistachio nuts. Artifacts uncovered at the site include flint sickles and grinding stones for harvesting and milling cereal grains as well as clay figurines, woven baskets, and rugs.

The Far East

It is apparent that agriculture arose at several locations in the Far East from excavations of dozens of sites in Southeast Asia, including Spirit Cave in Thailand and sites in both the Yellow River (fig. 11.4) and Yangtze River Valleys in China. Among the earliest plants domesticated in the Far East are rice, foxtail millet, broomcorn millet, rape, and hemp, with evidence of domesticated cattle, pigs, dogs, and poultry as well.

Recent excavations in China provide clear evidence that agriculture was established between 7,000 and 8,000 years ago. In the settlements along the Yellow River, foxtail millet was the first cereal to be domesticated and become a dietary staple; it remained the dominant crop in North China until the historical period. Broomcorn millet was also cultivated but not as extensively. Rice (both long-grained and short-grained varieties) cultivation began in southern China approximately 7,000 years ago, about 1,000 years after millet had been domesticated. The archaeological evidence indicates that one center of origin for cultivated rice was along the middle and lower reaches of the Yangtze River. Tilling tools, harvesting tools, and grain-processing tools made of stone, bone, shell, or wood have also been recovered from these sites, along with bones of domesticated animals.

The New World

In contrast to the Old World, the **Neolithic** (New Stone Age or agricultural society) cultures of the New World had domesticated an impressive array of plants but

Figure 11.5 *Preserved corn cobs from the Tehuacan Valley, Mexico. The oldest cob, on the left, is approximately one inch long.*

comparatively few animals. Among the earliest crops domesticated in the New World were corn, squash, chili peppers, amaranth, avocado, gourds, beans, and both white and sweet potatoes, with only dogs and llamas as domesticated animals. Most of the archaeological evidence for early agriculture in the New World has been obtained from the highlands of Mexico and Peru.

Probably the most thoroughly documented site is a group of caves in the Tehuacan Valley of central Mexico (fig. 11.4). Research in this area was initiated to obtain information on the ancestry of corn. Working with ancient corn cobs, archaeologists determined that corn was domesticated in this region by 5,500 years ago (fig. 11.5). Originally it was thought that these corn cobs were much older (7,000 years old), but new dating techniques have recently advanced the time frame. The Tehuacan Valley is one of the few sites where the transition from foraging to farming can be thoroughly documented. For thousands of years, people inhabited the caves seasonally as they foraged for plants and hunted animals. At first, there was a shift from hunting to a more intensified foraging of plants, along with the domestication of squash and avocado. Later, the list of domesticated plants expanded to include corn, bottle gourd, two species of squash, amaranth, three species of bean, and chili peppers; however, people still relied on foraging for the majority of their diet. Over the next few thousand years, agriculture became even more important as additional plants (tomato, peanut, guava), the dog, and later the turkey were domesticated. Artifacts such as stone tools, textiles, and pottery were also found. From this evidence, archaeologists have been able to reconstruct a picture of the life-styles of the inhabitants of these caves over a 12,000-year period.

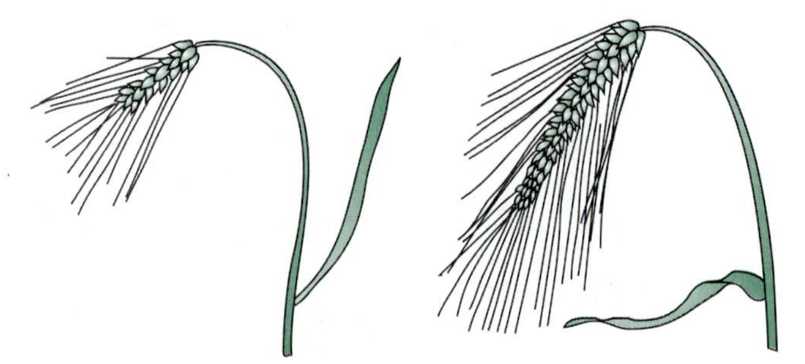

Figure 11.6 *Two-rowed wild barley (left) is contrasted with six-rowed domesticated barley (right).*

Characteristics of Domesticated Plants

Plants that have been **domesticated** are genetically distinct from their wild progenitors. Through the process of **natural selection,** wild plants have evolved mechanisms that ensure their survival in the environment, but once a plant is domesticated, traits are **artificially selected** to suit human needs and do not necessarily have a survival value. In fact, some of these traits might be detrimental to survival in the natural environment. For example, modern corn with its ensheathing husks cannot disperse its seeds; also, domesticated wheat and other cereals have fruiting heads that are nonshattering, which limits seed dissemination.

Most wild grasses have **shattering** fruiting heads, which will break apart at a slight touch or breeze and scatter their seeds over a wide area. A recessive gene is responsible for a tough spike with a **nonshattering** head. It would be natural for early foragers to gather those seeds attached to the tougher spikes. When agriculture began, the seeds most easily gathered would be planted and so pass on the nonshattering trait. Likewise, early foragers would select for larger seeds, fruits, or tubers and over time the domesticated varieties become larger than their wild counterparts. For example, wild barley has two rows of grains, while the domesticated varieties have six rows (fig. 11.6). In fact, archaeological evidence of six-row barley is indicative of agriculture at that excavation. Plant breeders continue to select for desired traits today using traditional, as well as more sophisticated, genetic manipulations (*see* Chapter 15).

Centers of Plant Domestication

Within each area of the world where agriculture evolved, the native peoples developed indigenous crops for a staple food supply. Crops that were particularly suitable for agriculture slowly spread to surrounding regions as people traded with others or migrated to new areas, bringing their crops with them. This diffusion led to the emergence of

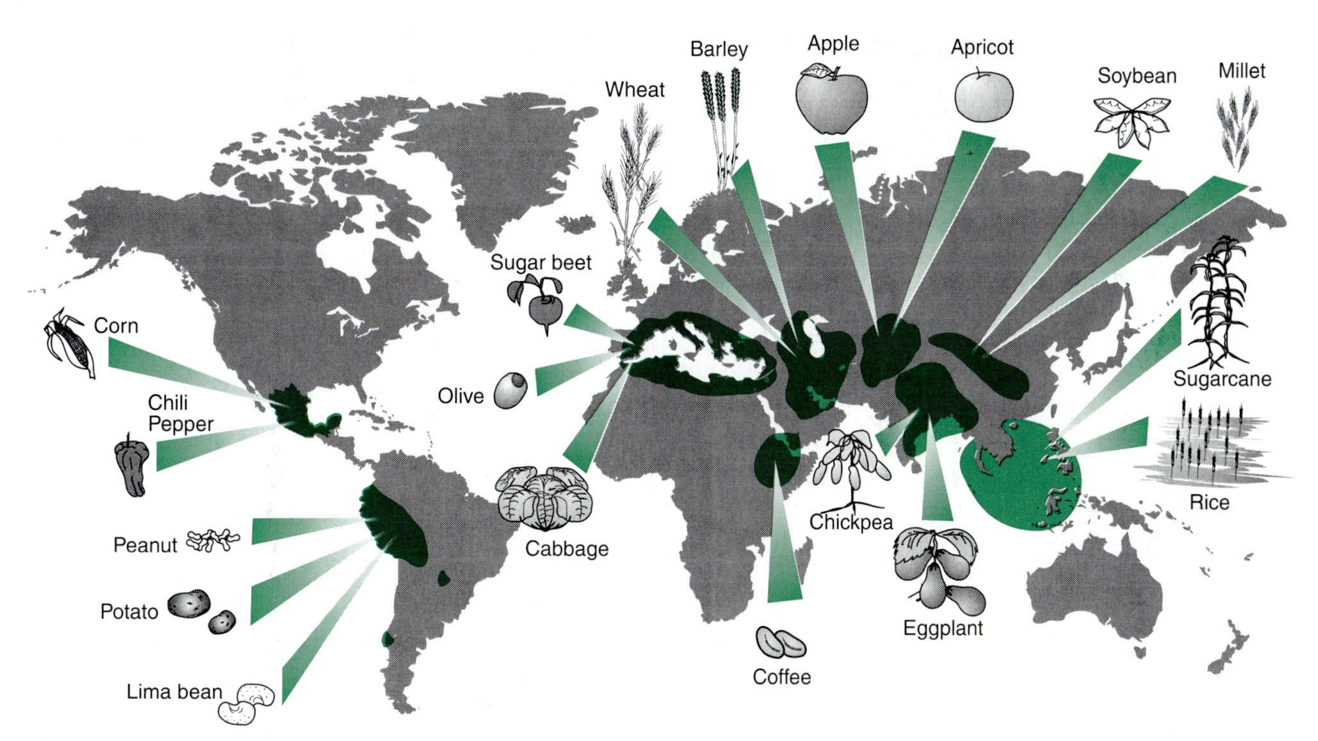

Figure 11.7 *Centers of origins as first defined by N. I. Vavilov. Examples of domesticated crops that originated from each center are indicated.*

principle crops associated with major centers of the world. In the Near East, wheat and barley were the dietary staples; in the Far East, rice; in Africa, it was sorghum and millet; in Mesoamerica, corn; and in South America, it was the potato and other root crops. As civilization continued to develop, trading and migration expanded the range of crops far from their origin, and today many crops are even more successful outside their native range. Potatoes, which became associated with the Irish, are actually native to the highlands of Peru; coffee, actually native to the mountains of Ethiopia, is most frequently linked to Colombia and Brazil; and the tomato, so essential to Italian cuisine, was first domesticated in the New World. The agricultural harvest of the United States would be meager if limited to commercial crops of native origin such as blueberries, cranberries, sunflower, pecan, and maple syrup.

Pinpointing the exact origin of important crops has intrigued scientists for many years. The name most frequently associated with this endeavor is Nikolai I. Vavilov, a Russian botanist. Vavilov directed plant collecting expeditions around the world and examined thousands of plants, looking for patterns of variation in crop plants and their wild relatives. He reasoned that areas that had the greatest diversity of a particular crop would most likely be the center of origin for that crop. Based on his research from 1916 to 1936, Vavilov proposed eight centers of origin for the major domesticated plants, six in the Old World and two in the New World. Examples of crops known to have originated in these centers are indicated in Figure 11.7. Vavilov's life ended tragically in a Soviet Gulag. Biology in the Soviet Union under Stalin was dominated by Trofim Lysenko, who rejected established genetic theory in favor of his own outdated views. Vavilov's adherence to Mendelian genetics clashed with the established policies of the State and he was sentenced to a Soviet prison camp, where he died in 1943, a martyr to the cause of scientific freedom.

Recent work has expanded the number of centers and questioned Vavilov's conclusions. Evidence suggests that while some crops have been domesticated more than once in different places, others did not originate where Vavilov indicated, and still others were developed over vast regions. For example, certain New World crops such as corn and cassava appear to have been independently domesticated in both Mesoamerica and South America. This search for the origin of certain crops is even more important today as plant geneticists strive to improve the gene pool of domesticated plants by tapping the genetic resources of wild strains (*see* Chapter 15).

Chapter Summary

1. The first human societies were based on a foraging life-style, gathering wild plants and hunting animals for food. Archaeological investigations have determined that the diet of Stone Age foragers was varied, especially in the variety of plants consumed. Studies on the diet and life-style of extant foragers, especially the !Kung San of the Kalahari Desert, reinforce the viewpoint that the foraging life-style more than satisfies the nutritional requirements yet allows time for activities not directed to food gathering and preparation.
2. The archaeological record indicates that, at least in some parts of the world, certain groups began to shift from the nomadic, foraging life-style to the sedentary one of agriculture. Many theories to explain this shift have been presented and discarded over the years but the currently held view believes that the switch to agriculture was gradual. A prolonged transitional stage ensued in which groups formed settlements but sent out members to hunt and gather.
3. The earliest agricultural settlements, nearly 10,000 years ago, have been found in the Near East, in the area known as the Fertile Crescent. Sites in the Far East in southeast Asia also document early agriculture. The New World dates from the Tehuacan Valley of Mexico are not as early.
4. Domesticated plants and animals are genetically different from their wild relatives since they have been shaped by artificial selection. Many of the traits, such as a nonshattering fruiting head, do not enhance a plant's survival value but have been selected to suit humanity's needs.
5. Since domesticated plants have been modified greatly from their wild ancestors after thousands of years of artificial selection, it has often been difficult to pinpoint their area of origin. Nikolai I. Vavilov laid the foundation for detecting the centers of origin of domesticated plants when he proposed eight centers, six in the Old World and two in the New.

Review Questions

1. What has been learned about foraging societies of the past by studying the !Kung?
2. Describe an early agricultural community in the Near East.
3. How do archaeologists reconstruct diets of prehistoric peoples?
4. How have the theories about the origin of agriculture changed in recent decades?
5. What crops were domesticated in the New World?
6. What is artificial selection and how has it shaped the world's crops?
7. Why is the Tehuacan Valley an important archeological site for the origin of agriculture in the New World?

Further Reading

Braidwood, R. J. 1960. The Agricultural Revolution. *Scientific American,* 203 (3): 130–148.

Bryant, V. M. and G. Williams-Dean. The Coprolites of Man. *Scientific American,* 232 (1): 100–109.

Fritz, Gayle J. 1995. New Dates and Data on Early Agriculture: The Legacy of Complex Hunter-Gatherers. *Annals of the Missouri Botanical Garden.* 82 (1): 3–15.

Gilbert, R. I. and J. H. Mielke. 1985. *The Analysis of Prehistoric Diets.* Academic Press, Orlando, FL.

Harlan, J. R. 1976. The Plants and Animals That Nourish Man. *Scientific American,* 235 (3): 88–97.

Harris, D. R. and G. C. Hillman (eds). 1989. *Foraging and Farming: The Evolution of Plant Exploitation.* Unwin Hyman, London.

Henry, D. O. 1989. *From Foraging to Agriculture: The Levant at the End of the Ice Age.* University of Pennsylvania Press, Philadelphia, PA.

Lane, M. A., L. C. Anderson, T. M. Barkley, J. H. Bock, E. M. Gifford, D. W. Hall, D. O. Norris, T. L. Rost, and W. L. Stern. 1990. Forensic Botany. *BioScience,* 40 (1): 34–39.

MacNeish, R. S. 1964. The Origins of New World Civilization. *Scientific American,* 211 (5): 29–37.

Yellen, J. E. 1990. The Transformation of the Kalahari !Kung. *Scientific American,* 262 (4): 96–105.

CHAPTER

12

THE GRASSES

Chapter Outline

Chapter Concepts

1. Grasses are members of the monocot family, Poaceae, whose characteristic grains are a vital food source.
2. Whole grains with the bran and germ intact are nutritionally superior to their refined counterparts, which contain only endosperm.
3. Wheat, corn, and rice, the major cereals, outrank all other plants as food sources for human consumption.
4. Grasses are also indispensable components of forage crops and landscaping designs.

The grass family is of greater importance than any other family of flowering plants. The edible grains of cultivated grasses, or **cereals,** are the basic foods of civilization, with **wheat, rice,** and **corn** the most extensively grown of all food crops. Other important cereals are **barley, sorghum, oats, millet,** and **rye;** all are among the top 25 food crops (fig. 12.1).

Characteristics of the Grass Family

There are approximately 600 genera and 8,500 species of grasses found throughout the world, making the grass family one of the largest and most widely distributed plant families. Grasses are the dominant plants in prairies and savannahs but can also be found wherever plants can grow, under a wide variety of environmental conditions from arctic marshes to tropical swamps. In fact, 25% of the world's vegetation belongs to the grass family.

Vegetative Characteristics

Members of the grass family, the Poaceae, are usually herbaceous, having linear leaves with parallel venation typical of monocots (fig. 12.2). Leaves usually have an alternate arrangement, and the base of each leaf forms a sheath that wraps around the stem. The stems, or **culms,** are often hollow between the nodes and usually unbranched. Many species may also have horizontal stems (either aboveground **stolons** or underground **rhizomes**) that can propagate the plant vegetatively by giving rise to new shoots (*see* Chapter 14). Both annual and perennial species of grasses occur, with most cereals being annuals and most pasture and lawn grasses being perennials. The primary root system is fibrous, and adventitious roots may also form from either of the lower nodes of erect stems (as **prop roots**) or from rhizomes or stolons.

The Flower

The flowers of grasses are borne in inflorescences, typically spikes, racemes, or panicles. The tassels of corn and the heads of wheat are common examples of grass inflorescences. The individual flowers are small, inconspicuous, and incomplete, with sepals and petals lacking completely or replaced by small structures called lodicules (fig. 12.3). Normally each flower has three stamens and one carpel, with a single ovule in the ovary but two styles and stigmas. The stigmas are enlarged and feathery and the

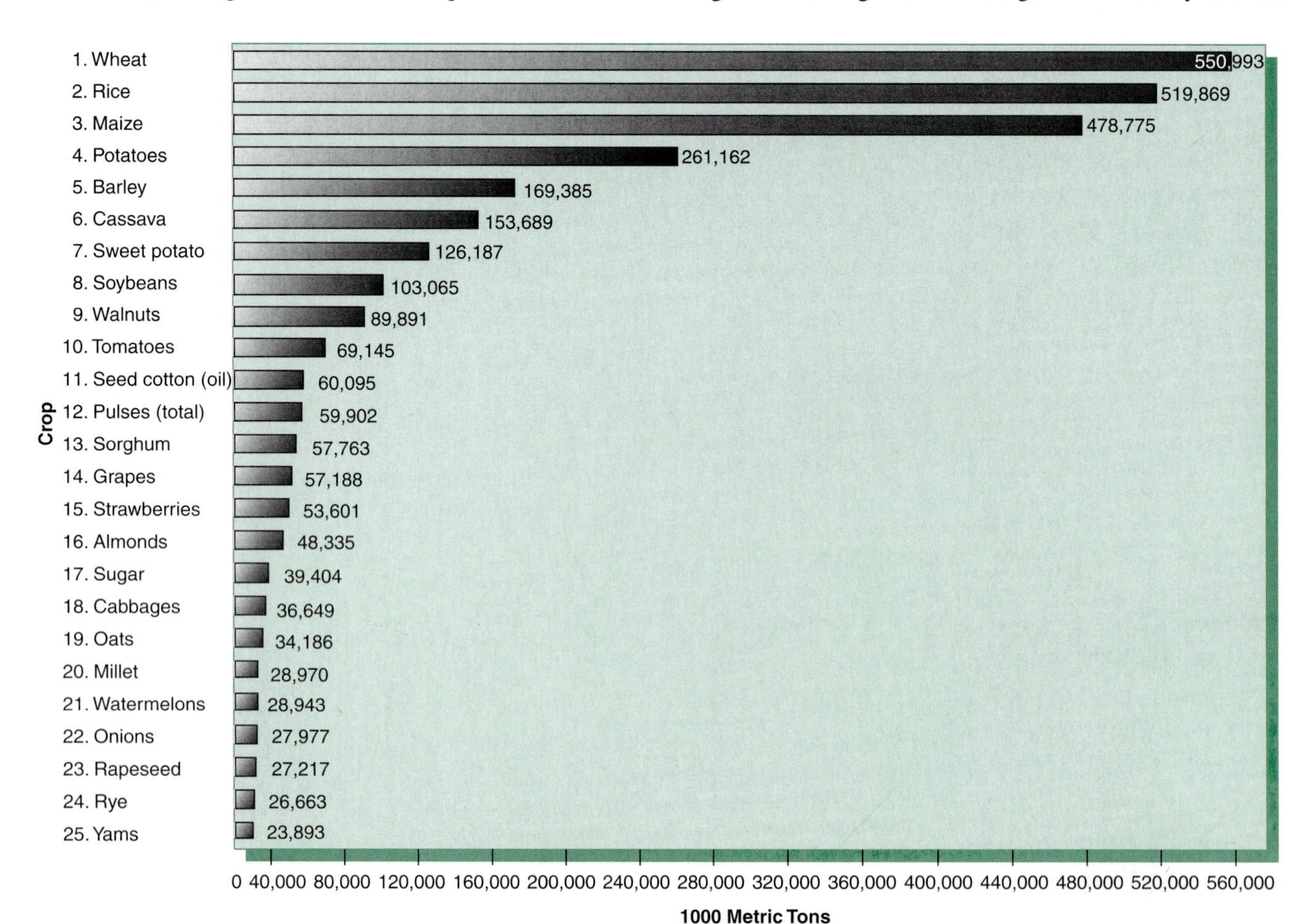

Figure 12.1 *Annual world crop production figures reveal that many of the top 25 crops are cereals.*
Source: FAO Production Yearbook, Vol. 45, 1991.

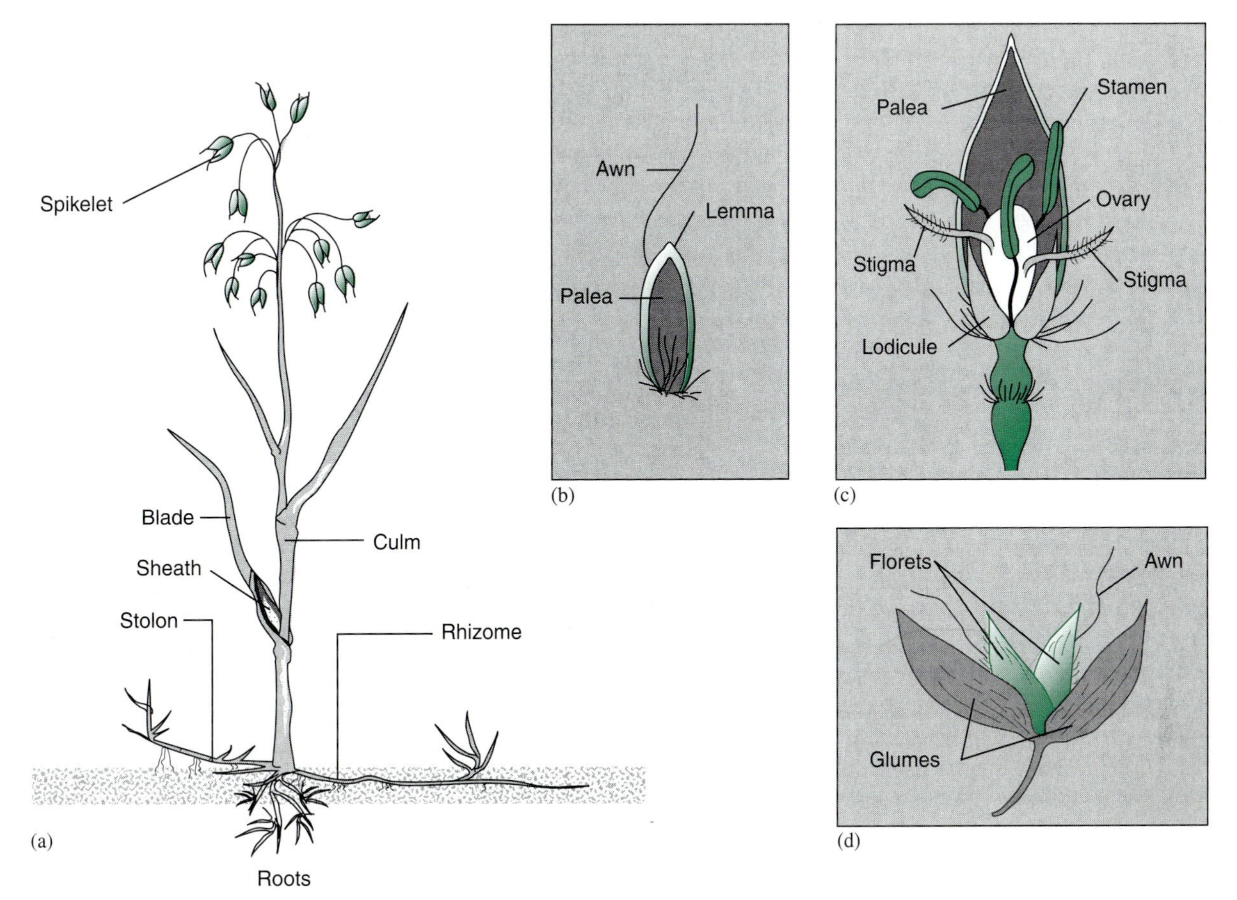

Figure 12.2 *Typical grass plant. (a) Whole plant. Some grasses can reproduce vegetatively through stolons or rhizomes; however, the major cereal crops lack these reproductive structures. (b) Each flower or floret is surrounded by a lemma and palea. (c) The grass flower consists of three stamens and one carpel with two separate styles and stigmas. (d) Glumes subtend each spikelet that bears one or more florets.*

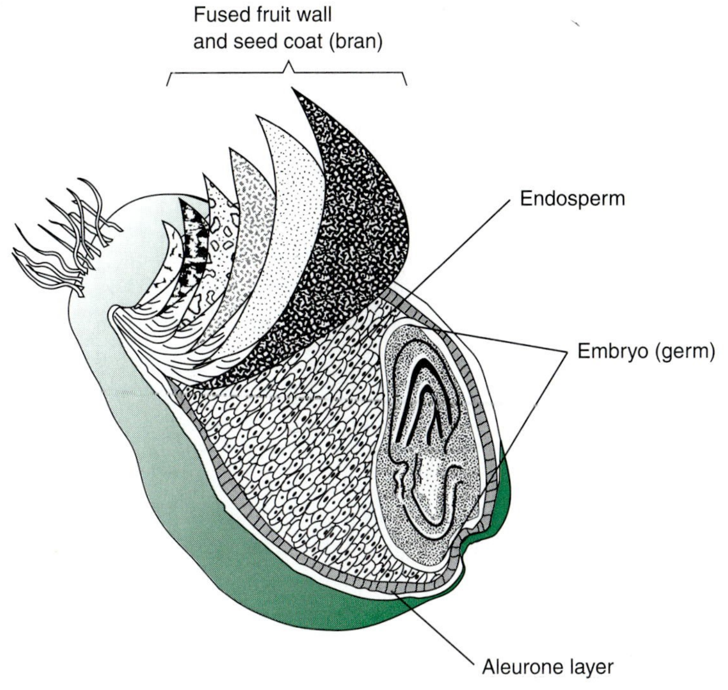

Figure 12.3 *A grain, the typical fruit of plants in the grass family.*

mature stamens pendant; these features facilitate wind pollination. Surrounding each flower are two bracts, the outer **lemma** and inner **palea;** the flower and the two bracts together make up a **floret.** One to 12 florets are arranged on a **spikelet,** which also may be subtended by two bracts called **glumes.** Often a slender bristle can be seen extending from either a glume or lemma (occasionally the palea); this structure is known as an **awn.**

The Grain

The typical fruits for the grass family are **grains,** which are dry, single-seeded indehiscent fruits (fig. 12.3). The bracts that surrounded the flower (or even the spikelet) now surround the grain and are called **chaff.** The outer wall of the grain, consisting of the pericarp fused to the testa, is known as the **bran.** Interior to the bran is a layer of enlarged cells known as the **aleurone layer,** which is normally high in protein. If the seed is allowed to germinate, this layer provides the enzymes that break down stored food for the growing embryo. The majority of the seed is occupied by endosperm, which contains stored food mainly in the form of starch. The scutellum transfers food to the embryo, which is surrounded by the coleoptile and coleorhiza sheaths. The embryo, with its sheaths, is often referred to as the **germ.**

The large amount of stored food in the grain makes this family valuable as a food crop. In the economically important cereals, the endosperm is mostly starch and in the refining process, the chaff, germ, and bran (usually with the aleurone layer attached) are removed, leaving only the starchy endosperm. Commercially, this refined product is available as white flour, corn starch, and white rice. In whole grain products, only the chaff is removed and the entire grain is used, providing certain nutritional advantages over the refined material. The bran provides fiber as well as some protein from the aleurone layer, and the germ is a source of vitamins, proteins, and some oils. Thus, brown rice, whole wheat flour, and even popcorn are more nutritious than refined grains. Their nutrient status is important because these three grains, whether whole or refined, are the dietary staples for the world's population, providing 50% of the calories consumed and the chief sources of both carbohydrates and proteins.

Wheat: The Staff of Life

Wheat is the most widely cultivated crop in the world and supplies a major percentage of the nutrient needs of the human population. Bread is literally the "staff of life" in many cultures. The vegetative appearance of the wheat plant is typical of most grasses, but the spikes are tightly packed with grains, which usually have long awns giving the fruiting head a bearded appearance (fig. 12.4). Wheat does best in temperate grassland biomes that receive about 30–90 cm (12–36 inches) of rain per year and have relatively cool temperatures. Some of the top wheat-producing countries are the Ukraine, United States, Canada, China, India, Argentina, France, and South Africa. It is one of the oldest domesticated plants, and it laid the foundation for western civilization. Domesticated wheat had its origins in the Near East at least 9,000 years ago, and wild species of wheat can still be found in northern Iraq, Iran, and Turkey.

Origin and Evolution of Wheat

The origin of domesticated wheat is a story of hybridization and polyploidy involving several species of wheat (*Triticum* spp.) and species of the closely related goat grass (*Aegilops* spp.). In contemporary wheat species, three groups can be identified on the basis of chromosome number: one group has 14 chromosomes, the second group has 28, and the remaining group has 42 chromosomes. This forms a polyploid series with a base chromosome number of 7 ($n = 7$) including diploid species (14 chromosomes), tetraploid species (28 chromosomes), and hexaploid species (42 chromosomes).

Figure 12.4 *Wheat, the most widely cultivated crop in the world.*

Triticum monococcum is one of the diploid species of wheat known as einkorn wheat (fig. 12.5). It was one of the first cultivated species of wheat and wild forms are still found. The domesticated form differs from the wild type in that the grains are nonshattering (they tend to stay on the stalk) in domesticated einkorn. Most of the other species of *Triticum* are polyploid hybrids that arose naturally through crosses with species of *Aegilops*, which also have a diploid chromosome number of 14.

It is believed that einkorn wheat, hybridizing with one of the goat grasses, *Aegilops* sp., gave rise to the tetraploid emmer wheat, *Triticum turgidum* (fig. 12.5). To fully understand the development of this tetraploid species, consider the diploid chromosome status of einkorn to be represented by AA and that for the goat grass BB (fig. 12.5). A hybrid between them would be AB with 14 separate chromosomes, and if that hybrid doubled its chromosomes, it would be AABB, the chromosomal makeup of tetraploid wheat. Tetraploid wheat evolved naturally before the origin of agriculture, and early societies in the Near East were cultivating emmer along with einkorn wheat. Domesticated emmer has nonshattering grains that are covered by clinging bracts (hulled) but other, later varieties of tetraploid wheats, such as durum, which is widely grown for pasta flour, have naked grains. Naked grains separate easily from the surrounding bracts, a trait referred to as free threshing.

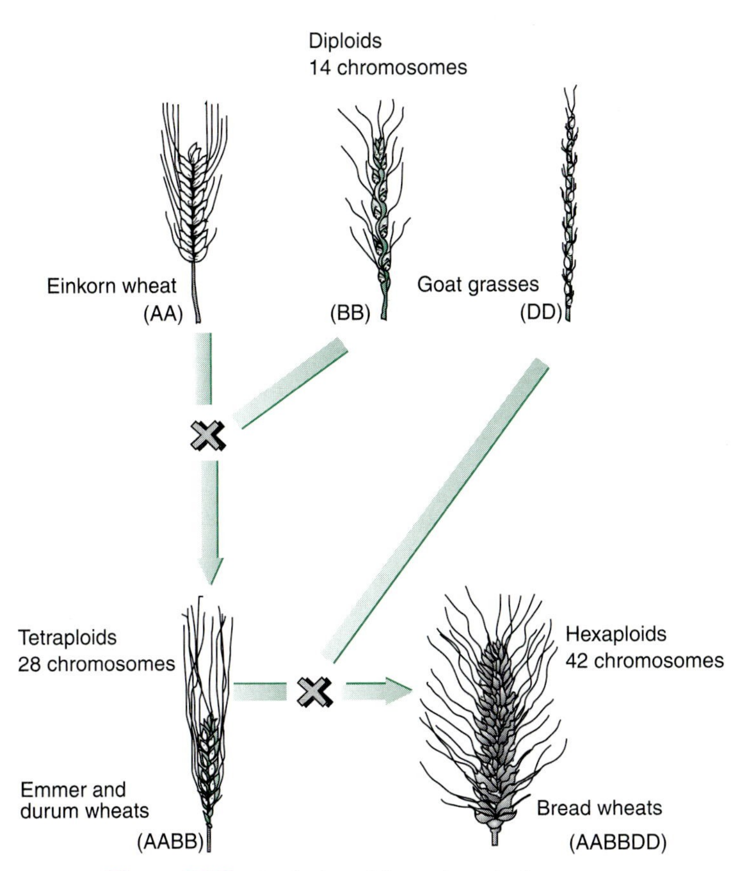

Figure 12.5 *Evolution of domesticated wheat.*

The next stage in the evolution of wheat was the result of another hybrid cross, this time between an emmer wheat and a goat grass (*Aegilops squarrosa*). Crossing the tetraploid AABB with the diploid goat grass DD produced the hybrid ABD. This hybrid doubled its chromosomes, thereby forming the hexaploid AABBDD, which first appeared 8,000 years ago in the Near East (fig. 12.5) and today is known as the bread wheat, *T. aestivum.* The genetic contributions of *A. squarrosa* were a higher protein content in the endosperm and greater tolerance to environmental conditions. Among the proteins in the wheat grain is **gluten,** an important component of flour for the elasticity it provides. This elasticity in wheat flour, coupled with a leavening agent (yeast, baking soda, or baking powder), allows dough to rise, making it suitable for breads and cakes (*see* A Closer Look: The Rise of Bread).

Modern Cultivars

The two types of wheat that are widely cultivated today are durum and bread wheat. Durum wheat (*T. turgidum*) is grown in the northern United States (especially North Dakota), Canada, southern Europe, and parts of India. It has a high gluten content and yields semolina flour, which is used to make spaghetti, macaroni, and noodles. *Triticum aestivum,* bread wheat, is the dominant type of wheat grown, making up approximately 90% of the world production. The flour from bread wheat is used for bread, pastries, and breakfast cereals. There are thousands of cultivars of wheat, enabling the crop to be grown under a variety of environmental conditions.

Wheat cultivars can be categorized by their growing conditions and their protein content. Hard wheat has a higher protein content (higher gluten) and their flour is usually used in bread making. Soft wheat has lower protein, yielding a soft flour that is better for pastries. Hard wheats are generally grown in areas with limited rainfall, while greater moisture is needed to grow soft wheat. Planting time is another criterion. Spring wheat is planted in the spring and is harvested in the fall; winter wheat is planted in the fall, it overwinters (requiring a cold period before flowering), and then is harvested the following spring or summer. In areas where the winter is very severe, spring wheat is grown. In North America this includes North and South Dakota, Minnesota, Montana, and in Canada as far north as the Arctic Circle. South of this area winter wheat is planted, thereby, dividing the wheat-growing regions of the continent into winter and spring wheat belts.

New cultivars are constantly being developed to improve characteristics such as disease resistance or yield. Wheat is highly susceptible to many diseases, particularly stem rust of wheat caused by the fungal pathogen, *Puccinia graminis* (*see* Chapter 22). A major aim of wheat breeding programs is to develop varieties resistant to the latest strain of the fungus. High-yielding varieties developed over the past 30 years have made a major impact on world food supplies and turned some wheat-importing countries into wheat-exporting ones (*see* Chapter 15).

Nutrition

Compared to the other major cereals, wheat is a nutrient-rich food. When considering the whole grain, wheat is missing only four of the known essential nutrients; those absent are vitamins A, B_{12}, C, and iodine. The protein content is good depending on the variety, with an average of 12.9%, but recall that cereal grains have incomplete proteins since the lysine and tryptophan contents are low. The nutrients, however, are not evenly distributed in the grain, with many of the nutrients concentrated in the bran and germ. Although eating the whole grain provides the greatest nutritional benefits, most of the wheat now consumed in the United States is refined; with bran and germ gone, many of the nutrients are lost (tables 12.1 and 12.2). To compensate somewhat for this nutritional deficit, refined white flour is often enriched with iron and three B vitamins (thiamin, riboflavin, and niacin). However, the consumption of whole wheat breads and cereals is on the increase, a good sign that Americans are becoming more aware of the nutritional benefits of whole grains.

TABLE 12.1

Nutrients in Whole, Refined, and Enriched Wheat Flours Per Cup

	Whole Wheat (120g)	White Unenriched (125g)	White Enriched (125g)
Calories	400	455	455
Protein (g)	16.0	13.1	13.1
Fat (g)	2.4	1.3	1.3
Carbohydrates (g)	85.2	95.1	95.1
Calcium (mg)	49	20	20
Phosphorus (mg)	446	109	109
Iron (mg)	4.0	1.0	3.6
Potassium (mg)	444	119	119
Thiamin (mg)	0.66	0.08	0.55
Riboflavin (mg)	0.14	0.06	0.33
Niacin (mg)	5.2	1.1	4.4

Source: USDA Handbook No. 456

TABLE 12.2

Vitamins and Minerals Lost during Refining

Nutrient	% Lost
Cobalt	88.5
Vitamin E	86.3
Manganese	85.8
Magnesium	84.7
Niacin	80.8
Riboflavin	80.0
Sodium	78.3
Zinc	77.7
Thiamin	77.1
Potassium	77.0
Iron	75.6
Vitamin B_6	71.8
Phosphorus	70.9
Copper	67.9
Calcium	60.0
Panthothenic acid	50.0
Molybdenum	48.0
Chromium	40.0
Selenium	15.9

Source: Henry A. Schroeder. 1971. Losses of Vitamins and Trace Minerals Resulting from Processing and Preservation of Foods. *American Journal of Clinical Nutrition,* Vol. 24, pp. 562–573.

CORN: INDIAN MAIZE

When the Europeans discovered the New World late in the fifteenth century, they found that the dietary staple was a cereal unknown to them and called **maize** by the Indians. Later to be named *Zea mays* by Linnaeus, this crop was grown from southern Canada to southern South America and formed the basis of the New World civilizations. (Although commonly called corn in the United States, this name is ambiguous. In many countries corn refers to the most commonly grown cereal; i.e., in England corn refers to wheat and in Scotland it refers to oats. In North America, the terms "corn" and "maize" have been used interchangeably; this practice will continue in this text.)

An Unusual Cereal

European herbalists of the sixteenth century were puzzled by the appearance of this plant when they compared it to their more familiar cereals. Corn is much larger than other cereals and is unusual for having separate staminate and pistillate inflorescences (fig. 12.6). The tassel at the apex of the stalk is the staminate inflorescence arranged in a panicle, with each floret consisting of three stamens surrounded by bracts. The pistillate inflorescence, a thickened spike, is borne on a lateral stalk and gives rise to the familiar ear of corn. The grains (kernels), which can be of various colors, are naked (they lack bracts around each grain), but the entire ear is tightly covered with specialized bracts known as husks. The silks can be seen beneath the husk; each silk is

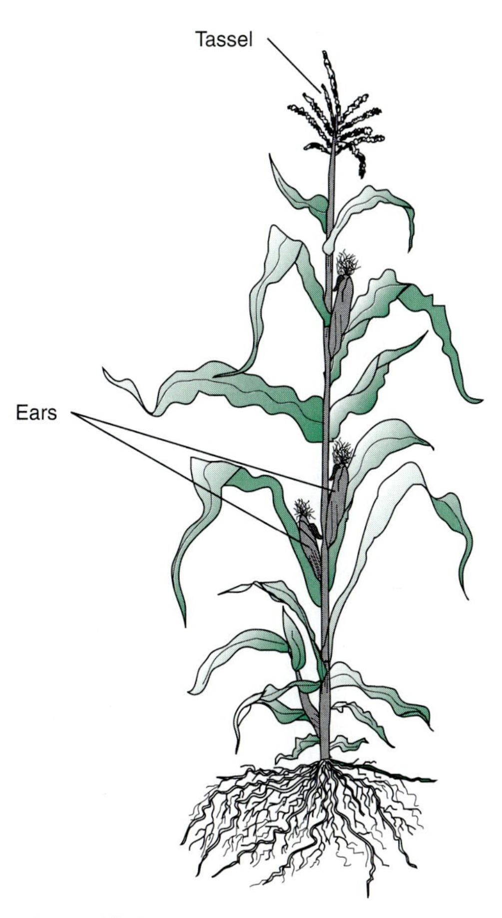

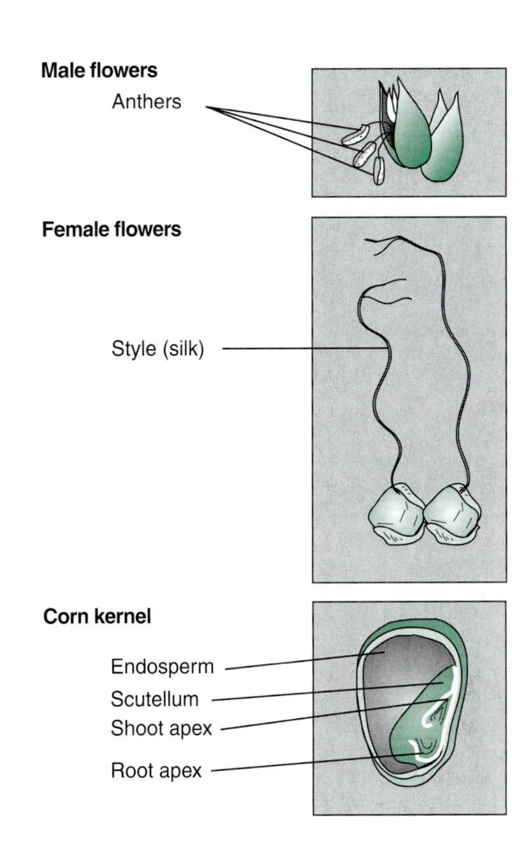

Figure 12.6 Zea mays, *corn.*

actually the style and stigma of an individual pistillate flower. Corn is poorly adapted for survival under natural conditions since its ensheathing husks totally prevent seed dispersal. The closely packed kernels in a fallen ear (even if they germinate) would produce seedlings under such intense competition that few would survive. Modern corn literally could not survive without human intervention.

Types of Corn

Several different types of corn are grown today, and can be characterized mainly by the nature of the starch present in the endosperm. Starch has two components, **amylose** (an unbranched chain of glucose molecules) and **amylopectin** (a highly branched chain of glucose molecules). Hard starch has a higher percentage of amylose than soft starch.

The types of corn are classified as popcorn, flint, flour, dent, sweet, waxy, and pod (fig. 12.7). Most of these were

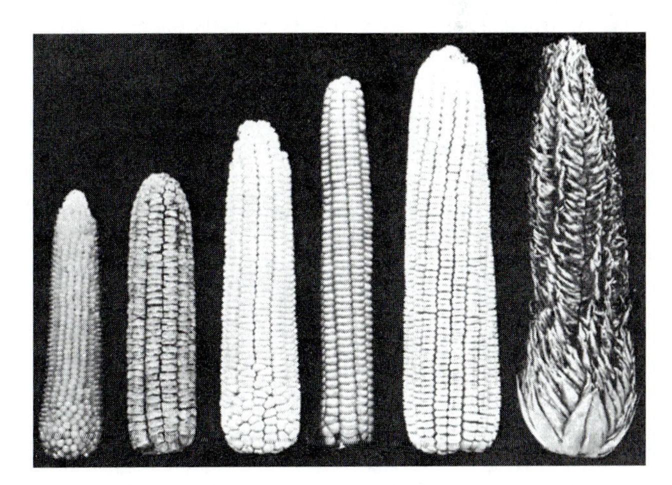
Figure 12.7 *Principal varieties of corn from left to right: popcorn, sweet corn, flour corn, flint corn, dent corn, and pod corn.*

A Closer Look 12.1

The Rise of Bread

Bread is the basic food for many different cultures and often supplies more than half of the dietary calories. Breads can be as different as the cultures that produce them: the corn tortillas of Mexico, chapatis of India, Scandinavian crisp breads, croissants of France, pumpernickels of Germany, and Jewish bagels. These are among the thousands of different types of breads that are basically made from mixing flour with water to make a dough. Any type of starchy meal can be used to prepare a dough that can be baked into a bread; however, the cereals are the foremost source of bread flours, and wheat flour is most commonly used for leavened bread (box fig. 12.1).

To make a leavened bread, flour must contain sufficient quantities of gluten. Gluten is a complex of proteins consisting largely of gliadin and glutenin. If flour containing gluten is mixed with water, the dough becomes elastic. When yeast is added and undergoes **fermentation** (anaerobic respiration), the resulting CO_2 is trapped as small gas bubbles, stretching and expanding the elastic dough. The dough rises and, when baked, results in leavened bread. Other leavening agents include baking powder and baking soda, which chemically produce the CO_2 bubbles. Of all the cereal grains, only wheat and rye have sufficient gluten to produce a leavened bread, and wheat is preferred for its higher gluten content.

The Egyptians are credited with discovering leavened bread using wheat flour almost 4,000 years ago. Of all the grains used by the Egyptians, only wheat flour had the potential to produce a leavened bread (since rye was unknown to the ancient Egyptians). At this time grains had to be parched or toasted in a fire prior to threshing. Applying heat to the grain made the glumes easy to remove but also changed the gluten, making it inelastic and unable to trap any CO_2. The resulting meal produced only flat breads. A new free-threshing form of wheat that could be threshed without heat set the stage for the discovery of leavened bread. It is assumed that yeast (from the air or possibly even from some beer added for flavor) was accidentally introduced into a dough prepared from these unparched grains (*see* Chapter 23). If the dough was set aside for a time before baking, it would rise and so produce a lighter, tastier bread.

Box Figure 12.1 *Bread is the staff of life.*

These early breads were prepared from whole grains that were coarsely ground or milled on grinding stones. Flour milled in this way was coarse with chaff, bran, germ, and even small pieces of grinding stone. As milling and sifting techniques improved over the centuries, the flour became more and more refined, and by the nineteenth century, iron roller mills replaced stone mills. The roller mills further refined the flour by removing the nutrient-rich germ and more of the bran. Thus, the refined flour was mainly starchy endosperm. The advantage of refined flour was a longer shelf life, since the oils from the germ in stone-ground flour became rancid within a few weeks. Unfortunately, the lower nutritional value of the resulting bread was detrimental for those relying on bread for a large portion of their diet.

actually cultivated by the Native Americans before the discovery of the New World. One of the oldest types, and possibly the most primitive, is popcorn, whose kernels swell and burst when heated. This was a trait useful for primitive peoples since the grains could become edible by heating without the need for arduous grinding.

Popcorn has extremely hard kernels with hard starch surrounding a core of soft starch. The endosperm cells in the center contain a large percentage of water; upon heating, the water turns to steam, building up pressure in the kernel. At a certain point, the whole kernel explodes and turns inside out. The moisture level inside the kernel is critical for successful popping, with the optimum moisture content between 13%–14.5%.

During the colonial period, New Englanders often ate popcorn for breakfast served with milk and maple sugar, and it has been suggested that popcorn was served at the first Thanksgiving. By the mid-nineteenth century, popcorn was considered more of a snack food and remains popular to this day. In fact, in recent decades "gourmet popcorn" has become fashionable, with varieties ranging from chocolate to taco flavored. With the high fiber of a whole grain, popcorn is healthier than other snack foods; however, butter or other toppings can increase the fat content and calories.

Flint corn also has hard starch near the outer part of the kernel and was the predominant corn grown by the Indians in northern areas of North America when the European settlers arrived. Flour corn, also known to the Indians, is similar in appearance to flint corn but has a softer endosperm that makes it easier to grind and prepare a dough by mixing with water. Unfortunately, the softer endosperm makes the kernels more susceptible to insect damage. Both flint and flour corn are no longer grown extensively, having been replaced by dent corn. Dent corn has both hard and soft starch with the hard starch along the sides and the soft restricted to the top and center. Upon drying, the kernel shows a characteristic dent as the soft starch shrinks. Today, dent corn is the most widely grown type in the Corn Belt, which encompasses most of middle America, and is primarily used for animal feed, corn starch, and corn meal.

Sweet corn contains a high concentration of sugar instead of starch in the cells of the endosperm. The sweet-tasting kernels are a popular vegetable, either fresh on the cob, canned, or frozen. Waxy corn is a relatively new kind of corn that first appeared in China and was introduced into the United States early in the twentieth century. It is named for the glossy, waxlike appearance of the cut kernels that is due to the presence of only amylopectin in the endosperm. Pod corn is a rare type of corn that occasionally appears in a field of corn. Unlike the other forms of corn, each kernel in pod corn is covered with glumes (fig. 12.7). It is a botanical curiosity and is not grown commercially. Some botanists believe that pod corn is a primitive type of corn that may be ancestral to the modern ones.

Corn is a summer annual that grows best under moderate conditions of temperature and moisture, but it is cultivated under more diverse environmental conditions than any other crop. An average temperature of 23° C (73° F) or above, with lots of sunshine, and 37 to 50 cm (15–20 in) of rain spread over a three-to-five month growing season are ideal. Corn requires a nutrient rich soil, and fields under cultivation for several years become depleted. The most frequently grown types of corn have thousands of cultivars, each with specific traits. For example, varieties of corn could be selected for early maturation, color of the kernels, size of the kernels, size of the ears, dwarf stalks, extra sweet taste (for sweet corn varieties), resistance to certain pests, etc.

The color variety of the kernels was a trait that impressed the European settlers. Kernel color is a complex characteristic that depends on pigmentation of the endosperm, aleurone layer, and pericarp. The familiar yellow kernels result only from yellow pigment in the endosperm; this pigment is apparently lacking in the white varieties. The multitude of colors and patterns visible in Indian corn used for autumn decorations result from pigments in the aleurone layer and pericarp (*see* A Closer Look: *Barbara McClintock and Jumping Genes in Corn*). Although both of these layers can be colorless, the pericarp may be red, orange, brown, or variegated, while the aleurone layer can be various shades of red, blue, purple, bronze, and brown. The mixture of colors from the endosperm, aleurone, and pericarp results in the kernel appearance. A black kernel results from a dark red pericarp over a blue or brown aleurone, and a greenish kernel results from a blue aleurone over yellowish endosperm.

Hybrid Corn

Corn has always been an important crop in the New World and is the most widely grown crop in the United States today. Modern hybrids of dent corn constitute the majority of the corn grown in the Corn Belt. The major characteristics introduced into the hybrids include two to three ears per stalk (as opposed to one ear, which was typical of most varieties at the beginning of the twentieth century) for greater productivity, and stronger stalks with standard positioning of the ears for easier mechanical harvesting. These improvements have been so valuable that the planting of hybrid corn increased from only 1% in 1930 to virtually 100% today.

Attempts to increase corn production early in the twentieth century led to the development of **hybrid** corn. Hybrid corn results from crossing inbred lines; these hybrids are hardier because of hybrid vigor or heterosis. Inbred lines consist of genetically homozygous plants that are produced by self-fertilization for several generations. Each inbred line reliably produces certain desired traits but are, unfortunately, weaker and less productive with each generation. By crossing inbred lines, the resulting offspring will have the desirable traits of both parents, as well as restored hybrid vigor (fig. 12.8). Because the original hybrid seeds were produced on inbred lines with small ears and,

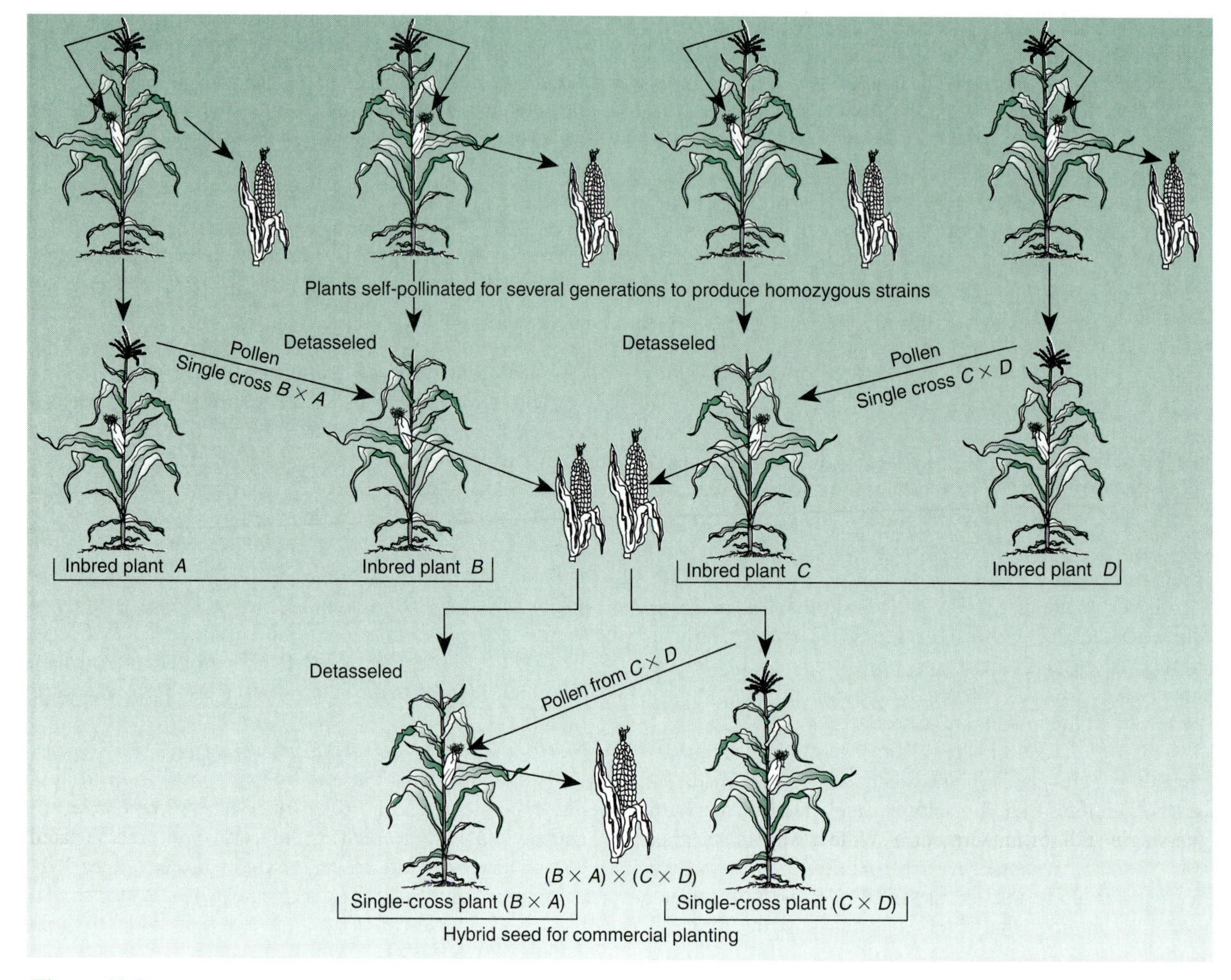

Figure 12.8 *Hybrid corn is essential for commercial seed production. Often the double-cross method is used, incorporating genes from four inbred strains.*

therefore, a small number of seeds, commercial production was impractical. To improve seed production, crosses between two single hybrids were made to produce a vigorous double hybrid that had a large number of kernels and still possessed the desired traits. Today the inbred lines are improved and it is possible to produce abundant seeds from a single hybrid cross. A disadvantage of planting hybrid corn is the need to purchase new hybrid seed each year. On the other hand, a new enterprise was created, as some farmers have specialized in planting the inbred lines to produce the hybrid seeds.

In the production of hybrid seeds, a row of one inbred line (to serve as the female parent or seed parent) is planted next to a different inbred line (that serves as the male or pollen parent). To prevent self-pollination of the female line, the staminate inflorescence must be removed or detasseled. This labor-intensive practice was the standard for many years; however, male-sterile lines were developed later that made detasseling unnecessary. A drawback to the male sterility method became apparent in 1970 when a new strain of a fungal pathogen, *Bipolaris maydis* (*Helminthosporium maydis*) that causes Southern leaf blight, appeared. Linked to male sterility was an increased susceptibility to this disease. By this time, 70% to 90% of the corn grown in the United States developed from seeds that had a male sterile parent and carried the increased susceptibility. Disaster followed when the blight struck and destroyed approximately 15% of the U.S. corn crop. In some states the devastation was even greater, with 50% loss reported (*see* Chapter 15). Since that time, other male-sterile lines have been developed that are not as susceptible. Also, detasseling has made a comeback with some seed companies.

Ancestry of Corn

Unlike wheat whose ancestors are fairly well identified, the origin of corn is a botanical mystery that has been the subject of speculation and controversy for a great many years.

Figure 12.9 *Teosinte,* Zea mexicana *is believed to be the ancestor of modern corn.*

While wild wheat can still be found in the Near East, no obvious equivalent exists for corn anywhere in the natural environment of central Mexico, where it was first domesticated over 5,500 years ago.

One school of thought, first proposed by George W. Beadle in 1928, holds that teosinte is the ancestor of modern corn. Teosinte, *Zea mexicana,* is a wild grass native to Mexico and Central America that shares some characteristics with *Zea mays* (fig. 12.9). Both species have terminal staminate inflorescences and lateral pistillate spikes. Teosinte, a much smaller plant, has multiple stalks with a tassel at the apex of each stalk, in contrast to the single stalk of modern corn. While a single corn plant normally forms only a few large ears each with multiple rows of kernels, teosinte produces numerous small ears each with six to ten kernels. The kernels are triangular in outline and have a very hard outer fruit case. This fruit case surrounds the kernels in teosinte, but is reduced to a cupule found at the base of each kernel in modern corn and remains attached to the corn cob. The spike of teosinte shatters easily at maturity to disseminate the kernels, very different than *Z. mays.*

In addition to sharing some vegetative similarities, corn and teosinte are closely related species that are able to hybridize and form fertile offspring. They have the same number of chromosomes ($2n = 20$), and chromosomes in the hybrids pair normally during meiosis. Beadle's hypothesis states that only two mutations would have been necessary to change teosinte into modern corn:

(1) a mutation to a nonshattering spike and
(2) a mutation to a soft or reduced fruit case.

In fact, an occasionally seen teosinte mutant (tunicate form) actually has a reduced fruit case with the kernels covered by soft glumes. This tunicate form also has less of a tendency to shatter. In breeding experiments carried out by Beadle and coworkers, crosses between teosinte and modern corn produced plants with small primitive ears similar to ones found in 7,000- year-old archaeological specimens.

The opposing school of thought has been led by Paul C. Mangelsdorf, who first suggested an alternative ancestry in 1939. Mangelsdorf now believes that both modern corn and the annual teosinte are descended from a cross between an ancestral wild corn, a primitive pod popcorn now extinct, and *Z. diploperennis,* a diploid perennial teosinte discovered in 1979. (Although another perennial teosinte has been known much longer, it is a tetraploid and crosses with corn result in sterile triploid hybrids.) In the F_2 generation of experimental crosses between *Z. diploperennis* and a primitive Mexican popcorn, there were a variety of fertile hybrids, some of which resembled annual teosinte and others modern corn. Mangelsdorf sees this as proof of his hypothesis. (In a previous version of this hypothesis, Mangelsdorf had suggested another grass, *Tripsacum,* as the genus that crossed with the wild pod popcorn; however, genetic studies and pollen analysis discredited this view.)

The most imaginative hypothesis was developed by Hugh Iltis in 1983, who suggested that the ear of corn evolved not from the slender pistillate spike of teosinte, but rather from the central spike of the staminate tassel. Iltis suggests that this morphological change came about by means of a "catastrophic sexual transmutation." Despite the decades of research on all these hypotheses, the ancestors of corn have still not been definitively identified.

Value of Corn

The majority of corn grown in the United States is used as animal feed. Only a small part of the corn harvest is eaten directly as a vegetable, with most processed commercially for food products or industrial applications (fig. 12.10). As food for either humans or livestock, corn is a good nutrient source of carbohydrates, fats, and proteins; however, the protein content is lower than wheat (7%–10%) and, like other cereals, is low in lysine and tryptophan. Corn geneticists have been trying to improve the protein balance in corn and have produced some varieties with higher lysine content. Corn is also deficient in niacin and what little niacin is present is in an unavailable form. Human diets based largely on corn can lead to the deficiency condition pellagra (*see* Chapter 10).

Corn starch, corn meal, corn flour, corn oil, and corn syrup are all processed from corn kernels. These products make their way into literally thousands of prepared foods, so that the average American is consuming corn one way or another in almost every meal. Corn-based breakfast cereals and snacks are prevalent in the American marketplace, with the snack foods alone accounting for almost a one billion dollar share annually. Corn oil, largely polyunsaturated, is obtained from the germ and used as salad oils, cooking oils, salad dressing, and margarine. Besides its use directly in food, corn starch is used as the base for producing corn syrup, which has become one of the most popular sweeteners in prepared foods (*see* Chapter 14). Industrial uses of corn starch are almost limitless, with laundry starch, pharmaceutical fillers, glues, lubricants, ethanol for use in gasohol, and biodegradable plastics and packing materials

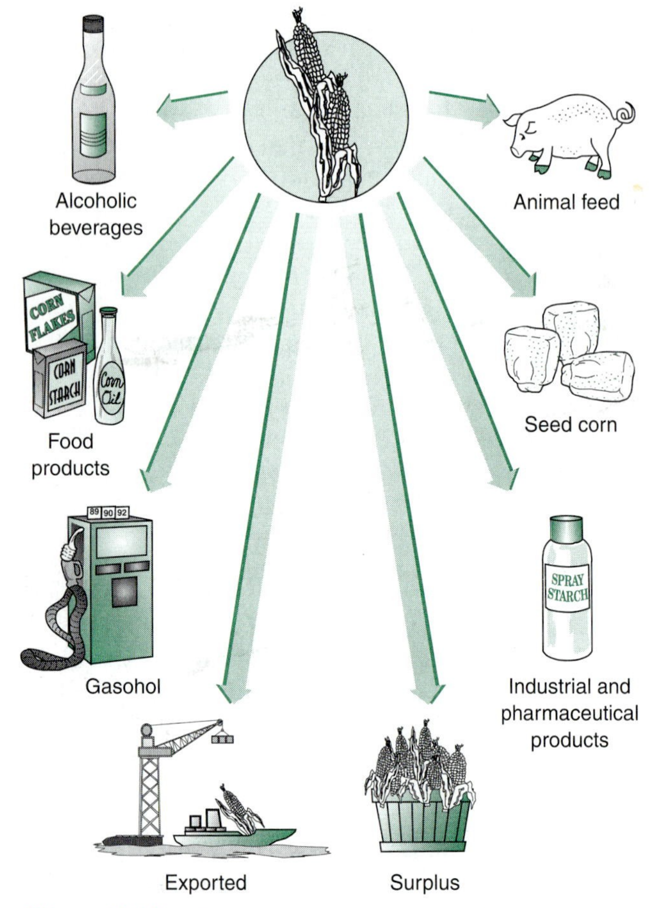

Figure 12.10 *The multiple uses of the United States corn crop.*

among the many products. During the 1980s, the industrial uses of corn increased by 160%, creating many new markets for corn growers. Additionally, corn is the base for fermented beverages (*see* Chapter 23) including *chicha* (a South American beer) and bourbon (corn mash whiskey).

Rice: Food for Billions

Rice feeds more people worldwide than any other crop, and it is the only crop grown exclusively for human food. It is estimated that over two billion people, mainly in Asia, rely on rice as a dietary staple. The oldest evidence of rice cultivation dates back 7,000 years and has been found in both eastern China and northern India. *Oryza sativa* is the main cultivated rice, although over 20 species in the genus are known. The only other cultivated species of note is *O. glaberrima,* which is grown in West Africa. (The grasses known as wild rice in North America, *Zizania aquatica* and *Z. texana,* are unrelated to cultivated rice and are, in fact, still mostly harvested from the wild by Native Americans.)

Oryza sativa is believed to have originated in lowland tropical areas that were subject to periodic flooding, and it is under these conditions that rice is still most productive. By contrast, rice is now a worldwide crop, able to grow under many different environmental conditions, with 11% of the world's arable land devoted to rice cultivation. It was introduced into North America early in the colonial period and soon became an important crop in the Carolinas. For about two hundred years it was confined to southeastern states, spreading to Louisiana, Texas, Arkansas, and California in the late nineteenth and early twentieth centuries. Although the United States produces less than 2% of the rice grown, it is one of the world's largest rice exporters.

A Plant for Flooded Fields

The rice plant is a large multi-stalked annual growing to an average height of approximately 1 meter (3 ft), with the typical vegetative appearance of grasses (fig. 12.11a). Each stalk is terminated by a panicle bearing grains surrounded by bracts. A distinctive characteristic of rice is the presence of air chambers in the stem that permit the diffusion of air from stomata in the leaves through the stem and eventually down to the roots. This adaptation, seen in many aquatic plants, permits rice to survive in flooded or waterlogged soils.

Although rice can be grown like other cereals without flooding (upland rice), in most areas of the world it is cultivated in flooded fields or paddies with 5 to 10 cm of standing water (lowland rice). This method of growing rice is an ancient one, dating back several thousand years. In fact, rice farmers in some parts of Asia are known as "farmers of 50 centuries." The paddies are diked with earthen dams and filled by rain or irrigation. Young seedlings started in seedbeds are transplanted, usually by hand, into the paddies (fig. 12.11b).

One weed of the rice paddies is known to be beneficial for cultivation. This weed is *Azolla,* a small aquatic fern that is inhabited by a symbiotic nitrogen-fixing cyanobacteria, *Anabaena azolla* (fig. 12.12). **Symbiosis** is the intimate association of two species, living together in a relationship that can be beneficial to one or both organisms. Nitrogen-fixing species are able to convert atmospheric nitrogen into a form of nitrogen that can be utilized by plants (*see* Chapter 13). Since all plants require nitrogen compounds for growth, this natural source of nitrogen reduces the need for costly fertilizers. Chinese and Vietnamese rice farmers observed that the rice crop improved when this weed was present and used it as a "green manure" for centuries, even though they were unaware of the nitrogen-fixation that was occurring.

When the grains are almost mature, the fields are drained to prepare for harvesting. Traditionally, harvesting has been done by hand using sickles. Threshing follows to free the grain from the outer bracts or chaff; **winnowing** then separates the grain from the fragments of chaff. In the United States and other industrialized countries, all stages of rice production from seeding (by airplane in some farms) through processing the grain, are highly mechanized, as opposed to the labor-intensive method still employed in developing nations.

(a)

(b)

Figure 12.11 *Rice,* Oryza sativa, *is the dietary staple for over two billion people. (a) Close up of fruiting stalks; (b) Rice paddy in Indonesia.*

Brown rice is the dehulled whole grain and is nutritionally superior to white rice, which has been milled and polished to remove the bran and germ. Brown rice contains more protein (8.5%–9.5%) than white rice (5.2%–7.6%) and more vitamins, especially thiamine and vitamin B_1. As pointed out in Chapter 10, a diet based on polished white rice can result in beriberi. Unfortunately, polished white rice is preferred for its taste, quicker cooking, and longer shelf life.

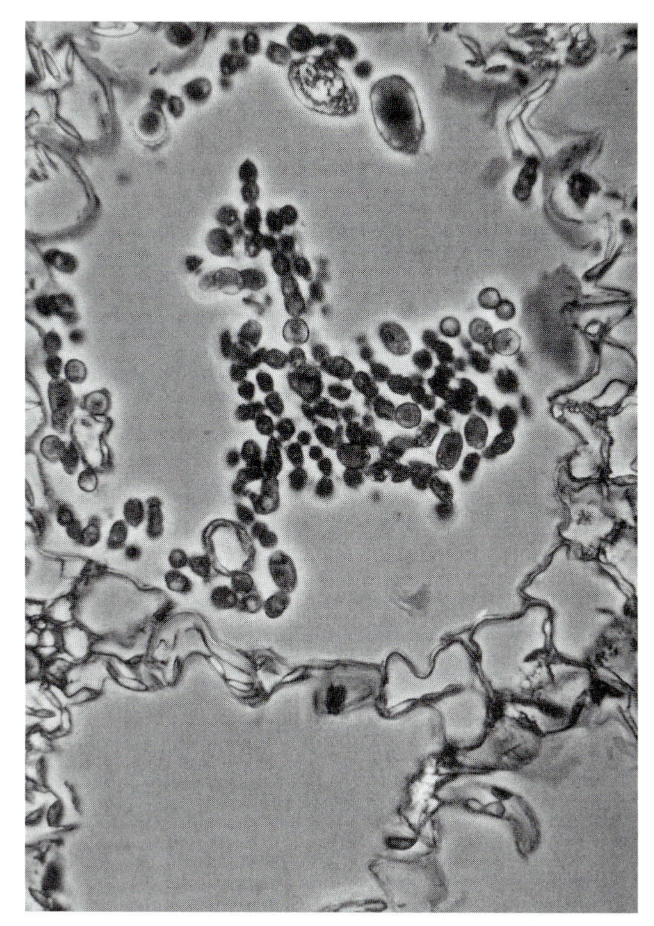

(a)

(b)

Figure 12.12 *The nitrogen-fixing cyanobacterium,* Anabaena azolla, *(a) has a symbiotic relationship within the leaves of the aquatic fern* Azolla *sp. (b)* Azolla *was originally noticed as a weed in the rice paddies, but today it is deliberately introduced to cut down on the need for expensive fertilizer.*

For many reasons corn has been a major research tool for geneticists:

Corn is easily grown and displays a great deal of variability;

Each kernel represents a different genotype so that a whole population exists on a single ear;

Corn has very large chromosomes; and

Chromosomal mutations can be easily studied with a light microscope.

Even in beginning biology classes, the inheritance of kernel color can illustrate basic genetic principles. One of the most intriguing aspects of corn genetics was the discovery of jumping genes by Dr. Barbara McClintock (box fig. 12.2).

Barbara McClintock was born in Hartford, Connecticut in 1902 and received her doctorate in botany from Cornell University in 1927. Most of her professional life was spent at Cold Spring Harbor Laboratory on Long Island, New York. This research center is famous for its significant work in genetics, molecular biology, and biochemistry, as well as virology and cancer studies. Most of Dr. McClintock's research involved corn genetics and the behavior of chromosomes. She was particularly interested in the inheritance of complex color patterns in Indian corn. McClintock concluded that the color patterns she observed were only possible if genes could move around from chromosome to chromosome. These jumping genes, called transposable elements or transposons, are fragments of chromosomes that move at random from one chromosomal position to another. When a transposon inserts on a chromosome, it alters or controls the normal expression of the neighboring genes.

She reported preliminary results of her work on these transposable genes in the late 1940s and published major articles in 1950, 1951, and 1953. Her work was viewed with a great deal of skepticism since it contradicted the prevailing theory that chromosomes consisted of genes in fixed positions. In addition, many scientists had difficulty accepting the controlling or regulatory nature of these genes. Vindication came with reports on regulatory genes in bacteria in the 1960s and transposable genes in bacteria in the 1970s. Widespread recogni-

Varieties

There are thousands of varieties of *Oryza sativa,* which differ in growing conditions or in the color, shape, size, aroma, flavor, and cooking characteristics of the grain. These varieties can be grouped into two major subspecies, *indica* and *japonica,* with a third, *javanica,* which is not as widely cultivated. The oldest varieties are *indica* varieties, primarily grown in the tropics. They produce long grains that do not stick together when cooked. By contrast, *japonica* varieties, which are grown in cooler subtropical to temperate regions, have short grains that are sticky when cooked. The stickiness of the cooked grains is dependent on the composition of starch in the endosperm; drier grains have more amylose. *Javanica* varieties are cultivated in Indonesia and are characterized by tall, thick stems and large grains. In the United States, *japonica* varieties are grown in California, while southern states grow varieties that are intermediate between *japonica* and *indica* (medium grain rice). New varieties have been developed that are high-yielding, disease-resistant, and early-maturing (*see* Chapter 15).

Other Important Grains

Rye and Triticale

It is believed that rye first came to human attention as a weed of cultivated wheat and barley fields about 5,000 years ago in southwestern Asia. It has always been noted for its hardiness, especially to cold and drought. Its ability to thrive in marginal areas was even noted by the Greek botanist Theophrastus, who reported the commonly held belief that wheat grown on poor soil would turn to rye. As wheat cultivation spread through Europe, weedy rye thrived in the colder regions and eventually became cultivated instead of wheat. *Secale cereale* is the only cultivated species of rye, although several wild species of *Secale* are known. The plant resembles wheat, but the spikes are slender and more elongate (fig. 12.13a). The grains yield a flour that is suitable for making leavened bread, although the gluten content is lower than wheat and the bread is somewhat soggy and heavy. During the Middle Ages cultivation of rye became widespread in northern Europe and gave rise to

(a)

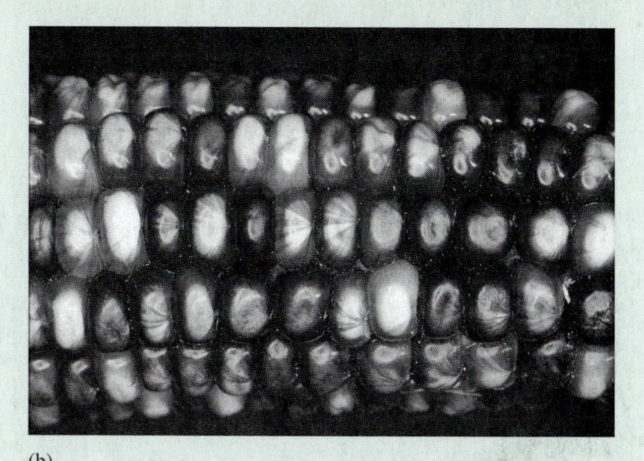

(b)

Box Figure 12.2 *Barbara McClintock (a) received the 1983 Nobel Prize in Physiology and Medicine for her work on moveable genetic elements, called jumping genes, which are responsible for (b) the complex color patterns in Indian corn.*

tion finally came in the 1980s with several awards, culminating in the Nobel Prize in Physiology and Medicine for 1983. Barbara McClintock was the first woman to receive the Nobel prize in this category for work done on her own. The Swedish institute that names the Nobel laureates in medicine compared her research to the significance of Gregor Mendel's. They said her studies

> "reveal a whole world of previously unknown genetic phenomenon . . . She was far ahead of the development in other fields of genetics . . . Her most important results were published before the structure of the DNA double helix and the genetic code had been discovered."

the familiar black bread of the peasants. Rye breads are still very popular in Sweden, Poland, Germany, and Russia. Because of the lower gluten content, rye flour is commonly mixed with wheat flour to prepare commercial "rye breads" sold in the United States today.

Crosses between wheat and rye have been made for over one hundred years to produce an intergeneric hybrid known as *Tritosecale,* or triticale (fig. 12.13b). Although the hybrids produced in the nineteenth century were sterile, techniques developed since the 1930s have permitted the breeding of triticale varieties that produce viable seeds. Much interest has centered on these hybrids, which combine desirable characteristics of each parent: the hardiness, disease resistance, and better protein quality of rye, and the higher yield of wheat. Varieties of triticale have been developed that have protein content equal to that of wheat (average 12.9%), but higher than rye (average 10.75%). Although rye has a lower protein content than wheat, it has a better biological value, since the lysine content is higher, and the lysine content of triticale approaches that of rye. One disappointment of triticale has been its poor performance as a bread flour. Although gluten is present, it tends to break up when the dough is kneaded, producing a bread that is heavy and not springy. Consequently, triticale flour is usually mixed with wheat flour for breads.

Oats

The inflorescence of the cultivated oat plant, *Avena sativa,* has a distinctive appearance when compared to other cereals. It is a branched panicle with an open, delicate appearance (fig. 12.13c); the spikelets are long, with up to seven florets (two is common). Since the oat grain is eaten unrefined, it is highly nutritious, with close to 15% protein and a good mix of vitamins, minerals, and oil. While the exact place and time of domestication is unknown, it was being cultivated in Europe around 4,500 years ago. Oats do well in moist temperate climates and possess some degree of cold-hardiness. In addition to *A. sativa,* there are several other cultivated and wild species of the genus.

Oats have always been considered a good food for horses, but historically have had mixed acceptance as human food. While some societies, like the ancient Romans, considered it only a food for animals, others developed many

(a)

(b)

(c)

(d)

(e)

(f)

Figure 12.13 *Other commercially important grains include (a) Rye, (b) Triticale, (c) Oats, (d) Barley, (e) Finger millet, and (f) Sorghum.*

dishes based on this grain. This is especially true in Scotland, where every celebration calls for an oat-based food. In the United States oats were primarily eaten as a breakfast food until the mid 1980s, when it was suggested that oat bran had cholesterol-lowering properties. The soluble fiber that makes oatmeal so gummy is the effective component in the bran (*see* Chapter 10). To cash in on this health claim, food companies began adding oat bran or oats to many processed foods, and even creating novel oat bran products.

Barley

Barley is one of the oldest domesticated crops and was brought into cultivation, along with wheat, approximately 9,000 years ago in the Near East. The cultivated barley plant, *Hordeum vulgare,* is similar in appearance to wheat, with long awns that give barley a bearded appearance (fig. 12.13d). Although other species of *Hordeum* exist, they are of minor importance for cultivation. While the spikes of some barleys have only two rows of grains, most cultivars today are the six-row type, with spikelets in threes on alternate sides. Barley can grow under a wide range of environmental conditions, tolerating cold temperatures, high altitude, low humidity, and saline soils.

Although barley was an important food for the ancient peoples of the Mediterranean region, today most barley is used as animal feed, with about one-third of the crop used in the production of malt for brewing beer (*see* Chapter 23). A small percent of barley is polished to make pearl barley, a common ingredient in vegetable soups. Whole barley grains contain 13% protein, but the protein level drops to 8% during the refining of pearl barley. Although direct human consumption of barley is almost insignificant, the importance of barley for livestock and in the brewing industry accounts for its fifth place among the world's crops.

Sorghum and Millets

Sorghum and millets are cereal grains that are seldom used as human food in North America, but are important staples in other parts of the world (fig. 12.13e,f). There are many cultivated varieties of sorghum, with most belonging to the species *Sorghum bicolor.* The vegetative appearance of sorghum is similar to corn, but with perfect flowers that are borne in a terminal inflorescence. Varieties of sorghum can be grown for their grain, their syrup, or as forage. The grain, which is most often used as animal feed in the United States, is ground and eaten as mush or baked into flat cakes for human consumption in Africa and India. The sweet sorghums, or sorgos, are used either as forage or for syrup (sorghum molasses). A special type of sorghum known as broomcorn is grown for its stiff inflorescence branches that are used to make brooms.

In the United States, millet is commonly used as a forage grass and as birdseed, but it is grown extensively in parts of India, Africa, and China as a staple cereal. The term *millet* actually refers to several different genera of grasses that were originally domesticated in the Old World and are exceptionally tolerant of drought conditions. Pearl millet, *Pennisetum glaucum,* is widely cultivated in India and Egypt, where it is ground to make a flour for bread. Millet grains can also be roasted or boiled and eaten like rice; finger millet, *Eleusine coracana* is prepared this way. The nutritional value of the millets is comparable to the other cereal grains; however, recent studies indicate that the lysine content is higher.

Other Grasses

This chapter has mainly focused on cereals that feed people, but grasses play other important roles as well. Wild and domesticated herbivores rely on both the grains and vegetative parts of grasses as food. The landscapes of lawns, parks, and playing fields are dominated by grasses. Other grasses provide society with both sugar from sugar cane (*see* Chapter 4) and building materials from bamboo (*see* Chapter 18).

Forage Grasses

Not only do grasses provide food for humanity, they also provide nourishment in the form of cereal grains (as discussed previously) or **forage** for livestock. There is actually more land dedicated to forage crops than to all other crops cultivated; but much of this land is marginal and either too hilly, rocky, wet, or dry for other crops. Forage, any vegetation consumed by domestic herbivorous animals, includes grasses and legumes (*see* Chapter 13). The majority of forage plants are herbaceous perennials and grown for their vegetative structures (leaves and stems), not their grains or seeds. The nutritive value of the forage grasses is not in the form of stored starch (as in the grains) but cellulose and hemicelluloses from the vegetative cell walls. Herbivores are able to digest these compounds through the action of symbiotic microorganisms within their digestive tracts. Forage can be consumed directly during grazing; the crop can also be cut, dried, and baled as hay; or it can be harvested and fermented by bacteria to produce silage. Hay and silage have value as stored food reserves when climatic conditions prevent grazing.

In many parts of the world natural grasslands provide pasture for grazing animals; however, in other areas large acreages are planted with forage crops. For example, Kentucky bluegrass, *Poa pratensis,* mixed with clover is grown as forage on many horse farms. Although native to Europe, this grass was introduced into North America during the colonial period and adapted well to cool, humid climates. It is considered one of the best forage grasses and is also one of the foremost lawn grasses. Many other introduced grasses native to Europe, such as timothy grass (*Phleum pratense*) and fescues (*Festuca* spp.), are also planted as forage in the eastern half of the United States. However, in the central United States many native grasses, such as big bluestem (*Andropogon*

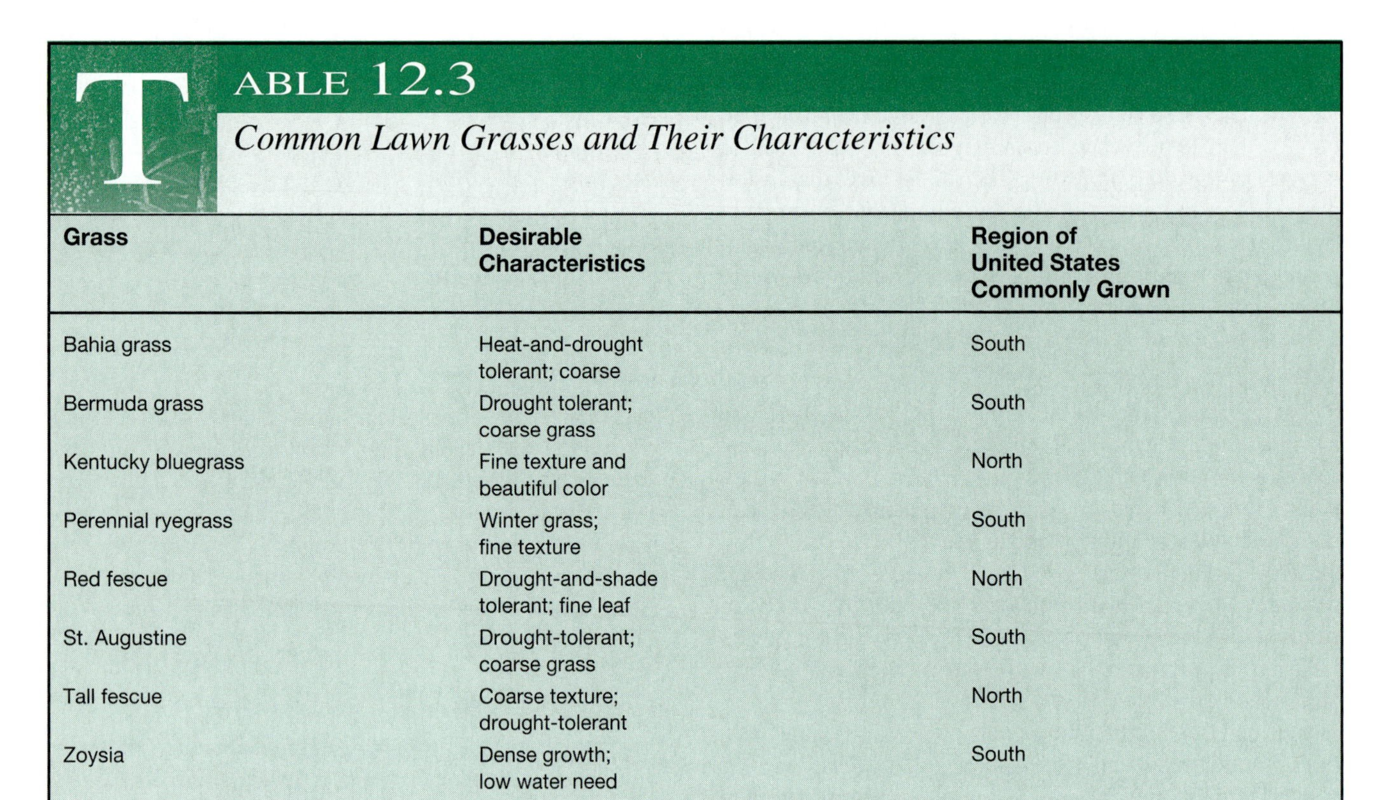

TABLE 12.3

Common Lawn Grasses and Their Characteristics

Grass	Desirable Characteristics	Region of United States Commonly Grown
Bahia grass	Heat-and-drought tolerant; coarse	South
Bermuda grass	Drought tolerant; coarse grass	South
Kentucky bluegrass	Fine texture and beautiful color	North
Perennial ryegrass	Winter grass; fine texture	South
Red fescue	Drought-and-shade tolerant; fine leaf	North
St. Augustine	Drought-tolerant; coarse grass	South
Tall fescue	Coarse texture; drought-tolerant	North
Zoysia	Dense growth; low water need	South

gerardi), little bluestem (*Andropogon scoparius*), and blue grama (*Bouteloua gracilis*), are important forage species. Unfortunately, overgrazing in some native pasture lands has decimated the native species to such an extent that inferior species have supplanted them.

Lawn Grasses

Imagine the ideal golf course, an emerald green rolling turf manicured to perfection; this would be a lawn that any homeowner would envy. In addition to its beauty and recreational uses, a lawn cuts down on mud and dust and cools the surface. These lawn plants, of course, are grasses, but unlike other crops, the harvest (grass clippings) is discarded. The cultivation also differs in that the plants are not spaced out but closely crowded together. This crowding maximizes competition, and nutrients and water are often supplemented to ensure luxurious growth. To maintain these "green carpets," billions of dollars are spent annually in the United States on sprinkling systems, maintenance, fertilizers, herbicides, etc. Lawn grasses differ in tolerance to drought, temperature extremes, shade, humidity, and salinity. These perennial grass species also vary in ability to grow in certain soil types and spread vegetatively, and, in appearance, there are differences in blade texture, color, and density (table 12.3). For the humid North, Kentucky bluegrass is the most common lawn grass, while Bermuda grass (*Cynodon dactylon*) is preferred in the dry southern states.

To maintain any lawn, frequent mowing is essential. All too often grass clippings end up in landfill areas, filling up the sites. One alternative to bagging clippings is the use of mulching mowers, which shred clippings so finely they decompose readily and return nutrients to the soil. **Composting** of grass clippings, along with other plant refuse, is another alternative that produces a nutrient-rich humus through microbial decomposition. This organic compost can be used later to enrich the soil.

1. Grasses are members of the family Poaceae. The characteristic grains with large amounts of stored food are a vital food source, providing 50% of the calories consumed by the world's population. Wheat, rice, and corn outrank all other plants as food sources for human consumption. Whole grains, with the bran and germ intact, are nutritionally superior to their refined counterparts, which contain only endosperm.
2. Wheat is the most widely cultivated crop in the world and is one of the oldest domesticated plants. Domesticated wheat had its origins in the Near East at least 9,000 years ago, and wild species of wheat can still be found in northern Iraq, Iran, and Turkey. Contemporary wheat species

include diploid, tetraploid, and hexaploid species. The polyploid species arose through hybridization between diploid wheat and goat grasses. Wheat is nutritionally superior to other grains but still lacks some essential nutrients.

3. Corn, a New World native, was the dietary staple of many Native American tribes prior to the discovery of America by European explorers. Corn is an unusual grass, having separate male and female spikes. Many different types of corn are cultivated, but all varieties are members of a single species, *Zea mays.* Corn was first domesticated in an area of central Mexico at least 5,500 years ago, but the origin of corn is a botanical mystery because no wild corn exists today. Although many botanists believe that teosinte is the ancestor of corn, other theories have been advanced.
4. Rice feeds more people worldwide than any other crop, and it is the only crop grown exclusively for human food. *Oryza sativa* is the main cultivated rice species, but over 20 species in the genus are known. Rice cultivation dates back 7,000 years and has been found in both eastern China and northern India. In most areas of the world, rice is cultivated in flooded fields or paddies, although it can be grown without flooding.
5. Other economically important grains include rye, triticale, oats, barley, sorghum, and millet. Triticale is a hybrid between wheat and rye that was developed to combine the hardiness and disease-resistance of rye, with the higher yield of wheat. Oats have always been considered a good food for horses, but historically have had mixed acceptance as human food. In the United States oats were primarily eaten as only a breakfast food until the mid-1980s, when it was suggested that oat bran had cholesterol-lowering properties. Direct human consumption of barley is almost insignificant, but the importance of barley for livestock and in the brewing industry accounts for its fifth place among the world's crops.
6. Grasses also provide nourishment for livestock in the form of forage. There is actually more land dedicated to forage crops than to all other crops cultivated; but much of this land is marginal and either too hilly, rocky, wet, or dry for other crops. Lawn grasses are indispensable components of landscaping designs and differ from crops in cultivation because plants are closely crowded together, maximizing competition between plants. Various species differ in tolerance to drought, temperature extremes, shade, humidity, and salinity and are, therefore, adapted to different climatic regions of the country.

Review Questions

1. Describe the vegetative and reproductive characteristics of a member of the Poaceae.
2. List the structural components of a cereal grain and the nutritional value of each part.
3. Trace the evolution of modern day bread wheat from its wild ancestors.
4. What are the different types of corn, what accounts for the differences, and which is thought to be the most primitive?
5. How is the cultivation of rice different than the cultivation of other crops? What is the value of *Azolla* in the cultivation of rice?
6. What is triticale? What advantages does it have over wheat or rye?
7. Beside cereal crops, what are other important economic uses of grasses?

Further Reading

Beadle, George W. 1980. The Ancestry of Corn. *Scientific American,* 242(1):112–119.

Heiser, Charles B. 1990. *Seed to Civilization: The Story of Food.* Harvard University Press, Cambridge, MA.

Hodgson, Harlow J. 1976. Forage Crops. *Scientific American,* 234(2): 61–75.

Hulse, J. H. and D. Spurgeon. 1974. Triticale. *Scientific American,* 231(2): 72–80.

Mangelsdorf, Paul C. 1986. The Origin of Corn. *Scientific American,* 255(2): 80–86.

Rhoades, Robert E. 1993. Corn, the Golden Grain. *National Geographic,* 143(6): 92–117.

Rutger, J. Neil and D. Marlin Brandon. 1981. California Rice Culture. *Scientific American,* 244(2): 42–51.

Swaminathan, M. S. 1984. Rice. *Scientific American,* 250(1): 80–93.

CHAPTER

13

LEGUMES

Chapter Outline

Chapter Concepts

1. The legumes are second only to the cereals in their importance in human nutrition and are an excellent source of high quality protein.
2. Nitrogen fixation is important for generating nitrogen compounds that can be used by plants in both natural and agricultural ecosystems.
3. Due to the wonders of chemistry, the soybean has been transformed into a variety of food products and has an ever-increasing role in the western diet.

Legumes are members of the bean family, Fabaceae, which includes all types of beans and peas as well as soybeans, peanuts, alfalfa, and clover. This large, widely distributed family also includes various trees and ornamentals such as black locust, wisteria, lupine, and the Texas bluebonnet (fig. 13.1a).

Characteristics of the Legume Family

Most members of this dicot family share a very similar flower and fruit structure. The five-petalled flower is irregular, with bilateral symmetry, and has been described as either butterfly-shaped or boat-shaped. The fruit is a **pod** or legume with one row of seeds; the seeds contain two prominent food-storing **cotyledons.** The two halves of a peanut are clear examples of cotyledons. Although the leaves of some legumes are simple, most are pinnately or palmately compound (fig. 13.1 a3).

The seeds of many legumes are an important food staple worldwide because they are rich in both oil and protein. They are higher in protein than any other food plant and are close to animal meat in quality. In fact, they are often called "poor man's meat" because they are an inexpensive source of high quality protein. The high protein content of legumes is correlated with the presence of **root nodules** (fig. 13.2), which contain **nitrogen-fixing bacteria.** These bacteria are able to convert free atmospheric nitrogen into a form that can be used by plants in the making of protein and other nitrogen-containing compounds.

Because of the presence of nitrogen-fixing bacteria, the cultivation of legumes enriches the soil. For this reason farmers often rotate legumes with crops that deplete soil nitrogen. Soybeans are often rotated with corn in the cornbelt region of the United States. Sometimes leguminous crops may even be plowed under as a "green manure" instead of being harvested. Before the advent of commercial fertilizers, these practices were more common than they are today, but they may gain renewed

(1)

(2)

(3)

(4)

Figure 13.1a *(1) Flower of garden pea; (2) Flower of false indigo; (3) Pod of chick-pea; (4) Pod of red bud.*

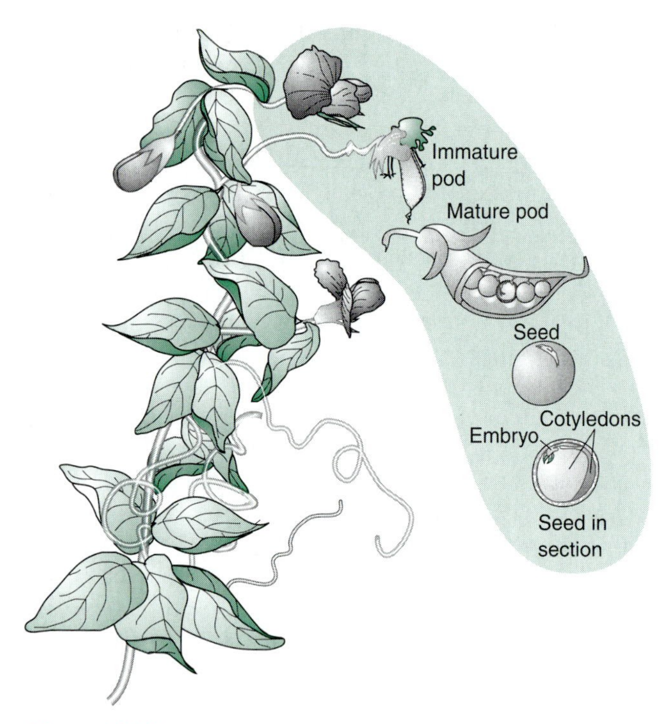

Figure 13.1b *Legumes are readily recognized by their distinctive flowers and pods.*

importance as fertilizer costs continue to rise and environmental awareness increases. Without the need for massive application of fertilizers, legumes can be cultivated worldwide, in even the poorest soils. Ecologists have even recommended planting fast-growing leguminous trees to reclaim eroded or barren areas.

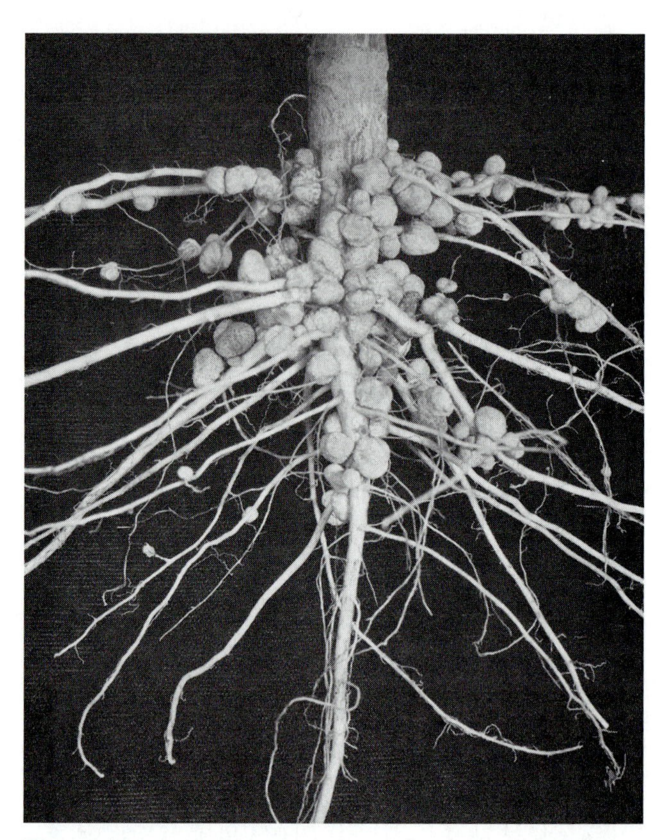

Figure 13.2 *Nitrogen-fixing bacteria,* Rhizobium spp., *inhabit nodules on the roots of many legumes.*

Important Legume Food Crops

As you recall from the discussion on the origin of agriculture, legumes have been cultivated for thousands of years in both the Old and New Worlds (*see* Chapter 11). A possible reason for their long history as food crops may be due to their seeds, which are easily harvested, have a low water content, and, when dry, are easily stored for long periods of time. These features, plus their high protein content and ease in growing, make them ideal crops.

Beans and Peas

To most people the word *legume* brings to mind beans and peas; these are, in fact, some of the oldest and most common food crops (table 13.1). Beans come in all shapes, sizes, and colors: kidney beans, lima beans, pinto beans, navy beans, green beans, wax beans, and butter beans are just a few of the many types familiar to us in the United States (fig. 13.3). Add to this the hundreds of varieties found in other parts of the world and you can see the diversity implicit in the term *bean.*

Beans are a good source of protein, with values ranging from 17% to 31%, and the average about 25%.

Figure 13.3 *Beans, peas, and lentils are the seeds of legumes. They come in a wide variety of shapes, sizes, and colors.*

Although the dry seeds were considered the only edible part for thousands of years, some of the most popular varieties today have edible pods such as green beans and wax beans. Beans are warm season annuals requiring a modest amount of rainfall. Like all legumes, they can tolerate most types of soil and can be grown worldwide.,

One bean of particular interest is *Vicia faba,* the broad bean or Windsor bean, which is an Old World species. It has been cultivated and eaten for several thousand years in the Mediterranean region; however, a disease called **favism**

TABLE 13.1
Common Edible Beans and Peas

Common Name	Scientific Name	Value
Black-eyed peas	*Vigna unguiculata*	Popular favorite in the South—a MUST on New Year's for good luck
Chick-peas	*Cicer arietinum*	Also known as garbanzo and ceci; common in Middle Eastern and Mediterranean foods
Kidney beans	*Phaseolus vulgaris*	Best known in chili; most-consumed legume in United States
Lentils	*Lens culinaris*	Used in soups and stews; most important legume in India
Lima beans	*Phaseolus lunatus*	A New World crop native to South America and named after Peruvian capital.
Mung beans	*Vigna radiata*	Widely cultivated in India and China. Best known as bean sprouts in oriental cooking.
Pinto beans	*Phaseolus vulgaris*	Mottled pink and brown beans; used in refried beans and other Tex-Mex dishes.

is associated with its consumption. In susceptible individuals, eating broad beans or even inhaling the pollen can produce favism, technically hemolytic anemia (the lysis of red blood cells). The disease is actually caused by an inherited enzyme deficiency common in Mediterranean people, but is aggravated by the type of alkaloids found in broad beans.

The term *pea,* like *bean,* denotes dozens of different kinds of edible seeds that have been cultivated for millennia. Those most familiar to us include green peas, split peas, black-eyed peas, lentils, chick-peas (garbanzos), and snow peas (fig. 13.3). Again the varieties, like snow peas with edible pods, and green peas with fresh (not dry) seeds, are of more recent origin. Nutritionally, peas are also a good source of protein, with averages about 21%. Unlike beans, peas are grown during the cooler seasons of the year in temperate zones. Biologically, the most famous pea is the garden pea, *Pisum sativum,* which Gregor Mendel used for his famous genetics experiments (fig. 13.1 a1).

Peanuts

Peanuts, also known as goobers and groundnuts, are originally native to South America, but are grown more extensively today in other parts of the world. Although the exact date of domestication is unknown, finely crafted gold and silver peanut-shaped jewelry was recently unearthed in Peru in the tomb of a Moche warrior priest (fig. 13.4). Carbon dating indicates that the tomb was from A.D. 290. This archaeological discovery shows that the peanut played a prominent role in the ancient Moche civilization.

In the sixteenth century, Spanish explorers discovered peanuts growing in South America and brought them back

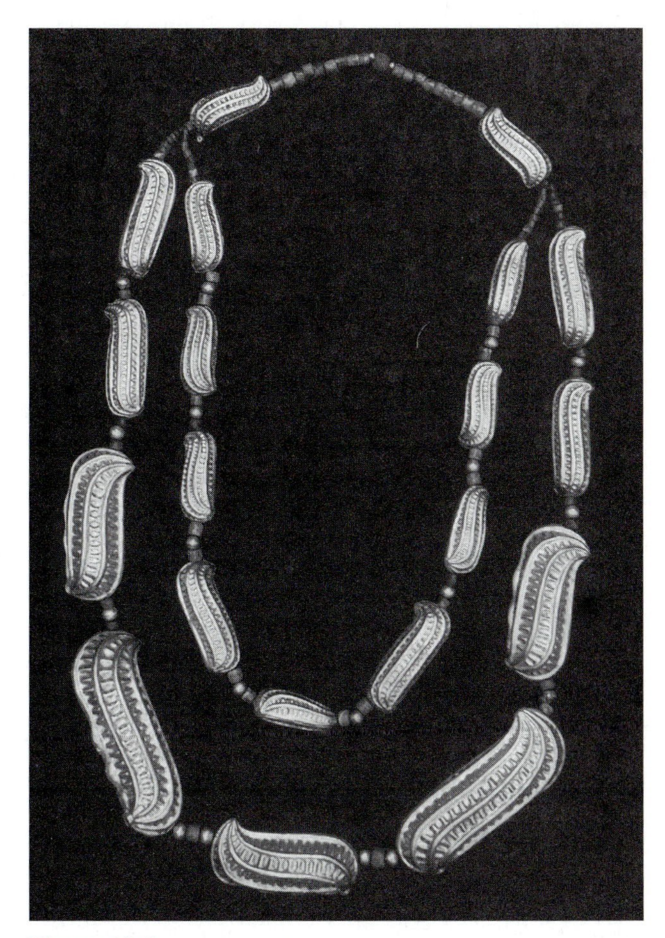

Figure 13.4 *A gold and lapis necklace reveals the prominence of the peanut to the agriculture of the Moche, a people who inhabited northern Peru 1800–1100 years ago.*

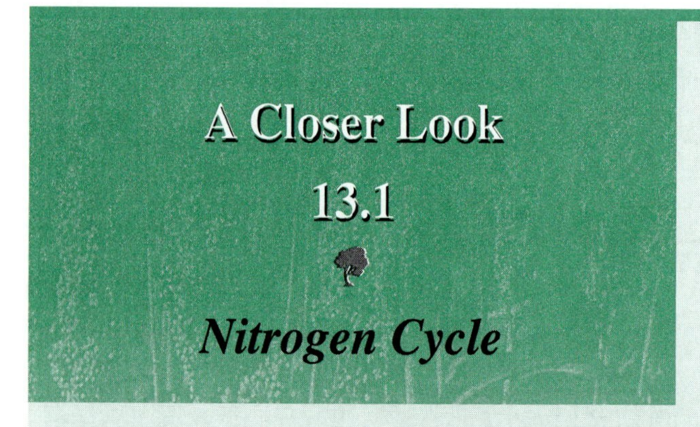

Nitrogen, one of the essential elements for all living organisms, is a major component of amino acids, proteins, nucleic acids, and other organic compounds. Nitrogen gas (N_2) makes up approximately 79% of the air we breathe, but unfortunately, most living organisms cannot use this form of nitrogen to make these cellular components. Certain bacteria and cyanobacteria have the enzymatic ability to reduce nitrogen gas to ammonium (NH_4^+), which cells can convert to other nitrogen-containing compounds. This process is called **nitrogen-fixation** and the organisms are called nitrogen-fixing. Some species of nitrogen-fixing organisms can live freely in the soil, while others are found in symbiotic associations with higher plants, as in the root nodules of legumes. The small water fern *Azolla* is known to have a symbiotic association with a nitrogen-fixing cyanobacterium (*see* Chapter 12).

Plants lacking a symbiotic nitrogen-fixing partner must rely on the nitrogen compounds present in the soil (box fig. 13.1). During the decomposition of dead plants and animals and their waste products, microorganisms break down the proteins and other complex nitrogen-containing organic molecules into ammonium. Although some plants can uptake ammonium directly, nitrifying bacteria present in the soil quickly convert ammonium to nitrite (NO_2^-) and then nitrate (NO_3^-). Nitrate is the form of nitrogen usually absorbed by plants. Most commercial fertilizers contain a mixture of both ammonium and nitrate.

Other sources of nitrogen compounds originate from the burning of fossil fuels, from volcanic activity, and from lightning. These atmospheric compounds are washed into the soil by rain and contribute to the cycling of nitrogen. Not all bacterial conversions make nitrogen available to plants; denitrifying bacteria found in wet soils actually break down ammonium and nitrates, returning nitrogen gas to the atmosphere.

to Europe. From there, trading introduced the peanut to Africa, where it soon became widely cultivated. The slave trade returned the peanut to the New World, but this time to North America. In the United States, the peanut is a staple crop of the South, growing best in light sandy soils and mainly cultivated in Georgia, North Carolina, Texas, Alabama, Virginia, and Oklahoma (fig. 13.5).

The peanut, *Arachis hypogea,* is one of nature's more unusual plants. After pollination, the flower stalk elongates downward, pushing the developing fruit into the soil. It is here, underground, that the fruit matures into a pod, characteristically with two seeds (peanuts) in a shell (fig. 13.6). The whole growing cycle takes about five months. Two varieties commonly grown in the United States are the larger-seeded Virginia peanut and the smaller Spanish peanut, which has a slightly higher oil content.

With 45%–50% oil and 25%–30% protein, the peanut is a highly nutritious seed that is used in many ways. It is a favorite in the United States, with over one billion pounds per year consumed mainly as a snack food, in candy, and in peanut butter. In fact, half of the U.S. crop is used to make peanut butter. (Peanut butter is a uniquely American food first developed by a St. Louis physician in the 1890s as a nutritious and easily digested food for invalids who had difficulty chewing.)

Today, peanut oil is found in margarine, shortening, salad dressing, cooking oil, in certain soaps, and in a variety of cosmetic and industrial products such as shaving cream, plastics, and paints. Even after the extraction of oil, the pressed cake that remains is used as a livestock feed that is rich in protein. Hogs in particular have such a fondness for peanuts that they will uproot them in fields if given the opportunity.

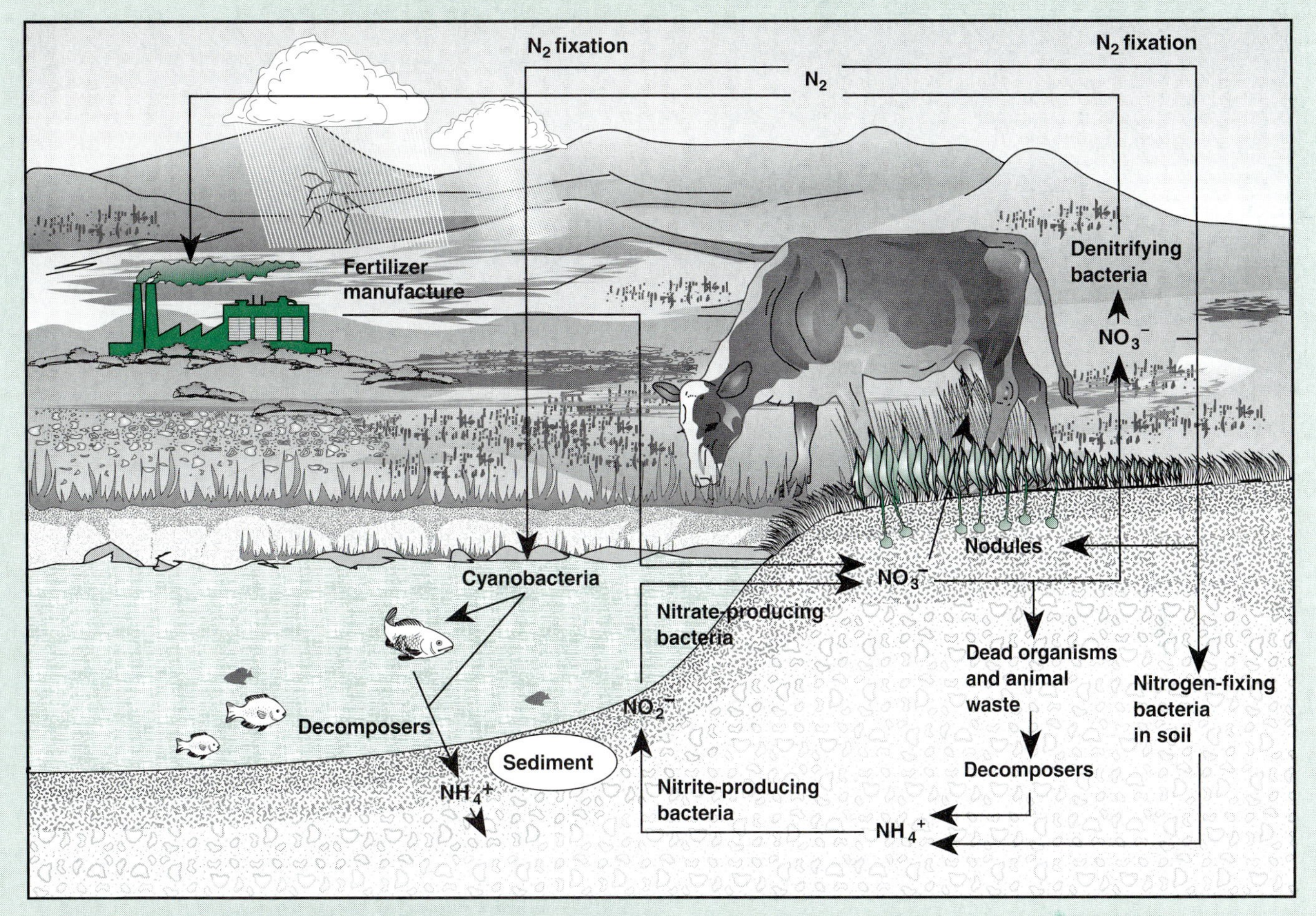

Box Figure 13.1 *The nitrogen cycle. Nitrogen is cycled through the environment via the actions of microorganisms. Nitrogen-fixing cyanobacteria and bacteria, free-living or associated with legumes, convert nitrogen gas into ammonium (NH_4^+). Bacteria and other microorganisms decompose dead organisms and animal wastes, releasing ammonium. Other bacteria convert ammonium into nitrite (NO_2^-) and nitrate (NO_3^-).*

Figure 13.5 *Oklahoma is one of several southern states where the peanut is extensively grown, a fact celebrated by a statue of the peanut in Durant, Oklahoma.*

Figure 13.6 *The peanut pod develops underground and must be dug up during harvesting.*

Figure 13.7 *George Washington Carver introduced and promoted the cultivation of the peanut to the Civil War-ravaged South.*

The versatility of the peanut is due in large part to the work of George Washington Carver (1864–1943), who developed over 300 food and industrial uses and encouraged its cultivation in the South (fig. 13.7). Carver, the son of a slave, developed an interest in agriculture during his college years and spent his whole teaching and research career at Tuskegee Institute in Alabama. He sought to revitalize Southern agriculture, since the soil had been exhausted during years of cotton cultivation, by introducing soil-enriching legumes like the peanut and the soybean. Carver changed the scope of agriculture in the South. When he began his work, the peanut was not even recognized as a crop; today, it is the second leading cash crop in the region.

Soybeans

Although peanuts will never lose their place in the American diet, roasted soy nuts are gaining popularity in health food stores. The soybean, *Glycine max,* is relatively new to the West but has been esteemed in the Orient for centuries (fig. 13.8). It was considered one of the sacred crops of the ancient Chinese, and evidence suggests that it was domesticated in northern China at least 3,000 years ago. Soybeans were first brought to Europe in the seventeenth century by the German botanist Engelbert Kaempfer.

Although introduced into North America in 1804, there was very little interest in growing the crop until the 1920s. Since then soybean production has risen dramatically, making the United States the world's leading producer. Because of this spectacular rise from a second rate crop, it has often been referred to as the "Cinderella crop." Reasons for the success story are the versatility of the soybean and its suitability for growing in the corn belt region of the Midwest; it does best in warm temperate climates with moderate amounts of rainfall.

(a)

(b)

Figure 13.8 *(a) The soybean plant* (Glycine max) *and pods with soybeans. (b) The ingredient common to all these products is the versatile soybean.*

Since ancient times in Asia, soybeans have been consumed in hundreds of different ways. Although the beans are unpalatable raw, when cooked they can be eaten whole. Most often, however, they are modified into a paste, curd, or "milk." One familiar soybean product is soy sauce; although many American brands are made synthetically, soy sauce traditionally is made by fermenting soybeans in brine.

A soybean product new to the American consumer is tofu (fig. 13.8), made from the curds of soy milk. (In making soy milk, beans are soaked in water and then pureed; this mixture is heated and the liquid, soy milk, poured off.) Tofu is extremely versatile and used in main dishes in Japanese and Chinese meals. In the United States, tofu is the basis for a low-fat, low-cholesterol "ice cream" dessert called Tofutti.

ABLE 13.2

Other Legumes of Interest and Their Uses

Plant	Scientific Name	Use
Carob	*Ceratonia siliqua*	Chocolate substitute
Fenugreek	*Trigonella foenum-graecum*	Spice
Kudzu	*Pueraria thunbergiana*	Ground cover
Indigo	*Indigofera tinctoria*	Dye
Licorice	*Glycyrrhiza glabra*	Extract
Mesquite	*Prosopis glandulosa*	Charcoal
Copaifera	*Copaifera officinalis*	Resin for paints, lacquers
Rosewood	*Dalbergia spp.*	Timber
Rosary pea	*Abrus precatorius*	Jewelry
Senna pods	*Cassia fistula*	Laxative
Tamarind	*Tamarindus indica*	Seasoning
Tuba-root	*Derris elliptica*	Insecticide

Other common soybean foods in the Orient are miso and tempeh. Miso is a fermented food of Japan prepared from soybeans, salt, and rice; the mixture is fermented by fungi for several months and then ground into a paste and used as a spread. Tempeh, fermented soybean cake native to Indonesia, is prepared by inoculating parboiled soybeans with mold and allowing them to ferment for a few days. The fungal mycelium binds the soybeans together into a cake, which can be sliced and cooked in various ways.

Soybeans are among the richest foods known, with 13%–25% oil and 30%–50% protein, depending on the variety. Overall, they have a higher protein content than lean beef. Although originally used solely as animal feed, the soy protein is used more and more in the human diet. After the extraction of oil, the soy meal that remains is made into a flour and can be included with wheat flour in a variety of breads, pasta, baked goods, and breakfast foods. By replacing a small fraction of the wheat flour with soy flour, the protein content is significantly improved.

Another product is textured vegetable protein (TVP), produced by spinning the soy protein into long slender fibers. TVP can pick up flavors from other substances and can, therefore, be used as meat extenders. By adding artificial flavorings and colors, TVP can be made into cholesterol-free imitation meats. Imitation bacon bits are made this way.

Soy oil is used extensively as cooking oil, salad oil, and in the manufacture of margarine, shortening, and prepared salad dressings. It is such a widespread ingredient in so many foods that the average American consumes almost 6 gallons of soy oil per year. Today, many manufacturers are replacing unhealthy saturated fats with soybean oil in commercially prepared food.

Industrially, soy oil can be used in dozens of processes for the manufacturing of paints, inks, soaps, insecticides, and cosmetics. Probably the most imaginative use of soy oil was the manufacture of a soybean-based "plastic" car by Henry Ford in 1940. Ford's commitment to the soybean was so great that at one point he stated that his goal was to "grow cars rather than mine them."

Lecithin, a common food additive, is a lipid extracted from soybeans. Added to many packaged foods such as cake mixes, instant beverages, whipped toppings, and salad dressings, it stabilizes and extends their shelf life. The use of soybeans should increase even more in the future. One possible market is the Third World. Attempts to improve the protein-deficient diets in these countries have included using soy products to enhance the nutritional value of the native foods.

Other Legumes of Interest

Legumes have many and varied uses; in terms of sheer numbers, legumes are by far the most utilized plants. Above and beyond their worth as a food source, legumes are valued as timber, forage, spices, ornamentals, and sources of medicines, insecticides, resins, and dyes (table 13.2). Following are a few noteworthy legumes.

A Super Tree for Forestry

Leucaena leucocephala is a tropical tree that has gained fame as one of the fastest growing species of woody plants. Some varieties have been reported to grow as much as 9 meters (30 ft) per year, earning it the nickname "Jack's beanstalk." In many poor countries, where firewood is the main source of

Figure 13.9 *Alfalfa* (Medicago sativa) *has been used as a forage since the days of the ancient Greeks and Romans.*

Figure 13.10 *All parts of the winged bean are edible; it may become an important food crop in the future.*

fuel, leucaena's remarkable growth makes it a quickly renewable energy resource. Likewise, the tree can be grown for pulp to make paper and paper products.

Like many legumes, leucaena has a symbiotic relationship with nitrogen-fixing bacteria. Nitrogen compounds produced by these bacteria accumulate in the leaves. Later, when the leaves fall they quickly decay, releasing nitrogen compounds that enrich the soil. Undecayed leaves can even be packaged and sold as an inexpensive fertilizer. Six bags of dried leaves contain the same nitrogen as one bag of commercially produced ammonium sulfate. Because of the soil-enriching properties, leucaena has been used as a "nurse" tree, providing nutrients and shade for coffee, pepper, vanilla, and other shade-loving crops.

Forage Crops

Worldwide, many legumes are planted and grown exclusively as pasture or forage crops. Their high protein content, which makes them ideal as a human food source, also makes them desirable as animal fodder. Combinations of carbohydrate-rich grasses and protein-rich legumes are grown in most pastures for direct consumption or for hay.

Alfalfa is probably the best known and most widely grown of these forage legumes (fig. 13.9). Commonly called lucerne in other countries, alfalfa has been cultivated as a forage crop since ancient times. It was introduced into Greece from Persia about 500 B.C. by the invading Medes. Linnaeus used this historical reference in assigning the scientific name *Medicago sativa.* The Romans recognized this crop as a superior feed for their horses; later the Spanish introduced alfalfa to the New World for the same purpose. Because of its high protein content, there has been recent interest in developing alfalfa products for human consumption. Alfalfa pills and extracts are a common commodity in health food stores. Other forage legumes include the clovers, sweet clovers, vetches, lespedezas, and bird's-foot trefoil.

Beans of the Future

Because of their nutritional value, there has been considerable interest and research in discovering "new" varieties of legumes. Ethnobotanists, such as Noel Vietmeyer from the National Academy of Sciences, have spent years searching the globe, in particular Third World countries, for little-known crops that have potential to become major food sources (*see* Chapter 15).

One of the interesting finds is the winged bean, *Psophocarpus tetragonolobus* (fig. 13.10). Native to Southeast Asia and New Guinea, this plant is valued because all parts of the plant are edible and highly nutritious. The pod, which has four extensions or "wings" running along its length, is edible either raw or cooked. The flowers, when cooked, are said to taste like mushrooms; the tendrils that support the vine taste like asparagus; and the leaves are eaten like spinach. The root is tuberous and can be eaten like the potato, but its protein content is much higher. Of greatest value are the seeds that are similar to soybeans, with a protein content of 37% and oil content of 20%. Some scientists believe that the winged bean may be as important to the future of tropical agriculture as soybean has proven to be in temperate agriculture.

Chapter Summary

1. Legumes are members of the Fabaceae or bean family and include several important food crops: peas, beans, soybeans, and peanuts. The highly nutritious seeds, and occasionally the pods, of legumes have become a food source second only to the cereals. In terms of nutritional quality, legumes are excellent sources of protein.
2. The high protein content of many legumes is due to their association with nitrogen-fixing bacteria in their roots. Nitrogen fixation converts the unusable N_2 gas into ammonium (NH_4^+), which the plants can incorporate in the synthesizing of proteins.
3. There are literally dozens of different types of peas and beans that have been valued for their high protein content.
4. The peanut (*Arachis hypogea*) has a South American origin and is unusual in that the plant itself literally "plants" its seed pods in the soil. The peanut is valued for both its oil and protein content. George Washington Carver was single-handedly responsible for developing the peanut as a major crop in the post-Civil War South.
5. The protein content of soybean (*Glycine max*) is one of the highest of all crops. The soybean has a long history of use in the Orient, where it has been modified by various treatments, making it an extremely versatile food.
6. Legumes are also valued as forage crops and as green manure to enhance soil fertility.
7. The search for new crops has uncovered several relatively unknown legumes, such as the winged bean, that may become the new Cinderella crops in the future.

Review Questions

1. What are the distinguishing characteristics of the legume flower and fruit?
2. What is the role of nitrogen-fixing organisms in the nitrogen cycle? How do legumes fit in?
3. Why is soybean called the "Cinderella crop"? List several products derived from soybean.
4. Trace the history of the peanut from its South American origins to the present day.
5. What is the nutritional value of legumes?
6. What legumes are being developed as economically important world crops?

Further Reading

Brody, Jane. 1985. *Jane Brody's Good Food Book.* Bantam Books, Toronto.

Hapgood, Fred. 1987. The Prodigious Soybean. *National Geographic,* Vol. 172 (1): 66–91.

Heiser, Charles B. 1990. *Seed to Civilization: The Story of Food,* Harvard University Press, Cambridge, MA.

National Research Council. 1989. *Lost Crops of the Incas: Little-Known Plants of the Andes with Promise for Worldwide Cultivation.* National Academy Press: Washington, DC.

Vietmeyer, Noel. July–August, 1987. Dazzling! *International Wildlife,* 31–35.

CHAPTER

14

STARCHY STAPLES

Chapter Outline

Chapter Concepts

1. Modified stems and storage roots can have several functions in plants: as food reserves, for asexual reproduction, and as the starting point for renewed growth after dormancy.
2. Starchy staples with their high carbohydrate content include some of the world's foremost crops and play major roles in the human diet.
3. Historically the potato has been pivotal to the development of several societies, from the ancient Incan civilization in South America to the preindustrial countries of Europe, especially nineteenth century Ireland.

Most plants store food reserves in the form of starch; often these reserves are stored in underground organs, either in some types of roots or modified stems. This is the case with all of the starchy staples considered in this chapter. All of them are of tropical origin, even though today some are extensively cultivated throughout the temperate world. These crops, all propagated asexually and not planted from seed, tend to be highly productive, yielding many tons per acre. An added characteristic of underground crops is their ability to provide food insurance against accidental or natural disasters such as fire, typhoons, or hail.

Since prehistoric days, humans have utilized these storage organs as valuable food sources and today some of these are significant world crops. The potato, sweet potato, and cassava are among the top ten crops in terms of tonnage produced annually. Nutritionally, the starchy staples are high in carbohydrates, mostly starch, but are low in protein and fat.

Storage Organs

Modified Stems

Stems are not always vertical and aboveground; many plants have modified stems that can be horizontal and even underground. These modified stems have a variety of functions. Some are specialized for asexual reproduction, some are specialized for food storage, and some can function in both capacities. In many herbaceous perennial plants, these modified stems serve as a protected food reserve. When the aboveground vegetative structures die back, these food reserves are available for renewed growth on the return of favorable weather conditions. These modified stems, like erect stems, have recognizable nodes and internodes.

Stolons or **runners** are aboveground horizontal stems that produce buds and roots at the nodes. These buds soon develop into plantlets; an entire area can be quickly invaded through this method of vegetative reproduction. One familiar example of plants that reproduce by stolons are strawberries; in fact, the current name is derived from *strayberry,* which refers to the stoloniferous habit (fig. 14.1a). The common lawn weed, crabgrass, and the familiar houseplant, *Chlorophytum* or spider plant, are two other stolon producers.

Horizontal stems that are underground, often just below the surface, are known as **rhizomes.** Reduced scale-like leaves are present on the surface of rhizomes and adventitious roots form all along the underside (fig. 14.1b). (Adventitious roots develop from organs other than roots.) Buds found at the nodes can give rise to new plants. In some, the rhizome may also be a food storage organ; ginger and iris are plants with this type of rhizome .

Tubers are the enlarged storage tips of a rhizome; the familiar white potato is a tuber. The "eyes" of the potato are actually buds located at the nodes, and each bud can give rise to a new plant (fig. 14.1c).

Bulbs and **corms** are modified stems found in monocots. Bulbs are erect underground stems with both fleshy and papery leaves; food is stored in the fleshy leaves. Onions, tulips, daffodils, hyacinths, and lilies are familiar bulbs. Not only will each bulb give rise to a new plant during the growing season, but the bulbs themselves can multiply. New bulbs can develop in the axil of the leaf scales; these eventually will separate from the parent bulb.

Although commonly confused with bulbs, corms store food reserves in the stem not the leaves (fig. 14.1d). These erect underground stems lack the fleshy food storage leaves and are covered only with dry papery leaves. Like bulbs, corms can multiply in number by producing small corms or cormels. Examples of plants that form corms are gladiolus, crocus, and taro.

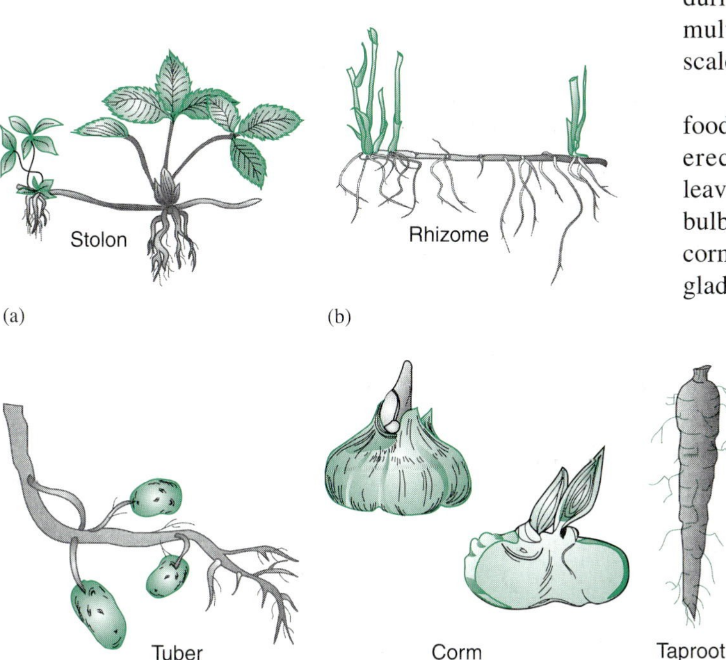

Figure 14.1 *Modified stems and roots: (a) stolon; (b) rhizome; (c) tuber; (d) corm; (e) taproot.*

Storage Roots

Tuberous roots are modified fibrous roots that have become fleshy and enlarged with food reserves. Like modified stems, tuberous roots can function in asexual reproduction, as well as food storage. Tuberous begonias, dahlias, and sweet potatoes all produce tuberous roots.

As described in Chapter 3, taproots are often food-storing organs for biennial plants such as carrots, rutabagas, and turnips (fig. 14.1e). They differ from tuberous roots in that taproots are enlarged primary roots. Because taproot crops are not

Figure 14.2 *The potato has been cultivated for approximately 8,000 years in the Andean highlands of South America. This ancient pottery vessel from South America is in the form of a potato.*

dietary staples (although they certainly supplement the diet with added nutrients), they are not considered in this chapter.

White Potato

South American Origins

Although the white potato, *Solanum tuberosum,* is often associated with Ireland or Idaho, its origins can be traced back to the Andean highlands of South America. As early as 8,000 years ago, Indians collected wild potatoes, the ancestors of today's cultivars, from the Andean plateau at an elevation of approximately 4,000 meters (12,000 ft). By the time the Spanish Conquistadors, led by Francisco Pizarro, arrived in the 1530s, the potato was the staple of the great Inca civilization that extended thousands of miles along the west coast of South America. Although the Spanish were attracted by the Incan gold, silver, and jewels, the real Incan treasure was later realized to be the potato (fig. 14.2).

The potato was introduced to Spain sometime during the middle to late sixteenth century. The exact date of introduction is lost to history, but records from sixteenth century herbalists suggest a 1580 arrival. Once introduced, cultivation of the potato slowly spread throughout Europe. Grown first as a curiosity and later as a food for livestock, the potato was only widely accepted as a food for humans in the eighteenth century. The slow acceptance by Europeans can be traced to misinformation about its edibility. Because European relatives of the potato (nightshade, mandrake, and henbane) were known to be poisonous or hallucinogenic, many assumed that the tubers would also be deadly. (In fact, the tuber is the only part of the plant that is safe to eat; all the aboveground parts are, indeed, poisonous.) Others feared that the potato could cause a variety of diseases such as leprosy, rickets, and consumption. Still others claimed that the potato was an aphrodisiac, stimulating lust and passion.

The Irish Famine

Nowhere else in Europe was the potato accepted as readily as in Ireland. It was an established crop as early as 1625, and a dietary staple for the Irish peasant throughout the eighteenth and the first half of the nineteenth century. The climate and soil in Ireland were ideal for the potato; even small plots of land could yield a large enough harvest to feed a family. The potato was so successful that historians link the subsequent population explosion (from 1.5 million to 8.5 million between 1760 and 1840) to the presence of this reliable food supply. The poor in Ireland subsisted on potatoes, some milk, and, occasionally, fish or meat; estimates are that the average adult consumed between 4 to 6 kg (8 to 12 pounds) of potatoes each day. This reliance on a single crop set the stage for disaster: the Irish potato famine of 1845–1849.

The most lethal pathogen of the potato is the fungus *Phytophthora infestans,* which causes the disease late blight of potato. The fungus attacks and destroys the leaves and stem, causing them to blacken and decay in a short time, thereby stopping tuber growth. In addition, the tubers themselves are attacked and rot in the ground, or even later in storage. In cool wet weather, the fungus can kill a plant within a week. The disease first appeared in continental Europe in 1844, most likely inadvertently carried with new varieties of potato from Central or South America. It first appeared in Ireland in August of 1845 and caused devastation during the next few years.

The disease struck several times during the period of 1845 to 1849, resulting in widespread destruction of the potato crop and devastating famines among the Irish peasantry. It is estimated that over one million died from starvation or from virulent diseases such as typhus and cholera that followed the famine. Another 1.5 million Irish emigrated to all parts of the world, but largely to the United States, resulting in a 25%–30% decline of the population in Ireland in less than a decade.

Continental Europe

The potato was also widely grown in Europe, where the cultivation was strongly encouraged by the aristocracy and leaders in several countries as a cheap, reliable food for the peasants. Fredrick Wilhelm, the Elector in Prussia, promoted the cultivation of potatoes as early as the 1650s. In 1744, his great grandson, Fredrick II (the Great) of Prussia, still attempting to gain acceptance for the potato, decreed that potatoes were to be cultivated in Pomerania and Silesia (now actually part of Poland), and distributed seed potatoes for planting. By 1770, the cultivation of the potato was extensive, partly due to crop failures of wheat and rye. During the War of Bavarian Succession (1778–1779), the opposing armies of Austria and Prussia faced off to a stalemate in Bohemia. Both armies consumed the local potato harvest until depletion of the food supply and cold weather forced them to retreat; today the war is also known as the *Kartoffelkrieg,* the Potato War.

In France, the champion of the potato was Antoine-Auguste Parmentier, who learned the value of the potato as a prisoner of war in Hanover during the Seven Years War (1756–1763). Apparently he survived only on potatoes during his time in prison. Upon his release, he returned to France and found widespread famine; he worked to popularize the potato as the solution to the famine. His efforts were rewarded when the French finally incorporated the potato into their daily diet following the French Revolution.

By the end of the eighteenth century the potato had at last gained widespread acceptance as food throughout Europe. Production continued to increase during the first half of the nineteenth century, until the potato blight devastated the crops throughout continental Europe. Although famine resulted, the effects were not as severe as the Irish famine, since the potato was not the sole dietary staple.

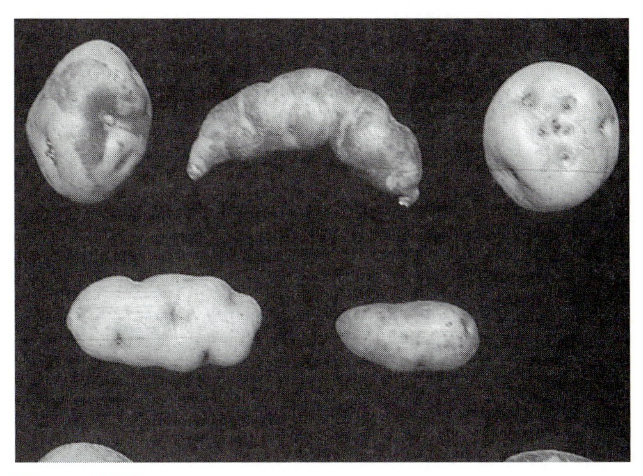

Figure 14.3 *Various potato cultivars available in the Andean highlands.*

The Potato in the United States

Today, Europe and the republics of the former Soviet Union produce over 70% of the world crop; China is the second leading producer, and Poland comes in third. The U.S. production is currently fourth, with only 5% of the world total. Although native to the New World, the potato made its appearance in North America through the European colonists. There is some doubt as to the exact date of introduction; some claim that the potato was introduced to Virginia as early as 1621, while others claim 1719 as the year of introduction to New England. Part of the uncertainty relates to confusion in historical records between the white potato and the sweet potato, which had been introduced to Europe around the same time. The word *potato* stems from the Arawak Indian word *batata,* which actually referred to the sweet potato, an edible root completely unrelated to the white potato.

In the United States today, potatoes are grown in virtually every state at some time of the year, but the top producing states are Idaho, Washington, and Maine. About one-third of the U.S. harvest is consumed fresh, and one-half is processed to make frozen French fries, potato chips, dehydrated flakes, and other products, including potato starch. Almost half of all processed potatoes are made into frozen French fries. The deep-frying of potato strips presumably originated in France, and tradition holds that they were introduced to the United States by Thomas Jefferson during his Presidency. Today, most French fries come prepackaged and frozen, a convenience food whose popularity skyrocketed in the latter half of the twentieth century with the availability of home freezers and the prominence of fast-food chains.

Another favorite processed potato product, the potato chip, is said to have been developed by George Crum, a short-order cook in Saratoga Springs, New York in the 1850s. The story goes that a customer complained about the thickness of his fried potatoes. In exasperation, Crum deep-fried superthin slices of potato, and the potato chip was born. These should not be confused with the British "chips," which are actually French fries; our potato chip is called a crisp in the United Kingdom.

Processed potatoes are nothing new in the history of the potato; Peruvian Indians from high in the Andes Mountains have prepared chuño, a kind of freeze-dried dehydrated potato, for the past 2,000 years. Potato tubers are spread on the ground when a heavy frost is expected. Following freezing, the potatoes are allowed to thaw during the day and then trampled underfoot to express water and bitter compounds, and also to remove the skins. Processing methods vary after these initial steps. One method uses extensive washings to leach additional bitter compounds; this is followed by drying. In contrast, another method continues the freezing-thawing-trampling cycle for several days prior to final drying. The resulting chuño, from either method, is used primarily as a basis for soups, stews, or desserts. This dehydrated product can be stored for several years without spoiling and provides a stable food reserve.

Solanum Tuberosum

The cultivated potato belongs to the genus *Solanum,* a large genus with over 2,000 species in the Solanaceae or nightshade family. Only 160 species in the genus are tuber-bearing, and only eight of these have been cultivated, with seven mainly confined to native regions in South America. *Solanum tuberosum* is the most widely cultivated potato; in fact, there are almost 6,000 cultivars (fig. 14.3) of this one species, but most commercial growers plant only a limited number of varieties. In the United States, of the 50 varieties that are grown commercially, 12 account for 85% of the potato harvest.

The potato plant is a bushy, herbaceous annual with an alternate arrangement of large, pinnately compound leaves. The white, pink, purple, blue, or yellow flowers are 5-merous, a typical dicot floral pattern. The fruit is a many-seeded berry, but normally the flower does not set seed because the precise temperature and daylength requirements are not met under

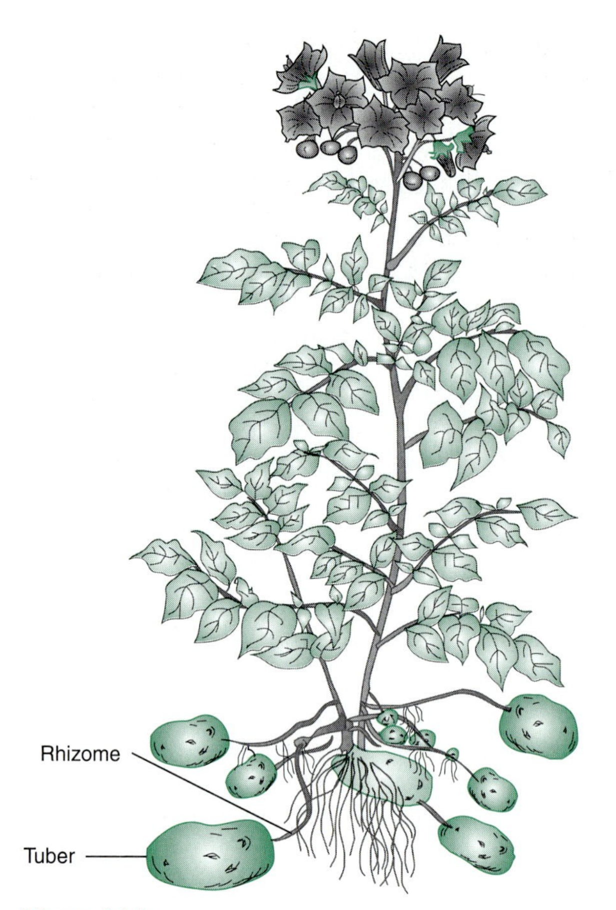

Figure 14.4 Solanum tuberosum, *cultivated potato.*

most northern agricultural conditions. Two types of stems are produced; the ordinary foliage-bearing stems and underground rhizomes that end in tubers (fig. 14.4). Anatomically, the tuber is a modified version of a dicot stem with an enlarged pith, a ring of vascular tissue, a narrow cortex, and a thin periderm. In some potato chips, the vascular tissue often appears as a darkened ring, with the narrow cortex on the outside and the large pith within the ring. The potato is a cool-season crop, with maximum tuber production at temperatures ranging from 15°–18°C (59°–64° F). Tuber formation is reduced at higher temperatures and inhibited at 29° C (84°F); this has limited the usefulness of the potato as a food crop in lowland tropical countries.

Potato cultivation involves propagation by seed potatoes, which are small whole tubers or cut tuber pieces containing at least one eye. Cultivation by seed potatoes is a method of asexual reproduction that produces plants genetically identical to the parent and maintains the desired traits within a cultivar. Seed potatoes are produced by certain growers who specialize in growing only seed potatoes; certified seed potatoes are ones that have been inspected and determined to be disease-free. This is an important consideration, since many diseases (in addition to late blight) can be transmitted through infected tubers. Although asexual reproduction is faster and produces plants with proscribed qualities, there are disadvantages as well. Genetically identical plants all share the same susceptibility to adverse environmental conditions and diseases. The devastation caused by *Phytophthora infestans* was compounded by the fact that most of the potatoes in Ireland were genetically identical, since they were derived from one or two plants originally introduced into the country.

Many of today's cultivars can be grouped into four familiar types: the round white, russet, round red, and long white. The round white is an all-purpose potato good for boiling, baking, or processing into chips, fries, or flakes; Kennebec and Katahdin are the most common round whites. Russets, also known as Idahos, are elongate, cylindrical tubers with a corky russet-colored skin and mealy texture that make excellent baking potatoes. Their shape also makes them ideal for French fries. Russet Burbank is the top-selling russet potato. The last two types, round reds and long whites, are usually sold as new potatoes, which are harvested earlier in the growing season and have a very thin skin. By contrast, late potatoes are harvested later and have a thicker, corky skin. Popular round reds, Red LaSoda and Norland, good for boiling, steaming, and roasting, are distinguished by their red skin and firm flesh. White Rose, a common long white, is a good all-around potato for home use; it has a tan smooth skin and is often even more elongate than a russet.

Potatoes are rich in carbohydrates (about 25% of the fresh weight); parenchyma cells within the pith are filled with starch grains. While not rich in proteins (only 2.5% of the fresh weight), the potato is actually a good source of high-quality protein because of the balance of essential amino acids. They are virtually fat-free (if not fried or topped with butter and sour cream), have no cholesterol, and are also good sources of potassium, iron, several B vitamins, and vitamin C. Much of the vitamins, minerals, and fiber of the potato occur in the periderm and cortex; therefore, the nutritional value is enhanced if potatoes are eaten with skins.

From its South American origins, the potato has traveled an interesting route to its present status as the world's fourth leading food crop.

Sweet Potato

Although the common name implies some relationship to the white potato, the sweet potato (*Ipomoea batatas*) is a storage root in the morning glory family, the Convolvulaceae. In fact, the flowers of the sweet potato resemble those of the familiar morning glory in gardens (fig. 14.5). The sweet potato plant is an attractive vine that is easy to grow at home. (A root placed in water will soon produce sprouts or slips, trailing shoots with adventitious roots.) Like the white potato, the sweet potato is propagated vegetatively. Slips, produced by the tuberous root in specially prepared beds, are transplanted into the field. The crop

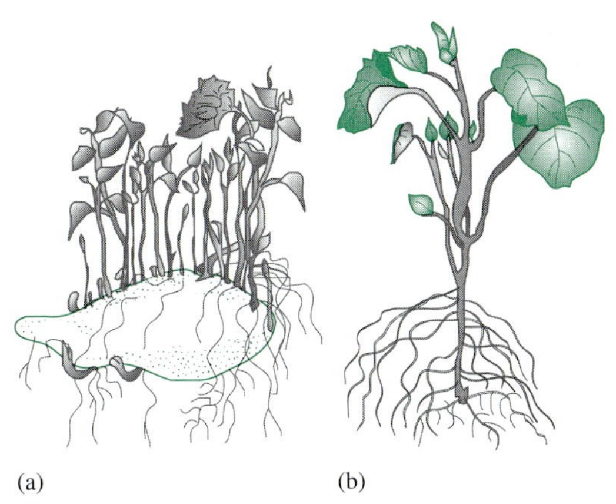

Figure 14.5 Ipomoea batatas, *sweet potato. (a) Vegetative propagation of sweet potato. Adventitious shoots, known as slips, develop from the tuberous root. (b) A single slip ready for planting.*

requires a long, warm growing season, since sweet potatoes are particularly susceptible to chilling injury.

Columbus discovered the sweet potato on his first voyage to the New World and introduced it to Spain on his return, at least 50 years earlier than the introduction of the white potato. There were several native names for the sweet potato in the New World, but in the Caribbean the Arawak Indians called it *batata,* which was eventually corrupted into the word potato. Originally the name potato was exclusive to *Ipomoea batatas,* but *Solanum tuberosum* also became known as a potato because it too was an underground crop. Following its introduction, the sweet potato became widely grown in Spain and other Mediterranean countries and was exported to northern Europe. The sweet potato was initially considered a delicacy in Europe; it was a favorite dish of England's King Henry VIII. It was also rumored to be an aphrodisiac, a claim that was later transferred to the white potato, along with the name.

The sweet potato is believed to be native to tropical South America; archaeological evidence dates its cultivation back several thousand years in Peru. It was widely grown as a staple crop in Central America and tropical South America prior to the European discovery of the New World. During this same period it was also cultivated in several Pacific Islands and New Zealand. Evidence of its history of cultivation in Polynesia suggests an earlier introduction. Was the sweet potato introduced to Polynesia by early seafaring natives or by the natural dispersal of seeds by birds or ocean currents? Thor Heyerdahl's voyage from Peru to Polynesia in the reed raft *Kon-Tiki* in 1947 demonstrated that a voyage by primitive peoples was a possibility. Its presence in the Pacific far from its South American origins still remains a mystery.

Today the sweet potato is a significant crop throughout the tropics and has expanded to warm temperate regions; in some countries, it is used as livestock feed as well as an important food staple. China dominates the world's production of the sweet potato, with several other Asian countries distant seconds. In Asian countries such as Japan, the sweet potato has uses beyond its consumption as a starchy vegetable. It is often sliced and dried, then ground into a meal that can be used in cooking, or fermented to make an alcoholic beverage. The sweet potato has also become an important component of the diet in several African countries. In the United States, it is primarily grown in the South, with North Carolina the leading producer. Two distinct varieties of *Ipomoea batatas* are recognized; a drier, starchier, and yellow-fleshed variety is favored in northern states while a sweeter, moister, and deep orange variety is preferred in the South. The latter variety is commonly called yams, but should not be confused with the true yam, *Dioscorea* spp., an important tropical tuber.

Like the white potato, the sweet potato is a nutritious vegetable rich in carbohydrates and certain vitamins and minerals. The sweet potato is accurately named since some of the carbohydrates are present in the form of sugar; they contain about 50% more calories than white potatoes but slightly less protein. Additionally, the carotene-rich root is an excellent source of vitamin A and is also a good source of vitamin C. Perhaps this nutrient-rich vegetable deserves a bigger place in the American diet than just at Thanksgiving.

Cassava

While unfamiliar to most North Americans except in the form of tapioca pudding, another tuberous root that is an important starchy staple is cassava, *Manihot esculenta.* This member of the spurge family (Euphorbiaceae) is a vital food to millions of people in the tropics. It is known by many different names including manioc, tapioca, yuca, and mandioca. Cassava ranks fourth behind rice, sugar, and corn as a source of calories for the human diet in tropical countries.

Origin and Spread of Cassava

Manihot esculenta, a New World crop, had its origins in South America, probably Brazil; there is speculation that it may have been independently domesticated in Central America as well. Cassava was a well-established crop in the New World tropics long before the arrival of the Europeans. In South America today, Brazil is the leading producer of cassava. The Portuguese, who ruled Brazil, introduced cassava into West Africa in the sixteenth century and it was introduced into East Africa in the middle of the eighteenth century. Extensive cultivation began in the twentieth century, and today Africa, particularly the countries of Zaire and Nigeria, is the leading producer of cassava. The Portuguese may also have introduced cassava into Asia via the port of Goa, on the Indian subcontinent, in the early eighteenth century. A later introduction from

Figure 14.6 Manihot esculenta, *cassava. Starch reserves are stored in the large tuberous roots visible in the photo.*

Mexico brought cassava to Indonesia and the Philippines. During the nineteenth century, cassava became a widely grown crop in areas of tropical Asia. Today Asia, especially Thailand and Indonesia, closely follows Africa in annual production, and South America is a distant third. As with many other crops, cassava cultivation has become greater outside the region of origin.

Botany and Cultivation

Manihot esculenta is a tall shrub with palmately compound leaves and numerous tuberous roots that are similar in appearance to sweet potatoes but usually much larger (fig. 14.6). Plants are propagated by stem cuttings that contain axillary buds; growth is fairly rapid and little care is needed following planting. The propagation is distinctive since none of the root, the economically valuable part, is used, unlike the other starchy staples where part of the harvest must be saved for cultivation. Although usually propagated from stem cuttings, cassava can also be cultivated from seed, which can be a source of new genetic varieties.

Many varieties of this tropical crop are tolerant to a wide range of moisture and soil conditions. Although predominantly grown in the hot lowland tropics, varieties exist that can survive the colder temperatures of tropical highlands. Although generally grown in areas that have moderate to heavy rainfall, cassava requires well-drained soils to prevent root rot. It can tolerate dry periods that last as long as six months and, therefore, does well in areas where rainfall is seasonal. This partly explains its popularity in the drier regions of Africa. Other advantages of cassava cultivation are its ability to grow well even in the nutrient-poor soils typical of the tropics, and its tolerance of acid soils, as well as those high in aluminum. Cassava is also resistant to many insects and fungal pathogens. Especially noteworthy is its resistance to locusts, which have periodically plagued parts of Africa. Even when the locusts eat the leaves, the roots are protected and the plant rapidly regenerates.

Depending on the variety, roots may be harvested anywhere from eight months to two years after planting; all the roots from a single plant may be harvested at one time or removed individually from the ground as needed. Flexible harvesting is possible because the tuberous roots remaining in the ground continue to grow and do not decay. Once harvested, however, the roots are subject to rapid decay and must be dried or processed in some manner within 24 hours.

Processing

Varieties are classified as either sweet or bitter, based on the concentration of poisonous hydrocyanic acid (HCN). If not removed, this toxin can cause death by cyanide poisoning. The HCN is liberated by the action of enzymes on **cyanogenic glycosides** present in cassava. There is no distinction between the sweet and bitter varieties other than the concentration of the toxins, and intermediate types exist. Environmental conditions are known to influence the production of cyanogenic glycosides, and what is a sweet variety in one locale can become a bitter variety when grown under different conditions. Sweet varieties with lower HCN levels can be eaten with little preparation; peeling followed by boiling, steaming, or frying will remove most of the toxins. Bitter varieties, which actually taste bitter due to the much higher levels of HCN, must undergo more extensive preparation to be detoxified before the roots are consumed. Traditional methods of treating the peeled bitter roots vary from country to country and include drying, soaking, boiling, grating, draining, and fermenting, or combinations of these. Although the sweet varieties are often cooked and consumed fresh, they can be processed similarly to the bitter varieties to make a meal or flour. The processed products all keep better than the quickly deteriorating fresh roots. Because of its bulk and perishability, most of the fresh cassava harvested is used locally and not exported. Worldwide, about 65% of the cassava grown is used for human food, with roughly half of that used as processed products.

In South America, the traditional preparation produces a meal called farinha. The peeled roots are grated (sometimes following a lengthy period of presoaking) and squeezed through a long cylindrical woven basket known as a tipiti (fig. 14.7). One end of the tipiti is tied to a tree while the other end is tied to a pole that is used to stretch the tipiti, thereby expressing juice from the grated pulp. This method eliminates much of the HCN from the pulp. The resulting mash is roasted until dried; in this form it can be stored for long periods and eaten many different ways. In the Caribbean Islands and the northeast coast of South America, the ground meal that results from processing is

Figure 14.7 *In traditional preparation of cassava meal, farinha, the peeled roots are grated and the juice containing HCN is expressed through a "tapiti," a woven basket, shown here.*

Figure 14.8 *True yams,* Dioscorea *spp., are important staples in many tropical areas.*

commonly made into a large circular flat bread. This bread is called *casabe,* from which the word *cassava* is derived.

In Africa, traditional methods of preparation to form gari involve one additional step—fermentation. The peeled roots are grated, with the resulting pulp placed in cloth sacks. Stones and logs are piled on top of the bags to squeeze out the juices containing HCN. This process takes several days, during which time the pulp naturally ferments, imparting a characteristically sour taste to gari. Following the fermentation, the pulp is roasted or fried until dry, and then stored. Gari, like farinha, can be eaten dry, used in soups or stews, or made into a doughy porridge.

One of the simplest techniques of processing cassava is the traditional method used in Indonesia, where peeled roots are sliced and dried in the sun, allowing the HCN to diffuse out. The resulting chips are called gaplek. They can be stored for long periods or ground into flour and used to make breads, cakes, cookies, noodles, and other products. These are just a few of the many traditional ways to prepare cassava. In many parts of the world today, mechanization is replacing these labor-intensive methods.

Starch is the main nutrient supplied by cassava (approximately 30% of the fresh weight); unfortunately the root is very low in protein (1% or less) and can result in kwashiorkor, protein deficiency, among peoples who rely on cassava exclusively. Improving the protein content of the root is a high priority among cassava breeding programs. In some African countries, cassava leaves are also consumed as a cooked vegetable and are a better source of protein, which can partly offset the low values in the root. Roots also contain moderate amounts of calcium and several B vitamins, and are a reasonably good source of vitamin C, but their main value to the human diet is as a source of energy.

Although in Africa almost all of the cassava grown is used for human consumption, in Asia and the Americas a considerable quantity is used for animal feed and for commercial starch production. Processing cassava roots into dried chips or pellets for animal feed has been increasing in recent years, so that today 20% of world production is fed to animals. Southeast Asia is the leading producer of animal feed for export, with much of it going to European markets. Cassava starch has many applications in the food, textile, paper, and pharmaceutical industries.

Although a staple in the tropics, the only form of cassava most North Americans have eaten is tapioca pudding, made by cooking tapioca pearls with milk, eggs, sugar, and vanilla. The pearls are partly gelatinized cassava starch made by heating moist cassava flour in shallow pans; this causes the wet granules of starch to stick together and form beads.

Other Underground Crops

Yams

Dioscorea spp. are mainly tropical tuber crops, true yams, that are important staples in many areas of the world, especially West Africa, southeast Asia, the Pacific Islands, and the Caribbean. This large genus in the Dioscoreaceae (yam family) is composed of several hundred species, of which ten are major food sources; evidence indicates that yams have been cultivated for over 5,000 years in tropical Africa. Tubers of these vines can vary from small ones the size of potatoes to massive ones often weighing over 40 kg (88 lbs) and 2–3 meters (6–9 ft) in length (fig. 14.8). Yams can be prepared in ways similar to the potato, with boiling and mashing the most common method. As much as 20% of the tuber is starch, with the protein content averaging about 2% and vitamins and minerals in negligible amounts.

Medically the tubers were an important source of **saponins** (sapogenic glycosides), a type of steroid, which had been used to make human sex hormones and cortisone. In fact, early research on the development of the birth control

A Closer Look 14.1

Banana Republics: The Story of The Starchy Fruit

Americans consider bananas a favorite dessert or snack food; however, for millions of people in tropical countries bananas are an important dietary staple. The familiar sweet bananas are usually eaten raw, but starchy plantains, which are more important as food in the tropics, are traditionally cooked and eaten as a vegetable. Cultivation of the sweet banana is greatest in Central America, while Africa leads the world in plantain production. Unlike other starchy staples, bananas are true fruits, even though modern cultivars are seedless.

Bananas are native to southeast Asia and are believed to be among the first cultivated plants in that area. It is thought that Polynesians spread the banana throughout the Pacific islands. Records from India indicate that bananas have been cultivated there for at least 2,500 years. About 2,000 years ago Arabian traders introduced bananas into parts of Africa, where the crop flourished. In fact, the word *banana* comes from West Africa. Portuguese and Spanish colonizers spread bananas throughout tropical regions, and early in the sixteenth century, they were introduced to the New World. Bananas became so well established that many early explorers thought they were native to the New World.

Four centuries after their introduction, bananas played a pivotal role in the history of Central America. Early in the twentieth century large U.S. corporations, such as United Fruit Company, developed extensive banana plantations in Central America, along with corporate-run railroads and steamships. Over the next 50 years, United Fruit exerted tremendous control over the economies and governments of several countries that became known as "Banana Republics." The rise of nationalism, starting in the 1950s, led to the decline in influence of United Fruit.

Bananas are produced by various species in the genus *Musa* in the Musaceae, the banana family. Most cultivars today are sterile triploids that probably arose as a cross between a diploid and a tetraploid species. Bananas require a tropical climate with constant moisture, and members of the genus are cultivated throughout the tropics for the fruit, the fiber (*see* Chapter 18), or even the foliage, which is often used to wrap foods as a type of natural waxed paper.

The banana plant, often called a tree, is a large herbaceous annual that may reach 6 meters (20 ft) or more in height (box fig. 14.1). The "trunk" of this monocot is not woody, but is actually a rosette of overlapping, tightly packed leaf bases that arise from an underground corm. The leaf blades themselves are also large, often about 2.5 meters (8 ft long) or more in length. Although the leaf blade is complete when it first develops, it is soon torn into strips by prevailing winds.

pill was based specifically on diosgenin extracted from yams and shown to inhibit ovulation. Today, synthetics have largely replaced the natural compounds.

Yams are planted on small hills, with their twining vines and cordate leaves supported by stakes. They grow best under humid and subhumid conditions, so their cultivation is mainly restricted to the wet lowland tropics, and as such, *Dioscorea* is a crop that does well in tropical rain forests. Propagated from eyes, the tubers of this monocot are harvested after a long growing season of 8 to 12 months. Harvesting tubers is a laborious task, since the tubers are generally deeply buried. For this reason, yam production has declined in recent years and been largely replaced by cassava.

Taro

Poi, the traditional dish of the native Hawaiians, is prepared from the corm of taro (*Colocasia esculenta*), a member of the Araceae or arum family (fig. 14.9). This plant is related to, and resembles, elephant's ear, a common landscaping plant in the southern United States. The large leaves of taro also play a role in Hawaiian traditions, since foods are wrapped and cooked in the leaves during a Hawaiian feast or luau. In fact, the word *luau* is the Hawaiian name for these leaves.

In preparing poi, the corms are steamed, crushed, made into a dough, and then allowed to ferment naturally by microorganisms. This doughy paste is then eaten with the fingers or rolled into small balls; this food was one of the main staples in the traditional diet, with large amounts consumed daily. The corm may also be cooked similar to potatoes (baked, steamed, roasted, or boiled) or processed into flour, chips, and breakfast foods. Nutritionally, the corm contains approximately 25% carbohydrate, which includes about 3% sugar, 2% protein, and very little fat. The majority of the carbohydrate present is a fine-grained, easily digested starch. The corms are also a good source of calcium due to the presence of calcium oxalate crystals; if the corm

Box Figure 14.1 *A banana plant.*

After approximately one year the apical meristem converts from vegetative growth to flowering, and the monoeocious inflorescence develops. The single flowering stalk contains 5 to 13 groups of flowers (often called hands or bunches), with each group covered by a large purple bract. Most groups contain pistillate flowers that develop parthenocarpic fruit, with staminate flowers confined to the end of the inflorescence. The fruit, which is technically a berry, requires about 80–120 days to mature. Bananas destined for market are picked green and shipped under controlled conditions to prevent ripening. Once yellow, bananas perish rather quickly.

Fruit production terminates the life of an individual plant, but new suckers or sideshoots develop from the corm. Since the fruits are seedless, these suckers are used in vegetative propagation. Suckers reach maturity in 9–12 months. Because of the asexual reproduction, all plants of one variety are genetically identical clones and share susceptibility to various pathogens. The worst threat to plantations is Panama disease, caused by a fungal pathogen, *Fusarium oxysporum,* that produces a vascular wilt. Growth of the fungus in the vessels of the xylem brings about the wilting; leaves turn yellow, then brown, and eventually die, even though water is available. Panama disease decimated Central American planatations early in the twentieth century and initiated the continuing search for resistant cultivars.

Nutritionally the fruits are good sources of energy and potassium and are easily digestible. Bananas are rich in starch; however, in the sweet varieties, some of the starch is converted to sugar as the fruit ripens.

(a)

(b)

Figure 14.9 Colocasia esculenta, *taro: (a) plant and (b) corms.*

A Closer Look 14.2

Starch: In Our Collars and In Our Colas

Starch is the most common carbohydrate storage product found in plants. Although the primary focus in this chapter has been the importance of starch in the diet, it is a versatile compound that has many commercial uses in both food and nonfood industries.

Sugars are the direct products of photosynthesis, but are generally not stored as such. Instead, accumulated sugars are stored in the form of starch. The starch forms as starch grains found in amyloplasts, an organelle abundant in parenchyma cells of storage organs. Chemically, starch is composed of thousands of glucose molecules bonded together in chains. There are two distinct components of starch, amylose and amylopectin, which differ in the structural configuration of their chains. Amylose is an unbranched chain of several hundred to a few thousand glucose units. On the other hand, amylopectin is a highly branched chain containing between 2,000 and 500,000 glucose units. Amylopectin is the major component (approximately 80%) of starch, with amylose making up the remaining 20%, but the ratio may vary with the species of plant. Amylose tends to be relatively soluble in water while amylopectin is insoluble. When potatoes are boiled, the amylose is extracted by the hot water, turning it cloudy. The same phenomenon happens when sticky varieties of rice are cooked, but here all the water is absorbed and the grains are clumped together due to the film of amylose adhering to the grains.

Starch can be extracted from almost any grain, tuber, or other storage organ, but commercially most starch is extracted from corn, potatoes, and cassava, with smaller contributions from wheat, rice, arrowroot, and sago. Although starch extraction varies from species to species, the following is a general description of the process. Basically, the plant material is ground or pulverized to liberate starch grains from the cells. The resulting pulp, a mixture of starch and plant debris, is sieved and washed to remove debris from the starch suspension. Finally, the starch is separated from the water (by filtration, settling, or centrifugation) and dried.

Commercial starch finds hundreds of industrial uses; among the foremost nonfood applications are the manufacture of adhesives and the production of sizings. Starch-based adhesives are used in the production of cardboard, paper bags, and remoistening gums for envelopes and stamps, to name a few. In the past, they had also been used in the lamination of veneers to make plywood, but today resins have largely replaced starch adhesives. Sizings are substances used as fillers or coatings, and starch is one of the most commonly used types of sizes. Steps in the manufacture of paper, cloth, thread, and yarn involve sizing; these steps can strengthen the material, impart a smooth finish, or prepare the surface for dyes. Starch finds other uses in the pharmaceutical industry as a binding and coating agent for medicines and in the petroleum industry as drilling compounds. Lastly, laundry starch is a familiar starch-based product with uses in the home as well as in industry.

One of the increasingly important uses of starch is in the production of sugar-based sweeteners. Starch is hydrolyzed with weak acids or enzymes to produce a glucose (dextrose) syrup, which may be marketed as such or enzymatically converted to fructose. High fructose syrups are in great demand in the food processing industry as a replacement for sucrose (table sugar), since fructose tastes sweeter than equivalent amounts of sucrose. Some starch is used directly in food processing for the manufacture of puddings, gravies, and sauces; in other cases, starch is blended with flour in the production of bakery goods.

The fermentation of starch by yeast produces alcohol that can be used for beverages, industrial solvents, or as a petroleum fuel substitute. In the past two decades replacement of gasoline by alcohol has received much media attention. As petroleum reserves dwindle and prices increase, the search for alternative fuels will intensify. Alcohol, produced from crops, has the potential to become an environmentally safe fuel substitute. Starch, as one of the world's renewable resources, will continue to have an expanding market.

is eaten raw, these crystals would cause intense burning or stinging in the mouth and throat. Cooking destroys the crystals. The crystals are characteristic of the arum family and are found in familiar plants such as *Philodendron, Dieffenbachia,* and Jack-in-the-pulpit (see Chapter 21).

Taro is believed to have originated in Southeast Asia and spread both east and west during prehistoric times. The eastward spread took it to Japan, the Pacific Islands, and as far east as Hawaii. It spread westward to India and then to the Mediterranean region during the time of the ancient Greek civilization. Eventually, it spread to Africa and from there it was brought to the West Indies and tropical America by African slaves. Today it is cultivated in the wet tropics, where it thrives under saturated soil conditions. It is propagated by planting either the tops of corms or cormels. Although not a major crop in the world market, taro remains locally important in some areas.

A similar member of the Araceae is yautia or cocoyam (*Xanthosoma* spp.), native to tropical Central and South America and also referred to as taro in some areas. Like taro, the underground storage organs can be prepared in various ways and have comparable nutritive value to the other starchy staples.

Current nutritional research advocates diets containing approximately 58% carbohydrates, 12% protein, and no more than 30% fat for good health. The starchy staples that are naturally low in fat and high in complex carbohydrates satisfy two of the major components of this recommendation and should be a major part of the human diet. This chapter has described the major starchy staples cultivated today; but with increased emphasis on the importance of carbohydrates in the diet, new starchy staples are likely to be developed and promoted for tomorrow's table.

Chapter Summary

1. Many plants store large quantities of starch in underground structures that are modified stems or roots. Botanically, these structures have several functions: as food reserves, for asexual reproduction, and as the starting point for renewed growth after dormancy. Some of these storage organs are starchy staples that include some of the world's foremost crops. In addition to their direct consumption, potatoes and cassava are also sources of commercial starch, which has many applications in both food and nonfood industries.
2. Native to the highlands of South America, the potato was the dietary staple of the ancient Incan civilization and introduced to Europe in the sixteenth century. Cultivation of these tubers spread slowly throughout Europe, becoming the food of peasants. The crop was so successful in Ireland that it led to a population explosion and, later, massive famines when late blight of potato destroyed the crops in the 1840s. Today, about one-half of the U.S. potato crop is processed to make French fries, potato chips, and dehydrated flakes. This continues a long tradition, since Peruvian Indians have prepared a processed potato product, chuño, for over 2,000 years. The potato plant is a member of the nightshade family, the Solanaceae. Many members of this family are poisonous; in fact, the potato contains toxins in all parts except the tuber. Potatoes are cultivated by seed potatoes, which are small tubers or cut tuber pieces containing at least one "eye." Although this method of asexual reproduction is fast and produces plants with desired characteristics, all the offspring are genetically identical and share susceptibility to adverse environmental conditions and diseases.
3. Sweet potatoes are storage roots of a plant in the morning glory family and unrelated to the white potato. The sweet potato is believed to be native to tropical South America; archaeological evidence dates its cultivation back several thousand years in Peru. It was also cultivated in several Pacific Islands and New Zealand. Evidence of its early cultivation in Polynesia suggests an earlier introduction by seafaring natives or by the natural dispersal of seeds by birds or ocean currents. While of relatively minor importance in the United States, the sweet potato today is a significant crop throughout the tropics and some warm temperate regions.
4. Cassava is a tuberous root that is another important starchy staple vital to the food supply of millions in the tropics. Cassava ranks fourth behind rice, sugar, and corn as a source of calories for the human diet in tropical countries. Although native to South America, cassava is now cultivated throughout the tropics. Cassava varieties are classified as either sweet or bitter, based on the concentration of hydrocyanic acid. Various methods of processing have been developed to detoxify cassava and make it safe to eat.
5. Other underground crops include yams and taro, which are important staples in tropical areas, especially the South Pacific islands. Unlike other starchy staples, bananas are fruits. While sweet bananas are often considered a dessert or snack food, starchy plantains are important as food in the tropics, where they are traditionally cooked and eaten as a vegetable.

Review Questions

1. Discuss the various plant organs modified for storage.
2. What has been the social impact of the white potato?
3. What is the difference between the bitter and sweet varieties of cassava? How does their processing differ?
4. How can the presence of the sweet potato in both South America and Polynesia be explained?
5. Contrast the nutritional value of the starchy staples to cereals and legumes.
6. Describe some of the commercial uses of starch.
7. What is the connection between yams and birth control pills?
8. Describe the cultivation of the potato, sweet potato, cassava, and taro.
9. In what ways is the banana different from other starchy staples? Consider the growth and storage organs.

Further Reading

Burton, W. G. 1989. *The Potato,* 3rd Edition. Longman Group UK Limited, Essex, England.

Cock, James H. 1985. *Cassava: New Potential for a Neglected Crop.* Westview Press, Boulder, CO.

Hobhouse, Henry, 1985. *Seeds of Change: Five Plants that Transformed Mankind.* Harper & Row, New York.

Hughes, Meredith Sayles. 1991. Potayto, Potahto—Either Way You Say it, They A'peel. *Smithsonian,* 22 (7): 138–149.

Rhoades, Robert E. 1982. The Incredible Potato. *National Geographic,* 161 (5): 668–694.

Salamon, Redcliffe N. 1949. *The History and Social Influence of the Potato.* Cambridge University Press, Cambridge, England.

Schumann, Gail L. 1991. *Plant Diseases: Their Biology and Social Impact.* APS Press, St. Paul, MN.

Sokolov, Raymond. 1991. *Why We Eat What We Eat: How the Encounter Between the New World and the Old Changed the Way Everyone on the Planet Eats.* Summit Books, New York.

Viola, Herman and Carolyn Margolis eds. 1991. *Seeds of Change: A Quincentennial Commemoration.* Smithsonian Institution Press, Washington, DC.

CHAPTER

15

FEEDING A HUNGRY WORLD

Chapter Outline

Chapter Concepts

1. The major challenge in agriculture today is producing enough food to feed the world's population.
2. Traditional breeding programs and biotechnology are being used to develop high-yielding and disease-resistant cultivars.
3. Germplasm, the genetic information encoded in plants, is a valuable asset to plant breeders and must be preserved.

In the 1990s, the effects of hunger and malnutrition are vividly seen on the nightly news in reports from Somalia and Bosnia. Yet, hunger extends far beyond these war-torn regions, with approximately 1.4 billion people, 25% of the world's population, receiving insufficient food to meet daily requirements. It is only when the situation reaches crisis proportions in a specific area that the media focuses on the horrible conditions. Even with the staggering rates of starvation today, it must be remembered that the human population is still growing, and the need for food will continue to increase far into the next century. By developing higher-yielding and disease-resistant crops, plant scientists will be at the forefront of the efforts to increase the global food supply.

Breeding for Crop Improvement

Since earliest times humans have practiced plant breeding by selecting for certain traits. For most present-day crops, native plants were the starting point for selection. Even before the beginnings of agriculture, early peoples were probably selecting for certain sizes of fruits or seeds. An example of artificial selection can be seen with the trait for nonshattering heads of grain. While people were still foraging (*see* Chapter 11), it was easier to gather grain from plants with nonshattering heads, a recessive trait. These plants retained the grains, while the wild-type plants scattered seeds for optimum dispersal. When human groups shifted from foraging to farming, they planted seeds that were on hand. Often a large percentage of the seeds would be from plants with the nonshattering trait. Soon the nonshattering trait dominated the population, thereby establishing this characteristic as standard for cultivation. At first the selection was not deliberate, but later farmers may have intentionally selected for traits such as large seeds or certain fruit or seed colors by carrying out various practices such as crosspollination (fig. 15.1).

Figure 15.1 *Relief carving from Nimrud (ninth century B.C.) showing an Assyrian deity pollinating female date palm flowers. This has been interpreted as showing the artificial pollination and breeding of date palm flowers as practiced at that time.*

Although crop improvement was well established by the nineteenth century, the scientific basis for plant breeding was not understood until the widespread acceptance of Mendel's work. The development of hybrid corn in the early twentieth century is an example of the early use of selection on a widespread scale in modern agriculture (*see* Chapter 12).

Human selection often circumvents evolution through natural selection, but it may increase the rate of evolution for some plants. In some cases, the evolutionary change has been so great that the ancestral plants are not known. The most notable example is, of course, corn (*see* Chapter 12). Although it is believed that teosinte is the wild grass from which corn was domesticated, there are striking differences between the two species. Corn is totally dependent on humans for survival since it has lost the ability to disperse seeds in the natural environment, has much larger ears encased in husks, and has naked grains. These traits are lacking in teosinte (fig. 15.2).

An interesting example of artificial selection is in the development of many different crops from a single species, *Brassica oleracea.* Cabbage, kale, broccoli, cauliflower, kohlrabi, and brussels sprouts were all developed from the wild type *Brassica* by selection techniques that emphasized different parts of the plant (fig. 15.3). Today, selection is still important in plant breeding programs; however, genetic engineering techniques often reduce the long time periods required to establish new varieties.

The Green Revolution

Improvements in crop yield and quality achieved through plant breeding helped produce a modern and efficient agricultural system in the United States by the early part of the twentieth century. Yields increased, and have continued to increase, throughout this century. In fact, over

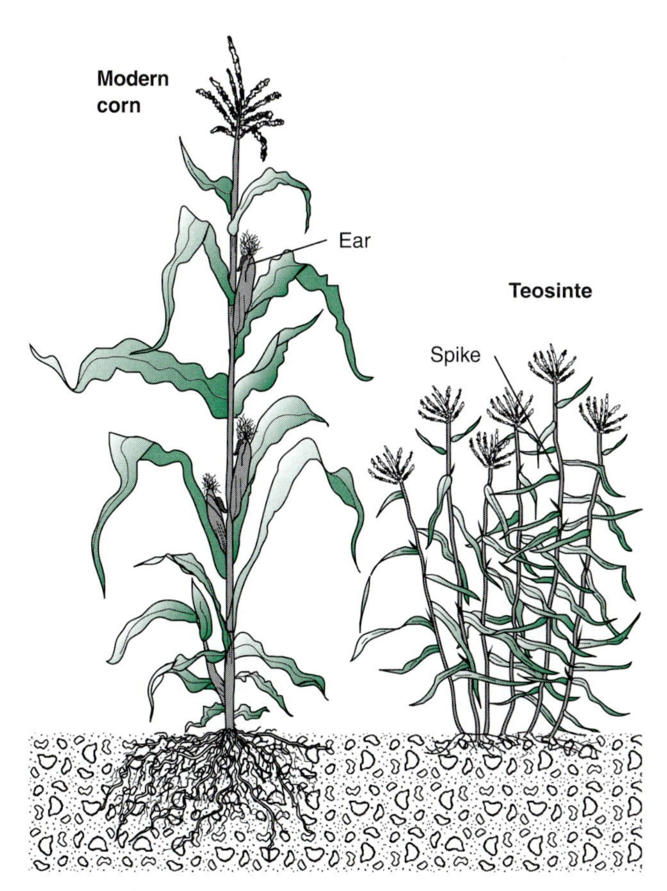

Figure 15.2 *The wild grass teosinte* (Zea mexicana) *is believed to have given rise to modern corn* (Zea mays). *Plants differ mainly in the spikes or ears. While modern corn has just two or three large ears that remain attached to the one main stalk, teosinte has numerous small brittle spikes (on many stalks) that shatter when the kernels are mature. Just a few genetic changes may have resulted in the evolution of modern corn.*

the past 50 years there has been a three-fold increase in agricultural production. The use of high-yielding varieties is the center of this accomplishment; however, these varieties need excellent growing conditions to be productive, and the use of mechanization, fertilizers, pesticides, and irrigation are an integral part of the higher yields. Over the past three decades, these technologies have been transferred to developing nations, where tremendous increases have also occurred. This achievement is known as the **Green Revolution.**

High Yield Varieties

The central thrust for solving the world food problems has focused on improvements of existing crops. During the past several decades major efforts have been made to increase the yields of wheat, rice, and corn, as well as other crops, especially in tropical and subtropical areas. Much of the research has been carried out at international agricultural research centers located in developing nations (table 15.1). These centers not only select and develop new varieties, but are also active in the collection and preservation of diverse species.

Wheat has been the principal crop behind the Green Revolution. Dr. Norman Borlaug (fig. 15.4), who is usually referred to as the "Father of the Green Revolution," developed high-yielding dwarf strains of wheat at the CIMMYT (Centro Internacional de Mejoramiento de Maiz y Trigo—International Center for Maize and Wheat Improvement) sponsored by the Rockefeller Foundation in Mexico City. The high-yielding varieties were produced from crosses, with a dwarf variety introduced from Japan following World War II. These varieties did well in Mexico and other areas like India and Pakistan. Within a 20-year period the Mexican wheat harvest doubled, and similar accomplishments were also realized in other countries. In India, the high-yielding varieties attracted so much attention that armed guards were needed to prevent people from stealing the seeds before they were ready for release. The rust-resistant cultivars that Borlaug developed have strong, stiff stems that can accept heavy applications of fertilizer without **lodging** (falling over during heavy winds or rain, especially near harvesting). The older, taller varieties tended to lodge from the weight of the grains when heavily fertilized, making harvesting very difficult. Similar success was achieved in the Philippines at the IRRI (International Rice Research Institute) where high-yielding dwarf rice cultivars were developed.

Starting in the late 1960s, food production increased dramatically in areas where these high-yielding crops were introduced, contributing to the chronicles of the Green Revolution. Breeding programs continue at these agricultural research centers, constantly developing and introducing new cultivars with improved resistance to pests and/or pathogens. Also, programs are now in place to improve crops such as cassava and sweet potato that are dietary staples in tropical countries. As long as the human population continues to grow, the need for increased food production will continue the search for improved crops.

Although the Nobel Prize Committee does not award a prize in agricultural sciences, Dr. Borlaug was awarded the Nobel Peace Prize in 1970 for his efforts to increase world food production. In making the award, the Nobel committee called the Green Revolution a

> "technological breakthrough which makes it possible to abolish hunger in the developing countries in the course of a few years."

The committee also stated that the Green Revolution contributed to the solution of the population explosion. Although the achievements were extraordinary, the Nobel committee may have been overly optimistic in assessing the early success of the Green Revolution, as we will discuss later.

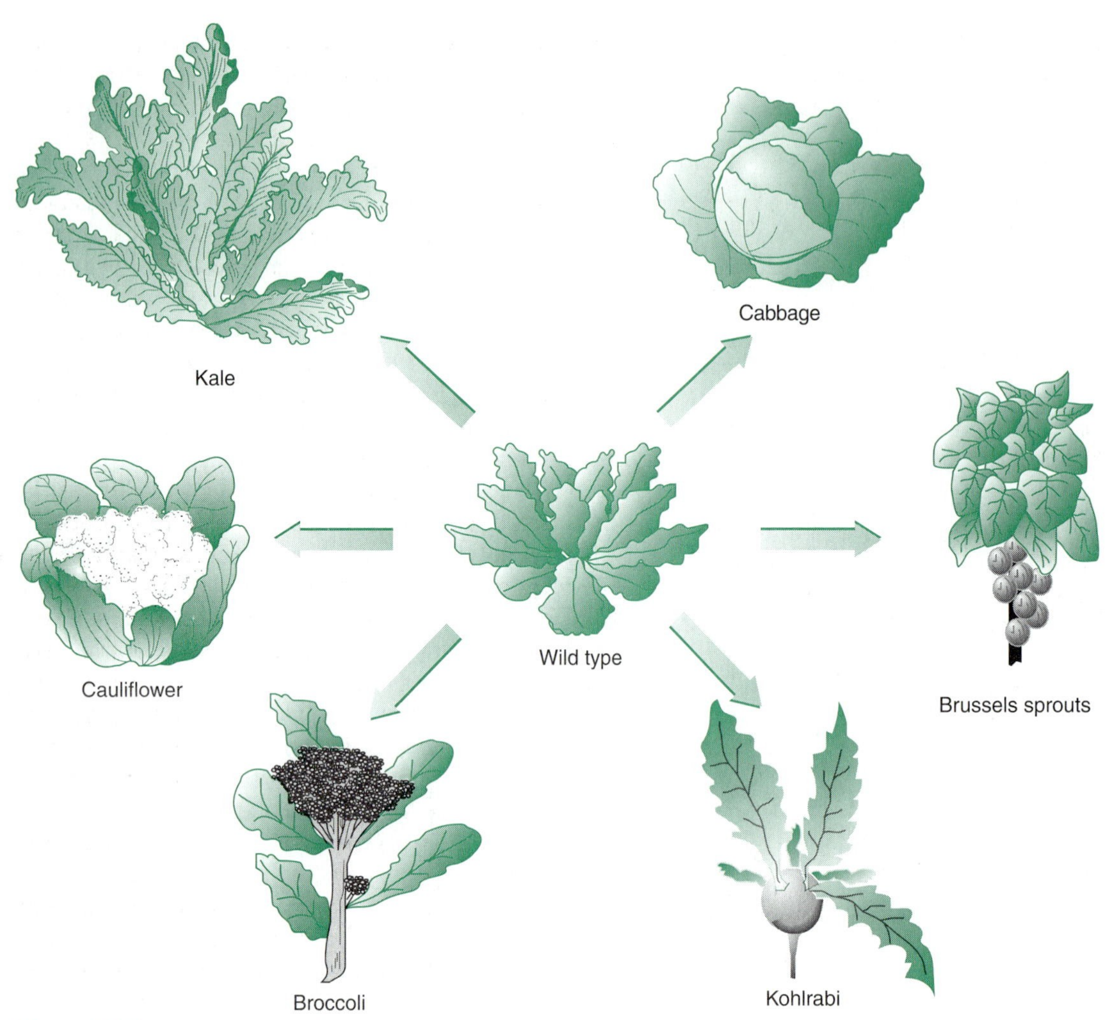

Figure 15.3 *Artificial selection has led to the development of many different crops from the wild type* Brassica oleracea.

Disease-Resistant Varieties

Crops are threatened by thousands of diseases that make the global food supply vulnerable. While some of the diseases may be noninfectious and caused by such factors as mineral deficiencies or pollutants, the majority are infectious diseases caused by a wide range of organisms including nematodes, bacteria, viruses, fungi, etc. Even with the heavy use of various pesticides, approximately 50% of the world's food crops are destroyed in the field or in storage. As a group, fungal pathogens make up the most serious threat and rank immediately behind insects as the chief competitors to human food (*see* Chapter 22).

The field of science that deals with the study of these diseases is plant pathology. While plant pathologists use many approaches to produce healthy plants, disease-resistant varieties of crop plants are the cheapest and most effective method. Breeding for disease-resistance must be coupled to yield, quality, climate, consumer acceptability, and a host of other traits the farmer demands. The primary efforts of plant breeders are naturally focused on the world's major food crops, especially wheat, rice, and corn, and disease-resistant varieties have been commercialized for many areas of the world. Unfortunately, some of the achievements in this area have been undermined by the continued evolution of the fungal pathogens.

Many instances exist where a single dominant gene in the host plant has been found to be the source of the resistance. Some parasites produce toxins or enzymes that enhance their ability to attack the host plant and thus contribute to their success. A resistant plant may produce a molecule that blocks the parasite enzyme or alters the toxin-binding site so that the parasite cannot be successful. With time, a mutation may arise in the parasite that results in a modified enzyme that cannot be blocked, or a modified toxin with improved binding ability. These genetic interactions between host and parasite constantly occur in the natural environment. Plant pathologists and breeders valiantly try to stay ahead of the newly evolving strains of pathogens that are capable of destroying various crops. This has been a problem especially for the development of rust-resistant varieties of wheat (*see* Chapters 12 and 22).

History of the Green Revolution

The earliest involvement of United States scientists with agriculture in developing nations came about in 1941 when Iowa State University set up an experimental station in

Figure 15.4 *Dr. Norman Borlaug, Father of the Green Revolution and recipient of the 1970 Nobel Peace Prize.*

Guatemala to study indigenous plants as sources of genetic material to improve U.S. crops. Also in 1941, the U.S. government sent four teams of agricultural scientists to travel throughout Latin America. The purpose was to determine how to strengthen the agriculture in the area, both for the benefit of Latin America and the United States, should the war in Europe continue to escalate. Rubber production was a major interest at this time because war with Japan would threaten the U.S. source of rubber. As a result of the inspection tour, experimental research stations were established in Peru, Ecuador, Nicaragua, and Guatemala. Agricultural programs received major emphasis, and U.S. foreign aid programs helped establish experimental stations within most non-Communist developing nations.

During the 1950s and 1960s, there was overwhelming enthusiasm for the agricultural programs, and tremendous confidence that science could solve the food problems in the hungry nations. One of the most acclaimed programs was CIMMYT in Mexico, where Norman Borlaug carried out his wheat-breeding experiments that led to the high-yielding dwarf cultivars.

William Gaud, the head of the U.S. foreign aid program, coined the term "Green Revolution" during a speech in 1968, and it quickly became the catch-word for both the media and the scientific community. The Green Revolution was credited with miraculous accomplishments. Two years later Norman Borlaug received the Nobel Peace Prize for his involvement in the Green Revolution. But the optimism about the Green Revolution during the 1960s, so eloquently expressed by the Nobel Committee when it honored Dr. Borlaug, met the cold realities of the 1970s.

TABLE 15.1
International Agricultural Research Centers

CIAT	Centro Internacional de Agricultura Tropical; Cali, Colombia
CIMMYT	Centro Internacional de Mejoramiento de Maiz y Trigo; Mexico City, Mexico
CIP	Centro Internacional de la Papa; Lima, Peru
IBPGR	International Board for Plant Genetic Resources; Rome, Italy
ICARDA	International Center for Agricultural Research in the Dry Areas; Aleppo, Syria
ICRISAT	International Crops Research Institute for the Semi-Arid Tropics; Hyderbad, India
IFPRI	International Food Policy Research Institute; Washington, D.C.
IITA	International Institute of Tropical Agriculture; Ibadan, Nigeria
ILCA	International Livestock Centre for Africa; Addis Ababa, Ethiopia
ILRAD	International Laboratory for Research on Animal Diseases; Nairobi, Kenya
IRRI	International Rice Research Institute; Manila, Philippines
ISNAR	International Service for National Agricultural Research, The Hague, Netherlands
WARDA	West Africa Rice Development Association; Monrovia, Liberia

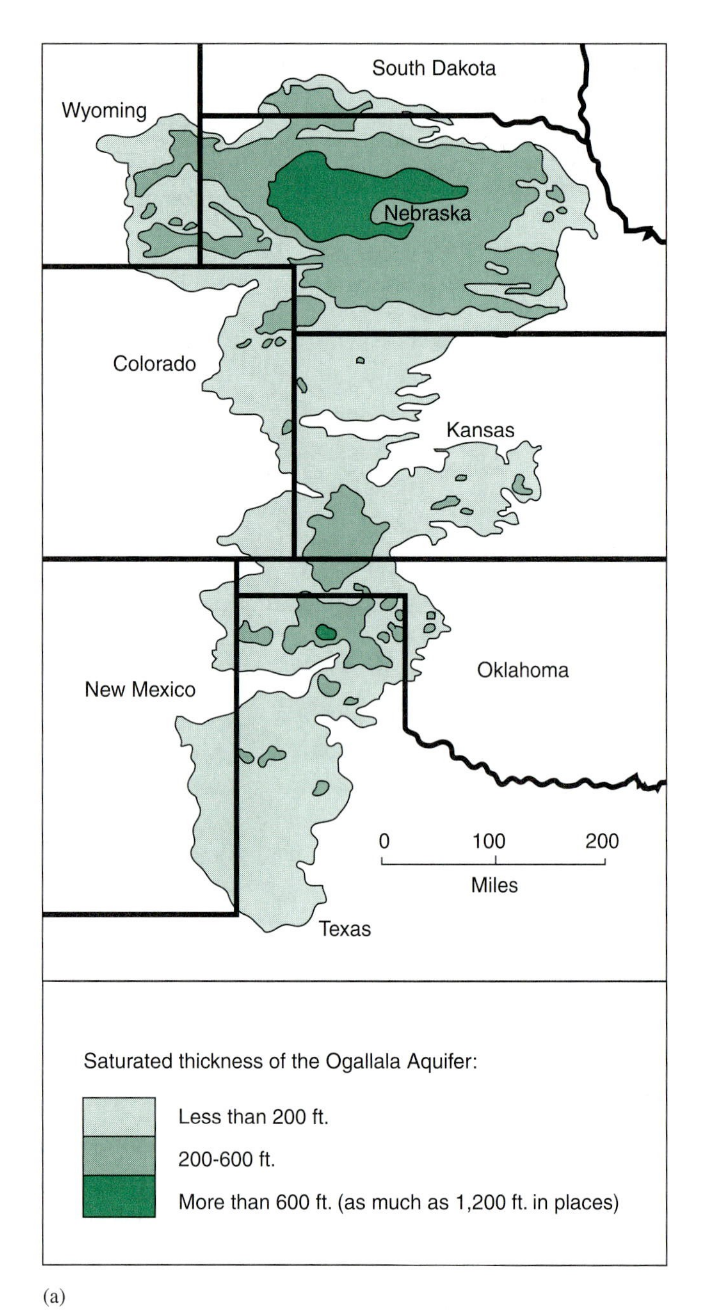

(a)

(b)

Figure 15.5 *Underground aquifers often supply much of the irrigation water used by farmers. (a) Ogallala Aquifer is the largest known body of fresh water. Yearly use of this water, mainly for irrigation, equals the total annual flow of the Colorado River, resulting in significant drops in the water table (as much as 50 m (162 ft) in some areas). (b) Sinkholes sometimes develop as land collapses after the removal of underground water. This home in Barstow, Florida was lost to such a sinkhole.*

(a) Source: U.S. Geological Survey.

Problems with the Green Revolution: What Went Wrong?

The Green Revolution is not without its problems and critics, nor is it accessible to everyone. Today's high-yielding and disease-resistant crop varieties are high impact crops; they are dependent on fertilizer and pesticide application, adequate water, and mechanized farming. Poor farmers cannot afford the seeds, fertilizers, pesticides, irrigation, farm equipment, and fuel needed to cultivate the high-yielding varieties.

Other problems are high energy costs, environmental damage, and loss of genetic diversity. The manufacture of inorganic fertilizers requires a great deal of energy (two barrels of oil per barrel of nitrogen fertilizer). In 1973, when the OPEC oil embargo resulted in major increases in the cost of petroleum products, the agricultural community suddenly realized the actual costs of high-impact agriculture. What had been profit soon became deficit. Because of their high cost, fertilizers are often reserved only for cash crops or export crops, not those grown for domestic consumption. Misuse of these compounds also causes serious environmental problems. Similarly, fossil fuels are required for the increasingly specialized and efficient farm machinery that has been developed to make planting and harvesting more efficient.

Application of pesticides for crop protection is also inherent in the Green Revolution. Insecticides, herbicides, and fungicides are all needed to maximize yield and were originally viewed as spectacular means to fight plant disease and increase yield. Today, the environmental and health effects of pesticide use and the problems with pest-resistance to these chemicals are growing concerns (*see* Chapter 25).

For maximum yield, the miracle crops also require ample water, either from optimum rainfall or irrigation. In the United States, where high-yielding crops are standard, 47% of the water supply goes for irrigation. Some have questioned the wisdom of massive irrigation in desert areas such as central Arizona for growing lettuce and pecan trees. Is this the most rational use of water? Often, irrigation water is pumped from underground **aquifers,** such as the Ogallala Aquifer stretching from Texas to South Dakota (fig. 15.5a), which had accumulated water for millions of years. The water from this and other aquifers is not being replaced as rapidly as it is used. In some areas, water is removed so rapidly that sinkholes have developed as land collapses into the spaces left after the water is pumped out (fig. 15.5b).

Some scientists feel that the whole premise behind the Green Revolution needs to be reexamined. Increasing food supplies in developing nations during the 1960s and 1970s has led to massive population increases in these countries. The situation has been described as similar to the Irish potato famine in the nineteenth century. The destruction of the potato by the fungus *Phytophthora infestans* directly caused the population of Ireland (8.5 million in 1840) to decrease by approximately 2.5 million, 1 million due to starvation and 1.5 million by emigration. The majority of the survivors were left in total poverty. It has been suggested that the catastrophe was actually due to the introduction of the potato centuries earlier. Prior to that, the population of Ireland was relatively stable at about 1.5 million, although most lived in total poverty. The potato was a cheap, reliable food source that allowed the population to grow. The Irish society, however, did not limit its birth rate, resulting in a population explosion. The only check on this population growth was starvation, which followed the famine.

Increased food production in areas where the population growth rate is uncontrolled is a dangerous situation. It ultimately means that more people will starve and more environmental degradation will occur. Even in the late 1960s, when the Green Revolution was being heralded, some scientists realized that population growth was destroying all the efforts being made to feed the hungry of the world.

Solutions?

In 1994, the human population reached 5.6 billion, with realistic estimates of 6 billion by the year 2000. At the current rate of growth, the earth's population is expected to double by 2025 and continue increasing until the middle of the twenty-second century (fig. 15.6). In order to feed the projected population, food production will need to more than double in the next 30 years and continue to increase well into the next century. Scientists debate whether the Earth can sustain this population growth.

The Earth's **carrying capacity** (the number of people the planet can adequately support) is not known. Estimates vary, and depend on the level of technology and on the lifestyles being projected. For example, estimates indicate that the present levels of food production, with an equal distribution of the food, could support 5.5 billion vegetarians, 3.7 billion people who get 15% of their food from animal products, or just 2.8 billion people who get 25% of their food from animal products. Other studies from the 1980s have estimated the population size supportable by the year 2000. The estimates were based on various levels of farming technology. With a low level of farming technology, the carrying capacity was 5.7 billion people, with an intermediate level of 14.4 billion, and with a high level of farming technology, 32.3 billion people. These projections have been viewed as too simplistic, with no economic or scientific analyses to back them up. Scientists must learn more about the interaction of populations, the environment, economy, and culture to determine an accurate assessment of the planet's carrying capacity.

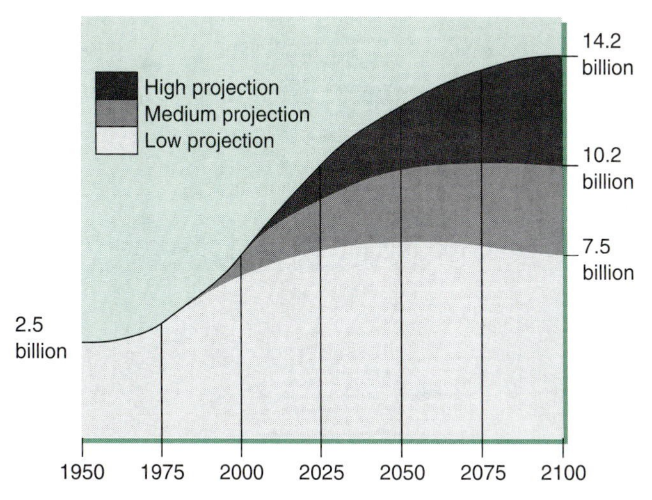

Figure 15.6 *The human population is currently showing exponential growth. Projections for when the rate will level off depend on the pace of development and use of birth control measures in less-developed regions of the world. The total populations, based on these projections, vary from a low of 7.5 billion people to 14.2 billion. The United Nations projection assumes a human population of approximately 10.2 billion by 2035.*

Source: 1988 World Population Data Sheet (Washington, D.C.: Population Reference Bureau, April, 1988).

The answer to feeding the world's population must be found in the regions where most of the people live and where most of the future population growth is anticipated—the developing nations. Innovative techniques must be encouraged to reduce the high-impact aspect of high-yielding cultivars. The use of reduced tillage and no-tillage (a new crop is planted without removing the debris from the previous crop) agricultural methods have greatly reduced soil erosion (fig. 15.7); however, the need for pesticides is often increased. These methods also increase the efficiency of water use, decrease the need for fertilizer, and reduce pollution from runoff and leaching. Biological controls such as predators (ladybugs, praying mantis) or pathogens (fungi, bacteria, and viruses) should be advanced to control agricultural pests and thereby reduce the use of pesticides. Organic fertilizers, such as manure and compost, can lower the need for expensive inorganic fertilizers, and nitrogen can be provided through biological fixation using improved strains of *Rhizobium* bacteria, free-living nitrogen-fixing bacteria, and also by genetically engineering the nitrogen-fixing process so that nonlegumes can benefit. New techniques in dryland agriculture may increase the efficiency of water use, and even permit sustained production in some areas when irrigation water becomes scarcer.

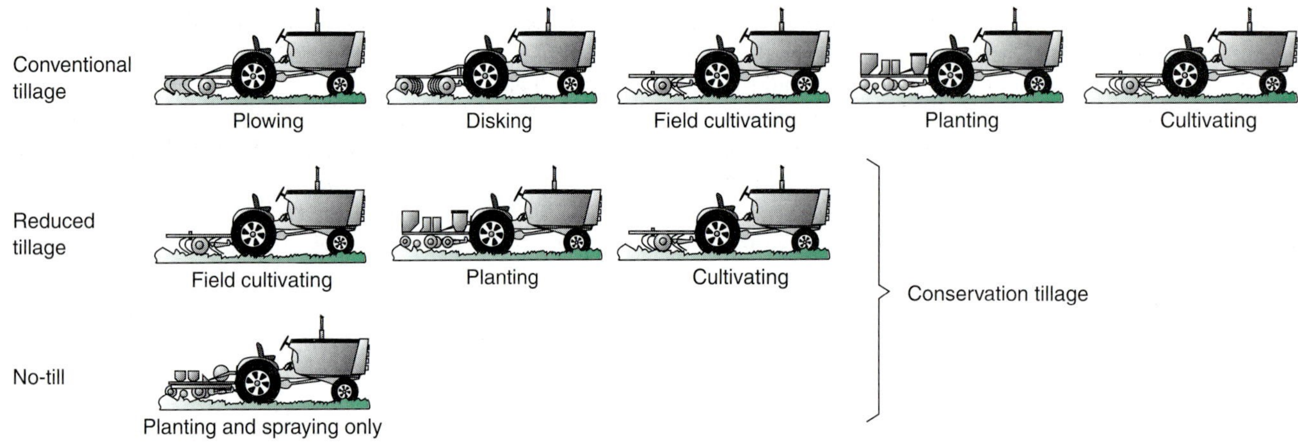

Figure 15.7 *Reduced tillage and no-tillage farming techniques have reduced soil erosion and energy use and also decreased the need for fertilizers. However, the need for pesticides is often increased.*

Genetic Diversity

Although the above methods will help improve productivity and reduce the environmental impact, there is still reason for concern. Some of the concerns center on the loss of genetic diversity in crop plants due to the predominance of monocultural practices.

Monoculture

Monoculture, growing the same crop year after year in a large region, has become a mainstay of modern agriculture (fig. 15.8). Several forces have increased the practice of monoculture in the twentieth century: mechanization, improvement of crop varieties, and chemicals to fertilize crops and control weeds and pests.

The mechanization of agriculture began with the steel plow and the steam engine. By the end of World War II, most commercial farms had switched to gasoline tractors. As machinery replaced draft animals, it was no longer necessary to produce as much livestock feed. Increasingly, more grain could be sold for profit and the increased profit return went into buying more farm machinery. With more sophisticated machinery, a single person could farm more land and farms became larger. Agricultural machinery became increasingly specialized to perform a single job, often for a single crop. The financial investment in specialized machinery proved to be a strong incentive for the farmer to grow only the crop for which the machinery was designed. The introduction of high-yielding varieties and new crops have also contributed to monocultural practices. By concentrating on the growing of a single, high-yielding variety, a farmer could maximize productivity. The large-scale growing of high-demand crops such as soybean and rapeseed (canola) further maximized profits and increased the risks associated with monoculture. The widespread use of chemical fertilizers and pesticides ensures that the crop will continue to produce year after year. Rotation of crops to encourage soil fertility or reduce pest populations is no longer necessary.

Figure 15.8 *Monoculture of wheat as far as the eye can see. The pattern is created by wheat fields lying next to fallow fields.*

The result of these practices is rampant monoculture. Often the major crop grown over a region is dependent on only a few varieties. For example, half of the wheat acreage in the United States is planted with just 9 varieties, 4 varieties account for 75% of the potato crop, 3 for 50% of the cotton, and 6 for 50% of the soybean. When genetic diversity is so limited, all individuals in the crop are essentially genetically identical and vulnerable to disaster. A single pest or disease can wipe out an entire crop since there is little genetic variation and, thus, no range in natural resistance. History has recorded time and again the dangers of monocultural practices: the Irish potato famine in the 1840s, the obliteration of Sri Lanka coffee plantations (*see* Chapter 16), the 1970 epidemic of corn leaf blight in the southern United States (*see* Chapter 12), and, in 1984, the destruction of 18,000,000 Florida citrus trees due to a bacterial infection.

Sustainable Agriculture

Some researchers believe that the future in agriculture is to move away from the instability and unnaturalness of annual, monocultural cultivation and instead model agriculture after natural systems. According to this concept, agriculture becomes sustainable; that is, crops can be harvested without the degradation of the environment (*see* Chapter 25). Agriculture would switch from a monocultural field planted annually to a polyculture of perennial crops. Warm-season and cool-season grasses and nitrogen-fixing legumes would be grown in close proximity. They would tend to mature at the same time to facilitate harvesting, but their life cycles would be different enough that they do not compete simultaneously for the same resources. Several perennials are being developed for polyculture planting. Illinois bundleflower (*Desmanthus illinoenis*) is a warm-season legume and a prodigious nitrogen-fixer. Its seeds are high in protein and borne on erect stalks that facilitate harvesting. Research is ongoing to improve the taste of the seeds to make them suitable for human consumption. A hybrid between Johnson grass (*Sorghum halapense*) and the African annual *Sorghum bicolor* has produced a perennial grain sorghum. Perhaps it will be possible to develop a modern perennial corn from the perennial wild relative, *Zea diploperennis.*

A field planted according to the suitable agricultural design would consist of mixtures of perennial crops that could be harvested for a number of years without replanting. This system has many advantages: soil erosion from yearly plowing is diminished; pest management is improved; and disease control is more effective. The model for this type of agriculture is nature itself. For example, the prairie is a sustainable ecosystem of grasses that lasted for millennia. The incredible fertility of prairie soil was the result of natural soil-building processes.

Germplasm

Today's food crops have their origins in the wild ancestors first domesticated by early farmers thousands of years ago. As agriculture developed over the centuries, artificial selection resulted in local varieties of a crop, so-called **land races,** or traditional varieties. Land races possessed unique and valuable genetic characteristics, traits that allowed them to survive cold, drought, or disease. **Germplasm** is the encoded information of a plant, the genetic instructions that dictate not only the type of plant but all of the traits unique to an individual plant. It is the raw material of the plant breeder and without it the creation of new, improved varieties is not possible. Unfortunately, as traditional varieties are abandoned and replaced with modern, high-yielding ones, and as habitat destruction brings about the extinction of wild relatives, valuable genetic heritage is lost forever. This irreversible loss of genetic diversity is known as **genetic erosion** and is of monumental concern.

Genetic erosion has accelerated since the introduction of high-yielding varieties during the Green Revolution in the 1960s. In 1959, farmers in Sri Lanka grew over 2,000 traditional varieties of rice. Today, that number has been reduced to 5. India has lost over 75% of its 30,000 land varieties of rice. It is estimated that the United States has lost 90% of the seed varieties brought here before 1950. In addition, human encroachment in natural habitats is expected to bring about the extinction of a quarter of the world's plants by the middle of the next century. What can be done to save this genetic diversity for future generations?

Seed Banks

One way nations and organizations have tried to offset genetic erosion is through the establishment of gene or **seed banks.** In a type of botanical Fort Knox, seeds of both domesticated plants and their wild relatives are collected from around the world and deposited. In the United States, the main seed bank is in the National Seed Storage Laboratory in Fort Collins, Colorado, established by the USDA in 1958. It is one of 19 such facilities located around the nation that make up the National Plant Germplasm System. More than 250,000 seed samples are carefully treated and housed here. To extend their viability, the seeds are dehydrated, sealed in aluminum foil bags, and placed in freezers. Other seed samples are preserved cryogenically, placed in tanks of liquid nitrogen at temperatures as low as -196° C (-321° F). Although these procedures may extend seed longevity for decades, inevitably some seeds will not survive prolonged periods of cold storage (fig. 15.9). It is necessary at certain points to take the seeds out of storage, induce them to germinate, and then grow the plants to refurbish the seed supply. The importance of this work is reflected by a growing network of seed banks throughout the world, in over 100 countries, all dedicated to the cause of preserving the genetic diversity of crop plants and their wild relatives. Some plants, like the white potato, do not readily produce seeds; these plants are usually propagated by cuttings that do not store well. For these plants, arboretums and conservation gardens have become "botanical zoos," maintaining representatives of old varieties and endangered wild plants.

Figure 15.9 *Seeds are cryogenically preserved in vats of liquid nitrogen at a seed bank.*

Conservation organizations such as Seed Savers Exchange and Native Seeds are devoted to saving the seeds of traditional varieties from extinction by encouraging their cultivation. Seed Savers stores more than 9,000 nonhybrid varieties of vegetable crops, and on their Heritage Farm in Iowa, they cultivate thousands of old-time varieties to supply seeds to those interested in growing them. These varieties are not grown on most farms because they do not ripen uniformly and are not suitable for mechanized harvesting. Native Seeds has set out to preserve the traditional seeds grown by Native Americans. This collection has been used to reintroduce Sonoran panic grass, *Panicum sonorum,* from Mexican samples. Extinct in the United States since the 1930s, Sonoran panic grass was a food staple to the Cocopah Indians and now, thanks to Native Seeds, is being cultivated by them once again. The value of preserving these local varieties is exemplified by the Australian interest in the rust-resistant sunflowers grown by Havasupai Indians of the Grand Canyon region. Accompanying the seed preservation is the preservation of the folklore and native knowledge about each plant, a legacy of knowledge to be passed on to future generations.

Germplasm is at the center of another international controversy, the question of ownership. Many developed nations on the cutting edge of agricultural biotechnology are home to relatively few major food crops. For example, the United States is the center of origin for only four: blueberries, cranberries, sunflowers, and pecans. Many developing nations, however, are rich in germplasm, since they are the centers of origin to over 96% of the world's food crops (*see* Chapter 11). Over the thousands of years that a plant population grows naturally in an area, a wealth of genetic variation develops. With accelerating genetic erosion, these genetic resources have become increasingly valuable to multinational seed companies.

Improving crops requires a constant infusion of fresh germplasm, since the average life of a new crop variety in the United States ranges only from five to nine years. Typically a new variety with improved disease or insect resistance makes its debut, performs well for a few years, and then is replaced by another more promising variety. Plant breeders must constantly search out traditional and wild relatives that possess desired traits. For example, plant breeders are interested in pig's weed (*Oryza ivara*), an endangered wild relative of rice in Sri Lanka. Some populations of this endangered weed are resistant to grassy stunt virus, a pathogen that is a constant threat to the rice crop in Asia. Given that most crops are grown outside of their centers of origin, a potential conflict arises when access to this germplasm is needed.

Ethiopia is the only primary center of genetic diversity for the coffee tree. Colombia is the leading coffee-producing nation, yet it relies entirely on germplasm from Ethiopian trees. Is it proper for Ethiopia to forbid the export of its germplasm to Colombia? What is the price tag for the genetic resistance to yellow dwarf virus that was transferred from Ethiopian barley to protect California's crop? Should developing nations be compensated for their germplasm? Recently, Occidental Petroleum compensated China for its rice germplasm, and other multinational corporations have followed suit. Should seed companies share their profits with a developing nation if new varieties are created by germplasm from that nation? Does a nation of origin have access rights to advanced varieties developed with their germplasm? Some authorities have suggested that a nation's local crop varieties be identified through molecular tags to ensure that royalties could be collected. An alternative suggestion is that a certain percentage of the profits be turned back to the country of origin for use in preserving natural habitats and maintaining germplasm centers. Others argue that germplasm, whether obtained from wild or high-yield varieties, is a natural heritage to the world and should be shared freely between nations.

Alternative Crops: The Search for New Foods

Thomas Jefferson once said,

> "the greatest service which can be rendered any country is to add a useful plant to its culture,"

a sentiment shared by those who search the world for alternative food crops. It is estimated that there are approximately 50,000 edible plant species but surprisingly, only 250–300 of these species are cultivated as food. The statistics are even more alarming when it is realized that only 22 domesticated plants are major crops and that just three crops, rice, corn, and wheat, provide 60% of the calories from plants in the human diet. Obviously, this reliance of the world's food supply on so few a number is a precarious practice. The United Nation's Food and Agriculture Organization and other groups have initiated searches to discover local crops that have the potential to be developed into alternative food sources for the world market. These alternative crops usually possess characteristics such as high nutritional quality, tolerance to harsh environmental conditions, or the ability to grow in poor soils, that make them attractive to plant breeders. The following is a discussion of several crops unknown to most of the world that may one day become staples in the world's larder.

Quinoa

Quinoa (*Chenopodium quinoa;* Chenopodiaceae, Goosefoot Family) has been a vital food crop in the high Andes of South America for centuries. In the languages of the Quechua and Aymara Indians, the descendents of the Incas, quinoa is referred to as the "mother grain" since quinoa and potatoes were the dietary staples of these highland communities. Quinoa has been described as resembling "a cross between sorghum and spinach," an appropriate description since both the leaves and fruits are edible. It is an annual, broad-leaved and 1–2 meters (3–6 ft) tall. Flowers are borne in terminal and axillary panicles. The fruits, called grains commercially, are actually achenes that have a hard, four-layered pericarp encasing each tiny seed. The seeds are high in protein, 12%–18%, when compared to most cereal grains. The protein is also of an exceptional quality for a plant source. It is high in the essential amino acids lysine and methionine; lysine is deficient in most cereals, while methionine is typically lacking in legumes. The carbohydrate content of the seeds is also high, 58%–68%, with approximately 5% sugar. Fat content ranges between 4%–9%, half of which is the essential fatty acid, linoleic acid. The mineral content of calcium and phosphorus is higher in quinoa than in other grains. Bitter saponins in the pericarp must be washed (traditional method) or milled out (commercial preparation) in the processing of the seeds. After the saponins are removed, the seeds can be cooked like rice and are often used as a high protein substitute for rice and other grains. Quinoa grains have a delicate taste that compares with wild rice. Grinding the seeds yields a flour that can be mixed with wheat flour to make bread. Since the 1980s, quinoa has been cultivated in the Colorado Rockies and has made its entrance into the U.S. marketplace as a gourmet or specialty food, used in soups, pasta, breakfast cereals, and desserts.

Amaranth

Another New World crop under development for the world market, is grain **amaranth** (fig. 15.10). Amaranths were an important staple, along with corn and beans, for many indigenous populations in the pre-Columbian era, but their use declined after the arrival of the Spanish. It seems that the amaranth seeds, with human blood used as glue, were shaped into figurines used in Aztec religious ceremonies. Consequently, the cultivation and eating of amaranth were banned by the Spanish conquistadors to crush what they considered pagan and heretical practices.

The genus *Amaranthus* includes at least 60 species, most of which are widely dispersed weeds. Amaranth can be brilliantly colored in shades of purple, orange, red, or gold; several varieties such as love-lies-bleeding and Prince of Wales feather are cultivated as ornamentals. The plants are tall, up to heights of approximately 2.5 m (8 ft) with broad, edible leaves that taste similar to spinach when cooked. The seeds of amaranth, however, are what attract the greatest attention as a potential nongrass cereal crop. Three especially promising species that have been targeted for development since the late 1970s by plant breeders are *A. cruentus, A. hypochondriacus,* and *A. caudatus.*

Unlike the major cereals (wheat, maize, and rice), amaranth, as a member of the Amaranthaceae, is only one of a handful of nongrass grains. Tiny and numerous (up to 50,000 per plant), the black, cream, tan, brown, or pink grainlike seeds are borne in elongate seedheads at the end of each stalk. The uncooked seed is indigestible because of a tough seed coat, but toasted, boiled, or popped like popcorn, it has a mild, nutty flavor. Amaranth seeds can also be ground into a flour that has been incorporated into baked goods.

Amaranth grains are easy to digest and have been traditionally given to those who are recovering from an illness or a fast. They have also been used by those allergic to other grains. Nutritionally, the protein content of the seeds ranges from 12.5%–17.6%, comparing favorably to the protein content of the three major cereals. Additionally, amaranth protein is rich in the essential amino acid lysine. Amaranth flour from toasted seeds enriches wheat flour or corn meal by improving both the protein quality and quantity. Six to 10% of the seed is an oil that can be extracted from the germ. It is unsaturated and high in linoleic acid. Amaranth oil also contains significant quantities of squalene, a high-priced hydrocarbon

(a)

(b)

(c)

Figure 15.10 *Alternative crops are being developed for the world's marketplace. (a) Grain amaranth. (b) Fields of tarwi in the Peruvian Andes. (c) Oca tubers.*

usually extracted from shark livers. Squalene is the raw material for synthetic steroids and also has uses in the cosmetic industry. Additionally, amaranth starch grains are extremely small and might have applications in non-allergenic aerosols, dusting compounds, or as a talcum powder substitute. Another desirable feature of the amaranth is its tolerance to unfavorable environmental conditions such as heat, drought, and poor soil conditions.

In 1977, research was begun in Pennsylvania by the Rodale Research Center to develop amaranth varieties suitable to the mechanized agriculture practiced in the United States. Varieties are being developed that are shorter and of uniform height, seedheads that do not shatter, and light-colored seeds that are preferred in cooking. Already, plant breeders have doubled amaranth productivity. It is used as forage for livestock and as a source of dyes. Popped amaranth grain can be eaten like popcorn. Breakfast cereals, breads, crackers, and pasta made from the flour can be found in health food stores throughout the United States and may soon move from the gourmet/specialty market to the corner supermarket if the successful development of this Aztec crop continues.

Tarwi

Much interest has been expressed in the developing **tarwi** (*Lupinus mutabilis*), a South American legume (fig. 15.10). The high protein (46%) and oil content (20%) of the seeds rivals that of the soybean. The quality of the protein is exceptionally high, as well as being rich in lysine, and when eaten with cereals, the ideal nutritional balance of essential amino acids is achieved. Despite this nutritional value, tarwi is largely unknown outside of the highland areas of Peru, Bolivia, and Ecuador. Pre-Incan people domesticated this lupine more than 1,500 years ago. As a significant source of protein, tarwi, along with corn, white potato, and quinoa, made up the staple diet of the peoples of the Andean highlands.

The plant is a branching annual 1–2.5 m (3–8 ft) tall with palmately compound leaves. The large, purplish-blue flowers emit a honeylike fragrance. Each fruiting pod contains 2–6 round seeds, usually white in color. The seeds must first be processed (traditionally soaked in running water) to remove bitter alkaloids, but it may be possible to develop sweet varieties with reduced levels of alkaloids that require little or no processing. The polyunsaturated oil from the seeds has been described as being similar to peanut oil in properties and uses. In the Andean highlands, tarwi seeds are most often used in soups, stews, and salads or eaten by the handful as a snack. Tarwi can also be grown as a livestock feed.

Although tropical, tarwi grows well in temperate regions, since it is adapted to the cooler conditions of high altitudes. Additionally, tarwi tolerates a wide range of environmental conditions including frost and drought. As a nitrogen-fixing legume, it can be grown in soils of poor quality and, in fact, is an excellent green manure. Currently, tarwi is being grown experimentally in Europe, Australia, South Africa, and Mexico to develop its potential as a world crop.

Tamarillo

A relative of the tomato, the tree tomato or **tamarillo** is another Andean crop making inroads into the world's marketplace. *Cyphomandra betacea* (Solanaceae) is a plant of the subtropics; it reaches heights of 1–5 m (3–16 ft) in height and has large, shiny heart-shaped leaves. Pinkish flowers hang on pendant branches. Fruits are egg-shaped and brilliantly colored; commercial varieties are typically bright red or golden yellow. The flesh may also be variously colored and contains numerous small, edible seeds. The taste of the tamarillo is described as tangy, although sweeter varieties are also available. Tamarillos are eaten peeled; they are often served as toppings on cakes or ice cream. They also can be eaten sliced and added to sandwiches and green salads. The fruits can be blended for juice and when mixed with milk, ice, and sugar make a delicious drink. Tamarillos are also good when cooked and are an ingredient in soups, stews, baked goods, and relishes. Tamarillo jams, jellies, preserves, and chutneys are another application for this fruit.

Nutritionally, tamarillos are high in several vitamins: A, B_6, C, and E. They are low in calories, less than 40 per fruit. Unlike many Andean crops, tamarillos have begun to find a world market. Introduced into New Zealand, tamarillos have been popular there for more than 50 years. In fact, New Zealand is the origin of the name "tamarillo." Commercial growers there have begun air shipping this fruit to North America, Japan, and Europe, hoping that tamarillos will duplicate the success of kiwi fruit, an earlier New Zealand import.

Oca

The **oca** is a tuber crop from the Andes that one day might rival another Andean tuber, the white potato. A member of the Oxalidaceae or wood sorrel family, oca (*Oxalis tuberosa*) is a herbaceous perennial with three-parted leaves similar to those of a shamrock. The tubers, actually tuberous rhizomes, resemble "stubby, wrinkled carrots" in colors ranging from white to red (fig. 15.10). The flesh of the tuber is white and, in some varieties, has a slightly acid taste that has been described as "potatoes that don't need sour cream." The acidity is due to the presence of oxalic acid, a substance also present in spinach. Other varieties are so sweet they are sold as "fruits." The tubers can be boiled, baked, or fried. Additionally, they can be eaten fresh in salads and/or pickled. Nutritionally, the oca is similar to that of the white potato but with higher protein levels. Also, a large percentage of the carbohydrate content is sugar. As a highland crop, it is tolerant of temperatures and high altitudes that are prohibitive to most other crops. Ocas have potential as a livestock feed, especially for pigs. The oca has also become popular in Mexico and New Zealand. Oca is sold in New Zealand as "New Zealand yam," a name bestowed by that country.

Biotechnology

Biotechnology can be broadly defined as the use of living organisms to provide products for humanity. Applications as simple as the use of yeast to make bread and culturing other fungi to produce antibiotics can be included in biotechnology. The definition is often focused to particular ends, and for a plant breeder, a working definition of biotechnology might be the use of **cell culture** or **genetic engineering** to create plants with new, useful characteristics. For the most part, the intent behind biotechnology is no different than traditional plant breeding. A promising trait is identified and bred or engineered into a valuable crop plant. For example, disease-resistance or herbicide-resistance might be introduced into an established variety, or large grain size bred into a variety that is already disease-resistant. It differs from traditional plant breeding in the speed of introducing new traits and the types of hybridization that are possible.

Cell and Tissue Culture

The ability to develop whole plants from single cells has triggered major interest in cell culture. Although this technique has been available to scientists since the mid-1950s, the applications have grown tremendously in the past decade. For cell or tissue cultures, small pieces of plant tissue are grown on a nutrient medium supplemented with plant hormones. After a few days, the cells begin dividing and produce a small undifferentiated mass of tissue known as a **callus.** The cells in the callus continue to grow and divide for several weeks and eventually produce tiny plantlets with stems and roots. When large enough, these plantlets can be replanted in soil and grown to maturity (fig. 15.11). Most of the time, plantlets produced from a single callus are genetically identical or clones; however, sometimes a mutation occurs giving rise to a plantlet with different characteristics. These mutations are known as **somaclonal variants,** variants that develop from a single somatic cell (the one that originally produced the callus). Although the appearance of mutations among offspring are relatively rare when produced through normal sexual reproduction (about one in a million), the rate of mutation in cell culture appears much greater.

Variants that develop by this technique may have useful traits. In one study at a New Jersey biotechnology company, 230 tomato plantlets were generated from callus that developed from tomato leaf tissue. When mature, 13 plants showed unique traits including larger fruit, tangerine-colored fruit and denser (fleshier) fruit. Mutant traits developed this way are passed on to the offspring of these plants and can lead to commercially successful new varieties.

(a)

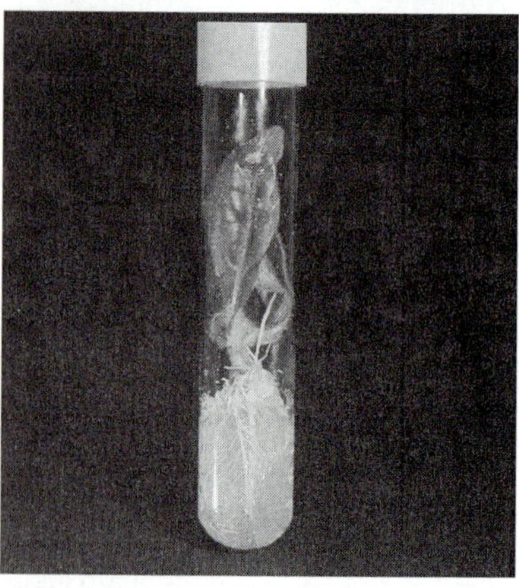

(b)

Figure 15.11 *Plant tissue culture can be used to regenerate tobacco plants. (a) Undifferentiated tobacco callus originally cultured from embryo tissue. When hormones and nutrients are added to the culture medium, tiny plantlets will develop. (b) When large enough, these plants can be repotted in soil and grown to maturity.*

Gametes, especially pollen, can also be cultured to produce callus; however, the cells are haploid since they are products of meiosis. Callus produced this way is often exposed to the drug **colchicine,** which blocks microtubule assembly and, thus, prevents the spindle from forming during mitosis. Although the chromosomes are duplicated, the chromatids are not pulled to opposite poles of the cell, thereby doubling the number of chromosomes. The resultant callus tissue becomes diploid and homozygous for every characteristic.

Another method that gives rise to plantlets from cell culture involves the use of **protoplasts,** plant cells whose walls have been removed by enzyme treatment. These protoplasts can also be grown in culture in a nutrient medium. Protoplasts from different species have been fused together using various chemical treatments, and the nuclei have even fused, giving rise to new plants. This technique may be useful for crossing plants that cannot be crossed through normal sexual reproduction.

In addition to searching for new traits, scientists can select for specific mutant types by exposing developing callus cells or protoplasts to certain poisonous substances. Those cells that survive have genes that provide resistance to the poison. The resulting plants and their offspring should also show resistance to this chemical. A useful application of this technique is in the development of varieties that are resistant to herbicides. This will permit widespread herbicide application to a cultivated crop without damage to the crop plant. Control of weeds is a significant method of increasing yields, since weeds compete with crops for nutrients, harbor pests, and clog irrigation systems. This same technique can be used to screen for cells resistant to a parasite toxin. Plants generated from these cells may be resistant to the parasite that produces the toxin.

Genetic Engineering and Transgenic Plants

The field of molecular biology has provided science with an understanding of how genetic information is stored, replicated, and translated into specific proteins for cellular development, functions, and control. The tools of genetic engineering, especially recombinant DNA technology (*see* Chapter 7), hold great promise in the field of plant breeding. While traditional methods of plant breeding permit the transfer of genes within a species or to a closely related species, genetic engineering allows the transfer of useful genes from one organism to a totally unrelated plant species. Organisms that contain a "foreign" gene in each of their cells are called **transgenic.** Transgenic plants express the transferred gene because all organisms use the same genetic code. Early recombinant DNA research used bacteria to express genes from higher organisms. For example, the gene for human insulin was transferred into the bacterium *Eschericia coli,* which could then produce this valuable protein. Today, it is possible to transfer foreign genes into eukaryotic plant or animal cells; the same gene for human insulin has been successfully transferred into tobacco plants, which are then able to synthesize the human protein.

The development of transgenic plants begins with the identification and isolation of the gene that controls a useful trait and the selection of an appropriate vector, which transports the gene into the plant cell. In many plants foreign genes are introduced, using a **Ti (tumor-inducing) plasmid.** A **plasmid** is a small circular strand of DNA found in bacterial cells. The Ti plasmid normally occurs in *Agrobacterium tumefaciens,* a bacterium responsible for crown gall disease in plants. It invades plant cells and causes tumorlike growths. The tumor-causing genes of the plasmid can be removed without interfering with its ability to enter the nucleus of a plant cell. The plasmid can then be used to introduce desired foreign genes into the host plant cell (fig. 15.12).

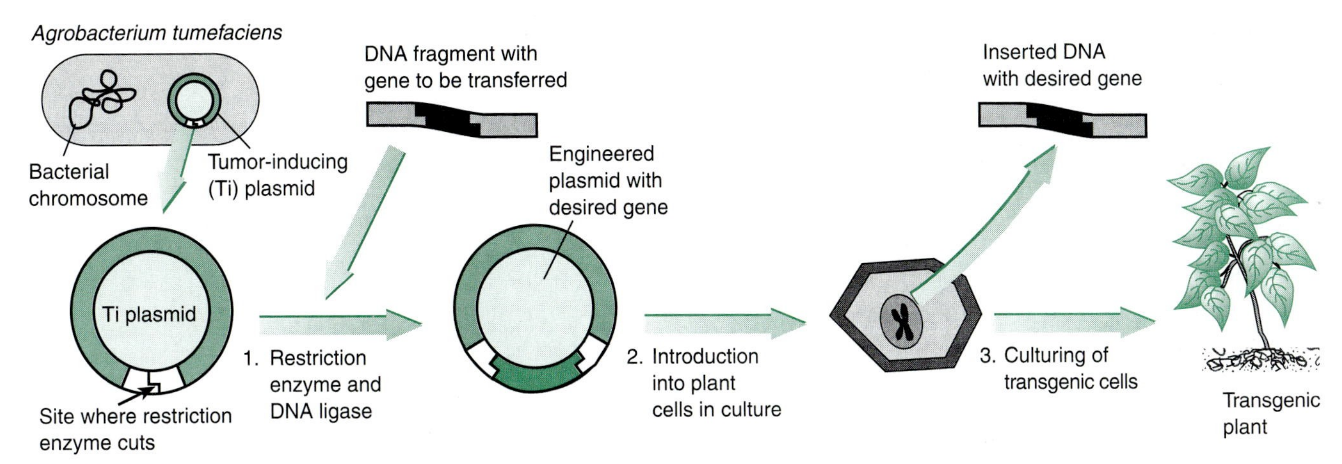

Figure 15.12 *Transgenic plants are often developed using the Ti plasmid from* Agrobacterium tumefaciens *as a vector.*

Genetic engineering of plants has already yielded some interesting results in situations where a useful trait is determined by a single gene. A bacterial gene that confers resistance to the herbicide glyphosate has been introduced into various crop plants. This simplifies the application of herbicide, since the crop plants are not threatened by the chemical.

The gene for a protein with insecticidal properties has been transferred from the bacterium, *Bacillus thuringiensis,* to corn, cotton, and potato. The plants express the gene and manufacture the protein. Field tests have shown that these plants are resistant to Colorado potato beetle, cotton bollworm, and corn borer, each a major pest on the respective crop plant. It is estimated that this technology can save 50%–70% of the cost of pesticides used to protect these crops and thus reduce environmental damage from pesticide use.

In a similar fashion, genetically engineered bacteria have also provided new solutions to agricultural problems, such as frost damage. Each year, approximately 1.5 billion dollars in crop losses occur due to frost damage, especially in the spring. Back in the late 1970s, Steven Lindow, then a graduate student, learned that ice forms on potato leaves around specific **ice-nucleating proteins.** Lindow found that the nucleation sites were cell wall proteins of *Pseudomonas syringae,* a common leaf surface bacterium. When many ice-nucleating bacteria are present, leaves freeze at -2° C (28° F); however, when bacteria are removed, plants can survive temperatures as cold as -5° C (23° F).

Although various measures were tried to eliminate or replace the bacteria to protect the potato plants, none of the measures were commercially feasible. Lindow then used genetic engineering to delete genes that controlled for the formation of the wall protein from *P. syringae.* In both laboratory and greenhouse experiments, plants sprayed with cultures of these altered bacteria lacked ice-nucleation sites and could survive lower temperatures without damage. Field tests of these bacteria were delayed many years while scientists and citizens debated the safety of releasing **genetically engineered microorganisms (GEMs)** into the environment. Finally, in the spring of 1987, field tests were conducted at Tulelake, California, on 3,000 potato seedlings. The altered bacteria did indeed protect the plants, but the debate over the use of GEMs has not diminished.

Environmental Considerations

Today the USDA, EPA, and FDA are all involved in approval of genetically engineered crops. To date, there have been over 1,000 field tests of altered crops in the United States alone. This is an amazing accomplishment when we realize that the first genetically engineered crop was developed at Washington University in St. Louis in 1982. In the spring of 1994 the first of these genetically engineered crops was approved for commercial marketing, the Flavr-Savr tomato, which was altered to slow the ripening process and prolong shelf life.

There are many issues and questions involved in the development and use of genetic engineering and transgenic plants. Some scientists hold that developing transgenic organisms is only an extension of traditional methods of hybridization and the products should be treated no differently. Others argue that transferring genes from one species to another can seriously impact the environment.

Among the concerns expressed are the length of time the altered organism will survive in the environment, how quickly it multiplies, how far can it travel, and what environmental impact could occur if the organism becomes established in the wild. Can the altered organism pass on its new traits to other organisms in the wild? Can the gene for herbicide resistance be passed on to weedy relatives of cultivated crops through natural hybridization? Although these questions largely remain unanswered, biotechnology will play an important role in increasing agricultural productivity.

A Closer Look 15.1

Mutiny on the HMS Bounty: *The Story of Breadfruit*

The introduction of new plant crops is always a risky proposition. People are generally conservative in their diet choices, choosing foods that are similar to the ones they know, and rejecting those that differ. The attempted introduction of the breadfruit tree to the Caribbean is an example of food crop introduction that went horribly awry, precipitating one of the most famous mutinies in naval history. Native to Malaysia, the breadfruit tree (*Artocarpus altilis*) is widespread throughout the South Pacific islands and its huge, multiple fruits are a traditional food staple of Polynesians. A member of the Moraceae (mulberry family), a single tree can bear up to 500 spherical, seedless green fruits, each weighing about 3 kilograms (6 lbs) (box fig. 15.1). The unripe fruits have a high starch content and can be fried, boiled, or baked. Ripened breadfruits are sweet and mushy and can be eaten raw or as an ingredient in pies,

(a)

Box Figure 15.1 *(a) The large, multiple fruits of breadfruit are a starchy staple in Tahiti.*

Chapter Summary

1. Approximately 25% of the world's population receives insufficient food to meet daily nutritional requirements. The major challenge in agriculture is producing enough food to feed the world's population. Dramatic improvements in crop yield have been achieved through breeding of high-yielding and disease-resistant varieties.
2. Over the past three decades, efforts have been made to increase the yields of major crops in developing nations. The central focus of the Green Revolution has been the use of high-yielding and disease-resistant varieties; however, these crops are high-impact crops. Yields depend on fertilizer and pesticide application, adequate water, and mechanized farming, making it economically impossible for poor farmers. Other problems include high energy cost, environmental damage, and loss of genetic diversity. Some scientists feel that the whole premise behind the Green Revolution needs to be reexamined. Increased food production is dangerous in areas where the population growth is unchecked, since it ultimately means more people will starve.
3. The loss of genetic diversity in crop plants and their wild ancestors due to monocultural practices is a serious concern to plant breeders. The cultivation of just one or two high-yielding varieties for most major crops has resulted in the loss of thousands of traditional varieties, and with them their genetic heritage. Monoculture is, in itself, an unstable system in that a single pest or disease can wipe out the crop of an entire region of essentially genetically identical and similarly susceptible individuals.
4. Sustainable agriculture is a movement away from the instability of current monocultural

(b)

(b) Bligh and his loyal crew were set adrift in a longboat by the mutineers of the H.M.S. Bounty

puddings, and other desserts. Breadfruit can also be fermented to produce a sour condiment used as a spice.

The British naturalist Joseph Banks, who accompanied Captain James Cook in his first expedition to Polynesia and Australia, was one of the first Europeans to observe groves of breadfruit trees in Tahiti. Impressed with the quantity and nutritional value of breadfruit, he suggested to Great Britain's King George III that it would be an ideal, inexpensive food for the slave population in the British West Indies. (Plantations in these British island colonies depended on slaves to grow cash crops of sugar, coffee, cacao, and indigo.) Banks' arguments were apparently persuasive; in 1789, Lieutenant William Bligh set sail to Tahiti on the HMS *Bounty* for the sole purpose of obtaining breadfruit trees for transport to the West Indies. During a six-month stay in Tahiti, Bligh secured nearly a thousand potted breadfruit saplings, and when he set sail for the return voyage, the *Bounty* was said to look like a "floating garden." The saplings were treated with special care; they were kept below deck at night to keep them warm and then shifted back to the deck in the morning for sunshine. The growing saplings required so much water that the crew's supply of drinking water was diminished. For a variety of reasons that have been the source of much speculation in the book *Mutiny on the Bounty* and three Hollywood movies, the crew rebelled. Captain Bligh and those loyal to him were set in an open boat, 4,000 miles from the Dutch island of Timor (box fig. 15.1). Miraculously, Bligh and his companions reached Timor unscathed. The breadfruit trees were not as lucky; they were unceremoniously dumped into the ocean by the mutinous crew.

Bligh was tenacious; after his triumphant return to England and the hanging of some of the captured mutineers, he went again to Tahiti and this time brought 1,200 trees safely to Jamaica in 1793 aboard the HMS *Providence.* Despite Bligh's travails, the breadfruit did not win acceptance by the slaves. They preferred the plantain or cooking banana over the exotic breadfruit. Or perhaps they resented, naturally enough, any food forced on them by their slave masters.

practices to a more environmentally sound approach in which crops are harvested without degradation to the environment. The monoculture of annuals would be replaced with a mixture of perennial crops more closely resembling a natural system.

5. Germplasm is the genetic information encoded within a plant; it is the raw material of the plant breeder and, without it, the creation of new varieties is impossible. Genetic erosion results when germplasm is lost to the plant breeder because traditional varieties are no longer cultivated or when wild ancestors become extinct due to habitat destruction. Seed banks preserve valuable germplasms by providing a storehouse for the seeds of domesticated plants and their wild relatives. An approach taken by other groups is to encourage the continued cultivation of traditional varieties. As the incalculable value of germplasm is realized, the question of ownership and proprietary rights of a country to this natural resource are being challenged and defined.
6. Although there are approximately 50,000 species of edible plants, only three domesticated crops provide 60% of the calories from plants in the human diet. The search for locally used crops that have the potential to be developed for expansion on the world market is an ongoing one. Some alternative crops that might play a bigger role in feeding a hungry world are quinoa, amaranth, tarwi, tamarillo, and oca.
7. Biotechnology is being used, along with traditional breeding, to develop high-yielding and disease-resistant cultivars. Using methods of biotechnology, including cell culture and genetic engineering, desired traits such as insect resistance from one species can be introduced into an established crop plant. Transgenic

plants contain one or more genes transferred from another species and express that gene by producing a foreign protein. The first genetically engineered plant was developed in 1982. By the spring of 1994, over 1,000 field tests had been carried out on various genetically engineered crops, and the first of these was approved for commercial marketing.

Review Questions

1. Can increasing crop yields solve the world's food problems?
2. Discuss the positive and negative effects of introducing a gene for herbicide resistance into a crop plant.
3. Describe the procedure to produce a transgenic plant.
4. Why have some scientists considered the Green Revolution flawed?
5. Many varieties of common crop plants may be disease-resistant. What other characteristics must the varieties have to be useful for widespread cultivation?
6. What is monoculture and how has modern agriculture encouraged its spread? What are the dangers of monoculture? Cite examples of agricultural disasters that were the result of monocultural practices.
7. What is the significance of germplasm to agriculture? Outline the controversies concerning the control of germplasm.
8. How might the practice of sustainable agriculture improve crop productivity?
9. What is the value of alternative crops to the world's food supply? What characteristics are important to consider in the search for alternative crops?

Further Reading

Cleveland, D. A., D. Soleri, and S. E. Smith. 1994. Do Folk Crop Varieties Have a Role in Sustainable Agriculture? *BioScience,* 44(11): 740–751.

Cohen, Joel E. 1992. How Many People Can Earth Hold? *Discover,* 13(11): 114–119.

Gasser, Charles S. and Robert T. Fraley. June, 1992. Transgenic Crops. *Scientific American,* 266(6): 62–69.

Gillis, Anna Maria. 1993. Keeping Traditions on the Menu. *BioScience,* 43(7): 425–429.

Heiser, Charles B. 1990. *Seed to Civilization—The Story of Food.* Harvard University Press, Cambridge, MA.

National Research Council. 1984. *Amaranth: Modern Prospects for an Ancient Crop.* National Academy Press: Washington, DC.

National Research Council. 1989. *Lost Crops of the Incas: Little-Known Plants of the Andes with Promise for Worldwide Cultivation.* National Academy Press: Washington, DC.

Oster, Gerald and Selmaree Oster. 1985. The Great Breadfruit Scheme. *Natural History,* 94(3): 34–41.

Paddock, William C. 1983. Healthy Plants—A Threat to Civilization (and a Challenge to the APS), In: *Challenging Problems in Plant Health* (Thor Kommedahl and Paul Williams, eds.) American Phytopathological Society, St. Paul, MN.

Power, J. F. and R. F. Follett. 1987. Monoculture. *Scientific American,* 256(3): 78–86.

Reisner, Marc. 1986. *Cadillac Desert, the American West and Its Disappearing Water,* Viking Press, New York, NY.

Rhoades, Robert E. 1991. The World's Food Supply at Risk. *National Geographic,* 179(4): 74–105.

Schumann, Gail. 1991. *Plant Diseases: Their Biology and Social Impact.* APS Press, St. Paul, MN.

Shell, Ellen Ruppel. 1990. Seeds in the Bank Could Stave Off Disaster on the Farm. *Smithsonian,* 20(10): 94–105.

Shulman, Seth. 1986. Seeds of Controversy. *BioScience,* 36(10): 647–651.

Sokolov, Raymond. 1991. *Why We Eat What We Eat: How the Encounter Between the New World and The Old Changed the Way Everyone on the Planet Eats.* Summit Books, New York.

Tangley, Laura. 1986. Agricultural Biotechnology: Who's Holding the Reins? *BioScience,* 36(10): 652–655.

Tucker, Jonathan B. 1986. Amaranth: The Once and Future Crop. *BioScience,* 36(1): 9–13.

Viola, Herman J. and Carolyn Margolis, eds. 1991. *Seeds of Change: A Quincentennial Commemoration.* Smithsonian Institution Press, Washington, DC.

UNIT

IV

Commercial Products Derived from Plants

Plants are the source of many useful materials, such as lumber for the home construction industry.

16

Stimulating Beverages

Chapter Outline

Chapter Concepts

1. Caffeine and caffeinelike alkaloids have a stimulating effect on the central nervous system.
2. *Coffea arabica, Thea sinensis,* and *Theobroma cacao,* plants naturally rich in caffeine, have long been used as sources of stimulating beverages, and historically have played an important role in human affairs.
3. Today coffee, tea, chocolate, and cola are consumed globally and are a mixed blessing to the world's population.

The human need for water is even more pressing than the need for food. From earliest times, we have desired to quench our thirst with drinks more flavorful than water alone. This desire, coupled with the fortuitous discoveries of caffeine-rich plants, led to the creation of stimulating beverages in many cultures. Our most familiar examples of stimulating beverages are coffee, tea, and cocoa, all rich in caffeine.

Physiological Effects of Caffeine

Caffeine and related stimulants belong to a class of chemicals called alkaloids, which are substances mainly produced by plants (*See* Chapter 19). Although their roles in plants are uncertain, it has been suggested that they may discourage grazing animals. Generally, alkaloids have diverse physiological and psychological effects on animals; specifically, caffeine and caffeinelike alkaloids are stimulants to the central nervous system. Caffeine is known to speed the heartbeat, increase blood pressure, stimulate respiration, and constrict blood vessels. It has long been used to alleviate fatigue and drowsiness, thereby promoting alertness and endurance.

Caffeine also improves athletic performance by drawing on fat reserves for energy and by increasing motor skills of conditioned reflexes. It enhances the pain-relieving effects of aspirin and acetaminophen and is found in many over-the-counter and prescription drugs. Many headaches are caused by dilated blood vessels; caffeine is known to constrict these blood vessels and so alleviate the pain. Caffeine is also included in many diet pills for several reasons, including its action as an appetite suppressant and its properties as a weak diuretic.

On the other hand, not all effects of caffeine are positive. In some individuals, caffeine can cause insomnia, nervousness, irritability, and rapid heartbeat. There is no set dosage that produces these effects; in some individuals even one cup of coffee induces this "caffeinism." Research has shown that there is a link between caffeine and birth defects; therefore, pregnant women should restrict their intake of caffeine. Recent studies also suggest that women trying to become pregnant should also limit caffeine consumption. The research, although preliminary, indicates that there may be a connection between caffeine intake and infertility in women. Heart patients should also limit caffeine intake because it affects the cardiovascular system.

Like many drugs, caffeine is addictive and can cause withdrawal symptoms. Without caffeine, users become irritable, nervous and restless, unable to work, and often develop severe headaches. Even if you are not a coffee or tea drinker, caffeine is a common additive in many soft drinks and medications (table 16.1). In fact, there is concern about the caffeine consumption of children who drink large quantities of soft drinks. Two cans of many soft drinks contain the equivalent amount of caffeine found in a cup of coffee. This concern has resulted in the greater availability of caffeine-free beverages.

Table 16.1
Caffeine Content of Common Products

Product	Caffeine Content
Coffee	
Drip (5-ounce cup)	115 mg (60–180 mg)*
Instant (5-ounce cup)	65 mg (30–120 mg)*
Decaf (5-ounce cup)	3 mg (1–5 mg)*
Tea	
Brewed (5-ounce cup)	40 mg (20–90 mg)*
Cocoa	13 mg
Soft Drinks (12-ounce serving)	
Coca-Cola	45.6 mg
Diet Coke	45.6 mg
Dr. Pepper	39.6 mg
Mr. Pibb	40.8 mg
Mountain Dew	54.0 mg
Mellow Yellow	52.8 mg
Pepsi-Cola	38.4 mg
RC-Cola	36.0 mg
Diet Rite	36.0 mg
Over-The-Counter Medications (per tablet or caplet)	
Excedrin, Extra Strength	65 mg
Anacin	32 mg
NoDoz	100 mg
Vivarin	200 mg
Midol, Maximum strength	600 mg

*A range in caffeine content occurs due to variation in brewing time and strength.

Coffee

Coffee is made primarily from the seeds of *Coffea arabica,* a tree native to the mountains of Ethiopia. Consumption of coffee has a long history, but its origins are lost in legends. One popular myth states that goats actually discovered the stimulating properties of the plant. Let out to graze, they came back one day friskier than normal. Investigating, the goatherder discovered the animals had been eating the berries of a nearby tree. Trying some himself, he enjoyed the same stimulating effect and introduced the fruit to others. The fame of this berry soon spread.

At first, coffee berries were eaten whole; later they were crushed and mixed with fat and eaten as a stimulating food. The practice of roasting the seeds and producing what we would recognize as coffee began in the thirteenth century in Yemen.

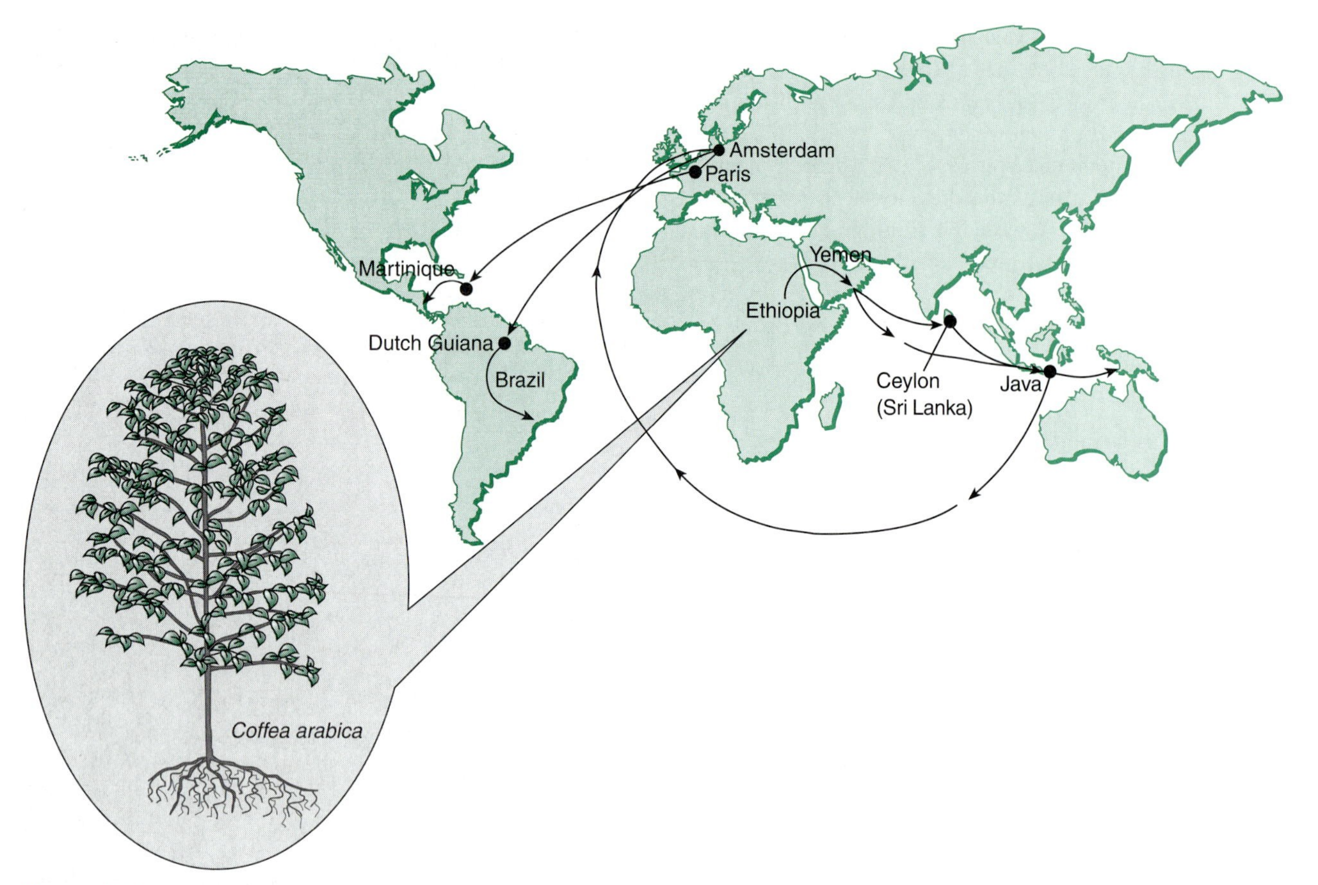

Figure 16.1 Coffea arabica, *native to the mountains of Ethiopia, was spread through many tropical areas. Plantations became well established in the New World early in the eighteenth century.*

An Arabian Drink

By A.D. 1500, coffee trees were widely cultivated in Yemen and coffee drinking had spread rapidly throughout the Arabian world (fig. 16.1). Although first used to keep worshippers awake through long vigils, coffee drinking soon acquired social aspects. Coffee houses were established to accommodate this rapidly spreading habit. From the beginning, coffee houses were controversial. Religious leaders felt that the time spent in the coffee houses should have been spent in the mosques. Political leaders also felt threatened by the political discussions common in coffee houses. Although at times efforts were made to outlaw coffee houses, they remained an integral part of the Arab culture. In fact, some terms we associate with coffee, like *mocha, kava,* and even *coffee* itself, are derived from Arabic terms.

Venetian traders introduced coffee to Europe in 1615, and by 1700 coffeehouses could be found throughout Europe. They were so popular in London that one could be found on every corner. The coffeehouses were not mere restaurants, but centers for commerce, the arts, intellectual discourse, and political debate (fig. 16.2); they were often nicknamed "penny universities." Tuition at these universities was a penny for a cup of coffee. Lloyd's of London, the famous insurance underwriters, began in a coffeehouse in 1688. Edward Lloyd opened a coffeehouse located near the wharves on the Thames River; it soon became a convenient meeting place for seafaring men. Overhearing the gossip about ships and their cargoes, underwriters who also frequented the place began insuring the shipping trade.

Figure 16.2 *Copper engraving of a seventeenth century coffeehouse.*

Edward Lloyd undertook no insurance business himself; he provided the facilities and shipping information for his customers to conduct their insurance business. After 300 years, Lloyd's of London, no longer a coffee house, is still operating strong. Although coffee houses or cafes are still popular in most European countries, by the end of the eighteenth century coffee houses had virtually disappeared from England, as tea became the caffeine beverage of choice.

The first of the North American coffee houses opened in Boston in 1669. They did not achieve the intellectual status of their European counterparts but were meeting places for businessmen and merchants. Three hundred years later, during the 1960s, coffee houses in the United States had a resurgence and became focal points of political thought and socially conscious folk music. Today, coffee house are popular again, specializing in gourmet and exotic blends.

Coffee drinking is still making inroads today; thousands of coffee houses have opened in Japan in the past 25 years. Coffee may soon become the most popular stimulating beverage in the world.

Plantations

Through the seventeenth century, coffee cultivation was limited to North Africa and the Arabian Peninsula; the city of Mocha was the center of this lucrative trade. The Dutch introduced coffee trees into their colonies in the East Indies and Ceylon (now called Sri Lanka). Java plantations soon became the major supplier of coffee beans to Western Europe, breaking the Arabian monopoly. From Java a tree was taken to the Amsterdam Botanical Garden in 1706 (*See* fig. 16.1). An offspring from this tree went to the Jardins de Plantes in Paris. A cutting from this Parisian tree was taken to the French colony of Martinique, eventually giving rise to the first Caribbean coffee plantation. The Dutch had already introduced coffee into the New World in their colony of Dutch Guiana (now the independent country Suriname) in 1718. Coffee was introduced into Brazil about 10 years later when a Portuguese envoy brought seeds from the Guianas. In Brazil and other areas of tropical Latin America, coffee plantations flourished. Brazil and Colombia are presently the world's leading coffee producers.

Today the fate of coffee is directed by the International Coffee Organization, which sets and regulates the production and price of coffee for its member nations.

Coffee rust, caused by the fungal pathogen *Hemileia vastatrix,* has had a major impact on the history of coffee cultivation. Introduced in the 1850s, coffee rust decimated the coffee plantations in Ceylon by 1892. The coffee industry of Ceylon never recovered; tea plantations replaced those of coffee. Other coffee-growing areas, such as Indonesia and Africa, were also ravaged by this disease. Even today coffee rust still limits the cultivation of *Coffea arabica* to uninfested areas of the world. Latin American growers are constantly vigilant to prevent the introduction of this destructive pathogen.

(a)

(b)

Figure 16.3 *The fruit of the coffee tree is a berry usually called a coffee "cherry." (a) A cluster of coffee cherries ready for picking. (b) Cherries are often picked by hand, especially in mountainous regions. This picker in Colombia carries a basket of cherries.*

The coffee tree is a small evergreen tree or shrub with shiny, simple leaves. The plant bears clusters of small, white, fragrant flowers in the axils. After the flower is fertilized, it develops a berry, usually called a coffee "cherry" because it is red at maturity (fig. 16.3). Within the cherry is a fleshy edible pulp surrounding two greyish seeds. Only the seeds are used in the production of coffee; the seeds are the coffee "beans" of commerce.

Coffee trees are cultivated in tropical and subtropical climates. They need about 150–250 (60–100 in) cm of rainfall per year and grow best in the cool highlands (1,000–2,000m/3,000–6,000 ft), where the temperature is a stable 20° C (68° F). The plants cannot tolerate frost at all; a killing frost can devastate the coffee industry. The plants start bearing fruit at 3–5 years of age and will continue bearing for about 35 years. A mature tree will yield 2.5–3.0 kg (5.5–6.6 lb) of berries per year. The berries are usually picked by hand, especially in the steep mountainous areas (fig. 16.3). Mechanical harvesters can be used in those plantations that are located on gently rolling hills. Each coffee-picking machine shakes off the cherries and performs the work of 100 laborers.

Once the cherries are picked, depulping follows to free the beans from the fruit. There are two methods of depulping: the wet method favored in Latin America and the dry method used predominantly in Africa. In the wet method, the cherries are floated in large tanks to remove debris and then mechanically depulped. The residual pulp clinging to the beans is allowed to ferment for up to 24 hours. This fermentation is not alcoholic but refers to natural enzymatic changes brought about during processing and should not be confused with anaerobic fermentation involved in the production of beer and wine. Once fermentation is completed, the beans are washed and dried, and the seed coats mechanically removed. The beans, green without the seed coat, are ready for roasting. In the dry method, the cherries are allowed to dry in the open for several days while the pulp ferments. Following drying, the beans are freed mechanically and can then be roasted.

From Bean to Brew

The flavor and aroma of coffee depend on the fine art of roasting. The temperature and timing of roasting are critical for producing the different varieties of coffee. Light roasts, which are generally preferred in North America, are accomplished at temperatures of 212°–218° C (414°–424° F). Dark roasts are produced at higher temperatures; the higher the temperature, the darker the beans. Dark brown Vienna Roasts are produced at 240° C (464° F) and black French Roasts at an even higher temperature of 250° C (482° F). As a general rule, the lighter the bean the milder and the sweeter the coffee; the darker the bean the stronger and less sweet the flavor. The strength of the brew relates only to the flavors and aromas and not the caffeine concentration.

The varieties result from the chemical reactions that take place when the bean is roasted. One reaction that occurs is the conversion of starches to sugars; this begins when the temperature reaches 207° C (405° F). At slightly higher temperatures, 212°–218° C (414°–424° F), the sugars begin to caramelize and the bean turns brown. At 238° C (460° F), carbonization begins as the sugar burns, leaving only the carbon that darkens the bean. Other reactions occurring during roasting involve the release of substances such as the essential oil caffeol that gives coffee its characteristic aroma. At roasting temperatures above 240° C (464° F), some oils are driven to the surface. Roasting also helps break down the cell walls, which aids in grinding.

Unroasted beans can be stored for long periods, but, once roasted, the beans rapidly deteriorate in flavor, noticeable to some in even a month's time. Ground beans have an even shorter shelf life, but vacuum packaging slows down the deterioration. Once opened, the ground coffee should be refrigerated to preserve the flavor. However, coffee connoisseurs insist that freshly roasted and freshly ground beans are the only ways to insure a good cup of coffee.

Varieties

The majority of the coffee beans that are processed are produced by *Coffea arabica* trees but it is not the only species in the genus. There are at least 60 other species of *Coffea*, but only two others are cultivated, *C. canephora* (*robusta*) and *C. liberica. Coffea arabica* is known for producing the choicest and mildest coffees. Approximately 85% of the coffee grown is arabica, although there are over 100 varieties being cultivated. In Latin America, it is the only species grown.

Coffea canephora, grown in Africa and Asia, has a stronger and harsher taste and is used primarily in instant and decaffeinated coffee. Even though it produces an inferior coffee, canephora is resistant to coffee rust and is grown in those regions where the fungus is endemic. Also, it is easier to harvest and produces much higher yields per acre than arabica. These factors account for its increased production in recent years. *Coffea liberica* produces a bitter coffee and its use is limited. Like canephora, liberica grows well at lower altitudes and is cultivated throughout Africa.

Most brand-name coffees are a blend of several varieties of beans. Select coffee beans are available at gourmet coffee shops so that connoisseurs can create their own blends. Varieties are so distinctive that professional coffee tasters can identify the country of origin and even the region where the bean was grown. Most coffee experts agree that high-mountain-grown coffees are the best and Jamaica Blue Mountain is one of the finest.

Instant coffee was created by a Japanese chemist in 1901, but did not gain popularity until years later. During World War I, instant coffee was shipped to U.S. troops overseas; this was the first widespread use of instant coffee. To manufacture instant coffee, brewed coffee is dehydrated by various means. The most common method is by spray drying; the hot coffee is sprayed through nozzles into a tall room. As the coffee falls, the water evaporates, leaving a dry powder. The powder can be made to appear granular, looking more like ground coffee, by mixing it with steam or water. Freeze-drying is the latest innovation in the dehydrating processing. Here freezing under a vacuum dehydrates the coffee and produces the coffee crystals.

With today's more health-conscious public, decaffeinated coffee has increased in popularity. No longer restricted to a few bland-tasting instant brands, decaffeinated coffee can also be found as roasted beans and ground coffee blends.

The decaffeination process is surrounded by controversy. Various solvents are used to extract up to 99.7% of the caffeine from green unroasted beans that have been softened by steam. The solvent is then thoroughly rinsed from the beans. The concerns involve the solvents used in the processing. Methylene chloride is one of those questionable compounds. In 1989, the FDA banned the use of this chemical in hair sprays, since laboratory animals who inhaled methylene chloride developed cancer. Its use in the decaffeination process has not yet been banned. Another method used to remove the caffeine involves the solubility of caffeine in water. This method is preferred to the use of chemicals, since water is the solvent. Although the benefits for decreasing caffeine intake are clear, there is some concern that drinking decaffeinated coffee might elevate blood cholesterol levels, but this has not been conclusively proven.

From the accidental discovery of a goatherd long ago, coffee has made a global impact on history, social customs, economics, international trade, and the human diet.

Tea

Oriental Origins

According to the Chinese, tea was discovered by the Emperor Shen Nung in 2737 B.C. when a tea leaf accidentally fell into water that was being boiled for drinking. A different legend comes from India. Dharma, a saintly Buddhist priest, vowed to spend seven years without sleep in contemplation of Buddha. After five years, he found it increasingly difficult to stay awake. In desperation he grabbed the leaves of a nearby bush and began to chew them. The leaves, from a tea plant, refreshed him and allowed him to complete his vow. Whatever the origin, today tea is the world's most popular beverage, next to water. Every day 800 million cups or glasses of tea are consumed around the world.

Cultivation and Processing

Tea is made from the dried tip leaves of the species *Camellia sinensis,* a small tree or shrub native to the area adjoining Tibet, India, China, and Burma. The plant, if left unpruned, uncut, and unpicked would grow to a height of 7.5–10.5 meters (24–34 ft). It can be grown from sea level to over 1800 m, but the finest teas come from plantations or estates at the higher elevations. The plants flourish in tropical or subtropical climates where there is abundant rainfall and there is no danger of frost (fig. 16.4).

Each tea plant is pruned to encourage shrubby growth. The plants are usually kept at approximately one meter (3–4 ft), with a flat top that facilitates easy plucking or hand harvesting of the leaves. For best quality teas, only the terminal bud and top two leaves of each branch are harvested. Plucking stimulates the plants to produce new shoots or flushes with tender young leaves and buds. New flushes will appear every one to two weeks (fig. 16.4).

Harvested tea leaves may be treated in one of three ways to produce black tea, green tea, or oolong tea. In the United States, 95% of the tea is black tea or fermented tea. The processing of black tea begins with withering. The freshly picked leaves are taken to racks where hot, dry air is passed over them for 12–24 hours. During this time the leaf loses much of its water content. After withering, the leaves are rolled, usually by machine; rolling breaks up the cells, releasing enzymes that start the next process, fermentation. The leaves are then spread out in cool, humid fermentation rooms for up to several hours.

(a)

(b)

Figure 16.4 *Tea plantation in India. (a) Field of shrubby tea plants with tea pickers visible in the background. (b) A single tea plant following plucking.*

Fermentation brings about chemical changes that turn the leaf a copper color. The last stage is the firing or drying of the leaves. The leaves are passed through hot air chambers that stop fermentation, reduce the moisture content, and turn the leaf black.

Green teas are not fermented. The leaves are not withered; after plucking, the leaves are steamed, rolled, and dried. These leaves remain green.

Oolong teas are semi-fermented; they are lightly withered to permit a partial ferment, resulting in leaves that are a greenish-brown color. After drying, the leaves are sorted into sizes or grades. The smallest tip leaves are called orange pekoe; larger leaves are pekoe and souchong. Broken leaf pieces are sieved out and also sorted according to size; these are used predominantly in tea bags.

The Flavor of Tea

The essentials oils and tannins in the leaf determine the flavor of the tea. Essential oils are volatile substances that contribute to the essence or aroma of certain species. The flavorful aroma of a steaming cup of tea is due to its essential oil, theol. Certain types of tea have additives that modify the flavor and aroma. Jasmine tea, a semifermented tea, has a delightful taste and fragrance due to the addition of jasmine blossoms during the withering process. Earl Grey tea, a traditional British favorite, has bergamot oil added, while other teas may have mint leaves or spices added.

Tannins are plant compounds found in a great many plants in addition to tea. In plants they are believed to discourage herbivores from eating the leaves, and have been widely utilized as stains, dyes, inks, or tanning agents for leather. The tannins in tea are responsible for staining the teapots and teeth of habitual tea drinkers. Black teas in particular, are very rich in tannins and are responsible for the color changes that occur during processing. During fermentation, the colorless tannins turn copper colored, then black when heated. Research suggests a link between certain forms of cancer and extremely high consumption or industrial exposure to tannins.

The stimulating effects of tea are due to the caffeine and theophylline present in the leaf. In fact, when caffeine is metabolized in the human body, a small amount of it is converted to theophylline. Although theophylline is structurally similar to caffeine, the medicinal properties of theophylline are better known than its stimulating properties. Theophylline has been used to treat asthma for many years by directly relaxing the smooth muscles of the bronchial airways and thus opening constricted air pathways. This bronchodilating action helps relieve wheezing, coughing, and other respiratory symptoms. In an emergency, asthmatics can get some relief by drinking a few cups of tea or coffee.

History

Scattered reports of this Chinese drink first reached Europe in the mid-fifteenth century, but tea did not actually arrive in Western Europe until 1610, when Dutch traders brought it to Holland. It was in the 1650s that it finally reached England, the European country most associated with tea. The London coffee houses were the first to bring the beverage to the public following their success in introducing coffee just a few years earlier. The British fondness for tea developed relatively quickly; in 1680, the British imported a mere 100 pounds of tea. By 1700, the imports jumped to one million pounds and by 1780, 14 million pounds, making tea the British national beverage.

Tea first made its appearance in North America about 1650, brought to New Amsterdam by the Dutch, and was soon introduced into the British colonies as well. At a cost of $30 to $50 per pound for dried tea leaves, the beverage first became popular only in the homes of the well-to-do. Despite the cost, tea drinking had spread throughout the colonies by the late 1600s, and, by the mid-1700s, the colonists were avid consumers. Once they developed the tea-drinking habit, the colonists were dependent on imports for their tea leaves.

Although the British East India Company was the official import company, the colonists usually preferred the lower and duty-free prices offered by smugglers. In 1773, the East India Company complained to Parliament that they had a large surplus of tea and were anxious to improve the colonial tea trade. Parliament responded with the Tea Act of 1773, allowing the East India Company to sell the tea directly to the colonies without paying any of the taxes imposed on the colonial merchants. Thus, the East India Company hoped to monopolize the colonial tea trade by underselling the colonial tea merchants. At the same time, this would make selling black market tea less lucrative and would drive the smugglers out of business.

The Tea Act aroused indignation in the colonies. The colonial merchants felt their livelihood was threatened, and it renewed resentment toward British taxation. At this time no tax, no matter how small, without representation was acceptable to the colonists. The most famous response was the Boston Tea Party on December 16, 1773. Colonists disguised as Mohawk Indians boarded the British ships and dumped the cargo of tea into Boston Harbor. The aftermath of this incident contributed to the events that led to the Declaration of Independence and the Revolutionary War.

One lasting influence of the Boston Tea Party and the Revolutionary War was to make the United States a nation of coffee drinkers, but iced tea has become one of America's favorite cold beverages. In 1904 at the St. Louis Exposition, a young Englishman named Richard Blechyden was introducing teas from the Far East. A sweltering heat wave dampened the crowd's enthusiasm for hot tea. Blechyden experimented by pouring the hot tea into tall glasses filled with ice; it was an instant hit. Since then iced tea has become the national summertime drink.

The year 1904 was a significant year in the history of tea for another reason as well—the invention of the tea bag. Thomas Sullivan, a New York tea importer, sent out samples of tea in small silk bags instead of the more expensive tins that were usually sent. To Sullivan's surprise, the orders came in not for the tea itself but for the tea packaged in silk bags. His customers had discovered that tea could be easily made by pouring boiling water over the silk bags. Silk is no longer used, but the tea bag industry has grown so much that tea made this way makes up more than half of all the tea drunk in the United States today.

Chocolate

Although coffee and tea are known only as beverages, the third major caffeine plant gives us a confection as well as a beverage. In fact, it is the confection that makes the cacao tree so widely cultivated today. The source of all the chocolate and cocoa in the world is the seed of *Theobroma cacao,* the cacao tree native to tropical Central and South America. (Note: The cacao tree should not be confused with the coca plant, which is the source of cocaine.)

A Closer Look 16.1

Tea Time: Ceremonies and Customs around the World

While the midmorning coffee break is a familiar tradition in the United States, other cultures have developed customs and ceremonies involving tea. The British High Tea is a late afternoon light meal that may include such delicacies as scones, crumpets, cakes, and finger sandwiches, accompanied by piping hot brewed tea, often served with cream.

For centuries, Russians brewed tea using a metal urn called a samovar to boil the water (box fig. 16.1). The tea was traditionally served in tall glasses, often with a slice of lemon; in fact, the Russians were the first to flavor tea with lemon. When lemons were not available, a spoonful of jam was added to the hot tea.

Box Figure 16.1 *Tea has been incorporated into many social customs throughout the world. An elegant Russian samovar used to prepare tea.*

In Japan, the Tea Ceremony is a highly stylized ritual that symbolizes the Zen Buddhist concept that universal truths lie in simple tasks. In some houses, the ceremony is held in a special room that is austere to focus attention on the details of the ceremony. Green tea is poured, whipped to a frothy consistency, and served from lacquered bowls. The tea ceremony has not changed in centuries and is considered a link to the traditions of the past.

Food of the Gods

The cacao tree had been cultivated by the native peoples from southern Mexico, through Central America, and into northern South America for centuries before the Spanish Conquest. According to Aztec mythology, it was the god Quetzalcoatl that gave cacao beans to the Aztec people (fig. 16.5). The cacao beans were offered as gifts to the gods and also used to make a beverage consumed by noblemen and priests on ceremonial occasions. The scientific name *Theobroma* reflects this ancient tradition, since it literally means "food of the gods."

In 1502, Christopher Columbus was the first European to be introduced to the cacao beans. Although he learned that the natives used these beans as money and also to prepare a spicy beverage, he did not appreciate their significance. In 1519, when the Spanish conquistador Hernan Cortés invaded Mexico, he found Montezuma, the Aztec Emperor, drinking a liquid called *chocolatl* from a golden goblet. In the mistaken belief that Cortés was the reincarnation of the god Quetzalcoatl, he was showered with riches and offered some of this esteemed beverage. *Chocolatl* was made from roasted and coarsely ground beans from the cacao tree, combined with various spices including chili peppers and vanilla beans. Boiling water was added and the mixture was whipped to a foamy consistency. The resulting spicy and bitter beverage had no resemblance to today's cocoa.

Figure 16.5 *The Aztec god Quetzalcoatl, envisioned as a plumed serpent, is believed to have given cacao beans to the Aztec people.*

Cortés introduced this beverage to Spain when he returned in 1528. The Spanish court modified the recipe and added sugar, making it more palatable to European tastes. The recipe was highly guarded and the Spanish had a

(a)

(b)

Figure 16.6 Theobroma cacao, *the source of cocoa and chocolate. (a) Cacao tree showing podlike fruits attached to the trunk. (b) A cacao pod split open to reveal the pulpy seeds or beans.*

monopoly on the cacao beans; these factors kept cocoa a Spanish beverage for many years. By 1650, a more recognizable cocoa was served throughout Europe and soon it was competing with coffee and tea. Although popular, cocoa never rivaled coffee or tea as a beverage. The high fat content in the cocoa bean made processing difficult and produced a greasy beverage that many found unappetizing. Some of these problems were solved in 1828 when a Dutch chemist, Conrad van Houten, developed a process to remove some of the fat or cocoa butter. In 1847, an English company, Fry and Sons, added cocoa butter and sugar to the ground beans to make chocolate. This was the creation of the first chocolate bar.

Cultivation and Processing

Theobroma cacao is a small tree in the understory of tropical forests. Optimum growing conditions require a wet climate and warm temperatures, restricting cultivation to a zone 20° north and south of the equator. Three commonly grown varieties are criollo, forastero, and trinitario. Although criollo is grown in Central America, Venezuela, and Colombia, its susceptibility to disease and modest productivity have limited its cultivation. Forastero is by far the most cultivated variety in Brazil and West Africa, and trinitario is a hybrid between the criollos of Trinidad and the Amazon forasteros. Today the Ivory Coast and Brazil lead the world in cocoa bean production, while other tropical countries in West Africa and South and Central America are also major contributors.

The tree is characterized by football-shaped pods that form directly on the main trunk from small white or pinkish flowers (fig. 16.6a). The pods take 4–6 months to mature; depending on the variety, the mature pods turn either yellow or orange. Inside these fruits are 20 to 40 ivory-colored seeds or beans, surrounded by a white, sweet, sticky pulp (fig. 16.6b). When the pods are ripe they are harvested by machete and split open. In some plantations, a mechanized pod opener has replaced the machete. The pulpy seeds are removed and allowed to ferment for up to one week, depending on the variety. The chocolate taste and aroma are not found in the fresh beans; they develop as the beans ferment. The beans, now chocolate brown, are dried either mechanically or in the sun and shipped to processing centers in Europe and North America.

The beans are classified according to quality and origin; the best come from Chuao Valley in Venezuela, where a special type of criollo bean is grown. These are often reserved for the finest quality chocolates.

The processing begins with the roasting of the beans at temperatures of 120°–140° C (248°–284°F) for 20–50 minutes, which develops the rich color and full flavor of chocolate. Following roasting, the seeds are cracked open,

A Closer Look 16.2

Candy Bars: For the Love of Chocolate

There have been many changes in chocolate candy since the first was created in 1847. In 1875 in Switzerland, Daniel Peter, in collaboration with Henri Nestlé, created milk chocolate by adding condensed milk. Milton Hershey, the foremost American chocolatier, modified the Peter process by adding whole milk. He not only produced chocolate bars but is credited with creating the first candy bar in the late 1890s. The candy bar is a conglomeration of nuts, fruit, caramels and/or other ingredients, usually with chocolate. Soon the market place was filled with competitor candy bars. One early competitor that is still sold today is the Goo Goo Cluster, concocted in 1912 in Nashville, Tennessee. World War I gave a further boost to the candy-bar habit, when chocolate bars were handed out to the doughboys for quick energy. Other early candy bars capitalized on the health craze that swept the nation in the 1920s, with claims of aiding digestion and names like Vegetable Sandwich.

At 500 calories per 100 grams (3.5 oz), chocolate is a high-energy food. Milk chocolate is about 57% carbohydrate, 32% fat, and 8% protein, and traces of vitamins and minerals. The nutritional value of candy bars improves when nuts and fruits are added.

Despite television commercials and the ubiquitous candy machines, Americans are not the greatest consumers of chocolate. We each consume a mere 4 kilograms (8.8 lbs) per year, not much compared to the 10 kilograms (22 lbs) in Switzerland. What is the mystique of chocolate? Perhaps it is the stimulating effects of the alkaloids. Chocolate contains both caffeine and theobromine, an alkaloid closely related to caffeine but milder in its effects. Perhaps it is the feeling of a love affair. Evidence suggests that phenylethylamine, a component of chocolate, and also found in the human brain, appears to increase in the brain when someone falls in love. So some chocoholics may be addicted to the feeling of "falling in love" rather than to the candy itself.

freeing the large cotyledons, or nibs, from the seed coat and embryo. The nibs are crushed to produce a dark brown oily paste, the chocolate liquor. This liquor can be solidified into squares of baking chocolate, or the cocoa butter can be removed from the liquor with heat and pressure to produce a brown cake, which is pulverized into cocoa powder. During this processing, alkali is added to neutralize the acidity of the cocoa. This step, called dutching, also increases the solubility and darkens the color of the cocoa powder.

Cocoa butter has many uses. In addition to its being added to the liquor to produce confectionery chocolate, it is the main ingredient for white chocolate. Interestingly enough, the FDA doesn't consider white chocolate "chocolate" because it contains no chocolate liquor. Cocoa butter is also used in a variety of suntan lotions, soaps, and cosmetics.

Today, the majority of cacao beans are processed to make chocolate. The recipe for chocolate starts with the chocolate liquor; sugar, cocoa butter, vanilla, and often milk are added during conching. The conching process, which may last for several days, involves a mechanical kneading and stirring that gives chocolate its velvety smoothness (fig. 16.7). After conching, the liquid chocolate is poured into molds and cooled (fig. 16.8).

Figure 16.7 *The conching process imparts smoothness to chocolate.*

Figure 16.8 *Far removed from the tropical cacao tree, these delectable chocolate confections gladden the heart of many "chocoholics."*

(a)

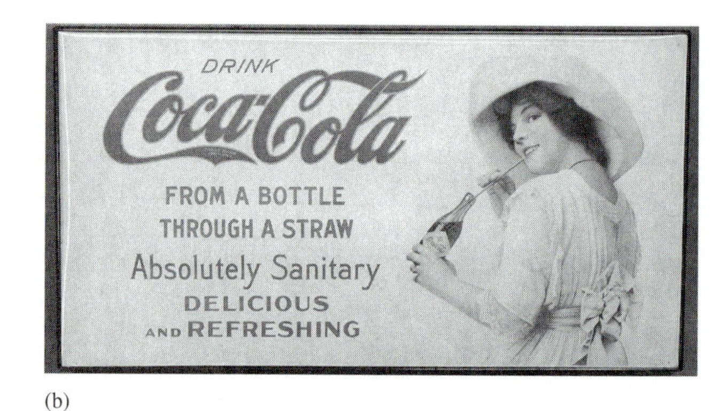

(b)

(c)

(d)

Figure 16.9 *Early bottles and advertisements for coca-cola.*

Coca-Cola: An "All-American" Drink

The drink that has become synonymous with American culture (fig. 16.9) begins with the seeds of the kola tree (*Cola nitida*) found in West Africa. The tree, a relative of the cacao tree, bears pods that usually contain eight seeds The fleshy seed coats are removed and the seeds are allowed to ferment before they are dried and pulverized. In Africa the seeds, with their high caffeine content, are used as a stimulant and as an appetite depressant. The seeds can be pulverized to prepare a tea, or can be chewed whole. In addition to the caffeine, small quantities of kolanin, which acts as a heart stimulant, are also present in the seed.

Coca-Cola first made its appearance on May 8, 1886 in Atlanta, Georgia. Dr. John Styth Pemberton, a pharmacist, concocted a beverage using carbonated water, caramel for coloring, an extract from coca leaves, and an extract from the powdered kola seeds. Other ingredients, including sugar, vanilla, cinnamon, and lime juice are also used but the exact formula is a highly guarded secret. The original recipe, in Dr. Pemberton's handwriting, is locked in an Atlanta bank vault. While coca extracts are still used, since 1903 the cocaine has been removed before the extracts are added.

Other Caffeine Beverages

Throughout the world other caffeine beverages are consumed by local populations. Mate and guarana are two well-known beverages in South America. Mate or Paraguay tea is made from the leaves of a holly, *Ilex paraguensis,* and is popular throughout Central and South America, while guarana, made from the pulverized seeds of *Paullinia cupana,* is primarily consumed in Brazil. Among the Jivaro Indians in Peru, a ritual morning hot drink is prepared from the caffeine-rich leaves (up to 7.5%) of *Ilex guayusa.* The leaves of other hollies are also known to contain small quantities of caffeine and have been used to make beverages. Most notable is *Ilex vomitoria,* the yaupon holly, whose leaves were sold for tea in the southern states during the Civil War blockade.

Chapter Summary

1. Caffeine and caffeinelike alkaloids have a stimulating effect on the central nervous system. Plants naturally rich in caffeine have a long history of use for alleviating fatigue and drowsiness. Historically, many of these plants have played an important role in human affairs.
2. Coffee is obtained from the seeds of *Coffea arabica,* a tree native to Ethiopia. Coffee drinking spread to Europe from the Middle East early in the seventeenth century, and by 1700 coffee houses could be found throughout Europe. Coffee plantations were established in South America before 1720. The trees flourished in the hospitable climate and soil, and today Brazil and Colombia are the world's leading coffee producers. Depulping, fermentation, and roasting are required to produce the characteristic flavor of coffee, with the roasting process the most critical and most variable. Instant coffee and decaffeinated coffee are two twentieth century innovations to the long history of coffee.
3. Tea, made from dried tip leaves of *Camellia sinensis,* is the world's most popular beverage. The tea plant is a shrub native to the area around Tibet, India, China, and Burma. It flourishes in tropical or subtropical areas where there is abundant rainfall. The essential oils and tannins in tea are responsible for its distinctive flavors and aromas but the stimulating effects are due to the caffeine and theophylline present. Tea was introduced to western Europe in the seventeenth century. Its popularity was most evident in England, where it became the national beverage. By the mid-1700s the British colonists in North America were also avid tea drinkers. The Tea Act imposed by Parliament laid the foundation for the Boston Tea Party and other events leading to the Revolutionary War.
4. The seeds of *Theobroma cacao* are the source of both a beverage and a confection, cocoa and chocolate. The cacao tree is native to tropical Central and South America and was used to prepare a spicy beverage consumed by Aztec noblemen and priests. In the sixteenth century, Spanish conquistadors introduced the beverage to the Spanish court, where it was modified by the addition of sugar. In Europe, cocoa never rivaled coffee or tea, but the popularity increased after the first chocolate bar was developed in the mid-nineteenth century.
5. The kola tree, native to West Africa, is the source of a caffeine beverage that has been consumed increasingly during the twentieth century. Although the widespread popularity of cola drinks is apparent in contemporary society, mate, guarana, and other caffeine beverages are consumed by local populations throughout the world.

Review Questions

1. Describe the physiological action and effects of caffeine in the human body.
2. What is the importance of caffeine in contemporary society? Do you think decaffeinated products will diminish this impact?

3. Contrast the geographic spread of coffee plantations with the history of coffee consumption.
4. Describe the processing of green, black, and oolong teas.
5. Trace the steps from freshly picked coffee cherries to a freshly brewed cup of coffee.
6. How have the products of the cacao tree changed from the time of Montezuma to the present?

Further Reading

Galvin, Ruth Mehrtens. 1986. Sybaritic to Some, Sinful to Others, But How Sweet It Is. *Smithsonian,* 16 (11): 54–65.

Hopley, Claire. 1991. Chocolate and Charity: A Quaker Legacy. *British Heritage,* 12 (3): 56–68.

Klein, Richard M. 1975. The Tea Mystique. *Natural History,* 84 (10): 12–29.

Roden, Claudia. 1977. *Coffee.* Faber & Faber, London.

Starbird, Ethel A. 1981. The Bonanza Bean Coffee. *National Geographic,* 159 (3): 388–405.

Young, Gordon. 1984. Chocolate: Food of the Gods. *National Geographic,* 166 (5): 663–687.

17

Herbs and Spices

Chapter Outline

Chapter Concepts

1. Essential oils are volatile substances that contribute to the essence, the aroma or flavor, of herbs and spices.
2. The desire for spices had a significant impact on the history of world exploration, colonization, and trade.
3. In temperate regions, the use of herbs goes back to prehistoric time, and four plant families provide the majority of herbs in use today.

The aroma of a freshly baked cinnamon bun, the zip of a peppery stew, the tang of oregano in pizza sauce, and the distinctive flavor of real vanilla ice cream, are all familiar sensory inputs that embellish life. Today, the herbs and spices responsible for these tastes and others are taken for granted, but in the past the desire for spices was a driving force that actually shaped human history. Although there is no clear-cut distinction between herbs and spices, **herbs** are generally aromatic leaves, or sometimes seeds, from plants of temperate origin, while **spices** are aromatic fruits, flowers, bark, or other plant parts of tropical origin. While herbs and spices are mainly associated with cooking, they have been used in herbal medicine, as natural dyes, and in the perfume and cosmetic industries.

Essential Oils

The characteristic scents of aromatic plants are due to the presence of **essential oils,** volatile substances that contribute to the essence or aroma of certain species. A simple distillation of plant parts can extract these volatile oils and thus capture the essence of a plant. Essential oils are widely distributed in plant organs, but are most commonly found in leaves, flowers, and fruits where they occur in specialized cells or glands. An essential oil is considered a type of **secondary plant product,** a compound that occurs in plants but is not critical for the plant's basic metabolic function. This sets secondary products apart from primary products such as sugars, amino acids, proteins, and nucleic acids, without which plants could not exist.

Chemically, most essential oils are classified as **terpenes,** a large group of unsaturated hydrocarbons with a common building block of $C_{10}H_{16}$. Essential oils are end products of metabolic pathways, but the exact role of these substances in plants varies. Clearly, the essential oils in flowers serve to attract pollinators by their alluring scents, while the function of other essential oils in different plant parts has been debated. At one time it was believed that many essential oils were merely waste products of metabolism that accumulated. Now it is thought that many essential oils may play a significant role in discouraging herbivores, particularly insects, and inhibiting bacterial and fungal pathogens.

History of Spices

There is no evidence of how primitive people first discovered herbs and spices, but it is reasonable to assume that they were attracted to the pleasant aromas of these plants and found many uses for them. The ancient Egyptians used herbs and spices extensively in medicine, cooking, embalming, and as perfumes and incense. The Ebers Papyrus, dated about 3,500 years ago, is a scroll that lists the medical uses of many plants and includes anise, caraway, mustard, and saffron, as well as many other familiar herbs and spices. Cinnamon and cassia are also mentioned in Egyptian records. These two spices, native to Southeast Asia and China, are evidence that an active spice trade was already in existence.

Ancient Trade

During the time of the Ancient Greek civilization, the spice trade was flourishing between the Mediterranean region and the Far East. Spices such as cinnamon and cassia, as well as black pepper and ginger from India, were sought by the Greeks. The middlemen for this trade were Arab merchants who brought the spices mainly by caravan from India, China, and Southeast Asia. To protect their monopoly, the Arabs invented misleading and fanciful stories about the source of the spices. For example, the Arabs claimed that pepper only grew under waterfalls protected by dragons.

When Alexander the Great conquered Egypt, he established the port city of Alexandria, which became the leading trading center for spices from the East and a meeting place for traders from Europe, Asia, and Africa. Alexandria remained one of the most important spice-trading centers for centuries.

Spices were even more prominent in the Roman Empire as their culinary use expanded, along with their application in medicine and luxury items (perfumes, bath oils, and lotions). The Romans went to great expense to procure these precious spices; large amounts of gold and silver flowed to the East in payment. When Nero's wife Poppaea died in Rome during the first century, a year's supply of cinnamon was burned at her funeral as a tribute; this spice was considered the most precious commodity in the imperial storehouse. After the first century, Rome began trading directly with India by ship, via the Red Sea, to the Indian Ocean, thus breaking the centuries-old Arab monopoly on spice trading. It is worth noting that both the Greeks and Romans made extensive use of native herbs, as well as the more exotic spices; in fact, almost all the herbs and spices that are in use today were known to these ancient civilizations.

As the Roman civilization spread its influence through Europe, they introduced exotic spices to the local tribes. Already familiar with many temperate herbs, these tribal people quickly developed a taste for these luxuries. When Alaric the Visigoth threatened Rome, he demanded and received 3,000 pounds of pepper, in addition to gold, silver, silks, and other valuable items. This highlights the value of the spice, and also illustrates the extent and volume of the spice trade. When Rome fell in A.D. 476, the trade between Europe and the East virtually disappeared; several centuries elapsed before the spice trade actively resumed.

Marco Polo

During the Dark Ages, exotic spices from the East were rare and the people of Europe had to rely, for the most part, on native temperate herbs. Many of these herbs were valued for medicinal as well as aromatic purposes and were grown in monastic and royal gardens throughout Europe. Merchant travelers during this period kept a limited supply of spices flowing from the Arab trading centers to Europe. Later the Crusades, beginning in 1095, increased the importation of spices and other goods from the Near East. Venice and Genoa, the great merchant cities of Italy, rose to prominence during this time. One Venetian trader in particular, who spurred on the European desire for spices and valuables from the East, was Marco Polo.

In 1271, Marco Polo began his travels at 17 years of age, in the company of his father and uncle who were making a return journey to the court of Kublai Khan in China. They spent 25 years in the Orient and witnessed firsthand many of the riches. A few years after his return to Venice, Marco Polo was taken prisoner during a war between Venice and Genoa. During this year of captivity in Genoa, he dictated the memoirs of his adventures, later published as *The Travels of Marco Polo.* In the book he described in striking detail the spice plantations of Java, the immense pepper stores in China, and the abundance of cinnamon, pepper, and ginger on the Malabar coast of India. His accounts whetted the European appetite for the riches of the Orient and lured more and more travelers eastward in search of these spices. New overland routes were established and soon explorers were searching for sea routes to the East.

Age of Exploration

None were more determined to find a sea route than Prince Henry of Portugal, better known as Henry the Navigator, who sought to break the Venetian-Muslim trade monopoly. He established a school of navigation in 1418, where he gathered the leading astronomers, cartographers, geographers, and navigators of the day. Although he died before a sea route was discovered, his efforts laid the groundwork for the Age of Exploration that followed. In 1486, Bartholomew Dias discovered the Cape of Good Hope at the southern tip of Africa, proving a sea route to India was possible; Vasco da Gama made the possibility a reality when he reached the west coast of India in 1497 (fig. 17.1).

While the Portuguese were exploring southern routes to reach the Orient, the Genoan Christopher Columbus, under the flag of Spain, sailed west in search of the spices of the East. From his first voyage in 1492, he was convinced that he had discovered the route to China and Japan. Although he did not return with valuable oriental spices, he still was able to persuade the Spanish monarchs, Queen Isabella and King Ferdinand, to eventually finance three more voyages. Christopher Columbus never did find the black pepper and cinnamon for which he so ardently searched, but he firmly established Spain's claim to the New World and introduced a great many plants, including yams, sweet potatoes, cassava, kidney beans, maize, capsicum peppers, and tobacco to Europe. Columbus died in obscurity in 1506, still believing that he had reached the East and never realizing that he had discovered the New World, a discovery worth far more than the spices he was seeking. Several years later the Portuguese Ferdinand Magellan, also sailing for Spain, led the expedition that circumnavigated the globe (1519–1522) and discovered a western route to the Spice Islands (now known as the Moluccas). Many of the important oriental spices, including cloves, nutmeg, mace, and pepper, are native to these Spice Islands (fig. 17.1).

Imperialism

During the sixteenth century, it was Portugal who monopolized the spice trade to Europe through its outposts in India, China, Japan, and the Spice Islands. The explorations of da Gama, Pedro Alvares Cabral (1460–1526), and Alfonso de Albuquerque (1453–1515), and the subsequent military victories, established Portuguese control over the major spice trading centers throughout the East. The Portuguese were ruthless in their control, often enslaving native populations to labor in the plantations. Without competition, the price of spices soared throughout Europe, and the revenues brought such tremendous wealth and power to Portugal that other European nations sought to break the Portuguese stranglehold and share the riches.

The Dutch and English eventually broke this control early in the seventeenth century. By 1621, the Dutch forced the Portuguese from the Spice Islands, securing control over nutmeg and cloves for The Netherlands. By the end of the seventeenth century, the Portuguese were only left with holdings in Goa (India) and Macao (a small island off the coast of China), and the British were only minor players in the spice trade. The Dutch were the dominant force in the East Indies, Ceylon (Sri Lanka), and the Persian Gulf spice markets. Their control over the spices was even harsher than the Portuguese. For example, to artificially create a scarcity and inflate the price of nutmeg and cloves, they uprooted 75% of those trees on the Spice Islands. Cinnamon from Ceylon, more than any other spice, brought huge profits to the Dutch East India Company, the officially sanctioned trading conglomerate. As with the Portuguese, the country that controlled the spices controlled great wealth and power.

In the latter half of the eighteenth century, the Dutch monopoly began to break down. There were many reasons for this: the British and French began spice plantations in their own colonies; the English East India Company gained a strong foothold on the Malabar coast of southwest India that produced pepper, ginger, and cinnamon; and a British blockade during the war between England and The Netherlands resulted in near bankruptcy for the Dutch East India Company.

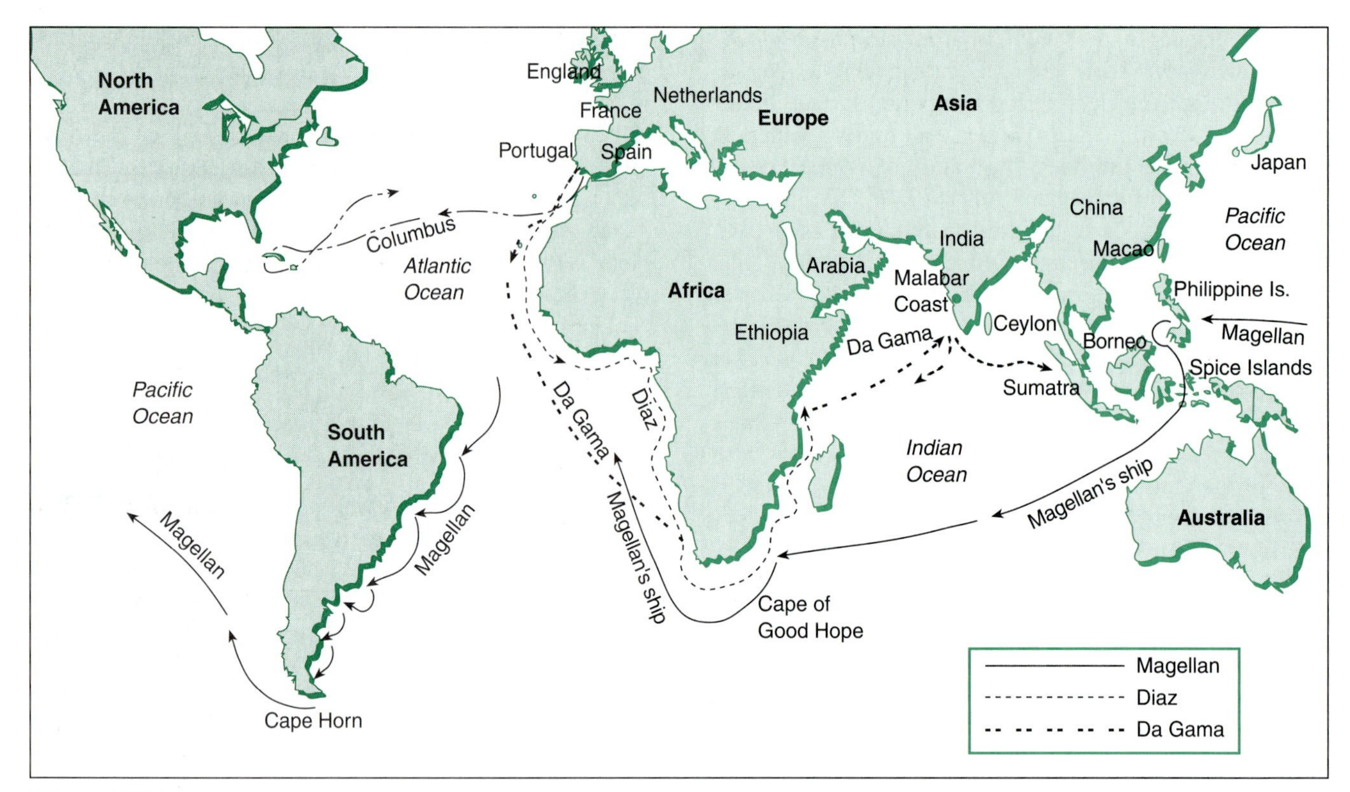

Figure 17.1 *European explorations during the fifteenth and sixteenth centuries were, in part, the result of the desire for exotic spices.*

By the end of the eighteenth century, the nearly 200 years of Dutch control had ended.

By the early nineteenth century the English East India Company had control over most of the spice-rich Orient, but the decentralization of the spice trade had begun and a spice monopoly would never occur again. In the late eighteenth and early nineteenth centuries, even the United States became involved, through clipper ships that sailed from New England to the island of Sumatra in Indonesia in search of spices, principally pepper. Many fortunes were made from this lucrative spice trade centered in Salem, Massachusetts.

New World Discoveries

During the sixteenth century, Spain greatly expanded its influence in the New World, conquering and subjugating native populations in Central and South America. The New World spices, introduced first to Spain, included allspice, vanilla, and several varieties of capsicum peppers such as chili peppers and paprika. The capsicum peppers could be grown in many parts of the temperate and tropical world and soon became important in the diets of many people in Europe, Africa, and Asia. Allspice still remains an exclusively New World spice since it has not been successfully grown elsewhere. However, since the mid-nineteenth century, vanilla has been cultivated in many tropical areas. Trade in these New World spices never had the allure, importance, or adventure associated with the spices from the Orient.

Spices

The spices of the Old World that dominated trade and were the impetus to exploration still remain prominent commodities throughout the world, as are the New World spices. The following discussion focuses on those spices that have the most widespread use (table 17.1).

Cinnamon: The Fragrant Bark

Cinnamon is one of the oldest and most valuable spices known; its use is documented in ancient Egyptian, Biblical, Greek, Roman, and Chinese accounts. It was one of the main spices sought in the early explorations. This spice comes from the bark of an evergreen tree, *Cinnamomum zeylanicum,* in the laurel family (Lauraceae) native to India and Sri Lanka, where it grows best under wet tropical conditions. The similar spice cassia, which may also be called cinnamon, comes from several related species, but primarily *Cinnamomum cassia,* native to Southeast Asia. Trees under cultivation are kept small and bushy, but in the wild could reach a height of 12 meters (39 ft). Two-year-old stems and twigs are cut and the bark removed with a special curved

TABLE 17.1
Common Spices, their Scientific Names, and the Plant Part Used

Spice	Scientific Name	Part Used
Allspice	*Pimenta dioica*	Fruit
Capsicum peppers	*Capsicum annuum*	Fruit
	Capsicum frutescens	
Cassia	*Cinnamomum cassia*	Bark
Cinnamon	*Cinnamomum zeylanicum*	Inner bark
Cloves	*Eugenia caryophyllata*	Flower
Ginger	*Zingiber officinale*	Rhizome
Mace	*Myristica fragrans*	Aril
Nutmeg	*Myristica fragrans*	Seed
Black Pepper	*Piper nigrum*	Fruit
Saffron	*Crocus sativus*	Stigma
Turmeric	*Curcuma longa*	Rhizome
Vanilla	*Vanilla planifolia*	Fruit

knife. The outer layer of the bark is scraped away and the inner bark curls into quills (cinnamon sticks) as it dries. Imperfect quills and trimmings are ground into powdered cinnamon. Cassia differs in that the entire bark is used to make the quills. In the United States, much of what is called cinnamon may actually be cassia (fig. 17.2). While today these spices are usually associated with baking, they have also been used in medicines, perfumes, and scents (fig. 17.3).

Black and White Pepper

Pepper, the most widely used spice today, was once a precious and most desired commodity. Both black and white pepper are obtained from the dried berries of *Piper nigrum,* a climbing vine native to India and the East Indies, where it thrives in a hot wet climate (fig. 17.4). For black pepper, berries are picked green just prior to ripening; the berries are allowed to dry for a few days and during this process they turn black and shrivel. They may be either sold in this form, as peppercorns, or ground. Since the biting flavor is due to volatile oils, peppercorns begin to lose their flavor after grinding; for this reason, the taste of freshly ground pepper is often preferred. To obtain white pepper, the berries are allowed to ripen on the vine; after harvesting, the outer hull is removed, leaving a greyish-white kernel that is ground. White pepper is slightly milder, lacking the pungency of the black.

Cloves

Cloves were valued in ancient China where they were used to sweeten the breath of court officials before they addressed the emperor. They are native to the Spice Islands. Centuries later, when the Dutch controlled these islands, they confined the production of cloves to a single island in order to drive up prices. A Frenchman managed to steal some clove seeds in 1770 and established plantations on French colonies. Cloves are the unopened flower buds of *Eugenia caryophyllata,* an evergreen tree in the myrtle family (fig. 17.5). The buds have to be picked with care because, once opened, they are useless as a spice. After picking, the buds are dried and marketed as whole cloves or ground and used in desserts, beverages, meats, pickling, sauces, and gravies. In Indonesia, cloves are mixed with tobacco for cigars and cigarettes. In addition to its use as a spice, extracted clove oil has been used in medicines, disinfectants, mouthwashes, toothpastes, soaps, and perfumes; however, synthetics are now replacing the natural oil.

Nutmeg and Mace

Nutmeg and mace are two spices obtained from a single plant, the nutmeg tree, *Myristica fragrans,* native to the Spice Islands. Nutmeg trees are dioecious, with pistillate trees bearing apricotlike fruit from which the spices are derived. The fruit is a drupe with a fleshy mesocarp that is removed, exposing the aril-covered endocarp (fig. 17.6). This **aril** (a thin, scarlet, netlike covering) is removed from the fruit, dried, and ground to become the aromatic mace (not to be confused with the aerosol chemical used in self-defense). After removing the aril, the pit, consisting of the stony endocarp and seed, is cured by drying until the seed rattles freely in the shell. The shell is cracked, releasing the seed or nutmeg, which is sold whole or ground. Both spices have similar properties, with a strong, spicy but slightly bitter, aromatic flavor and are used in baking sweets as well as meat and vegetable dishes.

(a)

(b)

Figure 17.2 *(a) Illustration from a sixteenth century French text depicts a cinnamon harvest. (b) Cassia quills (on the left) are thicker and darker than cinnamon quills. Unlike cassia, only the inner bark is used in cinnamon.*

Figure 17.3 *Ceremonial spice box or* b'samim *used in the Jewish Havdalah ceremony marking the end of the Sabbath. Spices (notably, cloves and cinnamon) are placed inside the box, which is passed around during the ceremony so that participants can enjoy the aromas. Symbolically, this promises that the fragrant memories of the Sabbath will be with them throughout the coming week. This silver filigree* b'samim *(Poland, early nineteenth century) is from the Gershon and Rebecca Fenster Museum of Jewish Art in Tulsa, Oklahoma.*

Both nutmeg and mace have received some notoriety as potential hallucinogens. To achieve a hallucinogenic state, very large quantities of either spice must be consumed. The essential oils are believed to be the hallucinogenic agents, but because of the toxicity of these compounds, the hallucinations are accompanied by many unpleasant side effects including nausea, vomiting, dizziness, and headaches.

Although nutmeg and mace were not known to the ancient western civilizations, they had reached Europe by the twelfth century and were two of the precious spices of the Middle Ages. Later the Portuguese, and then the Dutch, controlled production of these spices in the Spice Islands until the French smuggled seedlings to their island colonies. Yankee traders in the nineteenth century developed a profitable scam by producing fake wooden nutmegs, which they sold as the real thing. The nickname of Connecticut, the "Nutmeg State," reflects this historical anecdote.

Figure 17.4 *The berries of the vine,* Piper nigrum, *are the source of black pepper, the world's most widely used spice.*

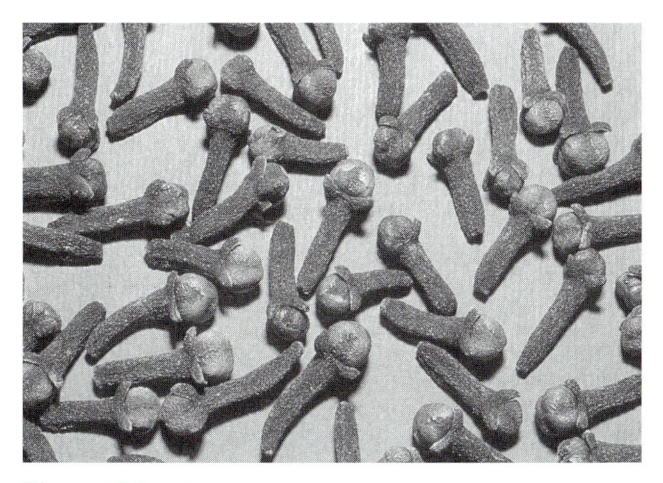

Figure 17.5 *The dried flower buds become the familiar clove "nail."*

Ginger and Turmeric

Two very different spices that come from the same family are ginger and turmeric. Ginger is obtained from the rhizomes of *Zingiber officinale,* a small herbaceous perennial native to tropical Asia but cultivated throughout the tropics. Not only do the rhizomes provide the spice, but small portions of the rhizome are used for vegetative reproduction and give rise to aboveground shoots (fig. 17.7). Nine

Figure 17.6 *The fruit of* Myristica fragrans, *a drupe, is the source of two spices, nutmeg and mace. Mace is derived from the netlike aril that is wrapped around the pit. Within the pit is a single seed, the source of nutmeg.*

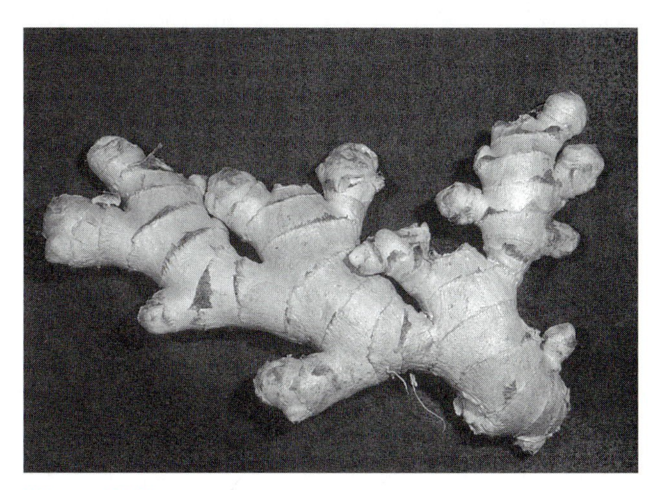

Figure 17.7 *Ginger rhizome.*

months to one year after planting, the greatly enlarged rhizomes are harvested and either cleaned and sold fresh, or peeled, dried, and ground. The aroma and taste of ginger are characteristically spicy, hot, and pungent, and the best ginger today is said to be from Jamaica. This spice was introduced into the New World by the Spanish, where it grew so successfully that by 1547 Jamaica was exporting ginger to Spain. Ginger is also a versatile spice used in baked goods, especially gingerbread, Oriental dishes, pickles, vegetables, meats, poultry, and ginger ale.

Also native to tropical Asia, *Curcuma longa* is the source of turmeric, another spice obtained from a dried rhizome. Turmeric is also used as a brilliant yellow dye to color both food and fabric. The propagation, harvesting, and processing of turmeric are similar to that described for ginger. Although turmeric is not as familiar as ginger, it is a common ingredient in prepared yellow mustard, the main spice in curry powder, and often a substitute for the more

(a)

(b)

Figure 17.8 *(a) The delicate stigmas of* Crocus sativa, *the source of saffron, must be separated carefully by hand. (b) In this wall painting from ancient Crete, the stigmas of saffron flowers are being harvested by a trained monkey.*

costly saffron. Turmeric is also used in Middle Eastern and East Indian cooking and even used medicinally and cosmetically in parts of Asia.

Saffron

The modern world's most expensive spice is obtained from the delicate stigmas of an autumn crocus, *Crocus sativus,* in the Iris family. The species is native to eastern Mediterranean countries and Asia Minor. While saffron is not one of the exotic tropical spices from the Far East, it is a spice that was much desired by the ancient civilizations of Egypt, Assyria, Phoenicia, Persia, Crete, Greece, and Rome (fig. 17.8). It was an important commodity in their spice trading and also valued as a yellow dye. In the East, the yellow dye symbolized the epitome of beauty and its aroma was considered sublime. The word *saffron* is from the Arabic word *zafaran,* which means yellow. It was the Arabs who introduced saffron to Spain in the tenth century, and today Spain dominates world production with 70% of the market.

Crocus sativus is propagated by corms; small cormlets are planted, giving rise to the familiar crocuses. Every autumn the appearance of the purple flowers signals the beginning of the saffron harvest. The blooming period is short, about two weeks, and the flowers must be picked in full bloom before wilting; often the critical time period for harvesting is limited to a few hours. Once picked, the flowers must be carefully stripped of their orange-red three-parted stigmas; again haste is important to remove the stigmas before the petals wilt (fig. 17.8). Most of this backbreaking and delicate work of harvesting and stripping has been traditionally done by hand, but mechanization is making inroads in Spain. After the stigmas are removed, they are dried by slow roasting and sold as either saffron threads (whole stigmas) or powdered. The stigmas from 150,000–200,000 flowers yield one kilogram (or 75,000–100,000 flowers for 1 lb) of the spice. In 1991, the retail price of saffron in parts of the United States was approximately $6.68 per gram (or $190.00 per ounce), validating its claim as the world's most costly spice. Hoping to cash in on the lucrative saffron market, unscrupulous merchants have been known to adulterate saffron with turmeric, marigold or safflower petals, or other substances.

The flavor and aroma of saffron are pungent, slightly bitter, and musky. A small amount of saffron imparts a delicate and enticing taste and color to many different foods. It is widely used in French, Spanish, Middle Eastern, and Indian cooking, being an essential ingredient in bouillabaisse, paella, arroz con pollo, saffron cakes and buns, and challah.

Hot Chilies and Other Capsicum Peppers

Not be confused with their namesake, black pepper, capsicum peppers from the New World have become equally important in international cuisine since their discovery by Columbus. When Columbus found the capsicum fruits to be as pungent as the black pepper from the Orient, he believed that his voyage west in search of spices had been justified, and it was with high hopes that he brought these New World fruits back to Spain. After their introduction to Spain, their cultivation and use spread throughout Europe, Asia, and Africa. Eventually they became as prominent in these regions as they were in the New World. Evidence indicates that capsicum peppers had been widely cultivated for thousands of years by the people of tropical America. Although the time of their domestication is not known, fragments of a 9,000-year-old chili pepper were discovered in a Mexican cave. When Cortés conquered Mexico, the native peoples considered capsicum peppers indispensable elements in their diet, along with maize, tomatoes, and beans, and were growing numerous varieties that were used with every meal.

Capsicum peppers are the fruits of plants belonging to a single genus, *Capsicum,* which includes several cultivated

Figure 17.9 *Capsicum peppers come in a wide variety of sizes, shapes, colors, and degree of hotness.*

species and hundreds of varieties (fig. 17.9). *Capsicum annuum* is the most widely cultivated of these species and includes mild, sweet bell peppers as well as many varieties of hot peppers such as cayenne. This species, a member of the nightshade family (Solanaceae), is a small, bushy herbaceous annual that produces small white flowers similar to those of tomato or potato. The fruits are berries that vary considerably in shape, size, and color among the hundreds of varieties; the immature fruits are green and the mature fruits vary from yellow to purple to bright red and from long and narrow to almost spherical. *Capsicum frutescens* is cultivated mainly in the tropics and warm temperate areas and generally has a more fiery taste, such as Tabasco peppers.

The biting taste of capsicum peppers is due to the amount of capsaicin present; this compound is mainly found in the seeds and placental area (where seeds attach to the ovary wall). The capsaicin content is negligible in the sweet bell peppers but found in such high concentrations in hot chili or jalapeño peppers that even handling or cutting the peppers can irritate the skin. Capsaicin is so potent that it can be tasted in concentrations as low as one part per million. The potency of capsaicin has been utilized recently in two completely different applications, by police as a pepper spray to subdue unruly persons and by physicians in creams that are applied to relieve the pain of arthritis and other ailments.

Figure 17.10 *Pods of the vanilla vine, the source of vanilla flavoring.*

The fruits are also excellent sources of vitamin C; even one pepper is more than enough to satisfy the daily requirement. The amount of vitamin C is actually higher in peppers than in citrus fruits; in fact, vitamin C was first chemically isolated from paprika in 1932 by Albert Szent-Györgyi, who won a Nobel Prize in 1937 for other work.

Many varieties of capsicum peppers are sold whole, either fresh or dried, while powders are prepared by grinding the dried fruits of several varieties. Included in this last group are paprika, red pepper (both ground and crushed are sold), cayenne, and chili powder, which is actually a blend of spices in addition to the ground chili peppers. The versatility of the capsicum peppers is reflected in the number of different cuisines that are associated with them. Many Hungarian, Italian, Mexican, Cajun, Indonesian, Indian, and Oriental dishes all utilize some type of capsicum pepper or spice.

Vanilla

Another important spice from the New World tropics is vanilla, the only spice obtained from an orchid. *Vanilla planifolia,* a perennial vine native to the humid tropical rain forests of Central America and Mexico, produces elongate pods that are processed into the vanilla beans of commerce (fig. 17.10). Today vanilla orchids are cultivated in Madagascar, and the islands of the Indian Ocean (Reunion and Seychelles), as well as Mexico. To insure good pod production, the flowers are hand pollinated, particularly in areas outside the native range where the natural pollinators (certain bees and hummingbirds) are not present.

The pods are picked while still green and then undergo a traditional curing for several months. The first stage of this process involves sweating, alternately heating the pods in the sun and then wrapping them in blankets throughout the night. A slow drying process follows, during which the final aroma is developed. Uncured or unfermented vanilla beans lack the characteristic vanilla flavor, which is due to vanillin, a crystalline compound synthesized during curing.

When properly cured, the pods are black with crystals of vanillin on the surface. Alternative methods of rapid curing are used in some areas but these do not always produce a satisfactory product. Although whole beans are available, most are processed into vanilla extract by percolating cut beans in a 35% alcohol-water solution. Pure vanilla extract contains more than just vanillin; other flavoring compounds are leached out of the bean and contribute to the final extract. These may vary according to the growing conditions of a region and account for taste and quality differences; many feel that Mexican vanilla is top quality.

For imitation vanilla extract, vanillin is synthesized chemically from a variety of different compounds such as clove oil, lignin from wood pulp, or coal tar. The chief advantage of imitation vanilla is the low cost; however, it lacks the subtle flavor of pure vanilla extract. In addition, extracts of tonka beans, *Dipteryx odorata,* have been passed off as vanilla extracts. This is a dangerous substitute, since the tonka beans contain coumarin, a blood thinner that could cause internal hemorrhaging.

Long before the discovery of America, vanilla was an important commodity among the Aztecs. The beans were used as flavoring, perfume, medicine, and even as a means of tribute. It is believed that Bernal Diaz, a Spanish conquistador with Cortés, was the first European to report on the use of vanilla in the preparation of *chocolatl* (*see* Chapter 16). Vanilla beans were brought back to Spain, and soon its use spread throughout Europe. Mexico continued to be the leading producer of vanilla beans for several centuries until it was discovered that the flowers could be hand pollinated. Today Madagascar leads the world in cultivation of vanilla beans.

Allspice

The third New World spice discovered by Spanish explorers was allspice, the dried berries of *Pimenta dioica,* an evergreen tree in the myrtle family. Although there is some evidence that Columbus may have come across the fragrant leaves of a *Pimenta* tree on his second voyage, the spice long used by the Mayan civilization was not discovered by Europeans until the 1570s. Allspice is named for its multi-faceted flavor, which is similar to a combination of cinnamon, nutmeg, and cloves. The berries, which are picked nearly ripe and then dried, resemble peppercorns; the other common names for the plant, Jamaica pepper, clove pepper, pimento, and Jamaica pimento reflect this resemblance (*pimenta* means pepper in Spanish). Unlike capsicum peppers and vanilla, this spice has never been successfully cultivated outside the Western Hemisphere; Jamaica controls the world production of allspice today. This versatile spice is used whole or ground. Ground it is an ingredient in baked goods, cooked fruits, sauces, and relishes, while whole it is best known for its use in pickling vegetables and meats.

Herbs

In recent years there has been a renaissance in the use of herbs in everyday life, going beyond the kitchen to find applications in shampoos, cosmetics, soaps, potpourris, and medicines. Herbs are usually the aromatic leaves or sometimes seeds of temperate plants; however, other plant organs are sometimes considered herbs as well (fig. 17.11a). Throughout the centuries, thousands of plants have been used as herbs, but the following discussion will focus only on four well-known families.

The Aromatic Mint

The mint family, Lamiaceae, is the source of many important and familiar herbs: spearmint, peppermint, marjoram, oregano, rosemary, sage (fig. 17.11b), sweet basil, thyme, savory, and others (table 17.2). Members of this cosmopolitan family include mainly herbaceous plants and small shrubs, characterized by square stems and aromatic simple leaves with numerous oil glands. The Mediterranean region is an important center of origin for the mint family, and a variety of these herbs have been used for thousands of years by the civilizations that developed in this area. With the long history of use, each herb has evolved a distinctive biography, folklore, and usage, but space limits the discussion to a brief overview of this family.

Typifying mint flavors are spearmint and peppermint, *Mentha spicata* and *M. piperita,* respectively. Either the dried leaves or the distilled oils of these herbaceous perennials find widespread use as flavorings and perfumes in gums, candies, cookies, cakes, cigarettes, toothpastes, mouthwashes, antacids, soaps, jellies, ice creams, teas, and other drinks. Mint flavorings have become indispensable in everyday life; it would be illuminating to count the number of times mint is encountered in just one day. Menthol is the most abundant constituent of peppermint oil; for large-scale commercial extraction of menthol, the related Japanese mint, *Mentha arvensis* is frequently used because of its higher menthol content.

Both marjoram and oregano were used by the ancient Egyptians, Greeks, and Romans who employed the herbs both for cooking and medicine. These perennial herbs are both members of the genus *Origanum* but taste differently, with marjoram, *O. majorana,* being mild, and oregano, *O. vulgare,* having a biting flavor. The dried leaves of both herbs are widely used with meats and vegetables; oregano has become especially popular through its use in pizza and spaghetti sauce.

Sweet basil, *Ocimum basilicum,* is one of the oldest herbs known. This native of India has been considered a sacred plant in the Hindu religion (fig. 17.11c). The Greeks referred to basil as the "herb of kings"; indeed the word *basil,* some believe, is derived from the Greek *basileus,* meaning king. The dried leaves of this annual, noted for its sweet aromatic flavor, are widely used in a variety of foods but are most commonly associated with tomato dishes.

TABLE 17.2

Common Herbs, their Scientific Names, and the Plant Part Used

Herb	Scientific Name	Part Used
Anise	*Pimpinella anisum*	Fruit
Basil	*Ocimum basilicum*	Leaves
Bay leaves	*Laurus nobilis*	Leaves
Caraway	*Carum carvi*	Fruit
Cardamom	*Elettaria cardamomum*	Seed
Celantro	*Coriandrum sativum*	Leaves
Celery	*Apium graveolens*	Fruit
Chervil	*Anthriscus cereifolium*	Leaves
Chives	*Allium schoenoprasum*	Leaves
Coriander	*Coriandrum sativum*	Fruit
Cumin	*Cuminum cyminum*	Fruit
Dill	*Anethum graveolens*	Fruit, leaves
Fennel	*Foeniculum vulgare*	Fruit
Fenugreek	*Trigonella foenumgraecum*	Seed
Garlic	*Allium sativum*	Bulbets
Horseradish	*Armoricana rusticana*	Root
Leek	*Allium porrum*	Leaves
Marjoram	*Origanum majorana*	Leaves
Mustard	*Brassica alba; Brassica nigra*	Seed
Onion	*Allium cepa*	Bulb
Oregano	*Origanum vulgare*	Leaves
Parsley	*Petroselinum crispum*	Leaves
Peppermint	*Mentha piperita*	Leaves
Rosemary	*Rosmarinus officinalis*	Leaves
Sage	*Salvia officinalis*	Leaves
Savory	*Satureja hortensis*	Leaves
Shallot	*Allium ascalonicum*	Bulb
Spearmint	*Mentha spicata*	Leaves
Star anise	*Illicium verum*	Seed
Tarragon	*Artemesia dracunculus*	Leaves
Thyme	*Thymus vulgaris*	Leaves

The Parsley Family

This temperate cosmopolitan family, the Apiaceae, which gives us carrots and parsnips, also provides many familiar herbs: parsley, caraway, dill, fennel, celery, anise, coriander, celantro, cumin, and chervil (*see* table 17.2). Annual, biennial, or perennial in habit, the members of this family are easily recognized by their umbels (flat-topped inflorescences) and alternate compound leaves. The characteristic fruit for the family is a dry indehiscent **schizocarp,** which splits into two one-seeded identical halves commercially referred to as seeds. For many plants in this family these fruits are used as the herb; however, for others the fresh or dried leaves are the desired parts. Like the mint family, the herbs in the Apiaceae have a long history of usage, but only a few will be considered.

Parsley, *Petroselinum crispum,* is probably the best known and most widely used member of this family, valued for its use as a garnish as well as its flavor. Native to the Mediterranean region, parsley was revered by the early Greeks as symbols of both victory and death, and as such was used in crowns for champions and wreaths for tombs. It was the Romans who first valued parsley as a culinary herb, and today it is an almost indispensable ingredient in many dishes. On the other hand, the attractive, dissected, and usually curly green leaves are too often used just to enhance the visual presentation of food and then discarded uneaten. This is unfortunate since the leaves are high in vitamins A and C, as well as several minerals and are said to sweeten the breath after a garlic-laden meal.

(a)

(b)

(c)

(d)

Figure 17.11 *(a) Many of the most flavorful and useful culinary herbs can be attractively grown in an herb garden; (b) Sage; (c) Sweet basil; (d) Dill.*

Dill, *Anethum graveolens,* provides both leaves (dill weed) and fruits (dill seeds) that are widely used as seasonings (fig. 17.11d). Individual plants are generally not grown for both, because when dill weed is desired the plant is harvested before flowering. Dill oil is another important commodity from this plant that finds its greatest use in the pickling industry. This annual, native to the Mediterranean area and Europe, has finely dissected, feathery leaves and umbels with bright yellow flowers, making it an attractive garden plant.

The distinctive flavor of caraway seeds in rye bread is familiar to most people. Native to most parts of Europe, Asia, North Africa, and India, the biennial caraway, *Carum carvi,* is one of the oldest herbs known. The use of these seeds is thought to have originated with the ancient Arabs who called the seeds *karawya,* which is the source of the English word. In addition to flavoring bread, caraway seeds are used in cheeses, soups, sausages, and a variety of meat or vegetable dishes. Anise, cumin, celery, coriander, and fennel are additional members of the Apiaceae also valued for their schizocarps.

(a)

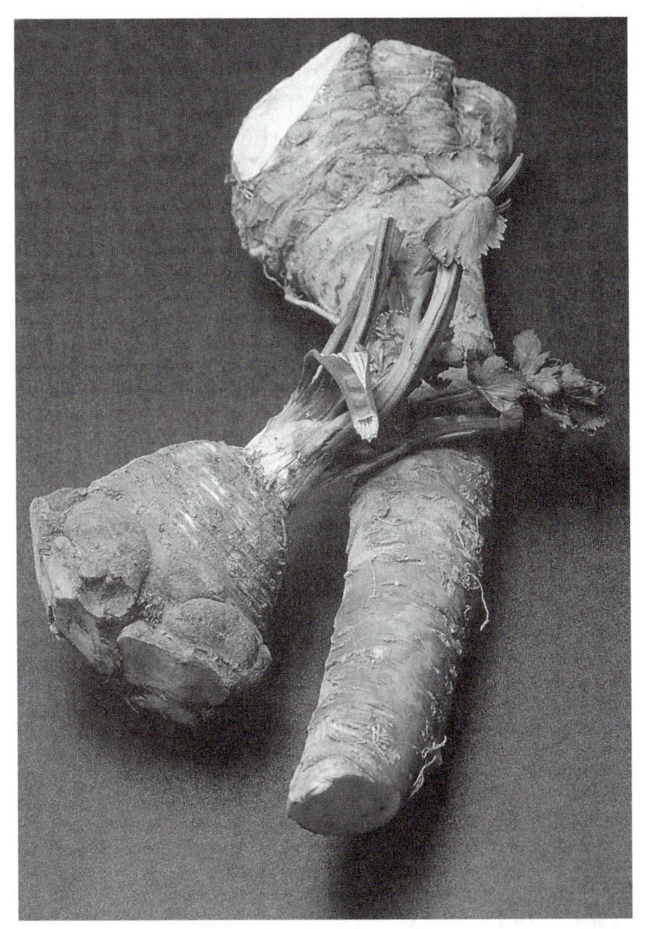

(b)

Figure 17.12 *(a) A field of mustard plants in flower. (b) The large taproots of horseradish are grated to make horseradish sauce.*

The Mustard Family

The Brassicaceae, mustard family, gives us many important vegetable crops such as cabbage, broccoli, cauliflower, brussel sprouts, turnips, and radishes, as well as two flavorful herbs or condiments, mustard and horseradish (table 17.2). This temperate cosmopolitan family is especially abundant in the Mediterranean area. Members of the family are easily recognized by their characteristic flowers, each with four petals arranged in a cross, which accounts for the old family name, the Cruciferae.

Seeds of *Brassica nigra, B. alba* and related species are the source of one of the most familiar seasonings, mustard (fig. 17.12a). Mustard produced from the seeds of *B. alba,* white mustard, is somewhat milder tasting than the more pungent product of *B. nigra,* black mustard. Both plants are annuals native to Europe and western Asia that have a long history of use both as condiments and medicinals. Forerunners of today's prepared mustard can be traced back to the late Middle Ages, when crushed mustard seed was mixed with vinegar to prepare a sauce. In many countries today, prepared mustard is the primary product of these species (recall that the bright yellow color of many brands of prepared mustard is due to turmeric). Whole seeds and ground seeds are two other ways that mustard is marketed; the whole seeds are primarily for pickling, while the ground mustard finds its way into hundreds of recipes. The sharp tangy taste of mustard is the result of reactions between sinigrin (in black mustard) or sinalbin (in white mustard) and myrosin, an enzyme. In the presence of water, these components react to produce volatile oils that give mustard its characteristic taste. Unless acidified (as with vinegar) the flavor of prepared mustard quickly deteriorates.

Horseradish, *Armoracia rusticana,* is an herbaceous perennial native to Europe that has been cultivated for centuries (fig. 17.12b). Although horseradish has a long history of use as a medicinal plant, its use as a condiment only dates from the Middle Ages in Denmark and Germany. Horseradish sauce is prepared from the taproots, which are white and faintly resemble a large misshapen carrot. The pungent aroma and hot biting taste are due to the interaction of two components, sinigrin and myrosin (identical to those found in black mustard), which combine to produce a volatile oil. In intact roots the volatile oil is not produced because the two components occur in separate cells; however, when the roots are scraped or grated, the components are liberated and free to interact. The volatile oil that is produced diffuses easily on exposure to air and the grated condiment quickly loses its pungency.

Sinigrin, which contributes to the potency of both mustard and horseradish, is similar to the chemical responsible for the caustic effects of mustard gas, a poisonous gas used in chemical warfare during World War I and during the Iran-Iraq war of the 1980s.

A Closer Look 17.1

Herbs to Dye For

For millenia, natural materials were the only source of dyes until the availability of synthetic aniline dyes from coal tar in the latter half of the nineteenth century. Even today, natural dyes are still important in many traditional societies. Dyes can be classified according to how easily a dye is affixed to the fiber. Direct dyes are soluble in water and readily picked up by the fiber. Turmeric and safflower are direct dyes that yield a yellow hue. In contrast, mordant dyes do not impart their color directly; the fiber must be treated with a chemical agent, the mordant. A mordant fixes the dye to the fabric; often, the color of the dye can be changed by using different mordants. Many lichens and club mosses contain alum, (potassium aluminum sulfate), a mordant used since antiquity. Another natural mordant is oak galls (box fig. 17.1), a source of tannins. (An oak gall is an abnormal growth on oak leaves and branches. It is caused by a female insect laying her eggs on the plant. The plant responds by producing a growth that encases the developing insects. The feeding activity of the insect stimulates the increased production of tannins in the gall.) A vat dye is insoluble and must be rendered soluble by the action of chemical agents or microorganisms. When dye-saturated fabric is exposed to air, the dye is oxidized. The color is permanent and does not fade in sunlight or after washing. Indigo blue is a vat dye discovered long ago in Asia.

Box Figure 17.1 *Oak galls.*

Natural dyes are obtained from both animal and plant sources; two of the best known animal dyes are Tyrian purple and cochineal red. Tyrian purple derives its name from the ancient Phoenician city of Tyre on the Mediterranean coast. The source of the dye is the mucous gland of several species of whelks, a type of shellfish. The dye was in much demand but extremely costly; the color of "royal purple" became synonymous with position and wealth in the ancient Greek and Roman societies.

The Aztecs obtained cochineal red from boiling the females of a type of scale insect that is commonly found on the *Opuntia* cactus.

The Lily Family: Pungent Alliums

The Liliaceae, which has a worldwide distribution, consists largely of herbaceous perennials that arise from rhizomes, bulbs, or corms. A single genus, *Allium* from central Asia, is the source of many familiar zesty herbs (table 17.2): onions (*A. cepa*), garlic (*A. sativum*), leeks (*A. porrum*), shallots (*A. ascalonicum*), and chives (*A. schoenoprasum*). The following discussion will focus on a brief consideration of the history, botany, chemistry, medicinal, and culinary use of onion and garlic. Onion, a biennial, produces a single large bulb, while the perennial garlic produces a composite bulb with each clove of garlic called a **bulblet** (fig. 17.13). Both species produce an inflorescence consisting of small lilylike flowers, which also make them attractive ornamentals.

Figure 17.13 *The pungent flavor and aroma of onion and garlic are due to sulfur compounds present in the bulbs and bulblets.*

ABLE 17.1

Some Common Household, Yard, and Garden Dye Plants

Plant	Scientific Name	Part Used	Color
Turmeric	*Curcuma longa*	Rhizome	Yellow
Yellow onion	*Allium cepa*	Papery brown outer layers	Burnt orange
Black walnut	*Juglans nigra*	Hulls	Dark brown, black
Red cabbage	*Brassica oleracea-capitata*	Outer leaves	Blue, lavender
Coreopsis	*Coreopsis* spp.	Flower heads	Orange
Lilac	*Syringa* spp.	Purple flowers	Green

Herbal dyes have also been used extensively throughout history. Blue is the rarest of all hues and, as the source of a deep, rich blue, indigo (*Indigofera tinctoria*) was one of the most valuable herbal dyes. A legume from the Old World tropics, the leaves are the source of the dyestuff. Different species of this plant were also used as a source of blue dye by the Aztecs. Indigo was one of the last of the herbal dyes to be replaced by synthetics. Woad (*Isatis tinctoria,* Mustard Family) is the temperate counterpart to indigo; in fact, the dye principle (indigotin) in the leaves of both plants is the same. Native to Europe, woad was the primary source of blue until the introduction of indigo in the eighteenth century. Woad, as with many dye plants, was used as a cosmetic. Julius Caesar's *The Gallic War* described the practice of the inhabitants of the British Isles of staining their bodies blue with woad before battle.

Red herbal dyes are equally as scarce as blue; the madder plant (*Rubia tinctorum,* Coffee Family) is an excellent and ancient source of this much-sought-after color. A mordant type dye is obtained from the slender roots of this herb. The famous "red coats," the uniforms that became synonymous with the British army, are a historical example of the importance of madder dye. Logwood, *Haematoxylon* spp., is one of many dye plants known to the Aztecs. A small tree in the Fabaceae, chips of the wood yield colors of blue, black, gray, brown, and purple. Logwood for dyeing purposes was one of the early exports to Europe, beginning in the sixteenth century. The dye haematoxylon is one of the few natural dyes that has not been replaced for commercial use by a synthetic. Haematoxylon is indispensable as a cell stain in the preparation of tissues for microscopic study.

Many common household, yard, and garden plants are sources of hidden colors; a few are listed in Box Table 17.1.

The pungent flavor and scent of onion and garlic are due to the presence of various volatile sulfur compounds that are released when the tissues are cut. In intact onions and garlic, these compounds are inactive and are only released through the action of the enzyme allinase. The main active ingredient released from garlic is allicin, while that from onion is known as the lacrimatory factor. These molecules are highly reactive and easily change into numerous other sulfur-containing compounds that have a wide range of biological effects. The most familiar effect is the tearing caused by the lacrimatory factor of onions. (Cutting onions under running water will wash away some of the lacrimatory factor, and cutting them cold will retard the enzyme activity.) Another undesirable effect of these sulfur compounds is the lasting aroma of onion and garlic on the breath after ingestion. After digestion, the sulfur compounds are transported by the bloodstream into the lungs, where they may diffuse into the exhaled air. The antibacterial and antifungal effects of garlic and onion extracts are also examples of the biological action of these sulfur compounds.

Both onion and garlic are two of the oldest cultivated plants used for both culinary and medicinal purposes, with a history even predating the ancient Egyptian civilization. The Ebers Papyrus, an Egyptian medical reference from 3,500 years ago, listed 22 uses of garlic as a treatment for various ailments. In the herbal medicine traditions of both India and China, onions and garlic have been prescribed for centuries as remedies for numerous conditions. Modern research has shown that these folk remedies have a sound scientific basis. The organic sulfur compounds from garlic do inhibit the growth of many disease-causing bacteria and fungi and also inhibit the formation of blood clots. (Blood clots have been linked to certain forms of cardiovascular disease as the cause of heart attacks and strokes.)

This chapter has focused on the history and culinary uses of herbs and spices; however, they also had widespread use as dyes (*see* A Closer Look: Herbs to Dye For), perfumes (*see* A Closer Look: *Alluring Scents* in Chapter 5), cosmetics, and most importantly, medicinals. In fact, the foundation of modern medicine is based on the herbal medicine practiced by many societies and this aspect of herbs will be examined in detail in Chapter 19.

Chapter Summary

1. The characteristic scents of aromatic plants, such as herbs and spices, are due to volatile substances called essential oils, which can be extracted to obtain the essence of the plant. Spices generally have a tropical origin and have been used for a variety of purposes since ancient times. The ancient Egyptian, Greek, and Roman civilizations all used spices extensively for cooking, embalming, medicine, and as perfumes. Trading routes for the purpose of obtaining these spices from India and China have a long history. Exotic spices became rare in Europe during the Dark Ages but trade routes to the East were reopened during the Crusades. Marco Polo's account of his travels to the court of Kublai Khan renewed the European desire for spices. The Portuguese, under the direction of Prince Henry the Navigator, were the first European power to travel by sea to India via the Cape of Good Hope. Christopher Columbus himself sailed west in an attempt to reach the spice-rich lands of the East.
2. The history of spices is the history of domination of the spice-rich lands by various European powers. Spice plantations were established in European colonies to cash in on the lucrative spice trade. Some of the most sought-after spices were cinnamon, black pepper, and cloves. Capsicum peppers, vanilla, and allspice from the New World were added to the list of spice treasures.
3. Herbs are usually the aromatic leaves of plants from temperate regions. The mint family is the source of many familiar herbs: peppermint, spearmint, marjoram, and oregano, to name a few. The parsley family is another important herb family; herbs in this family include parsley, dill, and caraway. Allicin is the sulfurous compound in onion, garlic, and related plants that is the source of tearing and unpleasant aromas but also medicinal effects.
4. Herbal dyes were one of the main sources of color in the ancient world, only recently replaced by synthetic dyes in the latter half of the nineteenth century.

Review Questions

1. What are secondary plant products? How have they been a benefit to society?
2. How did spices influence world history?
3. Distinguish between herbs and spices.
4. What plant parts are processed to prepare the following: cinnamon, nutmeg, cloves, black pepper, vanilla, parsley, mustard?
5. Discuss the medicinal value of herbs.
6. What familiar herbs are obtained from the mint and parsley families?
7. Although Columbus was searching for a source of black pepper, he introduced Europeans to the capsicum peppers. How have these measured up as source of valued seasonings?

Further Reading

Aikman, Lonelle. 1983. Herbs for All Seasons. *National Geographic,* 163 (3): 386-409.

Cannon, John and Margaret. 1994. *Dye Plants and Dyeing*. Timber Press: Portland, OR.

Grae, Ida. 1974. *Nature's Colors: Dyes from Plants.* Collier Books, London.

Parry, John W. 1953. *The Story of Spices.* Chemical Publishing Co., New York.

Robbins, Jim. 1992. Care for a Little Hellish Relish? Or Try a Hotsicle. *Smithsonian,* 22 (10): 42-51.

Rosengarten, F. 1969. *The Book of Spices.* Livingston, Philadelphia, PA.

Ward, Diane Raines. 1988. Flowers Are a Mine For a Spice More Precious than Gold. *Smithsonian,* 19 (5): 104-111.

Wernick, Robert. 1984. Men Launched 1,000 Ships in Search of the Dark Condiment. *Smithsonian,* 14 (11): 128-148.

18

Materials: Cloth, Paper, and Wood

Chapter Outline

Chapter Concepts

1. Many textiles are woven from plant fibers; one of the most important discoveries made by ancient peoples was to weave cloth from plant fibers.
2. Wood and wood products, which supply construction materials and fuel to much of the world population, rank right behind food plants in terms of their importance to humanity.
3. Paper, the chief medium of communication in contemporary society, is produced from the pulp of trees or other plants.

In the previous five chapters, the use of plants to provide sustenance has been examined, but plants clearly furnish society with more than food and drink. Since earliest times they have also been used to provide shelter, clothing, and fuel. Later, as civilizations developed, plants also supplied various media for written communications. This chapter will present a brief examination of some of the materials that plants contribute to society.

FIBERS

Plant fibers are one of the most useful plant materials; they have been used for millennia to make cloth, rope, paper, baskets, and numerous other articles. The term *fiber* actually has several meanings. Recall from Chapter 3 that anatomically, fiber cells are long and tapering, with extremely thick secondary cell walls. The composition of the fiber cell wall is chiefly cellulose, although lignin, as well as tannins, gums, pectins, and other polysaccharides, may also be present. The plant fibers of commerce often do not refer to individual fiber cells *per se* but stringy, elongated masses of plant material that are actually collections of fiber cells or entire vascular bundles.

The most valuable fibers are those that are nearly pure cellulose and white in color. Cellulose is an extremely strong material, with properties of tensile strength that rival that of steel. (Tensile strength measures the resistance of a fiber to tearing apart when subjected to tension.) Fibers that are high in lignin are generally of poorer quality, browner in color, and of lower mechanical strength.

Types of Fibers

Fibers can be classified according to their use. Those fibers that have been used to weave cloth are known as textile fibers; cordage fibers are used in making rope, and filling fibers are used as stuffing in upholstery or mattresses. Fibers may come from many sources. Natural fibers are mainly of plant or animal origin, although mineral fibers, such as asbestos, can also be used. Animal fibers, such as wool or silk, have a protein makeup, while plant fibers are composed primarily of cellulose. Some synthetic fibers are created using natural sources as a base. For example, cellulose from wood pulp is processed chemically into rayon.

Plant fibers can be classified according to where they are found on a plant. **Surface fibers** are found on the covering of seeds, leaves, or fruits; cotton cloth is made from seed hairs covering the surface of cotton seeds. Linen and ramie are made from **bast** or **soft fibers,** clusters of phloem fibers found in the inner bark of dicotyledonous stems. **Hard fibers** or **leaf fibers** are obtained from the vascular bundles or veins in leaves; these consist of both xylem and phloem, as well as the ensheathing fibers. Monocotyledonous leaves are the usual source of hard fibers; sisal and Manila hemp are examples of this type of fiber. Hard fibers usually have a higher lignin content than soft fibers (table 18.1).

Extracting the Fiber

Various methods are employed to extract fibers from the source material. Surface fibers are usually separated mechanically from plant material by **ginning,** in which machines tear the fibers loose. Many soft fibers are extracted from the stems through **retting,** a process that uses microbial action to degrade away the soft tissues, leaving the tough fiber strands intact and freed. The process of **decortication,** in which the unwanted tissues are scraped away by hand or machine, is the way hard fibers are commonly extracted.

Spinning into Yarn

Once fibers have been freed from the source material, they are cleaned of plant materials and dirt. Then the fibers are combed and laid parallel to each other to form a strand. Next, the strand is stretched or pulled with the fingers or from the weight of a spindle and the individual strands are twisted together to form the yarn or thread. The simplest method of spinning fibers into yarn is to roll the fibers together between the palms of the hands or against the thigh. The invention of the spindle made spinning easier. It is the rotation of the spindle that twists and holds the fibers together, forming the yarn. The finished yarn is then woven into cloth or used for other purposes.

Some time around the year A.D. 750 in India, the first spinning wheel was invented by mounting a spindle on a frame; the spindle rotates when the wheel is turned. The spinning wheel was faster than the hand spindle and made the yarn more uniform. Chinese inventors later added a treadle that further speeded up the process. One drawback to these earlier spinning wheels was that the spinner had to stop periodically to wind yarn on the spindle. The invention of the flyer in sixteenth century Europe overcame this problem. A U-shaped device affixed to the end of the hollow rod that holds the spindle, the flyer turns at a different rate than the spindle (fig. 18.1). In this manner, yarn can be wound and twisted on the spindle at the same time. During the time of the industrial revolution, spinning machines using multiple spindles mechanized the process of spinning.

King Cotton

The process of rendering cotton fibers into cloth was discovered independently during prehistoric times in both the Old and New Worlds. Archaeological evidence indicates that cotton was harvested from wild populations in coastal Peru as early as 10,000 years ago and domesticated by 4,500 years ago. It was also grown and used by native peoples of the American Southwest. In the Old World, cotton cloth dates back 5,000 years on the Indian subcontinent. Cotton spread westward from India to the nations of Assyria and Babylonia. The Romans and ancient Greeks knew of cotton but its use was limited, as they preferred linen. Cotton became a Muslim industry in the Near East in the ninth and tenth centuries. It was the Arabs who

TABLE 18.1

Important Plant Fibers and Their Uses

Plant	Scientific Name	Type of Fiber	Use
Coir	*Cocos nucifera*	Surface	Filling, cordage
Cotton	*Gossypium hirsutum*	Surface	Cotton cloth
	Gossypium barbadense		
Flax	*Linum usitatissimum*	Bast	Linen
Hemp	*Cannabis sativa*	Bast	Hemp cloth, canvas, cordage
Jute	*Corchorus* spp.	Bast	Burlap
Kapok	*Ceiba pentandra*	Surface	Filling
Manila hemp	*Musa textilis*	Leaf	Cordage
Piña	*Ananas comosus*	Leaf	Cloth
Ramie	*Boehmeria nivea*	Bast	Ramie cloth
Sisal	*Agave sisalana*	Leaf	Cordage, matting
	Agave fourcroydes		

Figure 18.1 *The Saxony wheel, a spinning wheel with a flyer, was introduced in Europe in the sixteenth century.*

introduced cotton cultivation to Muslim Spain. The Arabic influence is seen by the fact that the word cotton is derived from the Arabic *qutn.* Cotton was introduced into Florida in 1556, and by 1607 was grown as a commercial venture in Virginia. Less than 100 years later, cotton was a significant crop in the southern colonies. Cotton, for the most part, was a minor cloth in Europe until the introduction in the eighteenth century of New World varieties that, with their longer seed hairs, were more suitable for spinning. Currently, the top cotton-producing country in the world is the People's Republic of China. Other major cotton-producing countries are the United States, India, the republics of the former Soviet Union, Pakistan, Egypt, Brazil, and Turkey.

The Cotton Plant

Today, cotton is the most popular natural fiber, accounting for half of the world's textiles, and is one of the most economically important nonfood plants in the world. The cloth itself is woven from seed fibers obtained from species of *Gossypium.* The cotton plant is a member of the Malvaceae, or Mallow Family, of tropical and temperate distribution. There are 20–30 species in the genus *Gossypium,* native to parts of Asia, Africa, Central and South America, and Australia.

Cotton is a shrubby plant with palmately lobed leaves. In tropical climates, most species are perennials, but in temperate zones they are grown as annuals. Cotton grows best in a warm climate, with sufficient water during the growing season. Depending on the species, flower color ranges from white to purple. The fruit of cotton is a capsule, in commercial circles a boll (fig. 18.2), and when it splits open along five seams, reveals a white mass of fibers. The fibers are hairs that extend from the seed coats of each of the ten or so seeds in every fruit. As many as 20,000 seed hairs may grow from a single seed. Under natural conditions, the fibers enable seeds to become airborne and thus dispersed by wind currents. Each cottonseed hair, actually a single seed coat cell, is a twisted, hollow strand of cellulose up to 7.62 cm (3 in) in length that flattens at maturity (fig. 18.3). The hairs begin to develop from the integuments of the ovule after the fertilization of the flower. As the hair grows, the wall continuously thickens to the point that the lumen in a mature fiber is practically nonexistent. Two types of seed hairs cover the seed surface: the lint or staples, long slender fibers and linters, shorter, fuzzy hairs. High quality cotton is made from the longest of lints. The purity of the cellulose (90%) and its natural twist makes cotton an excellent fiber for spinning into yarn.

(a)

(b)

Figure 18.2 *(a) Cotton bolls mature 50 to 80 days after the flower is fertilized. (b) In an opened cotton boll, the fluffy white mass consists of seed fibers.*

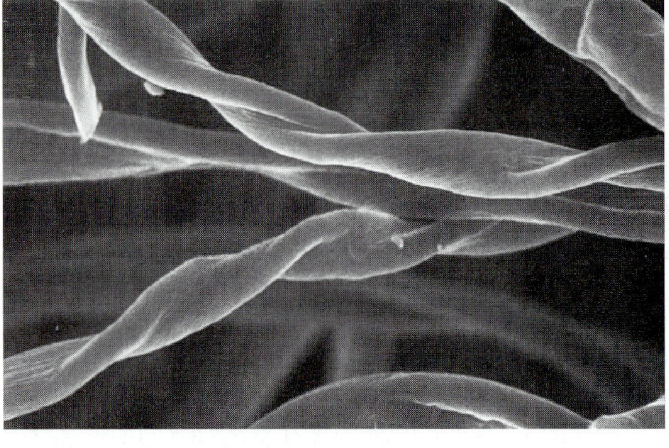

Figure 18.3 *Scanning electron micrograph (× 800) of cotton seed fibers reveals their natural twist, which makes them suitable for spinning.*

Old and New World Varieties

Commercially important species of cotton are *G. hirsutum* and *G. barbadense* from the New World, and *G. arboreum* and *G. herbaceum* from the Old. From these, numerous cultivars have been developed. The Old World cottons are diploid species ($2N = 26$) and produce a short lint, about 1–2 cm (0.4 – 0.8 in) in length. *Gossypium herbaceum* is a species that originated in southern Africa, eventually spreading to India and giving rise to *G. arboreum.*

Both New World species are tetraploids ($2N = 52$) and there is some evidence to indicate that they may be the result of a hybrid cross involving *G. herbaceum* and *G. raimondii* from northern Peru. *Gossypium hirsutum* is the predominant type grown in the world today. Also known as upland cotton, this species arose in Central America and Mexico and spread to South America. Lint from upland cotton varies from short to long, 2.2–3.0 cm (0.9–1.1 in). *Gossypium barbadense* is believed to have originated from the Andean region of South America, where its use dates back to as early as 10,000 years ago. The species includes the cultivars known as Sea Island, pima, and Egyptian cotton. The lint of *G. barbadense* is especially long and silky and its fibers are used to produce high quality, luxury cotton cloth.

Recently, commercial varieties of cotton that have naturally brown-colored lint suitable for machine spinning have been developed from native varieties developed by the Indians of Mexico before the Spanish conquest. Fibers from these varieties produce a naturally colored cotton cloth of green and browns, eliminating the need for chemical dyes.

The Cotton Gin

Cotton bolls mature 50 –80 days after fertilization; defoliants are sprayed on the plants, leaving only the bolls on the plant for picking. Picking cotton was formerly a labor-intensive process, but today it is performed primarily by machine. The seed hairs may be picked from the open bolls or the entire boll can be harvested. The harvested

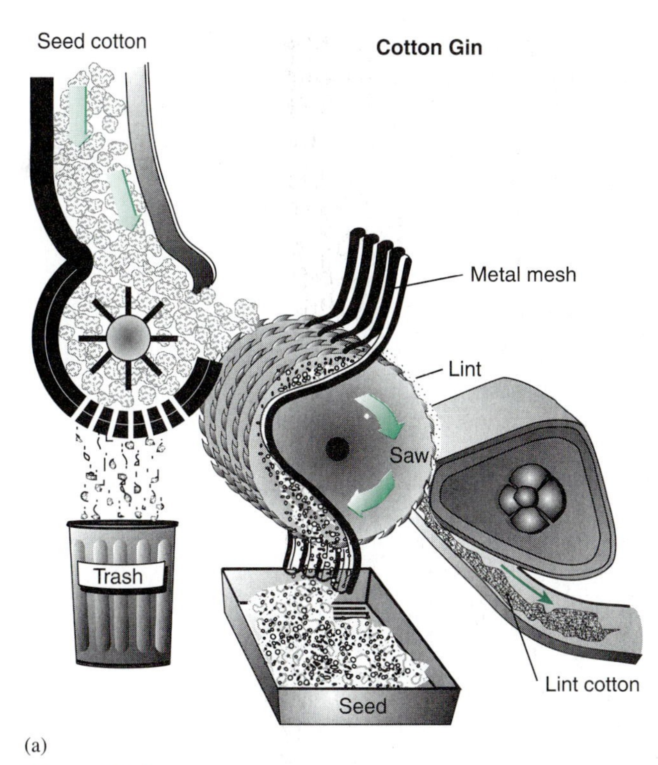

(a)

(b)

(c)

Figure 18.4 *(a) Operation of the cotton gin. The saw draws in the lint but the seeds are held back. (b) Cotton fiber can be seen as it falls onto the covered saws. (c) Ginned cotton fiber is baled and ready for shipment.*

bolls or fibers are next sent to a gin. During ginning, the lint is removed from the seeds. At first, ginning was painstakingly done by hand. The first type of gin invented, probably in India, was a roller gin. Lint was drawn between two closely set rollers; the lint passed through the rollers easily but the seeds did not. Unfortunately, this gin did not work with all species of cotton. Today, modern gins are larger versions of the saw gin, invented by Eli Whitney in the 1790s. The saw gin (fig. 18.4a, b) is essentially a roller studded with spikes and covered by a metal mesh. The spikes draw in the lint but the seeds do not pass through the mesh and are left behind.

Eli Whitney's invention had its greatest effect in the cotton-growing states of the South. Mechanized cotton production came of age at the same time Europe's demand for cotton grew. Consequently, more fields in the South were planted with cotton and, with each new acre of cotton planted, the demand for slave labor to grow and process the cotton also grew. Slavery, which had been dying out, underwent a resurgence as a direct result of the cotton boom. Immediately prior to the Civil War, cotton production was 4 million bales, making "King Cotton" the number one crop in the United States, a crop supported by enslaved labor.

Seeds can be used to make livestock feed; also, the seeds are the source of cottonseed oil used as a cooking oil. The ginned fibers are then packed into large bales (fig. 18.4c) and graded for quality. The graded bales are shipped to the appropriate yarn or cloth manufacturers, where the lint is straightened (carded) and then sorted into parallel bundles of similar size (combed) in preparation for spinning into yarn.

Finishing and Sizing

The finishing process may be applied to either the yarn or woven cloth; finishes may alter the appearance or modify the function of the textile. Most plant fibers are bleached to remove the natural color and cotton is no exception. Different methods of bleaching have been used throughout history, many of which sound bizarre by today's standards. The standard method in eighteenth century England was to soak the cloth or yarn in sour milk and cow's dung, followed by a steeping in lye. A bath in buttermilk followed and, after washing, the cloth was spread on grass until exposure to the sun bleached it white. A breakthrough in the process came about in 1774 when the bleaching effect of chlorine was discovered.

Mercerization is a finishing process for cotton that improves its strength, luster, and affinity for dyes. Named after John Mercer, who patented the process in the mid-1800s mercerization consists of passing cotton cloth through a bath of caustic soda (sodium hydroxide). The flat, cotton fibers swell into a round shape and the fiber becomes stronger and more lustrous. Mercerized cotton also has a greater affinity for dyes.

Permanent press is an example of a shape-retentive finish. Cellulose fibers like cotton wrinkle and crush easily. A permanent press finish involves the application of chemicals that cross-link the cellulose fibers. This gives the fabric a built-in memory so that the shape of the garment is retained even after laundering, with little or no ironing required.

Sizing is the application of materials to the yarn or fabric that produce stiffness or firmness. Sizing makes the fabric smoother and gives increased sheen; it also protects the fabric from soiling. Cotton is usually sized with starch, dextrin, or resin.

Linen: An Ancient Fabric

The oldest plant fiber used to make cloth may be flax (*Linum usitatissimum*). The stem fibers of this annual have been woven to make linen for at least 10,000 years. Flax fibers have been found at archeological sites of the Swiss Lake dwellers, a Stone Age people. It was also known to many ancient civilizations in Mesopotamia, Assyria, and Babylonia. But it was ancient Egypt that was known as the land of linen. Fragments of Egyptian linen have been found that are 6,500 years old. Egyptian linen was worn by priests and royalty, used as mummy cloths, and exported to other countries as the material of choice for sails. The ancient Greeks and Romans grew some flax for linen. Prehistoric textiles from the American Southwest also indicated the use of flax. In medieval Europe, Flanders became the first important linen center, as Belgium has an ideal climate for growing flax. Ireland is another country noted for the quality of its linens since the seventeenth century. Today, linen accounts for only 2% of the world's textiles; the leading areas of flax cultivation are the republics of the former Soviet Union, the People's Republic of China, and western Europe.

Figure 18.5 Linum usitatissimum, *the flax plant, is the source of a bast fiber that is woven into linen.*

The Flax Plant

Flax is delicate in appearance; its straight slender stems support sessile gray leaves and bell-shaped flowers of white or blue (fig. 18.5). The fruit or boll is a capsule containing 10 brown seeds. Flax is harvested after 100 days when the stalks are a golden color. Two types of flax are grown commercially: one for its seed oil (linseed oil) and one for its fibers. Linseed oil is used in the manufacture of paints, stains, varnishes, and linoleum. The oil cakes are used as livestock feed. The oil has been used medicinally as a laxative and poultice. Oilcloths, which predated plastic tablecloths, were made by applying linseed oil to canvas.

Unbranched varieties of flax are grown for linen production. The stems in these varieties reach a height of a meter (3 ft), with fibers as long. Flax is a bast fiber, composed of the phloem fiber cells found in the stem. Each flax stem contains between 15– 40 fiber bundles, with each bundle containing 10– 40 fiber cells, each 1.3–3.8 cm (0.5–1.5 in) long. To maintain the length of the fibers, the flax stems are pulled up by the roots from the ground, either by hand or machine (fig. 18.6). Gathered into bundles, the harvested stems are stood upright in the field to dry. Next, seed bolls are removed from the now-dried flax straw by rippling, pulling the heads of the plants through a comb or a special type of threshing machine (fig. 18.6).

(a)

(b)

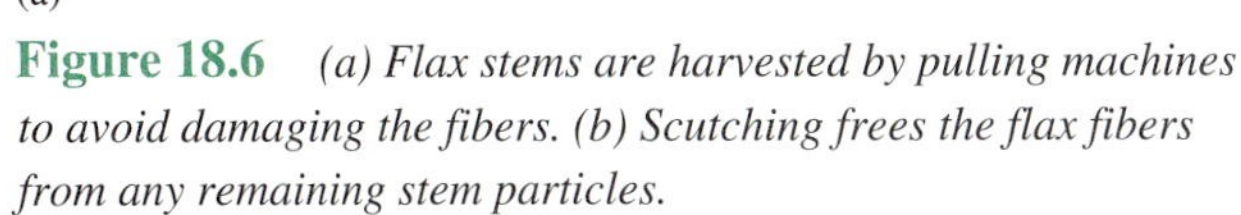

Figure 18.6 *(a) Flax stems are harvested by pulling machines to avoid damaging the fibers. (b) Scutching frees the flax fibers from any remaining stem particles.*

Processing Flax

Retting extracts the fibers from the stem by employing microbial decomposition to break down the outer part of the stem. The word *ret* is derived from the Dutch *roten,* which means "to rot." According to the traditional method of dew retting, the flax straw is spread over the fields to allow the interaction of dew or rain, sunlight, and microbes to rot away the soft, nonfibrous tissues of the stem. Dew retting usually takes four to six weeks, depending on the weather conditions. Immersing the flax straw in ponds or slow-moving rivers accomplishes the same purpose in one to two weeks. The retting process can be accelerated to a matter of days by soaking the straw in tanks of warm water. Chemicals can be added to the tanks to speed the retting process to a few hours. Most flax fibers are retted using a combination of dew and chemicals.

After the retted fibers are dried by the sun or in a kiln (oven), they undergo breaking with a flax break, a type of pounding used to free the fiber from the stem. Most flax breakes have been replaced by machines that crush the dried and retted flax as it passes through rollers. Any shives, or stem particles, that remain are scraped off during scutching. The now-freed fibers are hackled, or combed, to separate the short fibers (tow) from the long fibers (line). The flax fibers, which are naturally yellow in color (hence, the expression *flaxen-haired*), are bleached in the sun or by chemicals. The flax fibers are then spun, either wet or dry, into yarn; wet spinning is said to produce the best quality yarn.

Flax Fibers and Linen

Flax fibers consist of bundles of fiber cells held together by gums and pectins. Each fiber consists of about 70% cellulose, with the balance made up of waxes and pectins. The fibers are smooth and straight, not twisting as with cotton fibers. Line fibers are quite long, averaging 46–56 cm (18–22 in) in length. Tow fibers are always 30 cm (12 in) or shorter. The best flax fibers are a pale yellow color and require little or no bleaching. Flax fibers are naturally lustrous (due to the waxes) and extremely strong. Linen is durable and is used for both clothing and home furnishings such as drapes and upholstery. The flax fiber tends to draw moisture, a characteristic that has made linen a choice fabric for towels. On the down side, the fibers lack elasticity and resilience, making linen fabrics prone to wrinkling. In the process of cottonization, the individual fiber cells are separated out by dissolving the pectins between the cells. Flax fiber processed in this way is softer and can be blended with cotton, rayon, wool, or silk.

Other Bast Fibers

Ramie, hemp, and jute are, like flax, bast fibers and they have been used for a variety of purposes. Ramie (*Boehmeria nivea*) is a perennial shrub and member of the nettle family. Also called China grass, it has been cultivated for centuries in several Asian countries. After the ramie stalks have been cut, the bark is peeled or beaten away. Sun drying bleaches the fibers white; a bath of caustic soda removes pectins and waxes. Ramie fibers are some of the longest (45.7 cm or 18 in), strongest (8 times stronger than cotton), and most lustrous fibers. However, the fibers are somewhat brittle and removing the large amounts of pectins and gums in the stems is difficult. Ramie has recently made inroads into the Western world; its cultivation has been introduced into parts of Europe, and North and South America. Ramie fibers are often blended with cotton or rayon in the making of sweaters and other knitted fabrics.

Figure 18.7 *Most of the world's jute harvest comes from Bangladesh.*

Jute to Burlap

Obtained from the stems of *Corchorus* species in the Tiliaceae or linden family, jute fiber is used to make burlap, ropes, wall coverings, carpet backing, upholstery lining, and inexpensive clothing. The "sackcloth" worn by penitents in medieval Europe was made from jute. Jute is an annual plant, native to Asia, which thrives under the monsoon conditions of the wet tropics. The stems must be retted to free the fibers. The fibers vary in color from yellow to brown and are difficult to bleach; for that reason, many jute fabrics are in natural colors. The fibers are not as strong as other bast fibers; they range from 1.7– 6.7 m (5–20 ft) in length but are usually broken into smaller fragments during processing. About 90% of the world's supply of jute comes from Bangladesh and India. Jute is of particular importance to Bangladesh (fig. 18.7) as it is the main export crop of one of the world's poorest nations.

Another Use for Cannabis

The source of hemp is a plant more often associated with drugs than fabrics, *Cannabis sativa* (*see* Chapter 20). The processing of hemp fiber is similar to that of flax. The natural color of the fiber is dark brown and the length of the fiber varies widely. Staminate plants produce the longest fibers up to 2 m (6 ft) long; however, some fibers are as short as 2 cm (3/4 in). Hemp fibers are used primarily to make industrial fabrics like canvas (the word is derived from *Cannabis*), ropes, and twines. Although jeans today are made from 100% cotton, their inventor, Levi Strauss, originally produced his working clothes from hemp. The hemp cloth was imported from Nimes, France and was known as "serge de Nimes," which was later corrupted into "denim." The globally recognized term "jeans" is a derivative of the French pronunciation of the Italian city of Genoa, another city known for its manufacturing of hemp cloth.

Figure 18.8 *The traditional apparel of the Philippines is woven from piña leaf fibers, a close relative of the commercial pineapple.*

Miscellaneous Fibers

Hemp should not be confused with a leaf fiber, Manila hemp, much valued for making seaworthy ropes. Manila hemp, or abaca, is obtained from *Musa textilis,* a member of the banana family. The fibers are extracted from the huge petioles; the fibers themselves can be up to 5 m (15 ft) in length. Their natural color varies from off-white to dark brown; however, they can be bleached and dyed a wide variety of colors. In the Philippines where most Manila hemp is grown and produced, the fibers have been used to make lightweight clothing, cigarette filters, and teabags, but today most Manila hemp is made into marine rope.

Pineapple Cloth

The shirts, shawls, scarfs, and other traditional apparel of the Philippines are made of a gossamer-like material from the leaf fibers of the pineapple plant (fig. 18.8). Known as piña fiber, special varieties of pineapple (*Ananas comosus*) are grown for their leaves; in fact, fruit development is discouraged by pinching off young pineapples as they appear. Traditionally, the leaves are scraped with knives to uncover the fiber bundles. More scraping and beating cleans and separates the fibers. Machinery and chemical retting are modernizing the labor-intensive process. Piña fibers are 5 to 10 cm (2– 4 in), white in color, and soft and lustrous.

Sisal

Another one of the leaf fibers, sisal comes from the desert *Agave.* Two species are commonly used: *Agave sisalana* (sisal) and *A. fourcroydes* (henequen). Native to Central America, the fibers of these plants were originally used to make rope and coarse garments. Today the fiber is used primarily for making rope, string, and floor mats. Sisal fibers are approximately 1 m (3 ft) in length (fig. 18.9).

(a)

(b)

Figure 18.9 *(a)* Agave *leaves are decorticated to remove their fibers. (b) Sisal fibers are laid out to dry in the sun in Mexico.*

Kapok

This surface fiber has literally saved lives for it is the stuffing in life preservers. The kapok tree (*Ceiba pentandra*) is a majestic giant, native to the tropical rain forests of South and Central America. Grown in plantations in several tropical countries, the fibers develop within the large pods. The fibers grow inward from the ovary wall after fertilization, eventually completely covering the seeds. When the pods split open, the hairs enable the seeds to become airborne and thus dispersed. Workers gather the pods (fig. 18.10a) before they open; ginning is not necessary because the fibers separate easily from the seeds. A coating of cutin makes the fibers waterproof and a large lumen makes them lightweight; a life vest filled with kapok fibers does not waterlog and will support up to thirty times its weight. Kapok fibers are too fine and slippery to be spun but are used for padding and stuffing in upholstered furniture, mattresses, and pillows, since they are nonallergenic.

Coir From Coconuts

The familiar coconut is the source of coir, a seed fiber. This fibrous material makes up the mesocarp of the coconut fruit. The natural color of the coconut is a dark brown and the fiber is often left natural for it is difficult to bleach to lighter colors. The fibers are too coarse for garments but are valued for their durability in items such as ropes and door mats (fig. 18.10b).

Rayon: "Artificial Silk"

The idea of making synthetic fibers is an old one; Robert Hooke, the English scientist who coined the term "cell" (*see* Chapter 2), suggested the possibility in his *Micrographia,* published in 1664. Two hundred years passed before technology made the suggestion of cloth from artificial fiber a reality. The first synthetic fiber was rayon, originally called artificial silk. Rayon is a man-made fiber of pure cellulose. The viscous process for making rayon was invented in 1891 by British scientists, and their method, improved and modified over time, is still employed today.

The original source of cellulose in the production of viscose fibers was cotton linters; wood fibers are commonly used today. Wood (see next section) is processed into cellulose pulp and pressed into thin sheets about 6 m (2 ft) square. The sheets are soaked in an alkali solution, converting the pulp to alkali cellulose. The alkali cellulose is shredded into fine crumbs and aged; treatment with carbon disulphide produces a material called sodium cellulose xanthate (viscose). The xanthate is next dissolved in an alkali bath forming a thick, honey-colored solution. This spinning solution is next passed through the fine holes of a spinerette into a coagulating bath of salts and sulphuric acid that solidifies the liquid into solid fibers (fig. 18.11). The fibers have a high luster; sometimes delustering agents are added to reduce their shiny appearance. Viscose may be dyed in solution before the fibers have been spun out. Manufacturers can also put a crimp in viscose fibers to create fibers with better spinning quality.

Since rayon is a synthetic fiber, manufacturers can modify the size and shape of the fibers, producing rayon fibers that vary considerably. The strength and elasticity of rayon fibers are low; rayon tends to wrinkle and stretch easily. Like cotton, rayon has good moisture absorbency, which makes it easy to dye and comfortable to wear. Rayon blends well with cotton, giving the cloth a lustrous appearance. Adding rayon to ramie produces a linenlike fabric at a cheaper cost.

Wood and Wood Products

Wood and wood products rank right behind food plants in overall importance to society. Construction materials, furniture, paper, fuel, charcoal, and synthetic materials (i.e., rayon, cellophane, and cellulose acetates) are just a few of

(a)

(b)

Figure 18.10 *(a) The hairs encasing the seeds of the kapok tree are the source of a filling fiber. (b) Coir is the fibrous material from a coconut that is used to make floor mats.*

Figure 18.11 *Rayon filaments form when viscose is forced through a spinerette into a coagulating bath.*

the products that come directly or indirectly from trees. Although metals, concrete, and plastics have replaced many of the traditional uses, the strength of wood, its insulating properties, its versatility for construction, and its natural beauty ensure the continued need for lumber well into the future. Yet forest land is currently being lost at a rate that jeopardizes the future availability of wood.

Around one-third of the land surface is covered by forests that supply the wood and wood products used by humanity. The world's forests are being cut down at an alarming rate. Forest destruction is not new. It has been going on since humans learned to harness fire. However, it continues to get worse. Forestry scientists estimate that 30%–50% of the world's forests have already been destroyed, cleared for agriculture, firewood, or lumber. In some areas wood is being used at a faster rate than forests can regenerate it, a concern that has been expressed for many years. The Civilian Conservation Corps Edition of *The Forestry Primer,* published by the American Tree Association in 1933, stated the following:

> We find that our forests are going faster than they are being replenished. This is due to cutting for our needs and to destruction by forest fires, insect pests and diseases. Owing to our constant increase

in population we have ever new demands for what the forest yields us. Some forest uses change, such as when people build houses of concrete and steel instead of wood, but new uses are found so that the trees of the forest still remain tremendously important to the progress of our nation. We simply could not do without them.

While some reforestation programs are in place, many ecologists believe that new forestry management practices are needed to maintain forests as complex ecosystems. Only about 13% of the world's forestland is being managed, and only about 2% of the world's forests are protected in forest reserves. In Latin America only one tree is planted for every ten cut, and in Africa the percentage is even lower. With the world population continuing to grow, the demand for forest products also continues to grow, especially in developing nations. Better forest administration is needed to protect and sustain this valuable resource (*see* Chapter 25).

Hardwoods and Softwoods

Wood is secondary xylem consisting of largely dead cells involved in the transport of water and minerals as well as support. The overall value of wood resides in the strength and durability of the cell wall materials, principally cellulose and lignin.

Two categories of forest trees and lumber are recognized, **hardwoods** and **softwoods.** Hardwood refers to angiosperm trees, while softwood refers to gymnosperm trees or conifers. It is estimated that around 35% of the world's forest area is dominated by softwoods, and in North America, they cover more area than hardwood forests. In a literal sense, the terms are meaningless because some types of gymnosperm wood are actually harder (denser) than some hardwoods. However, as a group, angiosperm wood is generally denser than gymnosperm wood. The hardness indicates the sturdiness of the cell walls, which is largely a reflection of the amount of lignin. The harder the wood, the more resistant to wear. Some hardwoods are so dense that it is actually difficult to drive a nail into the wood, thus making it less desirable for construction.

In softwood, tracheids and some parenchyma cells are the only cell types present. The tracheids combine the functions of both support and conduction, while parenchyma cells carry out various metabolic activities. In hardwood, tracheids and vessels are principally involved in the conduction of water and minerals, while fibers are purely support elements. Xylem parenchyma occurs in longitudinal strands or in rays.

Recall from Chapter 3 that in woody plants the vascular cambium produces a ring of secondary xylem during each growing season. The centermost region of secondary xylem in a tree is known as **heartwood** (fig. 18.12). It is usually darker in color since the cells often contain tannins,

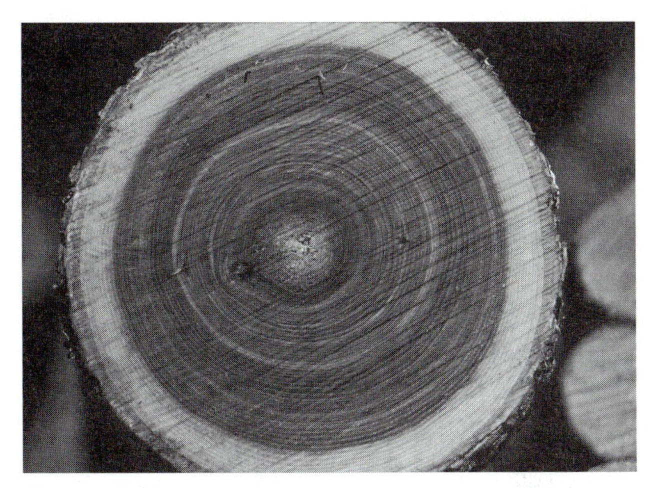

Figure 18.12 *Cross section of a tree trunk clearly shows the darker heartwood in the center surrounded by the lighter sapwood.*

gums, and resins, which accumulate in this older area and help prevent decay. The function of the heartwood is support, as this region is no longer involved in the transport of water and minerals. The region of secondary xylem outside the heartwood is **sapwood,** which functions in both support and conduction. The sapwood also stores carbohydrates and performs other vital functions. Since this region is actively involved in conduction, sapwood cells are normally wet. For lumber use, heartwood is usually preferred. Not only is it more resistant to decay, it is drier and less likely to shrink and warp.

Many characteristics of wood are determined by the thickness of the cell walls and proportion of vessels, tracheids, and fibers. The frequency and distribution of these components are distinctive for various species and contribute to the characteristic grain. The prominence of the annual rings and the direction of cutting also contribute to the grain. Tangential and radial cuts of the same species show variations in the pattern (fig. 18.13). Another important feature is the presence of **knots,** which are the bases of branches that have been covered over by subsequent lateral growth of the tree trunk.

Lumber, Veneer, and Plywood

The United States and Canada are the leading lumber-producing countries in the world. About half the wood harvested in the United States is used as lumber, primarily for construction, with a considerable amount used to make furniture. When the tree is living, about 50% of the weight of wood comes from water. Before the wood can be used the water content must be reduced to 10% or less. The wood can be dried (seasoned) in the open air or in specially built kilns.

The greatest use of softwood lumber is for home construction. Pines generally figure high on the list of important gymnosperms and are valued for their light but strong wood.

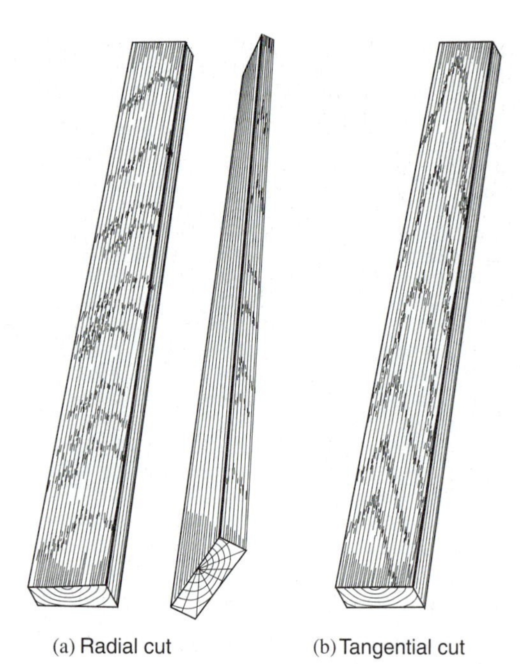

Figure 18.13 *The direction in which the logs are sawn determines the patterns on the finished wood. The radial cut (a), also known as quarter-sawn, produces parallel lines on the board, and the tangential cut (b), called plain-sawn, produces wavy bands.*

Worldwide there are 95 species in the genus *Pinus,* with 35 species native to North America. Historically white pine, *Pinus strobus,* had been the most important timber tree in the United States. During the colonial period, the tree was valued as a source of masts for British sailing ships. This species remained the most important pine species until the twentieth century, when overcutting and the introduction of a fungal disease, white pine blister rust, led to its decline. Today, longleaf, loblolly, and slash pines (often called yellow pines in reference to the wood color) are leading pines in the southern United States and are a major source of lumber, while ponderosa pine fills this role in the west.

Douglas fir (*Pseudotsuga menziesii*) is another one of the most desired timber trees in the world. This strong wood is used in the production of plywood and is a major source of large beams. Douglas fir grows throughout the Rocky Mountain belt but reaches its greatest development in the Pacific Northwest, where the species rivals redwoods in size and grandeur. The diameter and height of these trees enable the harvesting of many large knot-free boards. The species is in heavy demand by the lumber industry, and while large stands still occur, the species may be facing over-harvesting. Other important softwoods include spruce, hemlock, bald cypress, and red cedar (fig. 18.14).

The oaks are the most economically important hardwoods in the United States. The genus *Quercus* is a large one, with species placed into two general groups; the white oak group with rounded lobes on the leaves and the black or red oak group with pointed lobes. The white oak group is more important commercially and includes post, bur, and white oaks. White oak, *Quercus alba,* is the most prized of all species. The wood is very heavy, durable, and attractive. It is widely used in furniture, cabinets, flooring, trim, and whiskey barrels. Blackjack, scarlet, red, and pin oaks are among the more familiar trees in the black oak group, with *Quercus rubra,* northern red oak, the most commercially valuable. Although it is not as strong as white oak, it is used for general construction, flooring, furniture, posts, and railroad ties. Other hardwood trees of value are black walnuts, hickories, maples, sweetgums, tulip trees, and birches (fig. 18.14). Before the introduction of the chestnut blight fungus into North America, the American chestnut was also a widely used hardwood (*see* Chapter 6).

Furniture may be made from solid wood or constructed using a **veneer.** A veneer is a very thin sheet of a desired wood that is glued to a base of less expensive lumber. The veneer provides the look of fine wood at lower cost. Veneers are also used decoratively to produce exceptionally beautiful matched designs in fine furniture. Some of the most popular woods for veneers are black walnut, black cherry, bird's-eye maple, mahogany, and teak.

Plywood consists of three or more layers of thick veneer glued together. The grain of alternate layers is at right angles to each other. Since wood is strongest along the grain, the layering produces a sheet or board that is more uniformly strong than a comparable piece of solid wood. The result is a lightweight but strong building material for roof and wall sheeting, subflooring, shelves, cabinets, boxes, and signs. About 20% of harvested lumber is used in the production of plywood, with Douglas fir and various species of pine the most common sources.

Fuel

Throughout the history of human civilization, wood has been the chief source of fuel until relatively recent times. In many developing nations, the vast majority of harvested wood is still used as fuel. About 1.5 billion people depend on wood or charcoal for 90% of their energy needs for heating and cooking. Another billion people use wood for about 50% of their energy needs. Because of this dependence, it is estimated that each year 50% of the world's harvested wood goes to fuel.

Firewood gathering is hastening the demise of many tropical forests as well, since over one billion people in the tropics depend on wood for fuel. Local supplies are being cut faster than trees can regenerate. For example, in India, it has been estimated that the forests can sustain an annual harvest of approximately 39 million metric tons of wood; however, the demand in India is for 133 million metric tons annually. In addition, it is often necessary to travel long distances with heavy loads to meet fuel requirements. As wood becomes more scarce or the price of fuel rises, the poor will lose their ability to thoroughly cook their food. This may facilitate the spread of disease-causing organisms. This situation will only get worse as populations in developing countries continue to increase.

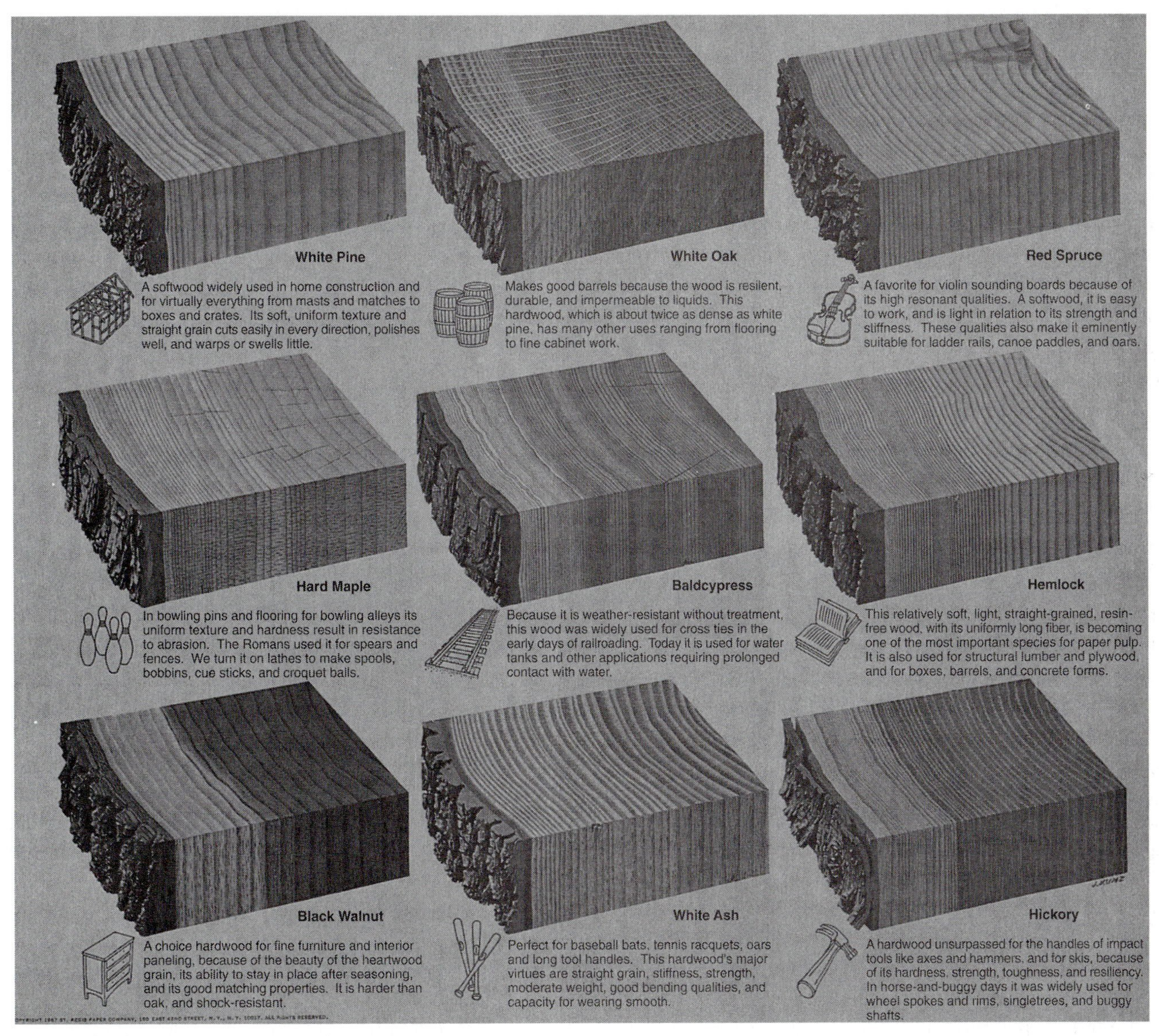

Figure 18.14 *The characteristics of several different types of wood make them ideal for specific purposes.*

In addition to being burned directly, wood can also be converted into **charcoal** by partial combustion in an oven or other enclosure that restricts airflow. Charcoal is almost pure carbon and burns at much higher temperatures than wood and can even be used for smelting ores into metals. Charcoal production was first developed over 7,000 years ago and ushered in the Age of Metals. During the Middle Ages the forests of southern Europe were decimated partly for the production of charcoal, which was used for smelting iron ore to make cannons.

Other Products from Trees

Resins include a broad collection of compounds that are composed of polymerized terpenes mixed with volatile oils (*see* Chapter 17). Resins are insoluble in water and apparently function naturally in furnishing protection to the tree. They discourage herbivores and also make the wood resistant to some decay-causing fungi. Fossilized resins produce amber, often considered a botanical jewel (*see* A Closer Look: *Amber: A Glimpse into the Past* in Chapter 9).

Although hardwood trees, especially members of the Fabaceae, produce resins, the best known commercial resins are extracted from conifers. Resin is produced in ducts or canals that occur throughout the tree. Resin oozes out when the tree is cut; it is easy to see and smell the sticky resin in a pine tree by just breaking off a cluster of needles. Commercial extraction is performed by slashing the bark, and the crude exudate collected is also known as pitch. When pitch is heated, some volatile components evaporate easily. These condense into turpentine, which is used as a thinner for oil-based paints and also as an organic solvent. After the volatile compounds are removed from the resin, the remainder is known as rosin, a material frequently used by musicians and baseball players. The bow of a string instrument is drawn across a block of rosin to make it slightly sticky. This increases the friction between the bow and the strings, resulting in more vibration and improved tone quality. The baseball player's bag of powdered rosin improves the pitcher's grip on the ball. Also, the stickiness of Band-Aids is due, in part, to rosin.

Pine pitch has been used for waterproofing since ancient times. It was commonly used to waterproof wooden ships; as a result, pitch, resin, and turpentine have been called naval stores. During the Colonial period in U.S. history, the British obtained large quantities of naval stores from North America to support England's large fleet. The extensive pine forests of North America were one of the reasons the British opposed independence for the American colonies.

Another nonwood product from trees is **cork.** As a tree increases in diameter, the epidermis is replaced by periderm (*see* Chapter 3). The major component of this tissue is cork, or phellem, produced by the cork cambium. Cork cell walls contain suberin. At maturity the protoplasm dies and the cells become air-filled. Cork is good insulation and possibly provides protection against fire damage in intact trees. The characteristics of cork were known for several thousand years. During early Greek and Roman times, cork was used as a seal for jars and casks, and for flotation devices when crossing rivers. Cork has some remarkable properties. Among other things, it is lightweight, good insulation, chemically inert, and long lasting. Although plastics and other synthetics have replaced some of the uses of cork, no synthetic substitutes exist with all its properties.

Commercial sources of cork are from the bark of the evergreen oak, *Quercus suber,* a tree native to the western Mediterranean region, with the greatest production from Portugal and Spain. The layers of cork may be several inches thick in *Q. suber,* which is one of the few species whose bark can be stripped repeatedly without significant damage. The outer bark can first be removed when a cork oak tree is about 20 to 25 years old, and subsequent strippings can then take place every 10 years for several hundred years (fig. 18.15). A large tree can yield up to one ton of cork at a single stripping. The initial cork is coarse, but subsequent strippings result in a closer-grained product. The stripped bark is seasoned briefly, then boiled to remove tannic acid and make it pliable. After drying, the cork can be trimmed, graded, and sent to market.

Figure 18.15 Quercus suber, *the commercial source of cork.*

Wood Pulp

Wood pulp is a watery suspension of pulverized wood containing tracheids, vessels, and fibers in hardwood pulp or just tracheids in softwood pulp. In industrialized nations, approximately 50% of the harvested wood goes into wood pulp, with the vast majority of pulp used in the manufacture of paper. About one-fourth of the wood pulp is produced mechanically by grinding the wood with water, making a slurry. The mechanical process produces the greatest yield, but paper produced by such pulp is weak and yellows quickly. Newsprint, catalogs, and paper towels are manufactured by this process.

A principal goal of processing wood for pulp production is to remove as much lignin as possible from the wood. Lignin, which constitutes 25%–35% of the wood, is brown in color and continues to darken with age, making it unsuitable for quality writing paper. Approximately three-fourths of the wood pulp is produced chemically by various methods that dissolve lignin and other noncellulose components. In one method, wood chips are dissolved in sodium hydroxide, while sulfites or sulfates are employed in other processing methods. Some of the dissolved lignin may be used in the production of other compounds such as vanillin, a vanilla substitute, and dimethyl sulfoxide (DMSO—*See* A Closer Look: *DMSO and the Lignin Connection* in Chapter 2). However, most of the separated lignin causes disposal problems, as do some of the chemicals used in the process. In addition, some pulp is also bleached to produce a white paper; unfortunately the discharged chemicals from this step also add to pollution problems.

In addition to paper, wood pulp is used in the manufacture of cardboard and fiberboard, as well as rayon, cellophane, and cellulose acetate. Eucalyptus trees are often planted specifically for rayon production, since these fast-growing trees can be harvested for pulp in seven to ten years. Cellophane and cellulose acetate are made by treating dissolved cellulose with acetic acid or acetic anhydride. The resulting material can be spun into fibers or rolled into sheets. Cellulose acetate is often used in the manufacturing of molded "plastics" such as glass frames, toothbrush handles, combs, car steering wheels, and pens.

The southern Atlantic states and the Northwest are the leading pulp-producing regions of the United States, with the southern pines the most important pulp species. Western softwoods, including spruce, fir, western hemlock, and Douglas fir, supply a significant proportion of pulp as well.

Paper

Although the "information superhighway" promises worldwide computer link-up and instantaneous electronic exchanges, paper is still the major medium of written communication in contemporary society. The United States accounts for over one-third of the world's production and use of paper and cardboard. Each year about one billion trees are cut down to satisfy the demand for paper and paper products, with each American directly or indirectly using approximately 273 kg (600 lbs) of paper. That amounts to 748 grams (1.6 lbs) of paper each day. Each Sunday edition of the *New York Times* consumes about 0.61 square kilometers (150 acres) of forest. With 44 million metric tons of paper produced each year in the United States alone (19 million tons for newsprint and 25 million tons for other types of paper), written communication links millions of people. What a contrast to its humble start as symbols on clay tablets!

Early Writing Surfaces

It is generally assumed by contemporary historians that the first written language was developed by the Sumerians around 5,000 years ago. Possibly because of economic or administrative needs, Sumerians came up with the idea of writing on clay. Although early examples were crude and totally pictographic, their system of writing became a conventional phonetic system over time. Clay tablets, prisms, and cylinders were the writing surfaces used by Sumerians. Angular sticks were used to make impressions in the soft clay, which was then baked, making the writing permanent.

Egyptian hieroglyphic writing is dated to about 100 years after the earliest Sumerian records. It has been suggested that the development of Egyptian writing was stimulated by the Sumerian example. The writing surface immortalized by the ancient Egyptians was papyrus.

Papyrus, *Cyperus papyrus,* is a sedge (fig. 18.16) that grows naturally in Egypt, Ethiopia, the Jordan River Valley, and Sicily. The papyrus writing surface was originally developed about 4,500 years ago; it was made of thin slices of the plant's cellular pith that was beaten and then laid lengthwise with other layers crosswise on it. The mat was moistened with water, then pressed and dried. In the final step, the papyrus sheet was rubbed smooth with ivory or a smooth shell. The finished sheets were made into rolls, often up to 9 meters (30 ft) in length.

The Greeks appeared to have learned about papyrus around 2,500 years ago. The use of papyrus by the Greeks, and later the Romans, continued until the fourth century, when it was superseded by parchment. After that, it was still used for official and private documents until the eighth or ninth century. Today, papyrus is only used for decorative pieces (fig. 18.16) and lives on semantically as the origin of the word "paper."

(a)

(b)

Figure 18.16 *Papyrus sheets were made from strips of pith from the stem of* Cyperus papyrus *(a) and used as a writing surface by ancient cultures. Today, papyrus is used only for decorative pieces (b) like this modern painting from Egypt.*

A Closer Look 18.1

Dry Rot and Other Wood Decay Fungi

Although the components of wood, especially lignin, are known for their durability and resistance to decay, certain specialized fungi are able to decompose cellulose and lignin. Many of these produce large fruiting bodies in the form of brackets, shelves, or crusts (*see* Chapter 22) on the trunks and branches of infected trees (box fig. 18.1). Saprobic species, which are essential for the decay of dead trees, produce similar fruiting bodies. The evident fruiting body reflects the vegetative mycelium within the tree, often extending 1–3 meters (3–10 ft) above and/or below the fruiting body. The mycelium of some fungi may grow within the tree for many years before the fruiting body develops on the bark or from the roots. Until the fungus is identified, it is usually not possible to know whether the fruiting body is that of a pathogen that just killed the tree or a saprobe that is merely hastening the decay.

Fungi can selectively attack either the heartwood or the sapwood. Sapwood rot and destruction eventually kill the tree. When the heartwood is attacked the result is heart rot, which seriously weakens the tree and makes it susceptible to being blown over. The timber value of the tree is also greatly reduced. Two types of heart rot are known: brown rot and white rot. Brown rot is caused by fungi that decompose the cellulose and leave the brownish lignin relatively untouched. White rot is caused by fungi that degrade

Box Figure 18.1 Ganoderma applanatum, *a widely distributed wood-rotting fungus.*

By the beginning of the first century, Roman writing implements were varied and included wax-coated wooden tablets for temporary writing and school materials. Letters were scratched on the waxed surface with a metal or bone stylus. For permanent records, writing was done on papyrus with a reed cut to a fine point and dipped in ink, and also on parchment or vellum with flat brushes and reeds.

Parchment is prepared from the skins of sheep, calves, or goats, while vellum is a finer quality parchment from kids, lambs, and young calves. In preparing parchment, the animal skin is cleaned and the hair removed. Both sides of the skin are scraped and smoothed and finally rubbed with powdered pumice. Parchment has been used for about 2,200 years. It gradually replaced papyrus and was itself replaced by paper after the development of printing. Today, parchment is still used for formal honorary documents and some diplomas.

The Art of Papermaking

Paper is prepared from pulp, a slurry of plant cells that are separated and dispersed in a watery suspension. While many plant materials can supply pulp, including straw, leaves, or rags, today most paper is prepared from wood pulp. The cells are matted into a thin layer that may be filled with clay or talc for added body, coated with sizing such as starch for smoothness, and then compressed. The cells must be long enough to matt well once the water is drained off. Tracheids, vessels, and fibers are the usual cells involved in the process; however, in the papermaking vernacular, these are all referred to as fibers. True papermaking can be traced back to China, early in the second century, where paper was made using a process similar to contemporary production. In the New World, the Mayans and Aztecs independently invented paper using fibers from native plants.

both the cellulose and the lignin, leaving the wood a whitish color. A few white rot species are even able to selectively degrade the lignin with little or no loss of cellulose.

There are at least 1,600 species of wood-rotting fungi responsible for the destruction of valuable timber and shade trees, as well as fallen and rotting trees. It is estimated that the damage caused by wood-rotting fungi amounts to $9 billion per year. In North America approximately 10% of the annual timber harvest is destroyed by wood-rotting fungi that cannot discriminate between rotting logs, valued timber, or stored lumber.

One economically important wood-rotting fungus is *Serpula lacrymans* (*Merulius lacrymans*), which is responsible for dry rot of timber. The name *dry rot* refers to the dry crumbly appearance of decayed wood, but it is somewhat misleading since the fungus needs moisture to become established. In fact, dry rot only occurs in wood exposed to moisture. The species breaks down cellulose, leaving the brown, crumbling wood with many cracks. Ships and older homes are especially susceptible to this destructive fungus, which often starts in poorly ventilated areas and spreads. The fungus is difficult to eradicate once it becomes established. When a timber or support beam is infected, it must be replaced, along with adjacent beams.

Dry rot was the blight of the British navy for hundreds of years. In 1684, Samuel Pepys, who was Secretary to the Admiralty Board, reported on the poor ventilation and dampness as well as the "toadstools" growing in the ships. Although, fungi were a well-recognized symptom of decay, at the time no one realized they were actually responsible for the decay. Inadequately seasoned timber and lack of ventilation created ideal conditions for fungal growth. British ships often rotted at anchor and the designation "Rotten Row" was used for the warships at the docks. During the American Revolution, so many British ships were damaged by fungi that it has been suggested the outcome of the war may be partially attributed to dry rot. This problem continued until the ironclad warships were introduced in the 1860s.

The presence of resins or tannins in the heartwood makes some trees more resistant to fungal decay than others. This is especially true for tannins, which sometimes may be present in amounts as high as 30% of certain woods. Redwood, cedar, bald cypress, and Douglas fir are the most resistant gymnosperms; black locust, black walnut, catalpa, chestnut, and red mulberry are angiosperms with this feature. Wood from these species are, therefore, useful for outdoor furniture, wooden decks, fences, and other articles exposed to the elements. Today, treatment of nonresistant wood with preservatives like creosote also prevents attack by wood-rotting fungi. Creosote is obtained from the distillation of coal and blast-furnace tar and is used for railroad ties, telephone poles, pilings, and other wood products.

The biotechnical potential of wood-rotting fungi has recently been highlighted. Testing is being done on the efficacy of pretreating wood chips with fungi that selectively degrade the lignin for pulp and paper production. Results indicate that the process improved paper strength while requiring 68% less energy than standard pulping techniques. Also, the ability of these fungi to bleach wood is being considered as a substitute for the pollution-causing chemicals now used to bleach pulp for paper production. As knowledge of these lignin-degrading fungi increases, it is likely that other industrial applications will develop.

Early Chinese scholars had traditionally written on strips of wood with a stylus. Later, woven silk and other cloth was used as a writing surface. According to tradition, paper was first made in the year A.D. 105 by Ts'ai Lun (A.D. 50–118), a eunuch attached to the Eastern Han court of the Chinese Emperor Ho Ti (A.D. 89–105). The material used by Ts'ai Lun was the inner bark of the paper mulberry tree, *Broussonetia papyrifera.* The bark was ground to a pulp with water and sieved through a mold made of bamboo strips with a cloth mesh. The art of papermaking expanded to include other materials such as bamboo, hemp, and rice straw, in addition to the paper mulberry bark. This produced papers of different quality and different textures.

For about 500 years, papermaking remained a Chinese property. It was introduced to Japan in 610, to Central Asia in 750, and to the Near East and Egypt around 800. The Moors introduced the use of paper to Europe, and the first European paper was made in Spain around 1150. Linen and cotton rags were the principal source of material for paper. The craft spread slowly through Europe over the next few centuries. The introduction of moveable type in the fifteenth century provided a major stimulus for the papermaking trade.

The increased use of paper in the seventeenth and eighteenth centuries resulted in shortages of linen rags and stimulated the search for cheaper substitutes. The first machinery to replace the hand-molding process was developed by the French inventor Nicholas Robert in 1798. Henry and Sealy Fourdrinier, British papermakers, improved on Robert's design in 1803 and papermaking machinery still bears the Fourdrinier name (fig. 18.17). (The Fourdrinier screen is a continuous belt of wire cloth onto which pulp is

Figure 18.17 *The Fourdrinier paper machine. Pulp is deposited onto a moving wire screen, shown here. The water drains through the screen, leaving a matt of fibers that comprise a sheet of paper.*

deposited.) The use of wood pulp was introduced around 1840, ending the search for an inexpensive and abundant raw material. The search for alternate sources of pulp is once again active because of the threats of deforestation.

Alternatives to Wood Pulp

Today, wood pulp is the major source of the world's paper supply even though many other sources of pulp exist. For hundreds of years, cotton and linen have been used to produce writing paper. Fine quality stationery and paper for permanent records still contain a large percentage of rags. Recently, Levi Strauss and Company started recycling denim scraps to produce paper for company stationery and are considering mass marketing this recycled product. Rice straw, a by-product of rice cultivation, is another possible source of pulp, as is bamboo, bagasse from sugarcane (*see* Chapter 4) and even hemp fiber.

One of the most promising alternatives is kenaf, *Hibiscus cannabinus,* a herbaceous plant in the Malvaceae, the mallow, family. It grows from seed to mature size, which is approximately 4 meters (12 ft) tall in about four to five months. By comparison, southern pine, a main source of wood pulp, takes seven to 15 years to reach harvesting size. Moreover, fiber yield from an acre of kenaf is three to five times the yield for an acre of pine. Kenaf is also relatively disease resistant and drought tolerant. The U.S. Department of Agriculture has suggested that kenaf is the most viable fiber plant for U.S. paper production (*see* Chapter 25).

Another source of pulp is recycled paper. Many types of paper can be soaked in water to release the component pulp fibers, which are reused to make new paper. Paper recycling can reduce deforestation and pollution, yet only 40% of the paper used in the United States is recycled. Americans use more and recycle less paper than any other industrialized country. If Americans recycled all the Sunday newspapers it could save over 500,000 trees each week or 26 million per year. The widespread development of one or more of these wood pulp alternatives is essential to reduce deforestation and ensure the availability of wood and wood products for future generations.

Bamboo

It has been said that "bamboo is all things to some men, and some things to all men." Over 1,000 applications for this treelike grass have been described; no other plant has so many uses. Large stems are used as posts and rafters in houses, and split sections form the side walls. Long sections of stem are ideal for water or drain pipes, while short sections are useful for containers or for musical instruments. Split bamboo is woven into hundreds of products such as baskets, screens, mats, fans, or carpets. Young shoots of small bamboo varieties are cooked, pickled, or even preserved in sugar. In some Asian countries, the farmer lives with bamboo from cradle to grave.

There are about 1,000 species of bamboo found in about 50 genera of the Poaceae, the grass family (fig. 18.18). Species occur in North and South America, Africa, Australia, and southern Asia. The greatest diversity occurs in China, with about 300 species in 26 genera; however, India has the largest reserves of bamboo, approximately 25 million acres. Although the species vary greatly in size, shape, and color, they all have a woody stalk or culm produced from an underground rhizome. In most species new stalks are produced every year. Generally the culms have a hollow pith that is solid at the nodes. The strong, lightweight stalks are the reason that bamboo is so useful. Unlike trees, there is no secondary growth. Bamboo does not increase in diameter after its initial growth. A bamboo culm will reach full size in 6 –12 weeks but cannot be cut until it is 3–5 years old. New culms contain high percentages of water and will shrink and crack if cut and dried.

Species range in size from small plants (just a few inches high) to giant grasses, 40 meters (120 ft) in height and 30 cm (12 in) in diameter. One striking feature of bamboo is the incredible rate of growth of new stalks; however, the most astounding feature relates to its flowering. Most species of bamboo only flower at long intervals of 30, 60, or even 120 years. All the plants of one species, wherever they are located, will come into flower around the same time. Specimens from a single culture have been taken to different

Figure 18.18 *Bamboo.*

parts of the world, such as the Royal Botanical Gardens at Kew (in England) and rain forests in Jamaica. The specimens continue to flower in synchrony with each other, suggesting an internal control mechanism. A Japanese scientist traced the flowering of black bamboo in historical records for more than 1,000 years. The first record was in the year A.D. 813; other documents indicated flowering every 120 years. After flowering, the aboveground parts of bamboo die back; it often takes five years or more for groves to become reestablished from either rhizomes or newly shed seeds.

Within China, the most useful bamboo is a large cane called *mao chu* or hairy bamboo (*Phyllostachys pubescens*), which accounts for two-thirds of the country's bamboo production. It is used to make furniture and even reinforcement rods in heavy construction. Tonkin, or tea stick, bamboo (*Arundinaria amabilis*) is highly prized and China exports 5,000 tons of it each year. It is widely used, especially in Europe, for garden stakes for fruits and vegetables, ski poles, fishing rods, and furniture.

Bamboo has been intricately linked with the giant panda. High in the mountains of the Sichuan Province of southwestern China, a native species of bamboo furnishes food for pandas; they consume both culms and leaves. After mass flowering and the resulting death of the culms, pandas often migrate to lower elevations, seeking other species of bamboo. Because of human pressures and widespread cultivation, alternate sources of food may be scarce. At these lower elevations, poaching becomes an additional threat to the survival of the panda.

Chinese bridges have traditionally been made from bamboo. Cables made of split bamboo strips woven or twisted together have incredible strength and have been used on hanging bridges for centuries. In fact, the bridge over the Min River in China hangs from bamboo cables that are about seven inches in diameter. This bridge, considered one of the engineering marvels of the world, has been in use for over 1,000 years.

Bamboo has a long history of use in papermaking, and today is the source of pulp for approximately two-thirds of all paper made in India. The products range from kraft paper (heavy brown) to fine writing stock. Although the yield from an acre of bamboo is not as great as pine, it is harvestable in three or four years, whereas a pine tree might take 15 years before it can be harvested. Expansion of this application of bamboo may be another viable alternative to wood pulp for a paper-hungry society.

Chapter Summary

1. In addition to food and drink, plants furnish society with the raw materials to provide shelter, clothing, and paper. Plant fibers have been extracted and treated to weave cloth and manufacture rope for millenia. Cotton fiber is actually elongated seed hairs and the techniques to render cotton fibers into cloth were discovered independently in both the Old and New Worlds. *Gossypium hirsutum* and *G. barbadense,* cotton species native to the New World, make the finest cotton cloth. These species are widely cultivated and have largely replaced Old World cultivars. The development of Eli Whitney's cotton gin brought about the mechanization of cotton production and with it an increased demand for slave labor, setting the stage for the American Civil War in the 1860s.
2. Linen, made from bast fibers of flax (*Linum usitatissimum*) was the cloth of choice for many ancient civilizations. Retting uses microbial action to free the flax fibers from the stem.
3. Other bast fibers that have been used to make cloth are ramie, jute (burlap), and hemp. Manila hemp, piña, and sisal are leaf fibers that have been used to make cloth, as well as a variety of other products such as rope and mats. Kapok is a surface fiber too fine to be spun but useful as padding and stuffing; coir is a seed fiber from the coconut valued for ropes and mats. Rayon was the first of many artificial fibers; it is made from extracting

cellulose from wood fibers and molding the cellulose into yarn.

4. Wood and wood products supply construction materials and fuel to much of the world population. The strength of wood, its insulating properties, its versatility, and its natural beauty sustain the demand for lumber. Yet forest land is being lost at a rate that jeopardizes the future availability of wood. Pine and Douglas fir are the leading softwoods produced in the United States. The greatest use of softwood lumber is for home construction. Oaks are the most economically important hardwoods in the United States with white oak the most prized oak species. The wood is heavy, durable, and attractive. It is widely used in furniture, cabinets, and flooring. Black walnut, hickories, maples, sweetgum, and birches are other important hardwoods. Each year approximately 50% of the world's wood harvest is used as fuel, especially in the developing nations. Over 2.5 billion people depend on wood or charcoal for the majority of their energy needs for heating and cooking. Other products from trees include resins and cork.
5. In industrialized nations, approximately 50% of the harvested wood goes into wood pulp, with the majority of pulp used in the manufacture of paper. Processing pulp removes the lignin, which is necessary to produce a quality writing paper. This has resulted in some environmental problems disposing of the residue, as well as processing chemicals. It has been suggested that wood-rotting fungi may be able to substitute for some of the pollution-causing chemicals. In the United States each year about one billion trees are cut down to satisfy the demand for paper and paper products. Although most paper today is prepared from wood pulp, this source of pulp has only been used since the mid-nineteenth century. Many plant materials can supply pulp including straw, leaves, or rags. Overuse of forest resources is renewing the search for alternative sources of pulp.
6. Bamboo is a treelike member of the grass family with a long history of use in construction, furniture, baskets, screens, carpets, paper, and food. The versatility of bamboo is due to the strong, lightweight stalks that have an astounding rate of growth. Bamboo species range from small plants to giant grasses and are found throughout the world. The greatest diversity of species occurs in China, producing about 300 species.

Review Questions

1. What are the various types of fibers in industry?
2. Describe the processing of cotton. How was Eli Whitney's invention of the cotton gin connected to the American Civil War?
3. List and briefly describe the steps in the processing of flax.
4. In what products would the fibers of sisal, agave, jute, hemp, piña, coir, or Manila hemp most likely be found?
5. What is pulp? What are possible sources of pulp? How is it used?
6. Why is bamboo considered a multipurpose grass?
7. Distinguish between the terms *hardwood* and *softwood,* and the terms *heartwood* and *sapwood.*
8. What are the major uses of wood in the world today? Why is there concern about the future?
9. Distinguish between papyrus sheets and true paper. How is each prepared?

Further Reading

Findlay, W. P. K. 1982. *Fungi: Folklore, Fiction, and Fact.* Mad River Press, Eureka, CA.

Gillis, Anna Maria. 1990. The New Forestry. *BioScience,* 40 (8):558–562.

Hall, F. Keith. 1974. Wood Pulp. *Scientific American,* 230 (4): 52–62.

Heinrich, Linda. 1992. *The Magic of Linen.* Orca Book Publishers, Victoria, British Columbia.

Janick, Jules, Robert Scherry, Frank Woods, and Vernon Ruttan. 1981. *Plant Science: An Introduction to World Crops,* 3rd Edition, W. H. Freeman and Co., San Francisco, CA.

Joseph, Marjory L. 1988. *Essentials of Textiles.* Holt, Rinehart and Winston, Inc., Fort Worth, TX.

Lewington, Anna. 1990. *Plants for People.* Oxford University Press, New York.

Marden, Louis, 1980. Bamboo, The Giant Grass. *National Geographic,* 158 (4): 502–529.

Margolis, Mac. November, 1991. Tomorrow's Trees. *World Monitor,* 34–40.

Northington, David and Edward Schneider. 1996 (in press). *The Botanical World.* Times Mirror/Wm. C. Brown Pub., Dubuque, IA.

Sandved, Kjell. 1993. *Bark: The Formation, Characteristics, and Uses of Bark Around the World.* Timber Press, Portland, OR.

Stevens, Jane. 1995. Bamboo is Back. *International Wildlife,* 25 (1): 38– 45.

Thompson, Jon. 1994. Cotton, King of Fibers. *National Geographic,* 185 (6): 60 –87.

Wilson, Kax. 1979. *A History of Textiles.* Westview Press, Boulder, CO.

UNIT

V

Plants and Human Health

Plants such as the purple foxglove (*Digitalis purpurea*) have long been valued for their healing properties.

CHAPTER

19

Medicinal Plants

Chapter Outline

Chapter Concepts

1. The connection between botany and medicine is an old established one; even today, 25% of prescription drugs are of plant origin.
2. Alkaloids and glycosides have been identified as the therapeutically important components of most medicinal plants.
3. Investigating folk remedies has led to important medical discoveries that have saved millions of lives.

No one knows where or when plants first began to be used in the treatment of disease, but the connection between plants and health has existed for thousands of years. Evidence of this early association has been found in the grave of a Neanderthal man buried 60,000 years ago. Pollen analysis indicated that the numerous plants buried with the corpse were all of medicinal value. An accidental discovery of some new plant food or juice that eased pain or relieved fever might have been the beginning of folk knowledge, which was passed down for generations and eventually became the foundation of medicine.

History of Plants in Medicine

The earliest known medical document is a 4,000-year-old Sumerian clay tablet that recorded plant remedies for various illnesses. By the time of the ancient Egyptian civilization, a great wealth of information already existed on medicinal plants. Among the many remedies prescribed were mandrake for pain relief, and garlic for the treatment of heart and circulatory disorders (*see* Chapter 17). This information, along with hundreds of other remedies, was preserved in the Ebers papyrus about 3,500 years ago.

Ancient China is also a source of information about the early medicinal uses of plants. The Pun-tsao, a pharmacopoeia that was actually published around 1600, contained thousands of herbal cures that are attributed to the works of Shen-nung, China's legendary emperor who lived over 4,500 years ago. In India, herbal medicine dates back several thousand years to the Rig-Veda, the collection of Hindu sacred verses. This has led to a system of health care known as Ayurvedic medicine. One useful plant from this body of knowledge is snakeroot, *Rauwolfia serpentina,* used for centuries for its sedative effects. Today the active components in snakeroot are widely used in Western medicine to treat high blood pressure. In all parts of the world, indigenous peoples discovered and developed medicinal uses of native plants, but it is from the herbal medicine of ancient Greece that the foundations of Western medicine were established.

(a)

(b)

Figure 19.1 *A wealth of knowledge about medicinal plants existed among ancient civilizations. Ancient Cyrenian coins (a) and (b) illustrating the silphium plant, which was widely used by women as an effective contraceptive but collected to extinction by the third or fourth century.*

Early Greeks and Romans

Western medicine can be traced back to the Greek physician Hippocrates (460–377 B.C.), known as the Father of Medicine. Hippocrates believed that disease had natural causes and used various herbal remedies in his treatments. Early Roman writings also influenced the development of Western medicine, especially the works of Dioscorides (1st century A.D.). Although Greek by birth, Dioscorides was a Roman military physician whose travels with the army brought him in contact with many useful plants. He compiled this information in *De Materia Medica,* which contained an account of over 600 species of plants with medicinal value. It included descriptions and illustrations of the plants, along with directions on the preparation, uses, and side effects of the drugs.

Many of the herbal remedies used by the Greeks and Romans were effective treatments that have become incorporated into modern medicine. For example, willow bark tea, the precursor to aspirin, was used to treat gout and other ailments. Unfortunately, some treatments from ancient times have been lost. Recently historians have discovered the use of silphium as an effective contraceptive by Greek and Roman women for over 1,000 years. This plant, a member of the genus *Ferula* and known as giant fennel, was the contraceptive of choice until the third or fourth century, when the plant was collected to extinction (fig. 19.1).

Dioscorides' work remained the standard medical reference in most of Europe for the next 1,500 years, since little new knowledge was added during the Middle Ages. Although medical botany was nearly at a standstill in Europe, progress was being made in the Islamic world. Most notable is the early eleventh century Persian, Avicenna, who wrote a *Canon of Medicine* that included new information on herbal medicine.

Age of Herbals

The beginning of the Renaissance in the early fifteenth century saw a renewal of all types of intellectual activity. Botanically, this was expressed in the revival of herbalism, the identification of medicinally useful plants. This, coupled with the invention of the printing press in 1440, ushered in the Age of Herbals (*see* Chapter 8). Some of the most richly illustrated herbals were those written by the four German "fathers of botany" in the sixteenth century: Otto Brunfels, Jerome Bock, Leonhart Fuchs, and Valerius Cordus. During this same period, the Englishman John Gerard published his famous work *The Herball or Generall Historie of Plantes* in 1597. Other English herbals published in the mid-seventeenth century were John Parkinson's *Theatrum Botanicum,* considered one of the best, and Nicholas Culpepper's *The Complete Herbal,* one of the most popular herbals of the day. All the herbals focused on the medicinal uses of plants but also included much misinformation and superstition (fig. 19.2). For example, Culpepper's herbal had a strong astrological influence and also revived the **Doctrine of Signatures.**

The most famous advocate of the Doctrine of Signatures was the early sixteenth century Swiss herbalist, Paracelsus. He believed that the medicinal use of plants could easily be ascertained by recognizing distinct "signatures" visible on the plant that corresponded to human anatomy. For example, the red juice of bloodwort should be used to treat blood disorders, and the lobed appearance of liverworts suggests their use in treating liver complaints. The belief behind the Doctrine of Signatures has been developed independently among many different cultures; however, there is no scientific basis to this concept (fig. 19.2).

During the eighteenth century, as scientific knowledge progressed, a dichotomy in medicine developed between practitioners of herbal medicine and regular physicians. At about this same time a similar split occurred between herbalism and scientific botany, the study of plants above and beyond their medical applications.

Modern Prescription Drugs

Although herbalism waned in the eighteenth and nineteenth centuries, many of the remedies employed by the herbalists provided effective treatment. Some of these became useful prescriptions as physicians began experiment-

(a)

(b)

Figure 19.2 *(a) Herbals from the Middle Ages contained richly illustrated descriptions of plants and often advanced the Doctrine of Signatures. (b) The "humanoid" form of this mandrake root suggested its use in promoting male virility and ensuring conception.*

ing with therapeutic agents. William Withering was the first in the medical field to scientifically investigate a folk remedy. His studies (1775–1785) of foxglove as a treatment for dropsy (congestive heart failure) set the standard for pharmaceutical chemistry.

In the nineteenth century, scientists began purifying the active extracts from medicinal plants. One breakthrough in pharmaceutical chemistry came when Friedrich Serturner isolated morphine from the opium poppy in 1806. Continuing this progress, Justus von Liebeg, a German scientist, became a leader in pioneering the field of pharmacology. With increased knowledge of the active chemical ingredients, the first purely synthetic drugs based on natural products were formulated in the middle of the nineteenth century. In 1839, salicylic acid was identified as the active ingredient in a number of plants known for their pain-relieving qualities and was first synthesized in 1853. This led to the development of aspirin, which is the most widely used synthetic drug today.

Although the direct use of plant extracts continued to decrease in the late nineteenth and early twentieth centuries, medicinal plants still contribute significantly to prescription drugs. It is estimated that 25% of prescriptions written in the United States contain plant-derived active ingredients (close to 50% if fungal products are included). An even greater percentage are based on semisynthetic or wholly synthetic ingredients originally isolated from plants. Today there is a renewed interest in investigating plants for medically useful compounds, with some of the leading pharmaceutical and research institutions involved in this search.

Herbal Medicine Today

While Western medicine strayed away from herbalism, 75%–90% of the rural population in the rest of the world still relies on herbal medicine as their only health care. The long tradition of herbal medicine continues to the present day in China, India, and many countries in Africa and South America. In many village marketplaces, medicinal herbs are sold alongside vegetables and other wares. Practitioners of herbal medicine often undergo a rigorous and extended training to learn the names, uses, and preparation of native plants.

The People's Republic of China is the leading country for incorporating traditional herbal medicine into a modern health care system. The resulting blend of herbal medicine, acupuncture, and Western medicine is China's unique answer to the health care needs of over one billion people. Plantations exist for the cultivation of medicinal plants and the training of doctors; active research programs also investigate potentially useful specimens. Thousands of species of medicinal herbs, some collected from all over the world, are thus available for the Chinese herbalist. Chinese apothecaries contain a dazzling array of dried plant specimens, and prescriptions are filled, not with prepackaged pills or ointments, but with measured amounts of specific herbs. These colorful apothecaries can also be found in U.S. cities such as San Francisco, with its large Chinese population (fig. 19.3).

Figure 19.3 *Chinese apothecary showing an amazing variety of dried herbs for medicinal use.*

While China melded traditional practices with Western medicine, in India traditional systems have remained quite separate from Western medicine. At Indian universities, medical students are trained in Western medicine; however, much of the populace puts its belief in the traditional systems. In addition to Ayurvedic medicine, which has a Hindu origin, Unani medicine with its Muslim and Greek roots is another widely practiced herbal tradition. Economics is also a factor in the reliance on indigenous cures, since the cost of manufactured pharmaceuticals is beyond reach for most of the population.

The renewed interest in medicinal plants has focused on herbal cures among indigenous populations around the world (fig. 19.4). This is especially true among indigenous peoples in the tropical rain forests, which have been studied by some of the world's leading ethnobotanists including Richard Schultes, Mark Plotkin, and Walter Lewis. They are spending time with local tribes to learn their medical lore. Hopefully, these investigations will add new medicinal plants to the world's pharmacopoeia before they are lost forever.

Tropical rain forests are of special concern since the widespread destruction of these ecosystems threatens to eliminate thousands of species that have never been scientifically investigated for medical potential. In addition to the destruction of the forests, the erosion of tribal cultures is also a threat to herbal practices. As younger members are drawn away from tribal life-styles, oral traditions are not passed on. Mark Plotkin has compared this loss of knowledge to the burning down of a library containing books that are one of a kind and irreplaceable. One step to preserving this knowledge was recently taken by the government of

Figure 19.4 *Statue representing contribution of ethnobotany to medicine at the Missouri Botanical Garden in St. Louis, Missouri.*

Belize, which established the Terra Nova Forest Reserve. This Central American reserve is a 6,000-acre sanctuary dedicated to the survival of medicinal plants and the traditional healers that use them.

Active Principles in Plants

The medicinal value of plants is directly connected to the vast array of chemical compounds manufactured by their various biochemical pathways. These compounds are considered **secondary plant products,** since they are not directly related to the plant's survival. Although scientists formerly believed that many of these compounds were simply waste products of metabolism, it is now believed that they do, in fact, have important functions. Some secondary products discourage herbivores, while others inhibit bacterial or fungal pathogens. One way of classifying medicinal plants is by the chemical makeup of the active principle. Two major categories of these compounds are **alkaloids** and **glycosides.**

Alkaloids

Alkaloids are a diverse group of compounds of which over 3,000 have been identified in 4,000 species of plants. Although widely distributed throughout the plant kingdom, most alkaloids occur in herbaceous dicots and also in the fungi. Three higher plant families particularly known for the occurrence of alkaloids are the Fabaceae (legume family), the Solanaceae (nightshade family), and the Rubiaceae (coffee family).

Although they vary greatly in chemical structure, alkaloids share several characteristics: they contain nitrogen, they are usually alkaline (basic), and they have a bitter taste. They affect the physiology of animals in several ways but their most pronounced actions are on the nervous system, where they can produce physiological and/or psychological results. Although some alkaloids are medicinally important, others are hallucinogenic or poisonous. It should also be noted that the difference between a medicinal or a toxic effect of many alkaloids (or any drug) is often the dosage. Common alkaloids include caffeine, nicotine, cocaine, morphine, quinine, and ephedrine. (Note that the names of most alkaloids end in *-ine.*) Although there are legitimate medical applications for morphine and cocaine, these alkaloids are discussed in Chapter 20.

Glycosides

Glycosides are widespread in the plant kingdom and are second in importance as medicines or toxins. They are so named because a sugar molecule (*glyco-*) is attached to the active component. The active portion of the molecule varies greatly, but the sugar is generally glucose. Glycosides are generally categorized by the nature of the nonsugar or active component, with the following types most common: **cyanogenic glycosides, cardioactive glycosides,** and **saponins.**

Cyanogenic glycosides release cyanide (HCN) upon breakdown. Recall that cassava contains cyanogenic glycosides, which must be removed before consumption. Also, the seeds, pits, and bark of many members of the rose family (apples, pears, almonds, apricots, cherries, peaches, and plums) contain amygdalin, the most abundant cyanogenic glycoside. The pits of apricots are a particularly rich source of amygdalin and are ground up in the preparation of **laetrile,** a controversial cancer treatment. Theoretically, laetrile releases HCN only in the presence of tumor cells and thus selectively destroys them. This has not been substantiated; therefore, laetrile has not been approved for cancer therapy in the United States.

Both cardioactive glycosides and saponins contain a **steroid** molecule as the active component. Cardioactive glycosides have their effect on the contraction of heart muscle and, in proper doses, some can be used to treat various forms of heart failure. The best known cardioactive glycoside used medicinally is digitalis, which will be discussed next. On the other hand, some of the deadliest plants, such as milkweed and oleander, contain toxic levels of cardioactive glycosides. Saponins are less useful medically but can be highly toxic, causing severe gastric irritation and hemolytic anemia. One useful saponin is diosgenin from yams (*Dioscorea* spp.), which can be used as a precursor for the synthesis of various hormones such as progesterone (one of the female sex hormones and an ingredient in birth control pills) and cortisone (*see* Chapter 14).

Medicinal Plants

Although hundreds of plant extracts are still an important part of the pharmaceutical industry, what follows is a limited selection from the many medicinal products that have had an impact on society either in the past or present. Additional compounds derived from plants are listed in Table 19.1.

TABLE 19.1
Plants of Known Medicinal Value

Scientific Name	Common Name	Family	Active Principle	Medicinal Use
Aloe vera	Burn plant	Liliaceae	Aloin	Skin injuries
Atropa belladonna	Belladonna	Solanaceae	Atropine	Diverse uses including relaxing muscles, dilating pupils, asthma
Cannabis sativa	Marijuana	Cannabaceae	Tetrahydro-cannabinol	Glaucoma; antinausea during chemotherapy
Catharanthus roseus	Madagascar periwinkle	Apocynaceae	Vinblastine; vincristine	Leukemia and lymphoma
Cephaelis ipecacuanha	Ipecac	Rubiaceae	Cephaeline emetine	Emetic for poisoning
Chondrodendron tomentosum	Pareira	Menispermiaceae	Tubocurarine	Muscle relaxant during surgery, cerebral palsy, and tetanus
Cinchona spp.	Feverbark tree	Rubiaceae	Quinine	Malaria
Colchicum autumnale	Autumn crocus	Liliaceae	Colchicine	Gout
Digitalis purpurea	Purple foxglove	Scrophulariaceae	Digitoxin	Congestive heart failure
Dioscorea spp.	Yam	Dioscoreaceae	Diosgenin	Steroid drugs for contraception and inflammation
Ephedra spp.	Ephedra	Ephedraceae	Ephedrine	Bronchial asthma; bronchitis
Erythoxyolon coca	Coca	Erythroxylaceae	Cocaine	Local anesthetic for eye surgery
Eucalyptus globulus	Eucalyptus	Myrtaceae	Eucalyptol	Cough suppressant
Hydnocarpus spp.	Chaulmoogra	Flacourtiaceae	Hydnocarpic acid	Leprosy
Papaver somniferum	Opium poppy	Papaveraceae	Morphine; codeine	Severe pain; cough suppression
Pilocarpus pennatifolius	Jaborandi	Rutaceae	Pilocarpine	Glaucoma
Podophyllum peltatum	Mayapple	Berberidaceae	Podophyllin; podophyllotoxin	Warts; cancer
Rauwolfia serpentina	Snakeroot	Apocynaceae	Reserpine	Hypertension
Salix alba	White willow	Salicaceae	Salicin	Pain; fever; inflammation
Sanguinaria canadensis	Bloodroot	Papaveraceae	Sanguinarine	Antiplaque mouthwash
Taxus brevifolia	Pacific yew	Taxaceae	Taxol	Ovarian and breast cancer

Foxglove and the Control of Heart Disease

Today in the United States, several million heart patients rely on digitalis as the primary treatment for their condition. Most are unaware of the plant source or the efforts of an eighteenth century English physician who was the first to scientifically investigate a folk remedy. William Withering was an English country doctor who had an extensive knowledge of the local flora and, in fact, wrote a book on the plants of the British Isles. Although he was aware of the flowering plant known as foxglove (*Digitalis purpurea*), his interest in it as a medicinal herb came about when he learned of a folk remedy for dropsy. As Withering later stated,

> "In the year 1775 my opinion was asked concerning a family receipt for the cure of dropsy. I was told that it had long been kept a secret by an old woman in Shropshire who had sometimes made cures after the more regular practitioners had failed."

Figure 19.5 *Foxglove,* Digitalis purpurea.

Dropsy is a condition characterized by severe bloating due to fluid accumulation in the lungs, abdomen, and extremities. Today it is known that the fluid retention is due to congestive heart failure, a failure of the heart to pump sufficiently. Although the folk remedy had 20 herbal ingredients, Withering was aware that foxglove leaves were the active component. He began treating dropsy patients with digitalis tea prepared from ground leaves and described one such patient

> "nearly in a state of suffocation, her pulse extremely weak and irregular, her breath very short and laborious, her countenance sunk, her arms of a leaden colour, clammy and cold. Her stomach, legs, and thighs were greatly swollen; her urine very small in quantity, not more than a spoonful at a time, and that very seldom."

Within a week she no longer showed any symptoms. For the next ten years he conducted careful experiments with powdered digitalis leaf to determine dosage, preparation, and effectiveness in treating dropsy. This culminated in the publication of *An Account of the Foxglove and Some of Its Medical Uses: With Practical Remarks on Dropsy and Other Diseases* in 1785. He believed that foxglove primarily acted as a diuretic in relieving dropsy. Although Withering did note that digitalis "has a power over the motion of the heart," he never positively made the connection between dropsy and heart failure.

Digitalis purpurea, the purple foxglove, is an attractive biennial in the snapdragon family (Scrophulariaceae) with a spike of large, purple, bell-shaped flowers (fig. 19.5). As with most biennials, it overwinters as a basal rosette of leaves, which gives rise during the next growing season to the flowering stalk that may be a meter (3 ft) in height. Because of the attractive flowers, foxglove is often used as a garden ornamental. The leaves contain over 30 glycosides with digoxin and digitoxin the most medically significant. In addition to *D. purpurea,* other *Digitalis* species, particularly *D. lanata,* also produce these cardioactive glycosides. The concentration of glycosides in the leaves is highest before flowering; after the leaves are dried and powdered, the glycosides can be extracted.

The cardioactive glycosides in digitalis slow the heart rate, while increasing the strength of each heartbeat so that more blood is pumped with each contraction. The resulting improvement in circulation decreases edema in the lungs and extremities and increases kidney output. While the digitalis glycosides are an effective treatment for congestive heart failure, they do not cure the underlying heart condition. It should also be noted that there is a fine line between a therapeutic and a toxic dose of digitalis. Even patients under a doctor's supervision can experience toxic side effects and must be carefully monitored for the correct dosage. The most dangerous toxic effect is a life-threatening arrhythmia (a rapid and irregular beating of the heart).

Aspirin: From Willow Bark to Bayer

Although aspirin is presently the most widely used synthetic drug, with Americans alone consuming 80 million pills a day, its origin is most definitely botanical. The bark of willow trees (*Salix* spp.) had long been known among many cultures as an effective treatment for reducing fever and relieving pain. The ancient Greeks used an infusion of bark from white willow (*Salix alba*) to treat gout, rheumatism, pain, and fever; this later became a well-known folk remedy throughout Europe. Many Native American tribes had independently discovered the healing powers of willow bark.

The evolution of the drug from willow bark tea into synthetic aspirin began in England in the eighteenth century. Reverend Edmund Stone experimented with willow bark tea for six years and found it beneficial for the treatment of fever and chills. In the early nineteenth century, French and German chemists sought to isolate the active compound from willow bark. In 1828, salicin was first isolated and, over the next decade, the extraction method was refined. Salicin is a glycoside of salicylic acid; salicylates occur widely in species of *Salix* (fig. 19.6) as well as meadowsweet (*Spirea ulmaria* or *Filipendula ulmaria*), poplars (*Populus* spp.), and wintergreen (*Gaultheria procumbens*).

Figure 19.6 *For hundreds of years, plants containing salicylates such as black willow,* Salix nigra, *were used in herbal remedies for the treatment of pain and fever.*

The next step was the laboratory synthesis of salicylic acid in the mid-nineteenth century by several German scientists. Now that salicylic acid was widely available, an inexpensive treatment existed for many ailments. It was used for rheumatic fever, gout, rheumatoid arthritis, and osteoarthritis. Today it is still used, but primarily applied topically to the skin for the removal of warts, corns, and various skin ailments. In 1898, while searching for a similar compound that caused less gastric distress to help his father who was afflicted with arthritis, Felix Hoffman, a chemist at Bayer Company, came across acetylsalicylic acid in the chemical literature. The new compound was more palatable and was soon given the name aspirin, a word with an interesting derivation. The "a" is from the acetylsalicylic acid and the "spirin" from *Spirea,* the plant from which salicylic acid was first isolated. (The names "salicin" and "salicylic acid," of course, reflect the *Salix* origin.)

Aspirin is valued for its three classic properties as an **antiinflammatory, antipyretic** (fever-reducing), and **analgesic** (pain-relieving), although the mode of action in the body is not completely understood. It has also found new uses in the prevention of heart attacks, strokes, and colon cancer. Aspirin may also delay the development of cataracts in the elderly and enhance the immune system in protecting the body against bacteria and viruses.

Of these new uses, probably the greatest attention has been given to the beneficial effects of an aspirin a day in the prevention of heart attacks. Various studies have shown that the administration of aspirin following a heart attack or stroke reduces the risk of a second heart attack or stroke. The subsequent death rate among the group treated with aspirin was significantly lower than the control group. More recent studies have shown that aspirin given to healthy middle-aged men reduces the incidence of a first heart attack by 44%. Studies are currently underway to see if aspirin has the same prophylactic action in women. Early results indicate that the use of one to six aspirin per week is effective in women as well.

In 1970, John R. Vane and others of the Royal College of Surgeons in London found that low doses of aspirin suppress the aggregation of blood platelets. This is a necessary step in the formation of blood clots that can block blood vessels and lead to heart attacks and strokes. (Vane later received a Nobel Prize for his work.) The mechanism behind this and other actions of aspirin is the suppression of **prostaglandins,** a group of local hormones that are widely produced throughout the body and have a number of regulatory activities. Prostaglandins are not stored in the body, but are released from injured cells or cells that have been stimulated by other hormones. Prostaglandins mediate physiological responses affecting functions from digestion to reproduction, circulation, and the immune system. An overproduction of prostaglandins can lead to headaches, fever, menstrual cramps, blood clots, inflammation, and other complaints.

Although aspirin is incredibly versatile, it is not without drawbacks. It has been known for quite some time that aspirin irritates the stomach. Scientists have found ways to coat it or buffer it; however, this is still one of its drawbacks. Again it is the suppression of prostaglandins that is responsible for the irritation. In the stomach, prostaglandins prevent the overproduction of acid and also promote the secretion of mucus that blocks self-digestion of the stomach lining.

A second drawback was discovered in the 1960s that resulted in limiting the use of aspirin for children and teenagers. Children who had taken aspirin while recovering from chicken pox or influenza occasionally developed unusual symptoms, including vomiting and change in mental alertness. This condition is known as Reye's syndrome and, although rare, can be fatal. It is unclear how aspirin is involved in this syndrome.

Malaria and the Fever Bark Tree

As the twentieth century draws to a close, malaria, known since antiquity, is still the world's most prevalent disease. Two to three million people die each year from malaria,

A Closer Look 19.1

Native American Medicine

Long before Europeans colonized the New World, Native American tribes had developed an impressive array of medicinal plants. Like other indigenous peoples, they had learned by trial and error which plants could heal or treat the diseases and injuries that afflicted them. Many tribes shared their knowledge with the settlers, with the result that some of these Indian remedies became well established herbal medicines. At the present time, much interest has been focused on identifying and saving medicinal plants from tropical rain forests; however, the wealth of medicinal plants from Native American tribes has not been fully explored. The potential of this resource is typified in the following plants.

Goldenseal (*Hydrastis canadensis*) was widely used among various woodland tribes. The ground-up yellow rhizome of this plant has been valued as a dye, an insect repellant, and an antiseptic/antibiotic wash for treating wounds, mouth sores, and eye inflammations. It was also used internally for stomach and liver ailments. Goldenseal's therapeutic action is due to two alkaloids, hydrastine and berberine. Studies have shown that hydrastine can reduce blood pressure, while berberine has antibacterial properties.

Witch hazel (*Hamamelis virginiana*) is a fall-flowering shrub or small tree native to the forests of eastern North America (box fig. 19.1a). Leaves, twigs, and bark have been used to prepare infusions to treat various aches and pains from sore muscles to sore throats. Today, a water or alcoholic extract known as *aqua hamamelis* is used as a topical astringent and antiseptic. In this form, it is probably more extensively used than any other herbal remedy from North America.

(a)

Box Figure 19.1 *Native American tribes have employed an impressive variety of plants for traditional medicines including (a) Witch hazel,* Hamamelis virginiana; *(b) Purple coneflower,* Echinacea *sp; and (c) Bloodroot,* Sanguinaria canadensis.

Purple coneflower (*Echinacea* spp.) was used medicinally by the Plains Indians as an antidote for snake bites, as well as bites and stings from other venomous animals (box fig. 19.1b). An extract of the rootstock was also applied topically for other skin ailments and, as a tea, drunk as treatment for various infectious diseases. Scientific studies have supported these claims; *Echinacea* stimulates the immune system, promotes wound healing and has mild antibiotic, antiviral, and antiinflammatory actions.

Bloodroot (*Sanguinaria canadensis*) is a spring-flowering perennial herb in the poppy family found in the eastern woodlands of North America (box fig. 19.1c). The blood-red sap from the rhizome contains several alkaloids, which account for its medicinal properties. Although toxic in large doses, a tea brewed from the rhizome was used to clear breathing passages in those suffering from coughs, laryngitis, and bronchitis. It was also used to treat rheumatism and is a powerful emetic. Perhaps the most acclaimed use has been as an anticancer agent. In the 1850s, a study by J.W. Fell showed breast cancer could be successfully treated with a procedure that included bloodroot sap. Later extracts found applications in the treatment of skin cancer. More recent uses of bloodroot have involved one of its alkaloids, sanguinarine, which is used as an antiplaque agent in mouthwashes.

(b)

(c)

and at least one million of these deaths are young children. Today malaria is largely confined to tropical and subtropical countries in Asia, Africa, and Central and South America (fig. 19.7a). In the past, however, malaria reached epidemic proportions in temperate areas of North America, Europe, and Asia as well.

Records of malaria can be found in ancient Egyptian papyri, which date back over 3,500 years, and in the work of the Greek physician Hippocrates. The Greeks also noted the higher incidence of the disease among people living near swamps or marshes. Much later, Italians believed that breathing bad air (*mal aria*) near swamps caused the disease, and their name for the disease eventually entered the English language. The correlation with the swamp was correct, but the mosquitos near the swamp, not the air, were the vectors of the disease. The details of how the mosquitos transmit malaria were discovered by Sir Ronald Ross, an English physician working in India during the late nineteenth century.

Four species of protozoans in the genus *Plasmodium* cause malaria in humans: *P. vivax, P. ovale, P. falciparum,* and *P. malariae.* The French physician Alphonse Laveran first observed the unicellular parasite in human blood in 1881. The four species each cause a slightly different form of malaria, varying in the recurring onset of fever and chills, the most characteristic symptoms of the disease. Other symptoms include anemia and an enlarged spleen due to destruction of red blood cells. *Plasmodium falciparum* is the species most often responsible for fatalities. Untreated, it can result in cerebral malaria, which is characterized by convulsions, seizures, and coma, and can lead to death.

The disease is initiated through the bite of a female *Anopheles* mosquito that has previously fed on the blood of an infected individual. The mosquito carries a stage of the parasite in its salivary glands and injects the parasite into the bloodstream with its bite. The parasite multiplies in the liver and eventually releases merozoites (the asexual stage in the life cycle of the parasite) into the bloodstream. The merozoites invade red blood cells and multiply; after a period of time the red blood cells rupture, releasing a new generation of merozoites that infect other red blood cells (fig. 19.7b). The cycle is synchronous with simultaneous rupture of red blood cells and release of merozoites and toxins, which cause the periodic fever and chills. The interval between symptoms is most commonly 48 hours. Male and female gametocytes are formed in some red blood cells. These do not develop any further in the victim, but, if they are ingested by an *Anopheles* mosquito, can complete sexual reproduction and begin the infection cycle anew.

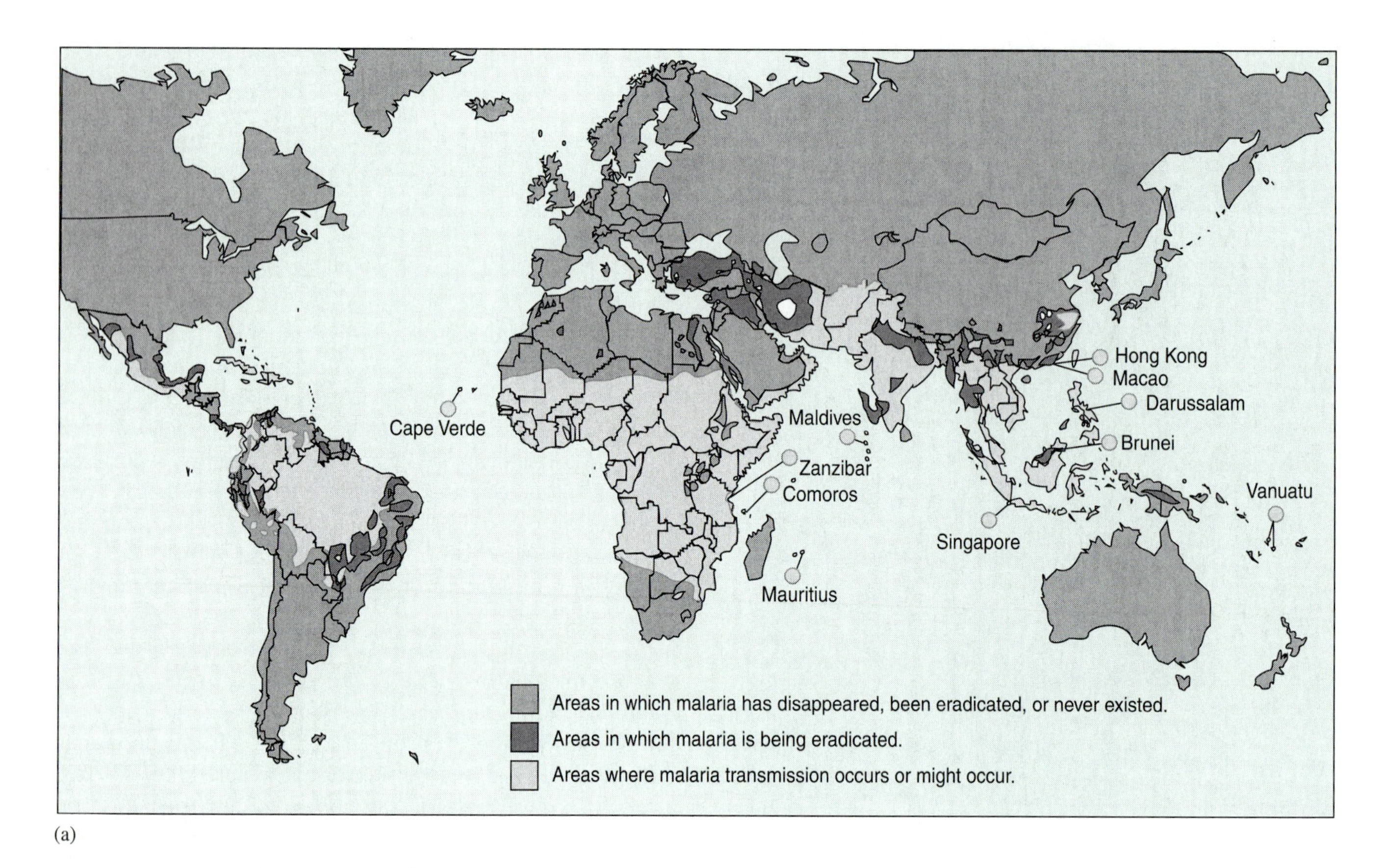

(a)

1. When a female *Anopheles* mosquito, harboring the malaria parasite in her salivary glands, bites an individual, the sporozoite stage of the parasite is injected. The sporozoites are carried to the liver in the bloodstream.

Salivary glands

2. In the liver, merozoites are produced and released into the bloodstream.

3. The merozoites invade red blood cells and multiply.

4. After a period of time the red blood cells rupture, releasing the new generation of merozoites, which then infect other red blood cells. The cycle is synchronous with simultaneous rupture of red blood cells and release of merozoites and toxins, causing the periodic fever and chills.

5. Male and female gametocytes are formed in some red blood cells. If the gametocytes are ingested when an *Anopheles* mosquito bites an infected individual, they can complete sexual reproduction and begin the infection cycle anew.

(b)

Figure 19.7 *Malaria is usually considered the world's most prevalent disease. (a) Currently, malaria is endemic in tropical and subtropical countries. (b) Life cycle of* Plasmodium vivax, *a protozoan responsible for one form of malaria.*

Source: *World Health Statistics Quarterly*, 41:69 (1988), World Health Organization.

Until the application of the fever bark tree, there was no effective treatment for malaria. The fever bark tree, called *quina-quina* (meaning bark or barks) by the Incas, is native to the eastern slopes of the Andes Mountains. This small evergreen tree with broad shiny leaves belongs to the Rubiaceae, the coffee family. The fever-reducing powers of the tree were well known to the Incas of Peru and they shared this knowledge with Jesuit missionaries. The Jesuits later used infusions from the bark to treat people infected with malaria. Legend has it that in 1638 one of the treated individuals was the Countess of Cinchon, wife of the Viceroy of Peru. Her miraculous recovery spread the reputation of the bark, which was soon introduced to Europe. (Later Linnaeus, believing the legend, named the genus *Cinchona* in honor of the countess.) A new trade developed, briefly under the monopoly of the Jesuits, in harvesting and shipping the "Jesuits' bark" to the Old World. By the end of the seventeenth century the powdered bark of the *quina-quina* tree was the standard treatment for malaria.

Figure 19.8 *The bark of* Cinchona *species is the source of quinine.*

The demand for the bark was tremendous since it took two pounds of powdered bark to treat one person with malaria. This resulted in exploitation of the wild stands of *Cinchona* trees, even threatening their survival. In 1820, two French scientists isolated the alkaloid quinine from the bark (fig. 19.8); within a few years the purified alkaloid was available commercially and replaced whole bark preparations. This increased the demand for the bark even more. Four out of a total of 36 alkaloids in *Cinchona* bark actually have antimalarial properties, but quinine is the most effective of these compounds. The concentration of quinine in the bark varies considerably among the 40 species within the genus, and even among different populations of the same species.

During the nineteenth century, the British and Dutch established plantations in India and Java, respectively. High-yielding strains of *Cinchona ledgeriana,* whose bark contained up to 13% quinine, were first developed in Bolivia. Seeds from these strains were purchased by the Dutch to start their plantations. The Dutch dominated the world quinine trade until sources were cut off during World War II, and synthetics filled the void. Today, the most widely used drug for the treatment of malaria is chloroquine, which is less toxic and more effective than quinine. Chloroquine also lacks some of the side effects often associated with quinine treatment: tinnitus (ringing in the ears), blurred vision, depression, and miscarriages. Unfortunately, the widespread use of this synthetic has fostered the development of chloroquine-resistant strains of *P. falciparum* in large areas of Central and South America, East Africa, and the Far East. Since these strains are not resistant to quinine, it is the drug of choice for these resistant infections, in combination with other drugs.

Quinine acts on the merozoite stage, killing the parasite in the bloodstream. It is also effective as a prophylactic, preventing the initial infection of red blood cells in travelers visiting malaria-infested areas. The prophylactic value of quinine brought about the development of tonic water used by the British in India. The British colonists made the quinine water more palatable by adding gin; this popularized "gin and tonic" as a favorite drink in the tropics. The greater reliance on synthetics, both for prophylaxis and treatment, has funnelled most of the quinine production today into flavoring for beverages.

Recently scientists have been investigating the antimalarial properties of a temperate weed *Artemesia annua,* wormwood. Long known in China to reduce fevers, this weed contains artemisinin, a terpene, which has had promising results in animal trials. Scientists are excited about this first natural antimalarial since quinine because it shows fewer side effects than quinine or synthetics and is effective against chloroquine-resistant strains of *P. falciparum.*

Snakeroot, Schizophrenia, and Hypertension

Snakeroot, *Rauwolfia serpentina,* exemplifies the belief of "doctrine of signatures." Because of the long coiled roots that resembled a snake, healers believed that the root could be used for treating snakebites (fig. 19.9). For over 4,000 years, Hindu healers in India employed the root of this rain forest shrub for the treatment of snakebites, insect stings, and even mental illness. Tea brewed from the leaves was known to have a soothing effect and was often used to induce a meditative state. Despite all these claims, *Rauwolfia serpentina* came to the attention of Western medicine less than 50 years ago. In 1952, the alkaloid reserpine was the first active principle isolated from the roots, but today dozens of alkaloids are known, with rescinnamine and deserpidine also important pharmaceuticals.

The sedative effects of reserpine (it depresses the central nervous system) made it valuable as one of the first tranquilizers prescribed for schizophrenia. A side effect observed during the administration of reserpine to mentally ill patients was a reduction in their blood pressure. Today, this side effect is the principal application of reserpine— a treatment for hypertension. In fact, the majority of the drugs prescribed for controlling high blood pressure contain *Rauwolfia* alkaloids. These alkaloids act on the nervous system by blocking neurotransmitters; this results in the dilation, or relaxation, of the blood vessels, causing the blood pressure to drop. The sedation and depression associated with their use as hypotensive agents are now considered negative side effects and must be monitored carefully.

Originally native to forests in India, *R. serpentina* is now cultivated in India as well as Thailand, Bangladesh, and Sri Lanka. Other species, including *R. vomitoria* from Africa, are also grown for their alkaloids and, in fact, also have a long history of use by native healers. Although the *Rauwolfia* alkaloids can be synthesized in the laboratory, it is much cheaper to extract them from natural sources.

Figure 19.9 Rauwolfia serpentina *is the source of the drug reserpine and other alkaloids that are used in the treatment of hypertension.*

Figure 19.10 *The glycosides found in the sap of* Aloe vera *are known to produce faster healing of burns and other skin ailments. Today, aloe sap is a common ingredient in many cosmetics, shampoos, and lotions.*

The Burn Plant

A folk remedy familiar to most people in contemporary society is the application of *Aloe vera* sap for minor burns and cuts. Members of the genus *Aloe* have been used for thousands of years as treatments for various skin ailments including rashes, sunburns, direct burns, scalds, and minor wounds. Members of the genus are succulent perennials in the Liliaceae (the lily family) originally native to Africa, and their medicinal use dates back to the ancient cultures in Africa and the Mediterranean area. After its introduction to the Western Hemisphere, it also became widely adopted by indigenous groups for its medicinal value. Aloes usually have basal rosettes of large, succulent, sword-shaped leaves that may give rise to a vertical inflorescence of brightly colored tubular flowers. *Aloe vera* (*A. barbadensis*), the burn plant or medicine plant, is probably the best known member of the genus but many other *Aloe* species are also used.

When cut, the succulent leaves yield a thick mucilaginous sap that can be soothing when applied to injured skin. The sap contains numerous compounds, including several anthraquinone glycosides, collectively referred to as aloin. Additionally, there is chrysophanic acid, the compound with possibly the greatest healing effect on skin. Studies have shown aloe sap promotes faster healing with less scarring by stimulating cell growth and inhibiting bacterial and fungal infection in injuries ranging from deep dermal burns to radiation burns. Other studies have shown that compounds in the sap inhibit pain, itching, and inflammation. The sap is also useful in treating skin and mouth ulcers, eczema, psoriasis, ringworm, athlete's foot, and poison ivy rashes.

Aloe sap is also used as a powerful purgative for the relief of constipation. The anthraquinones in the sap apparently irritate the gastrointestinal tract, resulting in its purgative action. The action is drastic, however, and is, therefore, often used in combination with other drugs to mollify the effect. Aloe sap has also been used internally as a traditional treatment for diabetes. In a recent investigation, dried aloe sap showed some promise in lowering blood glucose levels among a small group of noninsulin-dependent diabetics.

In recent years the cosmetic industry has capitalized on the moisturizing effects of the sap and it can be found in a variety of skin creams, shampoos, sunscreen lotions, and bath oils (fig. 19.10). Aloe has been cultivated since the fourth century B.C., when Alexander the Great reportedly conquered the island of Socotra in the Indian Ocean to obtain a supply of aloe. Today, there are large plantations in Texas and Florida to supply this time-proven folk remedy.

Cancer Therapy

Cancer is a diverse group of diseases characterized by uncontrolled growth of abnormal cells. Humans have suffered from cancer for thousands of years and treatments have

(a)

(b)

Figure 19.11 *Two important sources of anticancer drugs are (a) Periwinkle,* Catharanthus roseus, *and (b) Pacific yew,* Taxus brevifolia.

existed for just as long; even the Ebers papyrus gives instructions for the treatment of cancer. Today, in the United States, cancer is the second leading cause of death, and the search for cancer cures is relentless. Over the years plants have figured prominently in folk remedies and, as reported in the late 1960s, over 3,000 plant species have been used. In the United States, the search for plant sources of anticancer drugs began in the late 1950s under the National Cancer Institute, later joined in its efforts by the U.S. Department of Agriculture. Thousands of plants have been scientifically screened, and several have become part of standard chemotherapy for different forms of cancers (*see* table 19.1). But the search is not over, since only a small percentage of known plant species has been investigated. It is especially crucial for this work to continue in the tropics, where rain forest destruction is proceeding at an alarming rate and most species have not been described scientifically.

One remarkable story in the fight against cancer is the development of vinblastine and vincristine as treatments for various forms of leukemia and lymphoma. These alkaloids were isolated from the Madagascar periwinkle, *Catharanthus roseus* (*Vinca rosea*), a perennial herb native to the tropics (fig. 19.11a). This species had been used by traditional healers as a treatment for diabetes. Investigating this claim in the 1950s, scientists at the University of Western Ontario in Canada and Eli Lilly Pharmaceuticals in Indianapolis found no evidence of usefulness in treating diabetes. Extracts from the leaves, however, were found to be effective against leukemia cells and soon the alkaloids responsible were identified. The concentration of alkaloids in the leaves is very low; it takes 53 tons of leaves to produce 100 grams of vincristine.

Today, these alkaloids are widely used and are major chemotherapeutic agents. The alkaloids act on the microtubular level, where they interfere with spindle formation and thus prevent mitosis. Vincristine is especially effective for treating acute childhood leukemia and vinblastine is useful in treating Hodgkin's disease. Both alkaloids are also used in chemotherapy for other types of cancer.

Another anticancer drug that is currently the focus of extensive research is taxol, obtained from the bark of the Pacific yew, *Taxus brevifolia* (fig. 19.11b). Like the vinca alkaloids, taxol acts on the microtubules. The antitumor properties were first discovered in the 1960s during a screening program of the National Cancer Institute; however, taxol has only recently been approved for the treatment of cancer. Clinical trials showed the use of taxol to be especially promising in treating women with advanced ovarian and breast cancer. In one study, taxol halted or slowed the growth of tumors in more than half of 25 breast-cancer patients whose cancer had spread to other organs.

The supply of taxol was initially problematic since it had not been synthesized in the laboratory. The original source of the alkaloid was the bark from mature Pacific yew trees, slow-growing conifers, of old-growth forests in the Pacific Northwest. Concern about the destruction of these ancient forests has prompted alternative approaches to obtaining a continual supply of taxol. Other more common species of *Taxus,* including the common yew and Japanese yew, also contain taxol, with some occurring in the needles as well as the bark. Another promising direction is the establishment of taxol-producing tissue cultures of bark cells. One report indicated that cultures yielded seven times more taxol than naturally produced in the bark of the tree. Although still in the preliminary stages, this method may eventually become the primary source of taxol and other plant-derived medicines. Recently, however, taxol has been synthesized in the laboratory, ensuring a sufficient supply of this promising new drug.

The development of taxol from natural plant product to synthetic parallels that of many other medicinal compounds. Ethnobotanists today are searching for other promising folk remedies that will become the "taxol" of tomorrow.

Chapter Summary

1. From the earliest civilizations native plants have supplied medicinal compounds. Western medicine traces its traditions back to Greek and Roman herbal remedies that were used for hundreds of years. During the Renaissance, there was a renewal of interest in herbalism and a revival of the Doctrine of Signatures. Although herbalism waned in the eighteenth and nineteenth centuries, many of the remedies became incorporated into modern prescription drugs. Today 25% of prescription drugs are of plant origin and an even greater percentage is based on synthetic or semi-synthetic ingredients originally isolated from plants. While Western medicine strayed away from herbalism, 75%–90% of the rural population in the rest of the world still relies on herbal medicine as their only health care. Today there is renewed interest in studying the medicinal plants used by indigenous populations in remote areas of the world, especially the rain forests. Investigating these folk remedies may lead to important medical discoveries that may save millions of lives.
2. Alkaloids and glycosides have been identified as the therapeutically important components of most medicinal plants. Alkaloids are a diverse group of compounds that contain nitrogen, are alkaline, and have a bitter taste. Alkaloids have pronounced effects on the physiology of animals, especially affecting the nervous system. They are widely distributed among plants, with three families known to contain abundant alkaloids; the Fabaceae, Solanaceae, and Rubiaceae. Glycosides are another diverse group of compounds; the common feature they share is the presence of a sugar molecule attached to the active component. The physiological effects depend on the nature of the active component such as the cardioactive glycosides, which affect the contractions of heart muscle.
3. Digitalis, aspirin, quinine, reserpine, vinblastine and vincristine are just a few of the countless plant-based medicinal compounds that form the basis of modern medicine. Digitalis, the primary treatment for several heart ailments, is derived from foxglove, a biennial in the snapdragon family. The leaves contain over 30 glycosides, with digoxin and digitoxin the most medically significant compounds.
4. Today aspirin is purely synthetic, but it is based on traditional remedies from willow bark. Salicylic acid, the active principle in willow bark, was identified early in the nineteenth century and laboratory synthesis was achieved about 30 years later. In 1898, the Bayer Company developed acetylsalicylic acid, a modification they named "aspirin."
5. Quinine, the active alkaloid in the bark of *Cinchona* trees, has been used for centuries in the treatment of malaria. Although synthetic drugs are usually prescribed today, quinine is still an effective treatment and is the drug of choice for infections resistant to the synthetics.
6. In 1952, the alkaloid reserpine was isolated from the roots of *Rauwolfia serpentina,* a plant used by Hindu healers for thousands of years. Although first used as a tranquilizer, reserpine and other *Rauwolfia* alkaloids have their greatest use in the treatment of hypertension.
7. The sap of *Aloe vera* contains several glycosides that are used to treat minor burns and cuts. This herbal remedy, which has been used since ancient times, promotes faster healing and less scarring. The moisturizing effects of the sap has resulted in its inclusion in skin creams, shampoos, lotions, and other cosmetics.
8. Active compounds from several plants have become standard treatment for various forms of cancer, and researchers continue to look for and develop new anticancer compounds. The

alkaloids vinblastine and vincristine, isolated from the Madagascar periwinkle, are remarkably effective in the treatment of certain forms of leukemia. Recently, taxol, from the bark of the Pacific yew, has proven useful in the treatment of ovarian and breast cancer.

Review Questions

1. Trace the historical connection between medicine and botany. Is this still a valid connection today?
2. What parts of the world still rely on herbal medicine? Why? Do you think it is effective? Do you think more funds should be allocated to investigating herbal remedies?
3. What active principles in plants account for their medicinal properties?
4. Discuss the history and use of the following medicines: digitalis, quinine, and reserpine.
5. Trace the development of aspirin from an herbal remedy to the world's most widely used synthetic drug.
6. How has cancer therapy improved through the use of plant-derived medicines?

Further Reading

Aikman, Lonelle. Sept., 1974. Nature's Gifts to Medicine. *National Geographic,* 146 (3): 420 – 440.

Ayensu, E. Nov., 1981. A Worldwide Role for the Healing Power of Plants. *Smithsonian,* 12 (8): 87–97.

Blackwell, Will H. 1990. *Poisonous and Medicinal Plants.* Prentice Hall, Englewood Cliffs, NJ.

Cox, Paul Alan and Michael J. Balick. June, 1994. The Ethnobotanical Approach to Drug Discovery. *Scientific American,* 269 (12): 82–87.

Estes, J. Worth and Paul Dudley White. 1965. William Withering and the Purple Foxglove. *Scientific American,* 212 (6): 110 –117.

Hobhouse, Henry. 1987. *Seeds of Change.* Harper & Row, Publishers, New York.

Jackson, Donald D. 1989. Searching for Medicinal Wealth in Amazonia. *Smithsonian,* 19 (11): 94–103.

Klayman, Daniel L. 1989. Weeding Out Malaria. *Natural History,* 98 (10): 18–29.

Krochmal, Arnold and Connie Krochmal. 1973. *A Guide to the Medicinal Plants of the United States.* Quadrangle/The New York Times Book Co., New York.

Lewington, Anna. 1990. *Plants for People.* Oxford University Press, New York.

Lewis, Walter and Memory P. F. Elvin-Lewis. 1977. *Medical Botany: Plants Affecting Man's Health.* John Wiley & Sons, New York.

Mann, John. 1992. *Murder, Magic and Medicine.* Oxford University Press: Oxford.

Plotkin, Mark J. 1990. The Healing Forest: The Search for New Jungle Medicines. *The Futurist,* XXIV (1): 9–15.

Plotkin, Mark J. 1993. *Tales of a Shaman's Apprentice: An Ethobotanist Searches for New Medicines in the Amazon Rain Forest.* Viking, New York.

Riddle, John. M. and J. Worth Estes. 1992. Oral Contraceptives in Ancient and Medieval Times. *American Scientist,* 80 (3): 226–233.

Riddle, John. M., J. Worth Estes, and Josiah C. Russell. 1994. Ever Since Eve—Birth Control in the Ancient World. *Archaeology,* 47 (2): 29–35.

Weissmann, Gerald. 1991. Aspirin. *Scientific American,* 264 (1): 84 –90.

C H A P T E R

20

Psychoactive Plants

Chapter Outline

Chapter Concepts

1. Psychoactive compounds affect the central nervous system by acting as stimulants, hallucinogens, or depressants.
2. Most psychoactive drugs are alkaloids that have been employed by humanity for millennia.
3. Although many of these compounds have legitimate medicinal value in the proper dosage, the abuse of psychoactive drugs has led to devastating physiological, emotional, economic, and societal consequences.

The use of psychoactive plants dates back to the earliest societies when humans first stumbled on the mind-altering properties of unfamiliar plants. These plants were immediately valued for their ability to numb pain, relieve fatigue and hunger, or free the spirit from earthly concerns. Clearly these plants were used for millennia before written records appeared. Six-thousand-year-old Sumerian clay tablets refer to opium as the joy plant.

Psychoactive Drugs

Psychoactive drugs affect the central nervous system in various ways by influencing the release of neurotransmitters (chemical messengers within the nervous system) or mimicking their actions. Based on their effects they can be classified as **stimulants, hallucinogens,** or **depressants.** An important point to remember is that there may often be only subtle distinctions between medicinal, psychoactive, and toxic doses. For example, morphine and cocaine have legitimate medical uses for pain relief and local anesthesia, respectively. They can also be abused for their psychoactive effects and, when taken in overdose, result in death.

Stimulants excite and enhance mental alertness and physical activity; they reduce fatigue and suppress hunger. Cocaine, with its intensely potent effects, and the much milder caffeine, are well-known plant-derived stimulants. Hallucinogens such as peyote, marijuana, and LSD produce changes in perception, thought, and mood, often inducing a dreamlike state. Depressants dull mental awareness, reduce physical performance, and often induce sleep or a trance-like state. Opium and its derivatives, morphine and heroin, are classic depressants.

The term **narcotic** must also be considered when discussing psychoactive drugs. By strict definition, a narcotic drug is one that induces central nervous system depression resulting in numbness, lethargy, and sleep. Under this definition, narcotics would include opiates as well as alcoholic beverages. In contemporary use, the term "narcotic" is applied to any psychoactive compound that is dangerously **addictive;** therefore, the stimulant cocaine is classified as a narcotic. Addictive compounds generally elicit one or more of the following: **psychological dependence, physiological dependence,** and **tolerance.** Psychological dependence is a desire to reexperience the pleasure induced by the drug. In physiological dependence, the body has a need for the drug to avoid withdrawal symptoms. Finally, tolerance is the need for ever-increasing amounts of the drug to produce the same effect.

In the majority of psychoactive plants, the active compounds are alkaloids (although marijuana is a notable exception). As discussed in Chapter 19, alkaloids are distributed throughout the plant and fungal kingdoms but predominate in certain families. This chapter will focus on psychoactive compounds in angiosperms, with the fungal compounds discussed in a later chapter. Angiosperm families that are well-known for containing psychoactive alkaloids include the Solanaceae (nightshade family), Papaveraceae (poppy family), Rubiaceae (coffee family), Erythroxylaceae (coca family), Myristicaceae (nutmeg family), and Convolvulaceae (morning glory).

The Opium Poppy

The use of *Papaver somniferum,* opium poppy, for relieving pain was well-known by early societies as evidenced by their written documents. Babylonian and Egyptian writings, artifacts, and statuary attest to the fact that opium was revered for its analgesic and sleep-inducing properties. Hippocrates, Dioscorides, and Galen, leading contributors to early Western medicine, endorsed its use. Early peoples also valued its ability to allay the worries and sorrows of life. Surely, these ancient peoples were aware of the strongly addictive nature of compounds from this plant, and this curse has followed the opium poppy to contemporary society.

An Ancient Curse

The opium poppy is a large annual herb in the poppy family (Papaveraceae), with each stem topped by a showy flower of white, pink, red, or purple petals. After pollination, the ovary matures into a capsule. If the capsule is sliced when still green, it exudes a milky latex rich in potent alkaloids (fig. 20.1). The dried latex, which turns brown, can be peeled from the capsule and is known as crude opium. Originally native to the Middle East, it is now grown both legally and illegally throughout much of the world, with India the leading legal grower. In addition to the legal and illegal cultivation of drugs, poppy is also grown for its seeds and seed oil. Poppy seeds are common toppings for breads and rolls as well as ingredients for many muffins, cakes, and cookies. Poppy seed oil is a common cooking oil in some parts of the world. Although the quantity of alkaloids in the seed is negligible and, therefore, not at all addictive, it may be sufficient to register positive on some drug tests.

Opium has been eaten, drunk, and smoked for centuries. The usual method of preparation was to dissolve the crude opium in wine. Ancient pottery vessels for opium wine, pipes for smoking opium, and other artifacts decorated with poppy capsules have been dated back to ancient Cyprus and Knossos. Later Hippocrates and other early physicians advocated poppy wine as a medicine for many complaints. During the Middle Ages, a common method of preparing opium was to dissolve it in alcohol, and this tincture, later known as laudanum, became a popular medication for centuries.

Laudanum usage reached its peak in the ninteenth century, when it was consumed not only as a medicine but also as a mind-expanding drug by many in the so-called privileged class, including writers, artists, and intellectuals. The

Figure 20.1 *The flower and capsule of the opium poppy,* Papaver somniferum. *Crude opium is the dried, milky latex from the incised capsule.*

famous British poets Elizabeth Barrett Browning and Samuel Coleridge were both known laudanum addicts, as was Thomas De Quincey, who analyzed his addiction in *Confessions of an English Opium-Eater.* The French composer Hector Berlioz glorified the opium dream in his *Symphonie Fantastique.* The use of laudanum was so widespread during this time that little shame was attached to the addiction.

The Opium Wars

Opium use is directly tied to the history of China. The Chinese knew of opium for centuries but used it primarily for medicinal purposes. After A.D. 1600, opium use began to increase when tobacco smoking was introduced to China. Originally the tobacco was mixed with opium, but gradually the amount of tobacco was decreased until the opium was smoked alone. Opium smoking spread slowly and eventually many Chinese became addicted even though it was banned by the Emperor.

The addiction was fueled by the British desire for a favorable trade balance with China. The British wanted silk, tea, and porcelain from China; however, the Chinese demanded payment in silver since they had no interest in European goods. The drain of silver from Great Britain was intolerable to the British, and their solution was to meet the escalating Chinese need for opium. Opium was being widely cultivated in India under the auspices of the British East India Company. Merchant clipper ships from various countries, including the United States and Britain, brazenly smuggled the opium into China through the port of Canton. The trade was lucrative to everyone involved in the growing, shipping, and marketing of opium, and soon the flow of silver was reversed.

Alarmed by the rising addiction rate, the Chinese government sought to stop the opium trade. In 1839, the Chinese authorities confiscated and destroyed all the opium in Canton harbor. The British retaliated by sending warships, beginning the first Opium War, which lasted from 1839 to 1842. Chinese opposition was crushed and the British received major concessions, including the right to trade in opium, payment for the destroyed opium, the opening of more ports for foreign trade, and the establishment of Hong Kong as a British colony.

The second Opium War ten years later took a further toll on China when Britain, other European countries, and the United States were granted additional concessions. The British opium trade continued until 1913, when moral pressure inside both China and Britain brought an end to the trading that had cost China so much. By this time, however, China was cultivating its own poppies and about 25% of the population was addicted. The opium problems continued and were amplified during the Japanese occupation, only to end after the establishment of the communist People's Republic of China.

Opium Alkaloids

More than 20 alkaloids have been identified in opium, with morphine and codeine probably the most significant. Morphine was first isolated in 1806 by Frederic Serturner, a German scientist who named the compound after Morpheus, the Greek god of dreams. This, in fact, was the first active principle isolated from a plant. Morphine was soon recognized for its analgesic value and is still unsurpassed in its ability to deaden pain. Initially administered orally, morphine's full medical potential was not realized until after the development of the hypodermic syringe in the middle of the nineteenth century. It is known today that morphine is rapidly inactivated and excreted when taken in oral preparations. Although the exact mode of action is not known with certainty, opiate binding sites are found in the brain, spinal cord, and intestines. Morphine depresses those areas of the brain involved in the perception of pain and reduces the anxiety that accompanies pain. It is a general central nervous system depressant and, in overdose, can lead to death by completely suppressing the respiratory center in the brain.

Like opium, morphine is strongly addictive. This became evident when thousands of injured Civil War soldiers became dependent after injections of morphine. In fact, this was so widespread that morphine addiction became known as soldiers' disease. Physician-induced addiction became common because morphine was used to treat so many ailments during the last half of the nineteenth century. This was forcefully depicted in Eugene O'Neill's classic play "A Long Day's Journey into Night," which showed the ravages of his mother's morphine addiction. Among the many applications beyond the control of pain, morphine was frequently prescribed for diarrhea (it slows peristalsis in the digestive tract) and coughing (suppresses the cough reflex in the central nervous system).

With the growing awareness of the potential for addiction, morphine use declined around the turn of the century. Currently morphine is still the drug of choice for the control of intense pain from severe burns or visceral pain during the post-operative period, cases of terminal cancer, and kidney stones. Morphine is also administered in some cases to patients suffering from heart failure.

Medically, codeine is one of the most commonly used opiates. It has value as an oral analgesic but is only one-fifth as strong as morphine. It works well in combination with nonopiate analgesics such as aspirin and acetaminophen, and is especially useful in cough syrups since it suppresses the cough reflex. Since it is also addictive (but much less so than morphine), products containing codeine are available only by prescription in the United States.

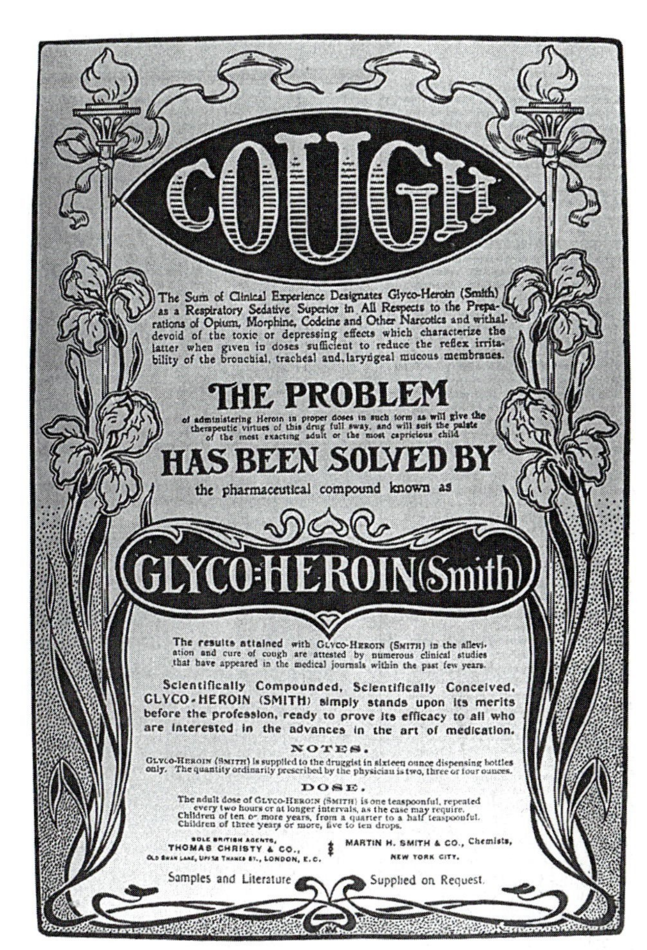

Figure 20.2 *As this advertisement from the turn of the century attests, heroin was an ingredient in many medications before its addictive properties were recognized.*

Heroin

In 1898, the Bayer company introduced heroin, which they believed was a nonaddictive opiate with analgesic properties superior to morphine and cough-suppressant properties superior to codeine. This semisynthetic derivative of morphine was widely dispensed in many over-the-counter medicines for about two decades. Cough syrups with heroin were promoted for their effectiveness in adults and children alike (fig. 20.2). Eventually physicians questioned the alleged nonaddictive nature of heroin, and by 1917 it was no longer available in cough syrups.

Contrary to the early promotion, heroin is actually six times more addictive than morphine, and its misuse has resulted in long-standing drug and crime problems. Heroin is not used medicinally in the United States nor is it manufactured here legally. However, heroin is used medicinally in other countries to control severe pain. India is the largest legal producer of both opium alkaloids and heroin for medical purposes. Most of the illegal opium, on the other hand, comes from Burma, Laos, and Thailand (The Golden Triangle) as well as Pakistan, Afghanistan, and Iran (The Golden Crescent), and recently Central and South America. Approximately 60% of all heroin coming into the United States originates from Burma, with Nigeria serving as a major trafficking center (fig. 20.3).

Heroin addiction became a serious problem in the United States and Europe after World War II. Today, heroin addiction is still a problem, with an estimated 500,000 addicts in the United States alone.

Withdrawal

Like any addictive substance, opiates will trigger withdrawal symptoms when the drug is no longer taken. The intensity of these symptoms depends on the type of opiate, dosage, and length of the habit. Withdrawal from heroin addiction produces the most severe symptoms, which begin to occur within 4 hours and reach a peak at 36 to 72 hours. Common symptoms include increased respiration, perspiration, runny nose, goose bumps, muscle twitches, insomnia, vomiting, and diarrhea. In fact, the abrupt cessation of a drug habit, or going "cold turkey," refers to the goose bumps prominent in a person undergoing opiate withdrawal. To alleviate the physiological trauma that the body suffers during withdrawal, some patients are gradually weaned from heroin addiction by substituting methadone, a synthetic opiate that is less addictive and has milder withdrawal symptoms.

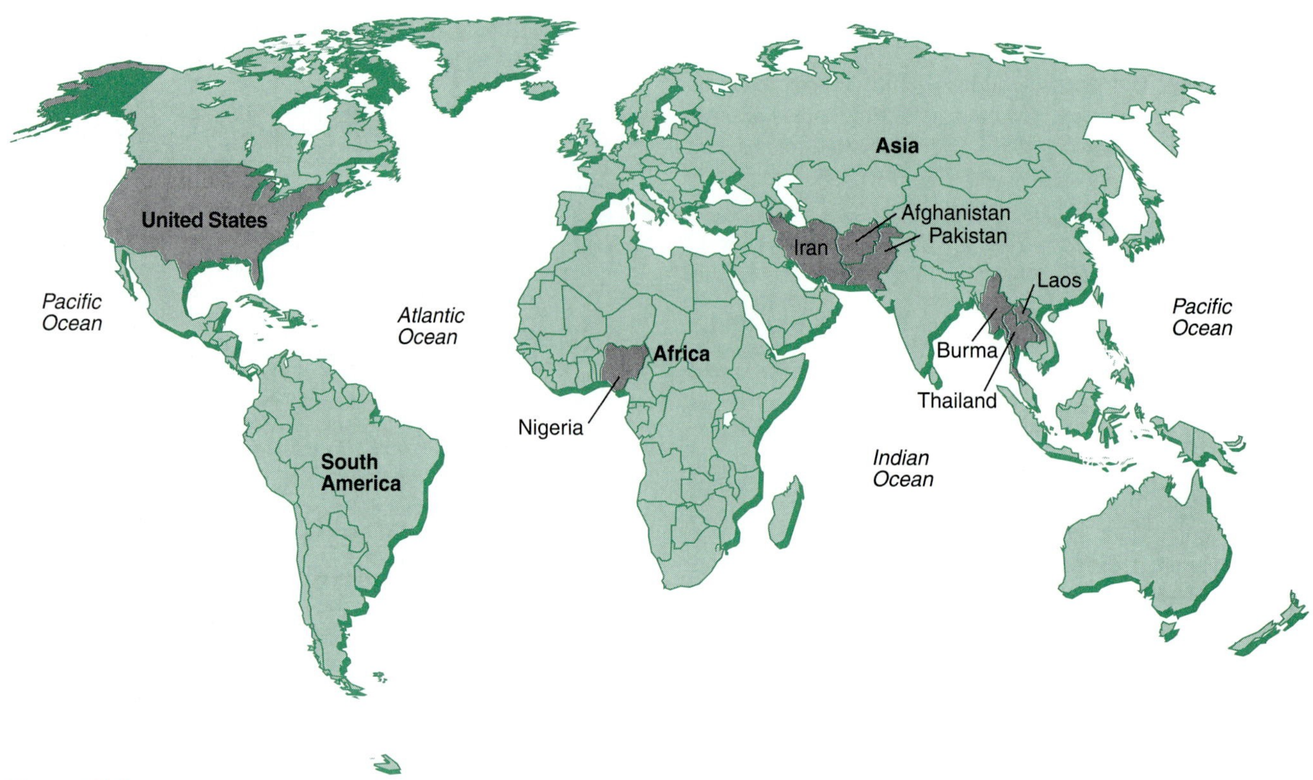

Figure 20.3 *In the illegal opium trade, cultivation of the opium poppy predominates in the countries of the Golden Triangle (Burma, Laos, and Thailand) and the Golden Crescent (Pakistan, Afghanistan, and Iran).*

Marijuana

Cannabis sativa is one of the oldest cultivated plants and has been used for varied purposes. In addition to its use as a hallucinogen, the plant has been used medicinally and for its fiber (*see* Chapter 18), its oil, and its seed. Marijuana is highly variable with some taxonomists believing it should be divided into at least three separate species, based on size and growth habit, from short branched forms to plants over 8 meters (25 ft) tall. Marijuana plants are dioecious annuals with inconspicuous staminate and pistillate flowers on separate individuals. Inflorescences occur in the axils of upper leaves; the palmately compound leaves are distinctive, with three to seven (usually five) toothed leaflets (fig. 20.4). The plants are known for their **resin** production by glandular trichomes, with the maximum amount of resin coating the unfertilized pistillate flowers and adjacent leaves. The golden resin is the source of the hallucinogenically active compounds, and the potency of resin varies greatly depending on genetic strains and growing conditions.

It is believed that the species originated in central Asia and today has a cosmopolitan distribution that is the greatest among all the psychoactive plants. Its wide cultivation has led to many common names for the species, including marijuana, hemp, grass, pot, hash, hashish, bhang, charas, ganja, ma, kif, and dagga.

Early History in China and India

The use of *Cannabis* can be traced back to ancient China where the hemp plant was valued for its fiber (for cloth and rope) and medicinal properties. About 5,000 years ago, the legendary Emperor Shen Nung recommended marijuana for the treatment of rheumatism, gout, malaria, beriberi, absentmindedness, female disorders, and constipation. There is also some evidence that the Chinese knew of and used the plant for its hallucinogenic properties, but this practice was not as prevalent here as it was in India.

Earliest documented records of marijuana's use as a hallucinogen can be traced to the Scythians (ancient nomadic Slav horsemen from Central Asia) who burned *Cannabis* and inhaled the hallucinogenic smoke. In a ritualistic ceremony, the Scythians piled the plants on burning coals in a small tent made of sheepskin pelts. When the vapors accumulated, the Scythians would lift the sheepskin and inhale. Pots of charcoal with remains of *Cannabis* have been found in Scythian excavations and dated to around 500 B.C.

Figure 20.4 Cannabis sativa *is dioecious; the pistillate plants and flowers (right) produce a higher concentration of the THC hallucinogen than the staminate plants (left).*

Marijuana use spread from its Central Asian homeland to Asia Minor, northern Africa, India, and elsewhere. The first written mention in India is found in the Sanskrit *Zend-Avesta* around 600 B.C. India has valued *Cannabis* for its medicinal properties, and folk medicine uses it extensively to treat a variety of ailments. It was the hallucinogenic properties, however, that gave *Cannabis* an integral role in religious ceremonies for achieving a contemplative state and placed it as one of the five sacred plants of ancient India. Three grades of *Cannabis* have been recognized in India: bhang, ganja, and charas. Bhang, the least potent of these, consists of the dried, cut tops that are then ground with spices to prepare a drink or candy. Even today, bhang is sold outside Hindu temples in India. Ganja is prepared from the resin-rich pistillate flowers and tops of specially-bred high-yielding strains of marijuana and is usually smoked. Charas, the most potent, consists of the pure resin itself (also known as hashish) from these special strains and is also smoked.

The use of *Cannabis* spread throughout the Muslim world, into the Middle East and Africa. Hookahs (water pipes) for smoking hashish were commonplace in bazaars and marketplaces, scenting the air with marijuana's sickly-sweet aroma (fig. 20.5). Arabic literature is replete with references to *Cannabis,* as can be seen in the famous *Thousand and One Nights,* where the potency of the drug was indicated by the expression

> "had an elephant smelt it, he would have slept
> from year to year."

Undoubtedly the most notorious legend about marijuana concerns the twelfth century Persian, Al-Hasan ibn-al-Sabbah, a leader of a Muslim sect. This fanatical religious sect, whose followers became known as Hashishins (after their leader), swore to kill all enemies of their faith. Legend has it that they were worked into a murderous frenzy by smoking *Cannabis* (the accuracy of this account is questionable). Nevertheless, the name of this sect, *hashishins,* left a legacy in the word *assassin,* as well as the name for the resin, *hashish.*

Spread to the West

Although Northern Europe had a long history of using hemp fiber, the psychoactive use of *Cannabis* was not well known before the nineteenth century. It is rumored that Napoleon's troops, returning from a campaign in North

Figure 20.5 *Alice, in the Lewis Carroll classic* Alice in Wonderland, *"stretched herself up on tiptoe, and peeped over the edge of the mushroom, and her eyes immediately met those of a large blue caterpillar that was sitting on the top with its arms folded, quietly smoking a long hookah, and taking not the slightest notice of her or anything else."*

Africa, popularized hashish smoking. By mid-century, Paris hashish clubs were frequent meeting places of the intellectuals, writers, poets, and artists. Members believed that the psychoactive properties of hashish enhanced their own creative abilities.

Marijuana was probably introduced to the United States around the turn of the twentieth century, possibly from Mexico or the Caribbean Islands. Mexican laborers brought the marijuana-smoking habit with them when they immigrated. In the 1920s it spread among the urban poor in the South, but was not in vogue until jazz musicians popularized smoking marijuana cigarettes. Not only was marijuana used recreationally, but it was also a popular ingredient in many over-the-counter medications.

During the 1930s, concerns about the dangers inherent in marijuana use led to the establishment of laws prohibiting its use. The Federal Bureau of Narcotics launched an "educational campaign" to make the public aware of the dangers of marijuana use. This campaign greatly distorted the problem and grossly exaggerated the dangers. These concerns culminated in the Federal Marijuana Tax Act of 1937, which controlled the legal sale of the plant and resulted in the virtual elimination of *Cannabis* from the nation's pharmacopoeia.

The dramatic increase in marijuana use came about during the social revolution of the 1960s. The youthful counterculture of this period embraced marijuana as the recreational drug of choice. In reaction to the pervasive usage of marijuana during the 1970s, many communities relaxed their laws concerning personal use and possession. The public demand for decriminalization of marijuana usage peaked in the late 1970s and is no longer a vocal issue. In fact, recent trends seem to be for stricter enforcement and harsher sentences for users and dealers.

THC and Its Psychoactive Effects

The chemistry of *Cannabis* is complex and includes a unique class of compounds called cannabinoids. These compounds, over 60 in number, are pure hydrocarbons, with 21 carbon atoms in an elaborate 2 to 3 ring structure. Although cannabinoids were known since the beginning of the twentieth century, it was not until 1964 that the psychoactive component of marijuana was identified as delta-9-tetrahydrocannabinol (THC), present in the resin. The concentration of THC in plants varies considerably, based on genetic strain, sex, climate, and growing conditions. Some of the most potent varieties of *Cannabis* are usually grown under hot dry conditions that seem to optimize THC production. Marijuana grown in Mexico, Colombia, and India generally has higher "street value" because of its higher THC concentrations. Sinsemilla, a seedless strain developed illicitly in California during the 1970s, however, is reported to have some of the highest levels of THC. Newer highly potent dwarf hybrids have also been illicitly developed for clandestine indoor cultivation.

The long-sought-after mind-altering effects of *Cannabis* include a sense of euphoria and calmness. Some of the most descriptive accounts of the intoxication were written by Walter Bromberg relating both personal experience and detailed observations.

> "Within a few minutes he begins to feel more calm and soon develops definite euphoria; he becomes talkative . . . is elated, exhilarated . . . begins to have . . . an astounding feeling of lightness of the limbs and body . . . laughs uncontrollably and explosively . . . without at times the slightest provocation . . . has the impression that his conversation is witty, brilliant . . ."

Perception of time and space may also be distorted and minutes may seem like hours.

Adverse effects of marijuana use have been studied since the 1930s, with research efforts intensifying after the identification of THC as the active component. Even moderate use of marijuana impairs learning, short-term memory, and reaction time. Because THC is fat-soluble, it accumulates in body tissues and measurable amounts may remain in the body for days after inhalation. The effects of marijuana on the male reproductive system show decreased sperm production and decreased testosterone levels. In pregnant women, THC is known to cross the placenta and might be potentially damaging to the fetus; however, mutagenicity studies thus far have been inconclusive. Marijuana or hashish is usually inhaled in smoke and, as such, can damage lung tissue much like cigarette smoking. More research is needed to determine if long-term harm can be traced to marijuana use.

Medical Uses in Chemotherapy and Glaucoma

Over the centuries in various cultures, marijuana has been employed to treat numerous ailments. Two uses in contemporary medicine include the treatment of glaucoma and as an aid to chemotherapy. Glaucoma is a group of eye diseases characterized by increased ocular pressure (pressure within the eye) and resulting damage to the optic nerve that may lead to blindness. In fact, it is the leading cause of blindness in the United States. Either smoking marijuana or ingesting preparations with THC have been shown to significantly reduce ocular pressure in patients with glaucoma. Patients undergoing chemotherapy for cancer often experience side effects of nausea, vomiting, and loss of appetite. These side effects have been reduced by the use of marijuana cigarettes as well as marinol, a synthetic form of THC. These and other medical applications have resulted in a storm of controversy over the medical dispensing of marijuana cigarettes. Despite evidence of its effectiveness, in early 1992 only 13 patients were legally allowed to smoke marijuana for medical purposes. Activists are pressuring the federal government to loosen controls over the use of marijuana to ease the suffering from AIDS, cancer, multiple sclerosis, and glaucoma.

Cocaine

Cocaine is the major alkaloid (1 of 14) of *Erythroxylum* (*Erythroxylon*) *coca* and *E. novogravatense,* the coca plant, a small tree or shrub with shiny evergreen leaves. The alkaloid occurs in the leaves, which can be harvested two or three times per year. A member of the Erythroxylaceae, the coca plant is native to the Andes Mountains of South America, where the leaves have been chewed for centuries.

South American Origins

Archaeological evidence, in the form of figurines depicting coca chewing, indicates that the plant has been used for at least 3,500 years in the Andes (fig. 20.6). To the Incas, the coca plant was important both socially and economically. According to myth, a god created the coca plant to alleviate hunger and thirst among the people. Since the great Incan civilization considered it sacred, coca chewing was mostly restricted to the ruling classes. Soldiers, workers, and runners were also permitted to chew the leaves for endurance. By the end of the fifteenth century, the use of coca was widespread among the Incas; however, casual chewing was considered a sacrilege. Economically, the coca leaves were used as a form of payment and could be given in exchange for potatoes, grains, furs, fruits, and other essential goods.

Figure 20.6 *Incan artifact depicts a coca user.*

The early Spanish explorers noted the practice of chewing leaves with added ashes. (One component of ash is calcium carbonate, or lime, which releases the cocaine from the leaves. This practice of adding lime is still used by the Indians of the Andes region today.) Following the Spanish Conquest, the native populations were enslaved and forced to work in mines under incredibly harsh conditions and little food. The Spanish soon learned the value of coca leaves; productivity and endurance of the enslaved Indians increased dramatically when they were given coca leaves to chew. King Phillip II of Spain declared coca leaves necessary to the well-being of the Indians and set aside land in the Andes for coca cultivation. Although specimens of coca leaves had been sent back to Spain, the leaves did not retain their potency in storage. As a result, coca chewing never gained popularity in Europe.

Freud, Holmes, and Coca-Cola

The contemporary social history of *E. coca* actually begins in the late 1850s when cocaine was isolated from coca leaves in a German laboratory. Although the anesthetic properties were quickly realized, it was the stimulating properties that brought it into the limelight. In 1884, Sigmund Freud became an enthusiastic advocate of the drug as a stimulant and as a means of combating morphine addiction. Freud published several papers on cocaine, which he described as a "magical drug." He used cocaine himself, recommended it to friends and family, and prescribed it for his patients, although years later he became disenchanted with the drug.

At this same time cocaine was gaining wide popularity in the United States and could be found in over-the-counter medicines, tonics, elixirs, and even beverages. One reason for cocaine's popularity in over-the-counter preparations for colds, asthma, and catarrh (hay fever) was its effectiveness in shrinking mucous membranes and draining sinuses. It was also promoted as a panacea for ailments from headaches to hysteria. Coca-wine, notably Vin Mariani, became one of the most popular beverages of late nineteenth century. Luminaries in both the United States and Europe, such as Jules Verne, Thomas Edison, and John Philip Sousa, praised its exhilarating properties. Another beverage that also incorporated cocaine was Coca-Cola, created in 1886 by an Atlanta pharmacist John Styth Pemberton who marketed his beverage as a "brain tonic." By the turn of the century, the negative effects of cocaine use were evident, and cocaine was eliminated from the Coca-Cola recipe in 1903 (*see* Chapter 16).

The widespread use of cocaine in the late nineteenth century is also evident from its appearance in literature of the day. One of the most familiar examples can be found in Arthur Conan Doyle's Sherlock Holmes stories, in which the legendary detective was a cocaine user.

By the beginning of the twentieth century, cocaine's image was tarnished and it was recognized as a dangerous drug. It was included under the Harrison Act of 1914, the first federal antinarcotic law that regulated the use of cocaine, opium, morphine, and heroin. Cocaine use gradually declined and was not a problem in the United States until the late 1960s.

Coke and Crack

By the mid-1970s, cocaine use dramatically increased, becoming the favorite drug of the middle and upper class by 1980. Under the misconception that consumption posed no long term danger, its use grew alarmingly during the 1980s. Cocaine-related deaths of celebrities and sports figures gradually began to change the public's perception. While today there is a realization that cocaine is one of the most dangerous drugs, cocaine abuse is still a serious national problem.

Cocaine makes its way into the United States from plantations in South America with Columbia, Ecuador, Peru, and Bolivia the major producers. Although cocaine production is illegal (except for legitimate medical applications), hidden coca plantations supply 400 tons of the drug annually. Coca plants grow best in the hot, humid highlands of these countries, where leaves can be harvested every 35 days. The leaves are put into large vats containing a dilute solution of sulfuric acid to extract the alkaloids and stirred several times a day for about four days. The liquid is removed and mixed with a variety of chemicals to produce cocaine base. Although the yield from this extraction is low (only about 400 grams per acre of leaves), the purity is high. The extract is 75% pure cocaine. Clandestine labs refine the base into cocaine hydrochloride, a white powder that is shipped north to Mexico and the Bahamas, and then on to the United States.

The cocaine hydrochloride is cut with various adulterants (additives to reduce the concentration); the street drug generally averages 12% cocaine hydrochloride. In this form the powder can be snorted and the alkaloid absorbed through the mucous membranes of the nose. Free basing and **crack** were modifications in the 1980s designed to produce quicker and stronger highs. Free basing purifies the powder, accomplished in part by boiling it in an ether solution to produce pure cocaine, the free base. (Ether is highly flammable and explosive. The use of any open flame or spark near ether is foolhardy.) The free base is then smoked in a water pipe to produce an intense high, which can reach the brain in 15 seconds. Crack is a form of free base prepared by heating a cocaine hydrochloride solution with baking soda. The resulting compound forms solid chunks, which can be broken into tiny "rocks," each costing a mere fraction of the cocaine powder. Like free base, crack is smoked and also produces a high within seconds. Crack is an especially addictive form of cocaine. Since the late 1980s, crack houses for the distribution and smoking of the drug have been a major epidemic of cities in the United States.

Medical Uses of Cocaine

Albert Niemann, who first isolated cocaine in the 1850s, is also credited with realizing its anesthetic properties. He noted that the purified crystal tasted bitter and soon numbed the tongue. Cocaine acts as a local anesthetic, temporarily blocking the transmission of nerve impulses at the site of application. Many of the widely used synthetic local anesthetics Novacain (chemically known as procaine) and Xylocaine (lidocaine) are structurally similar to cocaine. Another valuable property of cocaine is its ability to constrict blood vessels and therefore reduce blood flow when applied locally. This has made cocaine the anesthetic of choice for ear, nose, and throat surgery and was formerly used for eye surgery as well.

A Deadly Drug

Cocaine is a powerful stimulant to the central nervous system that produces a short-lived euphoric high. The high is accompanied by a burst of energy and alertness like an intense adrenaline rush. The resulting omnipotent feeling is the appeal that cocaine has for many users. The duration of the high varies with the method of administration: snorting—up to a few hours; injection—15 to 30 minutes; smoking—up to one hour. After the high, users experience depression and lethargy.

Physiologically cocaine has multiple effects on the body: increasing heart rate, respiration, blood pressure, and body temperature, and dilation of the pupils. Cocaine's varied effects are all related to its action on neurotransmitters. Cocaine blocks the reuptake of certain neurotransmitters; therefore, they continue to act, stimulating the nervous system and producing the hyperactive effects associated with cocaine use.

These same effects can be deadly. Cocaine abuse has resulted in sudden death due to heart attacks, cerebral hemorrhage, respiratory failure, and convulsions. It is impossible to predict a safe dose for a given individual. Some deaths occur with first-time users, others with chronic abusers. A safe dose one time may cause a fatality with the next use.

A less deadly but debilitating effect of chronic cocaine use is a psychosis that is often indistinguishable from schizophrenia, with accompanying paranoia and hallucinations. Heavy users develop insomnia and appetite loss. The destruction of nasal membranes is another side effect of chronically snorting cocaine. A constant runny nose, inflamed sinuses, and perforation of the nasal cartilage are symptoms of this destruction.

Cocaine is now considered one of the most addictive drugs. During the 1980s, medical opinion changed from the belief that cocaine was only psychologically addictive to the knowledge that it was physiologically addictive as well. Chronic users often develop tolerance, one classical criterion for addiction. Withdrawal symptoms may not be as obvious as in other drug addictions but include depression, irritability, and drug craving.

High in the Andes, coca chewing is still legal, and fresh coca leaves can be purchased at local markets at about one dollar per pound. Traditional ways of chewing coca with lime have not changed for centuries among the descendants of the Incans. Leaves are also brewed as herbal teas for a mild stimulating beverage and as a folk remedy for altitude sickness. The growing of coca for the illegal drug trade, however, has become a multibillion dollar business for the drug cartels. Some authorities believe that the only way to stop the drug traffic is to eradicate the source—the coca plant.

Cutting, burning, and herbicides have been used without much impact in the vast mountainous areas, where an untold number of farmers raise coca. Biological controls have been considered and researchers have studied the use of caterpillars that feed on coca leaves. Perhaps the control will come about naturally; a fungal disease that attacks the roots of the coca plant has reportedly been destroying many plants in Peru more effectively than eradication efforts.

Tobacco

Nicotine, the major alkaloid in tobacco, is the most addictive drug in widespread use and cigarette smoking is a notoriously difficult habit to break. The detrimental effects of tobacco smoking on health have been known for some time, yet the addictive powers of the drug compel many to continue this unhealthy practice. The rights of smokers versus the rights of nonsmokers is a hotly debated issue as more and more places are opting for a smoke-free environment. None of this was foreseen when Columbus and his crew encountered the inhabitants of Cuba smoking *tabacos.*

A New World Habit

Nicotiana is a genus of approximately 65 species, with the majority native to the New World and the remaining species found in Australia. The two species that have been widely cultivated are *N. tabacum,* which is one to three meters (3–9 ft) in height and *N. rustica,* a shrubby plant that is just over a meter in height. Both are tetraploid hybrids that originated in South America from natural hybridization of closely related *Nicotiana* species, followed by polyploidy. Members of the Solanaceae (the nightshade family), these *Nicotiana* species are herbaceous annuals in temperate areas and have extremely large leaves subtending a terminal inflorescence of white, yellow, or pink tubular flowers. Nicotine is synthesized in the roots and translocated to the leaves; it is the leaves that become tobacco after curing.

Natives of the Americas used tobacco for centuries, possibly even millennia, before the European discovery of the New World. Archaeological evidence indicates that tobacco was the first narcotic used in South America. Tribes throughout the continent utilized the leaves by chewing, smoking, as snuff (through the nose), or drinking infusions in various rituals and medicine. When Columbus set ashore on the island of Cuba in November 1492, his men encountered natives "drinking smoke" by inserting burning rolled-up leaves of *tabacos* into their nostrils and inhaling. Soon the plant was introduced to Spain, and within a short time thousands were addicted to the smoking habit.

Introduction to France followed soon after with credit generally given to Jean Nicot, French ambassador to Portugal. Nicot sent ground leaves and seeds of *N. rustica* to the French court for medicinal use as snuff. Nicot popularized the use of tobacco, and Linnaeus honored him in naming the genus 200 years later. John Hawkins, Francis Drake, and Walter Raleigh have all been credited with introducing and popularizing tobacco smoking in

(a)

(b)

Figure 20.7 *(a) Tobacco fields in Kentucky. (b) The flavor and aroma of tobacco develops during the curing process.*

England. Unlike Spain, where cigar smoking was preferred, pipe smoking became popular in England. Soon plantations in the colony of Virginia were established for growing tobacco, and, in 1629, over one and one-half million pounds of tobacco leaves were shipped to England. Tobacco became a major commodity in the colonies. Indentured servants from England and African slaves were used in the cultivation and processing of tobacco leaves on the large plantations in Maryland, Virginia, and the Carolinas. Many fortunes were made by the tobacco aristocracy and many of the founding fathers (George Washington, Thomas Jefferson, and Patrick Henry) were tobacco growers.

Cultivation Practices

Tobacco growing demands much care and attention. The plants are susceptible to many fungal, viral, and nematode pathogens, and conditions must be controlled to minimize spread of disease-causing agents. Seeds are planted in specially prepared seedbeds and seedlings are later transplanted to the fields (fig. 20.7). When the plants begin to flower, the inflorescence and any side branches are removed to promote maximum leaf expansion. The leaves are harvested and taken for curing as they mature.

The method of curing varies with the intended use of the tobacco, but curing is basically a fermentation-and-drying process during which the characteristic aroma and flavors of tobacco develop. The leaves are hung to dry at elevated temperatures, during which the chlorophylls disappear, resulting in a yellow-brown color (fig. 20.7). The leaves are sorted and graded, then aged for several months to several years to fully develop the aromas and flavors. The final step is cutting and blending different grades, as well as hundreds of flavoring additives, to produce the desired mixtures for cigarettes, cigars, pipe tobacco, and chewing tobacco.

Although there were prototypes during the Mayan and Aztec eras, the modern cigarette made its appearance in Europe after the Crimean War. The development of a cigarette manufacturing machine in 1881 produced a ready-made inexpensive tobacco product for the masses; however, it was not until after World War I that cigarette smoking dominated all other uses of tobacco in the United States. Per capita cigarette consumption in the United States reached a peak in 1963 at 42%. Although the number of smokers has declined to 25.5%, there is evidence that the number of cigarettes smoked per day has actually increased despite massive evidence on the negative health effects. Worldwide there are approximately one billion smokers (20% of the world population) who consume 5 trillion cigarettes a year.

Health Risks

When tobacco was first introduced to Europe, the Indian claim that tobacco had magical and curative powers was accepted by many. However, one vehement critic of tobacco use was King James I of England who wrote in *A Counterblaste to Tobacco* in 1604:

> A custome lothsome to the eye, hateful to the
> Nose, harmeful to the braine, dangerous to the
> Lungs, and the blacke stinking fume thereof,
> neerst resembling the horrible Stigian smoke of
> the pit that is bottomlesse.

Despite this royal condemnation, tobacco use increased exponentially in the subsequent decades. For the next 350 years claims about the healing powers (a panacea for most ailments) of tobacco persisted, and even in 1948 the *Journal of the American Medical Association* was stating that

> "more can be said in behalf of smoking as a form of escape from tension than against it . . . there does not seem to be any preponderance of evidence that would indicate the abolition of the use of tobacco as a substance contrary to the public health."

By the 1950s evidence was starting to accumulate that cigarette smoking was, indeed, harmful to health. But this was not widely accepted until the Surgeon General's report of 1964 that stated cigarette smoking was linked to lung cancer, heart disease, emphysema, and other diseases. It is now recognized that 85% of the lung cancer in men and 75% in women is due to cigarette smoking; in fact, 30% of all cancer deaths can be linked to smoking. (Chewing tobacco is strongly linked to oral cancers; long-term users have a 50-fold increase in the risk of developing oral cancer.) Each year 120,000 United States deaths due to coronary heart disease are attributed to smoking, and smokers have double the risk of developing heart disease. Pregnant women who smoke have a greater incidence of low-birth-weight babies and higher rates of premature and stillborn births. The risk of developing emphysema is 4–25 times greater than for nonsmokers, and smokers generally have 8.3 years shorter life expectancy.

Smokers are not the only ones affected by tobacco smoke. Much research has focused on the health effects in passive smokers, nonsmokers who are inhaling environmental tobacco smoke (secondary smoke). Children of smokers have an increased risk of respiratory disorders and reduced lung function, and in adults, passive smoking is a cause of lung cancer. In fact, it was the concern over the hazards of passive smoking in airplanes that prompted the ban on all smoking on domestic flights in the United States.

Tobacco smoke is a complex mixture of over 2,000 chemicals, plus extremely fine particulates, all of which can enter the lungs. Included in the mixture is nicotine, tars, carbon monoxide, hydrogen cyanide, aromatic hydrocarbons, and a host of other hazardous substances (table 20.1).

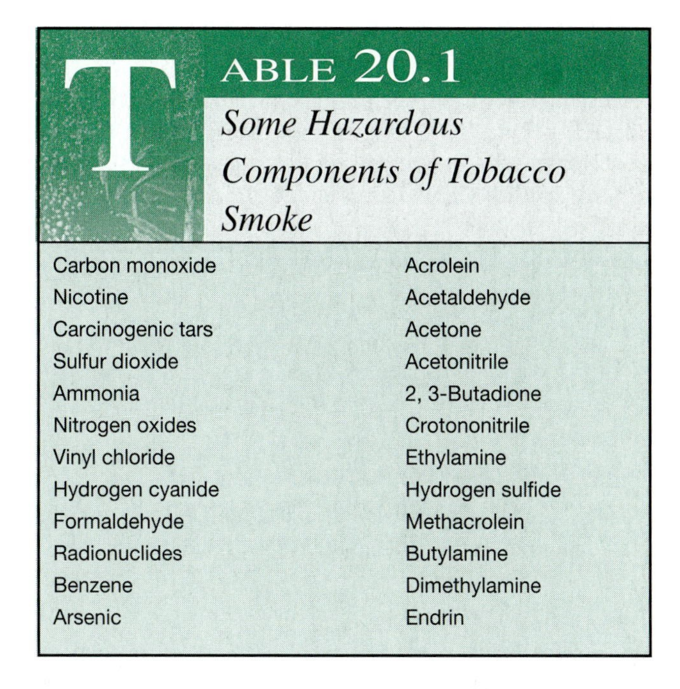

TABLE 20.1

Some Hazardous Components of Tobacco Smoke

Carbon monoxide	Acrolein
Nicotine	Acetaldehyde
Carcinogenic tars	Acetone
Sulfur dioxide	Acetonitrile
Ammonia	2, 3-Butadione
Nitrogen oxides	Crotononitrile
Vinyl chloride	Ethylamine
Hydrogen cyanide	Hydrogen sulfide
Formaldehyde	Methacrolein
Radionuclides	Butylamine
Benzene	Dimethylamine
Arsenic	Endrin

The alkaloid nicotine is addictive in any form of tobacco, and its addictiveness is as tenacious as other narcotics. Nicotine is volatilized during burning and absorbed upon inhalation. It exerts a stimulant action on the central nervous system, promoting arousal in certain areas of the brain. Nicotine also acts on the peripheral nervous system, first stimulating then blocking transmission of nerve signals. Sensations of heat, pain, and hunger are inhibited, while an increase in adrenaline causes increased heart rate and blood pressure, and constriction of blood vessels. Nicotine was isolated early in the nineteenth century and is known to be one of the most toxic plant poisons. In its pure form, ingesting just a few drops is lethal, causing death by respiratory paralysis. The poisonous properties have been put to use as an especially effective insecticide. Nicotine has also been implicated in various forms of cancer, with evidence indicating that nicotine is converted into carcinogenic products in the body.

Besides nicotine, other dangerous components of tobacco smoke include tars, organic substances produced during the burning of tobacco leaves. Tars are known carcinogens and cigarette manufacturers have attempted to reduce the tar, as well as the nicotine, content through filters. There is evidence that reducing the tar and nicotine content can somewhat decrease the negative health effects of smoking, although recent studies indicate that filters themselves might create health hazards. Poisonous gases are also found in cigarette smoke; carbon monoxide, hydrogen cyanide, and formaldehyde are just a few of the gaseous components. Carbon monoxide, in particular, has been shown to be associated with coronary heart disease and the retardation of fetal growth.

Like any addictive substance, withdrawal from nicotine is difficult and many approaches have been used with limited success. One recent innovation is the use of transdermal nicotine patches. Since nicotine can be absorbed through the skin, the patches provide a steady supply of the alkaloid, eliminating the other dangers associated with smoking. The concentration of nicotine in the patches is gradually reduced, weaning the patient from the addiction over several months. The obvious drawback to the patches is the continued use of a dangerously addictive substance, but many lives may be saved if it does, in fact, facilitate quitting smoking. Fifteen to 20 years after a smoker quits, the risks of developing lung cancer are 80% less than active smokers.

PEYOTE

As with many of the psychoactive plants discussed in this chapter, peyote has a long history of use, dating back at least 3,000 years in Mexico. Early accounts from sixteenth century Spanish missionaries described the plant and its use as a hallucinogen in religious rituals among many Indian tribes. Bernardino de Sahagun indicated

> "Those who eat or drink it see visions either frightful or mirthful; the intoxication lasts two or three days and then ceases . . . it sustains them and gives them courage to fight and not to feel hunger nor thirst; and they say that it protects them from all dangers."

Although the Spanish initially attempted to suppress the peyote rituals, the practices have continued to this day.

Mescal Buttons

Peyote (from the Aztec word, *peyotl*) is the cactus *Lophophora williamsii,* native to Mexico and southwestern Texas. The crown of the spineless gray-green cactus is sliced and dried into "mescal buttons" (fig. 20.8). Dried, the buttons can keep indefinitely, retaining their hallucinogenic properties, since the active ingredients are not volatile. The buttons are softened by saliva in the mouth and then swallowed; alternatively, the buttons may be soaked in water and the water consumed.

Peyote contains up to 30 alkaloids, with mescaline the most active hallucinogen in the group. Again the nervous system seems to be affected, since mescaline is similar in structure to a neurotransmitter. Investigations have shown that the effects of mescaline alone differ from the action of consuming the whole mescal button, where the alkaloids may interact synergistically. Intoxication induced by peyote may consist of strikingly brilliant visions, as described so vividly by Havelock Ellis in 1898:

> At first there was merely a vague play of light and shade which suggested pictures . . . Then, in the course of the evening, they became distinct, but still indescribable—mostly a vast field of golden jewels, studded with red and green stones, ever changing . . . I would see thick, glorious fields of jewels, solitary or clustered, sometimes brilliant and sparkling, sometimes with a dull rich glow. Then they would spring up into flowerlike shapes beneath my gaze, and then seem to turn into gorgeous butterfly forms or endless folds of glistening iridescent, fibrous wings of wonderful insects . . . Most usually there was a combination of rich sober color, with jewel-like points of brilliant hue . . .

Not all peyote experiences consist of these pleasant visions. Other writers describe anxiety and horribly frightening visual hallucinations of snakes and demons.

(a)

(b)

Figure 20.8 *(a) Peyote is a cactus (*Lophophora williamsii*) that is central to the dogma of the Native American Church. (b) The dried, cut tops of peyote are called mescal buttons and contain the alkaloid mescaline.*

The intoxication is often accompanied by physical effects as well. These include nausea, tremors, chills, and vomiting. Because of the interplay of alkaloids, peyote hallucinations are extremely complex, inducing a range of sensory experiences.

Native American Church

The use of peyote by members of the Native American Church has been a topic of controversy for over one hundred years. The origins of the Native American Church can be traced to the 1870s, when the Kiowa and Comanche Indians learned of peyote from tribes in Mexico and brought the plant back with them to Oklahoma (then Indian Territory). The tribes established a whole new cult, with a combination of Christianity and peyote rituals. The cult spread rapidly through the Southwest to the Plains states, and eventually even to Canada, incorporating elements from the cultures of various tribes. As the cult grew and spread, it met with much opposition, especially from missionaries and government officials. In an attempt to protect their

TABLE 20.2

Plants of Known Psychoactive Value

Scientific Name	Common Name	Family	Active Principle	Physiological Action
Atropa belladonna	Belladonna	Solanaceae	Atropine; scopolamine; hyoscyamine	Hallucinogen
Cannabis sativa	Marijuana	Cannabaceae	Tetrahydrocannabinol	Hallucinogen
Datura spp.	Jimsonweed	Solanaceae	Atropine; scopolamine; hyoscyamine	Hallucinogen
Erythoxylum coca	Coca	Erythroxylaceae	Cocaine	Stimulant
Hyoscyamus spp.	Henbane	Solanaceae	Atropine; scopolamine; hyoscyamine	Hallucinogen
Ipomoea spp.	Morning Glory	Convolvulaceae	Lysergic acid	Hallucinogen
Lophophora williamsii	Peyote	Cactaceae	Mescaline	Hallucinogen
Mandragora officinarum	Mandrake	Solanaceae	Atropine; scopolamine; hyoscyamine	Hallucinogen
Myristica fragrans	Nutmeg	Myristicaceae	Volatile oils	Hallucinogen
Nicotiana spp.	Tobacco	Solanaceae	Nicotine	Stimulant/depressant
Papaver somniferum	Opium poppy	Papaveraceae	Morphine; codeine; heroin	Depressant

religious rights, Native Americans formally incorporated their movement into the Native American Church in 1918. Even though officially recognized as a church and protected by the Bill of Rights, the opposition to peyote use continued. A long legal battle culminated in a Supreme Court ruling that gave the members of the Native American Church the right to use peyote as a sacrament. However, a more recent Supreme Court decision has restricted this right. In 1990, the Court upheld Oregon's right to outlaw peyote even for religious use, stating that the First Amendment does not permit people to break the law for the sake of freedom of religion. The controversy continues.

Lesser Known Psychoactive Plants

The previous discussion has focused on psychoactive plants that have become major problems in modern society. This has by no means exhausted the list of known psychoactive plants (table 20.2). Following is a description of some lesser-known plants that have potent effects on the central nervous system.

Certain members of the morning glory family, the Convolvulaceae, also figured prominently as hallucinogens and were revered as powerful medicines among the Aztecs; these morning glories continue to be used by tribes in present-day Mexico. The seeds of *Ipomoea violoaceae,* as well as other *Ipomoea* and *Rivea* species, contain amides of D-lysergic acid similar to, but far milder than, the potent hallucinogen LSD (*see* Chapter 24). Known as *ololiuqui* among the Aztecs, it was used in divination as well as religious and magical rituals. The shaman would consume a drink prepared from the seeds and, in the hallucinogenic state that followed, would divine the cause of a person's illness.

Nutmeg, *Myristica fragrans,* has a long history as a culinary spice, but has also been valued in some circles for its hallucinogenic properties. It was known to practitioners of Ayurvedic medicine in India as the "narcotic fruit." It was also mixed with betel nut or tobacco snuff for its intoxicating effects, and was occasionally used in other areas as a substitute hallucinogen when the drug of choice was unavailable. Since the active principle is volatile, the reaction to nutmeg is unpredictable and thus potency varies greatly. Side effects, however, are predictable and extremely unpleasant (headache, nausea, dizziness, vomiting, irregular heartbeat, etc.), making this an unpopular choice as a hallucinogen.

In addition to the potent psychoactive substances described in this chapter, caffeine and caffeinelike alkaloids should not be overlooked, since they too affect the central nervous system. Their impact on society and their physiological actions have been fully discussed in Chapter 16. Also, many other psychoactive plants have been used by cultures throughout the world, particularly in the tropical rain forests, which are only now being investigated. These studies should provide further evidence that will reinforce the pivotal role of psychoactive drugs in human society.

A group of alkaloids with similar structure and similar physiological action can be found predominantly in members of the Solanaceae (the nightshade family). These are known as the **tropane alkaloids** and include atropine, hyoscyamine and scopolamine. These alkaloids have a variety of physiological effects on the body:

they relax smooth muscles;

they dilate the pupils of the eye;

they dilate blood vessels;

they increase heart rate and body temperature;

they induce sleep and lessen pain;

they stimulate and then depress the central nervous system; and

some can even induce hallucinations.

As a group, the tropane alkaloids are extremely toxic, capable of inducing coma and death due to respiratory arrest. Because of the toxicity of these solanaceous plants, Europeans feared the whole plant family and were reluctant to accept the potato and tomato (see Chapters 6 and 14), the New World relatives. One unique property of the tropane alkaloids is their ability to be absorbed through the skin, and they have been administered this way. All three of these tropane alkaloids occur in varying concentration in deadly nightshade or belladonna (*Atropa belladonna*), henbane (*Hyoscyamus* spp.), mandrake (*Mandragora officinarum*), and Jimsonweed (*Datura* spp.).

Atropa belladonna, a branching herbaceous perennial native to Europe and Asia, has a long history of use as a medicinal, psychoactive, and poisonous plant. One use of the plant, reflected in its specific epithet "belladonna" was the practice by Mediterranean women of applying the plant's juice to the eyes. The result was dilation of the pupils to produce an alluring effect; hence "bella donna" or beautiful lady. This physiological response is due to the atropine in the plant and is used medicinally today by ophthalmologists. Atropine has multiple medicinal applications:

(1) as an antispasmodic for treating Parkinson's disease, epilepsy, and stomach cramps;

(2) as a bronchodilator for treating asthma;

(3) as a heart stimulant following cardiac arrest;

(4) as an antidote for various poisons or overdoses.

This last use is ironic, based on its toxic properties and its historical use as a poison itself. The application of botanical poisons probably reached its height during the Italian Renaissance, when special "schools" gave instructions in their preparation from deadly nightshade, henbane, mandrake, and other plants. These three plants were also used by witches and sorcerers of the Middle Ages to

Chapter Summary

1. The use of psychoactive plants in human society dates back millennia. The active principle in psychoactive plants works by affecting the release of neurotransmitters in the central nervous system. Depending on their effect, psychoactive drugs can be classified as stimulants, depressants, or hallucinogens. A psychoactive drug is classified as addictive if it causes one or more of the following actions: psychological dependence, physiological dependence, and/or tolerance. The active principles in most psychoactive drugs are classified as alkaloids.
2. The opium poppy (*Papaver somniferum*) is the source of a milky latex from the green capsules that, when dried, becomes crude opium. Opium has been eaten, drunk, and smoked for centuries. More than 20 alkaloids have been identified in opium, but the chief ones are morphine and codeine.
3. Heroin is a semisynthetic derivative of morphine and the most strongly addictive of all the opiates. Heroin addiction is a major drug problem in most western cities.
4. Marijuana (*Cannabis sativa*) has an equally long history as a hallucinogen in India. It is the resin found in the upper leaves and flowers of the pistillate plants

prepare magic potions. These decoctions would induce hallucinations and frenzies during witches' convocations. Images of witches flying through the air on broomsticks and transforming themselves into animals originated as delusions of their drug-induced state (box fig. 20.1).

Along with its witchcraft applications, henbane was used medicinally in Europe for centuries as a sedative and pain reliever, especially effective for toothaches. Mandrake is an herbaceous perennial with a distinctive branching taproot that resembles the human form (see fig. 19.2). Because of the doctrine of signatures, this plant was assumed to have magical powers to treat male and female sexual complaints. Malelike roots were thought to insure sexual prowess and femalelike roots would supposedly cure infertility.

Unlike the previous solanaceous plants, *Datura* spp. have a cosmopolitan distribution and have been extensively used by many indigenous peoples for both medicinal and hallucinogenic purposes. In the New World, there are several species of *Datura* that have an extensive history as sacred hallucinogens. The generic name is based on a Sanskrit word *dhatura,* meaning poison, and reflecting its toxic properties. *Datura stramonium* is probably the most widely distributed species and is cultivated for its scopolamine content, which is used today for motion sickness and for its sedative effects. The common name for this species, jimsonweed or Jamestown weed, refers to an incident of accidental poisoning of British sailors in colonial Virginia in 1676. They mistook *Datura* for an edible plant and suffered the consequences of consuming this potent hallucinogen,

> "for they turn'd Fools upon it for several Days . . .
> In this frantick Condition they were confined, lest
> they should in their Folly destroy themselves . . .
> and after Eleven Days, return'd to themselves again,
> not remembering anything that had pass'd."

Box Figure 20.1 *A witch is being prepared for the sabbat in this engraving from the eighteenth century. The broomstick was used to administer the salve of tropane alkaloids to the body.*

that contains the psychoactive and nonalkaloid principle, THC.

5. Cocaine is derived from the leaves of the coca plant (*Erythroxylum coca*), a shrub native to South America. Chewing a wad of the leaves to experience mild stimulating effects has been practiced for at least 3,500 years in the Andes and is still an ongoing practice among native groups today. The extraction of cocaine from the leaves in the 1850s encouraged the spread of cocaine use in Europe and the United States around the turn of the twentieth century. The enchantment with cocaine use faded as the negative effects became apparent. Unfortunately, cocaine use in the more dangerously addictive forms of free base and crack became popular in the 80s. The lessons of the dangers of cocaine had to be relearned.
6. The most addictive drug in widespread use is nicotine, the major alkaloid in tobacco (*Nicotiana* spp.). Native to the New World, the smoking of tobacco has long been associated with the rituals of many native tribes. It is now recognized that the habit of cigarette smoking increases the risk of certain cancers, especially lung, and coronary heart disease. Passive smokers, especially the children of smokers, have been shown to have higher incidences of respiratory disorders and even lung cancer.
7. The use of peyote (*Lophophora williamsi*) has long been

associated with sacred rituals of many Indian tribes in Mexico. In the United States, the eating of mescal buttons has been a practice and right of members of the Native American Church, protected under the auspices of religious freedom.

8. Certain herbs that contain the tropane alkaloids have long been associated with the practice of witchcraft in medieval Europe; many of these drugs have medicinal qualities that are used in modern medicine today.

Review Questions

1. Distinguish between stimulants, hallucinogens, and depressants.
2. What is meant by the term *addictive?* Which psychoactive plants are known to be addictive?
3. What active principle accounts for the psychoactivity in most plants?
4. Give the history and use of the following psychoactive plants: opium poppy, marijuana, coca shrub.
5. Give examples of how psychoactive plants have been used in religious, magical, and healing rituals.
6. Trace the history of tobacco smoking and describe the health effects of this habit.

Further Reading

Blackwell, Will H. 1990. *Poisonous and Medicinal Plants.* Prentice Hall, Englewood Cliffs, NJ.

Boucher, Douglas H. 1991. Cocaine and the Coca Plant. *BioScience,* 41(2): 72–76.

Emboden, William. 1979. *Narcotic Plants.* Macmillan Publishing, New York.

Lehane, Brendan. 1977. *The Power of Plants.* McGraw-Hill, New York.

Lewis, Walter, and Memory P. F. Elvin-Lewis. 1977. *Medical Botany: Plants Affecting Man's Health.* John Wiley & Sons, New York.

Mann, John. 1992. *Murder, Magic, and Medicine.* Oxford University Press, Oxford.

Meyer, John A. 1992. Cigarette Century. *American Heritage,* 43 (8): 72–80.

Musto, David. Summer, 1989. America's First Cocaine Epidemic. *Wilson Quarterly,* pp. 59–65.

Musto, David F. 1991. Opium, Cocaine, and Marijuana in American History. *Scientific American,* 265 (1): 40–47.

Polan, Michael. 1995. How Pot Has Grown. The New York Times Magazine, Feb 19, 30–57.

Schultes, Richard Evans. 1979. *Plants of the Gods: Origins of Hallucinogenic Use.* McGraw-Hill. New York.

Weil, Andrew. 1995. The New Politics of Coca. The New Yorker, May 15, 70–80.

White, Peter. February, 1985. The Poppy. *National Geographic,* 167 (2): 143–189.

White, Peter. 1989. An Ancient Indian Herb Turns Deadly: Coca. *National Geographic,* 175 (1): 3–47.

21

Poisonous and Allergy Plants

Chapter Outline

Chapter Concepts

1. Poisonous plants are everywhere in the environment, including some of the most common wild, yard, and houseplants, and many accidental poisonings, especially of young children, occur each year.
2. Many plants have evolved chemicals that protect them from insects and other pests; some of these chemicals are being used as natural insecticides.
3. Certain plants are capable of triggering allergic reactions ranging from respiratory allergies to contact dermatitis.

The focus of this chapter is on plants that adversely affect the health of humans and other animals. Plants are the source of toxins so deadly that fatalities can result from miniscule amounts, yet poisonous compounds have also been utilized in medicine and as natural insecticides. Other harmful effects include allergic reactions that can be debilitating or even fatal.

Notable Poisonous Plants

As discussed in earlier chapters, the ability of ancient peoples to identify edible plants was the first step in the development of foraging societies. Also poisonous plants would be recognized and the information passed on to avoid future calamities. There is also evidence that from the earliest times, poisonous plants were utilized as a method of capturing prey. In fact, the word *"toxic"* is derived from the ancient Greek word *toxikon,* meaning arrow poison.

At various times in history, plant poisons were also employed for the more nefarious purpose of disposing of one's enemies. The Ebers papyrus lists the many toxins known to the Egyptians of 3,500 years ago, while in ancient Athens a cup of poison hemlock was the standard method of capital punishment. Knowledge about toxic plants reached a peak in the Renaissance in Europe when so-called "succession powders" often ensured an untimely death and a calculated ascent to power. Members of the De Medici family of Italy, especially Catherine De Medici, were renowned poisoners of this time.

As modern society has distanced itself from its natural surroundings, knowledge of toxic plants has shrunk to a small circle of trained professionals. Most people would be surprised to learn that many common weeds, landscape plants, and houseplants contain deadly poisons. It is estimated that there are thousands of plants and fungi that produce toxic substances, with hundreds of these found in the Americas alone. Why are so many plants poisonous? The obvious answer is that by making at least some part of the plant toxic, it affords protection against grazing animals and herbivorous insects.

The following discussion of poisonous plants is restricted to plant poisons that harm the body when they are taken internally, either by consumption or injection. **Alkaloids** and **glycosides** play a prominent role as poisonous compounds; remember that many of the medicinal and psychoactive plants can also be fatal at certain doses (*see* Chapters 19 and 20).

Poisonous Plants in the Wild

Strychnine, an alkaloid obtained commercially from the seeds of the Asian-Indian tree *Strychnos nux-vomica* (Loganiaceae), is a stimulant of the central nervous system, especially the spinal cord. This powerful nerve toxin induces muscle spasms and convulsions and is unique in the way it magnifies sensations of sight, smell, touch, and hearing. In fact, any sudden stimulus, such as a loud noise, can trigger a seizure in an affected individual. Historically, strychnine has been used to treat a variety of ailments (constipation, impotence, barbiturate poisoning), but its medical use today is confined to neurologic research. It also has been used illegally to enhance athletic performance. Strychnine is still employed, however, as a rodent poison, particularly for moles, and is the main ingredient in the common poison "peanuts" for the garden.

Figure 21.1 *In many tribes of South America, hunters use darts poisoned with curare in their blowguns.*

Curare is the arrow poison employed by many South American tribes in hunting game. It is applied to the tip of the arrow and if the arrow finds its target, the Indians say "he to whom it comes falls" (fig. 21.1). Curare paralyzes quickly. Various methods of preparations, and over 70 different species, have been used by South American tribes to make curare. However, the bark of two woody vines of the lowland tropical rain forest, notably *Strychnos toxifera* and *Chondrodendron tomentosum* (Menispermaceae), figure most prominently. The French physiologist Claude Bernard was the first to study the action of curare in the nineteenth century, correctly concluding that it blocked nerve impulses at the junction of nerve and muscle. In 1942, H. R. Griffith was able to demonstrate the anesthetic value of an extract of curare. It was the prior isolation of the chemical tubocurarine from curare by British researchers that made this application possible since tubocurarine, by itself, produces only a partial, reversible paralysis. An injection of curare, or the alkaloid tubocurarine, produces immediate muscle relaxation by blocking nerve impulses. Movement is impossible because of skeletal muscle paralysis; swallowing and speaking become difficult; and if untreated, death by asphyxiation follows due to impairment of the diaphragm. The medicinal value of curare as a muscle relaxant is unsurpassed and has found its greatest application in surgery, where adequate muscle relaxation can be achieved without excessive general anesthesia. This muscle toxin also has value in the treatment of spastic cerebral palsy, myasthenia gravis, polio, and tetanus.

(a)

(b)

Figure 21.2 *Two poisonous members of the carrot family are (a) poison hemlock and (b) water hemlock.*

Two members of the Apiaceae, poison hemlock (*Conium maculatum*) and water hemlock (*Cicuta* spp.) found in many areas, share the distinction of being among the most poisonous wild plants in North America (fig. 21.2). (These plants should not be confused with hemlock, the common name of the conifer *Tsuga canadensis* of northern

Figure 21.3 *The larval stage of monarch butterflies feed exclusively on milkweeds* (Asclepias), *passing on cardioactive glycosides to the adult butterfly.*

forests, which is not known to be poisonous.) Both are large, perennial herbs with big white umbels and pinnately compound leaves. They are often confused with each other and with such edible relatives as wild carrot and parsnip. Poison hemlock is notorious as an agent of death in ancient Athens; Socrates was condemned to death nearly 2,400 years ago for allegedly corrupting the young with his teaching philosophy and forced to drink the juice of poison hemlock as punishment. The alkaloid coniine is a central nervous system stimulant that affects the body in a manner similar to a nicotine overdose; paralysis creeps from the lower limbs upward. Death is due to paralysis of the diaphragm and subsequent respiratory failure.

The water hemlock, as its name implies, grows in wet or swampy areas. It differs only slightly in appearance from the poison hemlock in that the leaves are more finely divided and there are no purple splotches on the lower stem. The toxin is an alcohol, cicutoxin, which is found in highest concentration in the yellow sap that exudes from cut roots. Cicutoxin produces violent convulsions and, unless treated promptly, death invariably follows.

Milkweeds (*Asclepias* spp.) are common weeds and easily recognized by opposite or whorled leaves, milky juice when the plant is cut, dense umbels of flowers, and follicles with silk-tufted seeds (fig. 21.3). The poisonous compounds in milkweeds include the resinous galitoxin, as well as **cardioactive glycosides** similar to the drug digitalis (*see* Chapter 19). Livestock and humans, especially children, have been poisoned by eating milkweed. The larvae of the monarch butterfly can eat milkweed without ill effects and in fact, these caterpillars feed exclusively on certain milkweeds. Not only are these cardioactive glycosides not harmful to the caterpillars, but they are passed on to the adult, or butterfly, stage intact. The presence of these cardioactive glycosides causes the unfortunate bird who eats a monarch to become ill and vomit for several hours. Birds quickly learn to avoid eating the distinctively colored butterflies, going so far as to also avoid viceroy butterflies with similar orange and black markings.

Poisonous Plants in the Backyard

Most homeowners are unaware that many of the commonplace landscape plants are poisonous and sometimes fatal, especially to young children. One of the deadliest is the evergreen shrub oleander, widely planted in the southern states and California for its showy red, pink, white, or yellow flowers. All parts of *Nerium oleander* (Apocynaceae) contain dangerous levels of over 50 toxic compounds, including two cardioactive glycosides, oleandroside and nerioside which are similar in action to digitalis. The toxins are found even in the smoke from burning leaves and branches, and breathing it can be injurious. Unwary picnickers have died just from the toxin absorbed by their hotdogs roasted on green oleander sticks.

Other common poisonous shrubs are the yews, *Taxus* spp., most often grown as hedges, shrubs, or small trees. These conifers present an attractive picture in the landscape, having dark green, needlelike leaves that contrast with the red cup-shaped arils surrounding each black seed. The poisonous taxine alkaloids are found in all parts of the plant except the aril. Taxine acts on the central nervous system and, after the initial symptoms of dizziness and dry mouth, the heart begins beating erratically and breathing becomes labored. Death is due to cardiac or respiratory failure. Young children are attracted to the bright red arils and it has been reported that eating only one or two seeds may be fatal to a small child.

Rhododendrons and azaleas are broad-leaved, flowering shrubs of the heath family (*Ericaceae*) that contain poisonous compounds so toxic that leaves, tea made from the leaves, flowers, poilen and nectar, and even the honey made from it, are deadly. Grayanotoxins (andromedotoxins) are responsible for the toxicity symptoms by first stimulating and then blocking the nervous regulation of the heart. Writings from ancient Greece record the poisoning of soldiers from wild rhododendron honey. Humans have not been the only casualties; bees have also been poisoned from rhododendron nectar. Wild relatives of rhododendron are the laurels (*Kalmia* spp.), which also contain grayanotoxins and have a history as poisonous plants among certain Native Americans groups. The Delaware Indians would make a tea from the mountain laurel for use as a suicide potion.

The Fabaceae, or bean family, is known as a major source of alkaloids. Therefore, it is not surprising that a number of poisonous plants are legumes. The rosary pea (*Abrus precatorius*), lupines (*Lupus* spp.), black locust (*Robinia pseudoacacia*), *Wisteria* spp., Vetch (*Vicia* spp.), and Golden Chain (*Laburnum anagyroides*) are just a few legumes implicated in human poisoning. *Sophora secundiflora,* a shrub or small tree with pinnately compound leaves, stands out from the other poisonous legumes because of its long history of use as a hallucinogen by Indians of the North American Southwest. The bright red seeds (mescal beans) were the focal point of the Red Bean Dance practiced by the Arapaho and Iowa tribes, as well as by Indians in northern Mexico. Ingesting the seeds produces visions that were believed to reveal the future. Eating mescal beans is a dangerous practice since the beans contain the highly toxic alkaloid cytisine. Cytisine is similar to nicotine and high doses cause death due to paralysis of the respiratory muscles.

Surprisingly, many of the herbaceous ornamentals planted around homes, parks, and office buildings and cherished for their beauty contain deadly poisons. Spring flowering members of the Liliaceae and Amaryllidaceae such as tulip, hyacinth, star-of-Bethlehem, daffodil, and narcissus fall into this category. Few would suspect that the delicate bell-shaped flowers of lily-of-the-valley (*Convallaria majalis*), and also the leaves, bulbs, and red berries, conceal toxic concentrations of a host of glycosides and saponins. The saponins are irritants to the digestive tract causing vomiting, diarrhea, and a burning sensation in the mouth and throat. But it is the cardiac glycosides that cause fatalities by slowing the heartbeat to the point of cessation. One of them, convallotoxin, is reportedly one of the most toxic of all the known cardioactive glycosides. In the past, lily-of-the-valley preparations have been used medicinally in the treatment of cardiac insufficiency, but this application has been abandoned in favor of the more reliable digitalis from foxglove.

The spurge family (Euphorbiaceae) is another family that is potentially harmful due to its characteristic milky latex or sap, which exudes when the plant is cut or bruised. Poinsettia (*Euphorbia pulcherrima*), crown-of-thorns (*E. milii*), pencil tree cactus (*E. tirucalli*) and snow-on-the-mountain (*E. marginata*) are all known to cause irritation to the skin and mucous membranes. Another member of this family, the castor bean (*Ricinis communis*), is deadly because of its large and colorful, yet highly toxic, seeds. The mottled red, white, or brown seeds (fig. 21.4a and b) are extremely poisonous; ingestion of just one seed may kill a child, while eating three seeds would be fatal for most adults. The toxic protein ricin, most concentrated in the seeds, inhibits protein synthesis in the intestinal wall and causes the clumping of red blood cells. It is said to be the most deadly natural poison known. Symptoms do not develop until several hours or even days after the seeds are eaten. (Note: The seed coat must be broken to release the toxins; swallowing seeds whole provides some protection since the extremely tough seed coat prevents release of the poisons.) Nausea, vomiting, diarrhea, and burning of the mouth and throat are the first symptoms to appear. These are followed by electrolyte imbalances, retinal hemorrhaging, internal hemorrhaging in the digestive system and lungs, and extensive damage to the liver and kidneys. Death results from kidney failure. Interestingly castor

(a)

(b)

(c)

(d)

Figure 21.4 *Many common yard and houseplants are poisonous. (a) Castor bean (*Ricinus communis*) plant, (b) Seeds of the castor bean, (c) Philodendron, (d) Dumbcane.*

bean oil has been a staple of folk medicine from ancient times and is still used today as a laxative. It is not poisonous since ricin is not present in the oil.

Poisonous Plants in the Home

Poisonous plants are everywhere in the environment, including the home. Most calls to poison control centers concerning children ingesting houseplants involve the aroids (Araceae), some of the most popular houseplants and almost all poisonous. The two aroids most frequently cited are philodendron (*Philodendron* spp. and *Monstera* spp.) and dumbcane (*Dieffenbachia* spp.). The philodendrons are either vines or erect plants admired for their heart-shaped or dissected leaves. The dumbcanes are distinguished by speckled leaves on a moderately stout erect stem (fig. 21.4c,d). All parts of both of these plants contain crystal needles (raphides) of calcium oxalate. If swallowed, these crystals cause painful burning and swelling of the lips, tongue, mouth, and throat, which may persist for several days. Talking, swallowing, and even breathing may become labored. In fact, the difficulty in speaking after ingesting *Dieffenbachia* accounts for its common name, *dumb*cane. There are also indications that these plants contain toxic proteins that may intensify the pain and edema. Calcium oxalate crystals are also the culprits in the poisonous wild aroids, jack-in-the-pulpit (*Arisaema triphyllum*) and skunk cabbage (*Symplocarpus foetidus*).

Plants Poisonous to Livestock

Poisonous plants are not only a concern in human health but can also affect the well-being and economics of livestock. Numerous plants have been identified as poisonous to sheep and cattle and, subsequently, to the humans who eat the flesh or drink the milk of poisoned animals. White snakeroot played a devastating role in the settlement of the frontier lands west of the Appalachian mountains in the early nineteenth century. White snakeroot (*Eupatorium rugosum*) is a white-flowered perennial herb of the Asteraceae found abundantly in the rich, moist woods in the eastern half of the United States (fig. 21.5). All parts of the plant contain the toxic alcohol tremetol, which is highly soluble in oils or fats.

Figure 21.5 White snakeroot (*Eupatorium rugosum*) is the cause of milk sickness.

Cattle that ingested white snakeroot developed a condition known as trembles, characterized by a foul breath smelling like acetone, sluggishness, and muscular weakness. In some areas, entire herds were wiped out by trembles. It was eventually observed that humans who ate the meat or drank the milk of afflicted cattle often developed a related condition with symptoms similar to trembles known as milk sickness. Milk sickness was epidemic in certain areas of the frontier. In one county of Ohio, 25% of the population died from this malady. In fact, Nancy Hanks Lincoln, the mother of Abraham Lincoln, was a casualty of this disease. Tremetol was isolated from snakeroot in the early twentieth century and the connection between trembles and milk sickness was fully explained. Tremetol affects the metabolism of the liver by blocking the breakdown of lactic acid, a temporary end-product of cellular respiration, and thereby preventing the Krebs cycle. The subsequent accumulation of lactic acid lowers the pH of the body causing acidosis, which accounts for the acetonelike breath. This blocking of the exergonic metabolic pathways also explains the lethargy. Tremetol's solubility in products high in lipids such as milk also explains two additional observations, that lactating cattle were less affected by trembles than their calves and that young children were often the primary victims of milk sickness. This disease is now, fortunately, only of historical importance because the practice of milk pooling from many cows and dairies dilutes any tremetol present to nontoxic doses and also because of the successful eradication of white snakeroot from pastures.

The leguminous locoweeds or milk vetches (*Astragalus* spp. and *Oxytropis* spp.) have been a bane to ranchers in the western half of the United States and are considered to be some of the most toxic plants for horses, sheep, goats, and cattle. *Loco* is the Spanish word for crazy and refers to the staggering and trembling behavior of poisoned animals who walk into things and react to unseen objects. They are finally overcome by paralysis. The poisonous compounds are an unusual group of alkaloids that affect certain cells of the central nervous system, explaining the behavioral changes observed.

Indian hellebore (*Veratrum* spp.) has been widely used externally as a local anesthetic and additionally it was used internally as a purgative in certain religious ceremonies by many Native American tribes of the Pacific Northwest. In the past, hellebore extracts have been employed in conventional medicine to treat hypertension. Despite these valuable applications, Indian hellebore is known to be a highly toxic plant both for humans and livestock. It is a known teratogen (agent that causes deformities in the fetus) in sheep. Ewes that feed on the western *Veratrum californicum* early in their pregnancy commonly give birth to so-called "monkey-faced" lambs, which have the facial deformity of a single eye. The active compounds in *Veratrum* are a number of alkaloids found throughout the plant but most highly concentrated in the rhizome.

(a)

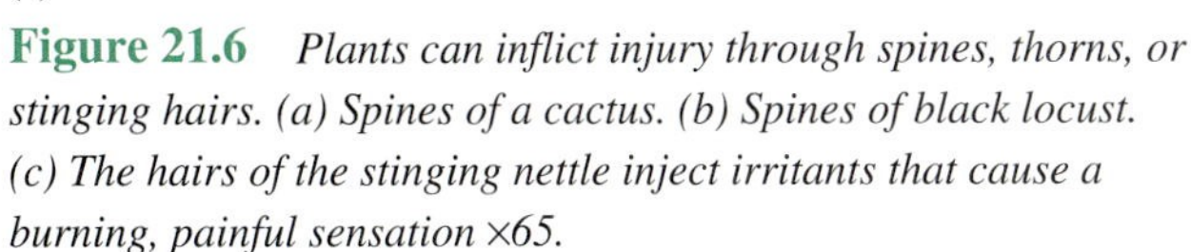

Figure 21.6 *Plants can inflict injury through spines, thorns, or stinging hairs. (a) Spines of a cactus. (b) Spines of black locust. (c) The hairs of the stinging nettle inject irritants that cause a burning, painful sensation ×65.*

(b)

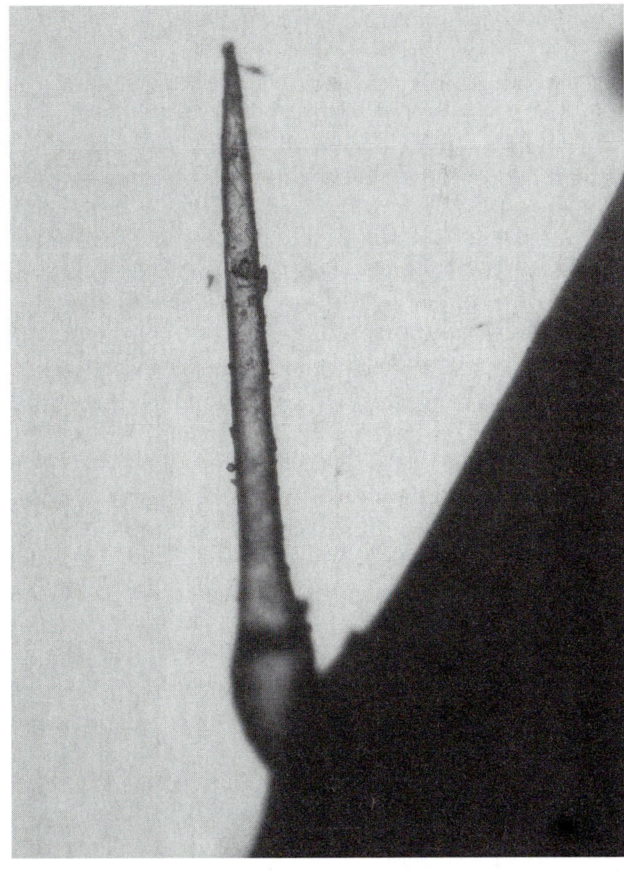

(c)

Plants Causing Mechanical Injury

Plants within this category cause injury usually by puncturing the skin via spines, thorns, prickles, burrs, or hairs. The foremost example of plants in this category would be the cacti. In the family Cactaceae, spines are modified leaves (fig. 21.6a) and seem to be a mechanism to ward off animal predation. Especially dangerous are the hardly noticeable glochids, minute barbed hairs, which, once embedded in the skin or mouth, are difficult to remove and may lead to infections.

Other plants possess stinging hairs that not only puncture the skin's surface but also inject chemical irritants, causing a painful burning sensation. The burning sensation is followed by itchiness, which may last for several hours. Leaves and stems in members of the stinging nettle family (Urticaceae) are covered with hollow hairs similar in operation to a hypodermic syringe (fig. 21.6c). When the tip of a hair penetrates the skin, chemicals such as a histaminelike compound are squeezed from the basal reservoir and injected into the skin, immediately producing a red swelling similar to a mosquito bite or hive (*see* Allergy and the Immune System later in this chapter).

Other plants produce compounds that can cause damage simply by contact when the plant is bruised, or are transported to the skin after the plant is digested. The end result is to make the skin extremely photosensitive and, upon subsequent exposure to ultraviolet light, symptoms vary from a mild sunburn to a blistered secondary burn.

(a)

(b)

Figure 21.7 *Both (a) St. John's wort and (b) snow-on-the-mountain are two plants that can injure through irritant juices.*

The weedy St. John's wort (*Hypericum perforatum*), which contains the specific phototoxin hypericin, was a serious problem for livestock in California during the first half of the twentieth century (fig. 21.7a). Sheep that foraged on St. John's wort became severely affected, with the skin in the face swelling to such an extent that the condition was commonly referred to as bighead. If the lips become affected, the animal is unable to feed and starves to death. Fortunately, a biological control program using beetles that feed exclusively on St. John's wort has successfully curtailed its spread.

Lastly, the milky latex of many members in the Euphorbiaceae have long been known to irritate the skin. In fact, unscrupulous beggars during the Middle Ages would burn their skin with spurge juice to fake leprosy and evoke more sympathy and charity from the public. The manacheel tree, (*Hippomane mancinella*) native to Florida and the Keys, produces a latex extremely caustic to skin and can even cause blindness if it gets in the eyes. Juice from snow-on-the-mountain (*Euphorbia marginata*) can burn skin on contact; in fact, it has been used to brand cattle (fig. 21.7b).

Insecticides from Plants

It is well known that plants produce a great diversity of secondary compounds; in fact, it has been estimated that the total approaches 400,000. Many of these reduce the palatability of the plant to insects; for example, the accumulation of alkaloids or saponins produces a repellent, bitter taste. The high concentration of tannins in oak leaves dissuades insect predation both by their bitter taste and their tendency to bind protein, reducing the digestibility of the leaves. In other cases, the secondary compounds may be outright toxic, such as cyanogens, or may affect the hormonal balance and life cycle of the insect, such as neem oil. Whatever the specific mechanism, secondary compounds afford the plant a means of chemical defense against predation. Research in this area has focused on the possibility of using these natural chemical defenses on a larger scale to protect agricultural crops from losses due to insect predation and to destroy those insects that are vectors of both human and livestock diseases. Botanical insecticides have the advantage of being nontoxic to humans and most livestock, and are biodegradable.

Two of the oldest and best known groups of botanical insecticides still in use today are pyrethrum and rotenone. Pyrethrum is a powder made from grinding the dried flower heads of *Chrysanthemum cinearifolium* commercially grown in Africa and South America. It was introduced into the United States in the 1800s. Later, the active ingredients were identified and named pyrethrin I, pyrethrin II, cinerin I and cinerin II; these are now used mainly in purified form, as well as forming the basis for the synthetic pyrethroids. Pyrethrins act as a nerve poison, quickly paralyzing many common household insects and are common ingredients of flea collars and quick-acting aerosol sprays. They are also available as a garden dust, which can be used safely on vegetables and fruits, since they are not known to pose any threat to human health (fig. 21.8a). (Some individuals, however, are strongly allergic to these compounds.) Pyrethrum, unlike some of the pyrethroids, breaks down readily in sunlight.

(a)

(b)

Figure 21.8 *Several insecticides have been developed from plants. (a) These products contain either the natural pyrethrins obtained from* Chrysanthemum cinearifolium *or the synthetic pyrethroids. (b) Pounded cubé root* (Lonochocarpus) *bark is placed in a net bag and dragged through the water to stupefy fish.*

Rotenone and related compounds called rotenoids are derived from the roots of several tropical legumes; for example, tuba root (*Derris*) grown in Malaysia, and cubé root (*Lonchocarpus*) from South America. Originally, the aqueous extracts of these roots were used by indigenous peoples to catch fish (fig. 21.8b). Rotenone causes nervous system paralysis in fish, making them surface in a helpless condition. The fact that rotenone is highly toxic to fish and might contaminate water resources has somewhat limited its use as a crop insecticide. Nonetheless, it is a widely used garden insecticide favored for its effectiveness against caterpillars and a host of common pests, and for its nonpersistence in the environment.

Figure 21.9 *The neem tree of India is the source of a seed oil that has been proven to be an effective insecticide.*

Recently, much excitement has focused on the potential of the neem tree or margosa (*Azadirachta indica;* family Meliaceae) as a source of natural insecticides. This tree is native to India and Burma (fig. 21.9) and is now widespread throughout the tropics. The medicinal applications of neem oil date back to Sanskrit records, and neem leaves and fruits have been used to repel insects in the home and in crop storage bins. As early as 1932, neem cake (the residue after the oil is extracted from the seeds) was utilized as an insect repellent in Indian rice paddies. Chemical investigation of the neem tree has identified a number of compounds acting as antifeedants or growth regulators against a variety of different pests. One of the most potent feeding deterrants isolated is azadirachtin from neem seeds.

Recent investigations have shown neem oil to be effective against more than 200 species of insect pests, as well as a number of viruses, bacteria, and fungi. It has been proven to fight tooth decay and a neem-based toothpaste is now on the market. There are even indications that neem oil can be used as a spermicide in birth control. Research is under way in many laboratories to identify further uses for this tree.

Allergy Plants

Certain plants or plant parts are capable of causing allergic conditions in sensitive individuals. Some plants may trigger respiratory allergies, due to the inhalation of airborne

A Closer Look 21.1

Allelopathy: Chemical Warfare in Plants

Certain plant species have been shown to produce and release chemical compounds into the environment that have deleterious effects on the growth and development of competing plants. This biochemical interaction between plants is known as **allelopathy.** The chemicals, typically volatile terpenes or phenolic compounds, are classified as secondary compounds and may be found in roots, stems, leaves, fruits, or seeds. The route of release is varied:

the allelopathic compounds may be leached from leaves and stems by rain;

released through microbial decomposition of litter;

volatilized into the air from leaves; or

exuded into the soil from roots.

Long before allelopathy was identified and confirmed, its consequences were noted in the natural environment. In *Natural History* Pliny (A.D. 23–79) observed that no plants grew beneath walnut trees, although plants thrived under the equally dense shade of alder trees. Much anecdotal information has also been gathered over the years about crops whose growth was inhibited by certain species. For example, A. P. de Candolle, a plant taxonomist of the nineteenth century, observed that thistles (*Cirsium*) growing in a field inhibited the growth of oats and that the presence of ryegrass (*Lolium*) had a similarly detrimental effect on wheat.

Research in the twentieth century provided the explanation for Pliny's observations. A chemical compound present in the leaves and green stems of black walnut trees (*Juglans nigra*) is leached by rainfall into the soil, where it is then hydrolyzed and oxidized to the allelopathic compound juglone. Juglone has been shown to be highly toxic to many plants, as well as a significant inhibitor of seed germination. These allelopathic effects result in a zone of inhibition around each walnut tree inhospitable to many plant species.

Box Figure 21.1 *The allelopathic effects of chaparral shrubs on herbs is visible as a bare zone.*

Allelopathic interactions play an important role in desert plant communities. Some desert shrubs, such as brittle bush (*Encelia farinosa*), release allelopathic chemicals that prohibit the growth of annuals to such an extent that a bare zone is noticeable beneath the leaf canopy and in the immediate surrounding area. Another desert shrub, guayule (*Parthenium argentatum*), releases a root exudate that inhibits guayule seedlings, as well as those of many other species. The subsequent spacing of desert vegetation resulting from these types of allelopathic interactions reduces competition for scarce water resources through the elimination of potential competitors.

Nowhere have the effects of allelopathy on vegetational patterns been more thoroughly investigated than in the Californian chaparral. A bare zone extending 1–2 meters (3–6 ft) clearly encircles and delineates the dominant shrubs, sagebrush (*Salvia leucophylla*) and artemesia (*Artemesia californica*). For 3–8 meters (10–25 ft) beyond the bare zone, the growth of herbs is inhibited and they are sparsely distributed and stunted (box fig. 21.1). Scientists have demonstrated that the leaves release volatile terpenes, predominantly camphor and cineole producing a vapor cloud around the shrubs. The terpenes are then adsorbed from the atmosphere by soil particles during the dry, hot summer. With the onset of winter rains, the terpenes are absorbed into the cuticle of seeds or roots, where they produce toxic effects inhibiting germination and growth. The fires that periodically sweep through the Californian chaparral destroy both the shrubs and their allelopathic terpenes present in the soil. Consequently, the soil is no longer contaminated and there is uninhibited growth of annual herbs and grasses for several years until the shrub stands reestablish themselves and the allelopathic pattern again develops.

pollen, while other plants cause contact dermatitis (skin rash) due to direct association with plant oils or resins.

Allergic reactions are considered hypersensitivity reactions. They are just one type of response produced by the immune system, the body system that differentiates "self" (the body's own cells or molecules) from "nonself" (foreign cells or molecules).

Allergy and the Immune System

The immune system is a complex system of specialized cells and organs that defend the body against attack by foreign invaders such as bacteria, viruses, fungi, and parasites. The basis of the immune system is the ability to distinguish between "self" and "nonself." Almost every cell within the body bears characteristic surface molecules that identify it as "self." When foreign substances are detected, the immune system responds with a proliferation of cells that either attack the foreign substance or produce proteins, called **antibodies,** that bind to the substance. Any substance capable of triggering an immune response is called an **antigen.** An antigen can be an invading pathogen or even a small portion or product of that organism.

In some people, however, harmless substances such as ragweed pollen, cat dander, or mold spores are perceived as a threat and stimulate the immune system to evoke an allergic response; in these cases the antigens are known as **allergens.** Allergies are inherited disorders. Although sensitivity to a particular substance is not inherited, the tendency to develop allergies is genetically controlled.

White blood cells are important components of the immune system; one class of these are **lymphocytes,** which are produced by stem cells in the bone marrow that give rise to all blood cells. One type of lymphocyte, **B-lymphocyte,** functions by producing antibodies, also known as **immunoglobins** (Ig). Antibodies are a major constituent of the blood, making up about 20% of the protein in the blood plasma. A human can produce 10 million different types of antibody molecules; however, an individual B-lymphocyte synthesizes only a single kind of antibody. Each type of antibody is capable of binding to only a single antigen.

All antibody molecules share the same structural features; each is composed of four polypeptide chains. Two chains are called light chains, which are identical to each other. The other two chains are called heavy chains, and are also identical to each other. In the antibody molecule, the two light chains and the two heavy chains are joined by bonds to make one Y-shaped antibody molecule (fig. 21.10). Both the light and heavy chains have a variable region that joins to form two antigen binding sites. The variable regions differentiate one type of antibody from another and also provide the antibody with the capability of binding to a certain antigen.

There are five classes of antibodies, **IgA, IgD, IgE, IgG,** and **IgM,** each with a different function in the immune system. The IgE class of antibodies is involved in allergic reactions, and people who suffer from allergies

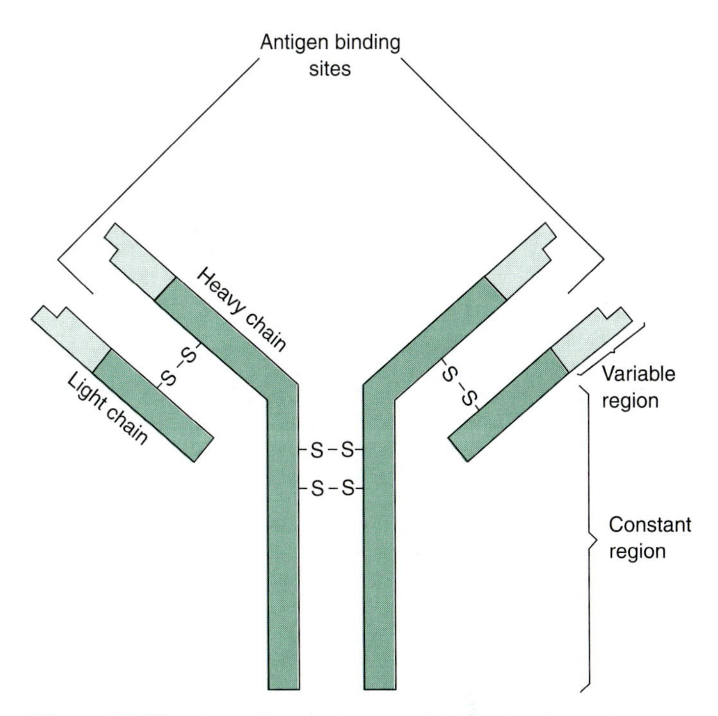

Figure 21.10 *An antibody or immunoglobin molecule consists of two identical light chains and two identical heavy chains joined to make a Y-shaped molecule.*

have elevated levels of these antibodies. When an allergic individual is first exposed to a particular allergen, lymphocytes produce specific IgE antibodies that are capable of binding to this substance. Unlike the other classes of antibodies, IgE antibodies do not circulate in the blood but are attached to the surface of basophils and mast cells (fig. 21.11). Basophils circulate in the bloodstream, while mast cells line the surface of the respiratory tract, intestines, and skin. When a particular allergen is encountered again, it will bind to the IgE antibodies on the surface of these cells, triggering a series of reactions inside the cells. Chemicals such as histamine, which are initially contained in vesicles (granules), are released from these cells and are responsible for the symptoms associated with allergies.

Respiratory Allergies

Approximately 20%–25% of the human population suffers from one or more significant allergies, with **hay fever** and **allergic asthma** the most common. These allergic diseases are major causes of illness and, although no complete data exist, it is believed that about 50 million people are affected in the United States today. Treatment of these conditions is a multibillion dollar a year business.

The name hay fever, which has little to do with hay and seldom produces a fever, was coined in the early nineteenth century when the condition was first recognized to coincide with the time of the year when hay was cut and baled. For this reason, the term **allergic rhinitis** is usually considered more appropriate. The typical symptoms of this condition are sneezing, runny nose, nasal congestion, and

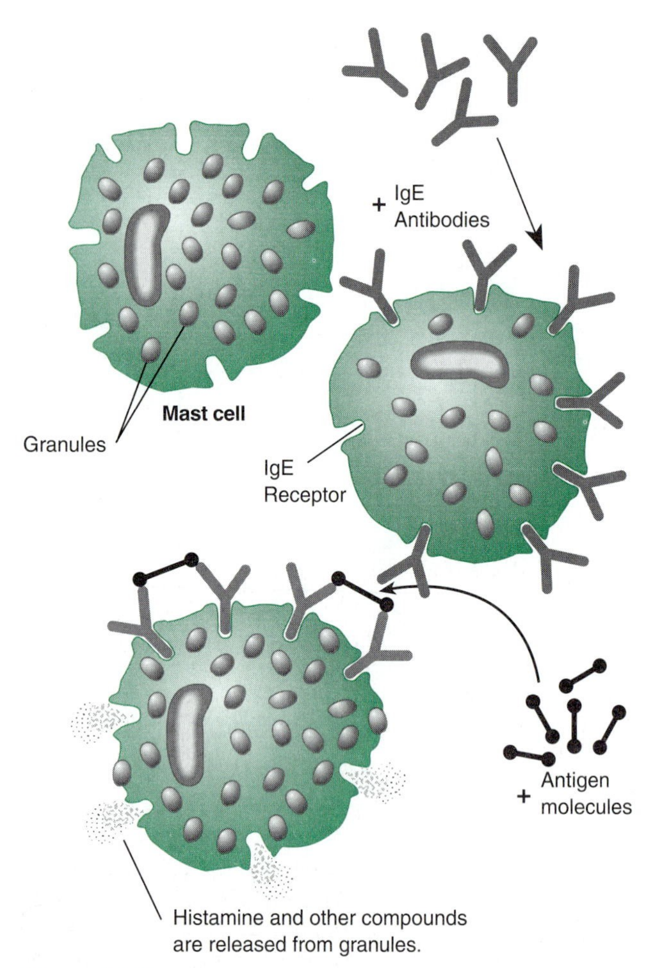

Figure 21.11 *Mechanism of an allergic reaction. After the first encounter with an allergen, IgE antibodies are produced and attach to the surface of mast cells lining the respiratory tract, skin, and intestines. When the allergen is encountered again, it will bind to the IgE molecules and cause the release of histamine and other chemicals that are responsible for allergy symptoms.*

postnasal drainage, along with red, itchy, puffy, and teary eyes. Although these symptoms do not seem serious, allergic rhinitis contributes to innumerable hours of suffering and loss of productivity. In addition, chronic or untreated hay fever can lead to asthma and other, more serious, long-term conditions. Approximately 40 million Americans suffer from hay fever and, in the United States, hay fever is responsible for more than 8 million visits to physicians, with over $2 billion spent on medication annually.

Asthma is a disease that is characterized by inflammation of the airways, bronchial constriction, and excessive mucous secretion, resulting in wheezing, coughing, and choking. This chronic breathing disorder is responsible for over 4,000 deaths per year in the United States. Asthma is an immunological-based disease and an allergen, viral infection, stress, or exercise can trigger an asthmatic episode. One disturbing trend is that the incidence and severity of asthma has been increasing throughout the world in recent years. During the decade of the 1980s there was a 29% increase in asthma prevalence. The exact reason for the increase is not known; in part, it is due to better diagnosis, but scientists believe that is not the full answer. Greater exposure to indoor allergens is thought to be a prime factor in this increase, especially in the inner city, where poverty may prevent proper medical treatment. In the United States alone, over 18 million people suffer from asthma, which accounts for $5 billion in medical costs each year.

Respiratory allergies can be triggered by exposure to pollen, mold spores, dust, animal dander, and insect allergens; however, the following discussion will focus on plants that produce allergenic pollen. Mold spore allergies will be considered in Chapter 24.

Hay Fever Plants

Pollen from plants that are considered important triggers of hay fever and asthma have several characteristics in common; the pollen is airborne, lightweight, abundantly produced, widely distributed, and allergenic.

With very few exceptions, plants that cause hay fever are wind pollinated. They produce abundant quantities of lightweight pollen that is passively dispersed into the air. (Recall that anemophilous or wind-pollinated plants generally invest metabolic energy into producing large quantities of pollen, while entimophilous or insect-pollinated plants produce showy petals and nectar to attract pollinators.)

Most allergenic pollen is 10–50 μm in diameter, a size range that facilitates the pollen remaining aloft for extended periods. Once airborne, it is possible for some pollen to be carried hundreds of kilometers. Such long-distance transport has been documented by aerobiologists, scientists who study airborne particles of biological origin. Eventually all pollen settles out, the distance transported dependent on pollen size, prevailing winds, and other atmospheric conditions. Large pollen grains tend to settle out very quickly, while smaller ones are carried greater distances.

Although some long-distance transport does occur, the majority of pollen grains settle out relatively close to their source, and allergy symptoms in susceptible individuals are closely correlated with proximity to the source. The most important hay fever plants are those that grow or are planted in large numbers close to human populations.

To be allergenic the pollen must contain actual allergens. These compounds are pollen wall proteins or **glycoproteins** (proteins with attached sugars). It is generally accepted that the role of these proteins in the pollen grain is to act as **recognition factors** for stimulating growth of the pollen tube when pollen lands on a receptive stigma. When a pollen grain is inhaled, it is usually deposited in the nasal passages or upper respiratory tract. The pollen wall proteins diffuse out of the pollen grain and into the respiratory tract, where they can trigger IgE-mediated responses in allergic individuals. These same molecules cause no reactions in people without allergies.

(a)

(b)

Figure 21.12 *(a) Short ragweed,* Ambrosia artemisiifolia, *and (b) giant ragweed,* Ambrosia trifida, *cause more hay fever than all other plants together.*

(a)

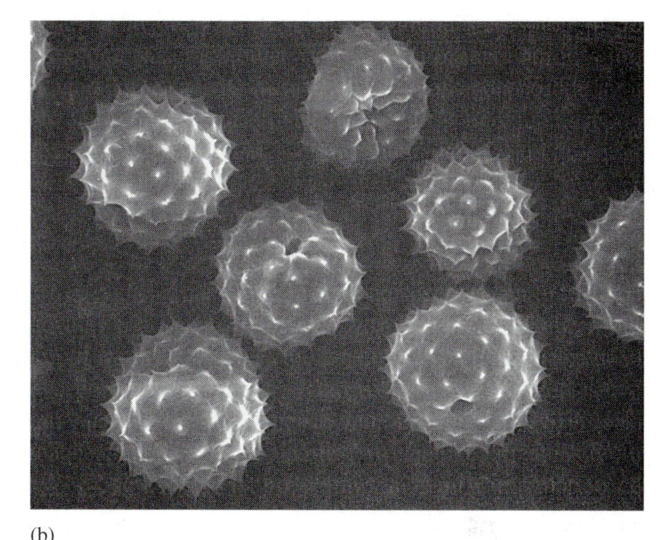

(b)

Figure 21.13 *(a) Staminate inflorescence of giant ragweed bearing clusters of florets. (b) Scanning electron micrograph of ragweed pollen ×1130.*

Ragweed

Undoubtedly the most notorious hay fever plant is ragweed, *Ambrosia* sp., a widespread genus in the Asteraceae, the aster or composite family. Interestingly, the generic name means "food of the gods" in Greek, surely an unlikely designation for the most important cause of allergic rhinitis in North America. Ragweeds are annual or perennial herbs, with lobed or divided leaves ranging from minute plants 15–20 cm (6–8 in) in height to giant ragweed, which can be 4 meters (12 ft) tall. Although 21 species of ragweed occur in North America, most allergy problems are caused by *Ambrosia artemisiifolia* (short ragweed) and *A. trifida* (giant ragweed), two species that account for more hay fever than all other plants together (fig. 21.12). Both are pioneer plants that are well adapted to invading disturbed soils. In natural environments these species are restricted; however, in areas where human intervention has cleared existing vegetation for farming, housing developments, shopping centers, or roadways, ragweed quickly becomes established.

Ragweed is monoecious, with staminate and pistillate flowers produced on the same plant. The small staminate flowers (each with five stamens) are grouped into inflorescences with 10–20 florets borne together in a cupule or involucre of bracts (fig. 21.13). Approximately 50 to 100 involucres occur on each of the many terminal flowering spikes. The thousands of staminate flowers on each plant result in the release of approximately one billion pollen grains.

In northern areas of the United States and in Canada, ragweed pollen release begins in early August, peaks by early September, and is completed early in October. An early, killing frost in late September often shortens the ragweed season in northern areas. In southern areas of the United States, the ragweed season begins later in August and continues until later in the fall. In Florida, as well as other areas of the Gulf Coast and the Southwest, the ragweed season may be much earlier or later, with some

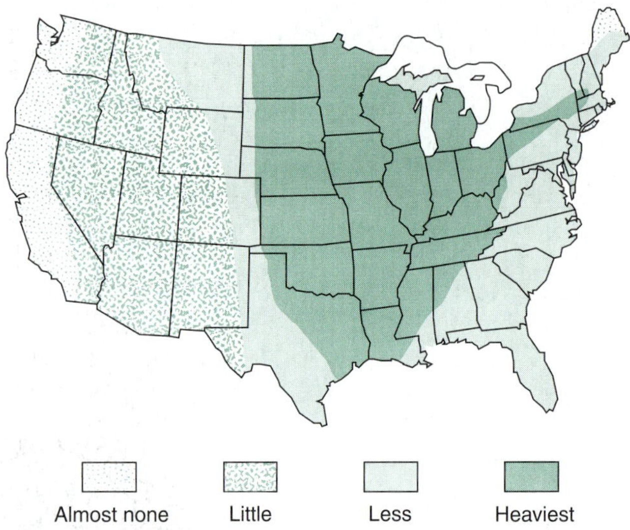

Figure 21.14 *Members of the genus* Ambrosia *are the most important allergy plants in the United States, with the heaviest distribution in the central states.*

species of ragweed actually flowering year-round. The prevalence of this genus in the United States results in the release of an estimated one million tons of pollen each year (fig. 21.14).

Pollen release normally occurs in mid-morning as the dew dries and humidities decrease. During the pollinating season, the daily atmospheric pollen concentration depends on local weather conditions, with warm, dry, windy conditions optimal for dispersal. On the other hand, heavy rainfall will wash pollen from the air and bring temporary relief to allergic individuals. Seasonal abundance of pollen often depends on growing conditions earlier in the year. Computer-generated pollen forecasting systems (based on weather and previous pollen levels) are being developed to predict the severity of the pollen season.

Other Hay Fever Plants

In addition to ragweed, the pollen from hundreds of seed plants are also known to trigger respiratory allergies. Allergists usually categorize pollen allergies into trees, grasses, and weeds, correlating symptoms with the season of pollen production.

Tree pollen is normally considered the leading cause of spring hay fever. In most cases, the inconspicuous flowers of wind-pollinated trees appear early in the season often before the leaves develop (fig. 21.15). Depending on climate and latitude, pollen release begins as early as late January for some species in the southern United States and extends through May for others. Pollen from oak, maple, elm, birch, pecans, walnut, and mulberry trees are among the many types recognized as major airborne allergens. Unfortunately, this list includes genera that are major components of the native vegetation, as well as popular shade and ornamental trees. Individuals allergic to tree pollen should avoid wooded areas in early spring and use species with showy (insect-pollinated) flowers for landscaping their yards.

Although angiosperm pollen is responsible for most cases of hay fever, pollen from certain conifers is a major cause of allergic rhinitis in some regions. One of the best-studied conifer allergens is the pollen produced by mountain cedar, *Juniperus asheii* a member of the Cupressaceae, the cypress family (fig. 21.15). Mountain cedar is a small-to-medium-sized evergreen tree native to the south central United States with the largest populations in central Texas where it is considered the most serious hay fever plant in the state. Additional small populations of the species occur in southern Oklahoma and northern Arkansas. An unusual feature of mountain cedar is that pollen release occurs in December and January, a time of the year when other pollen types are usually absent.

Grasses are the leading cause of early summer hay fever problems. Recall that the Poaceae, the grass family, is a large family with over 8,500 species (*see* Chapter 12). Although ragweed is considered the most important hay fever plant in North America, grass pollen allergy is the principal cause of allergic rhinitis throughout much of Europe and Central Asia.

There are approximately 1,200 grass species in the United States; however, only 35–40 species are allergenic and, of these, only 11 are important allergens. Bermuda grass, timothy, Kentucky bluegrass, Johnson grass, and orchard grass are the most important contributors to grass pollen allergies. While many native grasses pollinate in the late spring or summer, some species of lawn grass, such as Bermuda grass (*Cynodon dactylis*), will continue to flower throughout the growing season, especially when watered. Because of their proximity to people, lawn grasses present problems. Although these can be somewhat lessened by frequent mowing to prevent flowering, nearby lawns that are less well kept may produce pollen for six months or longer. In fact, in parts of Florida and other southern states, grass pollen may be present in the atmosphere year-round.

Weeds constitute a category of hay fever plants that include nongrass and nontree species; however, this is not a natural botanical category. As a category, "weeds" includes various types of vegetation that are not intentionally planted and are usually considered undesirable. These aggressive plants quickly invade disturbed sites associated with human activity. As such, they grow and pollinate close to populated areas. Although ragweed is the most important plant in this group, the pollen from many other common weeds are also well-known allergens. They often pollinate in late summer and fall, along with ragweed, and include marsh elder, lambs quarter, plantain and sage. Many of these cross-react with ragweed pollen; that is, they share common allergens so symptoms can be triggered by pollen

(a)

(b)

Figure 21.15 *During late winter and early spring, tree pollen allergens cause suffering for millions of people. (a) Staminate inflorescence of pin oak,* Quercus palustris *with inset showing the pollen. (b) Pollen cone of mountain cedar,* Juniperus asheii, *with inset showing the pollen. Pollen magnification ×800.*

from these plants in ragweed-sensitive people. In addition to the shared molecules, distinctive allergens are also present on the pollen grains.

Allergy Control

The best way to control respiratory allergies is to avoid the allergen. To someone allergic to cats or dogs, this is a possible solution, but to those with pollen allergies, avoidance is sometimes impossible. In the past, some cities have attempted to eradicate ragweed plants. This was very expensive and not terribly effective. Clearing a local area does not prevent the pollen from being carried from a more distant source, nor does it prevent new growth the following year from buried seeds. For many years, it was believed that moving to the desert would help allergy suffers; however, this was only a short-term solution. The growth of Phoenix and Tucson can be partly attributed to this belief. Allergic individuals soon became sensitive to the allergy plants of the area. In addition, many introduced landscaping plants such as mulberry, which grows fast but produces allergenic pollen. In fact, the problem became so bad in Tucson that an ordinance was passed against the further planting of mulberry trees.

Avoidance of pollen may necessitate a two-month ocean voyage (even here long-distance pollen transport may cause problems) or being sealed in an air-conditioned building (where efficient filters can remove the pollen from the incoming air) for the duration of the pollen season. For most allergy sufferers, therefore, total avoidance is not achievable since neither scenario is a feasible alternative.

There are many options available, however, for the treatment of respiratory allergies, including over-the-counter or prescription **antihistamines** (which can counteract the symptoms associated with the release of histamine from mast cells) and decongestants (which can relieve nasal and/or bronchial congestion). New antihistamines no longer result in drowsiness, an unwanted side effect commonly associated with antihistamines. Topical corticosteroids used as an aerosol or spray have also proven effective in reducing allergy symptoms. Immunotherapy or desensitization (commonly called allergy shots) is often indicated when symptoms cannot be adequately controlled by medication. Allergy patients are given a series of injections (often over several years) that contain weak extracts of the offending allergens. The extracts used in immunotherapy are prepared from actual pollen samples that have been collected from nature or from cultivated sources (fig. 21.16). In allergic individuals, these injections stimulate the production of IgG antibodies specific to the allergen. These block the binding to IgE antibodies and thus prevent the IgE-induced

Figure 21.16 *Commercial harvesting of ragweed pollen for the preparation of allergy extracts.*

response. Immunotherapy is often very effective in the treatment of pollen allergies (up to 90% effective in some studies) but has less success in the treatment of other allergies.

Contact Dermatitis

Contact dermatitis is an allergic reaction of the skin to something touched. While hay fever and allergic asthma are examples of immediate hypersensitivity, contact dermatitis is considered delayed hypersensitivity, since symptoms may take several hours to days before they appear. Like other allergies, it is triggered by the immune system responding to a harmless substance.

The most common plant that causes contact dermatitis is poison ivy, *Toxicodendron radicans,* a member of the Anacardiaceae, the cashew family. Poison ivy is a widespread weed native to the United States and southern Canada, where it is common in disturbed sites, floodplains, stream banks, lakeshores, and bordering woodland areas. It is extremely variable in form, growing as a woody shrub or vine that either spreads along the ground or climbs trees, fences, and poles. The leaves are alternate, each with three leaflets, but the leaf margin shows several forms (fig. 21.17). The margin may be smooth, toothed, or lobed, adding to the variability of the species. Clusters of white to whitish-yellow berries are found on all forms.

Related species include poison oak (*Toxicodendron quercifolium*), western poison oak (*T. diversilobum*), and poison sumac (*T. vernix*), which are equally potent allergens but not as widely distributed as poison ivy (fig. 21.17). Poison sumac and poison oak are found in the East, but occur as far west as Texas, while western poison ivy occurs in the Pacific coastal states. Some authorities consider all these species as the genus *Rhus* and consider poison oak to be a variant of poison ivy and not a separate species.

It is estimated that one out of every two people is allergic to poison ivy and related species, causing a serious skin rash on sensitive people. About 10% of the population is so sensitive that they require medical care even after a brief exposure. The appearance of the rash is normally delayed for 24–48 hours, or even longer, after contact with the plant. The rash usually appears as a red swollen area, followed by blisters and intense itching. The symptoms may last for a week or longer.

Urushiol, a resin that is present in all parts of the plant, is the actual allergen. Although the oily resin is produced within the plant, the slightest bruising of leaves, stems, or roots will release urushiol to the surface. It can be picked up by touching the plant or by touching clothing, animals, or gardening tools that were in contact with the plant. Urushiol rapidly bonds to proteins in the skin and may be spread to other parts of the body by rubbing, scratching, or just touching the rash or blisters. The compound is so reactive that a drop can cause dermatitis in 500 sensitive individuals. Urushiol is also long lasting; botanists have developed dermatitis after studying dried poison ivy plants that had been stored for over 100 years. Inhaling smoke from burning plants can also cause a massive skin rash, and more seriously affect eyes and lungs. This has caused serious problems for fire fighters battling forest fires, but can even be dangerous to the homeowner trying to eradicate plants by burning.

TABLE 21.1
Plant Foods Reported as Allergenic

Allspice	Gluten
Almond	Honey
Anise seed	Hops
Apple	Horseradish
Artichoke	Juniper berry
Baker's yeast	Lentil
Banana	Lima bean
Bay leaf	Mango
Beet	Millet
Black pepper	Mushroom
Brazil nut	Mustard
Brewer's yeast	Nutmeg
Buckwheat	Orange
Cantaloupe	Pea
Caraway seed	Peach
Cashew nut	Peanut
Castor bean	Pine nut
Celery	Pistachio
Chamomile	Poppy seed
Chestnut	Potato
Chicory	Psyllium seed
Chili pepper	Raspberry
Chocolate	Sage
Cinnamon	Sesame seed
Clove	Soybean
Coconut	Strawberry
Corn	Sunflower seed
Cottonseed	Sweet potato
Cumin seed	Tangerine
Dill seed	Tapioca
Fennel seed	Thyme
Filbert	Turmeric
Flaxseed	Vanilla
Garbanzo bean	Walnut
Garlic	Wheat
Ginger	

Food Allergies

Food allergies are sensitivities to certain foods that can cause a great diversity of symptoms in various body systems, including the gastrointestinal tract (abdominal pain, vomiting, and diarrhea), skin (hives, eczema, skin rash), and respiratory tract (rhinitis or asthma symptoms). The

Figure 21.17 *Poison ivy, poison oak, and poison sumac are common causes of contact dermatitis.*

most severe result of food sensitivity is anaphylaxis, a rare but sometimes fatal reaction with multiple symptoms, including swelling of the respiratory tissues, rapid drop in blood pressure, and cardiovascular collapse. (Anaphylactic reactions are not limited to food allergies; in fact, the most common triggers of anaphylaxis are bee sting and penicillin allergies.)

The diversity of symptoms often makes food allergies difficult to diagnose, with the offending food difficult to identify, since hundreds of foods are capable of triggering allergic reactions. Milk, egg, and fish allergies are some of the most thoroughly studied food allergies. Among plant-based foods, wheat, peanuts, tree nuts, soybeans, strawberries, and citrus fruits are considered leading allergens. The most common allergenic plant foods are listed in Table 21.1. The principal therapy for food allergies is avoidance, which makes individuals constantly vigilant about ingredients. While it is relatively easy to avoid a food when its presence is obvious, soy products, peanut products, and wheat are almost omnipresent staples in prepared foods. These hidden allergens, even in small quantities, can produce severe reactions in sensitive individuals.

Chapter Summary

1. Poisonous and allergy plants are known to adversely affect the health of humans and other animals. As foraging societies identified plants according to use, knowledge of poisonous plants accumulated. Poisonous plants have been employed as arrow poisons, as "succession powders," and as a form of capital punishment. Glycosides and alkaloids are usually the toxic compounds in most poisonous plants.
2. Many plants in the natural environment and those in the garden or house are poisonous. Some plant toxins, like curare, have been used for beneficial purposes in medicine. Most cases of plant poisoning in the home involve young children ingesting aroids.
3. White snakeroot, locoweed, and Indian hellebore are plants that have been implicated in the poisoning deaths of livestock.
4. Plants can cause mechanical injury by puncturing the skin via thorns, barbs, or spines. Stinging hairs not only penetrate the skin but inject histaminelike irritants. Other plants injure by releasing irritating resins or latexes that damage the skin and other organs.
5. Plant toxins have become an important source in the search for natural insecticides. Pyrethrum and rotenone are two examples of plant-derived insecticides; the neem tree of India is the origin of an oil that has been shown to be effective against insect pests, bacteria, viruses, and fungi.
6. Allelopathy is a type of chemical warfare waged by plants that affects the growth and development of competing plants. The effects of allelopathy on the spacing of vegetation are especially pronounced in desert and chaparral plant communities.
7. Allergies are hypersensitivity diseases caused by the immune system responding to harmless substances such as ragweed pollen or mold spores. Hay fever and asthma are the most common forms of respiratory allergies, affecting 20%–25% of the human population. Hay fever is characterized by sneezing, runny nose, nasal congestion, and watery eyes, while the symptoms of asthma include wheezing, coughing, and choking.
8. Pollen from ragweed plants is the most important cause of allergic rhinitis in North America. Ragweeds are pioneer species that are well adapted to invading disturbed soils. They occur abundantly through the eastern two-thirds of the continent, and in most parts of North America. Ragweed pollen occurs in the atmosphere from August through October. Approximately one million tons of pollen are produced each year in the United States.
9. In addition to ragweed, pollen from hundreds of seed plants are also known to trigger respiratory allergies. Like ragweed, hay fever plants produce abundant lightweight, airborne pollen. Pollen allergies are usually categorized into trees, grasses, and weeds, correlating the symptoms with the season of pollen production. Tree pollen is considered the leading cause of spring hay fever, grasses the leading cause of summer hay fever, and weeds the leading cause of fall hay fever.
10. The most common cause of contact dermatitis is poison ivy, a widespread climbing shrub in North America. The actual allergen is a resin, urushiol, present in all parts of the plant. The resin can be picked up by touching the plant directly or by touching clothing, animals, or even gardening tools that were in contact with the plant.
11. Food allergies can produce gastrointestinal symptoms, dermatitis, hives, or respiratory symptoms. A wide variety of plant foods are known to cause reactions in hypersensitive individuals.

Review Questions

1. How are some poisonous plants beneficial to society?
2. In terms of accidental plant poisonings, what dangers exist in the garden and home?
3. The medical term for hives (an allergic skin reaction) is urticaria. Can you explain the derivation of this word?
4. In what way are allergic responses immune system "mistakes"?
5. What are the characteristics of a "hay fever" plant and what common plants are major triggers of hay fever?
6. Avoidance is an effective treatment for what types of allergic reactions? Why is it not effective for other types of allergies?

Further Reading

Blackwell, Will H. 1990. *Poisonous and Medicinal Plants.* Prentice Hall, Englewood Cliffs, NJ.

Jelks, Mary. 1986. *Allergy Plants That Cause Sneezing and Wheezing.* World-Wide Publications, Tampa, FL.

Lampe, Kenneth and Mary Ann McCann. 1985. *AMA Handbook of Poisonous and Injurious Plants.* AMA: Chicago, IL.

Lawrence, Susan V. 1984. Recent Advances in Hay Fever Research are Nothing to Sneeze at. *Smithsonian,* 15(6):100–111.

Levy, Charles K. and Richard B. Primack. 1984. *A Field Guide to Poisonous Plants and Mushrooms of North America.* The Stephen Greene Press, Brattleboro, VT.

Lewis, Walter and Memory P. F. Elvin-Lewis. 1977. *Medical Botany: Plants Affecting Man's Health.* John Wiley & Sons, New York.

Lewis, Walter H., Prathibha Vinay, and Vincent E. Zenger. 1983. *Airborne and Allergenic Pollen of North America.* The Johns Hopkins University Press, Baltimore, MD.

Lichtenstein, Lawrence. 1993. Allergy and the Immune System. *Scientific American,* 269 (3):116–124.

National Research Council. 1992. *Neem: A Tree for Solving Global Problems.* National Academy Press, Washington, DC.

Turner, Nancy J. and Adam F. Szcawinski. 1991. *Common Poisonous Plants.* Timber Press, Portland, OR.

Wodehouse, Roger P. 1971. *Hay Fever Plants,* 2nd edition, Hafner Press, New York.

UNIT

VI

FUNGI

THE IMPACT OF FUNGI ON HUMAN AFFAIRS

Fungi provide a wealth of economic products including edible mushrooms, alcoholic beverages, and life-saving antibiotics.

22

Fungi in the Natural Environment

Chapter Outline

Chapter Concepts

1. Fungi are a diverse group of heterotrophic organisms with an absorptive type of nutrition; they generally have a threadlike body and reproduce by spores.
2. The majority of fungi are saprobes, actively recycling nutrients in the environment by decomposing organic matter; however, the ecological importance of the symbiotic fungi is also well-known.
3. Some plant diseases caused by fungi have had devastating influences on crop plants, resulting in major impacts on human society.

As a group, the fungi are among the most abundant organisms in the world, yet for most people they are probably the least understood group. The term *fungus* often elicits memories of moldy basements or grocery-store mushrooms without any awareness of their ecological importance or the tremendous impact these organisms have on society. Fungi are vital to the environment as decomposers or as partners in symbiotic relationships. They provide lifesaving antibiotics, fermented foods, and beverages; however, they are also responsible for devastating crop losses and debilitating human disease.

Characteristics of Fungi

Fungi are eukaryotic organisms that are neither plant nor animal but belong to a realm of their own, although recent molecular studies show fungi more closely related to the Animal Kingdom than the Plant Kingdom. Unlike plants that contain chlorophyll and carry out photosynthesis, fungi are **heterotrophic;** that is, they depend on external sources of organic material for food. Fungi have an **absorptive** type of nutrition, producing enzymes that are secreted into the environment and break down complex organic compounds into smaller molecules. These small molecules are then absorbed by the fungus.

Morphologically, fungi range from small, inconspicuous yeasts and molds to large, conspicuous mushrooms, puffballs, and bracket fungi. Microscopic observation of their structure shows that fungi exist as single cells such as yeast or, far more commonly, as threadlike **hyphae.** Hyphae usually branch extensively, and the collective mass of hyphal filaments is referred to as a **mycelium.**

Depending on the species, each hypha may be **septate,** divided by cross walls (**septa;** sing., septum) into many short cells or be **nonseptate** (lacking cross walls) with multiple nuclei existing in a common cytoplasm (fig. 22.1). Among those fungi with septate hyphae, each cell may contain one nucleus, or two genetically different nuclei, depending on the fungal group and stage of development. Hyphae with two genetically different nuclei within each cell are called **dikaryons.** While individual hyphae are microscopic, the whole mycelium is often visible to the naked eye. The mycelium may also develop highly specialized reproductive structures such as mushrooms, brackets, or puffballs. These are actually compact masses of tightly interwoven hyphae and are the visible portion of an extensive mycelium within the substrate. Mycelium may be growing in countless substrates such as underground, throughout a decaying log, or within a living tree.

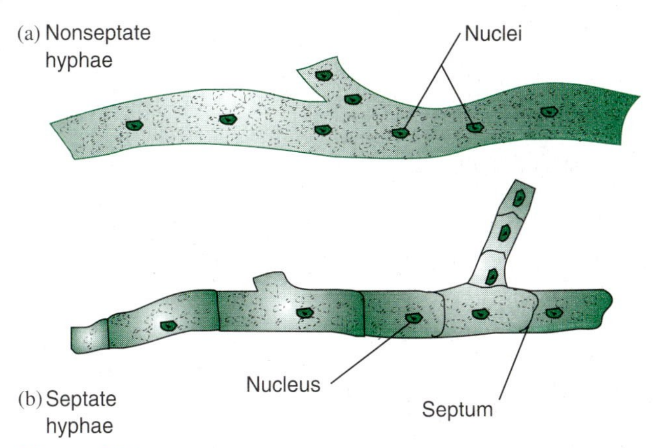

Figure 22.1 *Fungal hyphae can be (a) nonseptate with multiple nuclei existing in a common cytoplasm or (b) septate with cross-walls (septa) dividing the hyphae into many short cells.*

Reproduction in the fungi is usually carried out through the formation of **spores,** which may result from either sexual or asexual processes. Many fungi undergo both asexual and sexual reproduction at different stages in their life cycle, while others carry out just one of these reproductive processes. The sexual phase of the life cycle is known as the **perfect,** or **teleomorphic, state,** while the asexual phase is called the **imperfect,** or **anamorphic stage.**

Spores differ greatly in size, shape, color, and method of formation, but they are always microscopic, with unicellular to multicellular types. Spores may be formed from the breakup of hyphae, or on highly specialized hyphal branches often contained within **fruiting bodies.** Mushrooms, puffballs, brackets, and morels are well-known examples of these specialized fruiting bodies. In most groups of fungi, spore characteristics are important elements for identification and classification. The majority of fungal spores are dispersed by the wind, with various types of discharge mechanisms occurring within different groups of fungi. As a result, fungal spores are a normal component of the atmosphere and are present in large numbers any time that the ground is not covered with ice and snow. From spring through fall, it is not unusual to have spore concentrations exceeding 100,000 spores per cubic meter of air. This airborne transport is significant in the spread of plant disease, as well as in the suffering of people allergic to certain types of fungal spores.

Under adverse environmental conditions, some fungi are able to form **chlamydospores,** thick-walled dormant spores that develop from transformed vegetative hyphae. These are not usually dispersed and become free only when the adjacent hyphae decay. Another type of resistant or resting body is a **sclerotium,** which is a hardened mycelial mass that also enables the fungus to survive harsh conditions.

One feature the fungi share with plants is the presence of cell walls. The fungal wall, consisting of fibrils embedded in a matrix, is largely composed of polysaccharides (often over 90%) but also contains significant amounts of protein and lipid. In the majority of fungi, the fibrillar component of the cell wall is **chitin,** a nitrogen-containing polysaccharide; however, some fungi possess cellulose fibrils. The matrix, on the other hand, contains a variety of carbohydrates (glucose polymers) and proteins.

TABLE 22.1

Classification of the Fungi

I. **Kingdom Protista** Eukaryotic organisms with heterotrophic or autotrophic nutrition; includes protozoans, algae, and fungal-like organisms.
   A. **Division Myxomycota—**slime molds; plasmodial feeding stage; spores formed in a fruiting body.
   B. **Division Oomycota—**mycelial fungi with nonseptate hyphae; most form zoospores; sexual spores (oospores) are thick-walled resting spores.

II. **Kingdom Fungi** True Fungi (mycelial or yeast) with an absorptive, heterotrophic nutrition.
   A. **Division Chytridiomycota—**primitive fungi with flagellated asexual spores.
   B. **Division Zygomycota—**mycelial fungi with nonseptate hyphae; sexual spores (zygosporangia or zygospores) are thick-walled resting spores; asexual spores (sporangiospores) produced in a sporangium.
   C. **Division Ascomycota—**mycelial fungi or yeasts; sexual spores (ascospores) formed in an ascus; asexual spores (conidia) abundant in some groups.
   D. **Division Basidiomycota—**mycelial fungi; sexual spores (basidiospores) formed on a basidium; large fruiting bodies common; asexual spores abundant in some members.

Classification of Fungi

The fungi are a large heterogeneous group of organisms ranging from unicellular yeasts to large mushrooms. Over 70,000 species of fungi have been identified and estimates for the total number of species exceed one million. In general, classification of the fungi is based on the method of sexual reproduction. Sexual spores are the result of genetic recombination and normally follow **karyogamy** (fusion of haploid nuclei into a diploid zygote) and meiosis. These processes occur in specialized cells and, in some groups, the sexual spores form within specialized fruiting bodies. Asexual spores result from mitosis and may also be characteristic of certain groups. Asexual spores may form enclosed within a **sporangium** or be produced directly by the hyphae without any enclosing wall. These later spores are known as **conidia.**

In modern systems of classification, organisms that have traditionally been called *fungi* are found in two kingdoms, the Kingdom Protista and the Kingdom Fungi. The presence of chitin in the cell wall is one major distinguishing feature of those organisms in the Kingdom Fungi. Two major divisions are recognized in the Kingdom Protista, the Myxomycota and the Oomycota. Within the Kingdom Fungi, four divisions are recognized: the Chytridiomycota, the Zygomycota, the Ascomycota, and the Basidiomycota (table 22.1).

In addition to these six divisions, mycologists (scientists who study fungi) recognize a large group of asexual fungi, sometimes called the Deuteromycetes or the Fungi Imperfecti, which is included in the Kingdom Fungi. Although it is generally agreed that the vast majority of these are probably members of the Ascomycota, the sexual stage is not usually encountered and the exact classification is in doubt.

Kingdom Protista

Division Myxomycota

This division contains the **slime molds,** organisms that have affinities to both animals and fungi. The amoeboid feeding stage known as the **plasmodium** shows animal-like characteristics in actively engulfing organic matter and bacteria in the environment. The reproductive stage, however, aligns the slime molds with the fungi, since the spores have cell walls and are produced within fruiting bodies. Although several groups of organisms are included in the Myxomycota, this discussion will be limited to the true slime molds or the myxomycetes.

Slime molds produce unicellular spores that may remain dormant for long periods of time. When environmental conditions are favorable, the spores germinate and give rise to either a flagellated cell known as a **swarm cell** or an amoeboid cell known as a **myxamoeba.** The myxamoebae and swarm cells can ingest food, grow, and divide. Under unfavorable conditions, they can also encyst, forming spherical dormant cells that germinate when favorable conditions return. Eventually, the swarm cells or myxamoebae function as gametes. Compatible gametes

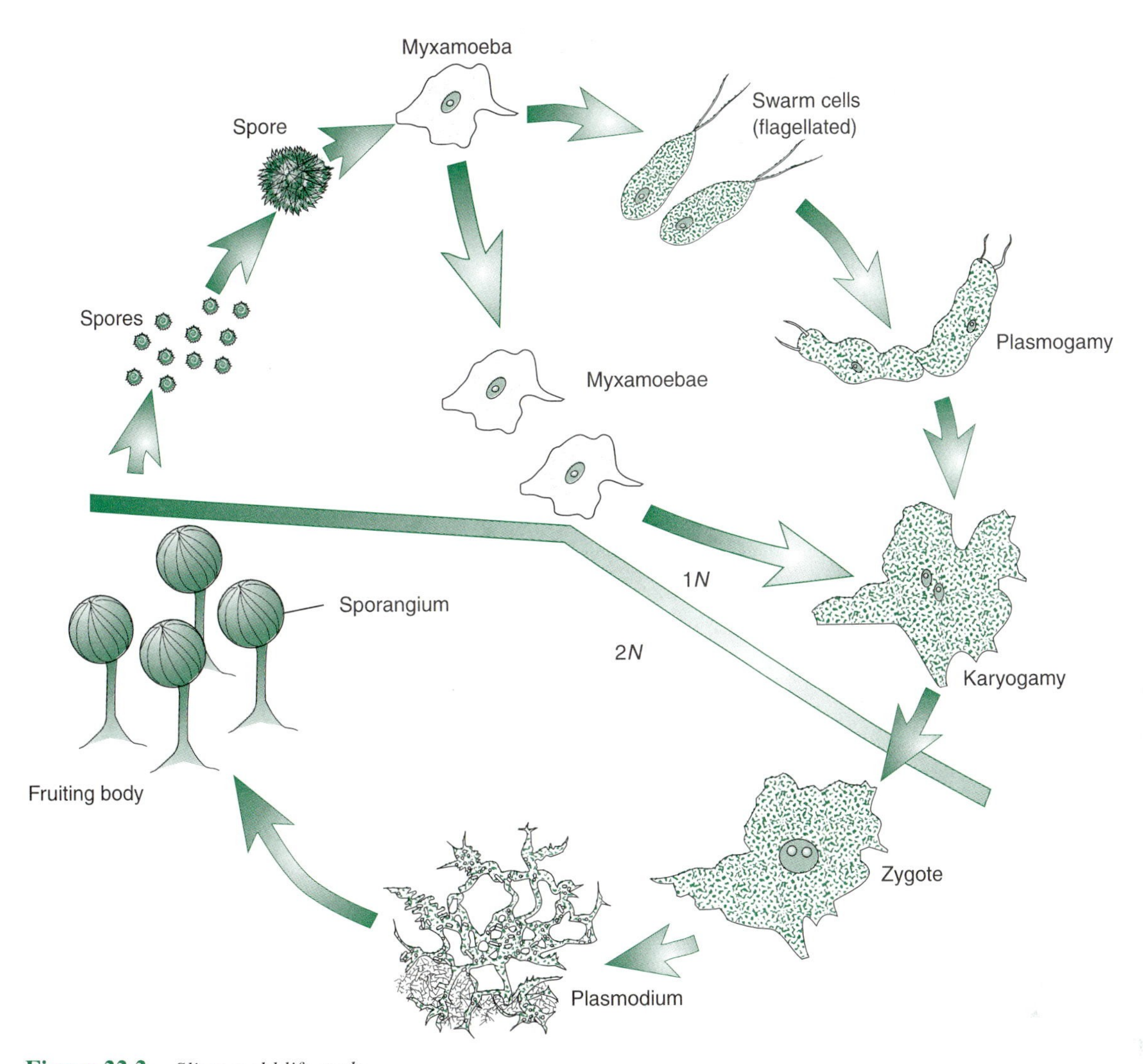

Figure 22.2 *Slime mold life cycle.*

fuse to form a zygote that is amoeboid. The diploid nucleus undergoes repeated nuclear divisions, forming a motile, multinucleate plasmodium. The plasmodium, which is frequently brightly colored, ingests food, grows, and crawls over the substrate. This is the main feeding stage of the organism, and its slimy appearance gives rise to the common name. Under unfavorable conditions, the plasmodium is capable of forming a hard, dry sclerotium, another resistant structure. Plasmodia eventually give rise to sporangia or similar fruiting bodies (fig. 22.2). Within the developing sporangia, meiosis and spore formation occur.

Slime molds frequently appear in cool, damp areas and can be found on moist soil and decaying vegetation, where they feed on bacteria, other microorganisms, and organic matter. Although a few species are weak plant pathogens, the slime molds as a whole are of little economic importance. They are, however, a curiosity. The sudden appearance of the miniature fruiting bodies occurring on decaying wood, or the plasmodium creeping over damp ground or grass, has surprised many woodland explorers unfamiliar with these fascinating organisms.

Division Oomycota

The characteristic sexual structure is the **oogonium,** which contains one or more **oospores** that result from fertilization. Another feature of this group is the presence of **zoospores,** motile asexual spores. These zoospores, which have two flagella, are produced in sporangia and are normally dependent on water for spore dispersal. Under certain conditions however, some **oomycetes** (common name for members of this division) are known to produce nonflagellated propagules (actually one-spored sporangia) that are dispersed by wind. These fungi can be found in aquatic habitats as well as damp soil.

The life cycle of this division will be represented by the genus *Saprolegnia,* a common saprobic water mold; a

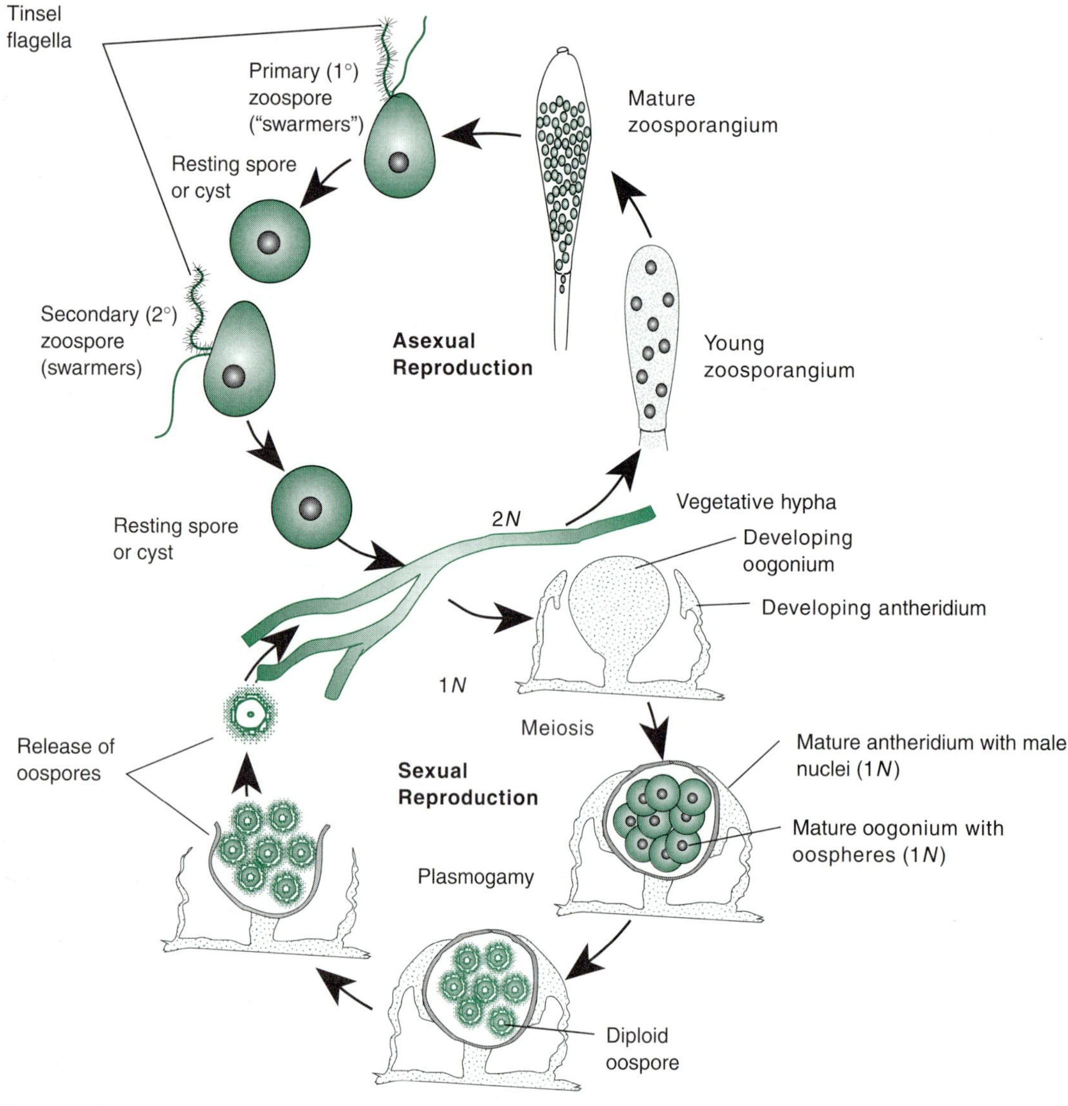

Figure 22.3 *Life cycle of* Saprolegnia *in the Division Oomycota.*

few species of *Saprolegnia* can be pathogenic to fish. *Saprolegnia* forms a branched, nonseptate mycelium and can complete its entire life cycle submerged (fig. 22.3). Biflagellate zoospores are produced in an elongated sporangium that has a cross-wall at the base. A pore develops at the tip through which the mature pear-shaped spores are released. After a short period of swimming, each zoospore encysts (withdraws its flagella and surrounds itself with a thin wall). The encysted state is maintained for several hours. When the zoospore re-emerges, it now appears bean-shaped, with laterally inserted flagella. Again, the zoospore swims about for a short time, then re-encysts. The cyst germinates immediately to produce a mycelium, thus completing the asexual phase of the life cycle.

The sexual stage of the *Saprolegnia* life cycle includes the production of male and female organs, or **gametangia,** (structures housing gametes). The female gametangium, or oogonium, is spherical, with a cross-wall at the base. Within the oogonium, the protoplasm divides into 5–10 uninucleate **oospheres** or eggs. Nearby hyphae develop that grow toward and appress themselves on the surface of each oogonium. The ends of these hyphae become delimited by cross-walls as multinucleate **antheridia.** Specialized threadlike hyphae develop from each antheridium and penetrate the oogonium and an oosphere. A male nucleus from the antheridium passes through this thread and enters each oosphere; the ensuing fertilization results in the production of thick-walled, resistant oospores, the characteristic structure of the class.

Included in the Oomycota are some well-known plant pathogens that have had a major impact on society, such as the fungi that cause late blight of potato, downy mildew of grape, and white rusts.

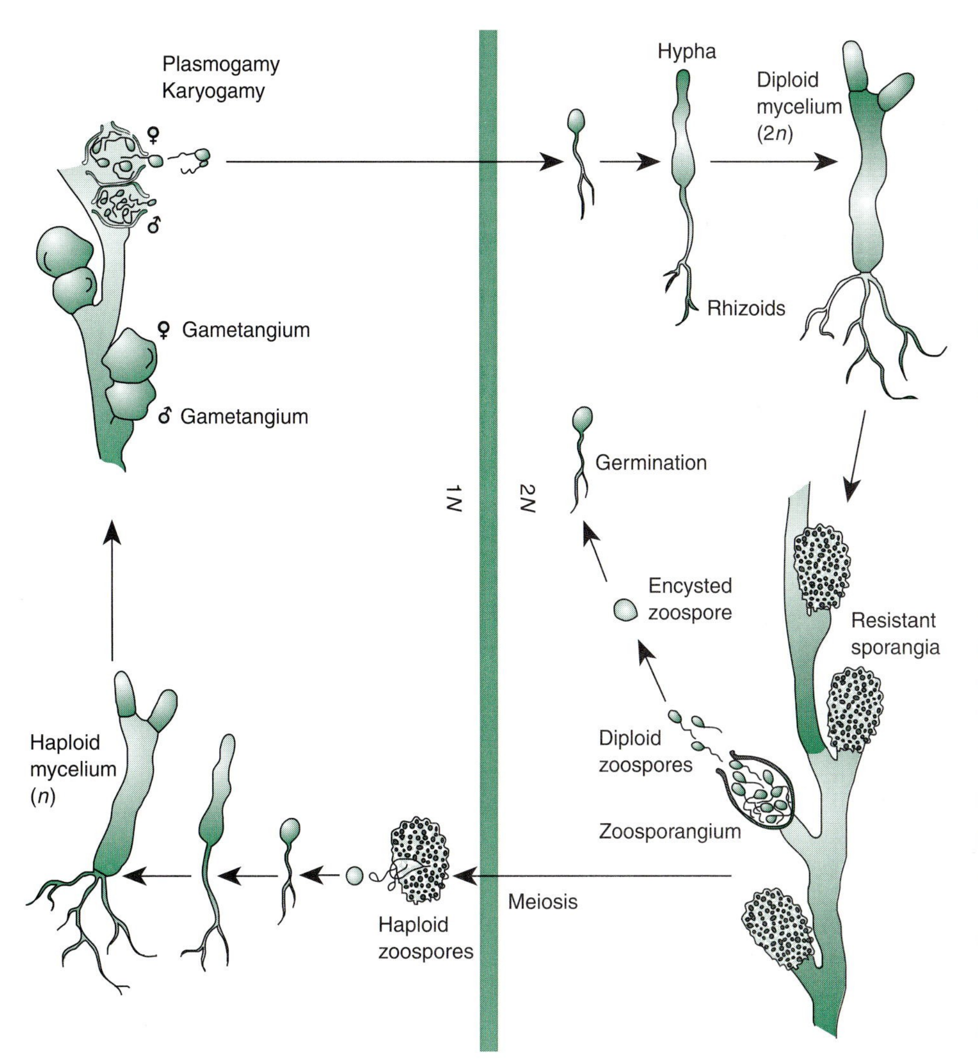

Figure 22.4 *Life cycle of* Allomyces arbusculus, *a member of the Division Chytridiomycota.*

Kingdom Fungi

Division Chytridiomycota

Both unicellular and mycelial fungi are found in this group, which occurs in aquatic and moist habitats. Many chytridiomycetes exist as parasites of algae, vascular plants, invertebrates, and even other fungi. Saprobic species are known for their ability to degrade cellulose, chitin, and keratin. Asexual reproduction is by zoospores with a single posterior flagellum. Sexual reproduction is by the fusion of gametes, and in many members of this division, both gametes are motile.

A well-studied genus of this division is *Allomyces,* a saprobe that shows an alternation of haploid and diploid generations (fig. 22.4). Both generations develop a sparsely branched mycelium. On the haploid stage, specialized gametangia produce motile gametes that are released into the environment. The male gamete is smaller and more active than the female gamete. After fertilization, the united gametes swim around briefly, then settle down. The nuclei fuse and the resulting zygote gives rise to the diploid generation. Zoospores are produced in sporangia on the diploid mycelium; these asexual spores repeat the diploid stage and allow for many generations of the diploid mycelium. The diploid generation also produces thick-walled, resistant sporangia that can remain dormant for many years. When environmental conditions are suitable, meiosis occurs in this structure and the resulting haploid zoospores give rise to the haploid mycelium.

Division Zygomycota

Members of this division are characterized by thick-walled sexual spores, called **zygosporangia** or **zygospores,** that develop from the fusion of compatible gametangia on hyphal tips. No distinct male and female gametangia are recognizable; however, opposite mating types are clearly present. These are usually designated as + and – strains. As compatible haploid hyphae grow toward each other, multinucleate gametangia differentiate at the tips of the hyphae. The end walls dissolve and the contents of the gametangia mix with compatible haploid nuclei fusing together. The resulting zygosporangium is a multinucleate, thick-walled, resistant structure.

Asexual spores, known as **sporangiospores,** are produced within a sporangium, which develops on a specialized erect stalk, the sporangiophore, that is separated from the hyphae by a cross-wall at the base. At maturity, the thin sporangial wall breaks down, releasing the thousands of spores that are normally dispersed by wind. These spores germinate to develop into the mycelium of either + or – strains.

Members of this subdivision are usually saprobes, which can occur in a variety of substrates. The genus *Rhizopus,* a common **zygomycete,** is frequently isolated from soil samples and can be an aggressive and bothersome laboratory contaminant (fig. 22.5).

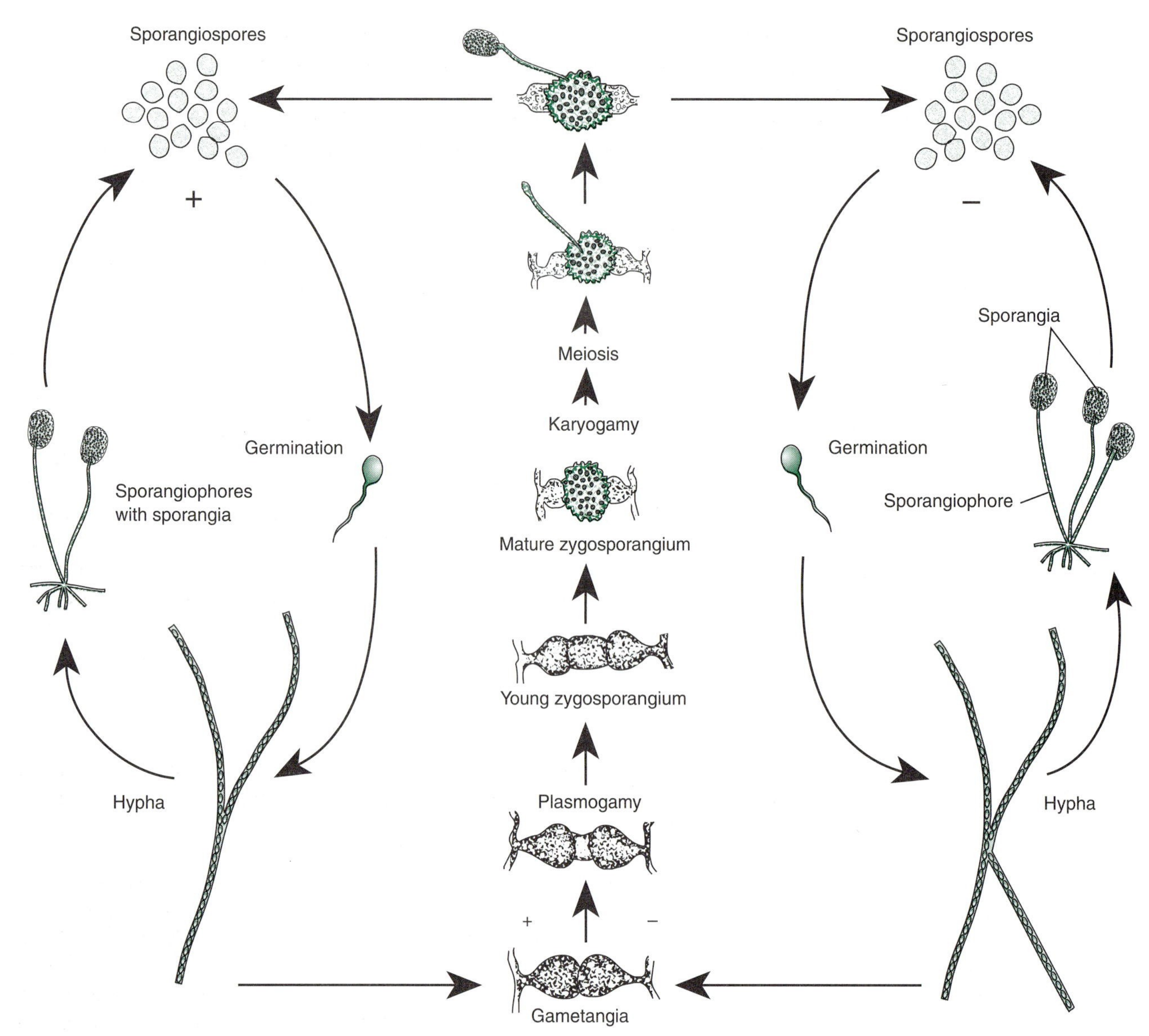

Figure 22.5 *Life cycle of* Rhizopus stolonifer *in the Division Zygomycota.*

One remarkable genus in this group is *Pilobolus,* commonly found on the dung of herbivores and possessing a unique dispersal mechanism for its sporangia (fig. 22.6). The sporangium is borne on top of a subsporangial swelling (an enlargement of the sporangiophore tip). When the spores are mature, turgor pressure builds up in the subsporangial swelling until the sporangium is explosively

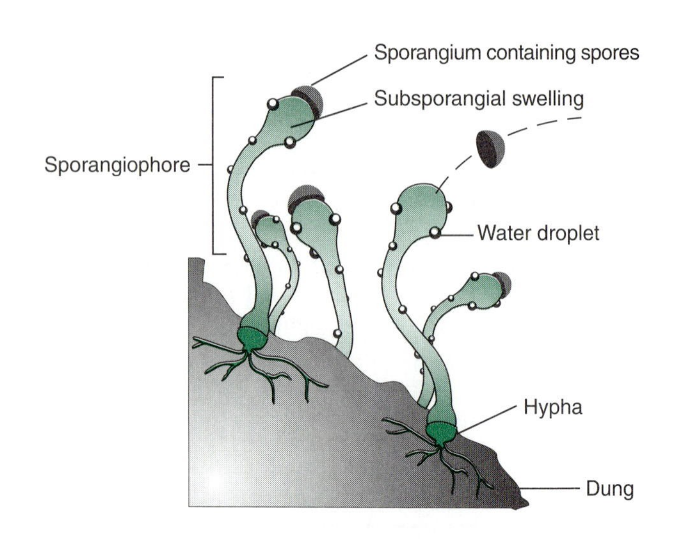

Figure 22.6 *Asexual reproduction in* Pilobolus *is highly specialized. The entire sporangium is "shot" away from the sporangiophore toward the light. Discharged sporangia adhere to nearby vegetation.*

shot away from the sporangiophore at a speed of about 14 meters (45 ft) per second. The sporangium travels up to 2 meters (6 ft) and adheres to nearby vegetation. The explosive discharge shoots the sporangium away from the "ring of repugnance," which usually keeps herbivores from feeding too close to their feces. When grazing animals eat the vegetation, the attached sporangia pass unharmed through the digestive tract and emerge with the feces. There, the spores germinate, the hyphae grow, and new sporangia develop, completing an unusual life cycle.

Division Ascomycota

Members of the Ascomycota are characterized by the production of sexual spores, **ascospores,** within a saclike structure called an **ascus** (plural, asci). Within the ascus, karyogamy and meiosis occur, followed by an additional mitotic division to produce eight haploid ascospores. Members of this division are commonly called **ascomycetes.** Classes within the division are determined by the morphology of the ascus and also the type of fruiting body **(ascocarp)**.

Ascomycetes range from simple yeasts, in which a naked ascus is produced by fusion of two compatible cells without a fruiting body, to the cup fungi and morels where asci line the depressions of the fruiting body (fig. 22.7a–c). The mycelium of these fungi is typically branched and septate, with individual cells initially uninucleate. When hyphae of different mating types become closely intertwined, male antheridia may form on one and female **ascogonia** (singular, ascogonium) on the other. Male nuclei pass from the antheridium into the ascogonium; however, karyogamy does not follow immediately. Ascogenous hyphae develop from the ascogonium with two compatible nuclei within each cell. The terminal cell of each dikaryotic hypha becomes an ascus, eventually giving rise to eight ascospores. In many ascomycetes the asci form in a layer, known as the **hymenium,** that lines the base of the ascocarps (fig. 22.7d).

In most ascomycetes the ascospores are shot from the ascus at maturity. High osmotic pressure develops within the ascus, causing it to swell. Soon, the ascus tip ruptures and the spores are explosively forced out into the atmosphere.

Asexual reproduction in mycelial ascomycetes is accomplished by the production of conidia, either singly or in chains. These spores, which are rapidly and abundantly produced, are generally dependent on airborne dispersal. In the yeasts, asexual reproduction is usually achieved by **budding:** a small outgrowth from the parent cell slowly balloons out and eventually becomes pinched off (*see* fig. 23.1).

Neurospora crassa, a common saprobic ascomycete has a long history of use as a genetic tool (fig. 22.8). The asexual stage produces conidia that develop in chains (this stage is often called red bread mold, a troublesome contaminant in bakeries), while sexual reproduction results in the production of flask-shaped ascocarps known as **perithecia** (singular, perithecium). Genetic studies on *Neurospora* by George Beadle and Edward Tatum in the 1930s established the one gene-one enzyme hypothesis. This states that the function of a gene is to specify the production of a specific protein. In 1958, Beadle and Tatum received the Nobel Prize in Physiology for this landmark work in molecular biology.

Dutch elm disease, chestnut blight, and powdery mildews are among the devastating plant diseases caused by members of the Ascomycota. Economically useful members of this division include the baking and brewing yeasts, as well as the morels and truffles often considered among the choicest edible fungi. While morels have an aboveground ascocarp (fig. 22.7b), truffles develop underground with no visible signs on the surface. In Europe these delicacies are collected using trained dogs and pigs that locate the scent of the underground fruiting body (fig. 22.9).

Division Basidiomycota

This group includes mushrooms, puffballs, and bracket fungi (fig. 22.10), the largest and most conspicuous fungi in the environment, as well as rusts and smuts, two groups of plant pathogens that lack fruiting bodies.

The fleshy fungi in this division are relatively familiar to the average person and have been extensively studied since antiquity because of the edible or poisonous nature of many mushrooms (*see* Chapters 23 and 24). These **basidiocarps** (fruiting bodies) are the visible portion of an extensive mycelium within the substrate, often extending for acres. In 1992, mushrooms made headlines across the United States with the discovery of a vast area on the Michigan-Wisconsin border where a population of *Armillaria bulbosa* that covered 37 acres was found to be a single organism. Studies showed that all the mushrooms in the population were genetically identical, indicating that they were all basidiocarps of a single mycelial colony. It was estimated that the colony weighed between 100 and 1,000 tons and was at least 1,500 years old based on observed growth rate, placing it among the oldest and largest living organisms. Shortly after this publication appeared, an even larger colony of *Armillaria ostoya* was found in Washington state. This colony covered 1,500 acres but was less than 1,000 years old.

The characteristic sexual spore in the basidiomycetes (the common name for this group) is the **basidiospore,**

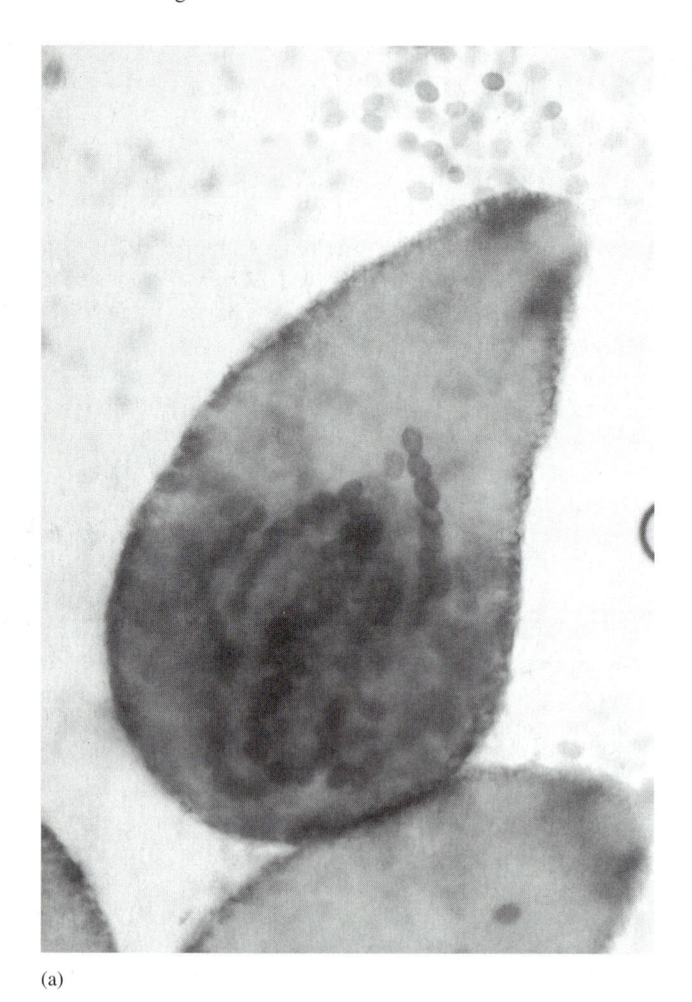

(a)

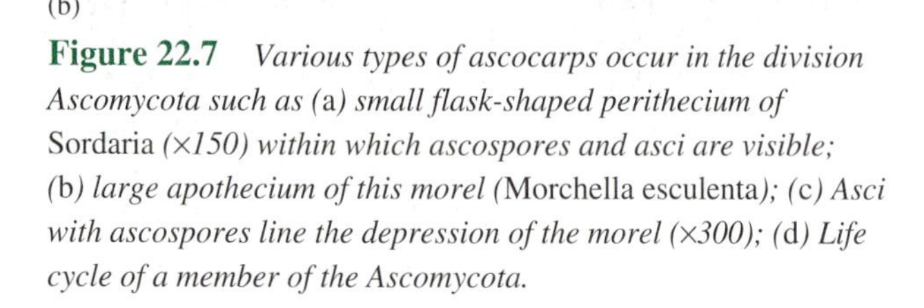

(b)

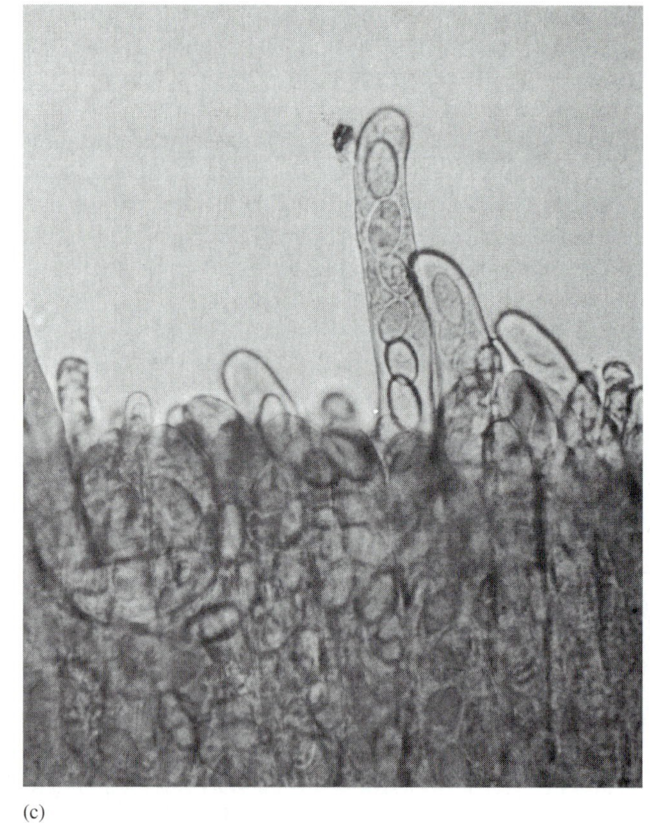

(c)

Figure 22.7 *Various types of ascocarps occur in the division Ascomycota such as* (a) *small flask-shaped perithecium of* Sordaria *(×150) within which ascospores and asci are visible;* (b) *large apothecium of this morel (*Morchella esculenta*);* (c) *Asci with ascospores line the depression of the morel (×300);* (d) *Life cycle of a member of the Ascomycota.*

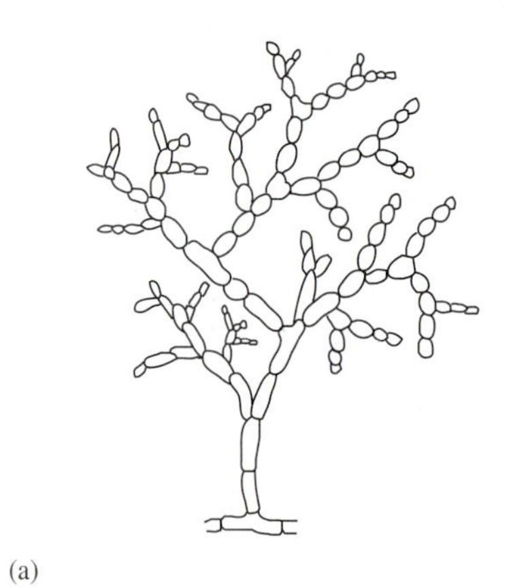

(a)

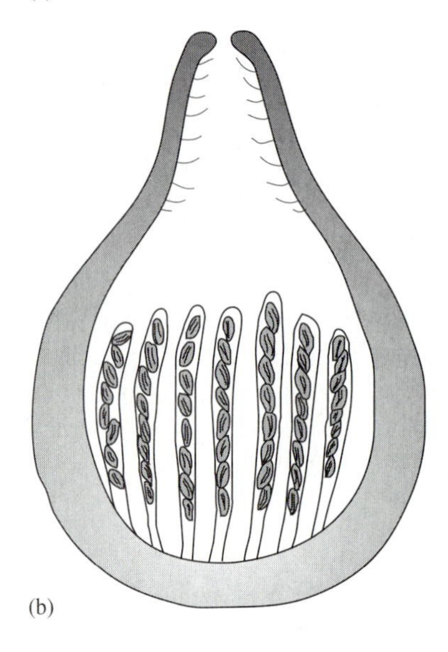

(b)

Figure 22.8
Neurospora crassa, *an ascomycete used in genetic studies.* (a) *Asexual stage produces chains of conidia;* (b) *Ascocarp showing asci and ascospores.*

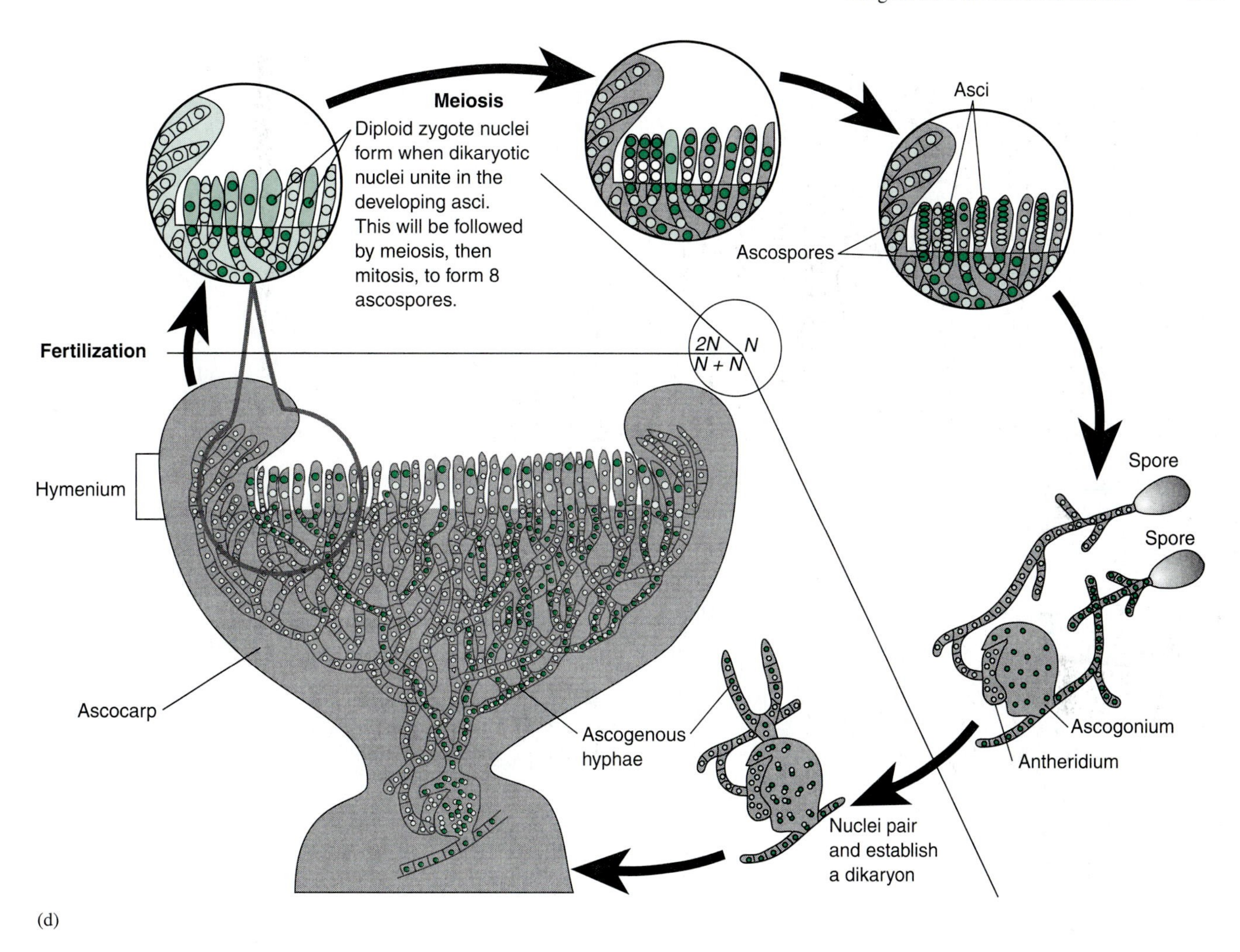

(d)

(a)

(b)

Figure 22.9 *Truffles are the most costly fungal delicacies.* (a) *In France, pigs are trained to sniff out the underground fruiting bodies.* (b) *A farmer admires the result of the hunt.*

(a) (b) (c) (d)

Figure 22.10 *Basidiocarps are the most conspicuous and familiar fungal fruiting bodies and include* (a) *mushrooms,* Armillariella tabescens, (b) *a bracket fungus,* Laetiporus sulfureus, (c) *a puffball,* Calvatia cyathiformis, *and* (d) *a stinkhorn,* Mutinus canis.

which forms externally on a **basidium** (plural, basidia). Each basidium bears four spores on peglike appendages (fig. 22.11). Karyogamy and meiosis occur within the basidium and the four haploid products of meiosis give rise to the four basidiospores.

Basidia line the gills of mushrooms and the pores of bracket fungi, with the basidiospores exposed to the atmosphere throughout development. When the spores are mature they are actively shot away from the basidium. Although the ballistics of discharge are not as dramatic as other fungi, this mechanism insures that the spores are freed from the gills or pores and reach the surrounding air currents for wind dispersal.

If the basidiospore lands on a suitable substrate, it will germinate, giving rise to a haploid mycelium. This stage of the life cycle is of limited duration, since the haploid mycelium must fuse with the haploid mycelium of a compatible mating type to establish a dikaryon before an extensive mycelium can develop. The dikaryon is the major stage in the life cycle and even includes the hyphae of the fruiting body. Fusion of the compatible nuclei occurs only in the basidium and is followed immediately by meiosis to form haploid basidiospores (fig. 22.11).

In puffballs, the basidia develop within the fruiting body, only exposing the basidiospores to the atmosphere when fully mature. These spores are passively dispersed in clouds or puffs as raindrops strike the fruiting body. Strong gusts of wind or small animals impacting the surface accomplish the same puffing action. The numbers of spores produced by puffballs, as well as by mushrooms and brackets, are astronomical (table 22.2), contributing to the atmospheric spore load.

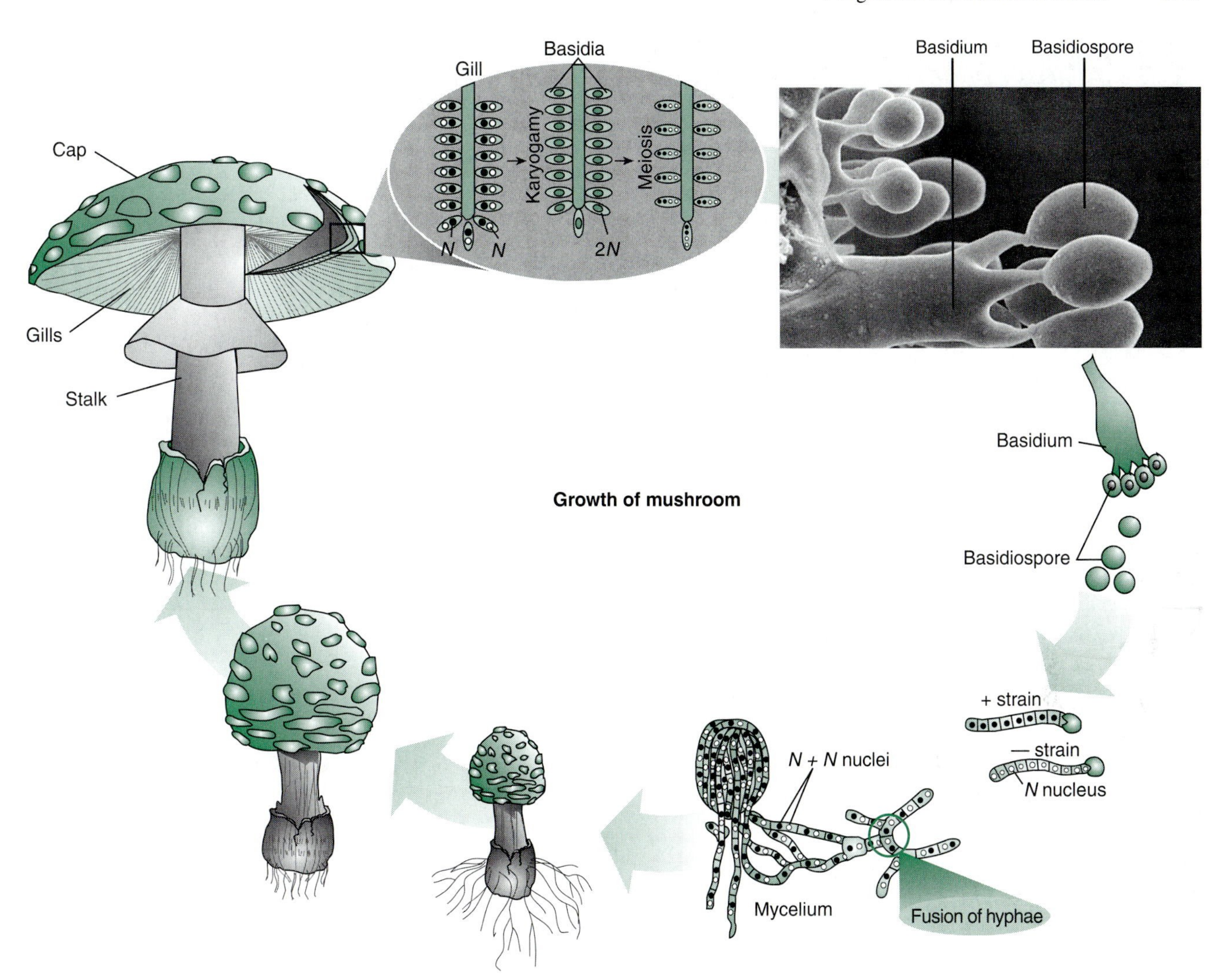

Figure 22.11 *Life cycle of a mushroom. Haploid basidiospores result from karyogamy and meiosis that occur in basidia that line the gills of the mushroom. On a suitable substrate, a basidiospore will germinate and give rise to a short-lived haploid mycelium. Compatible strains will pair to establish the dikaryon. The mycelium produced by the dikaryon is often very extensive, producing mushrooms when environmental conditions are right.*

In contrast to the airborne dispersal of spores from these fruiting bodies, basidiospores produced by stinkhorns are dispersed by flies. The mature spores are immersed in a slimy material with a foul odor that attracts flies. While visiting the fetid fruiting body, the flies eat some spores and pick up others on their body, thus playing an important role in the fungal life cycle.

Asexual spores are also formed by basidiomycetes, especially the rusts and smuts, two groups of plant pathogens in which asexual spores are the major dispersal units. The life cycles and impact of these pathogens is discussed below.

Asexual Fungi

The Deuteromycetes (also known as the Fungi Imperfecti) are an artificial grouping of fungi based on their asexual spores. These fungi are among the most abundant in the environment and can readily be isolated from many substrates, indoors or outdoors. At one time the sexual stages (teleomorphs) of all these fungi were unknown; however, many fungi classified as deuteromycetes have been determined to be the conidial (anamorphic) stages of ascomycetes (or in a few cases even basidiomycetes). For some members of this group, the teleomorph is therefore known; for others it has yet to be identified; while for still others the ability to undergo sexual reproduction has apparently been lost. Classification within this group is based on conidial morphology and development; various systems are in existence to identify and classify these fungi.

When the sexual and asexual stages of fungi were identified separately, they were given separate names. For example, it is now recognized that some species of the

Table 22.2
Basidiospore Numbers

Fungus	Total number of Spores	Duration of Spore Release	Spores per Day
Calvatia gigantea (Giant puffball)	7×10^{12}	—	—
Ganoderma applanatum (Artist's bracket)	5.5×10^{12}	6 months	3×10^{10}
Polyporus squamosus (Bracket fungus)	5×10^{11}	14 days	3.5×10^{10}
Agaricus bisporus (Commercial mushroom)	1.6×10^{10}	6 days	2.6×10^{9}
Coprinus comatus (Shaggy-mane mushroom)	5.2×10^{9}	2 days	2.6×10^{9}

Modified from A. H. R. Buller. 1922. *Researches on Fungi,* Vol. 2. Longmans, Green & Co., Ltd., London.

common mold genus *Aspergillus* are the conidial stages of an ascomycete genus known as *Eurotium,* which is not nearly as common as the conidial stage. The International Code of Botanical Nomenclature permits the use of both names, but rules that the name of the teleomorph stage should be used when referring to the whole fungus.

Several well-known fungi within this group are illustrated in Figure 22.12. Among the common molds included here is the genus *Penicillium,* a widespread indoor contaminant. Species of *Penicillium* can be readily isolated from moldy fruit or vegetables and often are the source of the "moldy" smell from basements, wet carpets, or old shoes. However, some species of this genus are quite beneficial and are the source of the antibiotic penicillin (*see* Chapter 24), while other species are used in the production of cheeses. *Penicillium roquefortii* is responsible for the unique taste, color, and flavor of blue, Roquefort, Stilton, and similar cheeses. *Penicillium camembertii* is used in production of Camembert and Brie, two additional varieties of cheese; enzymes produced by this fungus break down the curds of milk and develop the characteristic buttery consistency.

Role of Fungi in the Environment

Fungi are **heterotrophic,** obtaining organic material from their environment. As such, they can exist as **parasites** (obtaining nutrients from a living host and ultimately harming that host) or **mutualistic symbionts** (obtaining nutrients from a living host while providing some benefit to that host), but the majority of species and the most abundant in the environment are **saprobes.** Saprobes obtain nutrients from nonliving organic material or the remains and by-products of organisms.

Decomposers—Nature's Recyclers

Many saprobic species are found in the soil or as leaf surface microorganisms, where they exist on organic matter produced by the plant. Among the fungi, there are species that are capable of decomposing all the carbon compounds that occur in other organisms including cellulose, lignin, and keratin. In the environment, fungal saprobes play an essential role as decomposers, degrading organic materials and recycling nutrients.

The line between saprobes and parasites is not always clear-cut; some fungi are able to utilize both lifestyles under certain conditions. **Facultative parasites** (also known as opportunistic pathogens) can live as saprobes but will invade living tissue when suitable hosts are available. These fungi can be readily isolated from air and soil samples.

Soil is usually considered the most important environment for fungi; many species spend at least part of their life in or on the soil. It is known that one gram of soil may contain over 100,000 fungal spores.

When nutrients are available and environmental conditions are suitable, spores germinate and hyphae grow and branch within the substrate, typically producing a colony that eventually forms more spores. The types of fungi and their abundance in an area depend on the availability of

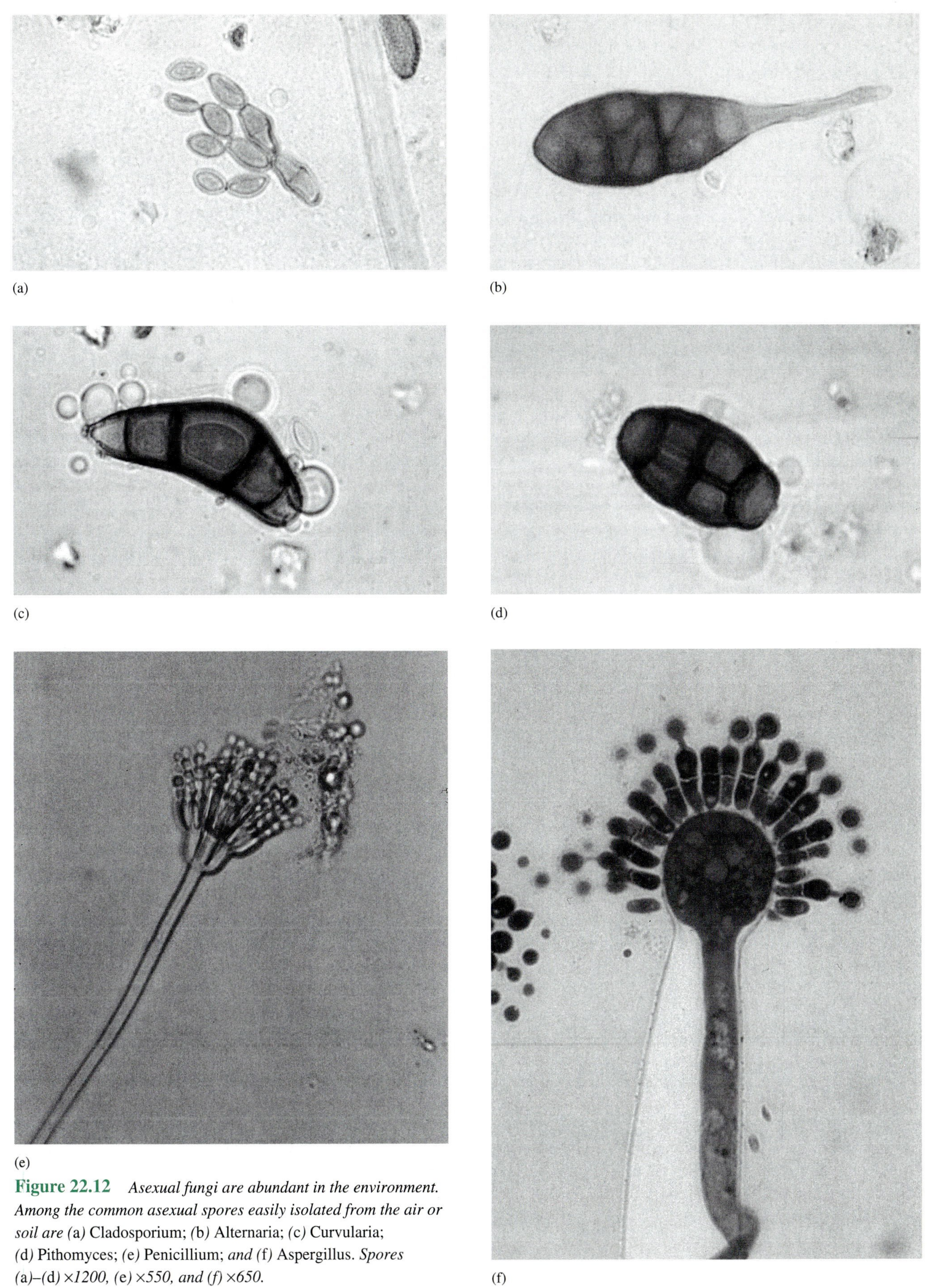

Figure 22.12 *Asexual fungi are abundant in the environment. Among the common asexual spores easily isolated from the air or soil are (a)* Cladosporium; *(b)* Alternaria; *(c)* Curvularia; *(d)* Pithomyces; *(e)* Penicillium; *and (f)* Aspergillus. *Spores (a)–(d) ×1200, (e) ×550, and (f) ×650.*

nutrients, water, and suitable temperatures. Saprobic fungi are especially versatile and able to utilize many different substrates for growth.

Mycorrhizae: Root-Fungal Partnership

Scientists estimate that 90% of all vascular plants possess fungi in mutualistic associations with their roots. This fungus-root association is termed **mycorrhizae** (literally, fungus root). The plant furnishes the fungus with sugars and amino acids, while the fungus aids in the absorption of water and minerals from the soil. This mutualistic association between plants and fungi has a long evolutionary history. Fossil evidence indicates that some of the earliest plants that colonized land about 400 million years ago had mycorrhizal fungi associated with them. Although early land plants were photosynthetic, they lacked extensive root systems. The mycorrhizal relationship was clearly an asset in obtaining water and minerals.

Different types of mycorrhizae occur. In one common type, the **ectomycorrhizae,** the fungus forms a sheath or mantle around the root, penetrating between the cells of the root epidermis and cortex (fig. 22.13). It is estimated that over 5,000 species of fungi form ectomycorrhizal relationships, typically on forest trees in temperate areas. Some trees such as pine, oak, and beech absolutely require the association for normal development. The fungal partner **(mycobiont)** of these ectomycorrhizae is usually a member of the Basidiomycota, and many woodland mushrooms are the reproductive structures of these fungi. In fact, some mushroom families are exclusively mycorrhizal.

Scientific interest in mycorrhizal fungi has increased in recent decades as their ecological role has become clearer. Their importance in forestry and agriculture has been recognized as attempts are made to revegetate areas abused by human activity. Even when a plant species can grow without mycorrhizae, the same species with mycorrhizae may be more tolerant of pollution, need less fertilizer, or grow in marginal soils (fig. 22.13). Extensive tests must be done to determine the most suitable mycobiont for each plant species in an ecosystem being reclaimed.

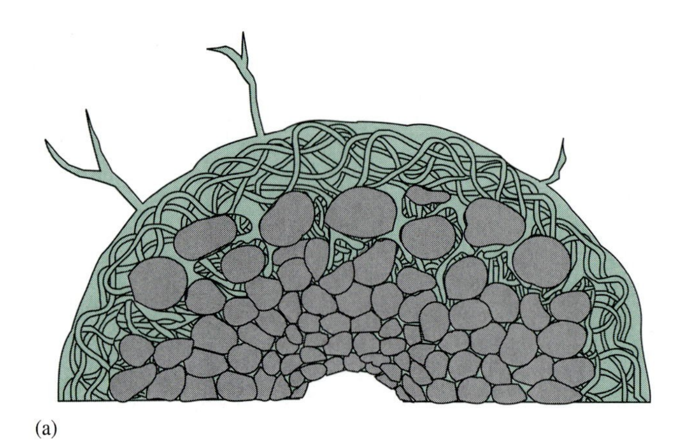

(a)

(b)

Figure 22.13 *Fungi can establish mutualistic relationships with the roots of plants. (a) In ectomycorrhizae, the fungus forms a sheath around the root, penetrating between the cells of the epidermis into the cortex. (b) Rough lemon plants with mycorrhizae grew better than those without mycorrhizae under all three nutrient regimes: no added nutrient solution, half-strength nutrient solution, and full-strength nutrient solution.*

Plant Diseases with Major Human Impact

There are many examples of plant diseases caused by fungi that have made a major impact on society and even changed human history. In fact, over 70% of all major crop diseases are caused by fungi. Recall the discussion of late blight of potato (*see* Chapter 14), when 25% of the population of Ireland was lost due to the ravages of *Phytophthora infestans.* During the 1840s, over one million people died from starvation or famine-related diseases, and over 1.5 million emigrated from Ireland.

Similar disasters have occurred at other times in human history. A more recent epidemic that resulted in large scale famine was caused by *Drechslera oryzae,* the fungus responsible for brown spot of rice. Symptoms of the disease include brown spots on the leaves and glumes of the rice plant. When the infection occurs at the seedling stage, the young foliage is severely damaged, the plants develop poorly, and the grain yield is significantly reduced. The fungus can also infect the seed during development, resulting in poor germination and weak seedlings. The most severe outbreak of this disease resulted in the great Bengal famine of 1942, when two million people died from starvation. A related fungus, *Bipolaris maydis,* (also known as *Helminthosporium maydis* and *Drechslera maydis*), which attacks corn and causes southern leaf blight, resulted in a widespread epidemic in the United States in 1970. About 15% of the total U.S. corn crop was lost, with yields in some states reduced 50% (*see* Chapter 12).

Although various plant diseases had been recognized since ancient times, they were not connected with pathogenic microorganisms until the nineteenth century. Early in that century, Benedict Prevost, a French scientist, reported that wheat bunt was caused by a smut fungus, but his findings were not accepted by the scientific community, who still believed in spontaneous generation. It had been generally accepted that fungi, other microorganisms, maggots, etc., appeared spontaneously on diseased and dying plants. They were considered the result of the condition, not the cause. During the investigations of the potato blight German botanist Anton de Bary, in 1861, proved experimentally that a fungus, *Phytophthora infestans,* was the cause of the disease. After de Bary's work, other plant pathogenic fungi were described (pathogenic bacteria and viruses were not identified until later in the nineteenth century), and plant pathology (the study of plant diseases) soon became an important field of study.

Even with the many advances in twentieth-century sciences, plant diseases still pose a major threat, and the world's food supply is vulnerable. Plant pathologists wage a constant battle to stay ahead of newly evolving strains of pathogenic fungi that are capable of destroying crop plants.

Rusts—Threat to the World's Breadbasket

There are about 6,000 species of rust fungi that attack a wide range of seed plants and cause destructive plant diseases. Some of the most important diseases of cereal crops are caused by these basidiomycetes, which have produced serious epidemics throughout history. Coffee, apple, and pine trees, and the economies depending on these crops, have also been devastated by rust fungi. Keep in mind that coffee rust destroyed the coffee plantations in Ceylon in the 1870s and 1880s (*see* Chapter 16) and today threatens production wherever coffee is grown.

The rust fungi, and also the smut fungi (*see* following text) lack a fruiting body, unlike other members of the Basidiomycota. Although sexual reproduction does occur, these fungi are recognized by the characteristic asexual spores formed on infected plant parts.

Millennia before scientists understood that pathogenic fungi could cause plant diseases, rust epidemics were studied and the reddish lesions noted on plants. These epidemics were recognized in ancient Greece and described in the writings of Aristotle and Theophrastus, who even noted that different plants varied in their susceptibility to these diseases. The Romans held a religious ceremony or festival, the *Robigalia,* to propitiate the gods, Robigo and Robigus, whom they believed responsible for the rust epidemics. In ancient times, wheat was the most important crop in these Mediterranean countries, and wheat rusts were major threats. Wherever wheat is grown today, rust fungi still remain serious pathogens.

In North America the most important rust pathogen is *Puccinia graminis,* subspecies (ssp.) *tritici,* the fungus responsible for stem rust of wheat. The organism has a complex life cycle that involves five spore stages (**basidiospores, spermatia, aeciospores, uredospores,** and **teliospores**) on two separate host plants, wheat and barberry (fig. 22.14). Basidiospores are produced in the spring when the overwintering teliospores germinate. The basidiospores, which are uninucleate and haploid, belong to either + or – mating types. If the basidiospores land on a young barberry leaf, infection is initiated, eventually giving rise to small flask-shaped structures, **spermagonia** (singular, spermagonium), on the upper surface of the leaf.

Each spermagonium produces masses of tiny unicellular spermatia that function as male gametes. These spores are formed in a sugary nectar or "honeydew" that oozes out of the neck of the spermagonium. Receptive hyphae also grow out of the spermagonium; these structures function as female gametes. Insects, which are attracted by the honeydew, will unknowingly transfer the spermatia from one spermagonium to another. When spermatia are carried to the receptive hyphae of the opposite mating type, successful fertilization is initiated. The nucleus of the spermatia enters the receptive hypha and a dikaryon is established, with two nuclei of opposite mating types within each cell. Within a short time, binucleate aeciospores are produced in cuplike structures known as **aecia** (singular, aecium) on the lower surface of the barberry leaf.

The aeciospores transfer the infection only to wheat plants. They cannot reinfect barberry. On wheat the aeciospores germinate, and hyphae enter the plant through stomata. The dikaryotic mycelium is now established in wheat and within two weeks the uredial stage develops, appearing as long narrow lesions on the stem. Dark red powdery masses of unicellular, binucleate uredospores occur in these lesions. Uredospores can reinfect wheat to produce several generations of **uredia** (singular, uredium). This stage in the rust life cycle is known as the "repeating stage." The uredospores are airborne and easily carried from one plant to another, giving rise to epidemics. In fact, the uredospores can be carried by prevailing winds for hundreds of miles. In Mexico and southern Texas, this stage can continue all winter (on winter wheat) and give rise to spring infections in northern states, since the uredospores are carried by prevailing southerly winds (fig. 22.15).

Near the end of the growing season, the uredia turn black as two-celled teliospores replace the uredospores. Teliospores are the overwintering stage of the fungus and also part of sexual reproduction. In the spring, the teliospore germinates, each cell giving rise to a short hyphae that becomes a basidium. The two haploid nuclei

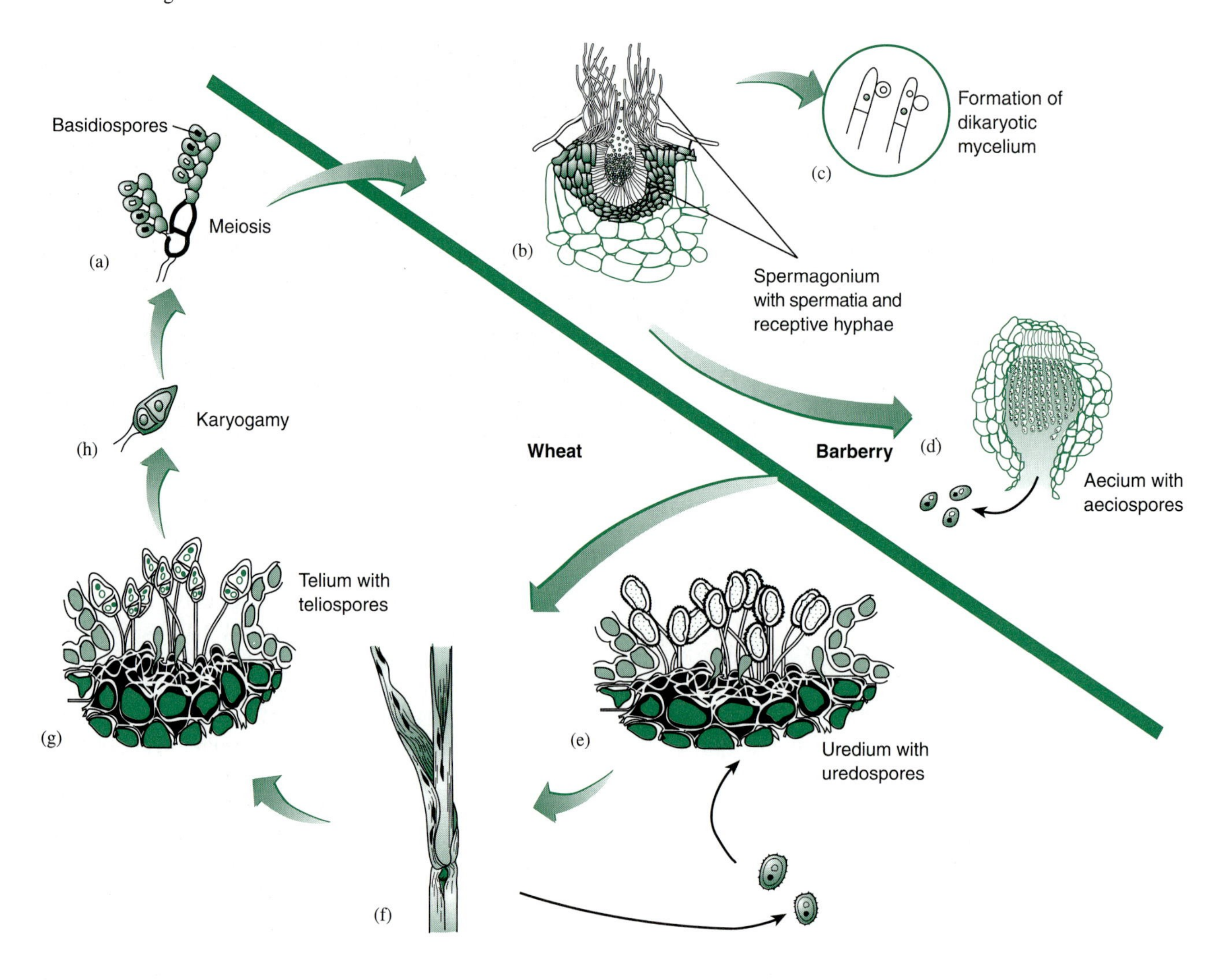

Figure 22.14 *Life cycle of* Puccinia graminis, *the cause of stem rust of wheat. The two hosts are barberry and wheat. Basidiospores* (a) *infect barberry and result in the development of spermagonia that produce spermatia and receptive hyphae* (b) *and* (c). *Aeciospores are also produced on barberry* (d). *These infect wheat, transferring the infection to the economically important host. Within 14 days uredia with uredospores* (e) *appear as long narrow lesions on the wheat stem* (f). *The uredospores can reinfect wheat throughout the growing season. Near the end of the season, uredospores are replaced by teliospores,* (g) *which overwinter. The following spring these give rise to basidia and basidiospores* (h) *and* (a), *thereby completing the life cycle.*

fuse, followed by meiosis, to produce four haploid nuclei, two of each mating type. Each nucleus migrates into each of the developing basidiospores, completing the life cycle.

Wheat plants infected by *P. graminis* are severely weakened but not destroyed, and the grain yield significantly reduced. It is estimated that worldwide over one million metric tons of wheat are lost annually to stem rust.

Although the connection between wheat and barberry (as alternate hosts for the pathogen) was not made until the 1860s, farmers had long before known that wheat rust was worse whenever barberry plants were nearby. In fact, there were laws enacted in France in 1660 requiring the removal of barberry plants from grain fields. Similar laws were enacted in the North American colonies of Connecticut in 1726 and Massachusetts in 1755. Eradication of barberry continues to be one of the major methods of control of stem rust of wheat. However, as noted, long-distance transport of uredospores can introduce wheat rust to areas where no barberry plants exist.

The most effective and cheapest control of wheat rust is through the use of resistant varieties of wheat. The ultimate goal of wheat-breeding programs is the development of a wheat variety resistant to all varieties of rust fungus. However, there are over 200 physiological races of *P. graminis* ssp. *tritici,* which differ in their ability to attack the many commercial varieties of wheat. New races of the

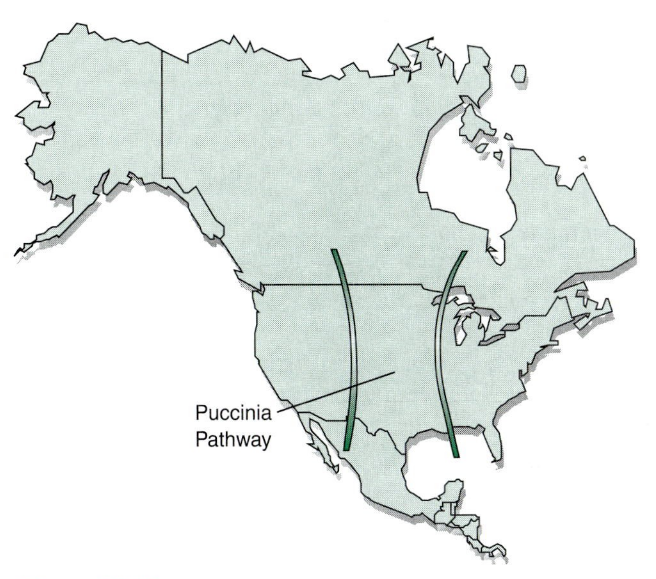

Figure 22.15 *Uredospores of* Puccinia graminis *are blown northward by prevailing winds from Mexico and the southern states to the northern plains and Canada.*

Figure 22.16 *Galls on eastern red cedar are produced by the cedar-apple rust fungus,* Gymnosporangium juniperi-virginianae.

fungus are discovered each year, so wheat breeders must be diligent to stay ahead of the fungus.

With five spore stages, *Puccinia graminis* is an example of a long-cycled rust. Some rust fungi have fewer spore stages and some complete their life cycle on a single host plant. Cedar-apple rust causes yellow spots on the leaves and fruit of apple and crab apple (and related species) and produces galls on red cedar (juniper) trees. This rust fungus lacks a uredial stage. Spermatia and aeciospores are produced on the apple trees and teliospores develop from the galls on the cedar (fig. 22.16). Because no repeating stage occurs, cedar-apple rust can be efficiently controlled by eliminating the alternate host from the immediate area.

Corn Smut—Blight or Delight

The smut fungi, like rusts, are plant pathogens that cause significant losses of grain crops. Unlike the rusts, only one host plant is involved and only two spore stages occur, teliospores and basidiospores. Incredible numbers of teliospores form from the mycelium, usually within galls. It is these masses of dark-brown or black spores, resembling soot or smut, that give the common name to these plant pathogens. Basidiospores form on germination of the teliospores in a situation similar to the rust. In each infected plant there is only one generation of teliospores. Many smuts overwinter as teliospores; however, other species overwinter as mycelium in infected grains.

Corn smut is a widespread disease, occurring wherever corn is grown, and is more common on sweet corn than other varieties. The fungus, *Ustilago maydis,* forms galls on any aboveground plant part but is most conspicuous when the galls form on the ear (fig. 22.17). The size and location of the galls reflect the degree of crop loss. Galls on the ear result in total loss, while those in other places reduce the yield or cause stunted growth.

The fungus overwinters as teliospores on plant debris and in the soil. In the spring, the teliospores germinate to produce a basidium that forms basidiospores. The haploid basidiospores are carried by wind or splashed by rain to the young tissues of corn plants, either young seedlings or the growing tissue of older plants. The basidiospore germinates and invades the plant tissue. For infection to continue, the dikaryon must be established by fusion of hyphae of compatible mating types. Alternatively, compatible basidiospores may fuse even before germination to establish dikaryotic mycelium directly. The mycelium grows prolifically in the infected tissue and stimulates the host cells to divide and form galls that may reach a size of 15 cm in diameter. The galls are initially covered by a membrane. As they mature, the interior darkens as the mycelium is converted into teliospores. Later, the membrane ruptures and releases the masses of dry spores.

One interesting feature of corn smut is the recent popularity in the United States of the young galls as a gourmet delicacy. The fungus has actually been cultivated for centuries in Mexico as a food item (fig. 22.17). At the present time farmers in Pennsylvania and several midwestern states are growing corn smut for the gourmet market, where it is known as "smoky maize mushroom" that sells for $50 per pound.

Dutch Elm Disease—Destruction in the Urban Landscape

The plant diseases just discussed have led to widespread problems by devastating crop plants; other pathogenic fungi have also made an impact on society by attacking

A Closer Look 22.1

Lichens: Algal-Fungal Partnership

Lichens are dual organisms, consisting of an alga (phycobiont) and fungus (mycobiont) in a mutualistic relationship. These organisms can survive in the harshest environments: on bare rocks, in hot deserts, in the frozen arctic. Through photosynthesis, the phycobiont produces food for both species, even though it only composes 5%–10% of the total lichen body. The mycobiont obtains water and minerals, forms a complex vegetative body, and develops both sexual and asexual reproductive structures.

The overall appearance of lichens is described as foliose (leaflike), fructicose (shrubby), or crustose (crustlike) (box fig. 22.1). Many lichens produce specialized asexual propagules known as soredia, which are small groups of algal cells intertwined with fungal hyphae. Wind, rain, and animals can all act as dispersal agents for these soredia, which eventually establish a new lichen colony.

The association of the two organisms is so complete that lichens are given scientific names as if they were a single organism. There are approximately 18,000 species of lichens. Interestingly, the phycobiont in more than 90% of all lichens belongs to just three genera of alga: *Trebouxia* and *Trentepohlia,* which are green algae, and *Nostoc,* a cyanobacterium. (In fact, *Trebouxia* is the algal component in 70% of all lichens.) By contrast, there are thousands of species of fungi that occur in lichens. These fungi only occur within lichens; they are never found without their phycobiont. Ninety-eight percent of these fungi are ascomycetes. An obvious reproductive structure visible on many lichens is a cup-shaped ascocarp typical of certain ascomycetes.

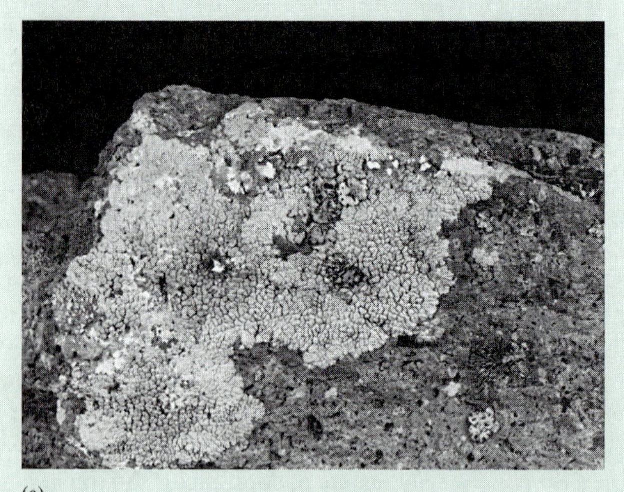

(a)

(b)

Box Figure 22.1 *Various growth forms common in lichens.* (a) *Crustose (crustlike) lichen growing on a rock;* (b) *Fructicose (shrubby) lichen growing on the ground;* (c) *Foliose (leafy) lichen on the bark of a tree.*

(c)

Lichens have incredibly slow growth rates, only expanding a few millimeters each year. Existing colonies often have great longevity, with some believed to be 4,500 years old. Although tolerant to extremes of temperature and drought, lichens are very sensitive to air pollution, especially sulfur dioxide. Scientists have even used the disappearance of certain lichen species as a means of measuring the extent of pollution within an area.

In the arctic, caribou and reindeer feed on a lichen species known as reindeer moss, which is the dominant vegetation in some areas. If other food is not available, these animals can exist on a purely lichen-based diet. Although lichens have not been used as human food, extracts from some lichens have been used medicinally as antibiotics.

Many lichens are brightly colored and for centuries were the source of dyes used by native peoples throughout the world (*see* A Closer Look: Herbs to Dye for in Chapter 17). While the use of lichen dyes has largely been replaced by synthetic dyes in industrial societies, lichens are still used in the manufacture of Scottish tweeds. Lichens are also used in the production of litmus paper, the acid-base indicator widely used in chemistry labs.

Figure 22.17 Ustilago maydis, *the cause of corn smut, produces galls that replace kernels. Although a devastating loss for the corn crop, the infected ears are actually considered a delicacy in Mexico (where they are known as* huitlacoche*) and are gaining in popularity in other markets as well.*

native forest and ornamental trees. During the first half of the twentieth century, the chestnut blight fungus caused the death of 80% of the native American chestnut trees throughout the eastern forests from Maine to Georgia (*see* Chapter 6). *Ophiostoma ulmi,* the cause of Dutch elm disease, has led to destruction of native American elm (*Ulmus americana*) trees and altered urban ecology by killing ornamental elms across the country. *Ulmus americana* had long been a popular shade tree throughout much of this country, valued for its overarching growth and broad canopy.

The disease first appeared in Europe around 1919, and the fungus was first described in Holland in 1921 by two Dutch plant pathologists. Although the name of the disease reflects these early studies on the fungus, it is believed that the disease originated in Asia. The fungus was accidently introduced to North America on contaminated logs and was first found in Ohio in 1930. Following its introduction, Dutch elm disease spread across the continent; and in the six decades since its appearance in the United States, it is believed that over 40 million elms have died. Although the aesthetic loss cannot be estimated, the cost of removing diseased or dead elms amounts to millions of dollars each year.

(a)

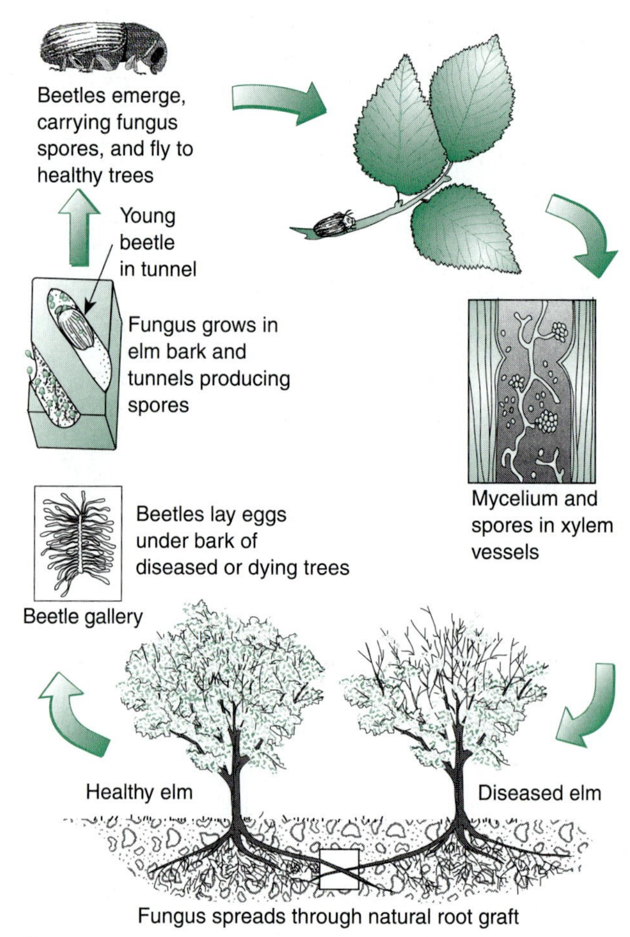

(b)

Figure 22.18 *Dutch elm disease, caused by* Ophiostoma ulmi, *has led to the destruction of native elm trees in North America. (a) Infected elm tree leafless on the left side; (b) Infection of elm trees is spread by bark beetles and even through natural root grafts.*

The disease affects all elm species, but it is most destructive to the American elm. The fungus causes a vascular wilt, and the first symptom is a sudden and permanent wilting of individual branches. The disease usually appears first on one or two branches, then slowly spreads to other branches, gradually killing the tree over a period of a few years (fig. 22.18a). In some cases, however, the entire tree develops the symptoms at once and dies within a few weeks.

The spread of this disease results from the activities of insects. Although the fungus *Ophiostoma ulmi* is responsible for the disease, elm bark beetles are the vectors that carry the fungal spores from infected elms to healthy trees (fig. 22.18b). These beetles feed on healthy trees or burrow into the bark and wood. Spores are deposited in the tree, germinate, and grow rapidly. The fungus grows in the vessels and produces spores that are carried upward in the xylem stream, thereby establishing new points of infection. Toxins are also formed by the growing fungus; these are actually responsible for the yellowing, wilting, and withering of the leaves.

Bark beetles lay their eggs in diseased or dying trees, forming breeding galleries or tunnels in the area between bark and wood (fig. 22.18). Mycelium and spores of *O. ulmi* are often found in the breeding tunnels. When adult beetles emerge from the galleries, they pick up the spores on their bodies and begin the infection cycle anew. The disease also spreads through natural root grafts between adjoining trees. The effects of this have been seen repeatedly on elm-lined streets, as one tree after another succumbs to the disease.

Since its first appearance in the United States, various measures have been employed to control the spread of the disease. For years, the major approaches were sanitation to remove infected trees and the use of insecticides to destroy the bark beetle. These measures were of limited effectiveness. Attempts at biological control of the beetle have recently focused on the use of another fungus that appears to disrupt the reproductive cycle of the beetles. Systemic fungicides have also been employed, as well as large-scale breeding programs to find resistant varieties of elms. Meanwhile, *Ulmus americana* continues its decline throughout the country.

Chapter Summary

1. Fungi are a diverse group of organisms ranging from small, inconspicuous yeasts and molds to large, conspicuous mushrooms, puffballs, and bracket fungi. The majority of fungi consist of a mycelium composed of highly branched hyphae. Fungi reproduce by spores that result from sexual or asexual processes. Fungi have an absorptive type of nutrition, producing enzymes that are secreted into the environment and break down complex organic compounds into smaller molecules. These small molecules are then absorbed by the hyphae.
2. Organisms that have traditionally been called fungi are found in both the Kingdom Protista and the Kingdom Fungi. Slime molds and water molds are included in the Protista, whereas all other groups are included in the Kingdom Fungi. The presence of chitin in the cell wall is one of the major distinguishing features. Classification in the Kingdom Fungi is normally based on the method of sexual reproduction, but a large group of organisms are considered asexual fungi, since a sexual stage seldom occurs.
3. The majority of fungi are saprobes, actively recycling nutrients in the environment by decomposing organic matter. Symbiotic relationships among the fungi are well-known and both mycorrhizal fungi and lichen fungi play important ecological roles. Other fungi exist as pathogens, obtaining nutrients from a living host.
4. The majority of plant diseases are caused by fungi. Some of these have devastating influences on crop plants and historically some have resulted in major impacts on human society. Rust fungi and smut fungi are especially damaging pathogens on cereal crops, while Dutch elm disease has resulted in the loss of millions of ornamental elm trees throughout North America.

Review Questions

1. What characteristics distinguish fungi from plants? What characteristics distinguish them from animals?
2. How does *Puccinia graminis* begin new infections on wheat when no barberry plants are in the area?
3. Why are scientists interested in mycorrhizal fungi?
4. Describe the life cycle of a mushroom.
5. Describe the ecological roles played by fungi.
6. Compare and contrast the rust and smut fungi.
7. Compare and contrast the sexual reproductive structures of the Ascomycota and Basidiomycota.
8. In what ways are the slime molds fungal-like and in what ways are they animal-like?

Further Reading

Agrios, George N. 1978. *Plant Pathology,* 2nd Edition, Academic Press, New York.

Barron, George. 1992. Jekyll-Hyde Mushrooms. *Natural History,* 101(3):47–53.

Buczacki, Stefan. 1989. *New Generation Guide to the Fungi of Britain and Europe.* University of Texas Press, Austin, TX.

Cooke, R. C. 1977. *Fungi, Man and His Environment.* Longman, London and New York.

Duran Sharnoff, Sylvia. 1984. Lowly Lichens Offer Beauty and Food, Drugs and Perfume. *Smithsonian,* 15(1):134–143.

Findlay, W. P. K. 1982. *Fungi—Folklore, Fiction, & Fact.* Mad River Press, Eureka, CA.

Kendrick, Bryce. 1992. *The Fifth Kingdom,* 2nd Edition. Focus Text, Newburyport, MA.

Lipske, Mike. 1994. A New Gold Rush Packs the Woods in Central Oregon. *Smithsonian,* 24(10):34–45.

Schumann, Gail L. 1991. *Plant Diseases: Their Biology and Social Impact.* APS Press, St. Paul, MN.

23

Beverages and Foods from Fungi

Chapter Outline

Chapter Concepts

1. The step common to both beer and wine making is the fermentation of sugar to alcohol, a conversion made possible through the action of yeast.
2. In distillation, the alcoholic content of a beverage is raised considerably through the principle of differential boiling points.
3. Fungi have a proven nutritional value when consumed directly as food or as agents of fermentation in fermented foods.

In 1810 Guy Lussac, a French biochemist, was the first to deduce the chemistry of fermentation. He recognized that each sugar molecule is converted into two molecules of ethanol and two molecules of carbon dioxide:

$$C_6H_{12}O_6 \longrightarrow 2\ C_2H_5OH + 2\ CO_2$$

It is known today that two ATPs are also formed from the chemical energy released when glucose is broken down during fermentation. The detailed steps of fermentation were not completely elucidated until the first half of the twentieth century and have been discussed in Chapter 4.

Yeast is a single-celled ascomycete (*see* Chapter 22). In an anaerobic environment, yeast will forego aerobic respiration in favor of fermentation. In the modern processing of beer and wine, most of the yeast fermenters are strains of *Saccharomyces cerevisiae* (fig. 23.1).

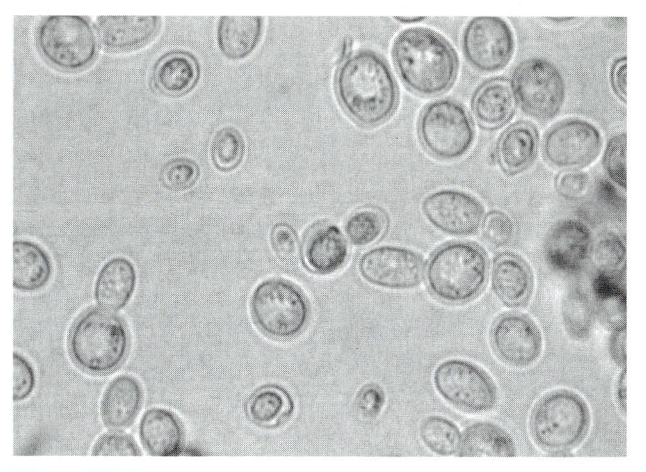

Figure 23.1 *Yeast,* Saccharomyces cerevisiae, *ferments sugar to ethanol and carbon dioxide during the production of wine and beer (×1200).*

Making Wine

The practice of growing grapes to make wine is an ancient one. Wine's first appearance was most likely an accident of nature when wild yeasts present on the grape skins fermented the juice. Whatever its exact origins, the skills in wine making were known to many ancient cultures. Five-thousand-year-old sealed wine jars have been found in Egyptian tombs. The Egyptians also developed fairly sophisticated practices in the cultivation of grapes for wine production, such as trellises for training vines, well-developed pruning methods, and irrigation of the vineyard (fig. 23.2). Both the ancient Greeks and Romans were skilled vintners. The Romans introduced many innovations to wine making such as classifying grape varieties and adding spices and herbs to flavor wines. They brought the wine grape and the practice of wine making to what is now France, Germany, and England. The skill was preserved and passed on in the monasteries of Europe during the Middle Ages and the Renaissance. Not until the renowned French microbiologist Louis Pasteur was called in for his services by the French wine industry in the mid-nineteenth century did the ancient art of wine making become the science of **enology.** It was Pasteur who discovered that yeast, through an anerobic process, converted grape juice into wine, and that bacterial contamination by *Acetobacter,* through an aerobic reaction, could turn wine into vinegar. He reasoned correctly that eliminating air (oxygen) from the wine barrel (primarily by keeping it full) could prevent the conversion to vinegar. He also discovered that the wine could be rid of bacterial contaminants by heating, a process now known as pasteurization, which is routinely used to sterilize many substances besides wine.

The wine grape was first introduced to California in 1769 by the Franciscan missionary, Father Junipero Serra, but wine making did not become established until the 1800s. The industry was almost destroyed by the ratification of the Eighteenth Amendment in 1919 and the nearly two-decades-long Prohibition Era that followed. Renewed growth in the 1960s and 1970s, established California as one of the major wine centers in the world. Today, California wines favorably compare with those from the finest wineries in Europe.

Figure 23.2 *The ancient Egyptians were skilled wine makers. This tomb painting depicts workers harvesting grapes and watering the vines.*

Next, we trace the making of wine from grape to bottle, according to the most commonly practiced methods of the California grape growers and vintners.

The Wine Grape

Although the sweet juice of other fruits, such as apricots or elderberries, can be fermented, wine is technically fermented grape juice. Wine begins with the wine grape *Vitis vinifera,* native to the area around the Caspian Sea between Europe and Asia. Members of the Vitaceae, or grape family, are found in warm temperate or tropical areas and are economically important as the sources of wine, table grapes, raisins, and grape juice. North America has many native species of grapes but most are not suitable for making wine. One well-known exception is *V. labrusca,*

Figure 23.3 *Wine is made from the fermented juice of grapes.*

TABLE 23.1

Some Varieties of the Wine Grape, Vitis vinifera

Variety	Color
Cabernet Sauvignon	red
Chardonnay	white
Gamay	red
Grenache	red
Muscat	white
Pinot Noir	red
Riesling	white
Sauvignon Blanc	white
Semillon	white
Zinfandel	red

(fig. 23.3) native to the eastern United States, and the base for several red wines. *Vitis vinifera* is a woody climbing vine with five-parted flowers borne in inflorescences opposite the palmately lobed leaves. Unlike some species of grape, the wine grape bears perfect flowers and is self-pollinated. The cultivation of grapes is known as **viticulture.** The wine grape is basically of two types, red or white, referring to the color of the grapes. Literally thousands of varieties of these two types exist (table 23.1). Shoots of the wine grape have been routinely grafted onto rootstocks of North American vines since the turn of the century. One North American species used for this purpose is *V. rupestris.* Grafting of *V. vinifera* to North American grapes came about because of the root aphid *Phylloxera,* an insect that destroys the grapevine by attacking its roots. In 1860, *Phylloxera* was accidentally introduced to European vineyards when grapevines, native to eastern North

Le Journal illustré

QUINZIÈME ANNÉE. — N° 39

Gravures :

DIMANCHE 22 SEPTEMBRE 1878

PRIX DU NUMÉRO : 15 CENTIMES

Texte

La Commission officielle visitant les vignobles suspects de phylloxera

Voir l'article, page 307.

Figure 23.4 *In a vain attempt to stop the spread of* Phylloxera, *infested European vineyards were uprooted and burned.*

America were sent to Europe. Native North American grapes are resistant to damage by the insect but can harbor this microscopic pest. Between 1870 and 1910, most French, German, Austrian, and even Californian (planted with the European wine grape) vineyards were devastated (fig. 23.4). In all areas of the world infested by *Phylloxera,* rootstocks of resistant American varieties are grafted to varieties of the wine grape.

The rootstock is planted in early spring. In late July or August, the desired grape variety is grafted to the rootstock. Alternatively, some growers plant the rootstock with the specific graft already in place. The graft will determine the variety of grape to be harvested. During the second season of growth, the vine is selectively pruned and trained to trellis wires for support and to maximize photosynthesis by spreading the leaves out for optimal exposure to the sun (fig. 23.5). These practices encourage shoot growth and more abundant fruit production. In the third year of growth, enough grapes are produced to harvest. Fruit production continues to increase during the next three years, usually leveling out in the sixth to eighth years. Vines are productive for 40–50 years. Grapevines are subject to several fungal diseases, as well as insect pests. Powdery mildew is

Figure 23.5 *Grape vines are pruned and trellised to maximize fruit production.*

a constant threat and a dusting of sulfur powder is applied annually to prevent the mildew from taking hold. The story of downy mildew, a more serious fungal disease, is presented in A Closer Look: Disaster in the French Vineyards.

Harvest

In the fall, grapes are harvested when their sugar content reaches a certain critical level. The grape, more than most other fruits, concentrates high levels of sugars. The grapevine produces sucrose in its leaves following photosynthesis, which is transported to the developing fruits and broken down into glucose and fructose. Approximately 20% of the juice of ripe grapes is a mixture of these two sugars. Organic acids, mainly tartaric and malic, are also found in the grapes in concentrations around 1%. To determine sugar content, a sample of grapes is analyzed. Determining the sugar content of the juice is necessary since growers are paid according to relative concentrations of sugars to acids. In colder regions, grapes tend to have higher acid levels, while in warmer areas the grapes are sweeter. Too sugary a grape produces a wine that is flat and more susceptible to bacterial spoilage. Too acidic, and the wine tastes green or hard. When the sugar concentration is optimal, the grapes are harvested either by hand or by machine into gondolas and then trucked to the winery. White grapes are harvested earlier than the red varieties.

Red or White

Grapes must be crushed, pressed, and fermented to become wine (fig. 23.6). Today, many grape varieties are crushed and destemmed mechanically. Crushing refers to breaking the skins to start the flow of juice, and mechanical crusher-stemmers also remove the stems at the same time. The resulting mix of skins, seeds, and juice is referred to as must. Pressing squeezes the skins to express the last bit of liquid in such a way as to free the juice of skins, pulp, and seeds. Leftover skins and seeds make up a cake, or pomace, usually returned to the vineyard as fertilizer. The expressed juice is transferred to stainless steel holding tanks or the traditional oak barrels in preparation for the fermentation process.

The juice of all wine grapes, whether red or white varieties, is initially white. Only through contact of the juice with broken red skins does the wine develop a red color. Red skins contain anthocyanin pigments and tannins that impart the rich color. In creating a white wine from red grapes, the skins are promptly separated from the juice in the press before fermentation. In the making of red wine, red skins are not removed from the juice until after fermentation. A rosé results when the pressing to remove the red skins takes place in the middle of the fermentation process, producing wine with a pink or pale red shade. To sum up, white wines, using either red or white grapes, are crushed, pressed, and then fermented. With red wines, the red grapes are crushed, fermented, and then pressed to remove the skins.

Fermentation

Sulfur dioxide is added to the juice to kill any unwanted, naturally occurring yeasts and bacteria brought in with the grapes (fig. 23.6). It also removes oxygen from the fermentation vat, a further insurance against any aerobic *Acetobacter* spp. present from turning the ethanol to vinegar. Specially developed wine yeast strains are then introduced into the grape juice. This contrasts with the age-old practice, called spontaneous fermentation, of relying on yeasts naturally present on the grapes. The yeast ferments the grape sugars, glucose and fructose, to ethanol and carbon dioxide gas. During this process, appreciable amounts of waste heat are produced that can negatively affect the wine and must be monitored carefully.

White wines are fermented at temperatures from 10°–15° C (50°–59° F) while red wines are fermented at higher temperatures, ranging from 25°–30° C (77°–86° F). Temperature control is extremely critical. If the fermentation temperature is too low, white wine develops an overly yeasty taste and red becomes too fruity and thin-bodied. If the temperature during fermentation is too high, the wine loses its fruitiness and tastes "cooked." Also, at higher temperatures, the yeast may be killed, stopping fermentation completely. Generally, controlling heat production during white wine fermentation is not as difficult as in the fermentation of red wines. In red fermentation, heat production is greater and more likely to reach unacceptably high temperatures above 30° C (86° F). In small wineries, temperature control is limited to keeping the winery cool. In large-scale operations, metal cooling jackets manage the temperature in each tank or barrel.

Red fermentation is very active because pressing takes place after fermentation. Consequently, a bubbling surface cap of skins and seeds forms on the top of the fermentation tank. This cap needs to be broken up several times a day during the fermentation process. This is done manually

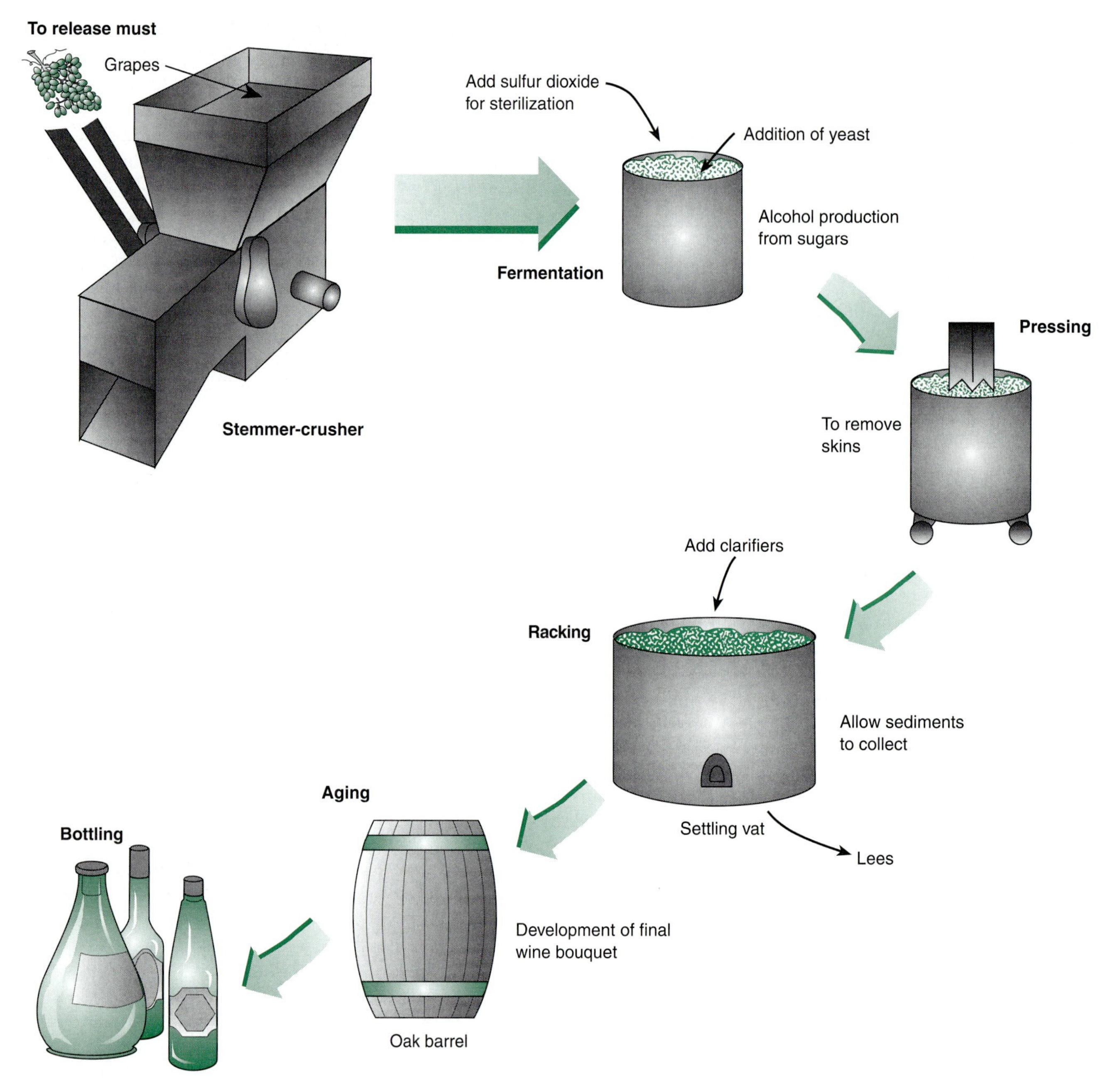

Figure 23.6 *Steps in the wine-making process. Wine production begins with the crushing of grapes to release the juice. In the processing of white wines, the grapes are pressed immediately to remove the skins. In the production of red wines, shown here, pressing is delayed until after fermentation. Before the yeast is added to begin the fermentation process, sulfur dioxide is added to eliminate any unwanted microorganisms. Fermentation may continue for days or weeks. Racking allows sediments to collect, clarifying the wine. In red wines, malolactic fermentation is often induced to reduce the acidity of the wine. The aroma and flavor of fine wines is obtained by aging in oak barrels. Bottling is the last step in wine production.*

with a wooden paddle or mechanically by pumping up a constant stream of juice from the bottom of the tank to the surface.

Fermentation goes on for days or weeks, depending on the temperature. The cooler the temperature, the longer the fermentation. Three to 14 days is about right for a red wine and 10 days to 6 weeks is usual for white wines. The rate of fermentation can be monitored in several ways. The most common method is to measure the decreasing sugar concentration. When there are no more fermentable sugars in the wine, fermentation is completed. In a dry wine, fermentation is complete and there is no residual sweetness. If a sweet wine is desired, the vintner must interrupt fermentation by lowering the temperature, filtering the wine of yeast, or adding sulfur dioxide to inhibit the yeast. The alcohol concentration of table wines averages 12%; yeast die at alcoholic concentrations above this. If higher alcoholic concentrations are desired, the wine must be either fortified or distilled. These options are discussed later. Most of the carbon dioxide produced during fermentation is allowed to escape, producing a still wine, a wine not highly carbonated.

Clarification

The newly fermented wine now undergoes the process of clarification, during which the wine loses its murky appearance and becomes crystal clear. Clarification begins with racking (fig. 23.6), when the wine is allowed to settle. Sediments, mostly used-up yeast known as lees, collect on the bottom and the wine is transferred to a second barrel or tank. To speed clearing, a clarifier is added, such as bentonite clay, egg whites, or gelatin. These substances act as adsorbents, collecting suspended particles and proteins as they move through the wine. Some wineries may even centrifuge the wine to remove sediments. Next, the wine is filtered. Cellulose pads, diatomaceous earth, or membrane filters are common methods. In premium quality wines, the wine is cold sterilized by passing it through millipore filters, with pores so tiny they exclude yeast and bacteria. Pasteurization, heating the wine to 81° C (180° F), or adding a preservative such as sorbic acid, are other methods that sterilize the wine by killing the yeast and any other microbes present.

Aging and Bottling

The aroma and flavor important to fine red wines and even a few whites, is obtained by aging the wine in oak barrels (usually white oak imported from France) or small casks for a period of several months to a year or two (fig. 23.6). A substitute but inferior method is to age the wine in tanks with oak chips. White wines, if aged at all, are generally not aged as long as red wines. For some red wines, a second fermentation may be induced during the aging process in the barrel. This second fermentation may be started by introducing lactic acid bacteria cultures (*Lactobacillus*), to break down malic acid still present in the wine. This type of fermentation is known as malolactic fermentation, as the malic acid is converted into lactic acid, a weaker acid, and is an effective way of reducing the acidity of a wine.

During bottling, the wine is placed in sterile bottles that are often flushed with nitrogen to minimize exposure to oxygen and the possibility of the aerobic conversion of ethanol to vinegar by *Acetobacter*. The bottles are precisely topped with wine to allow just enough room for the cork and to minimize the amount of air. Corked bottles are sealed with tinned lead and are stored bottoms up in order to wet the corks and keep air and bacteria out. Labels (fig. 23.7) reveal a great deal of information about the history of the grapes and wine. For example, the area where the grapes were cultivated and the year the wine was made, can be found by reading a label. If the vintage is given, it means that 95% of the wine was made during the indicated year.

Figure 23.7 *A wine label reveals important information about the bottled wine. The designation* Chardonnay *indicates that the wine contains at least 75% of that grape variety.* Napa Valley *indicates that 85% of grapes in the wine were grown and processed in this established viticultural area. The year* 1991 *is the vintage year and 95% of the wine must be of that year's growth.* Grown, produced, and bottled by *discloses that Burgess Cellars fermented and finished at least 75% of the wine on its premises.* St. Helena, California *is the bottling location and the alcoholic content of the wine is* 12.9%.

"Drinking Stars"

Whether champagne or Asti Spumante, these wines fizz due to an excess of carbon dioxide bubbles from a secondary fermentation. The origin of champagne was accidental, the result of an unfinished fermentation, and probably occurred in the colder wine-growing regions. Most likely, the primary fermentation of the wine began as usual after the harvest, but stopped before completion, as the weather cooled with the onset of winter. The wine maker assumed the fermentation was completed and the wine was subsequently bottled. When temperatures warmed up the following spring, the fermentation process restarted in the bottle since the wine still retained some residual sugar and yeast. The wine fermented to dryness and, when eventually opened, had a noticeable fizz from the abundant CO_2 bubbles.

Dom Perignon, the name most associated with champagne making, was a seventeenth-century Benedictine monk from the Champagne region in France. It was Dom Perignon who aptly described the sensation of champagne as "drinking stars." He is credited with the first careful study on how to deliberately induce the bubbles. He also perfected the art of blending different wines to come up with the base for the champagne and was responsible for developing the method of pressing red grapes immediately to achieve a white wine. The methodology was further improved in the eighteenth century and techniques developed during this time established the standards for the champagne industry today.

A Closer Look 23.1

Disaster in the French Vineyards

Box Figure 23.1 *Grapes infected by* Plasmopora viticola, *the fungus that causes downy mildew. The "downy" or cottony appearance is due to the growth of sporangiophores on the surface of the grapes.*

Although plant diseases can always cause serious problems, the mid-to-late nineteenth century was an especially adverse period for the wine industry in France. Starting in the 1860s, root-feeding aphids, *Phylloxera,* were attacking the grapevines. In an attempt to solve the problem, rootstocks of North American grape plants were introduced, and shoots of local French varieties were grafted onto the rootstocks. (North American species of grapes are resistant to damage by the aphids, even though roots are often infested with *Phylloxera.*) The grafted plants produced grapes of the desired variety while the roots tolerated the aphids. Unfortunately, the introduction of the North American plants introduced a far more lethal pathogen, *Plasmopara viticola,* the fungus that causes downy mildew of grape.

Plasmopara viticola is a member of the Oomycota and produces zoospores within a sporangium. However, unlike *Saprolegnia* (*see* Chapter 22), the sporangia themselves are airborne. Zoospores are released on landing if conditions are favorable. Infection occurs when a sporangium lands on a grape leaf. The zoospores will swim around for a short time, then settle down, encyst, and soon produce a germ tube. If the zoospore is on the undersurface of a leaf the germ tube may reach a stoma and start the infection cycle. The mycelium spreads through the leaf tissue, obtaining nutrients from mesophyll cells. Pale yellow spots typically appear on the upper surface of the leaf, while the lower surface of leaves or the grapes themselves appears cottony or "downy" due to the presence of branched sporangiophores growing out through stomata (box fig. 23.1). Sporangia developing on these stalks repeat the asexual stage of the fungus. Sexual reproduction often occurs at the end of the growing season with the production of oospores.

By the late 1870s, this fungus threatened vineyards throughout France, with Bordeaux vineyards especially hard hit. The wine industry was suffering major new losses and had not financially recovered from the *Phylloxera* epidemic. The turnaround came with the widespread use of a fungicide promoted by Alexis Millardet, Professor of Botany at the University of Bordeaux. Millardet's discovery of this fungicide came by chance in 1885 as he was walking by a local vineyard. Millardet noticed that the grape plants along the path were a blue-grey color, but were surprisingly healthy when compared to other plants in the vineyard. Upon investigation, Millardet learned that the farmer had sprayed the plants along the roadside with a poisonous-looking substance (a mixture of copper sulfate and lime) to discourage people from picking and eating his grapes. Millardet realized the significance of the healthy-looking plants and perfected the blend of chemicals, known as Bordeaux mixture, which solved the disaster in the vineyards. Bordeaux mixture became the first widely used fungicide and continues to be one of the most important ones in use.

Even today in the vineyards of Bordeaux, grape leaves look blue because of the application of copper fungicides used as a prophylactic to prevent infection by zoospores of *Plasmopara viticola.* Topical fungicides, such as Bordeaux mixture and other copper compounds, are surface protectants designed to prevent entry of the fungus into the plant. As such, the plants require complete coverage. Unfortunately the compounds are washed off by rain and require repeated applications throughout the growing season. However, once inside the plant, the fungus is not affected by the surface application of these compounds.

The constant use of copper fungicides can lead to a heavy buildup of the copper in the soil, which may eventually cause toxicity problems. Similar problems occur with the use of mercury compounds, which have also been used as fungicides. (Accumulation of toxic levels of heavy metals, such as copper or mercury, is a serious environmental problem in many areas. Heavy metals are known to denature many enzymes and are, therefore, a general biotoxic substance.)

The base wines (cuveé) are made separately from three principal grape varieties: the white Chardonnay and the reds, Pinot Noir and Pinot Meunier. The processing of these grapes into wine follows the same procedure as for any white wine, i.e., crushing followed by immediate pressing. The must undergoes the primary fermentation in stainless steel tanks, or in some companies, the more traditional oak barrels.

The three young wines thus produced are then blended according to the champagne producer's specifications to achieve a characteristic taste. In the traditional method, the blended wine is bottled and undergoes a second fermentation in the bottle. To enable the second fermentation to take place, sugar from any natural source (cane sugar, corn syrup, honey, or grape sugar) and yeast are added to the bottled wine. Some companies employ the tank method to induce the second fermentation. In this method, the cuveé is placed *en masse* in large pressurized tanks to which sugar and yeast have been added. In cheaper productions, the wine is artificially carbonated by cooling it to below freezing and pumping in CO_2.

Secondary fermentation is generally finished in two weeks but the wine stays in contact with the lees for several months to a year to develop aroma and flavor. Carbonation under pressure accumulates in the bottle. The champagne bottle is made of thicker glass than the typical wine bottle, since the high CO_2 content exerts a considerable amount of pressure, 75–90 lbs/inch^2. Even taking precautions, champagne bottles do sometimes explode. The bottle is sealed with a metal crown cap, the type seen on bottled soft drinks.

Riddling, the process of collecting sediments from the secondary fermentation in the bottle, is the next step. Madame Veuve Clicquot, an eighteenth-century champagne maker in France, invented riddling, which begins with placing the champagne bottles upended in a rack. Each day, the champagne is swirled and the bottle is gently uplifted and turned slightly (fig. 23.8). Although some riddling is done using a mechanical shaker, much of the industry still turns the bottles by hand. Over a period of time, sediments collect in the neck of the bottle. Formerly, the sediment collected against the cork. In modern practice, a tiny plastic bucket is inserted, during bottling, just inside the crown cap to collect the sediments.

Next, the sediments are removed from the bottle during the stage known as disgourgement. The most common method today is to immerse the neck of the inverted bottle in a subzero solution, freezing the collected sediment into a plug. When the crown cap is removed, either by hand with a bottle opener or mechanically from the now uprighted bottle, carbonation pressure blows out the plug. The dosage step adds wine or brandy to replace any liquid lost during disgourgement. Cane sugar may also be added, depending on the amount of sweetness desired. Brut champagne is the driest, while the designation doux indicates the sweetest. Although champagne is basically a white wine, pink champagne can be made by either blending a red wine in the cuveé or by permitting some skin contact with the red grapes before pressing. Although often ignored, champagne, strictly speaking, is a sparkling wine produced from grapes grown in the Champagne region of France, the northernmost grape growing region in that country. The chalky soils in the area are said to influence the flavor of the grapes in a unique way.

Figure 23.8 *Riddling is a step in the champagne-making process in which upended bottles are slowly and systematically turned to collect sediments in the neck.*

Fortified and Dessert Wines

Table wines have a final alcohol content of 9%–14%. Alcoholic concentrations much above 14% kill the yeast, ending fermentation. If a higher alcoholic content is desired, alcohol, usually in the form of brandy, must be added and the wine is said to be fortified. Fortified wines usually have an alcoholic content between 15%–21%. The higher alcohol content tends to act as a preservative, preventing bacterial spoilage and this, in fact, might have been the original intent of fortifying wines. In producing dessert wine, adding brandy brings the fermentation process to a halt and with it the conversion of sugars to alcohol. The natural sweetness of the wine is maintained yet the brandy brings the wine up to a sufficient alcoholic content. Examples of dessert wines are port and sweet Madeiras. Brandy added to fully fermented and relatively dry wine prevents spoilage during aging. Dry sherry is an example of this type of fortified wine, also called an aperitif.

Sherry begins as a dry white wine with an alcoholic concentration between 11%–14%. Sherry was probably the earliest fortified wine. The taste and aroma of sherry is due to the oxidation of some of the ethanol to acetaldehydes. It is this oxidation that imparts the "nutty" flavor of a sherry. This can be accomplished in several ways but the traditional method is to induce the oxidation by yeast, indigenous to the Jerez region of Spain (from which the word

Figure 23.9 *A Sumerian cylinder seal from the third millenium depicts beer drinkers. Note that the beer was drunk through straws, since early beers were not filtered to remove the spent grains.*

sherry is derived). The yeast forms a white surface on the wine in patterns that resemble flowers and is called *flor.* After it is fortified with brandy, sherry has an alcoholic content of 17%–20%.

The Brewing of Beer

Beer is basically fermented grain and second only to tea as one of the world's most popular drinks. It has a long history that dates back at least 6,000 years. Some anthropologists have argued that the origin of beer predates that of bread, and that agriculture developed to supply a ready source of cereal grains as much for the purpose of making alcoholic beverages as food. Records and artifacts indicate that ancient peoples such as the Sumerians, Egyptians, Hebrews, Incas, and the Chinese all knew and practiced the art of brewing. Recent discoveries of 5,000-year-old beer jars (pottery vessels used for fermentation or storage) verify that the Sumerians were one of the earliest brewers (fig. 23.9). A later Sumerian clay tablet not only praises Ninkasi, the goddess of brewing, but includes an ancient recipe for beer that uses *bappir,* a type of barley bread, as the start, and honey and dates for sweetening. Recently, Anchor Brewing of San Francisco followed this 3,000-year-old recipe to recreate what might have been one of the oldest beers.

Barley Malt

Beer can be and has been made from any starchy carbohydrate source, but barley, used by the ancients, is the basis of most modern beers. Since beer making begins not with sugars but with starch, there are several preliminary steps in brewing before the yeast is introduced and fermentation can begin. The first step is the preparation of the *malt* (fig. 23.10). Barley grains are moistened with water and then allowed to germinate for a short time, spread out on a malting floor. As the barley grains germinate, they produce enzymes that catalyze the breakdown of starch. The germinated grains are then dried in a kiln (oven) and crushed to make malt powder. (Malt shakes are flavored with this powder.) The color of the finished beer is determined by how the malt was roasted; pale or standard malts are dried at low temperatures, while specialty malts are roasted at higher temperatures for a longer period of time to carmelize the malt sugars and so impart a darker color.

Mash, Hops, and Wort

The malt is now added to the grain starch and water, and heated in a mash tun. This mixture is the mash and during this step, the malt enzymes break down the starch to fermentable sugars (fig. 23.10). The starch of any cereal grain, such as wheat, corn, rice, or barley, can be the start for the brewing process. Beers made with both barley malt and grain are known as malt liquors. German *weissbier* uses wheat both as the malt and the carbohydrate source. Beers in the United States typically use rice as the grain starch. Eventually, the mash is strained, producing a clear, amber liquid called wort (fig. 23.10). The wort is boiled with hops in a brew kettle. The hop, *Humulus lupulus* (fig. 23.11), is a vine in the hemp family (Cannabaceae), which is found in north temperate regions of Europe, Asia, and North America. It is actually the yellow-green pistillate flowers, whole or pelletized, that are added to the wort. It was discovered centuries ago by Bohemian brew masters that the hops add a source of desirable bitterness to the beer to counteract the sweetness of the malt. With all the sugars in beer, it would be sickeningly sweet without the tartness of hops. Other plants have been added to beer for this same purpose, such as spruce, ginger, ground ivy, sweet mary tansy, sage, wormwood, and sweet gale, but hops is the most widely used. Much of the flavor and aroma of beer is attributable to the hops. Also, hopping apparently has an antibacterial action, keeping the beer from spoiling, and hopped beers are said to retain the foamy head longer.

Fermentation and Lagering

Prior to the introduction of the yeast to begin the fermentation process, the wort is strained and cooled and placed in the fermentation tank (fig. 23.10). The main action of the yeast is to convert sugars into ethanol and carbon dioxide. Two types of yeast are used in the brewing process. Lager beers are fermented by yeast that settle to the bottom of the fermentation tank. (The word *lager* means "to store" in German and refers to the aging process of this type of beer.) Other beers are fermented by yeast that rise to the top of the fermentation tank. (Ale is produced by a top-fermenting yeast that ferments at higher temperatures than the cold-fermenting lager yeasts. It is one of the traditional English beverages, a robust, full-bodied, high-hopped brew. Today's beer, first developed by German monks, in comparison, is lighter and more delicate tasting.) Most beers in the United States are lager beers. Fermentation continues for up to a week, at which time the wort is now a beer. This

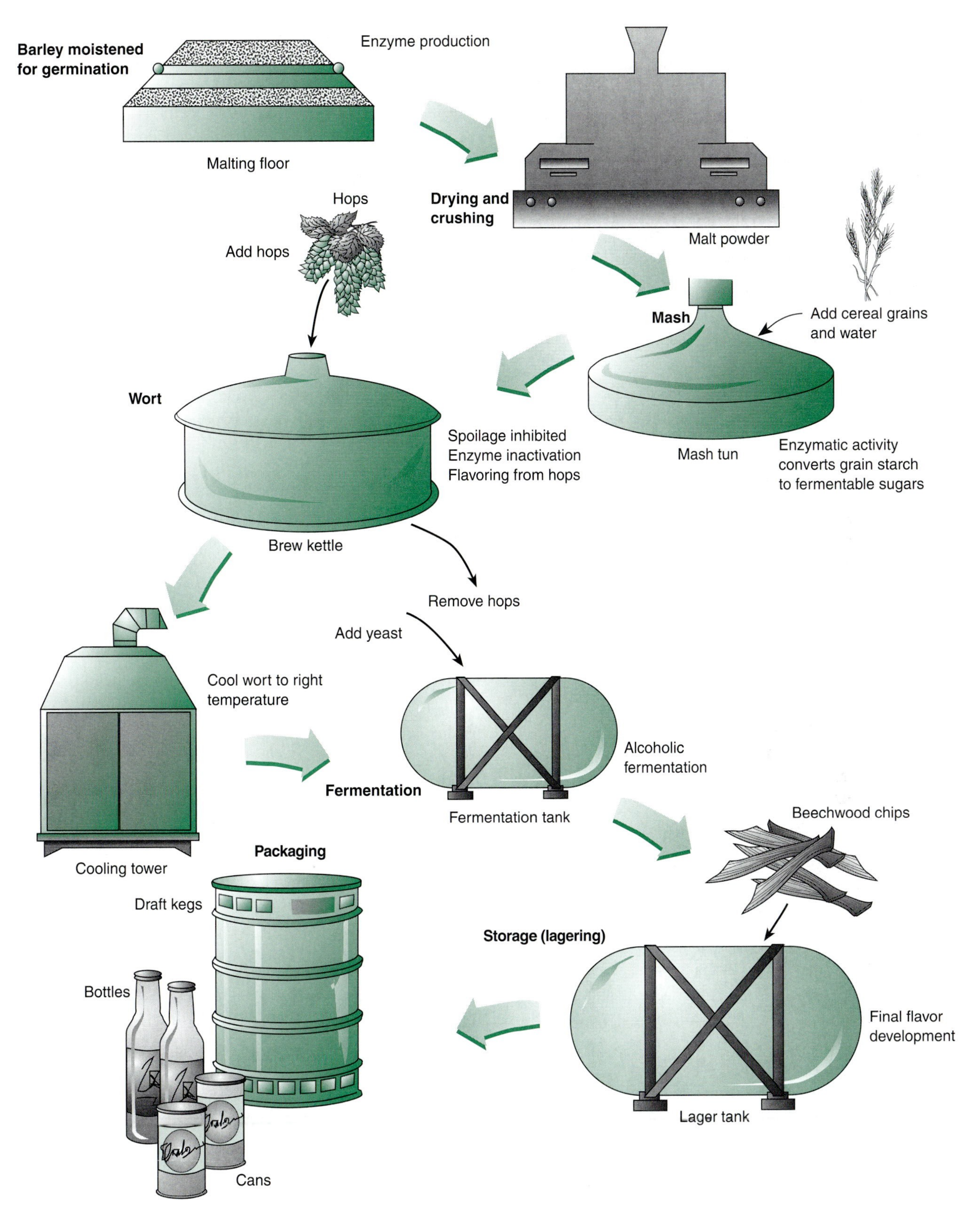

Figure 23.10 *Steps in the brewing of beer. Malting begins when barley grains are moistened in a mixing tank and spread out on the malting floor to induce germination. During germination, the grains produce enzymes that will break down starch. In the mash stage, malt and cereal grain are mixed with water, during which the malt enzymes convert grain starch to fermentable sugars. After several hours, the mash is strained, yielding a clear liquid, the wort. Wort and hops are added to the brew kettle and boiled. After straining off the spent hops, the wort must be cooled before the yeast can be added. With the addition of yeast, alcoholic fermentation begins. After several days, the green beer is transferred to lager tanks where it is aged for several weeks. Finally the beer is pasteurized or filtered before packaging in bottles, cans, or kegs.*

Figure 23.11 *The female inflorescences of hops,* Humulus lupulus, *are added to the wort for flavoring.*

new or green beer is then transferred to lager tanks where the beer is aged for three weeks (fig. 23.10). The flavor is often developed by adding beechwood chips to the tank. A second fermentation is started during aging by adding freshly yeasted wort, a process known as kraeusening. It produces the natural carbonation of the beer, although most mass-produced American beers are carbonated at bottling by adding CO_2 recovered from earlier fermentations. The lagering process lasts for several weeks. After lagering, the beer is clarified and packaged in bottles, cans, or kegs. Draft beers are sterilized by cold filtering, whereas most canned or bottled beers are pasteurized. The alcoholic content of beer is variable, from 3%–12%, depending on the type of beer, but most range from 4%–6%.

Two modern variants of brewing are light and nonalcoholic beers. By fermenting more of the carbohydrates in the mash, a glass of light beer has 95 Calories, as compared to the 150 in regular beer. Nonalcoholic beers are dealcoholized by brewing a beer normally, and then reducing its alcoholic content to less than 1.1% by evaporation or distillation. A variant of this process is the malt beverage, which is brewed at a low temperature to reduce the fermentation rate so as not to go above 1.1% alcohol. Dilution can also be used to reduce the alcoholic content of a beer to 0.5%. Nonalcoholic beers average about 55–82 Calories per glass.

Sake, A Rice "Beer"

Japanese sake or rice wine is produced by an intricate process that is more similar to beer making than wine making. As in the brewing process, a cereal, in this case rice, provides the grain starch that must be converted into fermentable sugars before yeast fermentation can begin. This conversion is accomplished by mixing steamed rice with the spores of the mold *Aspergillus oryzae.* The inoculated rice is then heated to a temperature of 35° C (95°F) for about a week. As the fungus grows on the rice, it produces enzymes that convert the starch into sugars. After several days, some of the resulting *koji* is mixed with more steamed rice and yeast is introduced to begin fermentation. The resultant *moto* serves as a starter culture and is mixed with the remaining batch of steamed rice (the *moromi*), continuing fermentation for up to three weeks. Sake has an alcoholic content of 20%, much higher than most beers and closer to what is found in distilled beverages.

Distillation

Distillation is the process by which a mixture of substances is separated by their differing boiling points. To make a **distilled spirit,** a beer or wine is boiled to leave the water behind and concentrate the alcohol. Since alcohol has a lower boiling point than water, heating an alcoholic beverage above the boiling point of alcohol, but less than the boiling point of water, will vaporize all of the alcohol and some of the water. The vapors are collected in a pipe; cool water running over the pipe recondenses the alcoholic vapors back, creating a liquid with a higher alcoholic content or proof than is possible through fermentation. (Proof number is equal to twice the alcoholic percentage.)

The Still

One type of distillation, known as the **pot-still** or batch method, employs a pot still (fig. 23.12). In this type of distillation, the beer or wine is heated in what is essentially a huge, covered copper kettle until the alcohol vaporizes. Rising from the pot is a neck or chimney to collect the alcoholic vapors. The vapors are passed on to cooling coils, worm-shaped tubes encased in cold water. As the alcoholic vapors pass through, they recondense back to liquid.

In the column, or continuous, still, the distiller has more control in separating out undesirable fractions from the alcoholic vapors. This type of still is shaped like a column, with a series of perforated plates on the inside. The beer or wine is introduced near the top of the column and percolates downward. Steam is introduced and, as the liquid moves down the column, it is repeatedly boiled and continually releases alcoholic vapors. By the time the liquid reaches the bottom of the still, only water and solids remain. The alcoholic content of the liquid has all been boiled off. The undesirable higher alcohols, essentially the extremely potent "heads" or acetaldehyde (the substance

Figure 23.12 *Copper pot stills are used in the distillation of beer to Scotch whiskey.*

that gives moonshine its raw "kick"), vaporize at the top of the column at the lower temperatures (21° C/70° F), while the equally unwanted "tails," the lower alcohols or fusel oils, are given off at the higher temperatures (84° C/183° F). Vapors from ethanol are given off in the middle temperatures (78° C/173° F). By fractioning, or separating out, the different alcoholic vapors, the undesirable higher and lower alcohols can be eliminated from the distilled product and the favorable alcoholic vapors from ethanol can be concentrated. This process of rectification or purification requires the skill of an expert distiller. In the pot still, the alcoholic beverage is heated for a certain amount of time to allow the higher alcohols to vaporize off and escape. The separation is not as fine as in a continuous still; some of the higher and lower alcohols are invariably mixed with the ethanol, producing a harsher and more potent product.

Distilled Spirits

The alcoholic content of distilled spirits ranges between 40%–50% or 80–100 proof. A grain spirit or neutral spirit is produced when the fermented beverage is highly distilled above 190 proof and is almost pure ethanol. It is colorless and tasteless and is used in the making of gin and vodka. Both are classified as unaged spirits, since they age for only a few weeks. Vodka is the distillation of a grain or potato beer and can either be unflavored or flavored with herbs. Gin is the distillate of a grain or cane beer (made from fermenting molasses). The characteristic taste of gin is from its steeping in the modified cones or "berries" of juniper and other herbs.

Most other distilled spirits are aged for a period of time in wooden barrels. Brandy, also known as cognac, is a twice-distilled grape wine that has been aged in oak casks. The color, aroma, and flavor associated with brandy is derived from the tannins, lignin, vanillin, and sugars that the alcohol picks up from the oak wood during the aging process. Different designations on the label of the bottle indicate the number of years the brandy has been aged. The most common designation is V.S.O.P., which stands for *very special old pale* and indicates that the hue of the brandy is not from adding caramel coloring but the result of aging in oak for at least 20 years. Some brandies are aged up to 50 years. The most distinctive brandies are produced in the Cognac region in southwestern France.

The term *whiskey* is derived from the Gaelic and means "water of life." Any whiskey is a distillate of a grain beer. The grain in Scotch whiskey is malted barley, kilned over a peat fire. The smoke from the peat gives this whiskey its distinctive flavor. After distillation, it is aged in oak for at least three years.

Bourbon is a corn whiskey whose development coincides with the colonial period and early years of the American republic. It is named after its place of origin, Bourbon County, Kentucky. The principal grain is corn, from 70%–90%, with small amounts of rye or wheat. The color, aroma, and flavor of bourbon is distinct from other corn whiskeys because of charred barrel aging. To legally be classified as bourbon, the whiskey must be aged in new, charred oak barrels for at least four years. Some are aged for 10–12 years. (The oak barrels must be new and cannot be used more than once. The used barrels are often sawed in half and sold as flower planters.) It is said that the Lewis brothers, Elijah and Craig, invented this distinctive aging process in 1789.

Fungi as Food

Edible Mushrooms

Mushrooms have long been a fixture in many European and Oriental dishes. As Americans have become familiar with their delicate flavors and aromas, a demand for a greater variety of these edible fungi has developed.

Most commonly eaten fungi are true mushrooms and members of the Basidiomycota or club fungi. The basidiocarp or fruiting body is the part that is eaten. Two popular exceptions to this general rule are the highly prized and costly morels and truffles, fruiting bodies of the ascomycetes *Morchella* and *Tuber,* respectively (*see* Chapter 22).

Until recently, most edible fungi had been collected from the wild, but the number cultivated to produce a greater and more reliable supply increases with every year. Cultivation eliminates the danger of mistaking a poisonous mushroom for an edible one, no small threat since every year hundreds of deaths are reported from accidental mushroom poisonings (*see* Chapter 24). Contrary to folk wisdom, there is no reliable test to distinguish a toadstool from an edible mushroom. The only sure way is to correctly identify the mushroom and, in some cases, that is a difficult task even for the expert.

A Closer Look 23.2

Alcohol and Health

Alcohol is a psychoactive drug affecting body organs, the brain, and the peripheral nerves. When a person drinks an alcoholic beverage, some alcohol is absorbed into the bloodstream directly through the stomach wall but most is absorbed through the wall of the small intestine. This explains why drinking on a full stomach slows the absorption of alcohol, while drinking on an empty stomach speeds up its absorption and hence its effects.

Alcohol is classified as a depressant of the CNS (central nervous system), inhibiting the centers in the brain dealing with speech, vision, balance, and judgment. In the case of an alcohol overdose, the level of alcohol is so high it depresses the respiratory control center in the brain stem and death results.

Alcohol eventually ends up in the liver, where the ethanol is metabolized at the rate of 10 ml/hour (1/3 ounce per hour). The enzyme alcohol dehydrogenase degrades ethanol first into acetaldehyde, then acetate, and eventually carbon dioxide and water.

There is some evidence to indicate that moderate drinking of alcohol, particularly of red wines, can reduce the risk of coronary heart disease by 40%, despite a diet high in saturated fats (14%–15% of the diet) and with serum cholesterol levels over 200 mg. This apparent contradiction is known as the French paradox, after the studies conducted on the people of France. Their high red wine consumption per day apparently prevents atherosclerosis, and thus lowers the risk of coronary heart disease. Various theories have been offered to explain this phenomenon. One suggestion is that somehow alcohol increases the level of HDL, the "good cholesterol" that removes cholesterol from blood vessels and returns it to

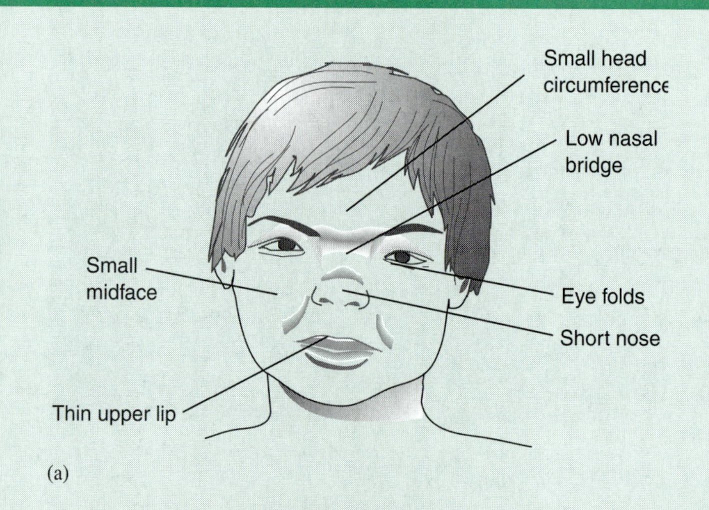

(a)

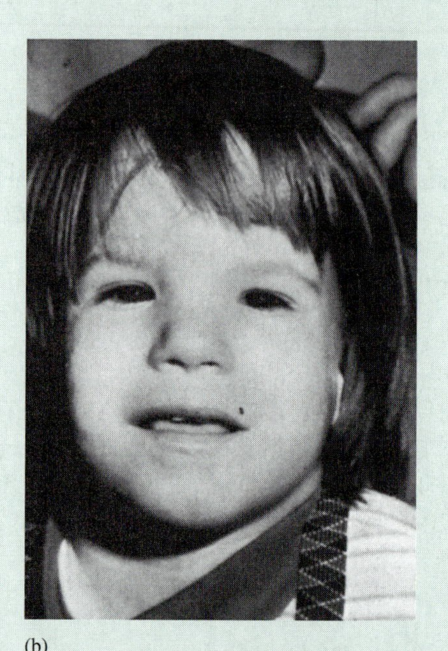

(b)

Box Figure 23.2 *Fetal alcohol syndrome occurs in children whose mothers drank heavily during pregnancy. These children exhibit telltale facial abnormalities and are often mentally retarded, learning disabled, or have behavioral problems.*

Two mushrooms, the button or field mushroom, *Agaricus bisporus,* and the Shiitake mushroom, *Lentinus edodes,* have been domesticated and cultivated for some time. The field mushroom is first in worldwide production and is the common white mushroom sold in western produce departments. Its cultivation dates back to 1650 in France. The Shiitake mushroom has recently been introduced to U.S. markets but is one of the most widely eaten mushrooms in the Orient, having been cultivated in China and Japan for several hundred years. Others widely cultivated are the straw mushroom (*Volvariella volvacea*) and the oyster mushroom (*Pleurotus* spp.) (fig. 23.13). Some examples of the edible fungi currently in cultivation are listed in Table 23.2.

Nutritionally, fresh mushrooms, as with most fresh fruits and vegetables, have a high water content, 85%–92% of the fresh weight. They are a source of complete protein the liver for degradation (*see* Chapter 10). Later studies found that this decreased risk of atherosclerosis only occurs when the amount of alcohol imbibed is excessive. Most alcoholics, for example, have no evidence of atherosclerosis. In the case of the French paradox, imbibing moderate amounts of red wine (approximately 20–30 g per day) apparently inhibits the aggregation of platelets, cellular fragments in the blood associated with the formation of blood clots. (Lowering the risk of blood clots decreases the incidence of a clot forming and blocking the coronary artery, and thus prevents a heart attack.) Recent evidence suggests that chemicals from red grape skin may be the active compounds preventing platelet aggregation.

Alcoholism affects 5% of drinkers. In the chronic alcoholic, the liver is one of the first of many organs to be seriously impaired. First, liver cells become fatty since alcohol disrupts the metabolic breakdown of fats. As fat accumulates in these cells, the cells enlarge and often rupture or form cysts, replacing normal cells. In the last stage, these are replaced by scarring, or cirrhosis, that can impede blood flow to the liver and disrupt normal functioning of this vital organ.

Chronic heavy drinking has adverse effects on many of the body's vital organs. The brains of alcoholics show atrophy due to death of nerve cells. An alcoholic's heart is usually larger than normal because one indirect effect of alcohol is to elevate the blood pressure. Scarring of the heart muscle results in arrythmia. The immune system is depressed and malnutrition is common.

Perhaps the most tragic consequence of heavy drinking is fetal alcohol syndrome (box fig. 23.2). First identified in 1970, fetal alcohol syndrome is the damage done to unborn children whose mothers drank during pregnancy, especially during the first trimester. When a pregnant woman drinks, the fetus is also exposed to the alcohol since it can cross the placenta. Children with this syndrome exhibit a range of symptoms, such as facial abnormalities, mental retardation, and/or behavioral problems. The recommendation from the medical community is to avoid alcohol completely during pregnancy, since even infrequent drinking might cause some damage to the fetus.

Figure 23.13 *The oyster mushroom,* Pleurotus ostreatus, *can be found on dead or dying trees. Although still sought after from the wild by many collectors, it is now widely cultivated and available in many supermarkets. World production is approximately 20,000 tons annually.*

Figure 23.14 *Commercial cultivation of* Agaricus bisporus *produces approximately one million tons of mushrooms annually, outpacing all other cultivated mushroom species.*

and in addition, have appreciable amounts of vitamins C, D, and some of the Bs. They are also naturally low in calories and high in fiber.

Cultivated mushrooms are grown on compost (fig. 23.14). Agricultural and industrial wastes are the raw materials for the compost. The exact nature of the compost differs with the specific fungus under cultivation, but cereal straws are the main ingredient of most. An important consideration in preparing the compost is to have the correct carbon-to-nitrogen ratio. Cellulose, hemicellulose, and lignin present in the straw are the carbon sources. Nitrogen is derived from proteins and amino acids and the compost is sometimes enriched with organic (such as horse manure) and inorganic nutrients to meet the necessary nitrogen requirements. Even so, it has been difficult to meet the specific nutrient requirements and environmental conditions of some edible fungi. For example, deciphering the specific requirements for cultivation of the delectable morel has been possible only recently.

Fermented Foods

Fungi, as well as bacteria, have also been utilized to modify foods through fermentations. Some common fermented foods well-known in the United States are cheese, yogurt, sausage, pickles, sauerkraut, soy sauce, and, of course, bread. Other fermented foods are little known outside their culture of origin (table 23.3).

Fermenting foods is a widespread and time-honored practice, predating historical records, and can be found in cultures throughout the world. In east Asian countries, soybeans sometimes mixed with cereals have been fermented by various bacteria and fungi to yield an impressive collection of flavoring agents and protein sources such as soy sauce, miso, and tempeh (*see* Chapter 12). In southern Asia, the Middle East, and Africa, the traditional fermented foods are the result of bacterial and yeast action on cereals to which a protein source has been added. In the Middle East, the protein source is usually milk; in the Indian subcontinent it is legumes.

Fermented foods have several inherent advantages regarding nutritional and keeping qualities. Microbial enzymes released during fermentation improve flavor and appearance, increase digestibility, and destroy undesirable components of the unfermented food. Protein values are enhanced in fermented foods through the enzymatic release of amino acids, and many of the microbial agents synthesize vitamins, further improving nutritional quality. Additionally, fermented foods are less subject to spoilage, an important factor in countries where refrigeration is uncommon.

TABLE 23.2
Edible Mushrooms and Fungi Grown in Cultivation

Common Name	Scientific Name
Cultivated white mushroom	*Agaricus bisporus*
Enoki	*Flammulina velutipes*
Inky cap	*Coprinus comatus*
Mu-erh	*Auricularia polytricha*
Nameko	*Pholiota nameko*
Périgord truffle	*Tuber melanosporum*
Oyster mushroom	*Pleurotus ostreatus*
Silver ear	*Tremella fusiformis*
Shiitake	*Lentinus edodes*
Straw mushroom	*Volvariella volvacea*
Strophaire	*Stropharia rugoso-annulata*
Yellow morel	*Morchella esculenta*

TABLE 23.3
Examples of Fermented Foods throughout the World

Food	Origin	Base	Use
Ang-kak	China, Philippines	Rice	Coloring agent (red)
Bongkrek	Indonesia	Coconut presscake	Meat substitute; snack
Doza	India	Rice, blackgram	Pancake
Gari	West Africa	Cassava	Starchy entree with sour taste
Ogi	Dahomey, Nigeria	Maize	Fermented cake
Pozol	Mexico	Maize	Fermented dough diluted with water and drunk
Sierra rice	Ecuador	Unhusked rice	Fermented rice
Soy sauce	Orient	Soybeans	Flavoring
Sufu	China	Soybean curd	Condiment
Tarhana	Turkey	Wheat meal and yogurt	Used in soup
Tempeh	Indonesia	Soybeans	Meat substitute; snack

Chapter Summary

1. The fermentation of sugar to alcohol through the action of yeast is the crucial step in the making of both wine and beer. Wine begins with the wine grape (*Vitis vinifera*). The wine grape is basically of two types, red or white, depending on the color of the grapes. The processing of a red or white wine differs in many respects. White wines, using either red or white grapes, are crushed, pressed, and then fermented. In the making of red wines, the red grapes are crushed, fermented, and then pressed. Newly fermented wines are racked to remove impurities, aged in oak barrels, and then bottled.
2. Champagne is essentially a blending of three young white wines that undergo fermentation in the bottle to produce the excessive carbonation associated with champagne. Table wines have a final alcohol content ranging from 9%–14%; in fortified wines, brandy has been added to raise the alcoholic concentration to levels of 15%–21%.
3. Beer is essentially a fermented grain. Since beer making begins not with sugars but with starch, preparation of a malt, mash, and wort are the steps necessary to provide a source of fermentable sugars. The alcoholic content of beer averages around 4%–6%.
4. In the production of distilled spirits, beer or wine is boiled and the alcoholic vapors collected and condensed to raise the alcoholic concentration to the 40%–50% range. Brandy is a twice distilled white wine; whiskey, gin, and vodka are distilled beers.
5. As a psychoactive drug, alcohol is classed as a depressant of the CNS; it inhibits the brain centers that control speech, vision, balance, and judgment. The French paradox is the acknowledgement that moderate drinking of alcohol can significantly reduce the incidence of coronary heart disease despite a diet high in saturated fats and cholesterol. Drinking alcohol during pregnancy is not recommended because of the risk of fetal alcohol syndrome to the unborn child.
6. Many fungi are edible and several cultures have a tradition of collecting wild mushrooms. More recently, it has been possible to cultivate a greater variety of edible mushrooms and other fungi, eliminating the risk of accidental poisoning by mistaking a poisonous mushroom for an edible one. Fungi have been employed to modify foods through the fermentative process. Fermented foods are often enhanced nutritionally and have improved flavor and appearance through the action of enzymes released during fermentation.

Review Questions

1. Contrast the steps of wine making to those required in the brewing process.
2. How does the making of champagne or a sparkling wine differ from the making of a still wine?
3. Contrast the operation of a pot still to that of a continuous still. Identify the following distilled beverages: grain neutral spirits, vodka, gin, whiskey, bourbon, and brandy.
4. Describe the effects, both positive and negative, of alcohol on health.
5. What is the nutritional value of edible mushrooms? How are mushrooms cultivated?
6. What are the advantages associated with fermented foods?

Further Reading

Chang, S. T. and P. G. Miles. 1984. A New Look at Cultivated Mushrooms. *BioScience,* 34 (6): 358–362.

Gibbons, Boyd. 1992. "Alcohol: the Legal Drug." *National Geographic* 181 (2): 2–39.

Hesseltine, C. W. and Hwa L. Wang. 1980. The Importance of Traditional Fermented Foods. *BioScience,* 30 (6): 402–404.

Katz, Solomon. H. and Fritz Maytag. 1991. "Brewing An Ancient Beer." *Archaeology,* 44 (4): 24–33.

Rose, Anthony H. 1981. "The Microbial Production of Food and Drink." *Scientific American,* 245 (1): 127–138.

Sokolov, Raymond. 1992. "Flower of the Vine." *Natural History,* 101 (2): 80–83.

Webb, A. Dinsmoor. 1984. "The Science of Making Wine." *American Scientist,* 72 (4): 360–367.

CHAPTER

24

Fungi that Affect Human Health

Chapter Outline

Chapter Concepts

1. Like plants, fungi produce a large number of secondary products, and some have been put to use as medicinal compounds.
2. Some fungal metabolites are toxic to humans and other animals; these include mycotoxins and mushroom toxins, both of which can cause serious or fatal reactions.
3. Some fungi affect humans directly by causing allergic reactions, as well as diseases ranging from mild skin conditions to fatal systemic infections.

Fungi produce a remarkable diversity of organic compounds. In addition to all the carbohydrates, lipids, and proteins that are components of cell structure and metabolic pathways, fungi produce a vast number of **secondary products,** organic compounds that have no direct role in major metabolic pathways but may serve to discourage predators or suppress competition. The formation of secondary compounds is quite specific. Often a certain compound is confined to one species or sometimes one strain of a species.

Thousands of secondary metabolites from fungi have been analyzed and characterized; many of them have widespread commercial importance while others have well-known health effects. Included in this group are alkaloids, as well as other compounds that may function as **antibiotics** or **toxins**. In many respects these secondary metabolites have made a significant impact on society. The value of the antibiotics (and even some of the fungal alkaloids) to medicine and society are clearly recognized, as are some of the harmful effects caused by toxic compounds.

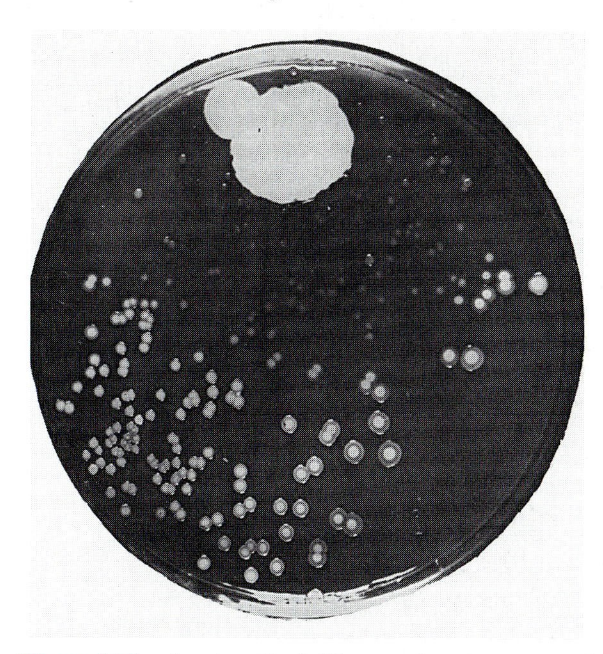

Figure 24.1 *Species of* Penicillium *produce the compound penicillin that has antibacterial properties. Fleming's discovery was made from this petri dish on which a colony of* Penicillium notatum *killed the bacteria in the culture.*

Antibiotics and Other Wonder Drugs

Antibiotics are compounds that are toxic to microorganisms. In the natural environment, these substances give the producing organism advantages over competing microorganisms for available nutrients and space. Production of antibiotics is one of the mainstays of the pharmaceutical industry and the primary weapon in the physician's arsenal for fighting bacterial infections. While it is hard to imagine modern medicine without antibiotics, large-scale production of these compounds is barely 50 years old. It all started with the accidental discovery of a contaminated bacterial culture by a British scientist.

Fleming's Discovery of Penicillin

Penicillin is a by-product of certain *Penicillium* species that inhibits the growth of Gram-positive bacteria (a group of bacteria separated on the basis of the chemical composition of the cell walls as identified with the Gram stain). The antibiotic acts by blocking wall synthesis in the bacterium and results in death of the bacterial cell by lysis.

Penicillin was first discovered in 1928 by British physician Alexander Fleming. But its potential was not realized for a number of years and mass production was not achieved until World War II. Although the antibacterial properties of other compounds had been seen previously, penicillin surpassed known therapeutic agents by suppressing bacterial growth without being toxic to animals or humans.

Fleming's discovery occurred while he was examining some old bacterial cultures before throwing them away. He found that a mold had contaminated his cultures while he was away on vacation (fig. 24.1). The fungus that Fleming identified as *Penicillium notatum* killed the culture of *Staphylococcus aureus* bacteria growing in the petri dish. He conducted some additional experiments on the bacteria-killing substance that he called penicillin and published his findings. Fleming's paper attracted little attention at the time, and his own experiments at purifying penicillin failed. Eleven years passed before research on this compound advanced.

In 1939, Howard Florey and Ernst Chain at Oxford University began an investigation of naturally occurring antibacterial compounds and came across Fleming's report on penicillin. Within a year, the team at Oxford had chemically analyzed the compound and demonstrated that it could destroy certain types of bacteria in test tubes. With World War II intensifying, Florey and Chain realized the potential of the drug for treating war wounds. Tests on infected animals were successful, and by 1941 the first human tests were conducted. Because of the escalating war in England, much of the penicillin research was moved to various sites in the United States. Miraculous cures were reported in human tests, and mass production was finally achieved.

One of the most dramatic cures was the recovery of Anne Miller of New Haven, Connecticut. Mrs. Miller fell ill with a massive streptococcal infection on February 14, 1942. For four weeks she barely clung to life as her temperature spiked to 106° (41° C) and 107° F (42° C). Doctors tried everything from sulfa drugs to transfusions, but nothing could stop the bacteria multiplying in her bloodstream. There was virtually no likelihood of survival. By chance, her physician knew someone working on penicillin research in New Jersey and he was able to obtain a tiny amount of the drug. It was administered during the afternoon of March 14, 1942. By 9:00 A.M. the next morning, Anne Miller's fever was gone, her vital signs were stable, and she was able to eat for the first time in four weeks. By afternoon, she was acting as if she were never ill; however, she was very weak and had lost 40 pounds. The miracle of her recovery electrified the scientific community. Shortly after Anne Miller's recovery, Alexander Flemming visited her in New Haven and acknowledged her contribution to the development of the miracle drug, stating

> "Thank you Mrs. Miller, you are my most important patient."

At the North Regional Research Lab of the U.S. Department of Agriculture, one team of researchers was looking for more high-yielding sources for the drug. Moldy fruits and vegetables were routinely collected from local grocery stores; then the fungi were isolated and tested for antibiotic production. In the summer of 1943, a cantaloupe was found contaminated with *Penicillium chrysogenum.* The isolated fungus produced 200 times more penicillin than Fleming's isolate. This species was used in the industrial production of the drug and continues to be used today.

By D-Day in 1944, there was enough penicillin to treat all the British and American casualties of the European invasion. Many even credit the development of penicillin as one of the reasons behind the Allied victory. By the time World War II ended, sufficient penicillin was available for civilian use. In 1945, Florey, Chain, and Fleming received the Nobel Prize for their work in developing the first "miracle" drug.

Soon after World War II, the pharmaceutical industry developed chemically altered versions of the penicillin molecule. These modified penicillins provided for greater stability, broader antibacterial activity, and also oral administration of the drug, which would permit home use of antibiotics.

Although other antibiotics have since been identified and developed, penicillin is the most widely used antibiotic and is still the drug of choice to treat many bacterial infections. Scientists have continued to improve the yield of the drug, and the present day strains of *P. chrysogenum* are biochemical mutants that produce 10,000 times more penicillin than Fleming's original isolate.

Although viewed as a miracle drug, the development of penicillin has had some drawbacks. Once wide-scale production of penicillin was achieved by the pharmaceutical industry and the drug was readily available, overprescribing by physicians and veterinarians commonly occurred. In addition, the antibiotics were incorporated into animal feed for use in feedlots. This widespread use led to the evolution of penicillin-resistant bacteria. (Since many species of bacteria reproduce every 20 minutes, the timetable for the evolution of new strains is much faster than evolutionary changes in other organisms.) By the early 1960s, penicillin resistance was evident among many types of bacteria, and, by the early 1990s, antibiotic resistance became a major cause for concern among the medical community.

A second drawback to penicillin is that a small percentage of the population is allergic to the drug, often resulting in severe or even fatal anaphylactic reactions. Anaphylaxis is a rapid and dramatic allergic reaction that can result in death through airway obstruction or irreversible vascular collapse. Penicillin is the most frequent cause of anaphylaxis in humans; it has been estimated that several hundred people die each year from anaphylaxis due to penicillin allergy.

Manufacture of Penicillin

Commercial production of penicillin is carried out in huge tanks with a high-yielding strain of *Penicillium chrysogenum.* The normal culture medium used is a corn steep liquor, a by-product of corn starch production. The main carbohydrate in the corn steep liquor is lactose, a disaccharide composed of glucose and galactose. About 30,000 liters (7,500 gallons) of the culture medium is sterilized and inoculated with spores of *P. chrysogenum.* During incubation, glucose is added periodically, the temperature is kept at about 24° C (75° F) and the pH adjusted to about 7.0. The culture is aerated with sterile air and agitated. The penicillin accumulates in the culture medium. After five or six days, the mycelium is filtered from the medium, which is then subjected to an extensive extraction and purification process to recover the penicillin.

Other Antibiotics

As early as 1943, while penicillin was still being developed, Selman Waksman of Rutgers University discovered **streptomycin,** a new antibiotic produced by *Streptomyces griseus,* which was effective against some bacteria that were immune to penicillin. Waksman's aim was to find an antibiotic that would inhibit the growth of *Mycobacterium tuberculosis,* the bacterium that causes tuberculosis. Although not produced by fungi, this antibiotic led to the development of a whole class of well-known drugs from **actinomycetes,** a type of bacteria. In addition to streptomycin, other well-known antibacterial drugs produced by actinomycetes are erythromycin, tetracycline, chloramphenicol, and nystatin. The majority of these antibiotics are produced by species of *Streptomyces.*

The development of penicillin led to the search for other fungi that produced antibiotic compounds. In 1948, Giuseppe Brotzu, an Italian microbiologist, had identified a compound produced by *Cephalosporium acremonium* that was an effective treatment for Gram-positive infections as well as some Gram-negative ones such as typhoid. Although Brotzu had achieved some success with his new antibiotic, the Italian authorities showed no interest in his work. Brotzu sent a culture of this fungus to Florey. The team at Oxford once again isolated the active compound, which they named cephalosporin. Since its initial isolation, a whole group of cephalosporins have been manufactured. These antibiotics have a broader spectrum than penicillins and are effective against many penicillin-resistant strains of bacteria. They are also much more expensive to produce; many of the newer cephalosporins are reserved for hospital use.

One additional antibiotic is griseofulvin, first isolated in 1939 from *Penicillium griseofulvum.* Unlike penicillin and the other antibiotics described, there are no antibacterial properties produced by this species. Griseofulvin is effective in treatment of many fungal diseases, especially dermatophytic mycoses that affect the skin, hair, or nails. Because the compound is only **fungistatic** (stopping growth) and not **fungicidal,** treatment must be continued until all the infected skin has been sloughed off.

Fungal Poisons and Toxins

Although antibiotics may be toxic to microorganisms, a compound is normally referred to as a fungal toxin when it affects humans or other animals. Fungal toxins broadly fall into two groupings, **mycotoxins,** formed by hyphae of common molds growing under a variety of conditions and mushroom toxins or poisons formed in the fleshy fruiting bodies of some fungi.

Mycotoxins

Mycotoxins are commonly produced by fungi growing in contaminated foods. These compounds can develop in grains or nuts in the field, but, more commonly, they develop in storage and remain within the food after processing and cooking. These toxins have been shown to have profound chronic and acute effects on both humans and livestock when the contaminated foods are eaten. In addition to direct toxic effects, mycotoxins are among the most potent known carcinogens.

The scientific community first became aware of the mycotoxins while investigating the cause of Turkey X Disease, which killed over 100,000 young turkeys in 1960 in England. The affected turkeys stopped eating, became lethargic, suffered hemorrhages under the skin, and died. Autopsies showed that their livers had undergone extensive necrosis (localized tissue death) and their kidneys had developed lesions. It was soon found that partridges, pheasants, ducklings, and other animals were also affected. Some "detective" work revealed that the only factor in common with all the poisonings was the use of Brazilian peanut meal as a feed supplement. A toxin was isolated from the feed and found to be associated with a common fungal contaminant, *Aspergillus flavus.*

The toxin was named **aflatoxin** to designate the fungus producing it, (A [*Aspergillus*] -fla [*flavus*] -toxin), although it was soon learned that four toxins were present, not just one. Within a few years it was shown that aflatoxins were not only toxic but also carcinogenic. In experiments it was found that even if the food did not contain toxic levels of aflatoxins, prolonged exposure caused liver cancer in laboratory animals. It is believed that these toxins are responsible for high rates of liver cancer among some groups of people in Asia and Africa, where contaminated food is often consumed. Although the association with

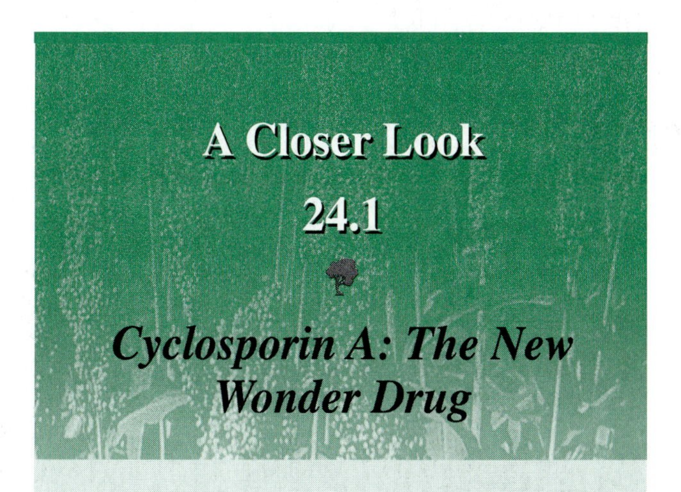

A Closer Look 24.1

Cyclosporin A: The New Wonder Drug

Cyclosporin A, a metabolite isolated from *Tolypocladium inflatum,* was called the "wonder drug of the 70s" for its ability to suppress the immune system and prevent rejection of transplants. Cyclosporin was first isolated when workers from Sandoz Pharmaceutical Company were searching for new antifungal agents from soil fungi. Although the extracts from *Tolypocladium inflatum* only showed weak fungicidal activity, there was unusually low toxicity. This prompted further examination and a pharmacological screening program. The screening showed that the extract had the ability to selectively suppress components of the immune system. With these results, the compound was purified and structurally analyzed. By 1980, the immunosuppressive drug, cyclosporin A, was ready for further clinical trials.

Cyclosporin is used for many organ transplants: liver, kidney, heart, and bone marrow. It has improved the success rate of organ transplants by preventing rejection reactions without completely suppressing the immune system. The advantage of cyclosporin A over other drugs is its level of effectiveness, combined with its low toxicity. Cyclosporin A is also the most cost effective immunosuppressant and is considered the prototype of a new generation of immunosuppressants.

Figure 24.2 *Aflatoxins are secondary metabolites produced by* Aspergillus flavus *(×1000).*

Figure 24.3 *When foods are not properly stored, fungi can cause spoilage.*

human cancer has not been proven experimentally, the toxic effects in humans were shown in India in 1974, when about 400 people were poisoned after eating corn containing aflatoxins; 106 died.

Today, foods that are most frequently contaminated by aflatoxins (such as peanut butter and grain products) are routinely screened before processing or sale. The permissible limit of mycotoxins is generally quite low (15–20 parts per billion); however, some scientists feel that no detectable levels of aflatoxins should be permitted because of the carcinogenic effects. Because cows and goats fed grain contaminated with aflatoxins produce milk with aflatoxins, limits also exist for livestock feed.

Aflatoxin was the first mycotoxin identified and studied; today over 200 mycotoxins have been identified from 150 species of fungi, with new ones discovered each year. Species of *Aspergillus, Penicillium, Fusarium, Alternaria, Cladosporium,* and *Stachybotrys* are among the common fungi known to form mycotoxins.

As indicated, *Aspergillus flavus* is a common saprobe that occurs on many grains and legumes in storage (fig. 24.2). Aflatoxins are produced under certain conditions by some strains of this species. *Aspergillus flavus* is also used to prepare fermented foods in the Orient. Fortunately, the strains of *A. flavus* that are used in fermentation do not develop toxins, or else the environmental conditions used in fermentation are not favorable for toxin production. Apparently, the ability to produce toxins may occur in one strain of a given species, while another strain of the same species is not toxigenic. Some environmental conditions also promote the development of toxins. Although much basic research still needs to be done to identify the precise conditions that cause common fungi to produce toxins, it is believed that warm temperatures are important environmental triggers for some species. These conditions more frequently occur in tropical and subtropical countries, where storage conditions are often less than ideal.

While all foods are subject to spoilage by fungi (fig. 24.3), the mold growing on forgotten food in the back of the cupboard might also be producing deadly mycotoxins. Knowledge of these toxins suggests caution about eating moldy food!

Ergot of Rye and Ergotism

St. Anthony's Fire

In many respects, the contamination of rye grains by the plant pathogen, *Claviceps purpurea,* could be considered another case of mycotoxin poisoning. The long history of this plant disease, the human and animal poisonings resulting from eating contaminated grain, and the medical uses of some of the compounds, warrants a fuller discussion.

Ergot, caused by the fungus *Claviceps purpurea,* affects rye and other grains. The fungus infects the flower, producing dark **sclerotia** (hardened mycelial masses) called ergots in place of the grain (fig. 24.4a). The sclerotia contain a complex mixture of secondary metabolites (over 100 different compounds), including numerous alkaloids. When infected rye plants are harvested, sclerotia can be collected and ground along with the healthy grain. The resulting flour and baked breads contain these alkaloids and cause the condition known today as **ergotism** in people consuming the contaminated bread. Animals grazing on infected grains in the field can also develop ergotism.

(a)

(b)

Figure 24.4 *Ergot of rye caused by* Claviceps purpurea *can lead to ergotism if infected flour is made into bread and eaten. (a) Infected head of rye showing ergots. (b) Sixteenth century woodcut depicts a peasant appealing to St. Anthony for relief from ergotism. The peasant has lost his right foot, and his left hand is surrounded by symbolic flames.*

(b) From G. Barger, "Ergot and Ergotism, " 1931

Historical records of epidemics caused by ergotism indicate that symptoms of ergotism followed two different patterns: chronic poisoning and acute poisoning. Chronic poisoning led to gangrenous ergotism, which began with cold prickling or burning sensations in the fingers and toes, gradually spreading to the entire limb. The limbs became swollen, and burning pains alternated with icy coldness; finally the limbs became numb. The affected parts turned black with a dry gangrene that gradually spread upward and necessitated amputation of the limbs (fig. 24.4b). Pregnant women frequently miscarried and fertility was generally lower during outbreaks of ergotism. Acute poisoning resulted in convulsive ergotism, which affected the central nervous system. Although itching, prickling under the skin, numbness, and muscle cramps or spasms were the initial symptoms, the condition quickly intensified into hallucinations and convulsions. Severe brain damage and fatalities commonly resulted from acute poisoning. Similar symptoms occurred in animals with ergotism; legs, hooves, and tails became gangrenous and spontaneous abortions of calves often occurred.

Today it is known that the symptoms can be attributed to the abundant alkaloids contained within the fungal sclerotia. Some of the alkaloids affect smooth muscles and cause vasoconstriction (constriction of blood vessels); these led to the muscle pain, spasms, miscarriages, and gangrene. Other alkaloids affect the central nervous system resulting in the convulsions, hallucinations, brain damage, and death.

Ergotism has been known for thousands of years, with historical records dating back to ancient Greece. During the Middle Ages, when rye bread was consumed by masses in Europe, epidemics of the plant disease and, consequently, outbreaks of ergotism, were common. Many epidemics occurred in France, where rye was the staple crop of the peasants.

In France during the eleventh century, the condition was known as "sacred fire" because of the burning sensations common in the extremities of afflicted individuals. In the twelfth century, the condition began to be associated with St. Anthony, a fourth century Christian monk who was thought to have power over fire. Many suffering from "sacred fire" made pilgrimages to the church of La Motte-au-Bois in central France, where the bones of St. Anthony rested. Many "miraculous cures" occurred and soon a hospital was built to treat the patients. Additional hospitals and

Figure 24.5 *Nineteenth century engraving by Howard Pyle depicts the Salem witch trials. Accusers seated are pointing at the alleged witches.*

houses were built for the monks to treat victims of the sacred fire, which eventually became known as St. Anthony's Fire. We know today that the "cures" brought about by the monks of St. Anthony can be attributed to the ergot-free diet patients received at the hospitals.

The historic impact of ergotism goes beyond the misery brought to the thousands of afflicted individuals. In 1722, Russian Czar Peter the Great was halted in his ambitions to invade Turkey and capture a warm-water port when both his soldiers and horses developed ergotism after eating contaminated bread and grain in the Volga Valley. The association of ergotism with other historical events has been somewhat more controversial. Several historians have implicated ergot poisoning with the hysteria that led to the Salem witch trials in 1692 (fig. 24.5). At that time rye was grown in the Massachusetts colony, and weather conditions the preceding year were ideal for the growth of *Claviceps*. It has been suggested that stored grain was contaminated with ergot. Common complaints by those who thought they were bewitched include muscle pain and spasms, tingling limbs, and convulsions, symptoms that are consistent with ergotism. Historian M. K. Matossian has also suggested that ergotism played a role in the panic of the French peasants, contributing to the French Revolution of 1789. Matossian claims that the panic, known to historians as the Great Fear, was caused by the consumption of contaminated bread rather than an insurrection against paying taxes.

Although the connection between the ergot (the fungal sclerotia) and ergotism was first made in 1676, it was not until the late eighteenth century that harvested grain was regularly checked for ergot. Removing the ergots from harvested grain put an end to the dreaded epidemics. Even though ergot poisoning of livestock still occurs, careful harvesting and milling make human poisonings rare. However, in the late summer of 1951 in Pont-St. Esprit, France a number of people were stricken with ergotism after eating contaminated bread, and 136 cases were reported in 1977 in Ethiopia. This last outbreak resulted in a number of deaths following the consumption of infected oats, also susceptible to *Claviceps purpurea.*

Documented medical uses of ergot go back to 1582, when it was discovered that preparations of ergot could cause uterine contractions and hasten birth. For centuries midwives employed ergot to induce abortions and as an aid in childbirth. Today the purified alkaloid ergometrine is used medicinally to reduce postpartum bleeding. The alkaloid ergotamine is an effective treatment for migraine headaches. By constricting the diameter of the cranial arteries, the pulsating pressure and resulting headache are relieved. Additional alkaloids are used medicinally for treating mental problems, high blood pressure, and other ailments. Fields of rye are deliberately inoculated with spores of *Claviceps purpurea* to produce the ergot needed for the pharmaceutical industry.

Another group of ergot compounds are the lysergic acid alkaloids; these are the toxins responsible for the hallucinations associated with convulsive ergotism. **Lysergic acid diethylamide (LSD)** was first synthesized from lysergic acid by the Swiss chemist Albert Hofmann in 1938. LSD is the most potent psychoactive drug known. The hallucinogenic effects were demonstrated in 1943 when Hoffman ingested 0.25 mg of the compound and described the dreamlike state and fantastic visions that followed. In his hallucinations, furniture took on grotesque forms and people appeared as witches. Hoffman wrote,

> "A demon had invaded me, had taken possession of my body, mind and soul. I jumped up and screamed, trying to free myself from him, but then sank down again and lay helpless on the sofa . . . I was seized by a dreadful fear of going insane."

LSD affects midbrain activity by interfering with the action of the **neurotransmitter** serotonin. In small amounts LSD mimics the action of the neurotransmitter, but in larger amounts it is antagonistic to the action of serotonin. The hallucinations and changes in sensory perception are due to disruptions in the normal pathways of sensory stimulation.

In addition to the hallucinations, LSD induces various physiological responses, including increases in blood pressure, respiration, and perspiration, often accompanied by heart palpitations. In some cases, the effects are more serious, causing unpredictable side effects and a schizophreniclike condition. These "bad trips" have resulted in many fatalities.

During the 1960s Timothy Leary, a professor at Harvard University, popularized the use of this hallucinogen and urged young people to

"drop out, turn on and tune in."

Leary helped usher in the hippie generation with its provocative music and psychedelic colors. One of the Beatles' popular songs of the period was "**L**ucy in the **S**ky with **D**iamonds," describing LSD-induced images of tangerine trees and marmalade skies.

In spite of the possibility of long-term mind-altering activity and unpredictable side effects, there has been a sustained interest in the medical uses of LSD. The drug has been used with limited success in psychotherapy for the treatment of schizophrenia.

Recreational use of LSD peaked during the late 1960s. Although the illegal use of this drug decreased during the 1970s, the street use has, unfortunately, been on the increase again in the 1990s.

The Destroying Angel and Toadstools

The common term *toadstool* refers to poisonous mushrooms; however, there is no easy way to find out if a mushroom is poisonous. In many instances, both poisonous and edible species occur within a single genus. Although folk tales suggest that cooking mushrooms with a silver spoon (or silver coin) will indicate poisonous species when the silver turns black, these fables have no scientific basis and can lead to death. No single test or rule exists to indicate the edibility of a mushroom. The only way to know for certain whether a wild mushroom is edible or poisonous is to learn mushroom identification from a trained mycologist. **NEVER** eat a wild mushroom unless the identification is certain.

Although stories of mushroom poisonings abound, only a small percentage of species are actually toxic to humans. The actual number of poisonous (or even edible) species is difficult to determine because the vast majority of mushrooms are not utilized as a food, nor have they been analyzed for the presence of poisons or toxins.

Mushroom toxins are broadly categorized by the physiological effects of the toxin on humans:

(1) protoplasmic poisons that destroy cells in the liver and kidneys and are often lethal;
(2) neurotoxins that affect either the central or autonomic nervous system and often lead to hallucinations; and
(3) gastrointestinal irritants that generally lead to nausea, vomiting, cramps, and diarrhea but seldom cause fatalities.

The following discussion focuses on the group of toxins that are responsible for most fatalities.

Amatoxins are a group of deadly protoplasmic toxins that are widely distributed in species of several genera including *Amanita, Conocybe, Galerina,* and *Lepiota.* The amatoxin levels in these genera are high enough to cause human fatalities; however, the toxin also occurs in minute levels in other mushrooms. Although a few species in the genus *Amanita* are actually edible, 50% of all mushroom poisonings and 95% of all fatalities are caused by members of the genus *Amanita.* Common warnings for wild mushroom collectors are geared to recognize and avoid members of this genus:

(1) Never eat white-capped mushrooms not positively identified;
(2) Beware of any mushroom with an **annulus** (ring) on the stalk;
(3) Beware of any mushroom that bears remnants of a **universal veil** such as a volva (cup or sac) at the base of the stalk, or warts, scales, or patches on the surface of the cap (fig. 24.6).

The universal veil is a membrane that totally surrounds some young mushrooms. When the mushroom expands, the veil ruptures, often leaving a volva at the base of the stalk and remnants on the cap. The **partial veil** is a membrane that covers the gills during development. After it ruptures, it remains as an annulus on the stalk. In some members of the genus *Amanita,* both volva and annulus are prominent structures that aid in identification. One difficulty of looking for these features is that the volva and annulus may weather away, leaving the poisonous mushroom without its most distinguishable characteristics. *Amanita virosa* (the destroying angel) and *Amanita phalloides* (the death cap) are the most notorious killers in this group (fig. 24.6).

The mechanism of amatoxin poisoning is at the level of gene expression by inhibiting mRNA synthesis. Without mRNA, the cells stop synthesizing proteins and eventually die. Usually there is a delay, often 12–24 hours, between eating the mushroom and the appearance of symptoms. During this delay, considerable cellular damage is taking place. The first symptoms include intense vomiting, severe diarrhea, and fierce abdominal pain. These symptoms are caused by damage to the cells of the gastrointestinal lining. The vomiting and diarrhea often result in severe dehydration. If the victim survives these initial symptoms, there is a

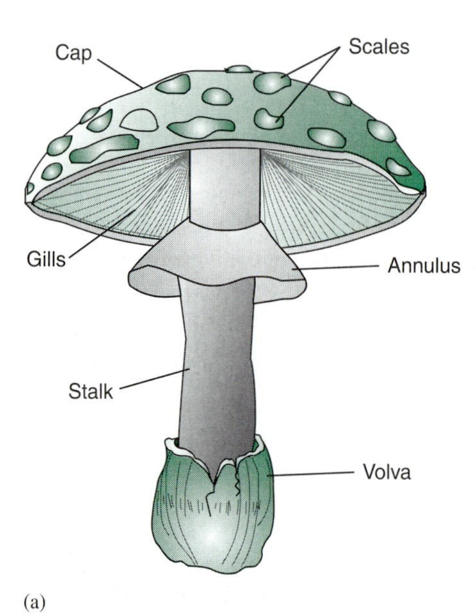

(a)

(b)

Figure 24.6 *The genus* Amanita *contains many toxic species. (a) Morphological features of mushrooms often found in the genus* Amanita *(b)* Amanita virosa, *the destroying angel.*

brief period of ostensible improvement. However, the toxins soon start affecting the liver, and severe liver damage develops. This stage is often fatal, but for those surviving the liver damage, kidney failure follows, with secondary effects on the heart and brain.

No antidote exists for amatoxin poisoning. Treatments consist of attempts to remove the toxin from the body, to increase the rate of its excretion, and to maintain electrolyte and blood sugar levels through supportive care. In addition, various measures to support the damaged liver and kidneys have been used, including intravenous administration of B vitamins, vitamin C, vitamin K, penicillin, and thioctic acid. The value of some of these measures has not been established, but thioctic acid has been used successfully in both the United States and Europe.

Soma and Hallucinogenic Fungi

Although the genus *Amanita* is well known for its deadly members, other species of this genus are noted for their psychoactive properties. *Amanita muscaria,* the fly agaric, is a widely recognized mushroom with a long history of use as an intoxicant. The morphological features of *A. muscaria,* with its orange-to-red cap and white scales, make it one of the most commonly photographed and widely illustrated mushroom species (fig. 24.7). The common name of this mushroom arises because flies are attracted to the mushroom and then killed from its insecticidal properties.

Ibotenic acid, the major toxin in this mushroom, is converted into muscimol when consumed, and it is muscimol that produces the psychoactive effects. Several species of *Amanita* contain ibotenic acid; however, *A. muscaria* is

Figure 24.7 Amanita muscaria, *the fly agaric.*

the best known of the group. Symptoms usually develop within 20 to 90 minutes and last for up to 12 hours. Many symptoms are similar to alcohol intoxication but may progress into far more serious effects, including epileptic-type seizures. Typical effects begin with dizziness, loss of coordination, and drowsiness, often accompanied by nausea and vomiting. These are followed by a period of sleep with vivid dreams. Later there is a period of hallucinations, with colored visions and distortions of size. In other cases, the medulla of the brain is affected, causing distortions in the motor control center and possibly resulting in seizures. Although fatalities are few from this type of mushroom toxin, they have occurred after a large dose of these mushrooms.

The properties of *Amanita muscaria* were discovered thousands of years ago by native peoples in various parts of the world. It is believed that this mushroom was used in ancient India to prepare the intoxicant Soma described 4,000 years ago in the *Rig Veda,* the book of sacred Hindu psalms. For centuries among some tribes in Siberia, the mushroom was dried and used as an intoxicant. The toxin is apparently excreted unaltered in the urine, which was collected and drunk for a second dose. In fact, it was known that intoxication would occur for up to four or five passages through the kidneys. *Amanita muscaria* was also used by several Native American tribes in the northern United States and Canada.

Among the native tribes of Mexico and Central America another group of mushrooms were revered for their hallucinogenic properties. These mushrooms have been used for thousands of years in religious and healing rituals. The Aztecs called these sacred mushrooms Teonanacatl, which means "flesh of god." It was believed that the mushrooms were a gift from god, who communicated with the people through the mushroom. Ancient mushroom-shaped statues found throughout the area attest to the reverence paid to these sacred mushrooms. Early Spanish explorers and missionaries described the rituals during which the mushrooms were consumed. Diego Duran described part of the ceremonies at the coronation of Montezuma in 1502:

> ". . . all the company proceeded to eat the toadstools, a food which deprived them of their reason and left them in a worse state than if they had drunk much wine . . . Others had visions during which the future was revealed to them. . . ."

Rituals involving the ingestion of sacred mushrooms have continued to the present time in parts of Mexico, especially among the Mazatec Indians. In recent decades scientists have observed and participated in these ceremonies and described the visual hallucinations and euphoria following the ingestion of various species of *Psilocybe* (fig. 24.8).

The major toxic compounds in these mushrooms are the alkaloids, psilocybin and psilocin. In the body, psilocybin is converted into psilocin, which is the biologically active molecule. Psilocin is structurally similar to the neurotransmitter serotonin and, like LSD, interferes with the action of this substance in the brain. Hallucinations usually begin within 30–60 minutes of ingestion and last for several hours. Generally, the results are visual hallucinations and distortions of time and space; however, sometimes the experience has caused vomiting, depression, and irritability. It has also been reported that excessive doses have resulted in paralysis, insanity, or suicide. These mushrooms are especially dangerous for young children; elevated temperatures, convulsions, and one fatality have been reported.

Figure 24.8 *Teonanacatl mushrooms of the genus* Psilocybe.

Psilocybin and psilocin also occur in certain species of *Panaeolus, Conocybe,* and *Galerina,* in addition to *Psilocybe.* During the 1970s, recreational use of these "magic mushrooms" gained considerable popularity in some parts of the United States. The mushrooms in this group are small and brown, often without distinctive features. It should be noted that similarity with amatoxin-containing species of *Conocybe* and *Galerina* can lead to misidentification and far more serious consequences.

Human Pathogens and Allergies

The connection between exposure to airborne fungi and disease was first made by Blackley in the 1870s when he described chest tightness and "bronchial catarrh" after inhalation of *Penicillium* spores. It has since been clearly demonstrated that exposure to fungal spores or even hyphal fragments can cause a variety of diseases ranging from allergic diseases (hay fever and asthma) to infectious diseases (such as histoplasmosis and aspergillosis).

Infectious diseases caused by fungi are loosely categorized into

(1) superficial mycoses (fungal diseases) that attack the skin, hair, and nails and
(2) deep or systemic mycoses that attack internal organs.

Only about 100 species of fungi are known to be human pathogens, and most cause relatively minor superficial mycoses. More serious fungal diseases occur in individuals whose immune system is compromised by disease, the prolonged use of antibiotics or immunosuppressive drugs, or other causes. Some of the organisms are only found as human or animal pathogens, while others can exist either as saprobes in the soil or as human pathogens. One final group of fungi that are responsible for human infections are opportunistic pathogens. These are common saprobic fungi such as *Aspergillus* and *Fusarium* (species of *Fusarium* also cause plant diseases), which can result in systemic infections in immune-compromised individuals. The majority of pathogens are asexual fungi, although teleomorphs of many have been identified in recent years.

Dermatophytes

Among the most common superficial fungal pathogens are dermatophytes that attack the outer layers of the body. These fungi utilize **keratin,** a protein found in skin, hair, and nails; most dermatophytes are species of *Microsporum, Trichophyton,* and *Epidermophyton.* Many cause a condition known as tinea or ringworm, which refers to the circular pattern on the skin produced by growth of the fungus (fig. 24.9a,b). Specialization often occurs, with some species only infecting certain parts of the body. For example, tinea capitis is a ringworm found mainly on the scalp, while tinea corporis may infect the whole body. Probably the most common ringworm infection, and, in fact, the most common mycotic infection, is tinea pedis, better known as athlete's foot. This well-known condition may persist indefinitely, causing occasional flare-ups, especially in warm weather (fig. 24.9c).

It should be emphasized that these fungi do not grow in living tissues; they invade only dead areas of the skin, hair, or nails. The symptoms caused by these infections are actually due to fungal metabolites. Transmission of the infection is generally by person-to-person contact in shared bathing facilities or locker rooms. Spores enter the skin through minor cuts or scratches to initiate the infection. Because these fungi utilize keratin, they cannot exist as soil saprobes. However, their conidia can survive in a dormant state for long periods of time in carpets, upholstery, or similar environments.

While some ringworm fungi only parasitize humans, other species infect domestic animals. A small number of dermatophytes are able to infect cats or dogs, as well as people. For many ringworm infections, topical applications of nystatin or oral griseofulvin is often the recommended treatment; however, other antifungal compounds are also used.

Yeast infections or candidiasis are another type of common superficial mycosis. Over 95% of these infections are caused by the fungus *Candida albicans,* a normal constituent of the microorganisms present on the skin as well as in the mouth, rectum, and vagina. Various conditions (such as excessive wetness) or weakened host defenses (caused by prolonged antibiotic therapy, steroidal therapy, diabetes, etc.) can trigger an overgrowth by this yeast, resulting in soreness and itching. This can cause diaper rash in babies or infection in the armpits, the vagina, or under the breasts in adults. Mucous membranes are particularly susceptible to yeast infections, and oral candidiasis, also known as thrush, is common in infants and young children (fig. 24.9d). In adults, thrush is frequently seen as an early mycotic infection in AIDS patients. Although candidiasis is usually limited to the skin, it can become systemic and cause a life-threatening illness in severely immune-compromised individuals.

Systemic Mycoses

All fungal diseases that infect tissues deeper than the skin are regarded as systemic mycoses. Most systemic mycoses are chronic conditions that develop slowly over long periods of time. Often months or years elapse before medical attention is sought. Again, it is immune-compromised individuals who are at risk. Many infections are seen in well-defined geographic areas; this is especially true for the most common mycoses in North America, histoplasmosis and coccidioidomycosis.

Histoplasmosis occurs in the East and Midwest and is especially common in the Ohio and Mississippi River valleys. The disease is caused by the fungus *Histoplasma capsulatum,* which is prevalent in the excrement of chickens, other birds, and bats. Anyone disturbing these droppings may inhale conidia, which can cause an infection in the lungs. In more than 95% of the cases, there are no symptoms. The individual is left with a small calcified lesion in the lung and resistance to further infection by the fungus. It is estimated that about 90% of the population living in the endemic area has had this mild infection. However, in a small percentage of the cases, the disease does stay localized and continues as a progressive lung disease, with symptoms similar to tuberculosis. Eventually the infection may become systemic and spread to other organs in the body. At this stage, the disease is usually fatal. *Histoplasma* is a dimorphic fungus. When growing in the soil or droppings, or in culture at room temperature, the fungus has a mycelial form with branching hyphae. However, in the body or when cultured at 37° C (98.6° F), it has a yeastlike morphology (fig. 24.10).

Coccidioidomycosis, also known as valley fever or San Joaquin fever, is prevalent in the hot dry deserts of the southwest, especially southern California and Arizona, where the species *Coccidioides immitis* exist saprobically in the soil. During the short rainy season, the fungus grows rapidly, producing thousands of spores that readily become

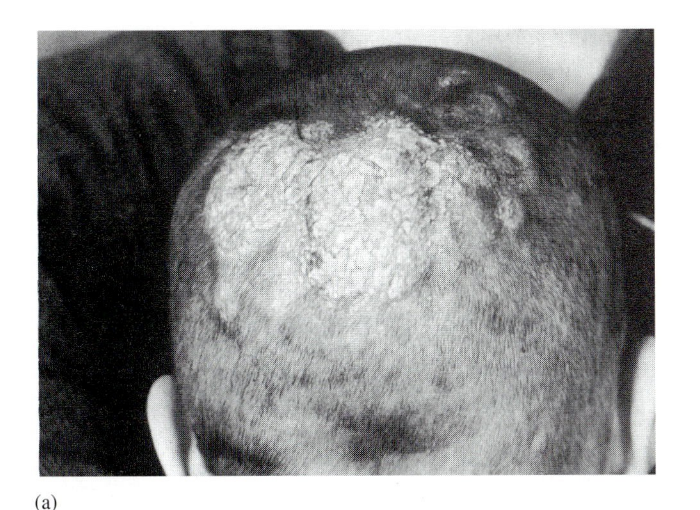

(a)

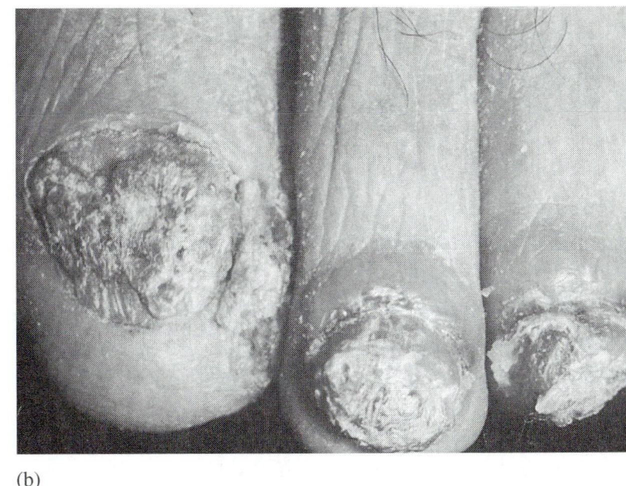

(b)

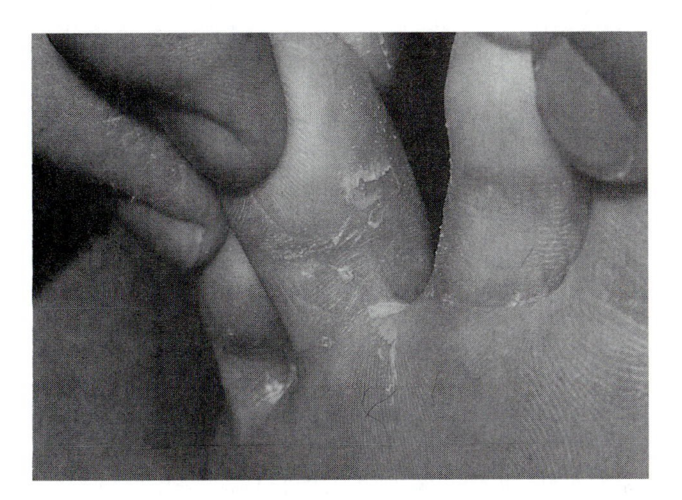

(c)

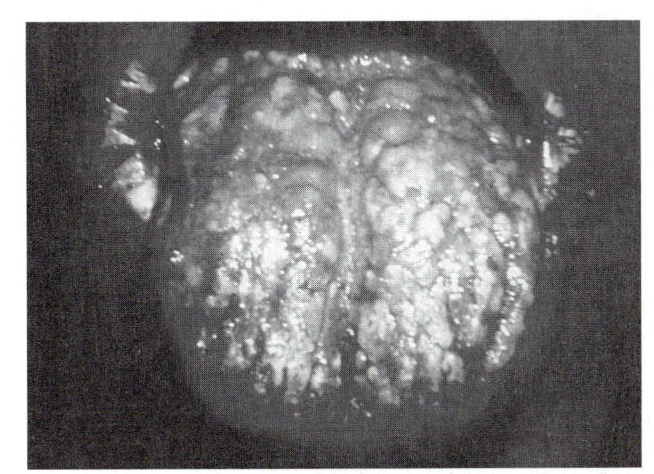

(d)

Figure 24.9 *Superficial mycoses. (a) Tinea capitis, ringworm of the scalp, caused by* Microsporum audouinii. *(b) Tinea unguium, ringworm of the nails, caused by* Trichophyton rubrum. *(c) Tinea pedis, ringworm of the foot (also known as athlete's foot), caused by several species of* Trichophyton *or* Epidermophyton floccosum. *(d) Oral candidiasis, thrush, caused by* Candida albicans.

airborne (fig. 24.11). Infection occurs when the spores are inhaled, and like histoplasmosis, it is centered in the lungs. Millions of people in the southwest have been infected by this fungus. In about 60% of the cases, there are no symptoms at all, and in about 40%, there are mild flulike symptoms with fever. However, in approximately 0.2%–0.5% of the infected individuals, the disease becomes systemic and may be fatal if not treated or is misdiagnosed. Incidence of valley fever in southern California increased dramatically after the January 1994 earthquake. Scientists speculate that the earthquake dislodged settled spores, which then became airborne.

Opportunistic infections are caused by a wide range of common saprobic fungi. These species are common and cause no infections when inhaled by healthy individuals. When the immune system is weakened by drug therapy or disease, infection by these fungi can occur. It has been estimated that invasive fungal infections occur in 25% of these high-risk individuals. The recent AIDS epidemic, as well as improved drug therapies for treating cancer patients and organ transplant patients, have resulted in a sharp increase in the incidence of opportunistic infections. Among the most common fungi causing these infections are species of *Aspergillus,* which cause aspergillosis, a lung disease with diverse symptoms. Often a pneumonia-like condition develops, whereas at other times pulmonary lesions occur. Formation of an aspergilloma, a fungus ball consisting of tightly interwoven hyphae within the lungs, is another possible symptom of aspergillosis. Allergic broncopulmonary aspergillosis is caused when the fungus, usually *A. fumigatus,* colonizes the mucus within the lungs, resulting in a severe allergic reaction. In many instances aspergillosis is fatal.

Treatment of systemic mycoses is difficult since there are few effective drugs. Amphotericin B produced by species of *Streptomyces* (one of the actinomycetes) is

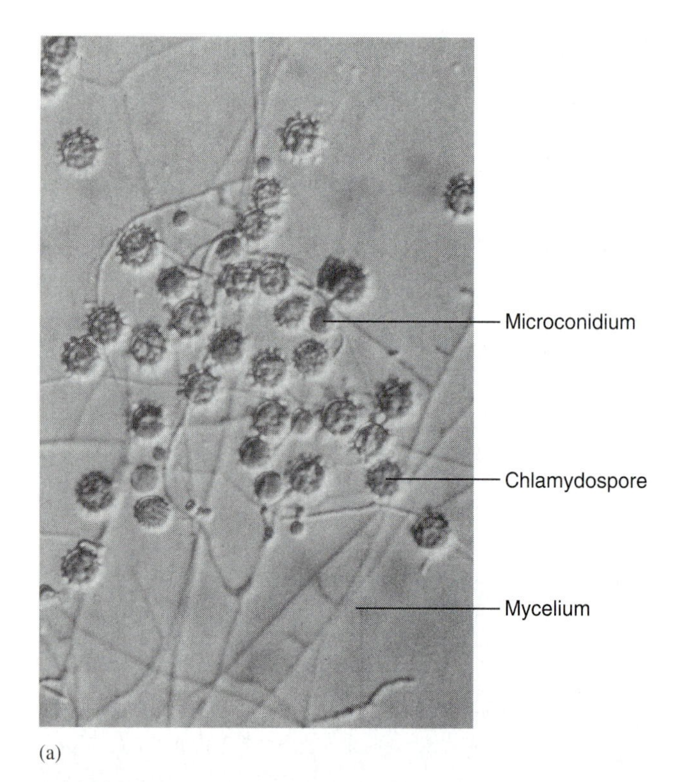

(a)

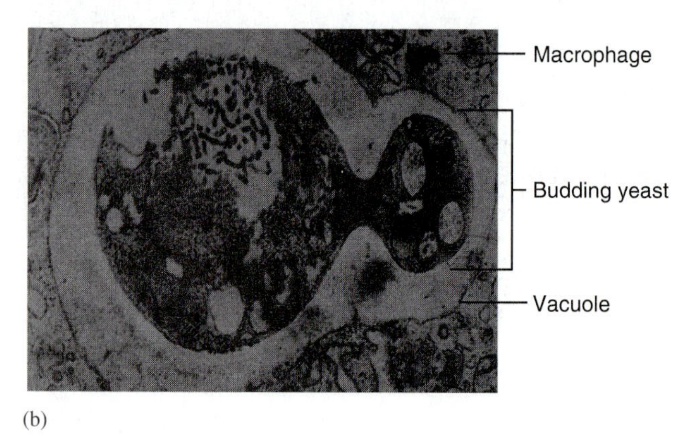

(b)

Figure 24.10 Histoplasma capsulatum, *cause of histoplasmosis, occurs in two growth forms. (a) Mycelium and spores as found in the soil. (b) Yeast-like cells are found in infected tissue.*

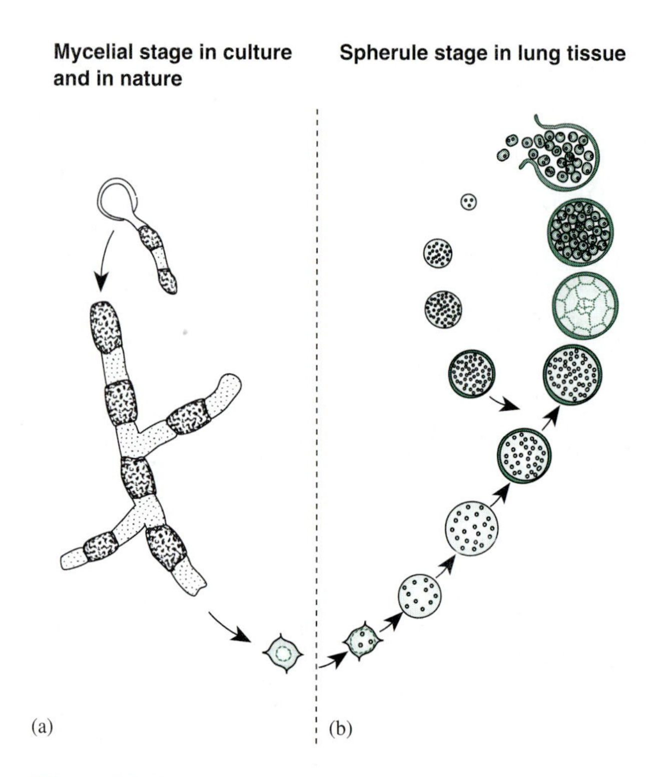

Figure 24.11 Coccidioides immitis, *cause of valley fever. (a) Mycelial stage occurs in culture and in the soil; (b) spherules occur in infected lung tissue.*

effective against many of the fungi when administered intravenously, but it is a very toxic compound with serious side effects. Miconazole and ketoconazole are also effective, but again show unpleasant and severe side effects.

Allergies

Many fungal spores contain **allergens** that are capable of causing respiratory allergies (both hay fever and asthma) in susceptible individuals (*see* Chapter 21). Allergens have been reported among every major group of fungi, but the majority of attention has been focused on the asexual fungi. Since the 1930s, spores of these fungi have been recognized as an important cause of both asthma and hay fever. Many of the important allergenic fungi include those whose spores are known to be most abundant in the atmosphere, *Cladosporium* and *Alternaria.* These are cosmopolitan genera that can be isolated from the atmosphere throughout the world (fig. 24.12). When conditions are right, atmospheric concentrations of *Cladosporium* spores frequently exceed 100,000 spores per cubic meter of air. Although *Alternaria* concentrations are far lower, its spores are considered much more allergenic.

Over the past decade, research has also focused on other groups of fungi as a source of allergens. Basidiospores from mushrooms, bracket fungi, and puffballs have been implicated as airborne allergens. These spores are produced at incredible levels and released into the atmosphere (*see* Chapter 22). From spring through fall the atmosphere may also contain high levels of other spores, including ascospores, smut spores, and rust spores. Far less is known

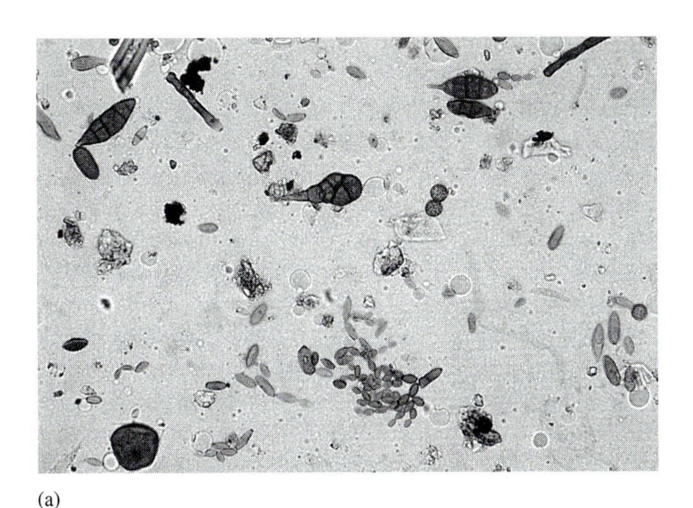

(a)

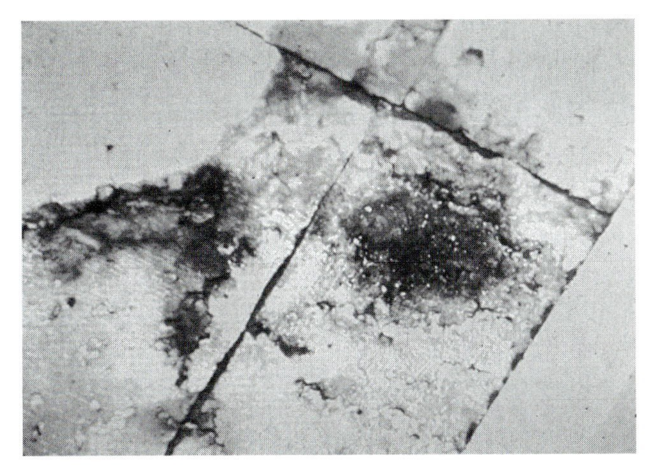

(b)

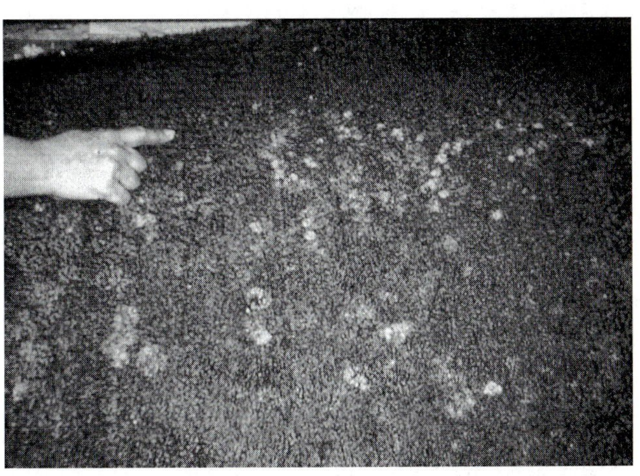

(c)

Figure 24.12 *Airborne fungal spores are a major cause of allergic diseases. (a) Outdoor air sample from Tulsa, Oklahoma showing abundant* Cladosporium *and* Alternaria *spores (×240). (b) and (c) Mold growth on ceiling tiles and carpets or other substrates can lead to high concentrations of airborne spores indoors.*

about the allergenic status of these spores, but they are also believed to be potentially important airborne allergens.

Unlike allergenic pollen, which occurs in definite seasons, allergenic fungi can be found in the atmosphere whenever the ground is not covered with ice or snow. This means that in many parts of the world, fungal allergens occur year-round. In addition, fungi can occur in the indoor environment in large numbers. Whenever unfiltered outside air is introduced either through ventilation systems or open doors or windows, pollen and spores will enter. Eventually, many of these will settle out. While the pollen cannot proliferate indoors, any time that moisture is available, saprobic fungi will colonize indoor substrates.

Common sites for indoor fungi include carpeting, upholstered furniture, showers, shower curtains and other bathroom fixtures, and potted plants and their soil. All of these contain abundant organic material for the growth of fungi (fig. 24.12). With moisture or even high humidity, spores can germinate and fungi can grow on these substrates and produce thousands of new spores. In buildings with central heating, ventilation, and air conditioning (HVAC) systems, in-duct filters should remove many of the spores present. Many instances, however, are known where the HVAC system itself served as an amplification and dissemination site for fungal spores. In these cases fungi have been found growing on the air filters as well as in the ducts. Among the common fungi that are found in indoor environments are species of *Cladosporium, Aspergillus,* and *Penicillium.* The latter two genera occur more frequently indoors than outdoors.

Over the past several decades, indoor mold contamination has become a more serious environmental and health problem as more and more people have insulated their homes and businesses for energy efficiency. The resulting amplified mold contamination has been considered one of the contributing factors in the increased incidence of asthma, as well as a possible cause of Sick Building Syndrome. This condition is usually characterized by numerous reports of symptoms such as burning eyes, headache, fatigue, dizziness, and breathing difficulties by workers in large, well-insulated office buildings. Sick Building Syndrome is often due to poor ventilation, with the accumulation of chemical and/or biological contaminants in the air.

Chapter Summary

1. Among the secondary products produced by some fungi are antibiotics, compounds that are toxic to microorganisms. In the natural environment, these substances provide advantages over competing microorganisms for available nutrients and space. In modern medicine, antibiotics are the primary weapons for fighting bacterial infections. Penicillin was discovered by Alexander Fleming when he examined a culture of bacteria contaminated with *Penicillium notatum.* Within the next 15 years, purification and mass production of penicillin was achieved. Although other antibiotics have since been identified and developed, penicillin is still the most widely used antibiotic and is the drug of choice to treat many bacterial infections. Today, antibiotic resistance has been found in many common bacteria, causing major concern among the medical community.
2. Some fungal metabolites are toxic to humans and other animals capable of causing serious or fatal reactions. These toxins include mycotoxins, which are formed by hyphae of common molds and mushroom toxins, formed in the fleshy fruiting bodies of basidiomycetes. Mycotoxins are commonly produced by fungi growing in contaminated foods, especially grains and nuts. Aflatoxin was the first mycotoxin identified and studied; today over 200 mycotoxins have been identified.
3. Toxic alkaloids produced by the ergot fungus, *Claviceps purpurea,* accumulate in the fungal sclerotium on contaminated rye plants. If the sclerotia are harvested and milled along with the grains of rye, contaminated bread results. Throughout the centuries, this has led to outbreaks of ergotism. Although some of the ergot alkaloids are known to have medicinal value, the lysergic acid alkaloids are responsible for hallucinations and other symptoms associated with convulsive ergotism. The hallucinogen LSD was first isolated from these alkaloids.
4. Mushroom toxins are categorized as protoplasmic poisons, neurotoxins, and gastrointestinal irritants. Most fatalities are caused by the protoplasmic poisons known as amatoxins that occur in various species of *Amanita* as well as other mushrooms. *Amanita muscaria,* which contains the neurotoxin ibotenic acid, has a long history of use as an intoxicant by native peoples in various parts of the world. Among the tribes of Mexico and Central America, various species of *Psilocybe* were revered for their hallucinogenic properties. These mushrooms, containing the alkaloids psilocybin and psilocin, have been used for thousands of years in religious and healing rituals.
5. Some fungi affect humans directly by causing superficial or systemic mycoses. Most fungal pathogens cause relatively minor skin infections such as ringworm and yeast infection, but more serious fungal diseases occur in individuals whose immune system is compromised. Histoplasmosis and coccidioidomycosis are two serious systemic infections that occur in North America. Opportunistic infections are caused by a wide range of common saprobic fungi. When the immune system is weakened by drug therapy or disease, infection by these fungi can occur. Invasive fungal infections occur in 25% of these high-risk individuals. The airborne spores of many fungi are known to cause respiratory allergies such as asthma and hay fever. Many of the important allergenic fungi include those whose spores are known to be most abundant in the atmosphere, *Cladosporium* and *Alternaria.* Recent concern has focused on the allergenic importance of fungi proliferating in indoor environments.

Review Questions

1. How did Fleming's discovery change the scope of medical science?
2. Differentiate between antibiotics and fungal toxins in terms of their human impact.
3. In what ways do mycotoxins threaten food safety in both developed and developing countries?
4. Discuss the positive and negative impacts of *Claviceps purpurea* on society.
5. Why are mushrooms in the genus *Amanita* among the most feared mushrooms? Discuss the various types of toxins found in this genus.

6. Describe the use of hallucinogenic mushrooms by native peoples in various parts of the world.
7. Why has there been an increase in the incidence of mycotic infections in recent decades? What are the most common serious mycoses?
8. Describe the importance of fungi as causes of allergic diseases.

Further Reading

Ammirati, Joseph F., James A. Traquair, and Paul A. Horgen. 1985. *Poisonous Mushrooms of the Northern United States and Canada.* University of Minnesota Press, Minneapolis, MN.

Blackwell, Will H. 1990. *Poisonous and Medicinal Plants.* Prentice-Hall. Englewood Cliffs, NJ.

Buczacki, Stefan. 1989. *New Generation Guide to the Fungi of Britain and Europe.* University of Texas Press, Austin, TX.

Findlay, W. P. K. 1982. *Fungi—Folklore, Fiction, & Fact.* Mad River Press, Eureka, CA.

Hill, Edwin. 1974. "That's Funny" said Fleming on Discovering Penicillin. *British History Illustrated,* 1 (3): 56–63.

Kendrick, Bryce. 1992. *The Fifth Kingdom,* 2nd Edition, Focus Text, Newburyport, MA.

Kiester, Edwin, Jr. 1990. A Curiosity Turned into the First Silver Bullet Against Death. *Smithsonian,* 21 (8): 173–187.

Lincoff, Gary and D. H. Mitchell. 1977. *Toxic and Hallucinogenic Mushroom Poisoning.* Van Nostrand Reinhold Co., New York.

Mann, John. 1992. *Murder, Magic, and Medicine.* Oxford University Press, Oxford, United Kingdom.

Matossian, M. K. 1989. *Poisons of the Past: Molds, Epidemics, and History.* Yale University Press, New Haven, CT.

Toutexis, Anastasia. June 6, 1988. Got That Stuffy, Run-Down Feeling? *Time,* 131 (23): 76.

Wasson, R. Gordon. 1969. *Soma Divine Mushroom of Immortality.* Harcourt Brace Javanovich, New York.

UNIT

VII

Plants and the Environment

Tropical rain forests with their tremendous diversity of plant and animal life are at risk. The strategy of extractive reserves makes use of the economically valuable and renewable products of an intact rain forest to preserve it.

CHAPTER

25

ECOLOGY

Chapter Outline

Chapter Concepts

1. An ecosystem is the functional unit of the environment; it is concerned with the interactions between organisms and the physical and chemical components of the environment.
2. Biomes are large terrestrial regions recognizable by their characteristic vegetation and shaped by climatic conditions; major biomes include grasslands, forests, and deserts.
3. The policy of extractive reserves is a strategy to maintain natural areas for the extraction of economically valuable, yet renewable, products with minimal environmental degradation.

Although ecological issues have become prominent in the latter half of the twentieth century, the concept of ecology, taken from the Greek *oikos* meaning "house," was first proposed by the German biologist Ernst Hackel in 1869. Ecology is the study of organisms in the environment, the surroundings in which organisms live. The areas on Earth in which organisms are found are collectively known as the **biosphere** (fig. 25.1). The biosphere is a 20 km (12 mi) zone extending from the depths of the ocean floor to the peaks of the world's highest mountains. But life is most abundant at the intersection of land (lithosphere), air (atmosphere), and water (hydrosphere), from 200 m (660 ft) below the surface of the ocean up to 6,000 m (20,000 ft) above sea level.

The Ecosystem

An **ecosystem** is the functional unit of study in the environment. It includes the study of **biotic factors,** organisms, and the interrelationships between organisms, as well as their connection to nonliving or **abiotic** components in the environment, such as energy flow and the cycling of minerals. The biosphere itself is an example of an ecosystem, although most ecologists study ecosystems on a smaller scale. A pond or an abandoned field is an example of an ecosystem. A terrarium or aquarium is an ecosystem in miniature for classroom or laboratory study.

The organization of life is well-structured and clearly defined, as depicted in Figure 25.2. A **population** is a collection of organisms of the same species in a given area. A **community** refers to all the populations of organisms occupying a given region. For example, the winged elm (*Ulmus alata*) is one of several populations of understory trees in an oak-hickory forest community in northeastern Oklahoma.

Ecological Niche

Identifying the **habitat** and **niche** is essential to understanding the role of an organism in the ecosystem. Habitat refers to the particular place in the ecosystem that an organism or a population occupies. It is the "address" of an organism, the specific location where the organism can be found. The habitat of cattails is a freshwater marsh; river birch is found bordering streams. The concept of niche is broader; it is defined as the functional role an organism

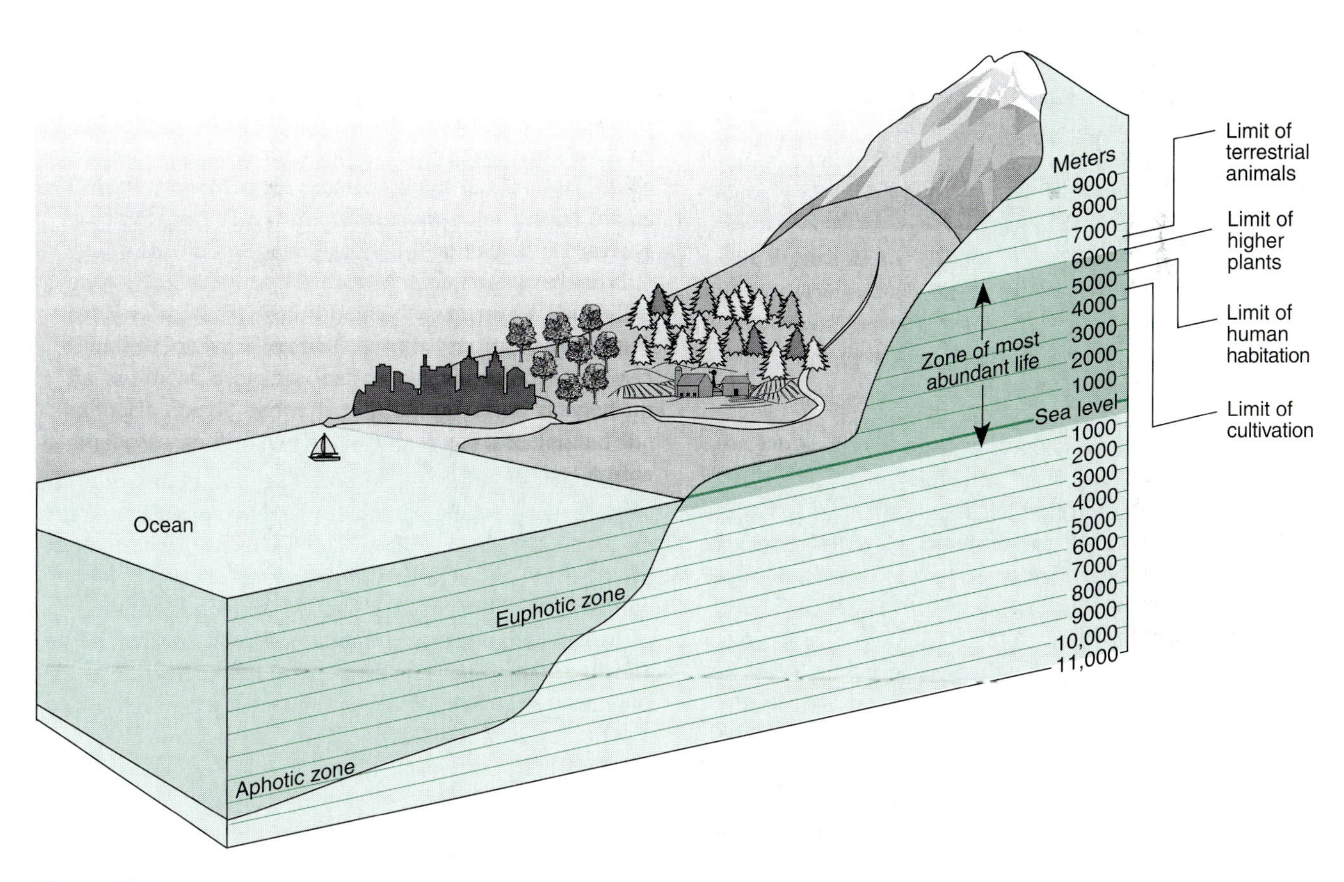

Figure 25.1 *The biosphere is the "thin skin of life" where organisms can be found on Earth. Life is most abundant in the intersection of the lithosphere, atmosphere, and hydrosphere. (No appreciable light penetrates the aphotic zone.)*

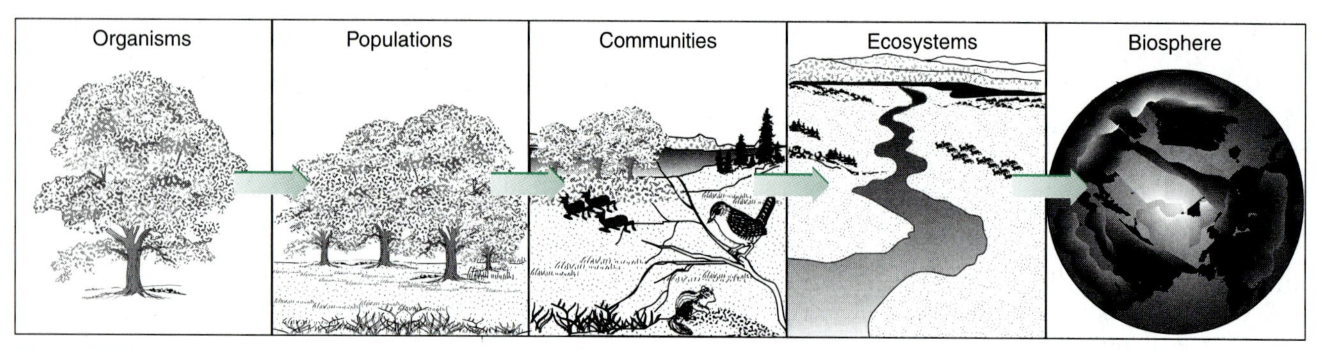

Figure 25.2 *The field of ecology deals with studies of organisms, populations, communities, ecosystems, and the biosphere.*

plays in the ecosystem. Niche encompasses not only where an organism lives but its relationships to other organisms in the ecosystem. Niche defines such factors as the nutritional requirements, temperature tolerance, or moisture conditions of a species.

Species that fulfill similar or identical niches in different geographical regions are known as **ecological equivalents.** For example, cacti in New World deserts fulfill a niche similar to that of the euphorbs in deserts of Africa. Species within these completely different plant groups even appear superficially similar. Both cacti and euphorbs possess photosynthetic stems that are armed with spines and contain water-storing tissues (fig. 25.3).

Under natural conditions, no two species may occupy the same niche indefinitely. One species will eventually displace the other as a result of **competitive exclusion,** outcompeting the other for nutrients, habitat, or other resources. This has been demonstrated by experimentally growing two species of clover together: white clover (*Trifolium repens*) and strawberry clover (*Trifolium castaneum*). The faster-growing white clover quickly established a leaf canopy that seemingly shaded out the strawberry clover. In time, however, the strawberry clover sent out shoots that overtopped the white clover, and the white clover was deprived of light. Displaced species, such as the white clover in this experiment, must occupy a different niche in order to survive. In some cases, the displaced species cannot adapt and becomes extinct. Introduction of alien species often results in displacement, and eventual extinction, of native species. This is the case with the Catalina mahogany (*Cercocarpus traskiae*), a tree native to Santa Catalina Island off the coast of California. The introduction of herbivores such as goats and sheep onto the island has caused severe overgrazing to the point that the population has been reduced to just a handful of trees. Another case in point is the introduction of the water hyacinth (*Eichhornia crassipes*). A species native to South America, it was deliberately introduced to Florida because of its beautiful flowers. Today, it is a major nuisance in southern waterways, where its population has grown to such an extent that dense mats crowd out native species.

Another factor in species survival is the type of niche characteristic of the species. Some species have rather broad tolerances for many environmental conditions and general requirements for survival. Species such as these are said to have a **generalized niche** and are less sensitive and more adaptable to environmental changes. Dandelions, coyotes, and human beings are examples of generalist species. In contrast, some species occupy **specialized niches,** those with a narrow range of tolerance or exacting requirements for survival. The giant panda of China is a specialist, feeding exclusively on bamboo, and is understandably more susceptible to extinction if its sole food source is threatened. Plants like the swamp pink (*Helonias bullata*) of the eastern United States have become endangered, since they are only found in freshwater wetlands, and many of these have been drained for development or agriculture.

Food Chains and Food Webs

Crucial to the functioning of an ecosystem is the relationship of organisms in a **food chain.** A food chain essentially asks "who eats whom?" A typical grazer food chain (fig. 25.4) begins with the **producers,** green plants or algae capable of producing complex organic compounds through the process of photosynthesis (*see* Chapter 4). These organisms are also classified as **autotrophs,** literally self-feeders. All other organisms in the grazer food chain fulfill their nutritional needs by using others as a food source and are known as **consumers** or **heterotrophs** (other feeders). All animals are consumers; those that feed exclusively on producers are called **herbivores.** Those that feed on other consumers are **carnivores.** (Some animals are classified as **omnivores,** since their diet includes eating a mix of plants and animals.) In sequential fashion, the grazer food chain begins with the producers, which are eaten by **primary consumers;** the primary consumers, in turn, are eaten by

(a)

(b)

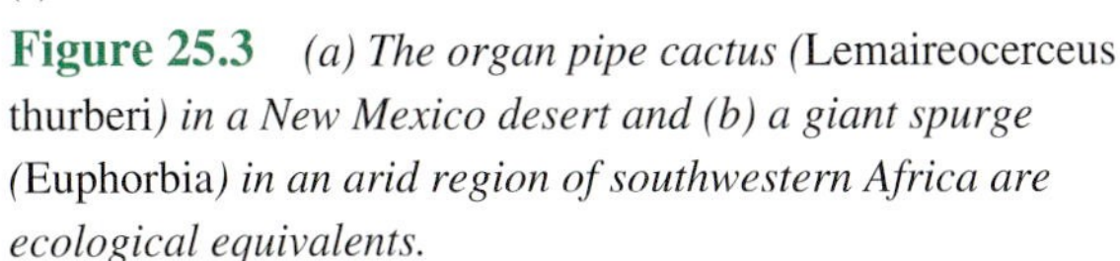

Figure 25.3 *(a) The organ pipe cactus (*Lemaireocerceus thurberi*) in a New Mexico desert and (b) a giant spurge (*Euphorbia*) in an arid region of southwestern Africa are ecological equivalents.*

secondary consumers; those eating secondary consumers are **tertiary consumers** and so forth. Each step in a food chain is referred to as a **trophic level;** most food chains have maximally four to five levels. In a **detritus** food chain (fig. 25.4), consumers (detritivores) meet their nutritional needs by degrading the remains of plants and animals and their waste products (detritus). Some larger animals, vultures and hyenas for example, are scavengers of the dead, and bacteria and fungi fulfill the role of **decomposers** or **saprobes,** degrading organic material. In most situations, a food chain is a simplification of the actual trophic interactions in an ecosystem. In reality, there are usually more than one type of producer, and most consumers have several alternative food sources. A **food web** such as that found in the salt marsh depicted in Figure 25.5, with multiple interactions between several food chains, is a more accurate representation.

Energy Flow and Ecological Pyramids

An important characteristic of ecosystems is the transfer of energy. The ultimate source of energy is the sun; solar energy is trapped by the chlorophyll present in algae and green plants for the purpose of photosynthesis. During photosynthesis, solar energy is transferred into chemical bond energy in the construction of organic compounds from the raw materials CO_2 and H_2O (*see* Chapter 4). Producers incorporate these organic compounds, or break them down to release energy during cellular respiration. Primary consumers eat the producers, obtaining the raw materials needed for the construction of their own tissues and energy. Secondary consumers eat herbivores to acquire energy, and energy is thus transferred along the food chain. Recall from Chapter 4 that energy transformations are governed by the first and second laws of

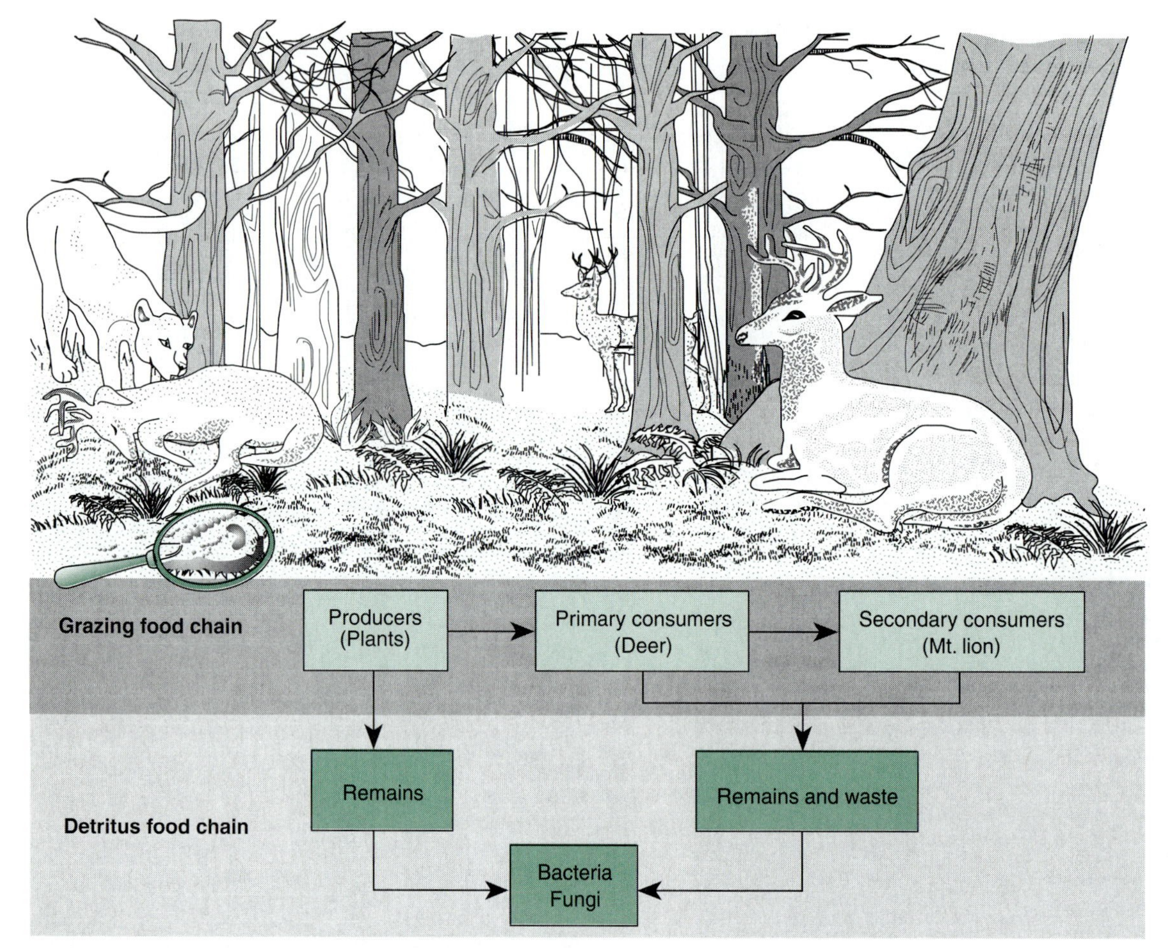

Figure 25.4 *The bottom level in a grazer food chain begins with producers, followed by succeeding levels of consumers. In a detritus food chain, both macroconsumers and microconsumers degrade the remains of dead plants, animals, and waste products as a source of nutrition.*

thermodynamics. According to the First Law, energy cannot be created or destroyed but can be transformed from one form or another. Photosynthesis transforms the radiant energy of the sun into the chemical bond energy within glucose. According to the Second Law, when energy is converted into another form, some of the energy is transformed into low quality heat energy that is lost into the environment. With each transformation, therefore, there is less high-quality energy available and the ability to do useful work is reduced. Much of the chemical bond energy in fuel molecules is dispersed as heat during the transformation into mechanical energy to drive an engine.

Energy flow in ecosystems can be depicted by **ecological pyramids** of **numbers, biomass,** and **energy**. A pyramid of numbers (fig. 25.6) indicates the number of organisms supported at each trophic level in a food chain. Fewer organisms are supported at successively higher trophic levels. In other words, it takes many small organisms to support the nutritional needs of one large organism. **Biomass** is the organic material comprising living organisms. It is usually represented as the dry weight (fresh weight minus the water content) of living matter. As with the pyramid of numbers, the total biomass for each trophic level usually declines with each step up the food chain.

Figure 25.5 *A food web in a salt marsh shows the complex interactions between food chains among organisms. The producers (1) are terrestrial and aquatic salt marsh plants. Primary consumers include herbivorous insects (2), invertebrates (3), and fish (4). Primary consumers are in turn eaten by carnivores such as the common egret (5), and shrew (6). Secondary carnivores (7), include the marsh hawk and short-eared owl. Mallard ducks (8), and song sparrows (9), sandpipers (10), Norway rat (11), and salt marsh harvest mouse (12), are omnivores feeding on both plants and animals.*

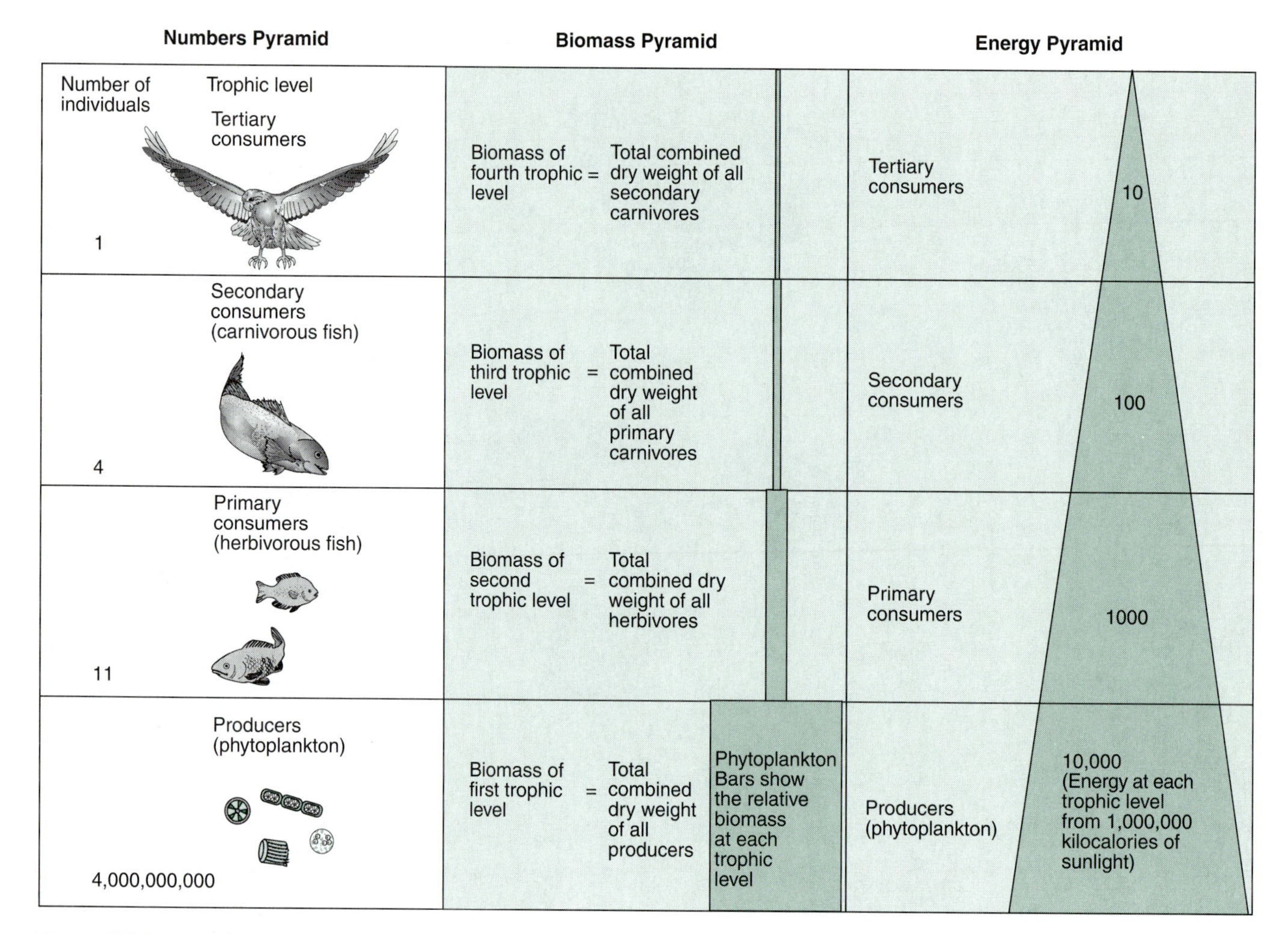

Figure 25.6 *Ecological pyramids of numbers (a), biomass (b) and energy (c). (a) A pyramid of numbers indicates the number of organisms in each trophic level that can be supported in the food chain. (b) Pyramids of biomass indicate the total biomass at each trophic level. (c) Pyramids of energy indicate the amount of energy transferred between ascending trophic levels.*

Likewise, the pyramid of energy illustrating the total energy content in kilocalories (Calories) for each trophic level follows the same pattern as the other ecological pyramids. It also shows the inefficiency of energy transfer between trophic levels and explains the shape of the ecological pyramids. Only 5%–20% of energy (averaging 10%) is passed between trophic levels. Consequently, most food chains consist of a maximum of only four to five levels. Beyond this number, the energy transferred is minimal and incapable of supporting life. This also explains why more herbivores can be supported than carnivores. One suggestion to conserve food resources for the burgeoning human population (5.6 billion in 1994) is to eat lower on the food chain. More human beings could be fed if they ate plant crops (producers) directly instead of feeding livestock with the crops and eating the meat produced. For example, 20,000 C of corn will feed 2,000 C of cattle to yield 200 C of meat, enough to feed only one human being. In contrast, the same amount of corn eaten directly would provide 10 people 200 C apiece.

Biological Magnification

In 1962, a book entitled *Silent Spring* by Rachel Carson brought to the attention of the scientific community and the public alike some disturbing observations about the use of pesticides. Since the end of World War II, numerous synthetic pesticides have been created and released in an attempt to control humanity's most serious pests. These pests, primarily insects, have been harbingers of human disease or destructive to crops and livestock. One of the new generation of pesticides that was widely used to control mosquitoes, as well as other harmful insects,

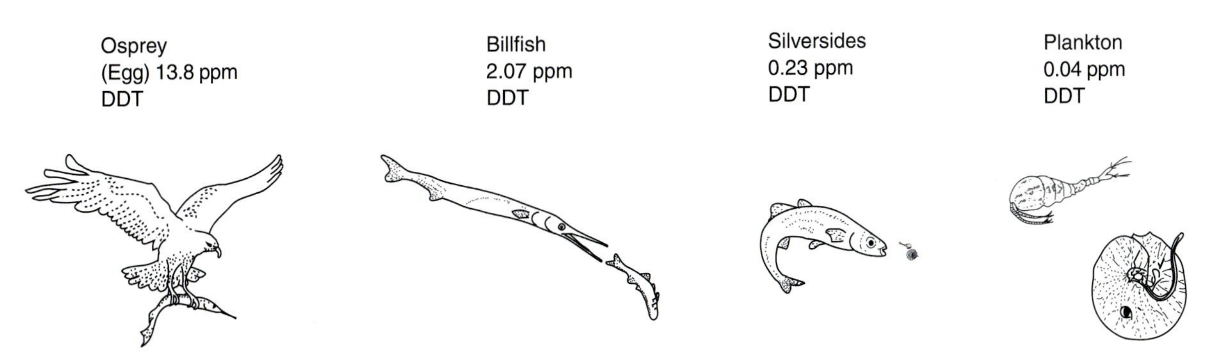

Figure 25.7 *The pesticide DDT is concentrated and passed along a food chain in a process called biological magnification. (DDT values are expressed in parts per million, wet weight, whole body basis.)*

was a chlorinated hydrocarbon known as DDT (dichloro-diphenyl-trichloroethane). DDT, as with many of the other synthetic pesticides produced during this time, is slow to degrade, persisting in the environment for several months or years. Also, it is known to be soluble in lipids, a major component of organisms. Stored in an organism's lipids, DDT was later shown to be passed up the food chain and, with each transfer to a higher level, became more concentrated in the fatty tissues. The transfer and concentration of a toxic material along a food chain is known as **biological magnification** (fig. 25.7). As *Silent Spring* pointed out, the biological magnification of DDT had dire consequences for many species of birds at the top level of food chains. For example, the American bald eagle, as a secondary consumer of fish-concentrated DDT (and its primary breakdown product DDE or dichloro-diphenyl-dichloroethylene), was exposed to such toxic levels that the amount of calcium deposited in the eggshells was reduced. Shells were so thin that the fragile shells broke during incubation, killing the developing embryo. Consequently, the numbers of bald eagles and other raptors (birds of prey), as well as some familiar songbirds, declined rapidly. DDT was also showing up in human tissue, even in the milk of lactating women. For these reasons, DDT has been banned in the United States since 1972. Unfortunately, this ban is not global; DDT is still used in many developing nations. Other toxic substances that can be biologically magnified have been identified and their use banned or restricted as well. Examples are mercury, PCBs (used as insulators in electrical systems), aldrin (a crop pesticide), and chlordane (used in termite control).

Biogeochemical Cycling

The natural world is the original recycler, cycling substances such as sulfur, phosphorus, and water through the abiotic and biotic components of the environment. The cycling of nitrogen has been discussed in Chapter 13. Here the carbon cycle, and how human activities have affected and upset this cycle, will be discussed.

The Carbon Cycle

All life on Earth is carbon-based; the primary organic compounds are constructed from carbon backbones. It is not surprising, therefore, that carbon makes up 40%–60% of the dry weight of all living matter. Carbon (fig. 25.8), in the form of gaseous carbon dioxide (CO_2), is removed from the atmosphere through the process of photosynthesis as carried out by green plants, algae, and cyanobacteria. Recall that the summary equation for photosynthesis is:

$$6CO_2 + 12H_2O \longrightarrow C_6H_{12}O_6 + 6O_2 + 6H_2O$$

Sunlight captured by chlorophyll provides the energy to drive the photosynthetic reaction. Carbon dioxide is fixed and reduced during photosynthesis to form glucose (*see* Chapter 4). Since they remove carbon dioxide from the atmosphere, the photosynthetic organisms are known as **carbon sinks.** The carbon in glucose can be used to make numerous organic compounds in the producers and later in the herbivores and carnivores along the food chain. Glucose is also utilized as fuel to yield energy for producers, consumers, and decomposers through cellular respiration. It is through cellular respiration that carbon dioxide is released back into the atmosphere:

$$C_6H_{12}O_6 + 6O_2 + 6H_2O \longrightarrow 6CO_2 + 12H_2O + 36ATP(\text{energy})$$

Likewise, the burning of any organic material, whether firewood or fossil fuels (coal, natural gas, or oil), also releases carbon dioxide to the atmosphere.

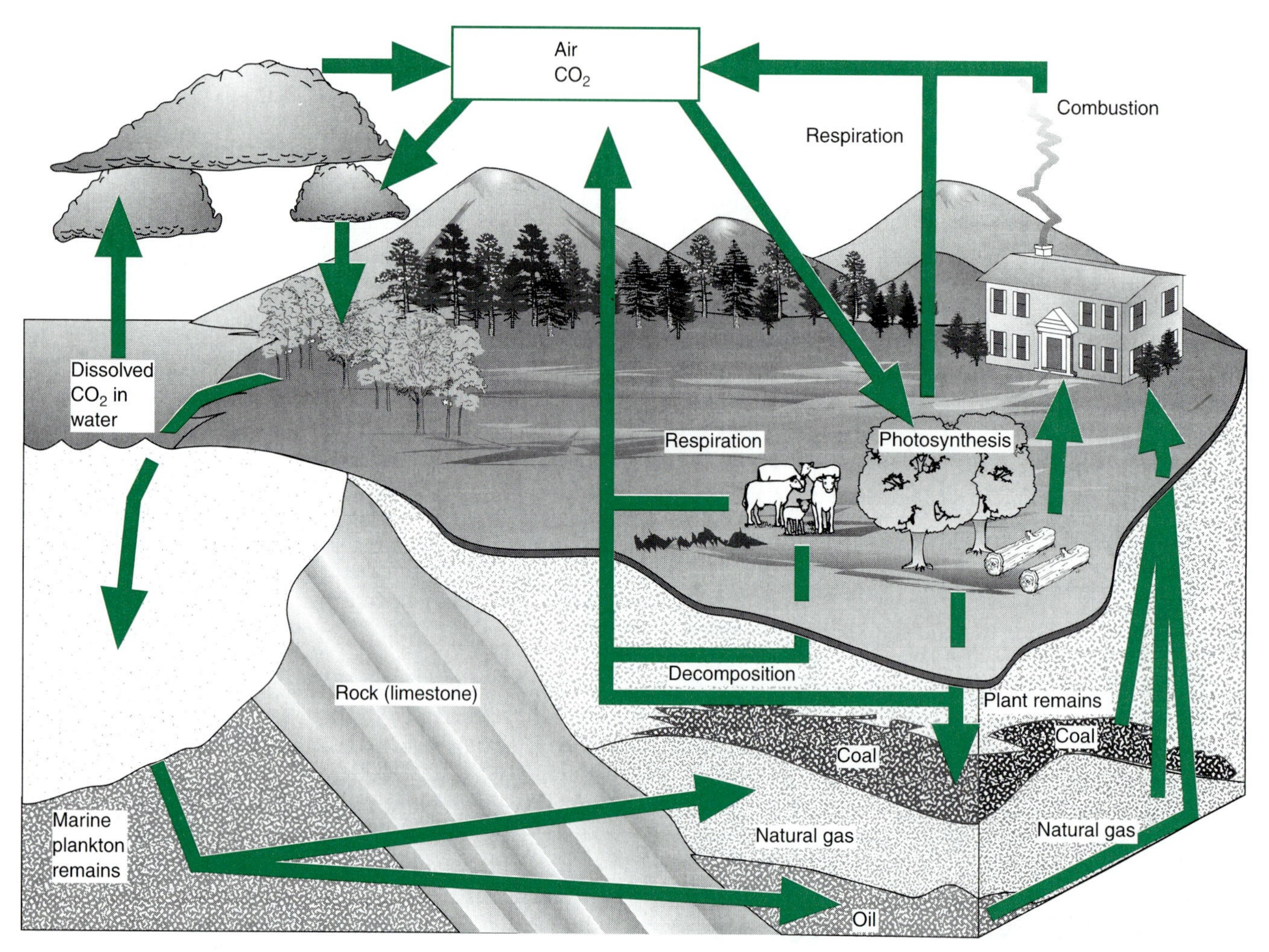

Figure 25.8 *The carbon cycle. Carbon dioxide enters organisms in the form of carbon dioxide when plants photosynthesize. Carbon is released back to the atmosphere when organisms respire carbon dioxide or during the combustion of carbon-based fossil fuels. Fossil fuels (oils, natural gas, coal) were formed from the remains of ancient organisms.*

The Greenhouse Effect

Humankind has upset the balance of the carbon cycle in several ways. Global deforestation, especially of the tropical rain forest, removes vast areas of vegetation, reducing the effectiveness of this carbon sink. At the same time, more CO_2 has been released into the atmosphere, primarily from the accelerated burning of fossil fuels. Global carbon dioxide has increased by 25% in the past century (fig. 25.9) and it is this increase in atmospheric CO_2 that has contributed to the **greenhouse effect.** Carbon dioxide, as well as some other greenhouse gases, acts as a heat trap. Much like the glass panes in a greenhouse, carbon dioxide in the atmosphere captures sunlight and converts it into heat, warming the atmosphere in the process. In fact, the temperature of the Earth has increased by 0.5° C (1° F) over the past 100 years as atmospheric CO_2 levels increased. It has been predicted that these levels will double over the next 50 years as fossil fuel consumption continues to increase. This will result in a 2°–5° C (4°–9° F) rise in temperature before the end of the twenty-first century. Further global warming is predicted to produce dramatic climatic changes. Agricultural regions will shift northward, as greater heat stress and more frequent droughts prevail in southern farming regions. Much of the world's best agricultural lands of today will be lost as the climate becomes too dry and hot to grow the major agricultural crops. Warmer climates will shift the distribution of plants and animals; some species undoubtedly will become extinct as the range of heat-tolerant species expands. Many of the eastern hardwood trees such as beech (*Fagus*) will die out in southeastern states, restricted only to the higher altitudes of a few mountains, and to New England and Canada, the northernmost realm of its current range. The polar ice caps will shrink and the oceans will take up more space due to thermal expansion of water molecules. A predicted 0.3–0.5 m

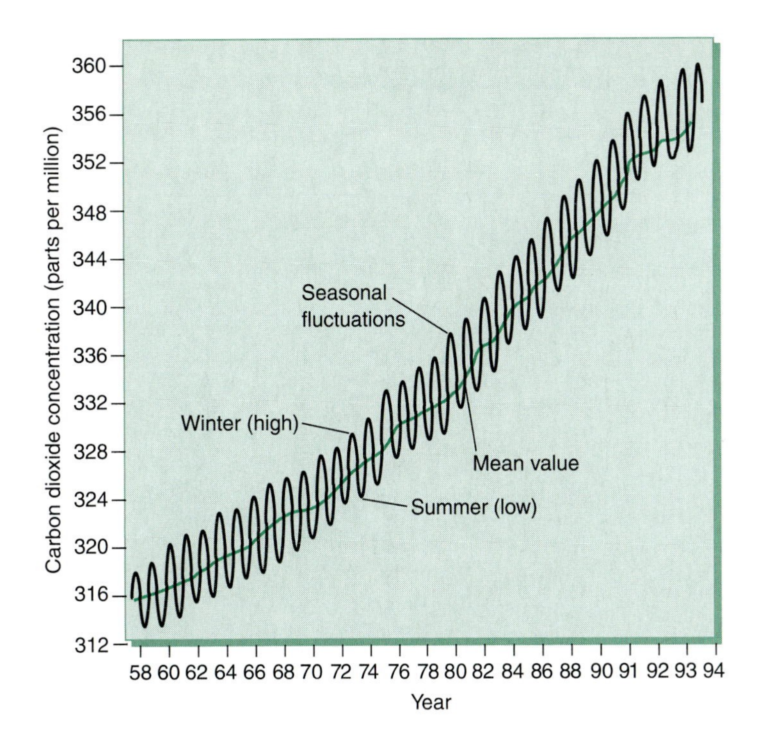

(a)

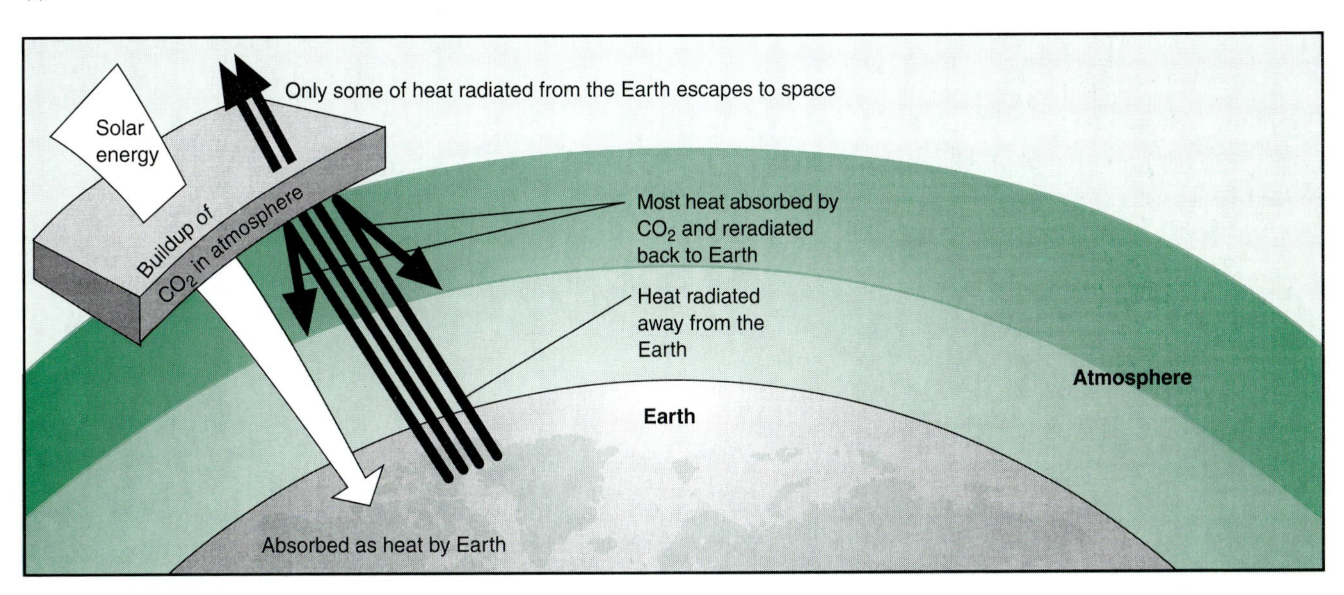

(b)

Figure 25.9 *(a) A steady rise of carbon dioxide in the atmosphere due to the burning of fossil fuels and the loss of carbon sinks has been documented. (Carbon dioxide levels are lower in the summer because photosynthetic rates are high during the summer season.) (b) Carbon dioxide and other greenhouse gases trap solar energy, reradiating it back to Earth as heat and, consequently, warming the atmosphere.*

(0.9 ft–1.5 ft) rise in sea levels by the mid-twenty-first century will flood the coastal areas of many nations, where half of the world's population resides. In 1992, representatives from nations around the world met in Rio de Janiero to discuss environmental issues, including the problem of global warming. As a result of this conference, 160 nations agreed to reduce carbon dioxide emissions in order to curb the greenhouse effect.

Succession

Ecosystems are not static; they may experience changes over time. **Ecological succession** is the gradual progression of orderly and predictable changes that occur in an ecosystem over time. These changes result from modifications of the physical environment through the actions of the biotic community. These modifications of the abiotic component of the environment, in turn, bring about changes in the biotic community. Two types of succession are recognized: **primary** and **secondary.**

Primary succession (fig. 25.10) begins in an area bare of vegetation. The colonizations of plants in a pond, a newly formed sandbar, and barren rocks are examples of primary succession. Pioneer species are those first inhabitants in the area; in the terrestrial environment, lichens are usually the first plant community. As pioneer species, they begin the process of soil formation by dissolving minerals from the rocks and, as they die, adding organic material. Also, the presence of

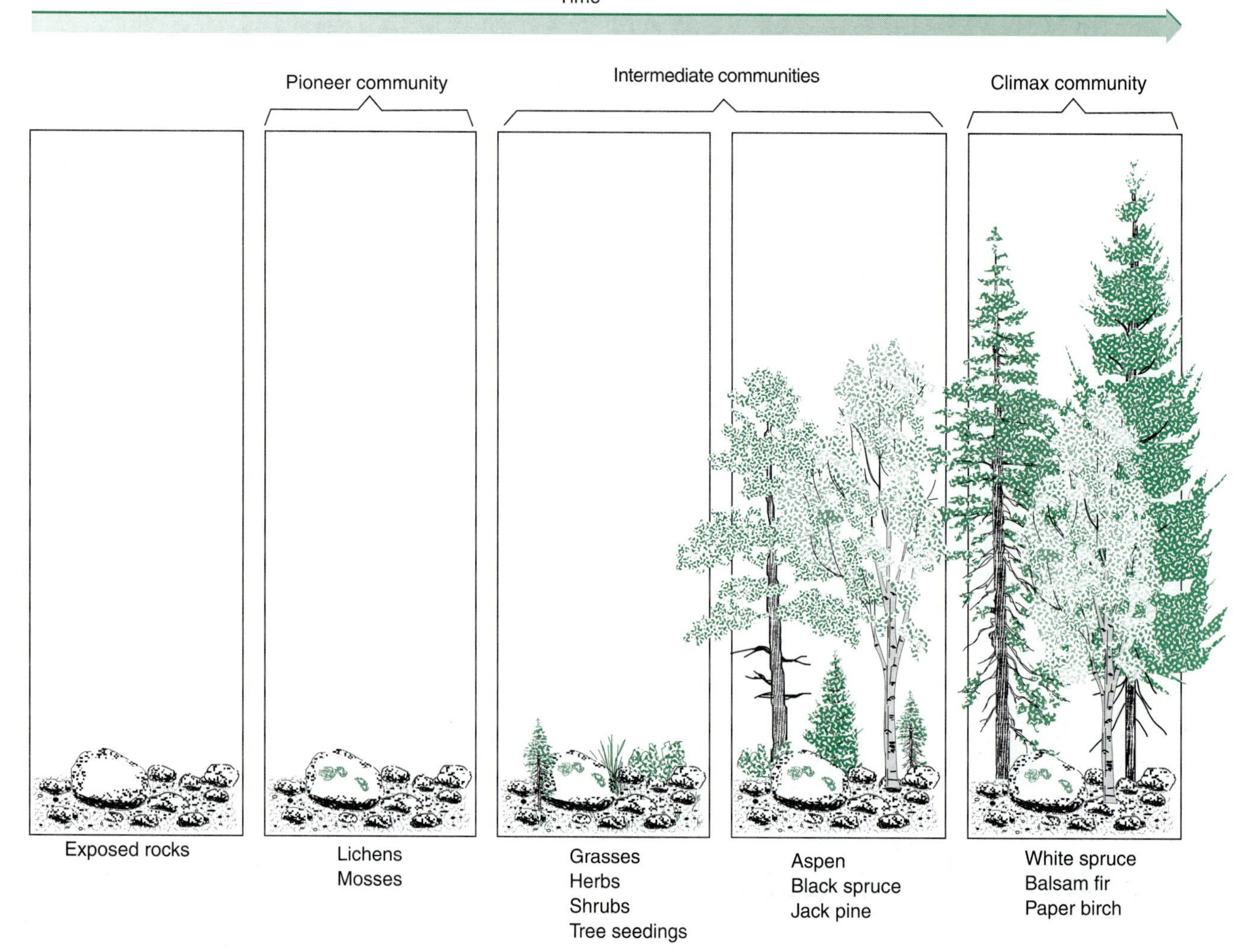

Figure 25.10 *Primary succession begins as a pioneer species of lichens and mosses invade bare rock. Several intermediate communities inhabit the area for a time before the self-perpetuating climax community of spruce-fir and birch forest forms.*

vegetation tends to trap and hold water. All of these modifications bring about changes in the environment, making it more amenable to succeeding communities of invading plants. Mosses, then grasses and herbs, follow and replace the pioneer lichens. Eventually the environment may be sufficiently modified to support a mature community of trees.

In secondary succession, natural forces or human intervention destroy the existing vegetation. Usually the soil remains, so the soil formation associated with primary succession does not occur. One of the best-studied examples of secondary succession is old field succession of abandoned farmland. Clearing the land of the natural vegetation is the first step in preparing the land for planting. If that field is abandoned and no longer farmed, the process of secondary succession begins (fig. 25.11), eventually returning the land to the original natural community. Other examples of secondary succession are the return of vegetation following the volcanic explosion of Mount St. Helens, May 18, 1980, and the recovery of Yellowstone Park after the devastating fires in 1988. The last successional stage in both primary and secondary succession is self-perpetuating and stable, and is known as a **climax** community.

The Green World: Biomes

Biomes are the largest terrestrial divisions of the biosphere. They are the climax communities for huge regions of the

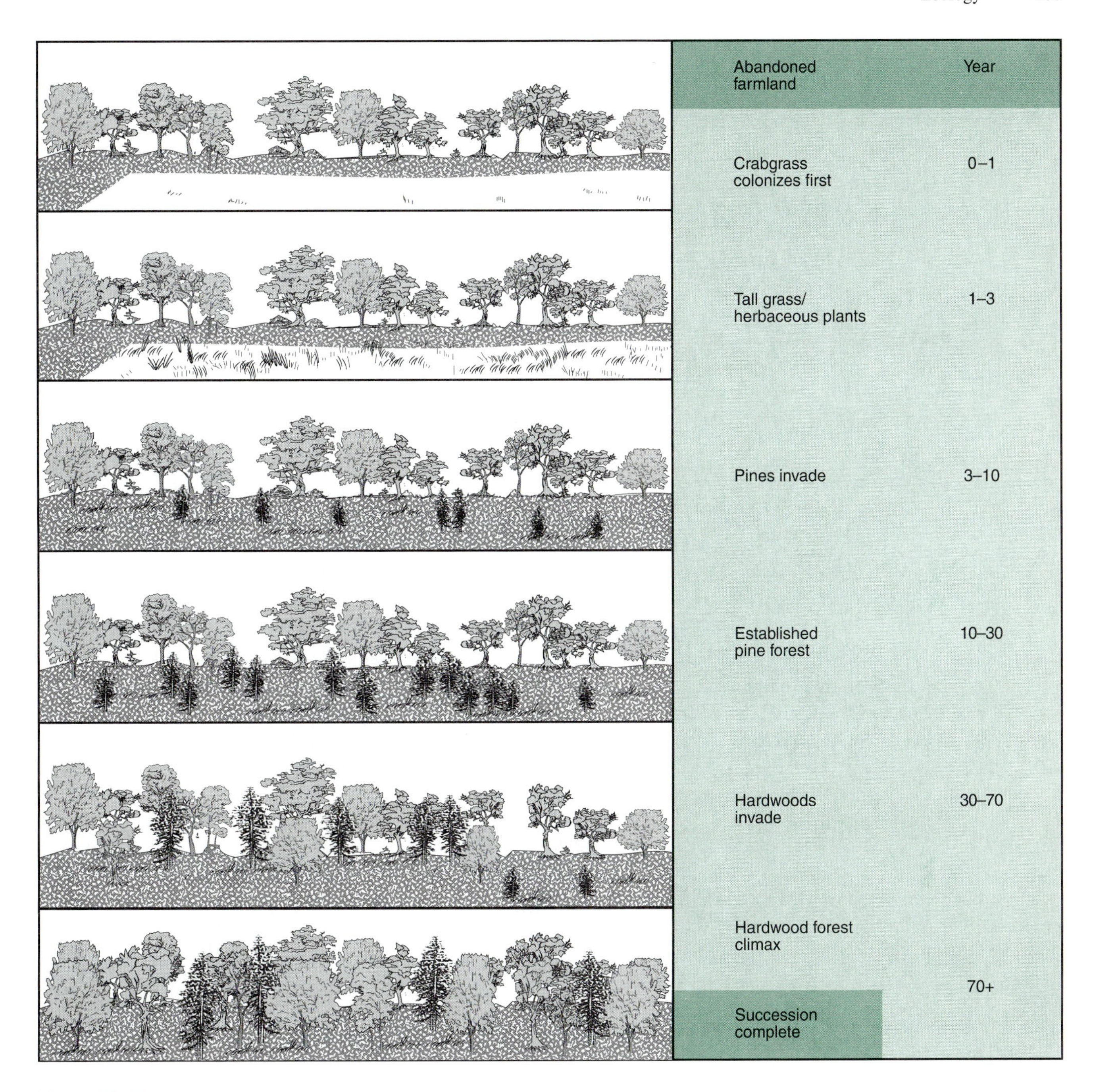

Figure 25.11 *The secondary succession of an abandoned farmland begins with the invasion of weedy species such as crabgrass that quickly colonize the area. Other herbaceous plants follow. Eventually sun-loving pine species invade and establish, in time, a pine forest. Hardwood saplings develop in the shade of the pine forest and form the climax hardwood forest 70 or more years after the successional process began.*

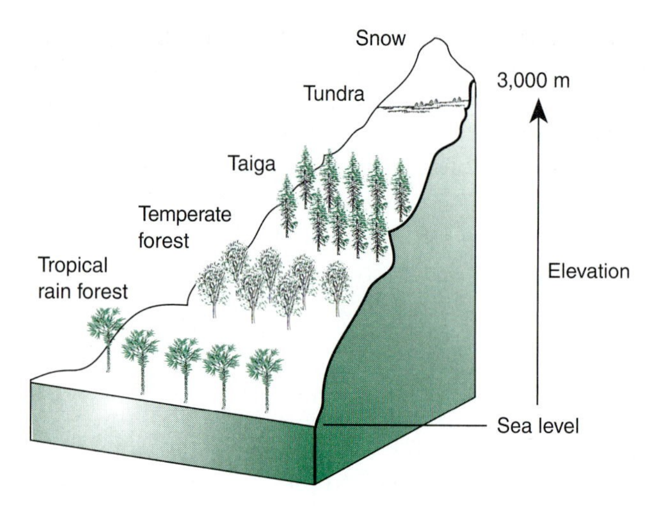

Figure 25.12 *Altitude also affects the distribution of biomes. The cooler temperatures at the higher elevations of a mountain, even one located on the equator, favor the development of biomes similar to those encountered when traveling a latitudinal journey from equator to North Pole.*

land and are recognized and defined by their distinctive vegetation and animal life. Climatic conditions, such as rainfall and temperature, largely determine the biome that develops in an area. Yearly precipitation is a key factor in determining whether a desert, grassland, or forest will develop at a given latitude. Deserts are found in areas with low yearly precipitation. If the yearly precipitation increases, grasslands develop. Still higher precipitation values favor the development of forest. Altitude also affects what biome develops (fig. 25.12). The cooler temperatures encountered at a mountain's higher altitudes promote the development of biomes normally seen at the higher latitudes. The major biomes of the world are depicted in Figure 25.13 and are discussed next.

Deserts

The **desert** biome develops when annual rainfall falls below 25 cm (10 in). Hot or low deserts form when evaporation and transpiration rates exceed rainfall. The Mojave desert in

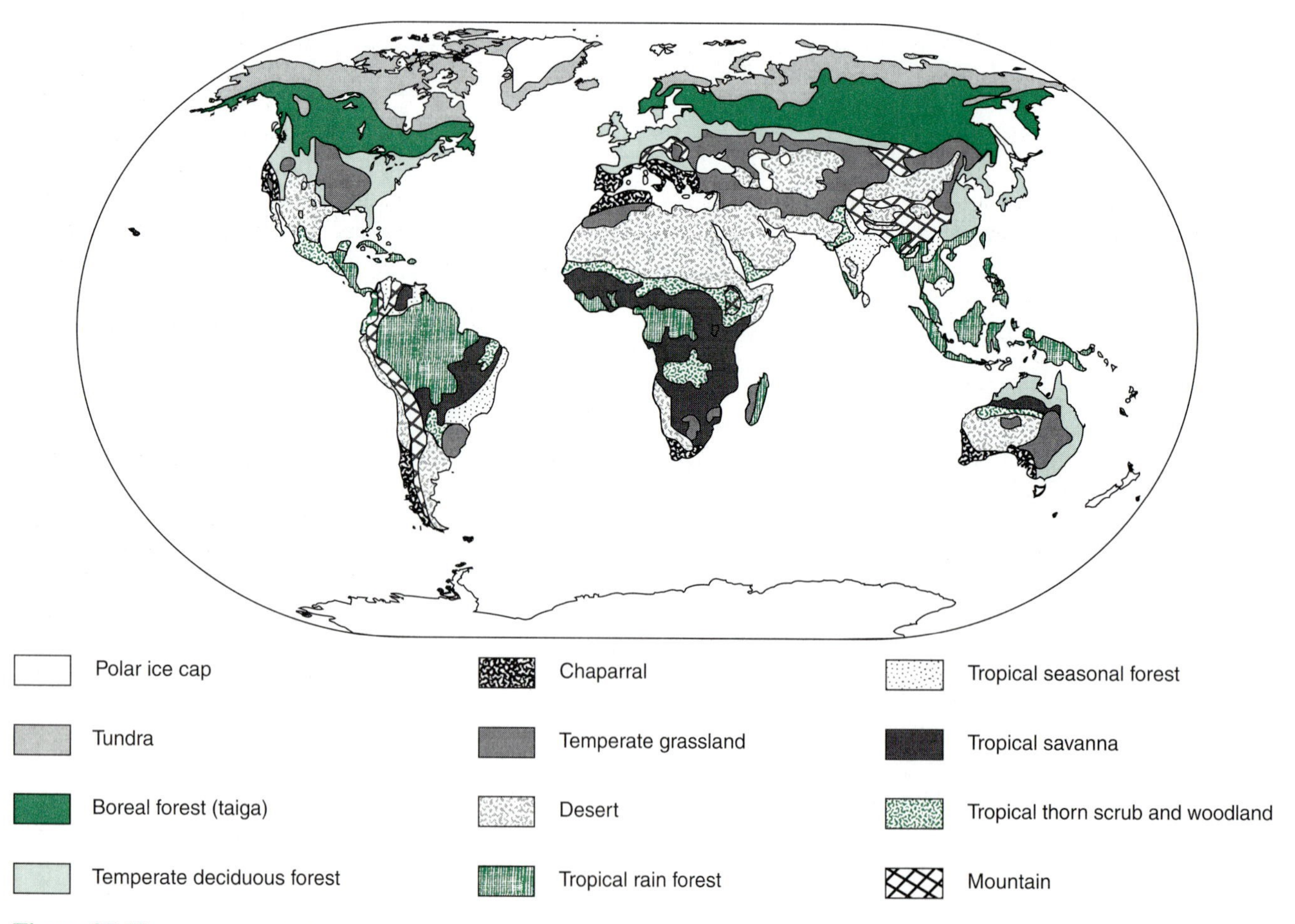

Figure 25.13 *Map of world biomes. Climatic conditions, particularly temperature and precipitation, determine the type of biome that forms in a region.*

California, and the Sonoran spanning New Mexico, Arizona, and southern California, are two low deserts in North America. Various cacti and drought-tolerant shrubs dominate the landscape of the low desert (fig. 25.14). Cold deserts form when a mountain range creates a rain shadow. An example of a cold or high desert is the Great Basin (comprising most of Nevada and parts of Utah, California, Oregon, and Idaho), formed because few precipitation clouds can pass over the Sierra Nevada mountain range. Sagebrush and other low shrubs characterize the high desert (fig. 25.14a). Desert soils are nutrient-poor, since there is not enough moisture for degradation of organic material. Some are high in minerals, but in levels that are toxic to most vegetation. Desert soils are also subject to wind erosion, since the vegetative cover is often sparse. Both the plants and animals that are the normal occupants of the desert exhibit adaptions to conserve water. Plants that have evolved adaptations to survive the xeric (dry) conditions of the desert are called **xerophytes.** One strategy to avoid the lack of moisture is seen in desert annuals, which germinate, grow, and reproduce only during the rainy season. Others, like the cacti, have lost their leaves to reduce water loss from transpiration and have developed stems that are both photosynthetic and succulent (water-storing). Other desert plants have extremely long tap roots that tap into underground water supplies. Native animals like the kangaroo rat have also adapted to the continual drought of the desert. The kangaroo rat is a burrower, only coming out in the cool of the evening. It also never needs to drink water; its highly efficient kidney system extracts sufficient water from its food alone.

Crops from the Desert

One way to help meet the food demands of the forecasted human population expansion in the next century is to convert marginal lands into agricultural lands. Many of these marginal lands are in arid and semiarid regions. Irrigation is one method of converting desert into farmland. But as water resources are already strained to meet agricultural, industrial, and residential demands, another solution has evolved: native desert plants already adapted to the arid conditions of high temperatures and water scarcity are developed into crops.

Natural rubber is typically thought of as a product of the tropical rain forest, but the desert shrub guayule (*Parthenium argentatum*) might change that perception (fig. 25.15a). Native to the Chihuahuan desert of Mexico and southwestern Texas, this plant produces a **latex** in the root that is stored in parenchyma cells of both root and stem. (Latex may be variously colored but is most often milky white. The chemical makeup of latex is also varied, but the terpenoids are among the most common constituents.) Wild stands of guayule were first used as a source of natural rubber in the late 1800s and early 1900s. During World War II, when supplies of natural rubber from the East were disrupted, the U.S. government initiated plantings of guayule as a domestic source of natural rubber. These plans were abandoned with the development of synthetic rubber and the end of the war. Although synthetic rubber is a suitable substitute in most cases, there is still a demand for the natural product. In fact, the United States imports nearly $1 billion in natural rubber per year. Considering the financial drain, there is renewed interest to develop guayule as a natural rubber crop; plantings and research to improve crop strains are ongoing in California and Arizona.

An evergreen shrub native to the Sonoran desert, jojoba (*Simmondsia chinensis*) (fig. 25.15b) has been nicknamed the botanical whale because its seed oil is similar in many ways to the oil obtained from sperm whales. The development of jojoba as an oil crop has been encouraged since 1969, when the sperm whale was placed on the endangered species list and the import of sperm whale oil was banned. The seed oil of jojoba is unique; most seed oils are triglycerides but jojoba oil is liquid wax, both colorless and odorless. Its unique chemical properties make it suitable for a wide range of applications, including high pressure lubricants, antifoaming agents in the fermentation industry, protective coatings, and even an edible cooking oil.

Another desert crop on the brink of worldwide importance is kenaf (*Hibiscus cannabinus*) (fig. 25.15c) which may replace trees as a source of pulp for papermaking. This annual member of the mallow family (Malvaceae) is native to east central Africa. Field trials indicate that kenaf would grow well in many of the southern states without irrigation. It is quick-growing and can tolerate saline soils. Paper made from kenaf is said to be equal in strength to that made from wood pulp and has the added advantages of not yellowing as quickly and holding ink better. Newsprint from kenaf is now a viable alternative, with greater productivity per hectare and half the cost of wood pulp.

Chaparral

Chaparral is found in only one location in North America—California. The chaparral has a Mediterranean climate of hot, dry summers and cool, rainy winters. Evergreen, broadleaf shrubs form dense thickets across the landscape (fig. 25.14b). Allelopathy (*see* Chapter 21) has been shown to influence the scattered spacing pattern of the vegetation and is the reason for the bare zones of inhibition around the shrubs. Many of the shrubs are **sclerophyllous,** having thick, leathery leaves because of an abundance of sclerenchyma cells. The leathery leaves cut down on transpiration rates during the long summers. Periodic fires sweep through the chaparral but the shrubs regenerate from underground organs. Mule deer, chipmunks, lizards, and several bird species are examples of the animals found in the chaparral. The chaparral soil is thin, relatively infertile, and unstable. Population pressures along the California coast have encroached on the chaparral. The ever-present threat of fire during the dry season, and the potential of mudslides from heavy winter rains, show the danger of unsuitable habitation in this biome.

(a)

(b)

(c)

(d)

Figure 25.14 *Major biomes in North America. (a) Low deserts of the American southwest are dominated by large cacti and drought-tolerant shrubs. (b) Broadleaf evergreen shrubs form thickets in the California chaparral. (c) Grasslands, like this short grass prairie with bison, are home to large grazing mammals. (d) The tundra is a treeless frozen plain located in the northernmost part of North America. (e) South of the tundra lies the boreal forest, a vast band of coniferous trees. (f) Much of the eastern half of North America is part of the deciduous forest of oak, maple, beech, hickory, and other hardwood species. (g) Epiphytes and vines festoon the many different tree species that abound in the tropical rain forest.*

Grasslands

The pampas, veld, steppes, and prairies are all examples of **grasslands.** The average rainfall is relatively low (25–75 cm/10–30 in), with warm summers and cool winters. Grasses are the dominant vegetation in this biome. **Tall grass prairies,** named for the 5 m (15 ft) heights typical of some grasses, are found in the eastern range of this temperate biome, where precipitation is greater. In the drier areas, **short grass prairies** (fig. 25.14c) are found. In past ages, these vast "seas of grass" attracted huge herds of grazing mammals. In the North American grasslands, bison and antelope thrived and ground-nesting birds like the prairie chicken and quail abounded. Periodic fires have also influenced the appearance of the grasslands, destroying trees and releasing nutrients from organic matter. Grasses, with their rhizomes belowground, can sprout back after a fire and quickly colonize a burned area. Many of the grass species are sod formers, producing compacted mats of soil, roots, and rhizomes that were cut into blocks by the pioneers and used to build homes and barns. The topsoil layer was extensive, reaching depths of 0.7 m (2 ft), and of a rich, dark color because of its high organic content. Few untouched grasslands still exist in the continental United States. These rich lands now support herds of cattle and farmland, where the native grasses have been replaced by the domesticated grasses, grains, or pasture crops such as wheat, corn, alfalfa, or fescue.

(e)

(f)

(g)

A variant of the grassland is the **savanna** (fig. 25.14). Trees are scattered throughout the grassy plain of the savanna. Located in the tropical latitudes, there is little seasonal temperature variation, but the annual rainfall of 85–150 cm (34–60 in) is distributed unevenly into pronounced dry and wet seasons. Giraffes, zebras, wildebeests, and antelopes are indigenous to the savanna. The herbivores in turn attract predators and scavengers like the lion and hyena.

Cold shapes the **tundra,** a polar grassland that is the northernmost biome bordering the ice cap. The tundra is an icy, rolling plain where grasses, sedges, mosses, and lichens dominate the landscape (fig. 25.14d). A few shallow-rooted shrubs, such as birch and willow, may be found but they are extremely stunted, reaching heights of only 30 cm (12 in). The soil is thin and nutrient-poor, inhibiting root growth. In the summer months, only the top 10 cm (4 in) of soil thaws. The rest is **permafrost,** permanently frozen the year around. Temperatures are cold throughout the year, averaging 10° C (50° F). The growing season in the summer is short, lasting only several weeks, but intense, for the days are 24 hours long. During the summer thaw, the tundra has a waterlogged appearance, dotted with many ponds. This appearance is deceiving, for the ponds are actually shallow puddles, since the permafrost prevents drainage and the soil is saturated. Yearly precipitation is low, usually between 10–25 cm (4–10 in). Herds of caribou are the predominant herbivores in the North American tundra. Smaller burrowing mammals such as lemmings, snowshoe hares, and weasels are also tundra inhabitants. Bird life includes ptarmigans, snowy owls, and migrant waterfowl that come to breed during the summer. Insects such as deerflies, mosquitoes, and blackflies are also abundant in swarms during the summer months. The tundra is a fragile biome that recovers slowly from disturbances. Damage from the building of the Alaskan pipeline in the 1970s is still evident.

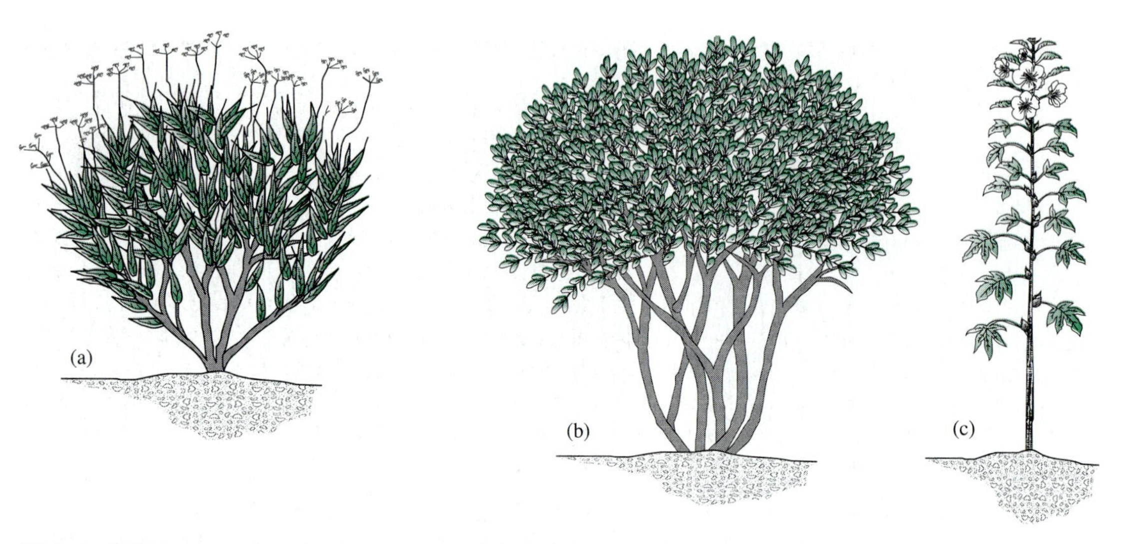

Figure 25.15 *Potential desert crops. (a) Guayule (*Parthenium argentatum*), produces a latex that is a source of natural rubber. (b) The seed oil of jojoba (*Simmondsia chinensis*) has many industrial uses. (c) The fibers of kenaf (*Hibiscus cannabinus*) can be used to make newspaper.*

Forests

Forests form when the annual precipitation is sufficient to support tree species. There are several types of forest biomes distinguished by their latitudinal distribution. Just south of the tundra, the **boreal forest** (also known as the **taiga** or **northern coniferous forest**) occurs as a gigantic band across North America, Europe, and Asia. Although the winters are long and cold, they are less severe than those in the tundra and, combined with the four-month growing season, allow for the complete thawing of the soil. An annual precipitation rate of 38–102 cm (15–40 in) also contributes to the milder conditions of the boreal forest. Conifers such as spruce, pine, hemlock, and fir dominate this biome in immense stands of only one or two species (fig. 25.14e). Leaf litter accumulates beneath the tree stands and decays slowly because of the cool temperatures. The soil is relatively infertile and acidic. Many species of large and small mammals are permanent residents or visitors to this biome. Caribou overwinter here and the moose, wolverine, bear, mink, lynx, and many rodents are other inhabitants of this biome. Many migratory species of birds also temporarily inhabit the forest. The pure stands of conifers in the boreal forest have been extensively harvested as a source of softwoods for lumber and pulp for paper.

The **moist coniferous forest** biome is limited to the Pacific northwest, where the northern climate is moderated by the influence of the Pacific ocean. The winters are mild and the summers cool. This biome is also referred to as the **temperate rain forest** because of the high yearly rainfall, 200–380 cm (80–152 in), the high humidity, and the prevalence of fog from the ocean. Conifers such as Douglas fir, northern arborvitae, and Sitka spruce are the dominant trees and often reach gigantic proportions. There is a well-developed understory of shrubs, ferns, club mosses, mosses, and lichens. Squirrels, deer, and many species of birds, such as the endangered spotted owl, inhabit this biome. The moist coniferous forest has been heavily exploited but tracts of old-growth forests (forests that are over 100 years old and have never been logged) still exist. The future of old-growth forests is at the center of a controversy between the timber industry and environmentalists.

Much of the eastern half of the United States was once covered by an extensive **deciduous forest** of oak, maple, beech, hickory, and other hardwoods (fig. 25.14f). (**Deciduous** means *to shed* and refers to the fact that these trees shed their leaves annually in the autumn, in contrast to conifers, which retain their leaves throughout the year.) The winters can be long and hard but the summers are hot. Precipitation is evenly distributed throughout the year and relatively high, from 75–152 cm (30–60 in) annually. The topsoil is characteristically rich in organic material. Animal life is abundant; reptiles, amphibians, insects, birds, and many familiar small and large mammals (white-tailed deer, wolves, bears, red foxes, cottontail rabbits, raccoons, and squirrels) abound. Most of the original deciduous forest is gone; less than 0.1% of the old growth forest remains. The forest has been cleared since colonial days for urban expansion, agriculture, and timber; however, as the agricultural center moved westward during the 1800s, from the eastern states to the central United States, much of the abandoned farmland has reverted back to forest.

The **tropical rain forest** has the greatest species diversity of any other biome. In one tenth of a hectare (1/4 of an acre) of tropical rain forest in South America, there are 208 tree species. Compare that to the approximately 25 tree species found in a similar area of the deciduous forest in the state of New Hampshire. The dominant vegetation is

A Closer Look 25.1

Buying Time for the Rain Forest

In 1985, a plan arose to protect the tropical rain forest from timber concerns yet maintain an economic base for the indigenous peoples who live in and depend on the forest for a livelihood. Nontimber forest products such as latex, resins, and nuts would be harvested and sold on the world market. These products are part of a renewable, sustainable harvest in that they could be extracted for the long term with minimal damage to the intact rain forest. Over thirty different products are sold, primarily from the tropical rain forest in South America.

Natural rubber and chicle are two latex products from tropical forests. Natural rubber, obtained from several tropical plant sources, was known to the indigenous peoples of the New World. Early explorers such as Columbus and Cortéz reported about the "bouncy" balls used in sport. The name rubber was later coined for its ability to rub out pencil marks. It remained a curiosity until 1823, the year Charles MacIntosh discovered that rubber-treated fabric became waterproof. In 1839, Charles Goodyear learned that treating rubber with sulfur and heat, a process called vulcanization, stabilized rubber's elastic properties. Vulcanization opened up numerous applications for rubber and accelerated the search for an abundant, quality source in tropical Africa and South America. *Hevea brasiliensis,* a tree native to the Amazon basin, yielded the purest and most elastic rubber of all and soon became the mainstay of the Brazilian rubber industry. Wild rubber trees are widely scattered in the tropical rain forest, usually only two to three trees per hectare. A single person collects rubber from 60–150 trees and much time and effort are spent in clearing trails from tree to tree. Rubber is found in latex vessels in the inner bark, close to the vascular cambium. Rubber tapping consists of wounding the bark with shallow, spiral incisions to release the latex (box fig. 25.1a). The incisions are made in the morning when latex flow is greatest. Small cups placed below the incisions catch the latex; it is collected in the afternoon. Tapping is usually done during the dry season, since the cups tend to fill with water in the rainy season and the trails to the trees are often impassable. The collected latex is dripped on a type of paddle or stick, which turns over a fire. Gradually, a large ball of solid rubber forms; the balls are then shipped to factories for further processing.

Box Figure 25.1a Seringuero *leader, Chico Mendes is shown tapping a rubber tree for latex; he was later assassinated by those who opposed the extractivism of the tropical rain forest.*

The *seringueiros,* Brazilian rubber tappers, have led the environmental movement to save the tropical rain forest and protect the indigenous people and their way of life. Their cause has been received by violent political opposition; one of the most visible leaders of the rubber tappers, Francisco (Chico) Alves Mendes Fihlo, was brutally murdered on December 22, 1988 by local ranchers who opposed the ban on clear-cutting the forest.

Chicle is used in the manufacture of chewing gums. Comparable to the extraction of rubber latex, workers called *chicleros* scar the bark of the chico tree (*Manilkara zapota,* Family Sapotaceae). The raw latex is solidified into 10 kg blocks or marquetas and stamped with the workers' initials. The latex is high quality but the chicle market has declined over the years since the availability of cheap synthetic gums. However, with the current interest in natural products it is beginning to rebound.

One of the most important economic plants of the South American rain forest is the Brazil nut (*Bertholletia excelsa*) in the Lecythidaceae family. The trees are found in forests with a prolonged dry season in the Guianas, Colombia, Venezuela, Peru, Bolivia, and Brazil. In fact, flowering begins shortly before the onset of the dry season. After fertilization of the flowers, the fruits take 15 months to develop, maturing during the rainy season. The large woody fruits fall to the ground with the seeds intact. The fruits weigh from 0.5–2.5 kg (1–5.5 lbs) and contain 10–25 seeds. A tree produces from 63 to 216 fruits. Various rodents gnaw through the woody bark to remove the seeds that also have an incredibly tough seed coat. Most seeds, shelled or unshelled, are sent to Europe and the United States (box fig. 25.1b). Over the long term, Brazil nut production appears to be more profitable than harvesting timber or clearing the forest for pasture.

Another potentially profitable product from the rain forest is tagua or vegetable ivory. Tagua is the very hard, cream-colored endosperm from the seeds of the *Phytelephas* genus of palms, with *P. macrocarpa* the most commercially important species. Tagua palms grow wild in the rain forests of northwestern South America. Tagua was once a very popular material in the manufacture of buttons and jewelry for European and North American markets, until it was replaced by the availability of cheap plastics in the 1940s and 1950s. Presently, the Tagua Initiative has sought to reestablish the tagua industry in countries like Ecuador and Colombia. Tagua buttons have already been introduced and sold in the United States on clothing by Esprit, L.L. Bean, etc. and tagua jewelry and carvings (chess pieces, animal figurines) are soon to follow (box fig. 25.1c).

Box Figure 25.1b *Brazil nuts are just one of the ingredients of* Rainforest Crunch, *a confection made from the sustainable harvest of the tropical rain forest.*

Box Figure 25.1c *Buttons and jewelry made from vegetable ivory,* Phytelephas *palm seeds.*

broadleaf, evergreen trees, which are divided into stories according to their height (fig. 25.14g). The upper story consists of the tallest trees, rising to heights of 80 m (260 ft); beneath lies the middle layer, where most trees are found, with an average height of 50 m (160 ft). There is little understory because of the dense shade. However, many epiphytes and lianas (vines) climb the trees to sunlight. Most animal life is arboreal: birds, insects, amphibians, monkeys, and reptiles. The tropical rain forest is so named because there is little seasonal variation in temperature and because of the high daily rainfall, 200–450 cm (80–180 in) per year. The soils of the tropical rain forest are extremely poor, since decay of organic material is rapid and nutrients are quickly taken up by existing vegetation. The tropical rain forest is currently being deforested at an alarming rate; it is estimated that half of the world's tropical rain forests have already been destroyed. Each year an area of forest equivalent to the state of Indiana is cleared; if this rate continues, the tropical rain forest will be gone in approximately 30 years. Several factors are contributing to this destruction.

Population pressures in many developing nations are causing governments to settle people in former wildernesses. With the introduction of people and roads, the forests are cleared in a slash-and-burn (swidden) type of agriculture. Because the forest soil is so nutrient-poor, crops can be grown in an area for only a few seasons before the land is abandoned and a new area has to be burned.

Land is also cleared to make pastures for cattle. Often, the cattle are grown for shipments to overseas markets in developed nations.

Harvesting the rain forest for highly prized tropical hardwoods such as mahogany or teak is often done with poor forest management practices. High-grading, when only commercially valuable trees are harvested, is all too often the common practice of timber companies. A road of destruction and waste leads from one desirable tree to the next. Also, in much of the world, the primary fuel for cooking and heat is firewood, another cause of deforestation. One strategy to save the tropical rain forest is the policy of extractive reserves, presented in A Closer Look: Buying Time for the Rain Forest.

Chapter Summary

1. The ecosystem is the functional unit of study in the environment. The environment is an organism's surroundings; it includes the study of biotic factors such as the interrelationships between organisms, as well as investigations into abiotic factors such as energy flow and the cycling of mineral nutrients.
2. Ecological niche is the functional role an organism plays in the ecosystem. Plants as producers are fundamental to most food chains, the sequential ordering of trophic relationships.
3. Biological magnification is the transfer and concentration of a toxic substance as it is passed along a food chain.
4. Human impact on the carbon cycle has resulted in the greenhouse effect in which the buildup of CO_2 acts as a heat trap warming the earth's lower atmosphere. Global warming is likely to cause a shift in agricultural regions and the range of many species of plants and animals.
5. Ecological succession is the gradual progression of orderly and predictable changes that may occur in ecosystems over time. The endpoint of ecological succession is the climax community.
6. Biomes are large, terrestrial regions recognizable by their distinctive vegetation and animal life. Climatic conditions, such as temperature and precipitation, largely determine the type of biome that develops in a region. Major biomes of the world include the desert, chaparral, grasslands, the savanna, the tundra, the boreal forest, the moist coniferous forest, the deciduous forest, and the tropical rain forest.
7. The strategy of extractive reserves is an attempt to preserve the tropical rain forest from destructive exploitation by developing economic markets for a renewable harvest of natural products such as natural rubber, chicle, Brazil nuts, and vegetable ivory.

Review Questions

1. What are the limits of the biosphere? What factors would limit the distribution of plants? What factors would limit the distribution of animals?
2. What are the components of an ecosystem? Detail the flow of energy in a food chain.
3. How do the concepts of habitat and niche overlap? How do they differ? Give examples.
4. It is clear that the greenhouse effect caused by rising levels of CO_2 will have many dire consequences in agriculture. Would there be any benefits to plants from higher CO_2 levels in the atmosphere?

5. List examples of primary and secondary succession and trace the transitional stages in each type of succession.
6. List and describe the major biomes found on the North American continent. Climbing up a mountain, the vegetational zones encountered are similar to the biomes passed on a journey north. Why? What is the altitudinal effect on temperature and precipitation?
7. Make a list of rain forest products obtained by extractive reserves available in local stores. Can these strategies be used to preserve natural areas other than the tropical rain forest?

Further Reading

Bennett, Bradley C. 1992. Plants and People of the Amazonian Rain Forests. *BioScience,* 42 (8): 599–607.

Carr, Thomas A., Heather L. Pedersen, and Sunder Ramaswamy. September, 1993. Rain Forest Entrepreneurs. *Environmen,t.* 12-15, 33-37.

Charas, Daniel D. *Environmental Science,* 4th Edition, The Benjamin Cummings Publishing Company, Redwood City, CA.

Dean, Warren. 1987. *Brazil and the Struggle for Rubber: A Study in Environmental History.* Cambridge University Press: Cambridge, MA.

Fearnside, Philip M. 1989. Extractive Reserves in Brazilian Amazonia. *BioScience,* 39 (6): 387–393.

Parfit, Michael. 1989. Whose Hands Will Shape the Future of the Amazon's Green Mansions? *Smithsonian,* 20 (8): 58–75.

Plotkin, Mark. 1990. The Healing Forest: The Search for New Jungle Medicines. *The Futurist,* 24 (1): 9–14.

Plotkin, Mark and Lisa Famolore, eds. 1992. *Sustainable Harvest* and *Marketing of Rain Forest Products.* Island Press: Washington, DC.

Raven, Peter H., Linda K. Berg, and George B. Johnson. 1993. *Environment.* Saunders College Publishing. Fort Worth, TX.

Rosenzweig, Cynthia and Daniel Hillel. 1993. Agriculture in a Greenhouse World. *National Geographic Research & Exploration,* 9 (2): 208–221.

Appendix A

Metric System: Scientific Measurement

Units of Length

Unit	Abbreviation	Equivalent
kilometer	km	1000 m
meter	m	1 m
centimeter	cm	0.01 m
millimeter	mm	0.001 m
micrometer	μm	0.000001 m
nanometer	nm	0.000000001 m
angstrom	Å	0.0000000001 m

Length Conversions

1 km = 0.6 mi	1 in = 2.54 cm
1 m = 39 in	1 ft = 30 cm
1 cm = 0.39 in	1 yd = 0.9 m
1 mm = 0.039 in	1 mi = 1.6 km

To Convert:	Multiply By:	To Obtain:
inches	2.54	centimeters
feet	30	centimeters
miles	1.6	kilometers

Units of Volume

Unit	Abbreviation	Equivalent
liter	L	
milliliter	ml	0.001 L

Volume Conversion

1 tsp = 5 ml	1 fl oz = 30 ml
1 tbsp = 15 ml	1 L = 1.06 qt
1 cup = 0.24 L	1 L = 0.26 gal
1 pt = 0.47 L	
1 qt = 0.95 L	
1 gal = 3.79 L	

To Convert:	Multiply By:	To Obtain:
fluid ounces	30	milliliters
quart	0.95	litcrs
milliliters	0.03	fluid ounces
liters	1.06	quarts

Units of Mass

Unit	Abbreviation	Equivalent
kilogram	kg	1000 g
gram	g	
milligram	mg	0.001 g
microgram	μg	0.000001 g

Mass Conversion

1 oz = 28.3 g	1 g = 0.035 oz
1 lb = 453.6 g	1 kg = 2.2 lb
1 lb = 0.45 kg	

To Convert:	Multiply By:	To Obtain:
ounces	28.3	grams
pounds	453.6	grams
pounds	0.45	kilograms
grams	0.035	ounces
kilograms	2.2	pounds

Energy Conversion

Calorie (cal) = energy required to raise the temperature of 1 g of water by 1° C

1 Calorie = 1000 cal = 1 Kilocalorie

1 Kilocalorie (Kcal) = 1000 cal

Temperatures Scales	Temperature Conversions
Celsius (Centigrade) ° C	$°C = \frac{(°F - 32) \times 5}{9}$
Fahrenheit = ° F	$°F = \frac{°C \times 9 + 32}{5}$

Temperature Conversion

	° F	° C
Boiling point of water	212	100
Human body temperature	98.6	37
Freezing point of water	32	0

Appendix B

Classification of Plants and Those Organisms Traditionally Classified as Plants Discussed in Plants and Society

Kingdom Monera—prokaryotes
- Cyanobacteria—also called the blue green algae

Kingdom Protista—unicellular or simple multicellular eukaryotes
- Division Pyrrophyta—dinoflagellates
- Division Chrysophyta—diatoms
- Division Euglenopyta—euglenoids
- Division Chlorophyta—green algae
- Division Rhodophyta—red algae
- Division Phaeophyta—brown algae
- Division Myxomycota—slime molds
- Division Oomycota—oomycetes

Kingdom Fungi—eukaryotic, mainly multicellular, absorptive heterotrophs
- Division Chytridiomycota—chytridiomycetes
- Division Zygomycota—zygomycetes
- Division Ascomycota—ascomycetes, sac fungi
- Division Basidiomycota—basidiomycetes, club fungi

Kingdom Plantae—eukaryotic, multicellular photosynthetic autotrophs
- Division Bryophyta—mosses and liverworts
- Division Psilophyta—whisk ferns
- Division Lycophyta—club mosses
- Division Sphenophyta—horsetails
- Division Pterophyta—ferns
- Division Cycadophyta—cycads
- Division Ginkgophyta—gingko
- Division Coniferophyta—conifers
- Division Gnetophyta—gnetophytes
- Division Anthophyta—flowering plants

Glossary

A

abiotic factor Nonliving component in the environment.
absorption spectrum Graph of absorbance values for different wavelengths of light.
absorptive heterotroph Mode of nutrition in which organisms secrete digestive enzymes into a substrate and then absorb the products of digestion; nutritional mode of fungi.
accessory fruit Most or part of the fruit is derived from tissue other than the ovary of a flower.
achene Simple, dry, indehiscent fruit with seed attached to pericarp at only a single point.
actinomorphic Regular or radially symmetric flowers.
active transport Movement of materials across a cell membrane and against a concentration gradient that requires the expenditure of cellular energy.
adaptation Trait that enhances the survival of an organism in its environment.
addictive drug Substance that induces an addiction by causing physiological dependence, psychological dependence, and/or tolerance.
adenine One of the purine bases found in nucleotides and nucleic acids.
adhesion Attraction between unlike molecules. The polar nature of water molecules causes them to adhere to a surface.
ADP (adenosine diphosphate) A nucleotide diphosphate that is often phosphorylated to form ATP.
adventitious Organ that forms in an unusual place; example: roots from leaves.
aeciospore Binucleate spore produced in an aecium.
aecium (pl. **aecia**) Cup-shaped structure in rust fungi composed of binucleate hyphae; aeciospores are formed here.
aflatoxin A complex of four mycotoxins produced by *Aspergillus flavus;* often found in peanut products.
agar Gelatinous product extracted from the walls of some red algae; used in growth media for microorganisms and in tissue culture.
aggregate fruit A fruit derived from a single flower with several separate ovaries; example: blackberry.
aleurone Protein-rich, outermost layer of endosperm in a cereal grain.
alginic acid Gelatinous material extracted from the wall of certain brown algae; used in a variety of commercial and industrial products.
alkaloid Nitrogenous substance, alkaline in solution, that produces physiological and/or psychological effects on the nervous system of animals.
allele Alternate expressions of a gene.
allelopathy The release of chemicals by certain plants that inhibit the growth of competing plants.
allergen Foreign substance that induces an allergic response.
alternate arrangement One leaf borne per node.
amatoxin A deadly protoplasmic toxin produced by several genera of fungi, including *Amanita, Conocybe, Galerina,* and *Lepiota.*
amino acid Nitrogen-containing building blocks of proteins.
amylopectin Component of starch consisting of highly branched chains of repeating glucose units; insoluble in water.
amyloplast Starch-storing plastid.
amylose Component of starch consisting of unbranched chains of repeating glucose units; soluble in water.
anabolic Chemical reactions that synthesize and require energy.
analgesic Pain-relieving remedy.
anamorphic The asexual or imperfect stage in a fungus life cycle.
anaphase Stage in mitosis in which sister chromatids separate and move to opposite poles of the cell.
androecium (pl. **androecia**) Collective term for all the stamens in a flower.
angiosperm A flowering plant whose seeds are contained within fruits.
annual ring Ring of xylem in woody stem composed of springwood and summerwood that corresponds in temperate regions to a chronological year.
annulus Remnant of the partial veil around the stipe of some mushrooms.
anther Pollen-producing part of a stamen.
antheridium (pl. **antheridia**) Male gametangium in which the male gametes, sperm, are produced.
antibiotic Substance that inhibits the growth of microorganisms, such as bacteria or fungi.
antibody Immunological protein produced by B lymphocytes in response to a specific antigen and capable of acting against said antigen.
anticodon The sequence of three nucleotides in a transfer RNA molecule that is complementary to a specific codon.
antigen Substance, usually a protein, that is foreign to the body and elicits an immunological reaction by antibodies.
antipodal One of several cells (usually three) of the embryo sac (female gametophyte) of angiosperms. They are located opposite the egg and synergids.
apical meristem Meristematic tissue located at the tips of roots and stems; responsible for primary growth of a plant.
apoplast Pathway of water entering a root that flows along intercellular spaces and cell walls.
aquifer Underground water reservoir.
archegonium (pl. **archegonia**) Female gametangium in which the female gamete or egg is produced and housed.
aril Thick, fleshy seed covering around some seeds.
artificial selection Selective breeding as practiced by humans on domesticated plants and animals.
ascocarp Fruiting body of an ascomycete.
ascogonium (pl. **ascogonia**) A female reproductive structure in some ascomycetes.
ascospore Sexual spore of an ascomycete fungus that is produced by meiosis, followed by mitosis.
ascus Saclike reproductive structure of ascomycetes in which meiosis, followed by mitosis, produces eight haploid cells called ascospores.
asexual reproduction Any type of reproduction not involving the union of gametes.
atom The smallest individual unit of an element that still retains those properties of an element.
atomic mass The total number of protons and neutrons in the atomic nucleus.
atomic nucleus Most of the mass of an atom; composed of protons and neutrons.
atomic number Number of protons in the atomic nucleus; there is a characteristic number for each type of element.
atomic weight unit The mass of one proton or neutron.
ATP (adenosine triphosphate) The energy compound of the cell.
ATP synthase A membrane-bound enzyme in mitochondria and chloroplasts that phosphorylates ADP to form ATP using energy from the passage of protons through the enzyme.
autotroph Self-feeder; an organism that can synthesize organic compounds from simple inorganic ones.
awn A slender bristle.
axil Upper angle between a leaf and stem.

B

basidiocarp Fruiting body of a basidiomycete.

basidiospore Sexual spores of basidiomycetes.

basidium (pl. **basidia**) Club-shaped reproductive cell in basidiomycetes undergoes meiosis to produce four haploid cells called basidiospores.

bast fiber Fibers located in phloem.

berry Simple, one-to-many seeded, fleshy fruit with thin exocarp; example: tomato.

beta-carotene Yellow to red pigment in plants; one of the most important of the carotenoids; converts to vitamin A in the body.

biennial Plant that completes its life cycle within two growing seasons; example: carrot.

binomial Two-word scientific name.

biological magnification Increase and concentration of toxins as they are passed along a food chain.

biomass Total dry weight of living organisms.

biome Large terrestrial communities recognized by characteristic vegetation.

biosphere The entire world of living organisms.

biotechnology The use of living organisms to provide products for humanity; using genetic engineering to create organisms with useful traits.

biotic factor Living component in the environment.

blade Flat, green part of a leaf; expanded or flattened portion of a brown alga.

B-lymphocytes Cells that manufacture antibodies involved in immunity.

boreal forest Northern coniferous forest biome, also called the taiga, located south of the tundra and dominated by conifers.

bract A floral leaf.

bran The husk of a cereal grain; including the pericarp and seed coat of a cereal grain. The bran is removed in the processing of refined grains.

budding Type of asexual reproduction in which a small blip of the parental body develops into a new individual.

bulb Vertical, underground stem with food-storing leaves covered by papery leaves.

bundle sheath Sheath of parenchyma or sclerenchyma cells surrounding the vascular bundles in leaves.

C

callus Mass of undifferentiated cells in tissue culture.

calorie The amount of heat needed to raise the temperature of 1 g of water 1° Celsius; one thousand calories = 1 kilocalorie or 1 Calorie.

Calvin Cycle Biochemical pathway in photosynthesis in which carbon from atmospheric CO_2 is fixed and reduced into carbohydrate.

calyx Collective term for sepals of a flower.

capsule Simple, dry, dehiscent fruit that opens along three or more seams or pores; example: cotton.

carbohydrate Organic compound containing carbon, hydrogen, and oxygen with the general formula of $C_n H_{2n} O_n$.

cardioactive glycoside Sugar-containing molecules with an active compound that affects the heartbeat.

carnivore Flesh-eating organism; animals that eat other animals.

carotenes Accessory photosynthetic plant pigments; protect chlorophyll from breakdown by light or oxygen.

carotenoids Class of plant pigments that includes carotenes and xanthophylls; most are yellow, orange, or red.

carpel The ovule-bearing part of a flower.

carrageenan Gelatinous material extracted from the walls of some red algae and used in a variety of commercial and industrial products.

carrying capacity Maximum number of a population that can be supported by the environment over a given period of time.

caryopsis Simple, dry, indehiscent fruit with a single seed that is fused to the ovary wall. Also called a grain. Example: wheat.

Casparian strip Water-impermeable strip of suberin found in the transverse and radial walls of endodermal cells.

catabolic Chemical reactions that release energy by degrading complex compounds into simpler ones.

catalyst An agent that speeds up a chemical reaction but is not used up in the reaction.

catkin Inflorescence of unisexual flowers.

cell culture (tissue culture) The culture of single cells or tissues to form callus and then to develop whole plants asexually.

cell plate Double membrane across the equator of a dividing cell that develops from the phragmoplast; it marks where the new cell walls will form.

cell wall Rigid outer layer in the cells of plants, fungi, bacteria, and certain other organisms.

cellular respiration Cellular pathway for the production of ATP.

cellulose Complex carbohydrate that occurs in the cell walls of plants.

central vacuole A membrane-enclosed sac that takes up most of the volume of a mature plant cell.

centromere Constricted portion of chromosomes to which spindle fibers attach; region joining sister chromatids.

cereal Edible grains of cultivated grasses; examples: wheat, rice, corn.

chaff Bracts surrounding a cereal grain; removed during threshing.

chalazal Ovule end opposite the micropyle.

chaparral Biome characterized by dense thickets of evergreen shrubs with mild rainy, winters and dry, hot summers.

chemiosmosis The coupling of ATP synthesis to electron transport by way of a proton gradient and ATP synthase.

chiasma (pl. **chiasmata**) The X-shaped configuration between the chromatids of homologous chromosomes that have exchanged genetic material during prophase I of meiosis.

chitin Complex polymer found in the cell walls of many fungi.

chlamydospore Thick-walled dormant spore in fungi.

chlorophyll Green pigment found in the chloroplasts of plants, algae, and cyanobacteria that is essential to photosynthesis.

chloroplast Membrane-bound organelle in plants and algae that contains chlorophyll and in which photosynthesis takes place.

cholesterol Steroid that is an integral part of cell membranes and a precursor to other steroidal compounds in animals.

chromatids The identical duplicated halves of a single chromosome seen during cell division.

chromatin Dispersed form of genetic material, DNA and protein, in the nucleus of a nondividing cell.

chromoplast Membrane-bound organelle containing pigments other than chlorophyll; the pigments may be yellow, orange, or red.

chromosome Condensed form of DNA and proteins that appear in a dividing cell.

Citric Acid Cycle Another name for Krebs Cycle.

class Taxonomic rank consisting of related orders.

climax community Stable, self-sustaining community that is the culmination of ecological succession.

clone The production of a genetically identical individual through asexual reproduction.

codominance A condition in which both alleles of a heterozygous pair are expressed independently.

codon Genetic code comprised of three nucleotide sequences.

coenzyme Organic molecule that is necessary for the proper functioning of an enzyme.

cofactor Inorganic molecule or ion that is necessary for the proper functioning of an enzyme.

cohesion Tendency of like molecules to stick together, usually due to hydrogen bonds.

coleoptile Sheath surrounding the embryonic shoot of monocotyledons.

coleorhiza Sheath surrounding the radicle (embryonic root) of monocotyledons.

collenchyma Ground tissue in plants with unevenly thickened primary cell walls; functions in support.

community All the populations in a given locale.

companion cell Phloem cell associated with a sieve tube member.

competitive exclusion No two species can occupy the same niche indefinitely.
complete flower A flower with all four floral whorls (sepals, petals, stamens, and carpels).
complete protein A protein that has all of the essential amino acids and in the correct proportions.
compound A substance formed by two or more elements in a definite ratio.
compound leaf A leaf with a blade divided into separate leaflets.
compound pistil A pistil composed of more than one carpel.
conidium (pl. **conidia**) Asexual reproductive spores of some ascomycetes and imperfect fungi.
consumer Organisms that cannot synthesize their food but must feed on other organisms.
coprolite Fossilized fecal material.
cork Suberized cells on the outer surface of woody stems and roots; produced by the cork cambium.
cork cambium Meristematic tissue that produces cork cells on its outer surface and phelloderm on its inner surface.
corm Underground, enlarged, food-storing stem covered by papery leaves.
corolla Collective term for the petals of a flower.
cortex Parenchymous tissue, the region between the epidermis and vascular tissue in a herbaceous root or stem.
cotyledon Seed leaf in the seed and seedling.
covalent bond Chemical bond formed when atoms share a pair of electrons.
crack Form of purified cocaine with widespread street use.
Crassulacean Acid Metabolism (CAM Pathway) A variation of the C_4 pathway that functions in a number of cacti and succulents; allows for the fixation of carbon dioxide during the night, then in the daytime the carbon dioxide is transferred to the Calvin Cycle.
crista (pl. **cristae**) Infoldings of the inner membrane of a mitochondrion.
crossing-over The exchange of genetic material between chromatids of homologous chromosomes during prophase I of meiosis.
cross-pollination Transfer of pollen from the stamen to the stigma of a flower of another plant.
culm Hollow, jointed stem of grasses.
cultivar Abbreviation for cultivated variety of plant.
cuticle Waterproof layer of cutin on leaves and nonwoody stems.
cutin Waxy material secreted by epidermal cells.
cyanogenic glycoside Glycoside that releases cyanide.
cytochromes Iron-containing enzymes that carry electrons in both photosynthesis and respiration.
cytokinesis The division of the cytoplasm to form daughter cells, usually accompanies mitosis and meiosis.
cytoplasm The entire contents of the cell exclusive of the nucleus.
cytosine A pyrimidine base in DNA and RNA.
cytoskeleton Cellular scaffolding of microtubules and microfilaments within the cytoplasm of the cell.
cytosol The more fluid portion of the cytoplasm.

D

dalton Atomic Weight Unit; the mass of one proton or one neutron.
dark reactions Biochemical reactions that make up the Calvin Cycle of photosynthesis.
deciduous Trees and shrubs that shed their leaves during the autumn.
deciduous forest Forest biome in which the dominant trees are deciduous.
decomposer Organism that obtains its nutrition by breaking down dead plants and animals or their waste products.
decongestant An agent that relieves nasal or respiratory congestion.
dehiscent fruit Fruit that splits open at maturity, facilitating seed dispersal.
deletion A mutation resulting from the loss of a small segment of DNA.
dendrochronology Study of the annual rings in trees to determine the timing of natural events in the past.
dendroclimatology Study of the annual rings of trees in order to interpret climatic changes in the past.
deoxyribonucleic acid (DNA) The genetic material of life.
deoxyribose Five-carbon sugar in DNA.
depressant Psychoactive drug that has a sedative effect on the central nervous system; actions include dulling mental awareness and inducing sleep.
dermal tissue Tissue that covers surfaces in plants.
desert Biome in which the annual precipitation is less than 25 cm (10 in).
detritus Wastes and remains of dead plants and animals.
dicotyledon A class of angiosperms in which the seedlings typically possess two cotyledons; commonly abbreviated to dicot.
differentially permeable membrane A membrane that allows the free passage of some materials but inhibits the passage of others.
diffusion The spontaneous movement of particles (ions, molecules, or atoms) from a region of higher concentration to one of lower concentration.
dihybrid cross A genetic cross between parents that differ for two traits.
dikaryon Mycelium of some fungi that have two separate haploid nuclei in each cell.
dioecious Plant species that has separate male and female plants; pollen-bearing and ovule-bearing flowers or cones are borne on different plants.
diploid Two complete sets of chromosomes in a cell; $2N$.
disaccharide Sugar consisting of two monosaccharides; example: sucrose is composed of glucose and fructose.
distillation Boiling of a liquid to evaporation and subsequent condensation of the vapors for the purposes of purification and concentration.
distilled spirit Alcoholic beverage with an alcoholic content between 80–100 proof, obtained by distillation of a beer or wine.
division Taxonomic rank that includes related classes; synonymous to phylum used in animal systematics.
domesticated plant A plant that has been genetically changed from the wild type due to artificial selection.
dominant The allele of a gene that masks or suppresses the expression of an alternate allele.
double fertilization The fusion of egg and sperm resulting in a zygote, and the simultaneous fusion of sperm with two polar nuclei resulting in the formation of endosperm that characterizes all angiosperms.
double helix The form of the DNA molecule; refers to the complementary nucleotide strands of the DNA molecule twisted into a helix.
drupe Simple, fleshy fruit with the seed enclosed in a hard endocarp (pit); example: cherry.
dry fruit Fruit in which the cells of the pericarp are dry (dead) at maturity.

E

ecological equivalent Species that fulfill similar niches in different geographical regions.
ecological pyramid of biomass Total biomass of all organisms at each trophic level in a food chain; typically biomass declines with successively higher trophic levels.
ecological pyramid of energy Total energy content of all organisms at each trophic level in a food chain; since only 5%–20% of energy is passed between trophic levels; energy content declines at successively higher trophic levels.
ecological pyramid of numbers Number of organisms supported at each trophic level in a food chain; typically, fewer organisms are supported at successively higher trophic levels.
ecological succession Orderly process of natural change in a community composition over time culminating in a self-perpetuating complex community; primary and secondary types are recognized.
ecosystem Community of living organisms interacting with the abiotic factors in the environment.

ectomycorrhizae A mycorrhizal relationship in which the fungus forms a sheath or mantle around the root, penetrating between the cells of the root epidermis and cortex.

egg Nonmotile female gamete.

egg apparatus Egg cell and adjacent synergids in the embryo sac (female gametophyte) of angiosperms.

electron Negatively charged particle of an atom.

Electron Transport System The final stage of respiration; a series of enzymes and coenzymes on the inner membrane of mitochondria functioning in the transfer of electrons and the resulting synthesis of ATP.

element Building blocks of matter that cannot be broken down into a simpler substance. Each element is composed of just one type of atom.

embryo Immature sporophyte that develops from a zygote.

embryo sac Female gametophyte of angiosperms; it is retained within the ovule.

endergonic reaction Chemical reaction that requires energy.

endocarp Innermost layer of the pericarp (fruit wall).

endodermis Innermost layer of the root cortex surrounding the stele; many of the endodermal cells have Casparian strips.

endoplasmic reticulum (ER) Membranous network of channels throughout the cytoplasm of a cell; some regions are studded with ribosomes (rough), while others are ribosome-free (smooth).

endosperm Nutrient tissue that forms by the fusion of a sperm nucleus with two polar nuclei during double fertilization in angiosperms.

energy The ability to perform work.

enzyme Proteins that act as catalysts to chemical reactions.

epicotyl Portion of the shoot of an angiosperm embryo or seedling above the cotyledons.

epidermis Outermost tissue in all young and nonwoody plant organs.

epigynous Floral parts (sepals, petals, and stamens) appear to arise from the top of an ovary; the ovary is said to be inferior.

epiphyte Plant that grows on top of another plant for support and position.

essential amino acid Amino acid that cannot be synthesized by an organism but must be obtained ready-made in the diet for proper health.

essential nutrient Nutrient that cannot be synthesized by an organism but must be obtained ready-made in the diet for proper health.

essential oil Volatile component that contributes to the scent and flavoring of aromatic plants.

ethylene Gaseous plant hormone involved in fruit ripening and other aspects of plant growth and development.

eukaryotic Cells with a nucleus and distinct membrane-bound organelles.

evolution Inherited changes in populations shaped by natural selection over time.

exergonic reaction Chemical reactions that release energy.

exine The outer layer of the pollen wall.

exocarp Outermost layer in the pericarp (fruit wall).

exon An expressed segment of a gene; exons are separated from each other by introns.

F

F_1 generation First filial generation; offspring from a genetic cross.

F_2 generation Second filial generation of a genetic cross.

FAD (flavin adenine dinucleotide) An electron receptor in cellular respiration.

family Taxonomic rank consisting of a group of related genera.

fat Triglyceride that is solid at room temperature; usually of animal origin.

fat-soluble vitamin Vitamins that can be stored in fatty tissues of the body; vitamins A, D, E, and K.

fatty acid Long chains of carbon and hydrogen; component of phospholipids and triglycerides.

fermentation Anaerobic cellular respiration; organic compounds are broken down to release energy without the use of oxygen as an electron acceptor.

Fertile Crescent Area between the Tigris and Euphrates Rivers in the Near East; some of the earliest documented sites of agriculture.

fiber Long and narrow sclerenchymous cell; functions in support; an important dietary component that provides bulk; component of fabrics, ropes, and paper.

fibrous roots Root system with several main roots; common to many monocots.

filament Part of the stamen in a flower that supports the anther.

flagellum (pl. **flagella**) Whiplike cellular structure of motility; eukaryotic flagella are composed of microtubules.

fleshy fruit Fruit in which the cells of the pericarp are alive at maturity.

floret One of the small flowers that make up the inflorescences in the composite and grass families.

fluid mosaic model Model of cell membrane structure composed of a lipid bilayer with scattered proteins; often described as a sea of lipids with protein icebergs.

follicle Single, dry, dehiscent fruit that splits along one seam; example: milkweed.

food chain Progression of organisms that feed on or decompose the preceding one.

food web Interrelationships between several food chains in an ecosystem.

forage crops Crops that are grown as food for domesticated herbivores.

forager Hunter-gatherer group.

frame-shift mutation A mutation caused by the insertion or deletion of nucleotides (less than three or a number not a multiple of three) resulting in the improper grouping into codons.

free-threshing grain Grain that separates easily from enclosing bracts.

fructose Six-carbon monosaccharide; often referred to as fruit sugar.

fruit A ripened ovary of an angiosperm flower.

fruiting body Reproductive structures in fungi.

fungicidal A compound that is able to kill a fungus and stop an infection.

fungistatic A compound that slows or stops the growth of a fungus.

G

G_1 Part of interphase known as Gap 1; time of active metabolism in the cell cycle.

G_2 Part of interphase after the synthesis of DNA and before the start of nuclear division and known as Gap 2.

gametangium (pl. **gametangia**) Structure in which gametes are produced.

gamete Sex cell.

gametophyte Haploid generation of plants that produce gametes.

gene Unit of hereditary information on chromosomes.

generalized niche Niche that has broad requirements; tolerates a range of conditions.

generative nucleus Nucleus of male gametophyte (pollen grain) that divides, producing two sperm.

genetic engineering The transfer of specific genes between organisms using techniques of molecular biology.

genetic erosion Irreversible loss of genetic diversity due to extinction of traditional varieties and wild ancestors of crop plants.

genetically engineered microorganisms (GEMs) Bacteria that have been genetically engineered by the insertion or deletion of DNA segments.

genotype Genetic makeup of an organism.

genus (pl. **genera**) Taxonomic rank consisting of a group of related species.

germ Embryo of a cereal grain.

germplasm Entire genetic makeup of an organism.

glucose Six-carbon monosaccharide; one of the most abundant simple sugars. It is the building block of both cellulose and starch and is important to several metabolic pathways.

glumes Pair of bracts at the base of a spikelet in a grass flower.

gluten Protein complex in endosperm of wheat and some other cereals that is essential in making a leavened bread.

glycogen Polysaccharide of glucose. Principal carbohydrate stored in animal and fungal cells.

Glycolysis Pathway in cellular respiration in which glucose is split into pyruvate.

glycoprotein A protein with attached sugars.

glycoside Physiologically active compound in plants that always contains a sugar group, although the active part of the molecule may differ.

glyoxysome A type of microbody involved in the enzymatic conversion of stored fats to sugars in some seeds.

Golgi body (apparatus) Organelle of membranous, hollow sacs arranged in a stack; functions in modification, storage, and packaging of secretion materials; may be called dictyosome in plants.

grafting The union of a part of one plant, the scion, to the root or stock of another plant.

grain Single, dry indehiscent fruit of a single seed that is fused to the ovary wall.

granum (pl. **grana**) Stacked thylakoid membrane within a chloroplast.

grassland Biome in which the annual moderate precipitation is enough to support the growth of grasses but insufficient to support a forest.

Green Revolution Introduction of scientifically developed food crops that can produce high yields under conditions of high inputs of water, fertilizers, and pesticides.

greenhouse effect The warming of the Earth due to the atmospheric accumulation of carbon dioxide and other gases, which trap heat and reradiate it back to the Earth's surface.

ground tissue Includes primary tissues of parenchyma, sclerenchyma, and collenchyma that make up much of the bulk of the primary plant body; function in support, photosynthesis, and storage; also known as fundamental tissue.

guanine Purine base present in both DNA and RNA.

guard cell One of a pair of specialized cells in the epidermis that regulates the opening and closing of a stoma.

gymnosperm Plants that bear naked or exposed seeds.

gynoecium (pl. **gynoecia**) Collective term for the carpels in a flower.

H

habitat Place or type of place where an organism lives.

hallucinogen Psychoactive drug capable of altering moods and perceptions of time and space.

haploid One set of chromosomes in a cell; 1*N*.

hardwood Angiospermous trees or the wood from angiosperms.

HDL (high density lipoprotein) Transport molecule that removes excess cholesterol from the body's tissues to the liver for degradation and elimination.

head Horizontal inflorescence of sessile flowers.

heartwood Core in woody stems, usually darker in color, and no longer functioning in water conduction.

hemicellulose Polysaccharide in plant cell walls that cross-link cellulose fibrils.

herb Nonwoody plant; aromatic plants whose leaves are used in seasoning.

herbaceous Nonwoody plants.

herbal A text that describes plants that are useful medicinally and in other ways.

herbarium A permanent collection of dried and pressed plants that provides information on the location and identification of the local flora.

herbivore Animals that eat plants.

hesperidium Simple, fleshy fruit with leathery exocarp; example: any citrus fruit.

heterotroph Other feeder; organism that is incapable of synthesizing its own food and must obtain its nutrition from other organisms.

heterozygous Having two different alleles for a given trait.

hilum Scar on a seed indicating where it was attached to the ovary.

holdfast Attachment organ or cell at the base of certain algae.

homologous chromosomes Chromosome pairs of the same size and shape that carry genes for the same traits.

homozygous Having two identical alleles for a given trait.

hunter-gatherers Human social group secures food sources from wild resources such as hunting animal prey or collecting edible plants from the wild.

hybrid Offspring of a cross between two species or between alternate homozygous conditions.

hydrogen bond Weak chemical bond formed when the slightly positive hydrogen atom of a polar covalent bond is attracted to the slightly negative atom of a polar covalent bond of another molecule.

hydrogenation Addition of one or more hydrogens to monunsaturated and polyunsaturated fatty acid chains.

hydroponics Growing plants without soil in liquid nutrient solutions.

hymenium The layer of fertile cells that produces spores in a fungal fruiting body.

hypertonic Solution with a greater solute concentration than that within a cell or reference solution.

hypha (pl. **hyphae**) Microscopic threads that make up the body of most fungi.

hypocotyl Region of stem in a plant embryo that is below the cotyledons.

hypogynous Floral whorls (sepals, petals, stamens) inserted below the ovary of a flower.

hypotonic Solution with a lesser solute concentration than that within a cell or reference solution.

I

IgE Immunoglobulin E; specific class of antibodies involved in allergic reactions; individuals who suffer from allergies have elevated levels of these antibodies.

immunoglobulin A category of protein known as an antibody.

imperfect flower Unisexual flower; either staminate or pistillate.

imperfect stage The asexual phase in a fungus life cycle characterized by the production of asexual spores.

incomplete dominance A type of inheritance in which the heterozygous phenotype is intermediate between the phenotypes of the dominant and recessive parents.

incomplete flower Flower lacking one or more floral whorls; typically either the sepals, petals, or both.

incomplete protein Protein that lacks the full complement of essential amino acids in the correct proportions.

indehiscent fruit Dry fruit that does not split open on maturity.

inferior ovary Ovary lies below the attachment of the sepals, petals, and stamens; an epigynous flower.

inflorescence A cluster of flowers.

insertion A mutation resulting from the addition of a small segment of DNA.

integral protein Protein that spans or penetrates the lipid bilayer of cell membranes.

integument Outermost layers of an ovule that typically develop into the seed coat.

internode Region on a stem between nodes.

interphase Stage in the cell cycle when a cell is not dividing.

intine The inner layer of the pollen wall.

intron An intervening or noncoding segment of a gene; introns separate exons.

involucre Whorl of bracts that subtend a flower or an inflorescence.

ion An atom or molecule that has gained or lost electrons, and has either a positive or negative charge.

ionic bond Chemical bond formed when ions of opposite charge attract.

isotonic Solution in which the solute concentration is equal to that within the cell or reference solution.

isotope Alternate form of an element with a different number of neutrons but the same number of protons and electrons.

K

karyogamy In sexual reproduction, the fusion of genetically distinct nuclei.

kinetic energy Energy of motion.

kingdom Largest taxonomic category consisting of related phyla or divisions.

Krebs Cycle The second stage of cellular respiration that occurs in the mitochondria; completes the breakdown of glucose into carbon dioxide.

L

land races Traditional varieties of plant crops.

lateral bud Bud found in the axil of a leaf, also called an axillary bud.

latex Milky juice exuded from some plants.

LDL (low density lipoprotein) Transports fats and cholesterol to the body cells, including the cells lining the bloodstream.
leaflet Subdivisions of a leaf blade.
legume Simple, dry, dehiscent fruit that splits along two seams, a pod. Member of the Fabaceae; a type of bean or pea.
lemma Bract in a grass flower.
lenticel Raised area in the bark of woody stems that permits the exchange of gases.
leucoplast Colorless plastid typically associated with starch formation and storage.
lichen Composite organism formed by the symbiotic association of a fungus and an alga.
light harvesting antenna A complex of several hundred chlorophyll and carotenoid molecules that form a part of each photosystem.
Light Reactions First steps in photosynthesis in which chlorophyll traps solar energy, driving the formation of ATP and NADPH; water is also lysed, releasing oxygen.
lignin Complex organic compound that strengthens the secondary cell walls of plants.
linkage Tendency of genes located on the same chromosome to be inherited together.
lipid Organic compounds that are insoluble in water; includes fats, oils, and steroids.
litter Partially decomposed plant material.
locus The location of a gene on a chromosome.
lumen Cavity bounded by secondary cell wall in dead plant cells.
lymphocytes A type of white blood cell; a component of the immune system produced by stem cells in the bone marrow.
lysergic acid diethylamide (LSD) A derivative of lysergic acid alkaloids, first isolated from ergot; strongly hallucinogenic.

M

macromolecule Complex organic molecule formed by joining smaller molecules; example: proteins are macromolecules formed by joining amino acids.
macronutrient Nutritional requirement needed in relatively large amounts.
major mineral Mineral requirement needed in relatively large amounts.
mass number Sum of the number of protons and neutrons in the nucleus of an atom; designated by a superscript to the upper left of the elemental symbol.
matrix Compartment of the mitochondrion enclosed by the inner membrane; site of the Krebs cycle.
megasporangium Sporangium that contains megaspores.
megaspore Spore that develops into the female gametophyte.
megaspore mother cell Diploid cell in megasporangium that, upon undergoing meiosis, yields megaspores.
meiosis Two successive nuclear divisions during which the chromosome number is halved; in plants, meiosis results in the formation of spores.
meristem Area of actively dividing cells in plants.
mesocarp Middle layer in the pericarp (fruit wall).
mesophyll Photosynthetic middle layer in the blade of a leaf; typically composed of palisade and spongy parenchyma.
messenger RNA (mRNA) Type of RNA created from DNA template that travels to ribosomes and directs protein synthesis.
metabolism Sum total of the chemical reactions in an organism.
metaphase Stage of mitosis in which the chromosomes are aligned along the equator.
microbody Membrane-bound organelles that are the site of certain enzymatic conversions; example: peroxisomes and glyoxisomes.
microfilament Solid rod of protein and part of the cytoskeleton.
micronutrient Nutrient required in relatively small amounts; vitamins and minerals.
micropyle The opening in an ovule through which the pollen tube enters during fertilization.
microsporangium Sporangium that contains microspores.
microspore Spore that develops into the male gametophyte.
microspore mother cell Diploid cell in microsporangium that undergoes meiosis to produce microspores.
microtubule Hollow protein rod found in cilia, flagella, the spindle, and the cytoskeleton.
middle lamella Layer of adhesive material (primarily pectins) found between adjacent cell walls.
mineral Inorganic, essential micronutrient.
mitochondrion (pl. **mitochondria**) Membrane-bound organelle that is the site of cellular respiration.
mitosis Nuclear division, usually accompanied by cytokinesis, in which the chromosomes are duplicated and divided up to form two identical daughter cells.
molecule Two or more atoms held together by chemical bonds that retain the properties of the compound.
monocotyledon A class of angiosperms in which the seedlings typically possess one cotyledon. Commonly abbreviated to monocot.
monoculture The cultivation of a single crop over a large region year after year.
monoecious Separate male and female reproductive structures are borne on the same plant; pollen-bearing and ovule-bearing flowers or cones are borne on the same plant.
monohybrid cross Genetic cross between parents that differ by a single trait.
monomer Building blocks of polymers.
monosaccharide The simplest carbohydrate, a simple sugar.
monounsaturated fat Composed of fatty acid chains in which there is only a single C-C double bond; examples: canola oil and olive oil.
multiple alleles A condition in which more than two alleles exist for a given trait.
multiple fruit A fruit derived from the fusion of the ovaries of several flowers in an inflorescence; example: pineapple.
mutation An inheritable change in genes or chromosomes.
mycelium (pl. mycelia) A network of fungal hyphae.
mycobiont The fungal partner in a mutualistic relationship such as mycorrhizae or lichens.
mycorrhiza (pl. **mycorrhizae**) Symbiotic association between a fungus and a plant root.
mycotoxin A toxic compound formed by the hyphae of common molds growing under a variety of conditions, especially in contaminated foods.

N

NAD (nicotinamide adenine dinucleotide) A molecule capable of being reduced that acts as an electron intermediate during cellular respiration.
NADP (nicotinamide adenine dinucleotide phosphate) A molecule that acts as an electron intermediate during photosynthesis. It is reduced during the light reactions and oxidized during the Calvin Cycle.
naked grain A grain that separates easily from the surrounding bracts.
narcotic Any psychoactive compound that is dangerously addictive; a compound that induces central nervous system depression resulting in numbness, lethargy, and/or sleep.
natural selection A guiding force of evolution in which organisms that are most fit to reproduce are selected.
nectar A sugary solution that attracts animals to plants.
nectar guide Color patterns present on petals that direct insects toward the nectar; often not visible to human eye.
nectary A gland that secretes nectar.
Neolithic The Stone Age period following the advent of agriculture.
net venation The netlike pattern of branching of veins on a leaf blade. Also known as reticulate venation; characteristic of most dicot leaves.
neurotransmitter A chemical responsible for the transmission of impulses across a neural synapse.
neutron A particle in the nucleus of an atom with no charge and a mass of approximately one atomic weight unit.
niche The particular role played by an organism in an ecosystem.

nitrogen-fixation The process of reducing nitrogen gas to ammonia and nitrates, forms of nitrogen that can be utilized by plants.
node The region of a stem where leaves or branches arise.
nonseptate hypha Fungal hyphae that lack septa or cross walls.
nonshattering Seeds and/or fruits do not split off and scatter from the fruiting head; a trait associated with domesticated plants.
nucellus Tissue in the ovule within which the embryo sac develops; integuments surround the nucellus.
nuclear envelope A double membrane with pores surrounding the nucleus.
nuclear pores Small openings in the nuclear membrane.
nucleic acid A molecule consisting of joined nucleotides; the two types are deoxyribonucleic acid (DNA) and ribonucleic acid (RNA).
nucleolus Spherical structure within the nucleus consisting of RNA and protein; assembly site for ribosomal subunits.
nucleotide A single unit of nucleic acid composed of a phosphate group, a five-carbon sugar (either ribose or deoxyribose), and a purine or pyrimidine base.
nucleus Membrane-bound organelle within eukaryotic cells; contains chromosomes and the nucleolus and is essential for the regulation of all cellular functions.
nut A one-seeded, dry, indehiscent fruit with a hard pericarp.

O

oil A triglyceride that is liquid at room temperature.
omnivore An organism that feeds on both plants and animals.
oogonium A female gametangium that occurs in some groups of algae and fungi; gives rise to oospores.
opposite arrangement Two leaves borne per node and arranged across the stem from each other.
order Taxonomic rank consisting of a group of related families.
organelle A body within the cytoplasm of eukaryotic cells; several different types of organelles occur, each with a specialized function such as the chloroplast, which functions in photosynthesis.
osmosis Diffusion of water (or other solvents) through a differentially permeable membrane.
ovary Enlarged basal portion of a single carpel or several fused carpels; contains one to many ovules.
ovule Structure that will become a seed following fertilization; literally an integumented megasporangium that contains the embryo sac before fertilization.
oxidization The loss of electrons or hydrogen from an atom or molecule.

P

P_{680} The reaction center for Photosystem II; a chlorophyll *a* molecule that is bound to a membrane protein and has a peak absorbance at 680 nm.
P_{700} The reaction center for Photosystem I; a chlorophyll *a* molecule that is bound to a membrane protein and has a peak absorbance at 700 nm.
paddy A flooded field used to cultivate lowland rice.
palea One of two bracts around the grass flower.
Paleolithic Old Stone Age; a cultural period during which early humans obtained food solely by foraging; ending in some areas approximately 10,000 years ago.
palisade parenchyma Parenchyma cells in the leaf mesophyll characterized by uniform rows of tightly packed cells with many chloroplasts beneath the upper epidermis.
palmately compound leaf Leaflets radiating from a common point.
panicle A branched inflorescence with the branches bearing loose flower clusters.
parallel venation Principal veins are parallel to one another; characteristic of monocot leaves.
parasite An organism that lives on or in the body of another living organism and derives nourishment from it.
parenchyma Ground tissue in plants with thin-walled cell varying in size and shape; the most abundant kind of cells in plants.
pectin A complex polysaccharide in the middle lamella and primary walls of plant cells.
pedicel An individual stalk of a flower that is part of an inflorescence.
peduncle The main stalk of an inflorescence or a single flower.
penicillin An antibiotic produced by various species in the genus *Penicillium.*
pepo A fleshy fruit with a tough outer rind that is composed of both receptacle tissue and exocarp such as cucumber, pumpkin, and melon.
perennial A plant that continues to live for an indefinite number of years.
perfect flower A flower having both stamens and carpels.
perfect stage The phase during the life cycle of a fungus when sexual fusion occurs, producing characteristic sexual spores.
perianth The petals and sepals together.
pericarp The fruit wall that develops from the ovary wall.
pericycle Root tissue sandwiched between the endodermis and the phloem; it is the outermost layer of the stele; meristematic region that gives rise to branch roots.
periderm Protective tissue that replaces the epidermis after secondary growth begins; it includes the cork, the cork cambium, and sometimes other cells.
perigynous A flower in which the base of the sepals, petals, and stamens form a cup around the ovary.
peripheral protein A protein on the surface of a biological membrane.
perithecium A flask-shaped ascocarp.
permafrost Soil that is permanently frozen.
peroxisome A microbody found in leaves and often associated with chloroplasts.
petal A floral organ that is leaflike and often brightly colored; a component of the corolla.
petiole The stalk of a leaf.
phenotype The physical appearance of an organism.
phloem The vascular tissue that conducts organic materials synthesized by the plant.
phospholipid A type of lipid molecule occurring in a bilayer in biological membranes; a lipid with two fatty acids and a phosphate group attached to glycerol.
phosphorylation The addition of a phosphate group to a molecule.
photon A unit of light energy.
photosynthesis The process that results in the conversion of light energy into the chemical energy of carbohydrates.
Photosystem I A light-harvesting unit located on the thylakoid membrane of the chloroplast, with P_{700} as the reaction center.
Photosystem II A light-harvesting unit located on the thylakoid membrane of the chloroplast, with P_{680} as the reaction center.
phragmoplast A system of microtubules and vesicles that arises between two daughter nuclei at telophase and forms the cell plate.
phylogeny The evolutionary history and relationship of a species.
physiological dependence The condition in which there is a physical need for a drug to avoid withdrawal symptoms.
pinnately compound leaf Leaflets attached on both sides of a common axis.
pistillate flower A flower having carpels but no stamens.
pit A pore in a secondary cell wall.
pith The central tissue of a dicot stem, consisting of parenchyma cells.
plasmid A small, circular DNA molecule found in bacterial cells.
plasmodesma (pl. **plasmodesmata**) A cytoplasmic strand that connects adjacent plant cells through pores in the cell wall.
plasmodium The vegetative stage of a slime mold.
plasmolysis Shrinking of protoplasm in a cell due to loss of water in a hypertonic environment.
plastid A class of organelles that includes chloroplasts, leucoplasts, and chromoplasts.
plywood A building material consisting of two or more thin sheets of wood bonded together.
pod A dry dehiscent fruit that splits along two seams; a legume.
point mutation The smallest mutation caused by the change of a single nucleotide.
polar nuclei Two nuclei found in the embryo sac that unite with a sperm to form the primary endosperm nucleus.

pollen An immature male gametophyte of seed plants.

pollen tube A tube that develops from the pollen grain and carries the sperm to the ovule.

pollination The transfer of pollen from an anther to a stigma.

polynomial A scientific name for an organism composed of more than two words.

polyploid Having more than two complete sets of chromosomes.

polysaccharide A polymer such as starch or cellulose composed of thousands of monosaccharides.

polyunsaturated fat A fat having several to many double bonds between carbon atoms.

pome A simple fleshy fruit; the outer portion formed by floral parts that surrounded the ovary; example: apple or pear.

population All the individuals of a species within a given area.

potential energy The energy stored in matter as a result of its location or chemical bonds.

Pressure Flow Hypothesis The theory that organic solutes move along a concentration gradient from source to sink through the phloem.

primary consumer Animals that feed directly on producers.

primary endosperm nucleus The product of the fusion of a sperm and two polar nuclei in the embryo sac of angiosperms; double fertilization.

primary growth Growth in length due to the activities of the apical meristems of shoot and root.

primary wall The wall layer of a plant cell deposited during cell expansion, generally thin and elastic.

producer An organism that manufactures food through photosynthesis.

prokaryotic cell A type of cell lacking a nucleus and membrane-bound organelles; found in the Kingdom Monera.

prophase The first stage of mitosis, characterized by the condensation of chromatin into chromosomes and the formation of the spindle.

prosthetic group Nonprotein groups that are attached to an enzyme or other protein and necessary for its function.

protein A macromolecule composed of one or more polypeptides, each composed of many amino acids.

proton A positively charged particle in the nucleus of an atom.

protoplast All of a plant cell excluding the wall.

psychoactive drug A drug that affects the central nervous system by influencing the release of neurotransmitters or mimicking their actions.

psychological dependence A condition marked by the strong desire to repeat the use of a drug to reexperience the feelings of well-being induced by the drug.

purine One type of nitrogen-containing base found in nucleotides; consisting of adenine and guanine.

pyrimidine One type of nitrogen-containing base found in nucleotides; consisting of cytosine, thymine, and uracil.

R

raceme A vertical inflorescence with stalked flowers.

radicle The embryonic root found in the seed.

receptacle The expanded tip of a pedicel or peduncle to which the floral organs are attached.

recessive An allele that is masked in the phenotype by a dominant allele.

recombinant DNA The introduction of genes from one organism into the DNA of a second organism.

redox reactions Oxidation-reduction reaction; a chemical reaction involving the transfer of electrons from one molecule to another.

reduction Gain of electrons or hydrogen from an atom or molecule.

resin An exudate common in conifers but also occurring in some angiosperms.

reticulate venation Netted venation; the arrangement of veins in a leaf that resembles a net; characteristic of dicot leaves.

retting The process that frees flax fibers by allowing microbial decomposition to break down the outer part of the stem.

rhizome A horizontal, underground stem.

ribonucleic acid (RNA) The nucleic acid formed from DNA and involved in protein synthesis; nucleotide of chain of phosphates, ribose sugars, and purine and pyrimidines.

ribose A pentose sugar present in RNA.

ribosomal RNA (rRNA) The type of RNA that is a component of ribosomes.

ribosome A particle composed of two subunits, each containing RNA and protein; functions in protein synthesis.

root cap A thimble-shaped group of cells found at the tip of roots; it functions to protect the meristem.

root hair A root epidermal cell that functions in water absorption.

root nodule Gall-like structures on the roots of legumes that contain symbiotic nitrogen-fixing bacteria.

rough ER A portion of the endoplasmic reticulum containing ribosomes.

runner A horizontal stem that grows along the surface of the ground; also known as a stolon.

S

samara A simple, dry indehiscent fruit with the pericarp bearing winglike outgrowths; winged fruit of maple.

saponin A glycoside with a steroid molecule as the active component, such as diosgenin from yams.

saprobe An organism deriving its food from the dead body or nonliving products of another organism.

sapwood Region of secondary xylem that actively transports water; light-colored wood immediately inside the vascular cambium.

saturated fat A fat in which all the carbons in the fatty acids are connected by single bonds, thereby having the maximum number of hygrogen atoms.

savanna A tropical grassland biome with scattered trees.

schizocarp A dry indehiscent fruit that splits into two one-seeded halves at maturity.

scion A small twig or bud that is grafted to a stock.

sclereid A sclerenchyma cell with a thick, lignified secondary wall having many pits; variable in form but not usually elongated.

sclerenchyma Tissue composed of cells with thick secondary walls; functioning in support or protection.

sclerophyllous Vegetation characterized by thick leathery leaves with abundant sclerenchyma cells; vegetation of a chaparral.

sclerotium (pl. **sclerotia**) A fungal resting body resistant to unfavorable conditions; a firm, hardened mass of hyphae (or a hardened plasmodium of a slime mold) that will germinate on the return of favorable conditions.

scutellum The single cotyledon in grass seeds.

secondary consumer An animal that feeds on other consumers.

secondary growth The increase in girth of stems and roots; produced by the activities of the vascular cambium and cork cambium.

secondary product Chemical compounds synthesized by plants or fungi but not critical for the basic metabolic functions of that organism; often functioning to deter predators or attract pollinators; a secondary metabolite.

secondary wall The innermost layer of a cell wall formed after cell elongation has ceased; often characterized by the deposition of lignin.

seed A matured ovule containing an embryo and food supply and covered by a seed coat.

seed bank Storage facilities for seeds of domesticated plants and wild relatives; facilities for preserving genetic diversity.

seed coat The outer layer of a seed that is developed from the integuments of the ovule; the testa.

seed plant Common term for gymnosperms and angiosperms.

self-pollination Transfer of pollen from stamen to stigma within the same plant.

sepal A leaflike floral organ that protects the unopened flower bud.

septate Divided by crosswalls into cells.

septum A dividing wall or partition; a cross wall in a fungal hyphae or algal filament.

shattering A trait found in wild plants in which the fruiting head breaks apart to scatter the seeds over a wide area.

short grass prairie A grassland biome characterized by short grass and low rainfall; also known as the plains.

sieve plate The perforated wall area in a sieve tube member.

sieve tube A long tube specialized for the conduction of food materials (products of photosynthesis) and consisting of several-to-many sieve tube members.

sieve tube member A phloem cell characterized by a sieve plate and enucleate condition; specialized for the conduction of products of photosynthesis.

simple fruit A fruit that develops from a single ovary.

simple leaf A leaf that is not divided into leaflets.

simple pistil A pistil that contains a single carpel.

smooth ER The portion of endoplasmic reticulum that lacks ribosomes.

softwood General term for the wood (secondary xylem) of conifers.

somaclonal variant A plant showing a mutation that developed asexually during the tissue culture of a single callus.

somatic mutation A mutation that occurs in cells of leaves, stems, or roots; a mutation occurring in any cells that are not involved in gamete formation.

sorus (pl. **sori**) A cluster of sporangia found on a fern leaf.

specialized niche A species occupying a niche with a narrow range of tolerance.

species A single kind of organism; often defined as a group of interbreeding populations reproductively isolated from any other such group.

sperm A male gamete.

spermagonium (pl. **spermagonia**) A structure that produces spermatia in the rust fungi.

spermatium (pl. **spermatia**) Minute, nonmotile male gametes that occur in the rust fungi.

spice A pungent, aromatic plant product derived from plants native to tropical regions and used to flavor foods.

spike An inflorescence in which the main axis is elongated and the flowers are sessile.

spikelet A small group of grass flowers; a unit of the inflorescence in grasses.

spindle The aggregation of microtubules that is involved in the movement and separation of chromosomes during mitosis and meiosis.

spongy parenchyma Part of the leaf mesophyll; cells are loosely arranged and contain chloroplasts.

sporangiospore A spore that develops within a sporangium.

sporangium (pl. **sporangia**) A structure in which spores are produced.

spore A reproductive unit (often unicellular) that is capable of developing into a new organism without fusion with another cell.

sporophyte A diploid plant that produces spores; the diploid phase of a life cycle that has an alternation of generations.

springwood The cells in the secondary xylem that are formed early in the season, usually with wide vessels (angiosperms) or wide tracheids (gymnosperms); also called early wood.

stamen The floral organ that produces pollen; consisting of an anther and filament.

staminate flower A flower having stamens but no carpels.

starch A polysaccharide composed of a thousand or more glucose molecules; the chief food storage material of most plants.

stele The vascular cylinder; vascular tissue making up the central cylinder of roots.

steroid A type of lipid containing four fused rings of carbon atoms with various side chains.

stigma The receptive portion of the carpel to which the pollen adheres.

stimulant A psychoactive compound that excites and enhances mental alertness and physical activity; often reduces fatigue and suppresses hunger.

stipe A supporting stalk; such as those in mushrooms and brown algae.

stipule A small appendage found in pairs at the base of leaves.

stolon A horizontal stem that grows along the ground surface; may form adventitious roots and plantlets; also known as a runner.

stoma (pl. **stomata**) A minute opening, bordered by guard cells, in the epidermis of leaves and stems.

stroma The ground substance of the chloroplasts where the reactions of the Calvin Cycle occur.

stroma thylakoid A thylakoid that does not occur in a granum; connects separate grana.

style The tissue that connects the stigma to the ovary in the carpel.

suberin A fatty material found in the cell walls of cork cells and the Casparian strip of the endodermis.

substrate The substance acted on by an enzyme; the surface on which a plant or fungus grows or is attached.

sucrose A disaccharide made from a molecule of glucose linked to a molecule of fructose; table sugar.

sugar A monosaccharide; a carbohydrate with the general formula $C_nH_{2n}O_n$.

summerwood The cells in the secondary xylem that are formed late in the season, usually with few vessels (angiosperms) or narrow tracheids (gymnosperms); also called late wood.

superior ovary An ovary located above the sepals, petals, and stamens.

symbiosis Two organisms that live in intimate association with each other.

symplast The interconnected protoplasm of all cells in a plant.

synapsis The pairing of homologous chromosomes that occurs in Prophase I of meiosis.

synergid One of a pair of short-lived cells that lay close to the egg in the mature embryo sac.

T

taiga A biome in the northern hemisphere dominated by conifers; the northern coniferous or boreal forest.

tall grass prairie A grassland biome characterized by many tall grasses up to 5 meters tall.

tannin A secondary product found in many plants that have been widely utilized as stains, dyes, inks, or tanning agents for leather; believed to function in plants by discouraging herbivores.

taproot A relatively large primary root that gives rise to smaller, lateral roots.

taxon (pl. **taxa**) A general term for any taxonomic rank such as species, genus, or order.

teleomorphic The sexual phase, or perfect stage, in a fungal life cycle.

teliospore A thick-walled spore found in the rust and smut fungi; karyogamy occurs within the teliospore and it gives rise to the basidium.

telophase The last stage of mitosis and meiosis during which the chromosomes become reorganized into daughter nuclei.

temperate rain forest A biome dominated by coniferous trees, high rainfall, and high humidity; moist coniferous forest.

tepal Members of the perianth that are not differentiated into sepals and petals.

terpene An unsaturated hydrocarbon formed from an isoprene building block; found in many plants in the form of essential oils.

testa Seed coat.

tetrad A group of four, such as the four haploid spores that form after meiosis, or the four chromatids in a bundle after homologous chromosomes pair.

thylakoid membrane A saclike photosynthetic membrane in chloroplasts; stacks of thylakoids form the grana.

thymine A pyrimidine base occurring in DNA but not in RNA.

Ti plasmid The tumor-inducing plasmid from the bacterium, *Agrobacterium tumefaciens;* commonly used as a vector for recombinant DNA studies in plants.

tissue A group of cells that perform a specific function.

toxin A poisonous substance.

trace element An inorganic element required in small amounts for plant growth.

trace mineral Dietary minerals that are required in minute quantities.

tracheid An elongated, tapering xylem cell that is specialized for conducting water and support with lignified pitted walls.

transcription The formation of RNA as a complementary copy of a portion of the DNA molecule.

transfer RNA (tRNA) A class of small RNA molecules that transfer amino acids to the correct position on the messenger RNA molecule at the ribosome for protein synthesis.

transgenic Cells or organisms that contain genes that were inserted into them from other organisms using the techniques of genetic engineering.

translation The synthesis of a polypeptide from a specific sequence of codons on a messenger RNA molecule; occurs at the ribosomes.

transpiration The loss of water vapor from leaves; occurs mostly through the stomata.

Transpiration-Cohesion Theory The theory that explains water movement in the xylem; the driving force is the pull of transpiration and the cohesion of water molecules.

trichome An epidermal appendage, such as a hair or scale.

triglyceride A type of lipid formed from three fatty acids bonded to a molecule of glycerol; a fat or oil.

triploid A cell or nucleus that contains three sets of chromosomes; common in endosperm.

trophic level A step in the movement of energy through an ecosystem; a step in a food chain.

tropical rain forest An endangered tropical biome with high rainfall and an exceptional diversity of species.

tube nucleus The nucleus or cell in the pollen grain that develops into the pollen tube.

tuber An enlarged, fleshy, underground stem tip, such as the potato.

tuberous roots Modified fibrous roots that have become fleshy and enlarged with food reserves.

tundra A treeless circumpolar biome with meadowlike vegetation above the Arctic Circle.

turgid A swollen, distended cell that is firm due to water uptake.

U

umbel An inflorescence in which the individual pedicels all arise from the apex of the peduncle.

universal veil A membrane that totally encloses some young mushrooms; after it breaks, its remnants appear as a volva at the base and scales on the cap.

unsaturated fat A fat containing one or more double bonds between carbon atoms.

uracil A pyrimidine found in RNA but not DNA.

uredium (pl. **uredia**) The structure that produces uredospores in rust fungi; sometimes called a uredinium.

uredospore A reddish, binucleate spore formed by rust fungi; often forms the repeating stage of the rust; also called a urediniospore.

V

vascular bundle A strand of tissue containing primary xylem and primary phloem, often surrounded by a bundle sheath.

vascular cambium Meristematic tissue that gives rise to secondary xylem and secondary phloem.

vascular cylinder The stele; vascular tissue making up the central cylinder of roots.

vascular plant A general name for any plant that has xylem and phloem.

vascular ray Sheet of parenchyma that extends radially through the wood, across the cambium and into the secondary phloem; rays are produced by the vascular cambium and function in lateral transport.

vascular tissue Tissue that is specialized for the long-distance transport of water or photosynthetic products; xylem and phloem.

vegan A pure vegetarian consuming no animal products at all.

vegetarian A person who does not consume animal flesh; some consume dairy products and eggs, while others are vegans.

vein A vascular bundle that forms part of the conducting and supporting tissue of a leaf.

veneer A thin sheet of wood, often with attractive grain, used to cover less expensive wood.

vessel A tubelike column of vessel elements that are connected by open end walls and are specialized for the conduction of water and minerals.

vessel element One of the cells forming a vessel and characterized by a perforation plate.

vitamin A naturally occurring organic compound that is necessary, in small amounts, for the normal metabolism of plants and animals.

W

water-soluble vitamin A vitamin that is not readily stored in the body, with excess eliminated in the urine; includes B vitamins and vitamin C.

weed A plant not valued for its use or beauty and not intentionally planted; a category of hay fever plants that includes nongrass and nontree species.

whorled arrangement Three or more leaves per node.

winnowing The process that separates the grain from the fragments of chaff.

wood Secondary xylem.

wood pulp A watery suspension of pulverized wood used in the production of paper, cardboard, fiberboard, rayon, cellophane, and other products.

X

xanthophylls A yellow carotenoid pigment found in chloroplasts.

xerophyte A plant adapted for growth in arid conditions.

xylem The vascular tissue specialized for the conduction of water and minerals; consisting of tracheids and vessel elements, fibers, and parenchyma cells.

Z

zoospore A motile spore.

zygomorphic flower A bilaterally symmetrical flower, capable of being divided into two symmetrical halves only by a single longitudinal plane passing through the axis.

zygosporangium (or **zygospore**) The thick-walled sexual spore formed by members of the zygomycetes.

zygote A diploid cell that is formed by the fusion of two gametes.

CREDITS

PHOTO CREDITS

Part Openers

Part I: © Brian Seed/Tony Stone Images; **Part III:** © Kunio Owaki/The Stock Market; **Part IV:** © Barbara Filet/Tony Stone Images; **Part V:** © Piotr Kapa/The Stock Market; **Part VI:** © Tony Stone Images; **Part VII:** © Robert Frerck/Tony Stone Images.

Chapter 1

1.2a: © Sherman Thomson/Visuals Unlimited; **1.2b:** © William J. Weber/Visuals Unlimited; **1.3,P.6a:** © Grant Heilman/Grant Heilman Photography; **p.6b:** © Dwight Kuhn; **1.4b:** © W.P. Wergin and E.H. Newcomb/Biological Photo Service; **1.9f:** © Cabisco/Visuals Unlimited.

Chapter 2

2.2a: Courtesy Armed Forces Institute of Pathology; **2.2b:** © Ed Reschke; **2.3a:** © W.P. Wergin & E.H. Newcomb/Biological Photo Service; **2.4:** © Cabisco/Visuals Unlimited; **2.5a:** © Ray F. Evert; **2.8b:** © Biophoto Associates/Science Source/Photo Researchers, Inc.

Chapter 3

3.2a: © Ed Reschke/Peter Arnold, Inc.; **3.2c:** © Biophoto Associates/Photo Researchers, Inc.; **3.3a:** © J. Robert Waaland/Biological Photo Service; **3.3b,c:** © Biophoto Associates/Photo Researchers, Inc.; **3.4a1:** © J. Robert Waaland/Biological Photo Service; **3.4b1:** © George J. Wilder/Visuals Unlimited; **3.5a(left):** © Runk/Schoenberger/Grant Heilman Photography; **3.5a(right):** © Ed Reschke; **3.5b(left):** © Carolina Biological Supply/Phototake, NY; **3.5b(right):** © Jack M. Bostrack/Visuals Unlimited; **3.6(right), 3.8b:** © Carolina Biological Supply/Phototake, NY; **3.9a:** © John D. Cunningham/Visuals Unlimited; **3.9b:** © Dwight Kuhn; **Box 3.2a:** © J.A.L. Cooke/OSF/Animals Animals/Earth Scenes; **Box 3.2b,c:** © Howard A. Miller, Sr./Visuals Unlimited; **Box 3.3:** © Times Mirror Higher Education Group, Inc./Jim Shaffer, photographer.

Chapter 4

Box 4.1a,b: Courtesy Dr. Mary E. Doohan; **4.3a:** © Dr. B.A. Meylan; **4.7a:** © Runk/Schoenberger/Grant Heilman Photography; **4.7b:** Courtesy of Herbert W. Israel, Cornell University; **Box 4.2b,c:** © Grant Heilman/Grant Heilman Photography; **4.13a:** © Keith Porter/Photo Researchers, Inc.

Chapter 5

5.3a: © Bill McMackins/Unicorn Stock Photos; **5.3b:** © Dede Gilman/Visions From Nature; **5.9a:** © Stephen Dalton/Photo Researchers, Inc.; **5.9b:** © Len Rue III/Irene Vandermolen/Animals Animals/Earth Scene; **5.9c:** © Len Rue III/Animals Animals/Earth Scene; **5.10:** © Wolfgang Kaehler; **Box 5.1a:** © Dr. Jeremy Burgess/SPL/Photo Researchers, Inc.; **Box 5.1B-1:** Courtesy of Joan Skvarla; **Box 5.1B-2:** Courtesy of Joan Nowicke, Smithsonian Institution; **Box 5.1c:** "Navajo woman and child gathering corn pollen," Harrison Begay, 0237.48, from the collection of The Gilcrease Museum, Tulsa; **Box 5.2:** © Michael Freeman.

Chapter 6

6.5: Earliest drawing of John Chapman, known as Johnny Appleseed, by an Oberlin College student; **6.6b:** © James Peter Stuart; **6.7:** © James Welgos/Photo Researchers, Inc.; **6.8a:** © Dr. William J. Jahoda/Photo Researchers, Inc.; **6.8b:** © C.P. Hickman/Visuals Unlimited; **Box 6.1a:** © Grant Heilman/Grant Heilman Photography; **Box 6.1b:** © William E. Ferguson.

Chapter 7

7.1: © William E. Ferguson.

Chapter 8

8.1, 8.2, 8.3a,b: Courtesy of the Harriet George Barclay Slide Collection, University of Tulsa; **8.5:** Hargrett Rare Book & Manuscript Library/University of Georgia Library; **8.6a,b:** Courtesy of Paul Buck **Box 8.1:** © Times Mirror Higher Education Group, Inc./Jim Shaffer, photographer; **8.9:** © Bridgeman/Art Resource, NY.

Chapter 9

9.3d: Courtesy of Paul Buck; **9.4a:** © T.E. Adams/Visuals Unlimited; **9.4c:** © David M. Phillips/Visuals Unlimited; **Box 9.1a,b:** © D.O. Hall, King's College, London; **Box 9.1c:** © Philip Sze; **Box 9.1d:** © D. Ogust/The Image Works, Inc.; **9.7a:** © John Shaw/Tom Stack & Associates; **9.7b,c:** © John D. Cunningham/Visuals Unlimited; **9.8:** © Kingsley R. Stern; **9.9:** The Field Museum, #75400, Chicago; **9.10a:** Courtesy of John Kimball; **9.10b:** © E.S. Ross; **9.11b:** Courtesy of A.P. Blair; **9.12a:** © Frank Pennington/Unicorn Stock Photos; **9.12b:** © Jeff Foott/Bruce Coleman, Inc.; **Box 9.2b:** © Ed Reschke/Peter Arnold, Inc.; **9.13a,** Courtesy of George S. Ellmore; **9.13b:** Courtesy of Paul Buck; **9.14a:** Courtesy of the Harriet George Barclay Slide Collection, University of Tulsa; **9.14b:** © Runk/Schoenberger/Grant Heilman Photography; **9.14c:** Courtesy of the Harriet George Barclay Slide Collection, University of Tulsa; **Box 9.3:** Courtesy of Frank M. Carpenter, Harvard University; **9.15:** © Ray Simons/Photo Researchers, Inc.; **9.16:** © John D. Cunningham/Visuals Unlimited.

Chapter 10

Box 10.b1: © Omikron/Photo Researchers, Inc.; **Box 10.b2:** © A. Vollan/WHO/Visuals Unlimited; **10.3a:** © Biophoto Associates/Science Source/Photo Researchers, Inc.; **10.3b:** © Ken Greer/Visuals Unlimited; **10.4:** © Lester Bergman.

Chapter 11

11.2: © Jack Harland; **Box 11.1b:** © David O. Norris; **11.3a,b:** Anthro-Photo File; **11.5:** © Robert S. Peabody Museum of Archeology, Phillips Academy, Andover, Massachusetts.

Chapter 12

12.4: © Max & Bea Hunn/Visuals Unlimited; **Box 12.1:** © Tony Craddock/SPL/Photo Researchers, Inc.; **12.7:** Courtesy of USDA; **12.9:** © Smithsonian/Antonio Montaner; **Box 12.2a:** AP/Wide World Photos; **Box 12.2b:** © John N.A. Lott/Biological Photo Service; **12.11a,b:** Courtesy of the Harriet George Barclay Slide Collection, University of Tulsa; **12.12a:** © George J. Wilder/Visuals Unlimited; **12.12b:** © Heather Angel/Biofotos; **12.13a:** © Hans Reinhard/Bruce Coleman, Inc.; **12.13b:** © Walter Hodge/Peter Arnold, Inc.; **12.13c-e:** © Grant Heilman/Grant Heilman Photography; **12.13f:** © Joe Munroe/Photo Researchers, Inc.

Chapter 13

13.1a, 1, 2, 4: Courtesy of Paul Buck and Karen McMahon; **13.2:** © The Nitragin Co., Milwaukee, WI; **13.3:** © Matt Meadows/Peter Arnold, Inc.; **13.4:** © Christopher B. Donnan, Courtesy Enrico Poli Collection; **13.6:** © Grant Heilman/Grant Heilman Photography; **13.8b:** © D. Ogust/The Image Works, Inc.; **13.10:** © Sigrid Salmela/Animals Animals/Earth Scenes.

Chapter 14

14.2: Courtesy of Idaho's World Photo Expo; **14.3:** Courtesy of H.D. Thurston; **14.6:** Courtesy of Kingsley Stern © Monice Emerson; **Box 14.1:** © Len Rue III/Irene Vandermolen/Animals Animals/Earth Scenes;

14.7: Courtesy of the Harriet G. Barclay Slide Collection at the University of Tulsa; **14.8, 14.9a:** Courtesy of Dr. Lamont Lindstrom.

Chapter 15

15.1: Courtesy of Allan Avery; **15.4:** The Bettmann Archives; **15.5b:** Courtesy of U.S. Geological Survey; **15.8:** © Grant Heilman/Grant Heilman Photography; **15.9:** Courtesy of the Agricultural Research Service, USDA; **15.10a:** © Grant Heilman/Grant Heilman Photography; **15.10b:** © Daniel W. Gade; **15.10c:** © Steven King/Peter Arnold, Inc.; **Box 15.1a:** © Max & Bea Hunn/Visuals Unlimited; **Box 15.1b:** © National Maritime Museum Greenwich; **15.11a,b:** © Carolina Biological Supply/Phototake, NY.

Chapter 16

16.2: © The Bettmann Archives; **16.3a,b, 16.4a,b:** Courtesy of the Harriet George Barclay Slide Collection, University of Tulsa; **Box 16.1:** © J. Greenberg/The Image Works, Inc.; **16.6a:** © Jane Thomas/Visuals Unlimited; **16.6b:** Courtesy of USDA; **16.8:** © James A. Hays/Unicorn Stock Photos; **16.7:** © Jim Pickerell/Tony Stone Images; **16.9a-d:** © Alan Weiner/Liaison Int'l.

Chapter 17

17.2a: National Library of Medicine; **17.2b:** © Times Mirror Higher Education Group, Inc./Bob Coyle, photographer; **17.3:** Courtesy of the Fenster Museum; **17.4:** © Noboru Komine/Photo Researchers, Inc.; **17.5:** © William E. Ferguson; **17.6:** © Verna R. Johnston/Photo Researchers, Inc.; **17.7:** © Robert & Linda Mitchell; **17.8a:** © DPA/The Image Works, Inc.; **17.8b:** © The Mansell Collection Ltd.; **17.9:** © D. Cavagnaro/Visuals Unlimited; **17.10:** © Wolfgang Kaehler; **17.12a:** © Jeff Gnass; **17.12b:** © M.D. Long/Visuals Unlimited; **17.13:** © D. Ogust/The Image Works, Inc.

Chapter 18

18.1: © Runk/Schoenberger/Grant Heilman Photography; **18.2a:** © Leonard Lee Rue III/Photo Researchers, Inc.; **18.2b:** © Joe Munroe/Photo Researchers, Inc.; **18.4b:** © Rod Heinrichs/Grant Heilman Photography; **18.4c:** Courtesy of the Harriet George Barclay Slide Collection, University of Tulsa; **18.6a,b:** Courtesy of Masters of Linen/International Linen Promotion Commission; **18.7:** © E.S. Ross; **18.8:** Courtesy of Lourdes R. Montinok from PINA, pg. 171, Amon Trading Corp. publishers; **18.9a,b:** Courtesy of the Harriet George Barclay Slide Collection, University of Tulsa; **18.10a:** © Norman Myers/Bruce Coleman, Inc.; **18.10b:** © Times Mirror Higher Education Group, Inc./Bob Coyle, photographer; **18.12:** © Gregory K. Scott/Photo Researchers, Inc.; **18.14:** Courtesy of St. Regis Paper Co.; **18.15:** © G. Kirtley-Perkins/Visuals Unlimited; **18.17:** Courtesy of Crane & Company, Inc.

Chapter 19

19.1a: © American Numismatic Society; **19.1b:** © British Museum; **19.2a:** The Bodleian Library, Oxford MS. Ashmole 1462 folio 15v.; **19.3:** © Barbara Rios/Photo Researchers, Inc.; **Box 19.1a-c:** Courtesy of Paul Buck; **19.8:** © Heather Angel/Biofotos; **19.10:** © Times Mirror Higher Education Group, Inc./Bob Coyle, photographer; **19.11a:** © Grant Heilman/Grant Heilman Photography; **19.11b:** © Christopher Joyce.

Chapter 20

20.1: © Scott Camazine/Photo Researchers, Inc.; **20.5:** From Lewis Carroll's, "Alice in Wonderland"; **20.6:** © Phoebe A. Hearst Museum of Anthropology, University of California at Berkeley; **20.7a:** © John Colwell/Grant Heilman Photography; **20.7b:** Courtesy of the Harriet George Barclay Slide Collection, University of Tulsa; **20.8a:** © Robert & Linda Mitchell; **20.8b:** Courtesy of Loran Anderson; **Box 20.1:** Wellcome Institute Library, London.

Chapter 21

21.1: © Dale Thompson/Bruce Coleman, Inc.; **21.7b:** Courtesy of James Stewart; **21.8a:** © Times Mirror Higher Education Group, Inc./Bob Coyle, photographer; **21.8b:** From "The Healing Forest" by R. Rauffauf & Richard Schuttes, with permission.; **21.9:** © Heather Angel/Biofotos; **Box 21.1:** © D. Cavagnaro/Visuals Unlimited; **21.12b:** Courtesy of Paul Buck; **21.13a:** Courtesy of Paul Buck; **21.16:** Courtesy of Ashland Farm Botanicals, Sedalia, Missouri.

Chapter 22

22.9a: © Jean-Yves Ruszniewski/Agence Vandystadt/Photo Researchers, Inc.; **22.9b:** Courtesy of the French Embassy Press & Information Division, New York; **22.11b:** © Biophoto Associates; **22.13b:** © Dr. J. Menge, UC Riverside; **Box 22.1a:** © Christine Case/Visuals Unlimited; **Box 22.1b:** © Jean T. Buermeyer/Photo/Nats; **Box 22.1c:** © Christine Case/Visuals Unlimited; **22.17:** © David Arora.

Chapter 23

23.2: From Lesko, L.H., King Tut's Wine Cellar, 1977. Reprinted with permission from B.C. Scribe Publishers.; **23.4:** From "Le Journel Illustre" 1878, Musee National Des Arts & Traditions Popwaires, Paris. © Reunion des Musees Nationaux.; **Box 23.1:** © Teaching Collection, PSU Dept. Plant Pathology, photographer unknown. Courtesy of William Merrill.; **23.8:** © Topham/The Image Works, Inc.; **23.9:** University of Pennsylvania Museum, Philadelphia (neg. # T35-13482).; **23.11:** © Grant Heilman/Grant Heilman Photography; **23.12:** © J. Creager/Visuals Unlimited; **Box 23.2b:** From Dr. Anne P. Streissguth, "Teratogenic Effects of Alcohol in Humans & Laboratory Animals," *Science* 209:353–61, Fig. 1, July 1980. © 1980 by the AAAS.

Chapter 24

24.1: The Bettmann Archives; **24.2:** © David M. Phillips/Visuals Unlimited; **24.3:** © Tom E. Adams/Peter Arnold, Inc.; **24.4a:** © Runk/Schoenberger/Grant Heilman Photography; **24.4b:** From G. Barger, "Ergot and Ergotism," 1931.; **24.5:** © The Granger Collection, New York; **24.7:** Courtesy of Harriet Burge; **24.8:** Courtesy of DEA; **24.9a:** © Everette S. Beneke/Visuals Unlimited; **24.9b,c:** © Carroll H. Weiss/Camera M.D. Studios; **24.9d:** © Everette S. Beneke/Visuals Unlimited; **24.10a:** © Arthur M. Siegelman/Visuals Unlimited; **24.10b:** Courtesy of Armed Forces Institute of Pathology. **24.12b, c:** Courtesy of Richard J. Shaughnessy; **21.15b (cone):** Courtesy of Paul Buck.

Chapter 25

25.3a: © Jeff Foott/Tom Stack & Associates; **25.3b:** © Kevin Schafer/Peter Arnold, Inc.; **25.14a:** © John D. Cunningham/Visuals Unlimited; **25.14b:** © William E. Ferguson; **25.14c:** © Grant Heilman/Grant Heilman Photography; **25.14d:** © Johnny Johnson/Animals Animals/Earth Scenes; **25.14e:** © Norman Tomalin/Bruce Coleman, Inc.; **25.14f:** © Joe McDonald/Visuals Unlimited; **25.14g:** © E.S. Ross; **Box 25.1a:** © Philip M. Fearnside, Nat'l Institute for Research in the Amazon-INPA; **Box 25.1c:** © Times Mirror Higher Education Group, Inc./Bob Coyle, photographer.

ART CREDITS

PC&F

1.1, 1.5, 1.6, 1.7, 1.8, 1.9B, 1.10, 1.11, 1.12, 1.13, 2.1, 2.3B, 2.5B, 2.6, 2.8A, 2.9, 3.1, 3.3A1, 3.3B1, 3.3C1, 3.4A-B, 3.6A, 3.7, 3.8A&C, 3.10, 3.11, 4.1A-B, 4.2, 4.3B, 4.4, 4.5, 4.6, 4.7C, 4.8, 4.9, 4.10, 4.11, 4.12, 4.13B, 4.14, 4.15, 4.16, 5.1, 5.2, 5.4, 5.5, 5.6, 5.7, 5.8, 6.1, 6.2, 6.3, 6.4, 7.2, 7.3, 7.4, 7.5, 7.6, 7.7, 7.8, 7.9, 7.10, 7.11, 7.12, 8.7, 8.8, 8.10, 9.1, 9.2, 9.4B, 10.1, 10.2, 10.5, 11.1, 11.4, 11.6, 11.7, 12.1, 12.2, 12.3, 12.5, 12.6, 12.8, 12.10, 13.1B, 13.8A, 13.9, 14.1A, 14.1B, 14.1C, 14.1D, 14.1E, 14.4, 14.5, 15.2, 15.3, 15.5A, 15.6, 15.7, 15.12, 16.1, 17.1, 18.4A, 18.5, 18.13, 19.2B, 19.7A, 19.7B, 19.9, 20.3, 20.4, 21.2A, 21.2B, 21.5, 21.10, 21.11A, 21.14, 21.17, 22.1, 22.2, 22.3, 22.4, 22.5, 22.6, 22.7B, 22.8, 22.11A, 22.13A, 22.14, 22.15, 22.18B, 23.6, 23.10, 24.6A, 24.11, 25.1, 25.2, 25.4, 25.5, 25.6A, 25.6B, 25.6C, 25.7, 25.8, 25.9A, 25.9B, 25.10, 25.11, 25.12, 25.13, 25.15A, 25.15B; Box art, pages 29, 43, 56, 102, 103, 138-141, 152, 199, 374; Text art, pages 99–104.

INDEX

D

E

F

G

H

I

J

K

L

M

N

O

P

Q

R